Modified Nucleosides

Edited by
Piet Herdewijn

Related Titles

S. Müller (Ed.)

Nucleic Acids from A to Z

A Concise Encyclopedia

2008

ISBN: 978-3-527-31211-5

C. Stan Tsai (Ed.)

Biomacromolecules

Introduction to Structure, Function and Informatics

2006

ISBN: 978-0-471-71397-5

Michael Wink (Ed.)

An Introduction to Molecular Biotechnology

Functional Oligonucleotides and Their Applications

2006

ISBN: 978-3-527-31412-6

S. Klussmann (Ed.)

The Aptamer Handbook

Functional Oligonucleotides and Their Applications

2006

ISBN: 978-3-527-31059-3

D. Crich (Ed.)

Reagents for Glycoside, Nucleotide, and Peptide Synthesis

2005

ISBN: 978-0-470-02304-4

Modified Nucleosides

in Biochemistry, Biotechnology and Medicine

Edited by
Piet Herdewijn

WILEY-
VCH

WILEY-VCH Verlag GmbH & Co. KGaA

The Editor

Prof. Dr. Piet Herdewijn
Rega Institute
Katholieke Universiteit
Minderbroedersstraat 10
003000 Leuven
Belgium

Library of Congress Card No.: applied for

British Library Cataloguing-in-Publication Data
A catalogue record for this book is available from the British Library.

Bibliographic information published by the Deutsche Nationalbibliothek
Die Deutsche Nationalbibliothek lists this publication in the Deutsche Nationalbibliografie; detailed bibliographic data are available in the Internet at http://dnb.d-nb.de

Cover illustration Adam Design, Weinheim
Typesetting Thomson Digital, Noida, India
Printing Strauss GmbH, Mörlenbach
Bookbinding Litges & Dopf GmbH, Heppenheim

Printed in the Federal Republic of Germany
Printed on acid-free paper

ISBN: 978-3-527-31820-9

Contents

Preface *XIX*
List of Contributors *XXI*

Part I **Biochemistry and Biophysics** *1*

1 **Investigations on Fluorine-Labeled Ribonucleic Acids by ^{19}F NMR Spectroscopy** *3*
Christoph Kreutz and Ronald Micura
1.1 Introduction *3*
1.1.1 NMR Spectroscopic Properties of the ^{19}F Nucleus *3*
1.1.1.1 General NMR Spectroscopic Properties *3*
1.1.1.2 ^{19}F versus ^{1}H NMR Spectroscopy *3*
1.1.1.3 Factors Affecting the ^{19}F Chemical Shift in Biomolecules *5*
1.1.1.4 Fluorine Relaxation in Biological Systems *6*
1.1.1.5 Solvent-Induced Isotope Shifts of ^{19}F NMR Resonances *7*
1.1.2 ^{19}F NMR Spectroscopy of Proteins *7*
1.1.2.1 Incorporation of Fluorinated Amino Acids into Proteins *7*
1.1.2.2 ^{19}F NMR Spectroscopic Studies of Proteins *8*
1.2 ^{19}F NMR Spectroscopy of Nucleic Acids *13*
1.2.1 Nucleic Acids with Fluorinated Nucleobases *14*
1.2.1.1 Transfer RNAs *14*
1.2.1.2 *HhaI* Methyltransferase in Complex with DNA Duplexes *14*
1.2.1.3 Minimal Hammerhead Ribozyme *15*
1.2.1.4 HIV TAR RNA *17*
1.2.2 Nucleic Acids with Fluorinated Ribose Units *19*
1.2.2.1 *R1inv* RNA *19*
1.2.2.2 RNA Secondary Structure Equilibria *21*
1.2.2.3 RNA Ligand Binding *22*
1.2.3 Influence of Fluorine Modifications on Nucleic Acid Structure *22*
1.3 Conclusions *24*
References *24*

Modified Nucleosides: in Biochemistry, Biotechnology and Medicine. Edited by Piet Herdewijn

ISBN: 978-3-527-31820-9

2 8-Oxo-7,8-Dihydro-2′-Deoxyguanosine: A Major DNA Oxidation Product *29*
Jean Cadet and Paolo Di Mascio
2.1 Introduction *29*
2.2 Formation of 8-Oxo-7,8-Dihydroguanine *30*
2.2.1 Single Lesion *30*
2.2.1.1 ·OH Radical *31*
2.2.1.2 One-Electron Oxidation *31*
2.2.1.3 Singlet Oxygen *32*
2.2.2 Tandem Lesions *34*
2.3 Reactivity of 8-Oxo-7,8-Dihydro-2′-Deoxyguanosine *35*
2.3.1 One-Electron Oxidation *35*
2.3.1.1 Secondary Oxidation Products *36*
2.3.1.2 DNA–Protein Crosslinks *37*
2.3.2 Singlet Oxygen *37*
2.3.2.1 Nucleoside *37*
2.3.2.2 Oligonucleotide *38*
2.4 Formation of 8-Oxo-7,8-Dihydro-2′-Deoxyguanosine in Cellular DNA *39*
2.4.1 Methods of Measurement *39*
2.4.1.1 HPLC Methods (HPLC-ECD and HPLC-MS/MS) *39*
2.4.1.2 Enzymic Assays *40*
2.4.2 Indirect Effects of Ionizing Radiation (·OH Radical) *41*
2.4.3 High-Intensity UV Laser Irradiation (One-Electron Oxidation) *41*
2.4.4 UVA Photosensitization ($^1O_2$) *42*
2.5 Synthesis of 8-OxodGuo and Insertion into Oligonucleotides *42*
2.6 Conclusions *43*
References *44*

3 Modified DNA Bases: Probing Base-Pair Recognition by Polymerases *49*
Eric T. Kool
3.1 Introduction *49*
3.1.1 The Importance of Understanding DNA Polymerases *49*
3.1.2 The Utility of Modified Nucleobases in Probing Mechanisms *50*
3.1.3 The Scope of this Chapter *50*
3.2 Basic Principles and Methods in Replication *51*
3.2.1 The Chemistry of Polymerases *51*
3.2.2 Different Classes of Polymerases *51*
3.2.3 Methods Used in Polymerase Studies *52*
3.3 Alternative Hydrogen-Bonding Schemes *53*
3.3.1 Thioguanine-Pyridone *53*
3.3.2 Benner Hydrogen-Bonding Variants *54*
3.4 Non-Polar Nucleoside Isosteres *56*
3.4.1 The Concept of Nucleobase Isosteres *56*

3.4.2 Synthesis, Structure, and Physical Properties *56*
3.4.3 Base-Pairing Properties *57*
3.4.4 Polymerase Behavior *58*
3.4.5 Other Classes of Polymerases *59*
3.4.6 Summary of Watson–Crick H-Bonding Effects in Polymerase Active Sites *60*
3.5 Non-Polar Steric Probes *60*
3.5.1 Isomers of Hydrocarbons Illustrate Hydrophobic "Packing" Effects *60*
3.5.2 Systematic Size Variants *62*
3.5.3 Systematic Shape Variants *63*
3.6 Minor Groove Hydrogen Bonds in Polymerases *64*
3.6.1 Base Analogues Testing Minor Groove Interactions *64*
3.7 Other Non-Polar Bases and Pairs *65*
3.7.1 The Quest for New Base Pairs *65*
3.7.2 A Broad Variety of Heterocycles and Hydrocarbons *66*
3.7.3 Benzimidazoles Continue the Debate on Steric Effects *67*
3.7.4 New Pairs of Hirao and Yokoyama *67*
3.8 Replication of Designed Bases in Living Cells *68*
3.8.1 Effects of Hydrogen Bonding in *E. coli* *68*
3.8.2 Effects of Nucleobase Size in *E. coli* *69*
3.9 Conclusions and Future Prospects *70*
3.9.1 What We Know About Replication *70*
3.9.2 And What Remains Unknown *71*
3.9.3 Future Directions *71*
References *71*

4 2′-Deoxyribose-Modified Nucleoside Triphosphates and their Recognition by DNA Polymerases *75*
Karl-Heinz Jung and Andreas Marx
4.1 Introduction *75*
4.2 Modified Nucleotides as Alternative Building Blocks to Natural Nucleic Acids *76*
4.2.1 Introduction *76*
4.2.2 Nucleotides with Downsized Residues: α-L-Threose-Derived Nucleotides *76*
4.2.2.1 Introduction *76*
4.2.2.2 Synthesis *77*
4.2.2.3 DNA Polymerase Studies *80*
4.2.3 Nucleotides with Downsized Residues: Glycerol-Derived Nucleotides *80*
4.2.3.1 Introduction *80*
4.2.3.2 Synthesis *81*
4.2.3.3 DNA Polymerase Studies *82*
4.2.4 Nucleotides with Expanded Sugar Residues: 1,5-Anhydrohexitol Nucleotides *82*

4.2.4.1 Introduction 82
4.2.4.2 Synthesis 83
4.2.4.3 Functional DNA Polymerase Studies 85
4.2.5 Nucleotides with Expanded Sugar Residues: Cyclohexenyl Nucleotides 85
4.2.5.1 Introduction 85
4.2.5.2 Synthesis 86
4.2.5.3 DNA Polymerase Studies 88
4.3 DNA Polymerase Selectivity: 4′-*C*-Modified Nucleotides 89
4.3.1 Introduction 89
4.3.2 Design and Synthesis 90
4.3.3 DNA Polymerase Studies 91
4.4 Concluding Remarks 93
References 94

5 Pyrimidine Dimers: UV-Induced DNA Damage 97
Shigenori Iwai
5.1 Introduction 97
5.2 Formation of Pyrimidine Dimers 98
5.2.1 Cyclobutane Pyrimidine Dimers 98
5.2.2 The (6–4) Photoproducts and their Dewar Valence Isomers 101
5.2.3 Other UV Lesions 101
5.3 Chemical Synthesis of Oligonucleotides Containing Pyrimidine Dimers 103
5.3.1 Oligonucleotides Containing a CPD 103
5.3.2 Oligonucleotides Containing the (6–4) or Dewar Photoproduct 105
5.4 Structure and Mutagenesis of Pyrimidine Dimer-Containing DNA 109
5.4.1 Tertiary Structures of Pyrimidine Dimer-Containing Duplexes 109
5.4.2 Base-Pair Formation by Pyrimidine Dimers 111
5.4.3 Mutations Induced by Pyrimidine Dimers 112
5.5 Repair of Pyrimidine Dimers in Cells 116
5.5.1 T4 Endonuclease V 116
5.5.2 Photolyases 118
5.5.3 Nucleotide Excision Repair (NER) 119
5.5.4 UV Damage Endonuclease (UVDE) 121
5.6 Bypass of Pyrimidine Dimers by DNA Polymerases 122
References 124

6 Locked Nucleic Acids: Properties, Applications, and Perspectives 133
Poul Nielsen and Jesper Wengel
6.1 Introduction 133
6.2 LNA in High-Affinity Hybridization: Designing Sequences 135
6.3 Structural Studies 139
6.4 Analogues of LNA and their Structural Impact 140
6.5 LNA as Potential Therapeutics 143
6.6 LNA-Probes 147

6.7 Concluding Remarks *148*
References *149*

7 Synthesis and Properties of Oligonucleotides Incorporating Modified Nucleobases Capable of Watson–Crick-Type Base-Pair Formation *153*
Mitsuo Sekine, Akio Ohkubo, Itaru Okamoto, and Kohji Seio
7.1 Introduction *153*
7.2 Natural, Enzyme-Assisted Sophisticated Devices for Maintaining Correct Base Recognition of Canonical Nucleobases *154*
7.3 Synthesis and Properties of Oligodeoxynucleotides Incorporating 4-*N*-Acylated Cytosine Derivatives *155*
7.4 Base-Recognition Ability of 4-*N*-Alkoxycarbonylcytosine Derivatives *157*
7.5 Synthesis and Properties of Oligonucleotides Incorporating 4-*N*-Carbamoylcytosine Derivatives *159*
7.6 2-Thiouracil as an Improved Nucleobase in Place of Thymine *160*
7.7 Modified Adenine Bases Capable of Recognizing the Thymine Base *161*
7.8 Design of Modified Guanine Bases Capable of Recognizing Cytosine *165*
7.9 Conclusions *168*
References *168*

8 The Properties of 4′-Thionucleosides *173*
Masataka Yokoyama
8.1 Introduction *173*
8.2 Synthesis of 4′-Thionucleosides *173*
8.3 Synthesis of Isothionucleosides *195*
8.4 Synthesis of L-Thionucleosides *196*
8.5 Synthesis of Thioxonucleosides *198*
8.6 Synthesis of Miscellaneous Thionucleosides *205*
8.7 Biological Activity of Thionucleosides *210*
8.8 Conclusions *219*
References *219*

9 *S*-Adenosyl-L-methionine and Related Compounds *223*
Christian Dalhoff and Elmar Weinhold
9.1 Introduction *223*
9.2 The Biochemistry of AdoMet *224*
9.2.1 AdoMet as a Methyl and Methylene Group Donor *224*
9.2.2 AdoMet as an Aminocarboxypropyl Group Donor *227*
9.2.3 AdoMet as an Adenosyl Group Donor *228*
9.2.4 AdoMet as a Ribosyl Group Donor *228*
9.2.5 AdoMet as a Radical Source *229*
9.2.6 AdoMet as an Amino Group Donor *229*
9.2.7 AdoMet-Dependent Riboswitches *230*
9.2.8 The Biosynthesis and Metabolism of AdoMet *230*

9.3 The Chemistry and Biochemistry of Modified AdoMet *231*
9.3.1 Synthetic Approaches to AdoMet Analogues *231*
9.3.2 Isotope-Labeled AdoMet *232*
9.3.3 Selenium and Tellurium Analogues of AdoMet *234*
9.3.4 Sulfoxide and Sulfone Analogues of AdoMet *236*
9.3.5 Sinefungin *237*
9.3.6 Nitrogen Analogues of AdoMet *237*
9.3.7 Aziridine Analogues of AdoMet *238*
9.3.8 AdoMet Analogues with Methyl Group Replacements *238*
9.4 AdoMet as a Pharmaceutical *240*
9.5 Concluding Remarks *241*
References *242*

Part II Biotechnology *249*

10 5-Substituted Nucleosides in Biochemistry and Biotechnology *251*
Mohammad Ahmadian and Donald E. Bergstrom
10.1 Introduction *251*
10.2 Synthesis *252*
10.2.1 Organopalladium Coupling Reactions *252*
10.2.2 Strategies for Post-Oligonucleotide-Synthesis Modification through Pyrimidine C-5 *253*
10.3 Incorporation of C-5-Substituted Pyrimidine Nucleotides into Nucleic Acids through Modified Nucleotide 5′-Triphosphates *255*
10.3.1 The Early Studies *255*
10.3.2 Incorporation of Diverse Functionality into DNA *257*
10.3.3 T7 RNA Polymerase-Mediated Synthesis of Modified RNA *262*
10.3.4 Incorporation of C-5-Appended Fluorophores *262*
10.4 C-5 Substituents that Stabilize DNA Duplexes *264*
10.5 Photochemistry *269*
10.6 Conclusions *271*
References *271*

11 Universal Base Analogues and their Applications to Biotechnology *277*
Kathleen Too and David Loakes
11.1 Introduction *277*
11.2 General Methods of Synthesis *278*
11.3 Properties of Universal Bases *282*
11.4 Structure, Stacking, and Stabilization *283*
11.5 Hydrogen-Bonding Universal Base Analogues *287*
11.6 Applications of Universal Base Analogues *290*
11.7 Triphosphate Derivatives *295*
11.8 Therapeutic Applications *298*
References *300*

Part III **Medicinal Chemistry** *305*

12 **The Properties of Locked Methanocarba Nucleosides in Biochemistry, Biotechnology, and Medicinal Chemistry** *307*
Victor E. Marquez
12.1 Introduction *307*
12.2 Structural Representation *308*
12.2.1 The Bicyclo[3.1.0]hexane Template *308*
12.2.2 Pseudoboat versus Pseudochair Conformations *309*
12.3 Synthesis of Locked Nucleosides *309*
12.3.1 North (*N*) Conformer Mimics *311*
12.3.1.1 Dideoxyribonucleoside Analogues *311*
12.3.1.2 2′-Deoxyribonucleoside Analogues *312*
12.3.1.3 Ribonucleoside Analogues *318*
12.3.2 South (*S*) Conformer Mimics *320*
12.3.2.1 2′-Deoxyribonucleoside Analogues *320*
12.3.2.2 Ribonucleoside Analogues *321*
12.3.3 Synthesis of *N*- and *S*-Methanocarba AZT Analogues *323*
12.3.4 Synthesis of Bicyclo[3.1.0]hexene Nucleosides *324*
12.3.5 Reshuffling of Groups on a Bicyclo[3.1.0]hexane Template *326*
12.3.6 Bicyclo[3.1.0]hexane Pseudosugars as Surrogates of Abasic Nucleosides *327*
12.4 Synthesis of Oligodeoxynucleotides (ODNs) Containing Locked Nucleosides *328*
12.4.1 The Dickerson–Drew (DD) Dodecamer *330*
12.5 Molecular Targets, Ligand Properties, and Binding Modes *331*
12.5.1 Kinases and Polymerases *331*
12.5.2 HIV Reverse Transcriptase *335*
12.5.3 DNA Methyltransferase *337*
12.6 Concluding Remarks *339*
References *339*

13 **Synthesis, Chemical Properties and Biological Activities of Cyclic Bis(3′–5′)diguanylic Acid (*c*-di-GMP) and its Analogues** *343*
Mamoru Hyodo and Yoshihiro Hayakawa
13.1 Introduction *343*
13.2 Synthesis of *c*-di-GMP and its Analogues *345*
13.2.1 Synthesis of *c*-di-GMP *345*
13.2.2 Synthesis of Artificial Analogues of *c*-di-GMP *347*
13.3 Chemical Properties of *c*-di-GMP and its Analogues *348*
13.3.1 Stability and Chemical Properties of *c*-di-GMP under Acidic, Basic, and Physiological Conditions *348*
13.3.2 Polymorphism of *c*-di-GMP in Aqueous Solutions *349*

13.4 Bioactivities of *c*-di-GMP and its Analogues *351*
13.4.1 Activity of *c*-di-GMP on Biofilm Formation *352*
13.4.1.1 Inhibition of Biofilm Formation and Prevention of Bacterial Infection of *S. aureus in vitro* *352*
13.4.1.2 Activity as an Immunostimulatory Molecule *353*
13.4.1.3 Activity on Biofilm Formation and Virulence Emergence of *P. aeruginosa* *357*
13.4.2 Inhibition of Proliferation of Human Colon Cancer Cells with *c*-di-GMP *357*
13.4.3 Biological Activity of *c*-dGpGp *358*
13.5 Conclusions *359*
References *361*

14 Siderophore Biosynthesis Inhibitors *365*
Courtney C. Aldrich and Ravindranadh V. Somu
14.1 Introduction *365*
14.2 Synthesis, Physico-Chemical Properties, Metabolism, Mechanism of Action, and Biological Activity *365*
14.2.1 Synthesis *365*
14.2.2 Physico-Chemical Properties *366*
14.2.3 Metabolism *367*
14.2.4 Toxicity *367*
14.2.5 Biochemical Target *368*
14.2.6 Mechanism of Action *369*
14.2.7 Biological Activity *369*
14.3 Background of Siderophores: Molecular Target and Rationale for Inhibitor Design *370*
14.4 Ligand Properties/Binding Mode *374*
14.4.1 Nature of the Linker *375*
14.4.2 Importance of the Aryl Ring *379*
14.4.3 Role of the Ribose *382*
14.4.4 Impact of the Nucleobase *385*
14.5 Conclusions *388*
References *388*

15 Synthesis and Biological Activity of Selected Carbocyclic Nucleosides *393*
Adam Mieczkowski and Luigi A. Agrofoglio
15.1 Introduction *393*
15.2 A-5021, Synguanol, and Cyclopropane Derivatives *395*
15.3 Lobucavir and Cyclobutane Nucleoside Derivatives *398*
15.4 Carbovir and 2′,3′-Unsaturated Nucleoside Derivatives *405*
15.5 Locked Nucleosides *413*
15.6 Conclusions *420*
References *420*

16 **4′-C-Ethynyl-2′-Deoxynucleosides** *425*
Hiroshi Ohrui
16.1 Introduction *425*
16.2 Murine Toxicity of Purine 4′EdNs *426*
16.3 4′EdA Derivatives Stable to Adenosine Deaminase, and their Biological Properties *427*
16.4 4′-*C*-Ethynylnucleosides without 3′-OH *429*
16.4.1 2′,3′-Dideoxy-4′-*C*-ethynylnucleoside *429*
16.4.2 2′,3′-Didehydrodideoxy-4′-*C*-ethynylnucleosides *429*
16.4.3 Carbocyclic and Other Heterocyclic Analogues of 4′-*C*-ethynylnucleoside *431*
References *431*

17 **Modified Nucleosides as Selective Modulators of Adenosine Receptors for Therapeutic Use** *433*
Kenneth A. Jacobson, Bhalchandra V. Joshi, Ben Wang, Athena Klutz, Yoonkyung Kim, Andrei A. Ivanov, Artem Melman, and Zhan-Guo Gao
17.1 Introduction *433*
17.2 Molecular Targets and Binding Modes *434*
17.3 AR Agonists as Clinical Candidates *434*
17.3.1 Modified Nucleosides as A_1 AR Agonists *435*
17.3.2 Modified Nucleosides as A_{2A} AR Agonists *437*
17.3.3 Modified Nucleosides as A_{2B} AR Agonists *439*
17.3.4 Modified nucleosides as A_3 AR Ligands *441*
17.4 Summary *444*
References *444*

18 **The Design of Forodesine HCl and Other Purine Nucleoside Phosphorylase Inhibitors** *451*
Philip E. Morris and Vivekanand P. Kamath
18.1 Introduction *451*
18.2 Purine Nucleoside Phosphorylase Enzyme Structure *452*
18.2.1 Purine-Binding Site *453*
18.2.2 Phosphate-Binding Site *454*
18.2.3 Sugar-Binding Site *454*
18.3 First-Generation PNP Inhibitors: Substrate Analogues *454*
18.3.1 Chemistry of First-Generation PNP Inhibitors *458*
18.4 Second-Generation PNP Inhibitors: Transition-State Inhibitors *460*
18.4.1 Chemistry of Second-Generation PNP Inhibitors *461*
18.4.2 Convergent Synthesis of Forodesine HCl *463*
18.5 Third-Generation PNP Inhibitors: Transition-State Inhibitors *467*
18.5.1 Chemistry of BCX-4208 *467*
References *469*

19 **Formycins and their Analogues: Purine Nucleoside Phosphorylase Inhibitors and their Potential Application in Immunosuppression and Cancer** *473*
Agnieszka Bzowska
19.1 Introduction *473*
19.2 Chemical Structure of Formycins and their Analogues *475*
19.2.1 Formycin A and B, Oxoformycin B *475*
19.2.2 Structural Modifications of Formycins *476*
19.2.2.1 N-Methyl and N-Substituted Analogues *476*
19.2.2.2 Other Base-Modified Analogues *478*
19.2.2.3 Sugar-Modified Analogues *478*
19.2.3 Formycin Phosphates and Polyformycin Phosphates *478*
19.3 Spectral Properties of Formycins *479*
19.4 Sources of Formycins *483*
19.4.1 Natural Sources and Biosynthesis *483*
19.4.2 Synthesis *483*
19.5 The Biological Activity of Formycins: A Brief Summary *485*
19.6 Formycins and Analogues as Purine Nucleoside Phosphorylase Inhibitors *488*
19.6.1 Molecular Target of Formycins: Purine Nucleoside Phosphorylase (PNP) *488*
19.6.2 Formycins and Analogues in Studies of the Molecular Mechanism of Catalysis: The 3-D Structure of PNPs *489*
19.6.2.1 Low-Molecular-Mass PNPs *490*
19.6.2.2 High-Molecular-Mass PNPs *492*
19.6.3 PNP Deficiency and the Potential Role of PNP Inhibitors *495*
19.6.4 Formycins and Analogues as Inhibitors of Mammalian PNPs *496*
19.7 Formycins as Inhibitors of Parasitic PNPs and Hydrolases *500*
19.8 Actual and Potential Applications of Formycins *501*
19.8.1 Formycin and Analogues in Assays of Enzyme Activity *501*
19.8.2 Formycin and Analogues as Protein Ligands for X-Ray Structural Studies *502*
19.8.3 Formycins and Analogues as Molecular Probes *502*
19.8.4 Formycins as Ligands in Affinity Chromatography *503*
19.8.5 Formycin B as a Tool to Study Nucleoside Transport *504*
References *504*

20 **1-(3-C-Ethynyl-β-D-*ribo*-pentofuranosyl)cytosine (ECyd)** *511*
Akira Matsuda
20.1 Introduction *511*
20.2 Synthesis of ECyd and its Analogues *511*
20.3 Cytotoxic Activity and Structure–Activity Relationships of ECyd Analogues *In Vitro*, and *In Vitro* Antitumor Activity *513*
20.4 Structural Features of ECyd and 4′-Thio-ECyd *513*
20.5 Metabolism and Mechanism of Action *517*

20.6 An Apoptotic Pathway Involving the Action of ECyd *519*
20.7 Combination of ECyd with Low-Dose X-Irradiation *519*
20.8 ECyd is Effective against Gemcitabine-Resistant Human Pancreatic Cancer Cells *520*
20.9 Conclusions *521*
References *521*

21 Syntheses and Biological Activity of Neplanocin and Analogues *525*
Dilip K. Tosh, Hea Ok Kim, Shantanu Pal, Jeong A. Lee, and Lak Shin Jeong
21.1 Introduction *525*
21.2 New Methodologies in the Synthesis of Neplanocin A *527*
21.3 Modifications on Neplanocin A and Aristeromycin *532*
21.3.1 C2′ Modification *533*
21.3.2 C3′ Modification *536*
21.3.3 C4′ Modification *538*
21.3.4 C5′ Modification *540*
21.3.5 C6′ Modification *550*
21.3.6 Base Modification *552*
21.3.7 Miscellaneous *559*
21.4 Conclusions *563*
References *563*

22 Clitocine and Its Analogues *567*
Hyunik Shin and Changhee Min
22.1 Clitocine: Isolation, Synthesis, and Biological Activity *567*
22.1.1 Isolation *567*
22.1.2 Synthesis *567*
22.1.3 Biological Activity *569*
22.2 Clitocine Analogues *571*
22.2.1 Aglycone Modifications *571*
22.2.2 Carbocyclic Analogues *573*
22.2.3 Acyclic Analogues *576*
22.2.4 5′-Amino Analogues *578*
References *583*

Part IV Antitumorals and Antivirals *585*

23 Capecitabine Preclinical Studies: From Discovery to Translational Research *587*
Hideo Ishitsuka and Nobuo Shimma
23.1 Introduction *587*
23.2 Drug Design and Discovery of Capecitabine *588*
23.2.1 5′-DFUR as a Lead Compound of Capecitabine *588*

23.2.2 5′-DFCR Derivatives 589
23.2.3 N^4-Acyl-5′-DFCR Derivatives 589
23.2.4 N^4-Alkoxycarbonyl-5′-DFCR Derivatives 590
23.3 Preclinical Studies 590
23.3.1 Tumor-Selective Delivery of the Active 5-FU 590
23.3.2 Anti-Tumor Activities 592
23.3.3 Dose Fractionation and Schedule 592
23.3.4 Safety (Dose Range and Mild Myelotoxicity) 593
23.4 Translational Research for Optimizing Capecitabine Efficacy 593
23.4.1 Factors that Influence Capecitabine Efficacy 593
23.4.2 Combination Therapy with Rational Partners 594
23.4.2.1 Combination with TP Up-Regulators 594
23.4.2.2 Combination with DPD Down-Regulators 594
23.4.3 Personalized Therapy of Rational Patient Populations 596
23.5 Conclusions 597
References 598

24 Tenofovir and Adefovir as Antiviral Agents *601*
Tomas Cihlar, William E. Delaney IV, and Richard Mackman
24.1 Introduction 601
24.2 Synthesis 602
24.2.1 Adefovir and Adefovir Dipivoxil 603
24.2.2 Tenofovir and Tenofovir Disoproxil Fumarate 605
24.3 Mechanism of Action 606
24.3.1 Membrane Transport and Intracellular Metabolism 606
24.3.2 Inhibition of Viral Polymerases 608
24.3.3 Spectrum of Antiviral Activity 609
24.4 Activity in Animal Models 612
24.4.1 Models for Retroviral Infections 612
24.4.2 Models for Hepadnavirus Infections 613
24.4.3 Herpes Models 614
24.5 Clinical Experience 614
24.5.1 Tenofovir Disoproxil Fumarate (Viread®) 614
24.5.2 Adefovir Dipivoxil (Hepsera®) 616
24.6 Drug Resistance 618
24.6.1 HIV Resistance 618
24.6.2 HBV Resistance 619
24.7 Novel Antiviral Nucleotides and Nucleotide Prodrugs 619
24.8 Conclusions 621
References 622

25 **Clofarabine: From Design to Approval** *631*
John A. Secrist III, Jaideep V. Thottassery, and William B. Parker
25.1 Introduction *631*
25.2 Clofarabine: The Background *632*
25.3 The Beginnings *632*
25.4 The Next Generation of Compounds *635*
25.5 Mechanism of Action of Clofarabine *639*
25.5.1 Transport and Metabolism to Active Metabolites *639*
25.5.2 Inhibition of DNA Synthesis *640*
25.5.3 Induction of Apoptosis *641*
25.5.4 Activity against Non-Proliferating Cancer Cells *642*
25.6 Clofarabine to the Clinic *643*
25.6.1 Clinical Trials and Approval *643*
25.7 Summary and Comments *644*
References *644*

Index *647*

Preface

The chemical modification of nucleosides has been – and will continue to remain – a major research topic in bioorganic and medicinal chemistry. The reason for this is obvious: nucleosides, oligonucleotides and nucleic acids are involved in all aspects of cellular life – from the storage of genetic information to metabolic regulation, catalysis, and energy supply. Indeed, research investigations in this area have led to the provision of many life-saving drugs in cancer and infectious disease, as well as invaluable diagnostic tools.

The study of the biological effect of modified nucleosides and oligonucleotides has provided us with a deeper insight into cellular functions. Their potential applications are continuing to explore new horizons, amongst others in the field of nanotechnology, microarrays, RNA interference, and small-molecule therapeutics.

Although, from the 1960s until the 1980s, the discovery of new biological active nucleosides and oligonucleotides was largely a matter of serendipity, this is no longer the case. Today, fundamental insights into the conformational behavior of nucleosides, enzyme mechanisms, the physico-chemistry of nucleosides and oligonucleotides, and the availability of increasing amounts of structural data on nucleosides and oligonucleotides has simplified the design process of biological active nucleosides and oligonucleotides. Likewise, whilst between the 1950s and 1970s the synthesis of a modified nucleoside was a difficult undertaking, the evolution of organic synthesis in general – and of nucleoside chemistry in particular – has led to modified nucleosides and oligonucleotides becoming much more easily available.

In this book, we have assembled different aspects of the chemistry and biology of modified nucleosides. In order to demonstrate the broad field of research that is covered by modified nucleosides, pure "chemical" chapters as well as more "biological" and even "clinically oriented" chapters are included. However, the major focus is on modified nucleosides, as it in this region that the applications are most clear, though new directions in the nucleotide and oligonucleotide fields have also briefly been discussed.

Clearly, it would be impossible to cover such as vast research field within a single volume, and consequently the contents of this book must be considered as a general introduction to the fascinating world of modified nucleosides.

Modified Nucleosides: in Biochemistry, Biotechnology and Medicine. Edited by Piet Herdewijn

ISBN: 978-3-527-31820-9

The editor cordially thanks all authors who have contributed to the book, and hopes that it will inspire scientists not only to study the chemistry and biology of modified nucleosides but also to explore new opportunities in this area.

Leuven, May 2008 *Piet Herdewijn*

List of Contributors

Luigi A. Agrofoglio
Université d'Orléans
UFR Sciences
Institut de Chimie Organique
et Analytique
ICOA UMR CNRS 6005
BP 6759
45067 Orléans Cedex 2
France

Mohammad Ahmadian
MDRNA Inc.
3350 Monte Villa Parkway
Bothell, WA 98021
USA

Courtney C. Aldrich
University of Minnesota
Academic Health Center
Center for Drug Design
516 Delaware St. S.E., 7–169 PWB
Minneapolis, MN 55455
USA

Donald E. Bergstrom
Purdue University
Birck Nanotechnology Center
Department of Medicinal Chemistry
and Molecular Pharmacology
1205 West State Street
West Lafayette, IN 47907-2057
USA

Agnieszka Bzowska
University of Warsaw
Institute of Experimental Physics
Department of Biophysics
Żwirki i Wigury 93
02089 Warsaw
Poland

Jean Cadet
Département de Recherche
Laboratoire 'Lésions des Acides
Nucléiques'
LCIB-UMR-E no. 3 CEA–UJF
Fondamentale sur la Matière Condensée
CEA/Grenoble
38054 Grenoble Cedex 9
France

Tomas Cihlar
Gilead Sciences, Inc.
Department of Biology
362 Lakeside Drive
Foster City, CA 94404
USA

Christian Dalhoff
Caprotec Bioanalytic GmbH
Volmerstrasse 5
12489 Berlin
Germany

Modified Nucleosides: in Biochemistry, Biotechnology and Medicine. Edited by Piet Herdewijn

ISBN: 978-3-527-31820-9

William E. Delaney, IV
Gilead Sciences, Inc.
Department of Biology
342 Lakeside Drive
Foster City, CA 94404
USA

Paolo Di Mascio
Universidade de São Paulo
Instituto de Química
Departamento de Bioquímica
CP 26077
CEP 05513-970
São Paulo, SP
Brazil

Zhan-Guo Gao
National Institutes of Health
National Institute of Diabetes,
Digestive and Kidney Diseases
Laboratory of Bioorganic Chemistry
Molecular Recognition Section
Bethesda, MD 20892-0810
USA

Yoshihiro Hayakawa
Nagoya University
Graduate School of Information
Science
Furo-cho, Chikusa
Nagoya
Japan

Mamoru Hyodo
Nagoya University
Graduate School of Information
Science
Furo-cho, Chikusa
Nagoya
Japan

Hideo Ishitsuka
Roche Diagnostics K.K.
Shiba 2-6-1
Minato city
Tokyo 105-0014
Japan

Andrei A. Ivanov
National Institutes of Health
National Institute of Diabetes,
Digestive and Kidney Diseases
Laboratory of Bioorganic Chemistry
Molecular Recognition Section
Bethesda, MD 20892-0810
USA

Shigenori Iwai
Osaka University
Graduate School of Engineering Science
Division of Chemistry
1-3 Machikaneyama
Toyonaka
Osaka 560-8531
Japan

Kenneth A. Jacobson
National Institutes of Health
National Institute of Diabetes,
Digestive and Kidney Diseases
Laboratory of Bioorganic Chemistry
Molecular Recognition Section
Bethesda, MD 20892-0810
USA

Lak Shin Jeong
Ewha Womans University
College of Pharmacy
11-1 Seodaemun-gu, Dae hyun-dong
Seoul 120-750
Korea

Bhalchandra V. Joshi
National Institutes of Health
National Institute of Diabetes,
Digestive and Kidney Diseases
Laboratory of Bioorganic Chemistry
Molecular Recognition Section
Bethesda, MD 20892-0810
USA

Karl-Heinz Jung
University of Konstanz
Department of Chemistry
Universitätsstrasse 10
78457 Konstanz
Germany

Vivekanand P. Kamath
BioCryst Pharmaceuticals
2190 Parkway Lake Drive
Birmingham, AL 35244
USA

Hea Ok Kim
Ewha Womans University
College of Pharmacy
11-1 Seodaemun-gu, Dae hyun-dong
Seoul 120-750
Korea

Yoonkyung Kim
Ewha Womans University
College of Pharmacy
11-1 Seodaemun-gu, Dae hyun-dong
Seoul 120-750
Korea

Athena Klutz
National Institutes of Health
National Institute of Diabetes,
Digestive and Kidney Diseases
Laboratory of Bioorganic Chemistry
Molecular Recognition Section
Bethesda, MD 20892-0810
USA

Eric T. Kool
Stanford University
Department of Chemistry
Stanford, CA 94305-5080
USA

Christoph Kreutz
University of Vienna
Department of Biomolecular
Structural Chemistry
Campus Vienna Biocenter 5
1030 Vienna
Austria

Jeong A. Lee
Ewha Womans University
College of Pharmacy
11-1 Seodaemun-gu, Dae hyun-dong
Seoul 120-750
Korea

David Loakes
Medical Research Council
Laboratory of Molecular Biology
Hills Road
Cambridge CB2 2QH
United Kingdom

Richard Mackman
Gilead Sciences, Inc.
Department of Biology
342 Lakeside Drive
Foster City, CA 94404
USA

Victor E. Marquez
National Institutes of Health
National Cancer Institute
Frederick Cancer Research and
Development Center
Laboratory of Medicinal Chemistry
376 Boyles St.
Frederick, MD 21702-1201
USA

Andreas Marx
University of Konstanz
Department of Chemistry
Universitätsstrasse 10
78457 Konstanz
Germany

Akira Matsuda
Hokkaido University
Faculty of Pharmaceutical Sciences
Kita-12, Nishi-6
Kita-ku
Sapporo 060-0812
Japan

Artem Melman
National Institutes of Health
National Institute of Diabetes,
Digestive and Kidney Diseases
Laboratory of Bioorganic Chemistry
Molecular Recognition Section
Bethesda, MD 20892-0810
USA

Ronald Micura
Leopold-Franzens-University
Institute of Organic Chemistry
Center for Molecular Biosciences
Innsbruck (CMBI)
Innrain 52a
6020 Innsbruck
Austria

Adam Mieczkowski
Université d'Orléans
UFR Sciences
Institut de Chimie Organique
et Analytique
ICOA UMR CNRS 6005
BP 6759
45067 Orléans Cedex 2
France

Changhee Min
LG Life Sciences Ltd./R&D
Drug Discovery Division
104-1, Moonji-dong
Yusong-gu
Daejeon 305-380
Korea

Philip E. Morris, Jr.
BioCryst Pharmaceuticals
2190 Parkway Lake Drive
Birmingham, AL 35244
USA

Poul Nielsen
University of Southern Denmark
Department of Chemistry and Physics
Campusvej 55
5230 Odense M
Denmark

Akio Ohkubo
Tokyo Institute of Technology
Department of Life Science
4259 Nagatsuta
Midoriku
Yokohama 226-8501
Japan

Hiroshi Ohrui
Yokohama College of Pharmacy
Department of Kampo Pharmaceutical
Sciences
601 Matanocho, Totsukaku
246-0066 Yokohama
Japan

Itaru Okamoto
Tokyo Institute of Technology
Department of Life Science
4259 Nagatsuta
Midoriku
Yokohama 226-8501
Japan

Shantanu Pal
Ewha Womans University
College of Pharmacy
11-1 Seodaemun-gu, Dae hyun-dong
Seoul 120-750
Korea

William B. Parker
Southern Research Institute
P.O. Box 55305
Birmingham, AL 35255-5305
USA

John A. Secrist III
Southern Research Institute
P.O. Box 55305
Birmingham, AL 35255-5305
USA

Kohji Seio
Tokyo Institute of Technology
Department of Life Science
4259 Nagatsuta
Midoriku
Yokohama 226-8501
Japan

Mitsuo Sekine
Tokyo Institute of Technology
Department of Life Science
4259 Nagatsuta
Midoriku
Yokohama 226-8501
Japan

Nobuo Shimma
Chugai Pharmaceutical Co., Ltd
Kamakura Research Institute
Kajiwara, Kamakura
Kanagawa 247-8530
Japan

Hyunik Shin
LG Life Sciences Ltd./R&D
Chemical Development Division
104-1, Moonji-dong
Yusong-gu
Daejeon 305-380
Korea

Ravindranadh V. Somu
University of Minnesota
Academic Health Center
Center for Drug Design
516 Delaware St. S.E., 7-169 PWB
Minneapolis, MN 55455
USA

Jaideep V. Thottassery
Southern Research Institute
P.O. Box 55305
Birmingham, AL 35255-5305
USA

Kathleen Too
Medical Research Council
Laboratory of Molecular Biology
Hills Road
Cambridge CB2 2QH
United Kingdom

Dilip K. Tosh
Ewha Womans University
College of Pharmacy
11-1 Seodaemun-gu, Dae hyun-dong
Seoul 120-750
Korea

Ben Wang
National Institutes of Health
National Institute of Diabetes,
Digestive and Kidney Diseases
Laboratory of Bioorganic Chemistry
Molecular Recognition Section
Bethesda, MD 20892-0810
USA

Elmar Weinhold
RWTH Aachen University
Institute of Organic Chemistry
Landoltweg 1
52056 Aachen
Germany

Jesper Wengel
University of Southern Denmark
Department of Chemistry and Physics
Campusvej 55
5230 Odense M
Denmark

Masataka Yokoyama
Emeritus Professor, Chiba University
Godo Shigen Sangyo Co. Ltd.
1365 Nanaido
Chosei-mura, Chosei-gun
Chiba Prefecture
Japan

Part I
Biochemistry and Biophysics

1
Investigations on Fluorine-Labeled Ribonucleic Acids by ^{19}F NMR Spectroscopy

Christoph Kreutz and Ronald Micura

1.1
Introduction

1.1.1
NMR Spectroscopic Properties of the ^{19}F Nucleus

1.1.1.1 General NMR Spectroscopic Properties

Fluorine, with its unique nuclear magnetic resonance (NMR) spectroscopic parameters, has 100% natural abundance, possesses an intrinsic NMR sensitivity almost as high as protons (83%), and offers a chemical shift dispersion that is about 100-fold that of protons (Table 1.1). These properties make fluorine an ideal candidate to be used as an alternative spin label complementary to ^{13}C- and ^{15}N-labeling. Thus, the introduction of fluorine produces a strong NMR signal that appears against a background devoid of signals from other nuclei.

Table 1.1 Gyromagnetic ratios (γ), NMR frequencies (ν) in a 9.4 T magnetic field, and natural abundances (a) of selected nuclides.

Nucleus	γ [rad.T^{-1} s^{-1} 10^{7}]	ν [MHz]	a [%]
^{1}H	26.75	400.0	99.985
^{13}C	6.73	100.6	1.108
^{15}N	−2.71	40.5	0.37
^{19}F	25.18	376.5	100.0
^{31}P	10.84	162.1	100.0

1.1.1.2 ^{19}F versus ^{1}H NMR Spectroscopy

A biopolymer exhibits a NMR spectrum, which is composed of the typical resonance frequencies (chemical shifts) of the different building units that comprise the

Modified Nucleosides: in Biochemistry, Biotechnology and Medicine. Edited by Piet Herdewijn

ISBN: 978-3-527-31820-9

biomolecule. In ^{1}H NMR spectroscopy, chemical shift values typically span a range of approximately 10 to 15 ppm. This leads to severe overlap and resonance degeneracy in the proton NMR spectrum, in particular, in the case of large biomolecules or if other phenomena such as chemical exchange are present. For example, resonances of the imino protons (protons that participate in Watson–Crick base pairing) are well separated for the 32 nt DNA/RNA hybrid depicted in Figure 1.1, whereas the residual resonances from sugar and nucleobase moieties show severe overlap. Multidimensional NMR methods and labeling techniques (^{13}C and ^{15}N) are required to overcome this hurdle.

In contrast, fluorine chemical shifts span a very wide range due to the intrinsic electronic situation of the fluorine nucleus (on average the ^{19}F nucleus is surrounded by nine electrons, whereas the hydrogen nucleus is only surrounded by a single electron). Thus, the problem of resonance degeneracy in ^{19}F NMR spectroscopy is almost non-existent (Figure 1.2). Chemical shift values are listed in Table 1.2 for fluorinated organic groups relative to various reference compounds.

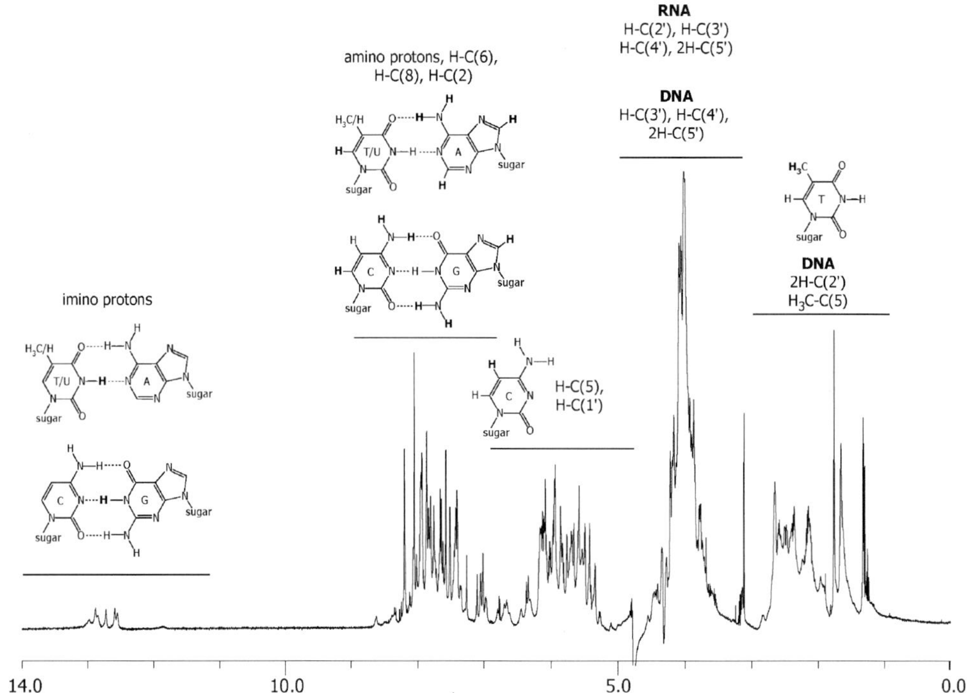

Figure 1.1 Typical one-dimensional proton NMR spectrum of a nucleic acid system (1.0 mM DNA/RNA hybrid, 50 mM sodium phosphate buffer, pH 7, 283 K, D_2O/H_2O ratio 1/9). Based on the local magnetic field induced by the chemical environment, resonances belonging to the different building units of the biopolymer (e.g., sugars, nucleobases) can be identified. However, with a chemical shift range of about 15 ppm, severe resonance overlap is encountered in the ^{1}H NMR spectrum.

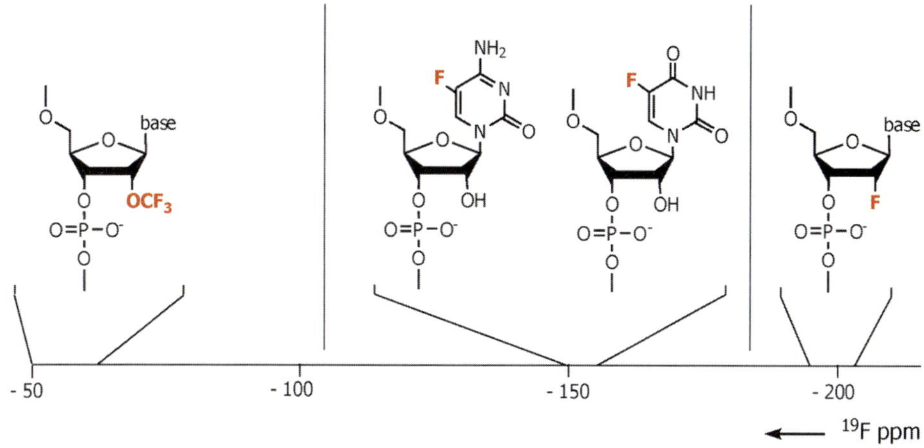

Figure 1.2 Typical chemical shift values that are found in fluorine-modified oligonucleotides (referenced to $CFCl_3$). With a chemical shift dispersion that is about 100-fold that of protons, resonance degeneracy is hardly encountered in ^{19}F NMR spectra.

1.1.1.3 Factors Affecting the ^{19}F Chemical Shift in Biomolecules

One of the most useful properties of the ^{19}F nucleus is the high sensitivity of the fluorine shielding parameter to changes in the local environment. This makes fluorine an ideal candidate for monitoring functional important transitions in biological systems via NMR spectroscopic methods. In the following, we will briefly describe two situations where changes in fluorine chemical shifts are expected: namely, the folding of a biomolecule; and the binding of a ligand at a receptor site.

Starting from a biomolecule containing a fluorinated reporter group in its native functional state, the fluorine spin label can be used to monitor transition to the denatured (unfolded) state. The local environment of the fluorine is supposed to be changed significantly by the unfolding process. In the unfolded state, the fluorine is exposed and thereby subjected to interactions that are predominantly due to solvent molecules. In contrast, during the folding process from an unfolded to the native state, secondary and tertiary structure elements are formed. Thus, the fluorine

Table 1.2 Chemical shift values of fluorinated functional groups.

Chemical group	Chemical shift relative to $CFCl_3$ [ppm]
$-C(F)H-$	−210
$-CF_2-$	−140
$R-C_6H_4-F$	−140
$-CF_2-C(O)-$	−125
$-CH-CF_3$	−75
$-C(O)-CF_3$	−81
$-SO_2F$	50

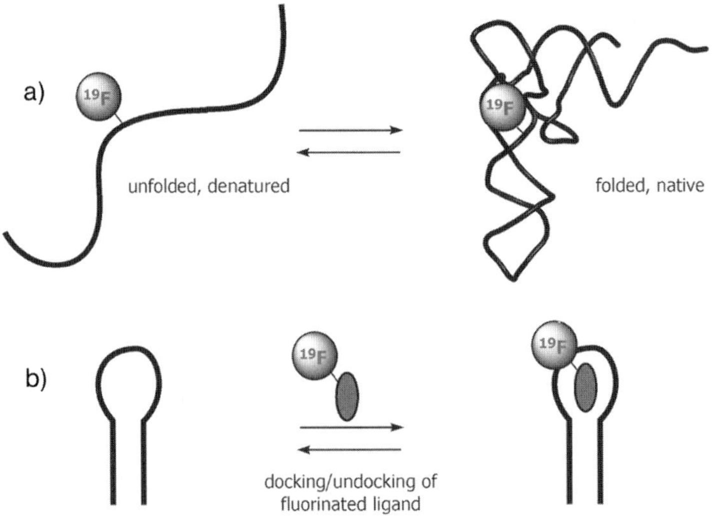

Figure 1.3 Functionally important transitions of biomolecules for ^{19}F NMR spectroscopic applications. (a) A ^{19}F spin label can be used to monitor the transition of a biomolecule from the unfolded to the folded, functional state. (b) Binding and release of a ligand at a receptor site can be detected by observing the ^{19}F resonance of the fluorinated ligand.

reporter nucleus becomes buried in the structural network made up by the biomolecule's architecture. Thereby, the chemical environment of the fluorine can be dramatically changed and interactions with solvent molecules are expected to be disrupted, resulting in a different chemical shift of the fluorine reporter (Figure 1.3a).

The second aspect of how ^{19}F NMR spectroscopy is involved in studies of biomolecules relates to interactions with a ligand. For a fluorinated small molecule that binds at a receptor site, the electric fields, short-range contacts and hydrogen-bonding possibilities experienced by the fluorine are different in the free versus receptor-bound states, and these will be reflected in a change of the chemical shift value (Figure 1.3b).

1.1.1.4 **Fluorine Relaxation in Biological Systems**

The fluorine spins relax mainly by two pathways. First, dipole–dipole interactions with the surrounding proton spins offer an effective relaxation pathway. Second, the anisotropy of the fluorine chemical shift leads to an enhanced relaxation rate.

Dipole–dipole interactions are very useful as they can lead to $^{19}F/^{19}F$ and $^{1}H/^{19}F$ nuclear Overhauser effects (NOEs) that provide interesting information about internuclear distances in the same way as do proton–proton NOEs. Fluorine relaxation can also provide a quantitative estimation of the degree of mobility in different parts of a large biomolecule. Furthermore, fluorine NMR experiments are also sensitive to the rates of processes which interchange the environments of the observed spins, and thus can produce quantitative data about the rates of processes, such as conformational change and ligand exchange. For a detailed discussion on fluorine relaxation, the reader is referred to the review by Gerig [1].

1.1.1.5 Solvent-Induced Isotope Shifts of ^{19}F NMR Resonances

Experiments have shown that a fluorine signal from a water-soluble molecule is deshielded by about 0.2 ppm when the solvent is changed from water (H_2O) to deuterium oxide (D_2O) [2, 3]. The magnitude of the effect thus reflects the extent to which a fluorine nucleus in a macromolecule is exposed to solvent. An established way of considering solvent exposure in fluorinated biomolecules is simply gradually to replace H_2O with D_2O, as an inverse correlation is found between the buriedness and the solvent-induced isotope shift (SIIS) effect [2]. The ^{19}F NMR resonances of the exposed regions of a biomacromolecule can be shifted by about 0.2 ppm by replacing H_2O with D_2O, whereas buried fluorinated residues do not experience any (or only a weak) SIIS. By replacing H_2O with D_2O, all exchangeable protons become replaced by deuterons; this leads to a slightly changed chemical environment, and may also bear influence on the ^{19}F H_2O/D_2O isotope shift. In other studies, $H_2{}^{18}O$ was used to induce an isotope shift of ^{19}F resonances of small fluorinated molecules and a 16 nt RNA containing a single 5-fluorouridine [4, 5].

1.1.2
^{19}F NMR Spectroscopy of Proteins

1.1.2.1 Incorporation of Fluorinated Amino Acids into Proteins

A wide variety of synthetic methods are available for the preparation of fluorinated derivatives of most of the common amino acids. The placement of these fluorinated materials into proteins has been accomplished by different strategies, including chemical synthesis and biosynthetic incorporation by organisms. Today, some fluorine-bearing aromatic amino acids, such as 4-fluorophenylalanine, 4-trifluoromethylphenylalanine, 6- and 5-fluorotryptophan, and 3-fluorotyrosine, are commercially available (Figure 1.4).

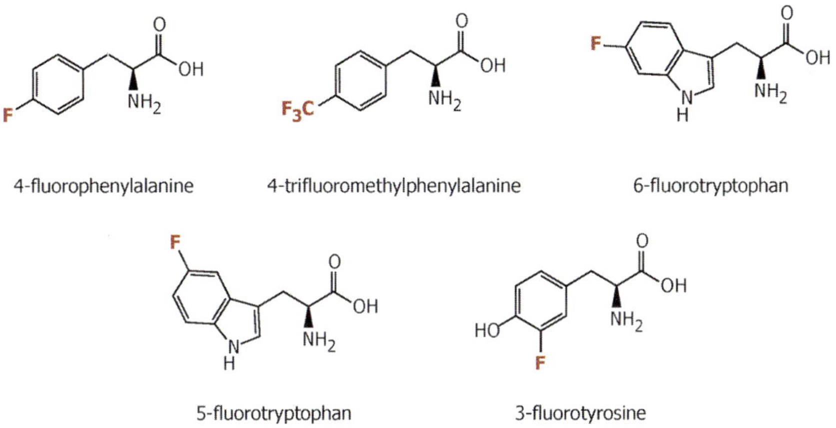

Figure 1.4 Fluorinated aromatic amino acids that can be incorporated into proteins by chemical or biosynthetic methods. In general, the fluorinated amino acids are well accepted in the native protein structure, where they serve as high-sensitivity spin labels for ^{19}F NMR spectroscopic applications.

By far the most widely used method for placing fluorinated amino acids into the primary sequence of proteins is biosynthesis of the protein by a living organism. Fluorinated bacterial proteins are usually obtained by using bacterial strains that are auxotrophic for tryptophan by adding fluorinated tryptophan to the growth medium. A variety of other approaches exist to incorporate aromatic fluorinated amino acids, including the use of glyphosphate or 3-β-indoleacrylic acid, both of which are inhibitors of aromatic amino acid synthesis [6, 7]. These biosynthetic approaches lead to the ubiquitous incorporation of fluorinated amino acids into the target protein, and the need to assign all observed ^{19}F resonances. If site-specific ^{19}F labeling of a protein is desired, then chemical synthesis of the peptide or protein is necessary. Alternative approaches to the site-specific incorporation of fluorinated aromatic amino acids by biosynthetic methods were also elaborated, for example by using an appropriately acylated suppressor tRNA that inserts the fluorinated amino acid in response to a stop codon substituted for the codon encoding the residue of interest [8]. In 1998, Furter *et al.* used such an approach to incorporate *p*-fluorophenylalanine in a site-directed manner into proteins [9]. The major advance here was the use of an *Escherichia coli* strain, which was equipped with a non-essential yeast aminoacyl-tRNA synthetase that charged its cognate yeast amber suppressor tRNA with the fluorinated amino acid analogue, which in turn was incorporated almost exclusively at a programmed stop codon. Although incorporation yields of up to 75% were obtained, the method also caused 3–7% of all phenylalanine residues to be replaced by the fluorinated counterpart.

Very recently, a further advance in site-specific labeling of proteins with ^{19}F modified amino acids was presented [10]. The site-specific introduction of trifluoromethyl-L-phenylalanine (tfm-Phe) was described by an orthogonal aminoacyl-tRNA synthetase/tRNA pair, capable of incorporating tfm-Phe into proteins. The synthetase/tRNA pair functions with high translational efficiency and fidelity for incorporating tfm-Phe using a nonsense codon in *E. coli.*

1.1.2.2 ^{19}F NMR Spectroscopic Studies of Proteins

Native and Denatured States of Green Fluorescent Protein Khan and coworkers successfully prepared uniformly 3-fluorotyrosine-labeled green fluorescent protein (GFP) [11], and subsequently assigned the observed ^{19}F resonances to all ten fluorotyrosines by successively replacing the fluorotyrosines by phenylalanines. Complete assignment was achieved with the additional aid of relaxation data and ^{19}F photochemically induced dynamic nuclear polarization (CIDNP). The ^{19}F resonances showed no overlaps, and were dispersed over a chemical shift range of 10 ppm. Most interestingly, two tyrosines (Tyr92 and Tyr143) exhibited a pair of signals that were interpreted by two ring-flip conformational states populated in the folded protein. The sensitivity enhancement by the photo-CIDNP mechanism revealed four ^{19}F tyrosine resonances which are solvent-exposed in the native state (Y39, Y151, Y182, and Y200).

Furthermore, the photo-CIDNP approach was used to characterize the denatured states of GFP. The photo-CIDNP spectra under unfolding conditions and using high

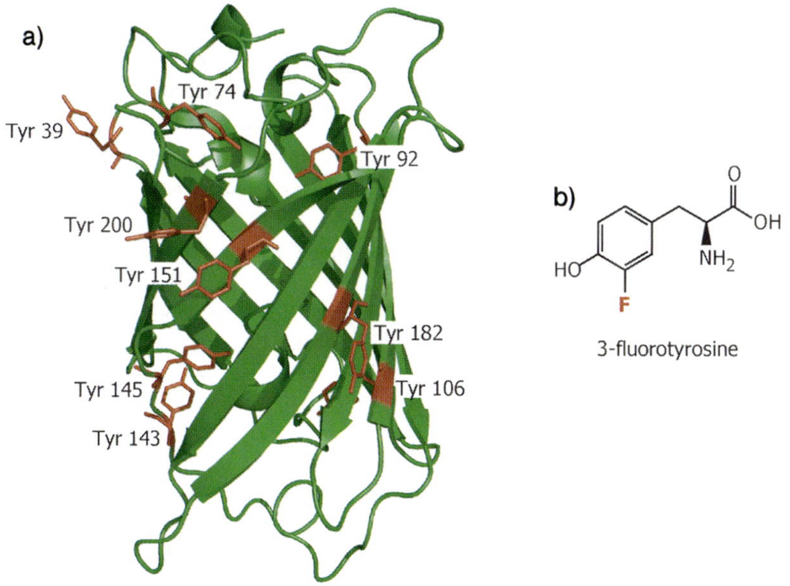

Figure 1.5 Green fluorescent protein (GFP) [11]. (a) The protein fold is depicted in ribbon presentation with the tyrosine residues used for ^{19}F labeling highlighted in red (PDB ID 1B9C). (b) The structural formula of 3-fluorotyrosine.

concentrations of chemical denaturants were qualitatively the same, showing positive enhancement, as all of the fluorotyrosine residues were solvent-accessible in the unfolded state. However, when a low pH value of 2.9 was applied instead, a positive and negative photo-CIDNP effect was found, indicating that folding intermediates with structured parts were still present, and meaning that a low pH is not sufficient to completely denature this protein (Figure 1.5).

Relation of Enzyme Activity to Local/Global Stability of Murine Adenosine Deaminase Uniformly 6-fluorotryptophan-labeled murine adenosine deaminase (mADA) was expressed in an *E. coli* strain, and by applying ^{19}F NMR spectroscopy, the activity of mADA was examined in the presence of denaturing agents [12]. By adding chemical denaturants, it was observed that mADA lost its enzymatic activity even before significant secondary and tertiary structure transitions had taken place. ^{19}F NMR spectroscopy revealed that the chemical shift change of the ^{19}F resonance of W161 close to the active site exhibited a correlation with the loss of enzymatic activity on the addition of urea. The urea-induced chemical shift change of another fluorinated tryptophan, namely W117, was correlated with the rate constant of the binding event of the transition state analogue inhibitor, deoxycoformycin. The two remaining fluorinated tryptophan residues, W264 and W272, showed hardly any significant chemical shift change upon the addition of small amounts of urea. This indicated that the parts of the mADA protein around the residues W264 and W272 were stable under slightly denaturing conditions. Taken together, the ^{19}F NMR spectroscopic

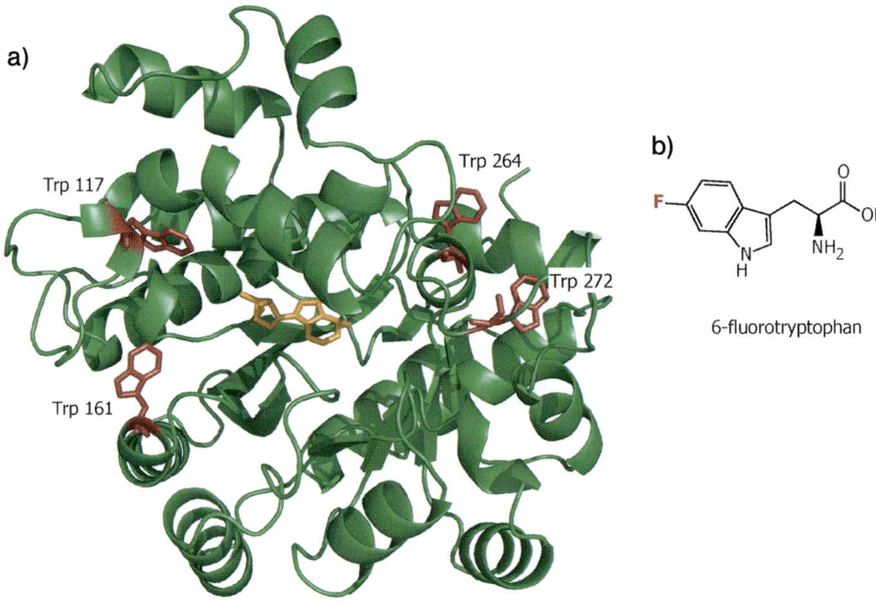

Figure 1.6 Murine adenosine deaminase (mADA) [12]. (a) The protein fold is depicted in ribbon presentation, with the tryptophan residues used for ^{19}F labeling highlighted in red (PDB ID 2ADA); the substrate analogue is highlighted in orange. (b) Structural formula of 6-fluorotryptophan.

results of the fluorinated mADA suggested that different regions of the protein exhibited different local stability upon the addition of urea, which in turn controlled the activity and stability of the protein (Figure 1.6).

Structural Studies of Bcl-xL/Ligand Complexes Fluorine atoms are frequently found in drugs and in drug-like molecules; for example 17% of the compounds listed in the MDDR (MDL drug data report) database contain a fluorine atom. Besides, the introduction of fluorine atoms often improves the pharmacokinetic properties of drug molecules. In a study conducted by Yu and coworkers, fluorinated ligands and the fluorinated protein Bcl-xL were used to derive structural restraints and information of the ligand–protein complex [13]. By using the method derived by Kim *et al.*, a uniformly ^{13}C and para-^{19}F-phenylalanine-labeled protein was expressed in *E. coli* by suppressing the biosynthetic pathway of the aromatic amino acids, phenylalanine, tyrosine and tryptophan, by the addition of the specific inhibitor glyphosphate [6]. Several multidimensional NMR spectra were recorded of the fluorinated Bcl-xL protein alone, and of Bcl-xL in complex with a series of fluorinated ligands (Figure 1.7a; compound **1**: K_D ca. 200 μM; compound **2**: K_D ca. 20 μM). NOEs between ligand **1** and several amino acids of Bcl-xL, A104, L108 and L130, could also be identified. Furthermore, a $^{19}F/^{19}F$ NOESY of the complex of fluorinated ligand **2** and the fluorinated Bcl-xL protein resulted in observable NOE contacts between Phe97 and the fluorine atoms of the ligand (Figure 1.7b) [13]. In conclusion, the

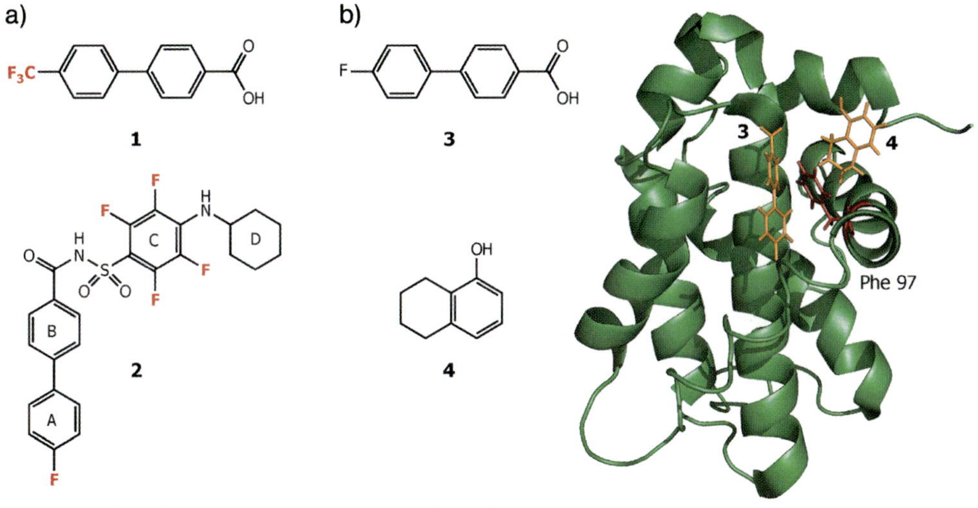

Figure 1.7 The Bcl-xL protein [13, 14]. (a) Typical high-affinity ligands, **1** and **2**. (b) NMR structure of the Bcl-xL protein in complex with 4′-fluoro-1,1′-biphenyl-4-carboxylic acid **3** (orange) and 5,6,7,8-tetrahydronaphthalen-1-ol **4** (orange). The phenylalanine residue 97 at the binding site is highlighted in red (PDB ID 1YSG) [14].

structural data obtained from ^{19}F NMR experiments were in good accordance with the known high-resolution structure of Bcl-xL and highly similar ligands, indicating that fluorine labeling can yield structural information additional to data obtained from traditional ^{13}C- and ^{15}N-labeling. Also in this study, fluorine was proven to be a non-invasive NMR probe nucleus, which did not alter the structural properties of proteins in the first order.

High-Throughput Screening Dalvit and coworkers presented several studies on ^{19}F NMR-based screening techniques for drug discovery [15–22]. NMR-based screening can be used to analyze ligand binding as well as to run functional assays. Thus, by establishing a structure–activity relationship (SAR), the dissociation constants and inhibitory activity of potential binders are accessible. Two ^{19}F NMR-based methodologies were introduced, namely "Fluorine chemical shift Anisotropy and eXchange for Screening" (FAXS) [18, 21] and "Three Fluorine Atoms for Biochemical Screening" (3-FABS) [16, 17].

FAXS is a ligand-based binding competition screening experiment which utilizes a weak affinity spy molecule bearing a CF or CF_3 group. The approach can be further extended by the use of a fluorinated control molecule, which exhibits no affinity against the protein target. These two molecules are selected from existing libraries, which is often an easy task as commercially available libraries contain fluorinated compounds at a rather high frequency (for example, 17% of the compounds listed in the MDL drug data report database). As fluorine atoms increase the lipophilicity of a compound, care must be taken that the "spy" and control substances are still highly soluble in aqueous solution.

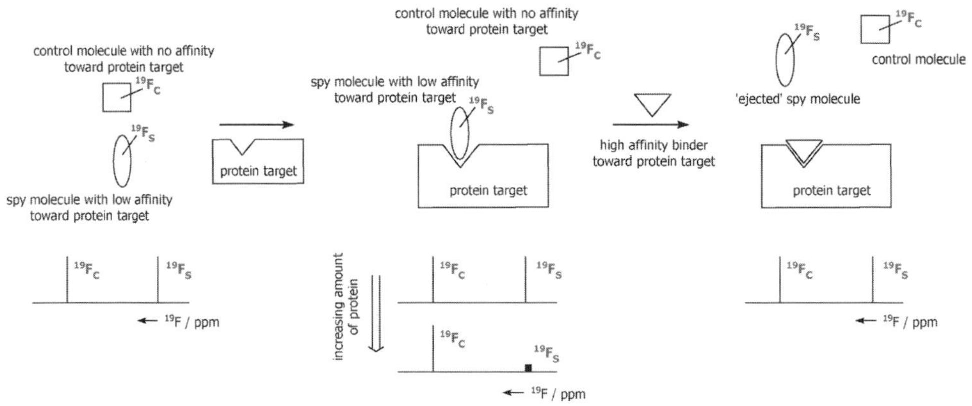

Figure 1.8 Schematics of the "Fluorine chemical shift Anisotropy and eXchange for Screening" (FAXS) approach for high-throughput screening [18, 21].

The spy molecule is replaced by a competing molecule during the screening process of chemical mixtures against the protein target. Thereby, the NMR spectroscopic parameters of the ^{19}F nucleus of the spy molecule are significantly changed (e.g., chemical shift, relaxation parameters). The screening is carried out by monitoring the intensities of the signals of the control and the spy molecule (Figure 1.8). If the K_D value of the spy molecule is known, the binding constants of the competing molecule towards the protein become accessible. It is the large chemical shift anisotropy (CSA) of fluorine that makes the difference in line-width for the spy molecule in the free versus bound state very large, in particular, at high magnetic fields. The FAXS approach, when performed with a weak-affinity fluorinated ligand (spy molecule) and a ^{19}F-labeled control molecule with no affinity toward the target, has proven to be very powerful and sensitive for the primary screening of ligands to a protein target of interest. Furthermore, current technological advances such as ^{19}F cryoprobes strengthen this NMR-based screening approach [15].

The 3-FABS approach (Figure 1.9) is a functional NMR-based assay which is used to study enzymatic reactions and to obtain IC_{50} values (the concentration of inhibitor at which 50% inhibition of the enzymatic reaction is reached) [16, 17]. The functional assay uses CF_3-tags on the enzyme substrates and ^{19}F NMR spectroscopy. Due to modification of the tagged substrate mediated by the enzyme, the chemical environment of the three fluorine atoms is changed, which in turn leads to a chemical shift change. The enzymatic reaction is quenched after a defined delay by the addition of a denaturant, which may be a chelating agent or a strong inhibitor. For screening purposes a reference sample without any test molecules is run, representing 0% inhibition (Figure 1.9a). Even multiple enzymes can be screened by the approach (Figure 1.9b), a point which is of special interest if the selectivity of an inhibitor for a target enzyme is tested in the presence of another enzyme of the same family. Recently, this approach has benefited from current technical advances, such as the introduction of ^{19}F cryoprobes [15].

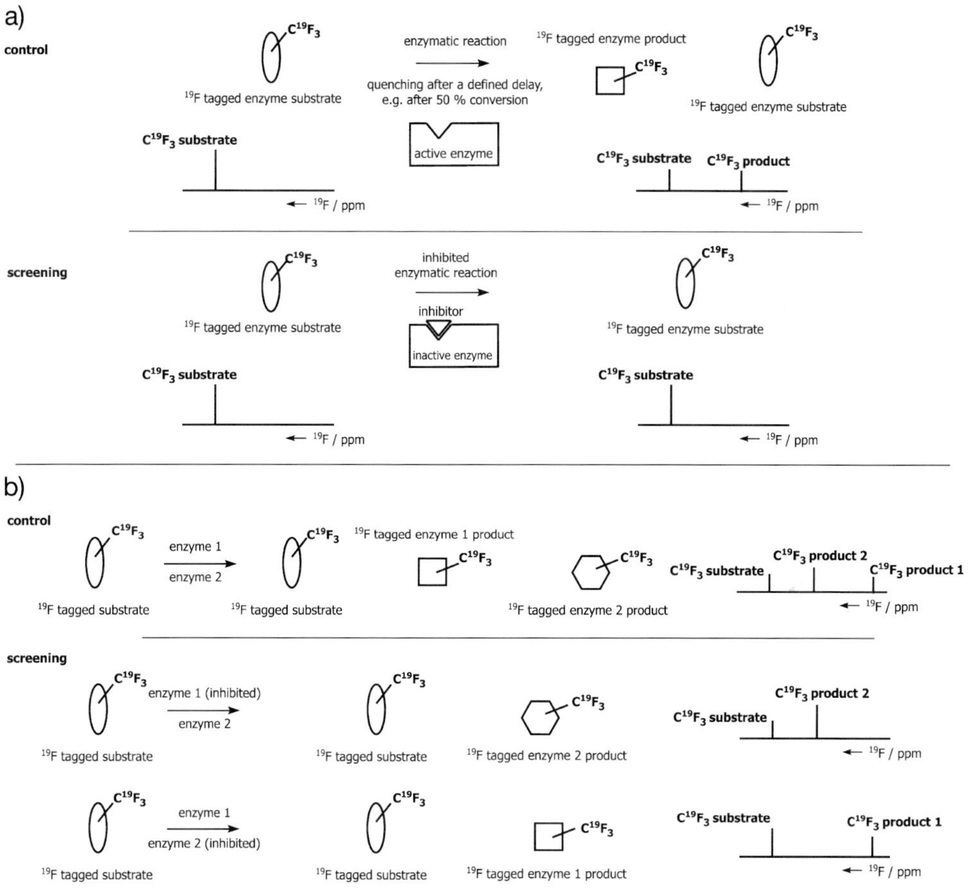

Figure 1.9 Schematic of the "3-*F*luorine *A*toms for *B*iochemical *S*creening" (3-FABS) approach for screening enzyme inhibitors. (a) The principle of 3-FABS screening. (b) Multiple enzyme reactions can be screened using the 3-FABS approach [16, 17].

1.2
^{19}F NMR Spectroscopy of Nucleic Acids

As described above, ^{19}F NMR spectroscopy not only provides important contributions towards the elucidation of protein structures and dynamics, but also contributes to the field of nucleic acids, where the number of interesting ^{19}F NMR spectroscopic approaches is currently increasing. For nucleic acids, multiple options exists to introduce a fluorine atom, with possible labeling sites being the ribose moieties and the nucleobases. In this respect, the fluorinated nucleoside analogues can be incorporated either by using biochemical methods, or by chemical solid-phase synthesis.

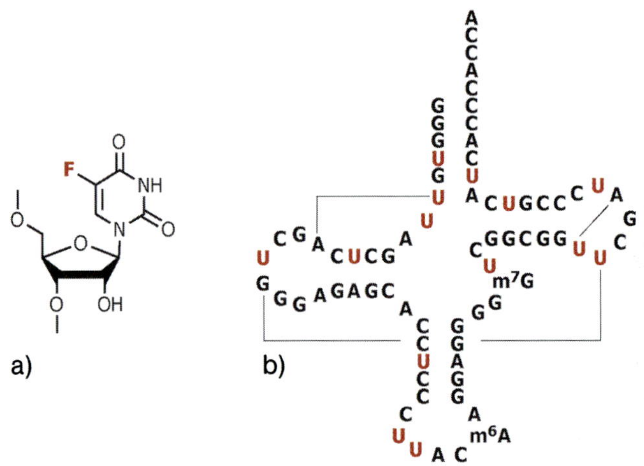

Figure 1.10 *E. coli* $tRNA^{Val}$. (a) Structural formula of a 5-fluorouridine unit. (b) Secondary structure with 5-fluoro-uridines shown in red. The *in-vitro* transcript lacks the modified nucleobases 7-methyl guanosine (m^7G) and N^6-methyl adenosine (m^6A). The solid lines indicate tertiary structure interactions of the modified uridines [32].

1.2.1 Nucleic Acids with Fluorinated Nucleobases

1.2.1.1 Transfer RNAs

Pioneering studies by Horowitz and coworkers on fluorinated tRNAs were commenced about 30 years ago [23–35]. In these experiments, tRNAs were *in-vitro* transcribed in the presence of 5-fluorouridine triphosphates, such that all uridine positions were replaced by the fluorinated analogue (Figure 1.10). Subsequent NMR spectroscopic characterization led to the assignment of all fluorine resonances. The fluorinated tRNAs were used in multiple studies, including the investigation of the solvent accessibility of a free tRNA and a tRNA in complex with its synthetase [29]. Furthermore, the tRNA constructs were used to assess the interactions with small molecules, such as ethidium bromide or psoralen [23, 25].

1.2.1.2 *HhaI* Methyltransferase in Complex with DNA Duplexes

A fluorinated DNA duplex was used to investigate the nucleotide flipping mechanism during nucleobase methylation mediated by the *HhaI* methyltransferases [36]. The flipping motion was studied by a 5-fluorocytidine placed at the methylation site; a second 5-fluorocytidine residing three nucleotides upstream served as an internal reference (Figure 1.11). The NMR spectroscopic and gel mobility data suggested that, in the binary (DNA/enzyme) and ternary complex (DNA/enzyme/cofactor), the flipping nucleotide (5-fluorocytidine) cycles through three states; namely, the cytidine remains stacked in the double helix, and populates an ensemble of extrahelical conformations and a locked external form in the enzyme active site. The addition of

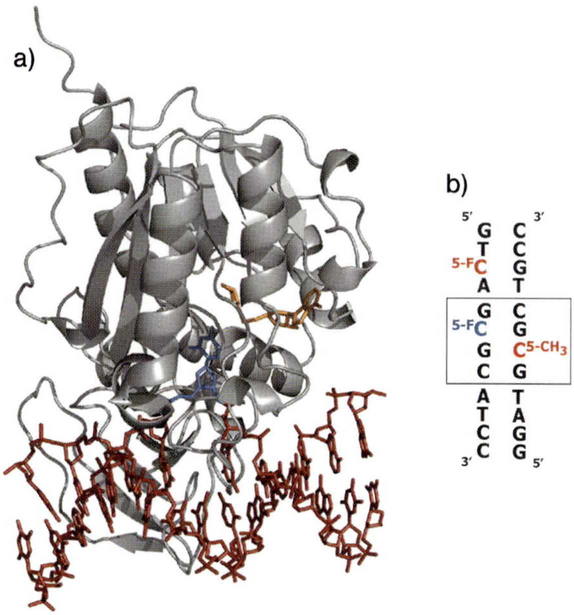

Figure 1.11 *HhaI* methyltransferase. (a) Presentation of the protein in complex with an unmodified DNA duplex (red) and the cofactor analogue AdoHyc (orange). The flipped cytidine residue is highlighted in blue (PDB ID 3MHT). (b) DNA duplex used in the ^{19}F NMR spectroscopic study [36]. The 5-fluorocytidine ($^{5\text{-F}}$C) in red served as internal reference, whereas the $^{5\text{-F}}$C residue in blue represented the target nucleotide for *HhaI* methyltransferase. On the opposite strand, the methyltransferase target nucleotide was replaced by an already methylated cytidine (red).

the cofactor analogue S-adenosyl-L-homocysteine (AdoHyc) led to an enhanced trapping of the target cytidine in the catalytic site of the enzyme. The study showed that the flipping mechanism was not exclusively associated with binding of the cytidine in the active site of the enzyme. Rather, the authors suggested an active role of the methyltransferase in the flipping process, which possibly occurs via the major groove.

1.2.1.3 **Minimal Hammerhead Ribozyme**

The fluorine labeling of nucleic acids has also proven to be very useful in folding studies, as exemplified by monitoring the metal ion-induced folding of the minimal hammerhead ribozyme [37, 38]. Two 2′-*O*-methyl-5-fluorouridines (5-F-U) were incorporated into the hammerhead ribozyme; one (5-F-U4) was located in domain 1, the other (5-F-U7) resided in domain 2, near the interface between the two domains (Figure 1.12) [39].

The fluorine labels responded to folding processes of the hammerhead ribozyme in a very sensitive manner (Figure 1.12). The 5-F-U7 label sensed two events, one of which occurred at a low Mg^{2+} concentration ($\sim$0.5 mM). This transition was attributed to the formation of domain 2, which builds the core structure of the

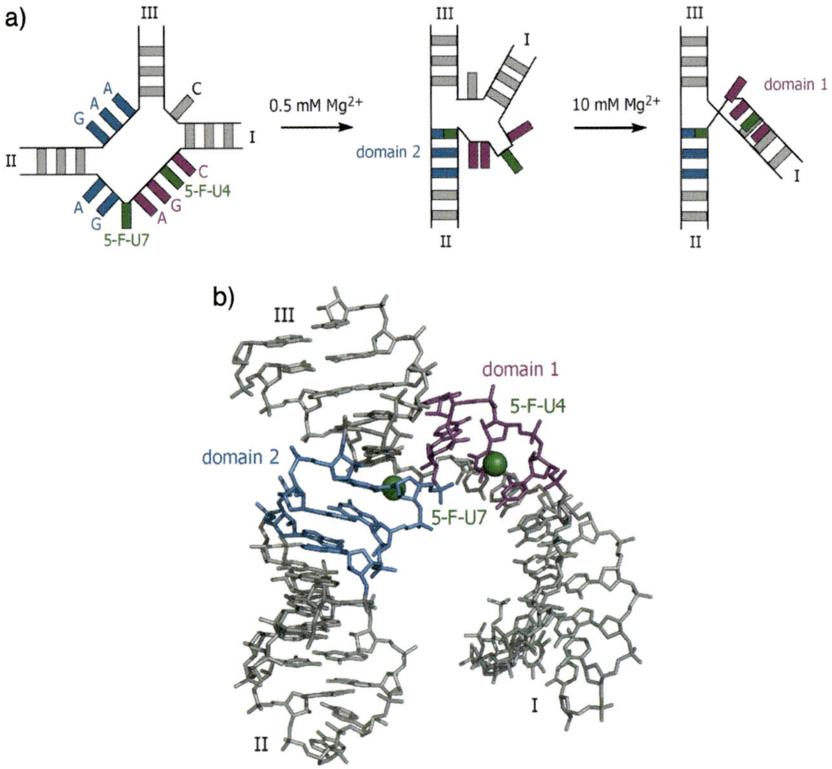

Figure 1.12 The folding process of the hammerhead ribozyme as deduced from ^{19}F NMR spectroscopic data by Lilley and coworkers [39]. (a) The two folding events sensed by the fluorine labels are shown schematically. At low Mg^{2+} concentration (0.5 mM), domain 2 is formed, which is indicated by a chemical shift change and a line width increase of the ^{19}F resonance of 5-F-U7. Above 1 mM Mg^{2+} concentration, domain 1 is formed which is sensed by both fluorine labels. (b) At the time of during the study, the crystal structure of the minimal hammerhead ribozyme (PDB ID 1HMH) [37] was used for structural considerations; the C5 atoms of uridine 4 and 7 are highlighted in green. (Figure adapted from Ref. [39]).

active ribozyme, but it is still not in its functional state (also manifested in the lack of cleaving activity at that magnesium ion concentration). The behavior of the 5-F-U7 label between 0 and 1 mM Mg^{2+} indicated that the fluorine nucleus sensed a reversible exchange process. The broadening of the resonance between 0 and 0.5 mM Mg^{2+} concentration and the subsequent narrowing between 0.5 mM and 1 mM Mg^{2+} probably arose from a fast exchange process between two folding states. By increasing the Mg^{2+} concentration, the exchange rate can be quickened, leading to a narrowing of the ^{19}F resonance at 1 mM Mg^{2+}. By analyzing the line width of the ^{19}F resonance, an approximation of the rate constant k of the exchange process was obtained, resulting in a k of 1000 s^{-1}. At millimolar Mg^{2+} concentrations, both fluorine labels 5-F-U7 and 5-F-U4, sensed the binding of Mg^{2+} which induced a structural rearrangement of the ribozyme towards the functional state. This second transition was attributed to the formation of domain 1, and followed a two-state model as the ^{19}F chemical shift changes and line width changes of both labels could be fitted to a two-site exchange process. The proposed folding model of the hammerhead ribozyme was later also assessed with fluorescence resonance energy transfer (FRET) studies [40].

1.2.1.4 HIV TAR RNA

^{19}F labeling was also used to study small-molecule binding and to identify the metal ion binding sites of the HIV-1 TAR RNA [41]. Multiple 5-F-U labels were introduced into the target RNA via the phosphoramidite solid-phase synthesis approach (Figure 1.13). In order to test the sensitivity and specificity of the fluorine labels on ligand binding, ^{19}F NMR titration experiments were conducted by adding increasing amounts of argininamide, which is known to bind at the UCU bulge region. The site-specific binding event was reflected via the fluorine labels, as only the resonance of the 5-F-U23 residue was strongly affected during the titration experiment. All other labels responded either moderately (5-F-U25 and 5-F-U38) or only very weakly (5-F-U31 and 5-F-U40) on argininamide binding. This can be easily rationalized by the NMR solution structures of HIV-2 TAR RNA in complex with argininamide [42], and of the ligand-free form of HIV-1 TAR RNA [43]. Whilst U23 is found in a bulged-out conformation for the free RNA, it participates in a base triple formed by U23-A27-U38, in the bound form. This leads to a drastic change in the chemical environment of the ^{19}F resonance of 5-F-U23, and in turn to the observed behavior. By fitting the chemical shift change data of the 5-F-U23 resonance, a dissociation constant for the argininamide ligand could be estimated (K_D ca. 300 μM).

The fluorinated TAR RNA was then probed for metal ion binding sites by titration experiments with Mg^{2+}, Ca^{2+}, and $Co(NH_3)_6{}^{3+}$. The ^{19}F NMR spectroscopic data showed that metal ions bind preferably at the bulge region and not in the 6-nucleotide loop region. Mg^{2+} and Ca^{2+} ions exhibited a similar affinity towards the bulge region with K_D values in the millimolar range. Co^{3+} ions also showed a site-specific affinity towards the bulge region. Unfortunately, concentrations in excess of five equivalents of $Co(NH_3)_6{}^{3+}$ led to irreversible aggregation of the RNA, and making determination of the K_D value impossible.

Hennig and coworkers recently reported on an enzymatic method to incorporate fluorinated nucleosides into a target RNA [44, 45]. These authors used 2-fluoroadenosine, 5-fluorocytidine, 5-fluorouridine triphosphates and T7 RNA polymerase for *in-vitro* transcription to obtain uniformly labeled HIV-2 TAR RNAs (Figure 1.14). It was shown that the ^{19}F resonances and imino proton resonances of 2-F-adenosine labeled RNA were easily assignable with a series of homonuclear and heteronuclear NOE experiments (Figure 1.14b) [44]. In a ^{1}H-^{19}F HOESY experiment, intense NOE correlation between the 2-F atoms and the imino protons of two (out of four) 2-F-adenosine residues were found (2-F-A20 and 2-F-A27). Based on a sequential heteronuclear NOE to the anomeric H1′ proton of the 3′ adjacent uridine 23, the ^{19}F resonance of the 2-F-A22 residue was assigned. The remaining ^{19}F resonance of

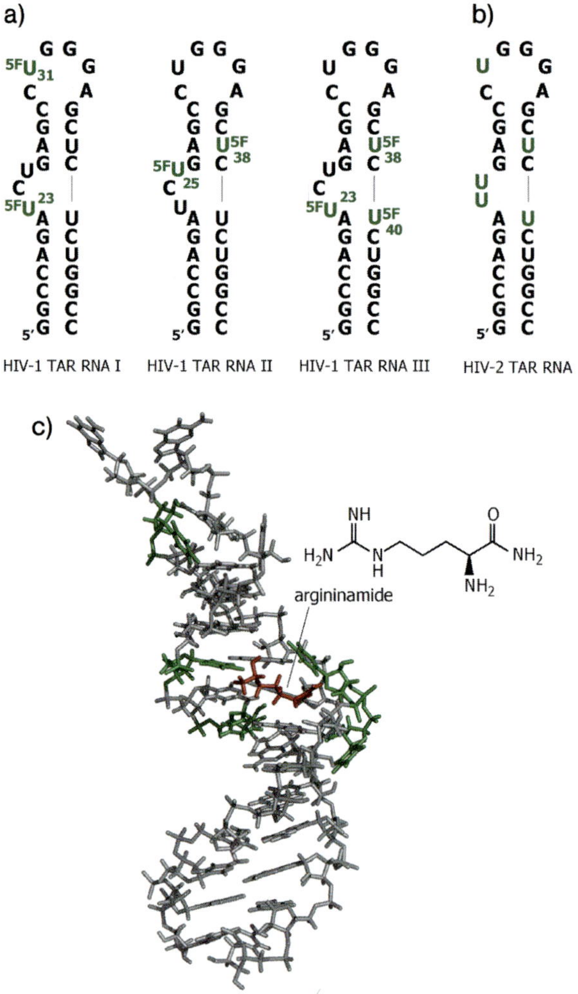

Figure 1.13 HIV TAR RNA. (a) Fluorinated derivatives of HIV-1 TAR RNA (I, II, III) used in the binding study of metals ions (Mg^{2+}, Ca^{2+},$Co(NH_3)_6{}^{3+}$) and argininamide [41]. (b) Structural formula of argininamide and secondary structure of the HIV-2 TAR RNA with a cytidine deletion in the bulge [42]. (c) NMR solution structure of the HIV-2 TAR RNA/argininamide complex (PDB ID 1AJU) [42]; the uridine residues which were replaced by fluorinated analogues are highlighted in green, the argininamide ligand is highlighted in red.

loop adenosine 35 was assigned by exclusion, as it is the only fluorinated residue in an unstructured RNA domain which exhibits a higher flexibility and no observable heteronuclear NOE correlation.

The study results illustrated that 2-F-adenosine represents a non-invasive NMR labeling nucleus, as the structural integrity of HIV-2 TAR RNA was not impaired and ^{1}H-^{19}F heteronuclear NOEs resembled that of the homonuclear proton NOEs in the unmodified RNA.

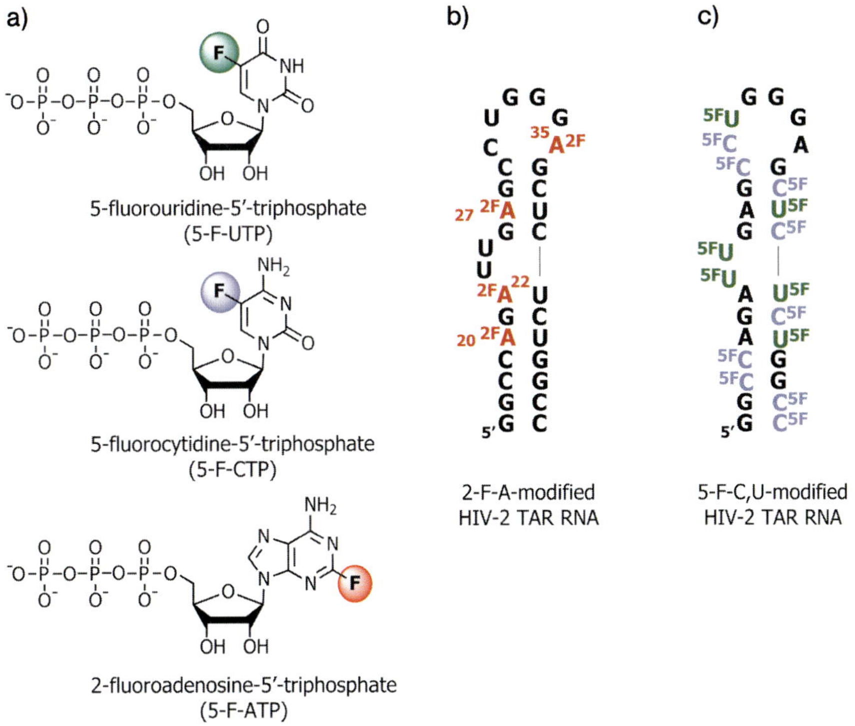

Figure 1.14 HIV TAR RNA. (a) Nucleoside triphosphates for the incorporation of 5-F-uridine (^{5F}U), 5-F-cytidine (^{5F}C), and 2-F-adenosine (^{2F}A) into the RNA targets by *in-vitro* transcription. (b) 2-F-adenosine-labeled HIV-2 TAR RNA [44]. (c) 5-F-pyrimidine-labeled HIV-2 TAR RNA [45].

Furthermore, Hennig and coworkers showed that from a fully 5-fluoropyrimidine-modified RNA, structural information about the nucleobase orientation can be obtained by the analysis of intraresidual ^5J(H1′,5F) coupling (Figure 1.14c) [45]. These couplings can also be used to facilitate NMR resonance assignment. The phenomenon of the rather unusual ^5J long-range coupling was qualitatively explained by a W-like conformational arrangement of the H1′-C1′-N1-C6-C5-F5 bond network. By applying a density functional theory (DFT) approach, an adequate description of the torsion angle (χ) dependence on the ^5J(H1′,5F)-constant was found and described by a generalized Karplus relationship.

1.2.2
Nucleic Acids with Fluorinated Ribose Units

1.2.2.1 *R1inv* RNA

Fewer examples are available of ^{19}F NMR studies on nucleic acids which utilize fluorine-modified riboses. In particular, 2′-deoxy-2′-fluoro (2′-F) nucleosides are

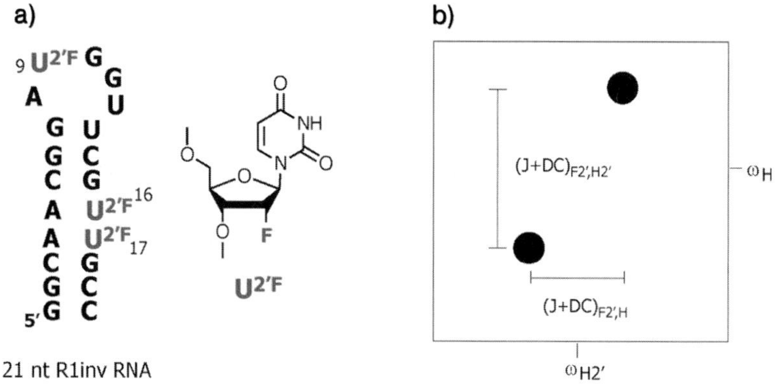

Figure 1.15 *R1inv* RNA. (a) Secondary structure of the 21-nt RNA target hairpin and structural formula of a 2′-F uridine unit [46]. (b) Schematic pattern from a X-filtered E.COSY experiment used for the determination of ^{19}F-^{1}H coupling. The large geminal F2′-H2′ coupling is evolved in the indirect dimension, whereas the long-range F2′-H1′/H3′/H6 couplings are detected in the direct dimension (frequencies ω, dipolar couplings DC).

interesting candidates for non-invasive spin labels, as this modification favors the C3′-endo ribose pucker, which is also the preferred conformation in double-helical A-form RNA. Luy and Marino introduced 2′-F-modified uridine residues at positions 9, 16, and 17 of the *R1inv* RNA hairpin (Figure 1.15a) [46]. The RNA was partially aligned by using filamentous bacteriophage Pf1 [47] and, by applying X-filtered-E.COSY-type methods, long-range dipolar coupling constants between $^{19}F/^{1}H$ spin pairs were obtained. In the experiment, the X-filter was tuned on a scalar coupling constant of approximately 50 Hz, selecting the $^{2}J(H2',F2')$ coupling constant. After the X-filter and evolution in t1 of the H2′ protons' chemical shifts and scalar couplings, the H2′ magnetization was correlated via a homonuclear mixing step to other protons. Finally, the long-range correlated protons (H3′, H1′ and H6 base protons) were detected without ^{19}F decoupling (Figure 1.15b).

The $^{19}F/^{1}H$-dipolar couplings were used in structure determination and refinement of the hairpin RNA. The obtained dipolar couplings of the fluorinated uridine residues placed in the double helix (positions 16 and 17) were in perfect agreement with a modeled UUG trinucleotide in A-form geometry. The authors further suggested that selective fluorine labeling of RNAs and the subsequent residual dipolar coupling (RDC) analysis might be a valuable tool in determining the interhelical orientations of large RNAs or RNA–protein complexes. However, care must be taken, if 2′-fluorinated nucleotide analogues are placed in non-canonical regions, where deviations of the standard A-form geometry are possible. In this study, the 2′-fluororuridine at position 9 was found to have a shifted C2′/C3′-endo population (towards C3′-endo), as compared to the C2′-/C3′-endo equilibrium position of uridine 9 found in the unmodified hairpin. To summarize, the selective labeling of double-helical RNA regions with fluorinated nucleotide analogues should provide a general method to obtain ^{1}H-^{19}F RDC restraints, which can be used to determine the relative orientation of helices in large RNAs, and in the structure refinement process of RNA structures.

1.2.2.2 RNA Secondary Structure Equilibria

A further concept relying on site-specific 2′-deoxy-2′-fluoro (2′-F) nucleosides in RNA, has been presented recently, the aim being to develop a simple tool to distinguish alternative RNA secondary structures of the same sequence [48]. This was achieved when a distinct 2′-F labeled nucleoside resided within a double helix of one fold, whereas it was part of a single-stranded region within the alternative fold [49]. Because the chemical environment of the 2′-F atom was significantly different for the two conformations, different chemical shifts were observed for the ^{19}F resonances (slow exchange regime). In this sense, a key feature of the approach relies on strategically "correct" positioning of single 2′-F labels within the RNA sequence of interest (Figure 1.16). Such a strategy also implies that the replacement of a 2′-OH group by a 2′-F atom only slightly alters the overall RNA structure and, in particular, the thermodynamic stability of an RNA double helix. In general, purine 2′-F nucleoside-labeled RNAs reflected the same equilibrium position when compared to their non-modified counterparts, while for pyrimidine 2′-F nucleoside-labeled RNAs the equilibrium position was shifted by at most 25% towards the fold containing the label within the double helix [49].

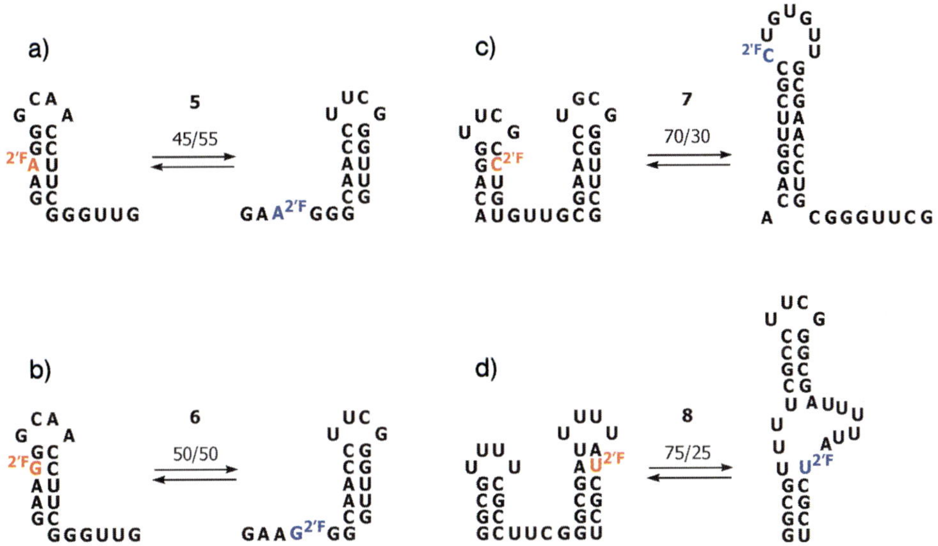

Figure 1.16 Bistable RNAs **5–8** containing site-specific 2′-F nucleoside labels. The equilibrium position between two respective secondary structures can be easily obtained based on the ratio of the ^{19}F resonances representing the individual folds [49].

1.2.2.3 RNA Ligand Binding

A novel concept for the verification of RNA binders via ^{19}F NMR spectroscopy has been introduced recently [50]. Since most of the RNA ligands known follow the concept of adaptive recognition [51], their binding alters the local RNA conformation to a certain extent. The RNA was therefore labeled with 2′-F nucleosides at selected positions. Because the chemical environment of the 2′-F atom was different for the free versus complexed RNA, different chemical shift values for the corresponding resonances were observed. A key feature of the approach relies on the strategic positioning of single site-specific 2′-F labels within the RNA target. This approach is particularly powerful because of simultaneous accessibility of an internal reference that can be represented by a fluorine label placed, for example, within a region where no binding occurs, and consequently, no resonance shift is anticipated.

The realization of the concept was demonstrated for the structurally well-characterized tobramycin–RNA aptamer (Figure 1.17) and additionally, for the flavin mononucleotide (FMN)–RNA aptamer. Both aptamers bind their ligands in a loop and bulge, respectively, with nanomolar to low micromolar dissociation constants, and thus two distinct signals were observed for the free RNA versus the complexed RNA (slow exchange mode). Moreover, a pronounced decrease of line width indicated a more rigid structure of the RNA in the bound state.

For weak to moderate binders, such as the interaction between streptomycin and the tobramycin–RNA aptamer, slight broadening and a defined shift of the loop 2′-F resonance reflected a well-behaved fast exchange ligand–RNA interaction, and allowed a straightforward determination of the dissociation constant K_D [50].

1.2.3
Influence of Fluorine Modifications on Nucleic Acid Structure

As fluorine modification is a non-native modification of a ribonucleic acid, the question remains as to whether the introduction of fluorine atoms results in structural perturbation of the target RNA, or not. Two studies should be mentioned at this point. In a detailed study on 5-fluorouridine substitutions, Gmeiner and coworkers found only marginal changes of the overall RNA structure and stability, indicating that single 5-fluorouridine replacements represent a structural non-invasive spin label option [53], as has been also implied by most other studies involving fluorine-labeled nucleobases, as described above.

In case of 2′-ribose labeling, the preference of ribose C3′-endo over C2′-endo conformation (2.2.1) was mentioned above [46, 54]. The shifted C2′-/C3′-endo equilibrium position of 2′-modified nucleosides may take influence on the RNA structure, when positioned in non-canonical regions, such as bulges and loops [46]. A defined influence on the structure of RNA was also found for the oligoribonucleotide r($C^FGC^F(U^FU^FC^FG)GC^FG$), wherein all pyrimidine nucleotides were replaced by their 2′-F-modified counterparts via the phosphoramidite solid-phase synthesis

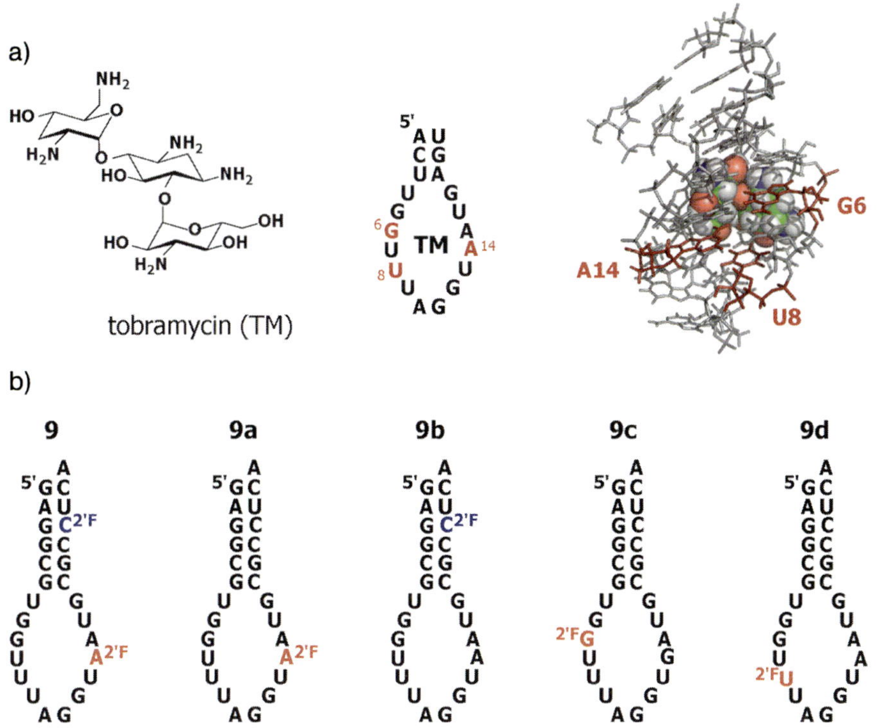

Figure 1.17 Tobramycin binding RNA aptamer. (a) Structural formula of tobramycin (TM), schematic representation of the tobramycin –RNA complex, and 3-D structure of the complex determined by NMR [52]. The nucleotides highlighted in red (G6, U8, and A14) were chosen as target sites for 2′-F modifications. (b) The approach to detect site-specific binders via ^{19}F-NMR was realized by fluorinated RNA construct **9** [50]. The 2′-F cytidine label (blue) in the double helix represented the internal reference, whereas the 2′-F adenosine label representing the "sensor" label (red) was placed in the binding region of the aptamer. The ^{19}F resonance of the sensor responded to the binding event of TM by a pronounced chemical shift and a decrease in linewidth, while the ^{19}F resonance of the reference remained unaffected. Assignment of the fluorine resonances was achieved by resonance comparison with the mono-fluorinated RNAs **9a** and **9b**. (c) RNAs **9c** and **9d** contained fluorine labels at alternative positions in binding region and further confirmed the concept of the novel binding assay.

approach [55]. A comparison of the structure of the fluorinated versus the unmodified RNA revealed that the fluorine-modified oligonucleotide existed in a 7 : 3 equilibrium between hairpin and duplex, whereas the unmodified RNA exclusively adopted the hairpin conformation. The interconversion rate $k_{ex}^{duplex\text{-}hairpin}$ was extracted from a ROESY experiment ($k_{ex}^{duplex\text{-}hairpin} = 0.28\ s^{-1}$). In the hairpin conformation, the loop 2′-F-uridine resonances seemed to underlie a chemical exchange process in the submillisecond time scale, probably due to a slow sugar pseudorotation between C2′- and C3′-endo sugar pucker.

1.3 Conclusions

In this chapter, we have summarized recent efforts in ^{19}F NMR spectroscopy of complex biological systems. The number of such studies dealing with nucleic acids is steadily growing and demonstrates – in particular for RNA – the power of site-specific fluoro-labeling in combination with NMR spectroscopic methods for the investigation of biophysical properties.

References

1 Gerig, J. T., Fluorine, N. M. R., *Chapter in On-line Textbook*, Biophysical Society, **2001**, (http://www.biophysics.org/img/jtg2001-2.pdf).

2 Hull, W. E., Sykes, B. D., Fluorine-19 nuclear magnetic resonance study of fluorotyrosine alkaline phosphatase: the influence of zinc on protein structure and a conformational change induced by phosphate binding. *Biochemistry* **1976**, *15*, 1535–1546.

3 Gerig, J. T., Fluorine NMR of proteins. *Prog. Nuclear Magn. Reson. Spectrosc.* **1994**, *26*(Part 4), 293–370.

4 Arnold, J. R. P., Fisher, J., The $H_2{}^{18}O$ solvent-induced isotope shift in ^{19}F NMR. *J. Magn. Reson.* **2000**, *142*, 1–10.

5 Arnold, J. R. P., Fisher, J., Observation of ^{18}O solvent-induced isotope shifts in ^{19}F NMR signals. *Chem. Commun.* **1998**, 1859–1860.

6 Kim, H.-W., Perez, J. A., Ferguson, S. J., Campbell, I. D., The specific incorporation of labelled aromatic amino acids into proteins through growth of bacteria in the presence of glyphosate: Application to fluorotryptophan labelling to the H^+-ATPase of *Escherichia coli* and NMR studies. *FEBS Lett.* **1990**, *272*, 34–36.

7 Leone, M., Rodriguez-Mias, R. A., Pellecchia, M., Selective incorporation of ^{19}F-labeled Trp side chains for NMR-spectroscopy-based ligand-protein interaction studies. *Chembiochem* **2003**, *4*, 649–650.

8 Noren, C. J., Anthony-Cahill, S. J., Griffith, M. C., Schultz, P. G., A general method for site-specific incorporation of unnatural amino acids into proteins. *Science* **1989**, *244*, 182–188.

9 Furter, R., Expansion of the genetic code: Site-directed *p*-fluoro-phenylalanine incorporation in *Escherichia coli*. *Protein Sci.* **1998**, *7*, 419–426.

10 Jackson, J. C., Hammill, J. T., Mehl, R. A., Site-specific incorporation of a ^{19}F-amino acid into proteins as an NMR probe for characterizing protein structure and reactivity. *J. Am. Chem. Soc.* **2007**, *129*, 1160–1166.

11 Khan, F., Kuprov, I., Craggs, T. D., Hore, P. J., Jackson, S. E., ^{19}F NMR studies of the native and denatured states of green fluorescent protein. *J. Am. Chem. Soc.* **2006**, *128*, 10729–10737.

12 Shu, Q., Frieden, C., Relation of enzyme activity to local/global stability of murine adenosine deaminase: ^{19}F NMR Studies. *J. Mol. Biol.* **2005**, *345*, 599–610.

13 Yu, L., Hajduk, P. J., Mack, J., Olejniczak, E.T., Structural studies of Bcl-xL/ligand complexes using ^{19}F NMR.*J. Biomol. NMR* **2006**, V*34*,221–227.

14 Oltersdorf, T., Elmore, S. W., Shoemaker, A. R., Armstrong, R. C., Augeri, D. J., Belli, B. A., Bruncko, M., Deckwerth, T. L., Dinges, J., Hajduk, P. J., Joseph, M. K., Kitada, S., Korsmeyer, S. J., Kunzer, A. R., Letai, A., Li, C., Mitten, M. J., Nettesheim, D. G., Ng, S., Nimmer, P. M., O'Connor, J. M., Oleksijew, A., Petros, A. M., Reed,

J. C., Shen, W., Tahir, S. K., Thompson, C. B., Tomaselli, K. J., Wang, B., Wendt, M. D., Zhang, H., Fesik, S. W., Rosenberg, S. H., An inhibitor of Bcl-2 family proteins induces regression of solid tumours. *Nature* **2005**, *435*, 677–681.

15 Dalvit, C., Mongelli, N., Papeo, G., Giordano, P., Veronesi, M., Moskau, D., Kummerle, R., Sensitivity improvement in ^{19}F NMR-based screening experiments: Theoretical considerations and experimental applications. *J. Am. Chem. Soc.* **2005**, *127*, 13380–13385.

16 Dalvit, C., Ardini, E., Flocco, M., Fogliatto, G. P., Mongelli, N., Veronesi, M., A general NMR method for rapid, efficient, and reliable biochemical screening. *J. Am. Chem. Soc.* **2003**, *125*, 14620–14625.

17 Dalvit, C., Ardini, E., Fogliatto, G. P., Mongelli, N., Veronesi, M., Reliable high-throughput functional screening with 3-FABS. *Drug Disc. Today* **2004**, *9*, 595–602.

18 Dalvit, C., Fagerness, P. E., Hadden, D. T. A., Sarver, R. W., Stockman, B. J., Fluorine-NMR experiments for high-throughput screening: Theoretical aspects, practical considerations, and range of applicability. *J. Am. Chem. Soc.*, **2003**, *125*, 7696–7703.

19 Dalvit, C., Fasolini, M., Flocco, M., Knapp, S., Pevarello, P., Veronesi, M., NMR-based screening with competition water-ligand observed via gradient spectroscopy experiments: Detection of high-affinity ligands. *J. Med. Chem.* **2002**, *45*, 2610–2614.

20 Dalvit, C., Flocco, M., Knapp, S., Mostardini, M., Perego, R., Stockman, B. J., Veronesi, M., Varasi, M., High-throughput NMR-based screening with competition binding experiments. *J. Am. Chem. Soc.* **2002**, *124*, 7702–7709.

21 Dalvit, C., Flocco, M., Veronesi, M., Stockman, B. J., Fluorine-NMR Competition binding experiments for high-throughput screening of large compound mixtures. *J. Comb. Chem. High Throughput Screening* **2002**, *5*, 605–611.

22 Dalvit, C., Papeo, G., Mongelli, N., Giordano, P., Saccardo, B., Costa, A., Veronesi, M., Ko, S. Y., Rapid NMR-based functional screening and IC50 measurements performed at unprecedentedly low enzyme concentration. *Drug Dev. Res.* **2005**, *64*, 105–113.

23 Hardin, C. C., Gollnick, P., Horowitz, J., Partial assignment of resonances in the fluorine-19 nuclear magnetic resonance spectra of 5-fluorouracil-substituted transfer RNAs. *Biochemistry* **1988**, *27*, 487–495.

24 Hills, D. C., Cotten, M. L., Horowitz, J., Isolation characterization of two 5-fluorouracil-substituted *Escherichia coli* initiator methionine transfer ribonucleic acids. *Biochemistry* **1983**, *22*, 1113–1122.

25 Chu, W. C., Liu, J. C., Horowitz, J., Localization of the major ethidium bromide binding site on tRNA. *Nucleic Acids Res.* **1997**, *25*, 3944–3949.

26 Chu, W.-C., Kintanar, A., Horowitz, J., Correlations between fluorine-19 nuclear magnetic resonance chemical shift and the secondary and tertiary structure of 5-fluorouracil-substituted tRNA. *J. Mol. Biol.* **1992**, *227*, 1173–1181.

27 Chu, W.-C., Feiz, V., Derrick, W. B., Horowitz, J., Fluorine-19 nuclear magnetic resonance as a probe of the solution structure of mutants of 5-fluorouracil-substituted *Escherichia coli* valine tRNA. *J. Mol. Biol.* **1992**, *227*, 1164–1172.

28 Chu, W.-C., Horowitz, J., Fluorine-19 NMR studies of the thermal unfolding of 5-fluorouracil-substituted *Escherichia coli* valine transfer RNA. *FEBS Lett.* **1991**, *295*, 159–162.

29 Chu, W. C., Horowitz, J., Recognition of *Escherichia coli* valine transfer RNA by its cognate synthetase: a fluorine-19 NMR study. *Biochemistry* **1991**, *30*, 1655–1663.

30 Hardin, C. C., Gollnick, P., Kallenbach, N. R., Cohn, M., Horowitz, J., Fluorine-19 nuclear magnetic resonance studies of the structure of 5-fluorouracil-substituted

Escherichia coli transfer RNA. *Biochemistry* **1986**, *25*, 5699–5709.

31 Gollnick, P., Hardin, C. C., Horowitz, J., Fluorine-19 nuclear magnetic resonance study of codon-anticodon interaction in 5-fluorouracil-substituted *E. coli* transfer RNAs. *Nucleic Acids Res.* **1986**, *14*, 4659–4672.

32 Horowitz, J., Ching-Nan, O., Ishaq, M., Ofengand, J., Bierbaum, J., Isolation and partial characterization of *Escherichia coli* valine transfer RNA with uridine and uridine-derived residues replaced by 5-fluorouridine. *J. Mol. Biol.* **1974**, *88*, 301–304.

33 Horowitz, J., Ofengand, J., Daniel, W. E. Jr., Cohn, M., ^{19}F nuclear magnetic resonance of 5-fluorouridine-substituted tRNA1Val from *Escherichia coli*. *J. Biol. Chem.* **1977**, *252*, 4418–4420.

34 Hardin, C. C., Horowitz, J., Mobility of individual 5-fluorouridine residues in 5-fluorouracil-substituted *Escherichia coli* valine transfer RNA: A ^{19}F nuclear magnetic resonance relaxation study. *J. Mol. Biol.* **1987**, *197*, 555–569.

35 Chu, W.-C., Horowitz, J., ^{19}F NMR of 5-fluorouracil-substituted transfer RNA transcribed *in vitro*: resonance assignment of fluorouracil-guanine base pairs. *Nucleic Acids Res.* **1989**, *17*, 7241–7252.

36 Klimasauskas, S., Szyperski, T., Serva, S., Wuthrich, K., Dynamic modes of the flipped-out cytosine during *Hha*I methyltransferase-DNA interactions in solution. *EMBO J.* **1998**, *17*, 317–324.

37 Scott, W. G., Finch, J. T., Klug, A., The crystal structure of an all-RNA hammerhead ribozyme: a proposed mechanism for RNA catalytic cleavage. *Cell* **1995**, *81*, 991–1002.

38 Martick, M., Scott, W. G., Tertiary contacts distant from the active site prime a ribozyme for catalysis. *Cell* **2007**, *126*, 309–320.

39 Hammann, C., Norman, D. G., Lilley, D. M. J., Dissection of the ion-induced folding of the hammerhead ribozyme using ^{19}F NMR. *Proc. Natl. Acad. Sci. USA* **2001** *98*, 5503–5508.

40 Nahas, M. K., Wilson, T. J., Hohng, S., Jarvie, K., Lilley, D. M., Ha, T., Observation of internal cleavage and ligation reactions of a ribozyme. *Nat. Struct. Mol. Biol.* **2004**, *11*, 1107–1113.

41 Olejniczak, M., Gdaniec, Z., Fischer, A., Grabarkiewicz, T., Bielecki, L., Adamiak, R. W., The bulge region of HIV-1 TAR RNA binds metal ions in solution. *Nucleic Acids Res.* **2002**, *30*, 4241–4249.

42 Brodsky, A. S., Williamson, J. R., Solution structure of the HIV-2 TAR-argininamide complex. *J. Mol. Biol.* **1997**, *267*, 624–639.

43 Aboul-ela, F., Karn, J., Varani, G., Structure of HIV-1 TAR RNA in the absence of ligands reveals a novel conformation of the trinucleotide bulge. *Nucleic Acids Res.* **1996**, *24*, 3974–3981.

44 Scott, L. G., Geierstanger, B. H., Williamson, J. R., Hennig, M., Enzymatic synthesis and ^{19}F NMR studies of 2-fluoroadenine-substituted RNA. *J. Am. Chem. Soc.* **2004**, *126*, 11776–11777.

45 Hennig, M., Munzarova, M. L., Bermel, W., Scott, L. G., Sklenar, V., Williamson, J. R., Measurement of long-range ^{1}H-^{19}F scalar coupling constants and their glycosidic torsion dependence in 5-fluoropyrimidine-substituted RNA. *J. Am. Chem. Soc.* **2006**, *128*, 5851–5858.

46 Luy, B., Marino, J. P., Measurement application of ^{1}H-^{19}F dipolar couplings in the structure determination of 2′-fluorolabeled RNA. *J. Biomol. NMR* **2001**, *20*, 39–47.

47 Hansen, M. R., Hanson, P., Pardi, A., Filamentous bacteriophage for aligning RNA, DNA, and proteins for measurement of nuclear magnetic resonance dipolar coupling interactions. *Methods Enzymol.* **2000**, *317*, 220–240.

48 Höbartner, C., Micura, R., Bistable Secondary structures of small RNAs and their structural probing by comparative imino proton NMR spectroscopy. *J. Mol. Biol.* **2003**, *325*, 421–431.

49 Kreutz, C., Kählig, H., Konrat, R., Micura, R., Ribose 2′-F labeling: A simple tool for the characterization of RNA secondary structure equilibria by ^{19}F NMR spectroscopy. *J. Am. Chem. Soc.* **2005**, *127*, 11558–11559.

50 Kreutz, C., Kählig, H., Konrat, R., Micura, R., A general approach for the identification of site-specific RNA binders by ^{19}F NMR spectroscopy: Proof of concept. *Angew. Chem. Int. Ed.* **2006**, *45*, 3450–3453.

51 Hermann, T., Patel, D. J., Adaptive recognition by nucleic acid aptamers. *Science* **2000**, *287*, 820–825.

52 Jiang, L., Patel, D. J., Solution structure of the tobramycin-RNA aptamer complex. *Nat. Struct. Biol.* **1998**, *5*, 769–774.

53 Sahasrabudhe, P. V., Gmeiner, W. H., Solution structures of 5-fluorouracil-substituted RNA duplexes containing G-U wobble base pairs. *Biochemistry* **1997**, *36*, 5981–5991.

54 Blandin, M., Tran Dinh, S., Catlin, J. C., Guschlbauer, W., Nucleoside conformations. 16. Nuclear magnetic resonance and circular dichroism studies on pyrimidine-2′-fluoro-2′-deoxyribonucleosides. *Biochim. Biophys. Acta* **1974**, *361*, 249–256.

55 Reif, B., Wittmann, V., Schwalbe, H., Griesinger, C., Wörner, K., Jahn-Hofmann, K. W. E. J., Bermel, W., Structural comparison of oligoribonucleotides and their 2′-deoxy-2′-fluoro analogs by heteronuclear NMR spectroscopy. *Helv. Chim. Acta* **1997**, *80*, 1952–1971.

2
8-Oxo-7,8-Dihydro-2′-Deoxyguanosine: A Major DNA Oxidation Product

Jean Cadet and Paolo Di Mascio

2.1
Introduction

It is now well documented that either exogenous or endogenous agents can modify cellular DNA, along with other cellular components. To ensure normal growth control and accurate DNA replication, cells have developed many strategies to manage stress. However, the failure of some of these defense mechanisms may lead to the development of pathologies such as cancer and neurodegenerative disorders. Reactive oxygen and nitrogen species might be produced by either endogenous sources, as cell aerobic metabolism and inflammation, or by the exposure to a variety of chemical and physical agents. Oxidation reactions occur within cells as the result of aerobic metabolism [1]. Thus, the superoxide radical ($O_2{\cdot}^-$) is generated as a side product of the incomplete reduction of molecular oxygen during electron transport in mitochondria and the endoplasmic reticulum [2]. Macrophages and neutrophils are also able to produce $O_2{\cdot}^-$ as part of the host defense system via the NADPH-mediated reduction of O_2 [3]. Other sources of rather unreactive $O_2{\cdot}^-$ include enzymatic reactions mediated by, for example xanthine oxidase, and metabolic activation of xenobiotics such as benz[*a*]pyrene [4]. The dismutation of $O_2{\cdot}^-$ radicals within cells – either enzymatically or chemically – leads to the generation of hydrogen peroxide (H_2O_2), another reactive oxygen species (ROS) which alone is also poorly reactive towards most biomolecules. However, in the presence of reduced transition metals – and particularly of Fe^{2+} – H_2O_2 may be involved in the so-called Fenton reaction, giving rise to either the strongly oxidizing hydroxyl radical (·OH) or a related reactive species [5]. Either X- or gamma-radiolysis of water molecules constitutes another possibility of generating ·OH radicals [6]. Ionizing radiation as the result of direct interaction, high-intensity ultra-violet (UV) laser pulses and type I photosensitizers are also able to generate radicals cations through one-electron oxidation of nucleobases and, eventually, of the 2-deoxyribose moiety [6, 7]. Singlet oxygen ($^1O_2$) – yet another ROS – may be generated intracellularly by the action of myeloperoxidase [8], an enzyme which is implicated in inflammation processes, and also by the UVA component [7] of solar radiation.

Modified Nucleosides: in Biochemistry, Biotechnology and Medicine. Edited by Piet Herdewijn

ISBN: 978-3-527-31820-9

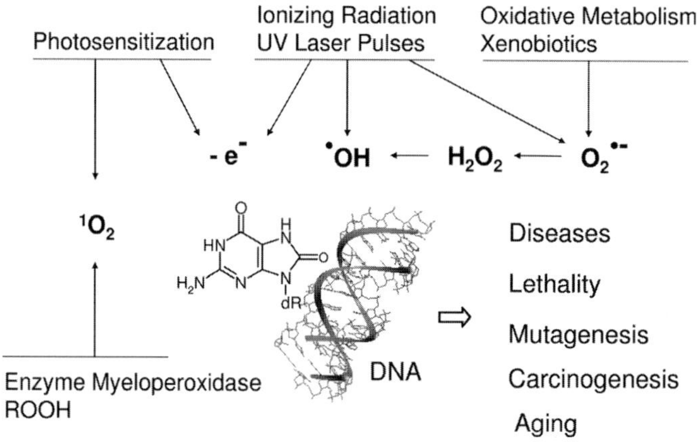

Figure 2.1 Oxidative reactions generating 8-oxo-7,8-dihydro-2′-deoxyguanosine, and the biological consequences.

During the past two decades, extensive progress has been achieved in the delineation of the mechanisms of DNA oxidation [9] upon exposure to various ROS and one-electron oxidants. Guanine, which exhibits the lowest ionization potential among the nucleic acid components, has recently received much attention (Figure 2.1). One ubiquitous guanine oxidation product that is often utilized as a exposure marker of oxidative stress is that of 8-oxo-7,8-dihydroguanine (8-oxoGua). Hence, within this chapter, emphasis is placed on describing the main oxidation reactions initiated by ·OH radical, one-electron oxidants and singlet oxygen, giving rise to 8-oxoGua in model compounds and in cellular DNA. In addition, the main one-electron and $^1O_2$ oxidation reactions of 8-oxo-7,8-dihydroguanine, which is more susceptible than the guanine precursor, are critically reviewed. Finally, methods aimed at synthesizing 8-oxo-7,8-dihydro-2′-deoxyguanosine (8-oxodGuo) synthons for their site-specific insertion into oligonucleotides are surveyed, together with the use of the latter probes for assessing DNA repair substrate specificity and DNA replication.

2.2 Formation of 8-Oxo-7,8-Dihydroguanine

2.2.1 Single Lesion

Guanine is a preferential DNA target to several oxidants, as it demonstrates the lowest ionization potential among the different purine and pyrimidine nucleobases [9a], and is the only nucleic acid component to exhibit significant reactivity toward singlet oxygen ($^1O_2$) at neutral pH [10].

2.2.1.1 ·OH Radical

The addition of ·OH to the purine ring at C8 of 2′-deoxyguanosine (**1**) leads to the formation of reducing 8-hydroxy-7,8-dihydro-7-yl radical (**2**) which, in the presence of oxidants such as O_2, give rise to 8-oxo-7,8-dihydroguanine (**3**). One competitive reaction of the 7-yl radical is a one-electron reduction that leads to the formation of 2,6-diamino-4-hydroxy-5-formamidopyrimidine (FapyGua) (**4**) through scission of the C8–N9 imidazole bond [11] [see Eq. (2.1)]. This was found to occur efficiently in model compounds with a high unimolecular rate ($k = 2 \times 10^5\ s^{-1}$), this being inferred from pulse radiolysis measurements [12]. The two overwhelming oxidation products of the purine moiety of **1** resulting from either the reaction with ·OH radical were isolated and identified as 2,2-diamino-4-[(2-deoxy-β-D-erythro-pentofuranosyl)amino]-5(2*H*)-oxazolone (**10**), and its precursor 2-amino-5-[(2-deoxy-β-D-erythro-pentofuranosyl)amino]-4*H*-imidazol-4-one (**9**) [13, 14] [see Eq. (2.1)]. The mechanism of their production may be rationalized in terms of transient formation of the oxidizing guaninyl radical **5**, which may arise either from dehydration of the ·OH adduct at C4 ($k = 6 \times 10^3\ s^{-1}$) or deprotonation of the guanine radical cation [12]. The addition of O_2 to the C5 carbon-centered radical **6** is at best a rather inefficient process, as the rate constant has been found to be lower than 10^3 $M^{-1}\ s^{-1}$. Evidence has been provided that $O_2{\cdot}^-$ reacts significantly with a rate constant that has been estimated as $3 \times 10^9\ M^{-1} s^{-1}$ for the 2′-deoxyribonucleoside [12]. A slightly lower reactivity has been assessed for the related dGMP nucleotide [15] and short oligonucleotides [16], the *k*-values being $1.3 \times 10^9\ M^{-1} s^{-1}$ and $0.47 \times 10^9\ M^{-1} s^{-1}$, respectively. This leads, after protonation, to the transient formation of a hydroperoxide **7** that is followed by a nucleophilic addition of a water molecule across the 7,8-ethylenic bond leading to **8**. Cleavage of the 1,6 bond is then accompanied by the release of formamide [17], and subsequent rearrangement leads to the formation of oxazolone **10** through the quantitative hydrolysis of unstable imidazolone **9** (half-life = 10 h in aqueous solution at 20°C) [13] [see Eq. (2.1)]. However, at this stage, the formation of 5-hydroperoxy-8-hydroxy-7,8-dihydro-2′-deoxyguanosine (**8**), the proposed key precursor of the imidazolone and oxazolone nucleosides (**9**, **10**), and which is likely to be highly unstable, remains to be established.

2.2.1.2 One-Electron Oxidation

The hydration reaction of the radical cation of guanine residues (**11**) produced by one-electron oxidants within double-stranded DNA also leads to the formation of the 8-hydroxy-7,8-dihydroguanyl radical (**2**) [18] [Eq. (2.1)]. Various processes and agents, including the direct effect of ionizing radiation, mono- and bi-photonic UV lasers [19], type I photosensitizers [20], Co(II) ion in the presence of benzoyl peroxide [21], peroxyl and oxyl radicals [22], together with radicals such as $CO_3{\cdot}^-$ [23], $Br_2{\cdot}^-$ $(SCN)_2{\cdot}^-$ [24], Tl^{2+} or $SO_4{\cdot}^-$ [25a], are able to promote the formation of the guanine radical cation (**11**) precursor of 8-oxodGuo **3** in isolated DNA. As discussed previously for reactions involving the OH radical, the adducts at C8 **2** of the guanine moiety, 8-oxoGua **3** and FapyGua **4**, are generated in competitive manner, by oxidation and reduction, respectively.

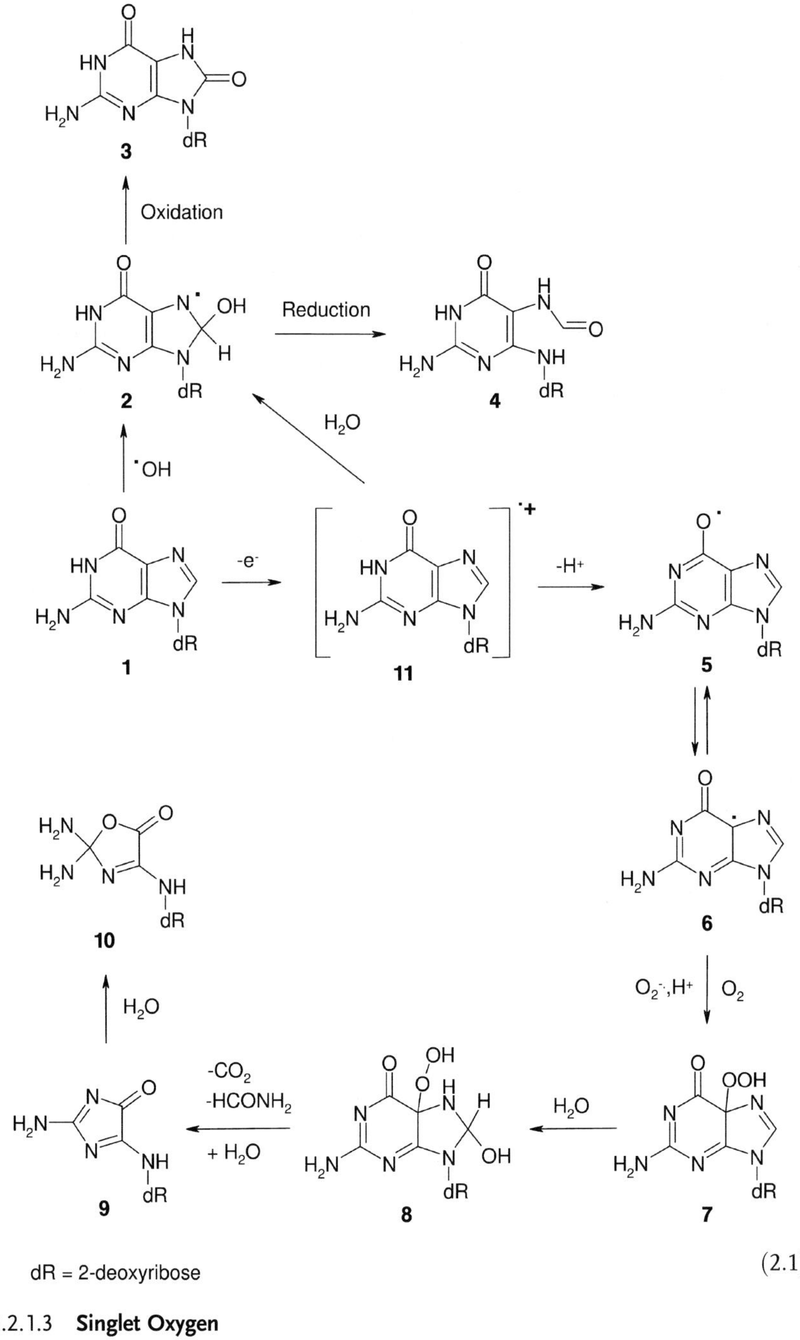

(2.1)

2.2.1.3 Singlet Oxygen

During the past 20 twenty years, much attention has been devoted to the elucidation of the mechanism of $^1O_2$ oxidation reactions of the guanine moiety of nucleosides and dinucleoside monophosphates in aqueous solutions [10]. As mentioned above, the guanine base is the only normal nucleic acid component that shows notable reactivity for $^1O_2$ in the $^1\Delta_g$ state in neutral aqueous solutions [26]. As an initially striking result, it was found that phthalocyanine and methylene blue sensitization of 2′-deoxyguanosine (**1**) to UVA radiation in aerated aqueous solutions led to the generation of the two main oxidized nucleosides, which initially (and tentatively) were identified as the (4*R**) and (4*S**)-diastereomers of 4-hydroxy-8-oxo-4,8-dihydro-2′-deoxyguanosine (**15**) [27, 28] [Eq. (2.2)]. Similar observations were made upon the exposure of thymidylyl (3′,5′)-2′-deoxyguanosine to photoexcited naphthalocyanines [29] that are able predominantly to generate $^1O_2$. More recently, it was proposed that the base moiety of the two related $^1O_2$-mediated oxidation products of guanosine in aerated aqueous solution exhibits in fact a spiroiminodihydantoin **17** structure [30] [Eq. (2.2)]. This finding is in agreement with previous reports which have established that the two latter diastereomers are formed as the result of one-electron oxidation of 8-oxo-7,8-dihydroguanosine. Most likely, this involves the formation of transient 5-hydroxy-8-oxo-7,8-dihydroguanosine which, upon rearrangement involving an acyl shift, leads to the 4*R** and 4*S** diastereomers of spiroiminodihydantoin ribonucleoside [31–33]. Further support for the spirocyclic connectivity of related 2′-deoxyribonucleosides (dSp) **17** was gained from highly relevant and unambiguous SELINQUATE ^{13}C NMR measurements [34]. Mechanistic insights into the formation of the two diastereomers of spiroiminodihydantoin nucleosides **17** were inferred from ^{18}O labeling isotopic studies [35, 36] and low-temperature ^{13}C NMR experiments [25b, 37]. Thus, the proposed mechanism of formation of dSp 17 [Eq. (2.2)] involves an initial [4 + 2] cycloaddition of $^1O_2$ across the 7,8 and 4,5-ethylenic bonds of the purine ring, and the successive conversion of several instable compounds. Thus, the 4,8-endoperoxides **12** formed initially may be converted into the 8-hydroperoxide **13** which, through dehydration, would give rise to an oxidized quinonoid intermediate **14** [Eq. (2.2)]. The subsequent addition of water to the reactive 5,7 double bond of the latter compound leads to 5-hydroxy-8-oxo-7,8-dihydroguanine (**16**), which is able to rearrange into spiroiminodihydantoin (**17**). Relevant chemical and conformational properties of dSp diastereomers **17**, either as free nucleosides or when inserted into DNA duplexes, were inferred from theoretical calculations [38, 39]. Recently, attempts have been made to assign the absolute configuration of the two diastereomers of dSp **17** using either a theoretical approach [40] or via the basis of NMR studies involving the consideration of key dipolar interactions within the modified 2′-deoxyribonucleosides [41]. This approach should allow a correlation to be made between the structural features and the enzymic processing of the lesions by DNA repair enzymes and replicative polymerases [40].

A relatively minor product of the $^1O_2$ oxidation of free 2′-deoxyguanosine has been identified as 8-oxo-7,8-dihydro-2′-deoxyguanosine (**3**), which plateaus to relatively low yields with increased time of exposure to singlet oxygen [41]. The formation of 8-oxodGuo **3** may be explained by the generation of diastereomeric 4,8-endoperoxides prior to rearrange into 8-hydroperoxy-2′-deoxyguanosine (**9**) [25b, 42]. Reduction of the hydroperoxide (**13**) [Eq. (2.2)], which is also a precursor of dSp, is the likely pathway giving rise to 8-oxodGuo **3**. Support for the latter mechanism was provided by the observation of an overwhelming formation of 8-oxo-7,8-dihydroguanosine at the expense of spiroiminodihydantoin nucleosides upon exposure of guanosine to $^1O_2$ in the presence of thiols [30]. Interestingly, the reduction pathway leading to the formation of 8-oxodGuo **3** is predominant in the $^1O_2$ oxidation of double-stranded DNA, even in the absence of any added reducing agent [26, 42]. It was also shown that FapyGua **4**, a degradation product that might be formed by the hydration of guanine radical cation **11** [Eq. (2.2)], followed by an opening of the imidazole ring according to a reductive pathway [43], is not generated in detectable amounts within isolated DNA. This was achieved using a thermolabile naphthalene endoperoxide derivative as a clean chemical source of $^1O_2$ [44]. This has led to a ruling out of the possibility for $^1O_2$ to act as a one-electron oxidant, in contrast to a previous proposal [45]. It should also be noted that the 4*R* and 4*S* diastereomers of dSp **17**, were found not to be generated in duplex DNA – at least under conditions where the extent of guanine oxidation remained less than 1%. Thus, it may be concluded that $^1O_2$ is a highly selective oxidant of double-stranded DNA, leading to the predominant formation of 8-oxodGuo **3**.

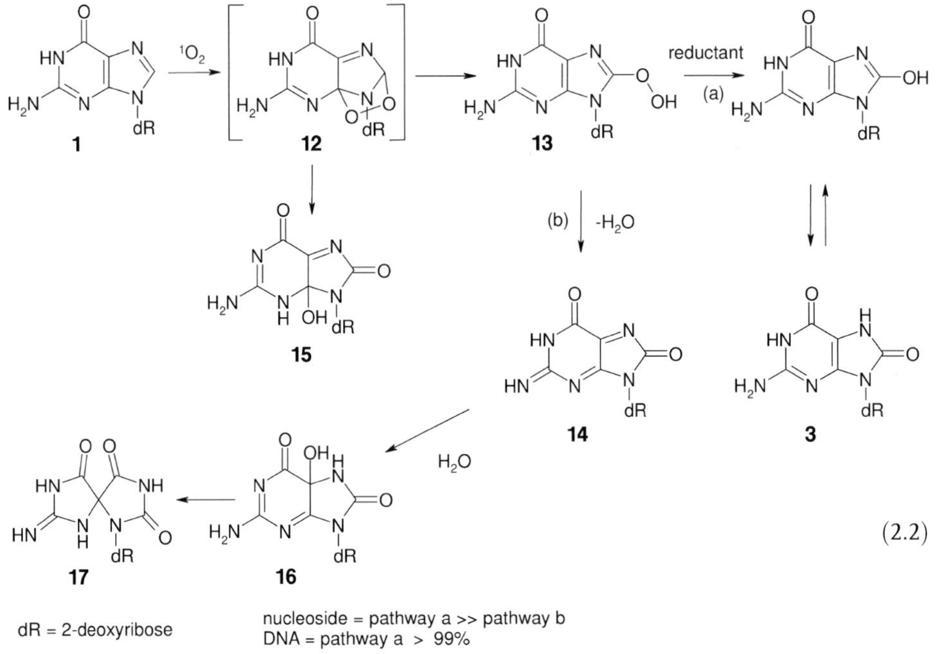

(2.2)

2.2.2
Tandem Lesions

New insights into the formation and the measurement of 8-oxoGua-formylamine (Fo) and the opposite sequence isomer (Fo-8-oxoGua) tandem lesions within DNA were recently gained from detailed studies that have involved the site-specific chemical insertion of the modified nucleosides into defined sequence oligonucleotides [46]. The accurate measurement of both sequence isomers of the vicinal

8-oxoGua and Fo damage was achieved using a recently designed high-performance liquid chromatography-tandem mass spectrometry (HPLC-MS/MS) assay [46b]. Interestingly, the formation of the two tandem lesions was shown to be linear within the low-dose range (5 to 100 Gy), thus confirming that these were generated from one radical hit. The respective radiolytic yields for 8-oxoGua-Fo, Fo-8-oxoGua and 8-oxoGua were 0.0001, 0.0013 and 0.0130 $\mu mol\,J^{-1}$. A reasonable mechanism of formation for the 8-oxoGua-Fo lesion involves, in the initial step, the addition of a ·OH radical at either C5 or C6 of any of the two pyrimidine bases, which is followed by the rapid reaction of molecular oxygen with the pyrimidyl radical thus generated [Eq. (2.3)]. The resulting peroxyl radical may at least partly react with a vicinal guanine by intramolecular addition at C8, as inferred from experiments using ^{18}O-labeled materials [47]. Subsequently, the adduct is able to rearrange, giving rise to 8-oxoGua on the one hand and an oxyl-type pyrimidine radical on the other hand; this is known, based on a β-scission mechanism, to lead to the formation of formylamine. Interestingly, the kinetic parameters of specific excision of the oxidized bases by bacterial DNA N-glycosylases – namely 8-oxoGua by Fpg and formylamine by endo III – was not greatly affected in oligonucleotides that contain both tandem base damage [46a]. The mutagenic potential of the 8-oxoGua/Fo tandem lesion has been assessed using a single-stranded DNA shuttle vector in which the tandem base lesion and each of the two oxidized bases was site-specifically inserted prior to being transfected into Simian COS7 cells [48]. The mutations induced after replication in mammalian cells were screened in bacteria. The presence of 8-oxoGua alone was not found to affect the survival (70% bypass), whereas the formylamine lesion was shown to be highly mutagenic when it arose from a cytosine residue. The mutagenic properties of the tandem 8-oxoGua/Fo damage appeared to be the result of a combination of the effects of both individual 8-oxoGua and Fo, with a high frequency of adenine insertion in front of the pyrimidine lesions [48].

2.3 Reactivity of 8-Oxo-7,8-Dihydro-2′-Deoxyguanosine

Interestingly, it was shown that 8-oxodGuo **3**, an ubiquitous exposure marker of DNA exposed to most oxidizing agents [9a] is a much better substrate than dGuo **1** to further oxidation by $^1O_2$ [49] and one-electron oxidants.

2.3.1 One-Electron Oxidation

There is a growing body of evidence showing that 8-oxodGuo **3**, the oxidation potential of which is about 0.5 eV lower than that of dGuo **1** [50], is a preferential target for numerous one-electron oxidizing agents. These include Na_2IrCl_6 [33, 51], $K_3Fe(CN)_6$, $CoCl_2/KHSO_5$ [31], a high-valent chromium complex [52], peroxyl radicals [53], triplet ketones, oxyl radicals [34, 54], ionizing radiation through the direct effect [55], and riboflavin as a type I photosensitizer [32].

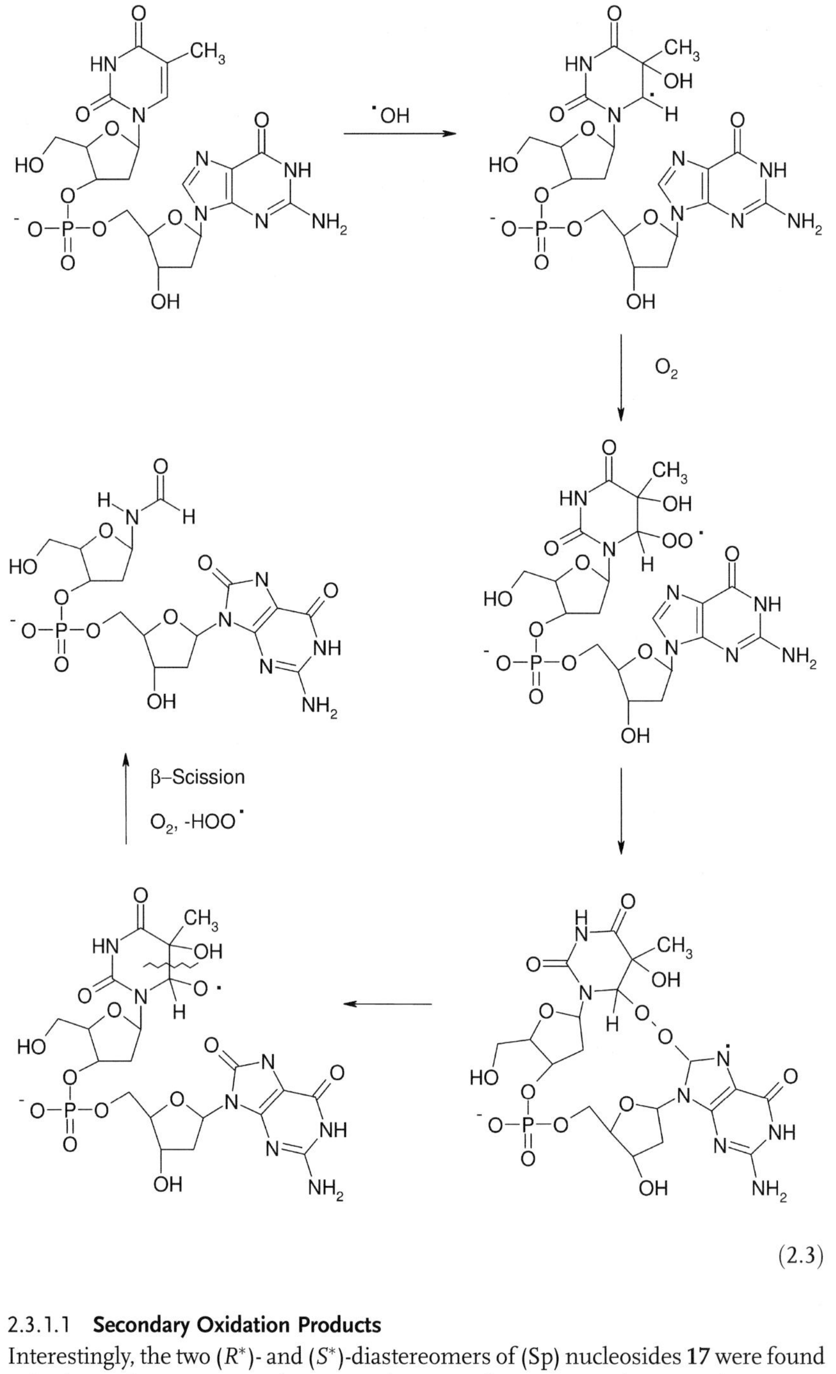

(2.3)

2.3.1.1 **Secondary Oxidation Products**

Interestingly, the two (R^*)- and (S^*)-diastereomers of (Sp) nucleosides **17** were found to be the predominant one-electron oxidation products of 8-oxodGuo **3** and 8-oxo-7,8-dihydroguanosine at neutral pH. Formation of the latter oxidized nucleosides was rationalized in terms of the transient generation of 5-hydroxy-8-oxo-7,8-dihydroguanine derivatives **16**, followed by rearrangement into **17** via an acyl shift [Eq. (2.4)]. The latter precursors were found to undergo a different decomposition pathway under slightly acidic conditions; this involves opening of the 5,6-pyrimidine ring followed by a decarboxylation reaction, with the subsequent formation of the two diastereomers of guanidinohydantoin (Gh) derivatives **18**. The oxazolone nucleoside **10**, together with its imidazolone **9** precursor, were also found to be one-electron oxidation products of 8-oxodGuo **3**, although generated in lower yields than spiroiminodihydantoin **17** and guanidinohydantoin nucleosides **18** [32].

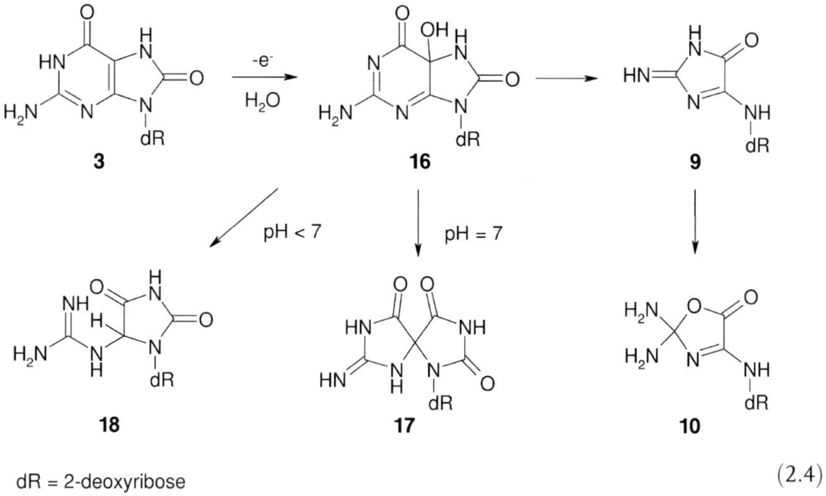

(2.4)

2.3.1.2 DNA–Protein Crosslinks

The one-electron oxidation of mixtures of 8-oxoGua containing 2′-deoxyoligonucleotides with proteins, including the MutY repair enzyme [56a] and single-stranded binding protein [56b] has been shown to give rise to crosslinks. This is likely to occur through nucleophilic addition of the free ε-amino group of a lysine residue at the C5 position of the oxidized purine moiety. The resulting C5 adduct was found, as was previously observed for the 5-hydroxy-8-oxo-7,8-dihydroguanine intermediate, to undergo competitive rearrangement leading to the formation of substituted guanidino and spiroiminodihydantoin derivatives. It should be noted that a similar rearrangement was found to occur for the KKK peptide adduct through the central lysine to the C8 of the guanine moiety of d(TpG) upon one-electron oxidation [57].

2.3.2 Singlet Oxygen

2.3.2.1 Nucleoside

The rate of reaction of $^1O_2$ with 8-oxodGuo **3** was found to be about two orders of magnitude higher than that with dGuo **1** [50]. It is likely that $^1O_2$ adds across the

4,5-ethylenic bond of 8-oxodGuo to generate a transient dioxetane **19** that decomposes according to two main processes [Eq. (2.5)]. A major degradation pathway that involves 1,2-bond cleavage of the dioxetanes would give rise to a nine-membered ring intermediate **20** which, upon intramolecular cyclization, is converted into 1,3,5-triazine-1(2*H*)-carboximidamide, 3-(2-deoxy-β-D-*erythro*-pentofuranosyl)-tetrahydro-2,4,6-trioxo (**21**). The latter six-membered ring nucleoside then slowly hydrolyzes, giving rise to 1-(2-deoxy-β-D-*erythro*-pentofuranosyl)-cyanuric acid (dCya) (**22**), with the concomitant release of urea [58]. The second pathway leads to the formation of 2,2,4-triamino-5-(2*H*)-oxazolone (**10**) and spiroiminodihydantoin nucleosides **17** [55] [Eq. (2.5)].

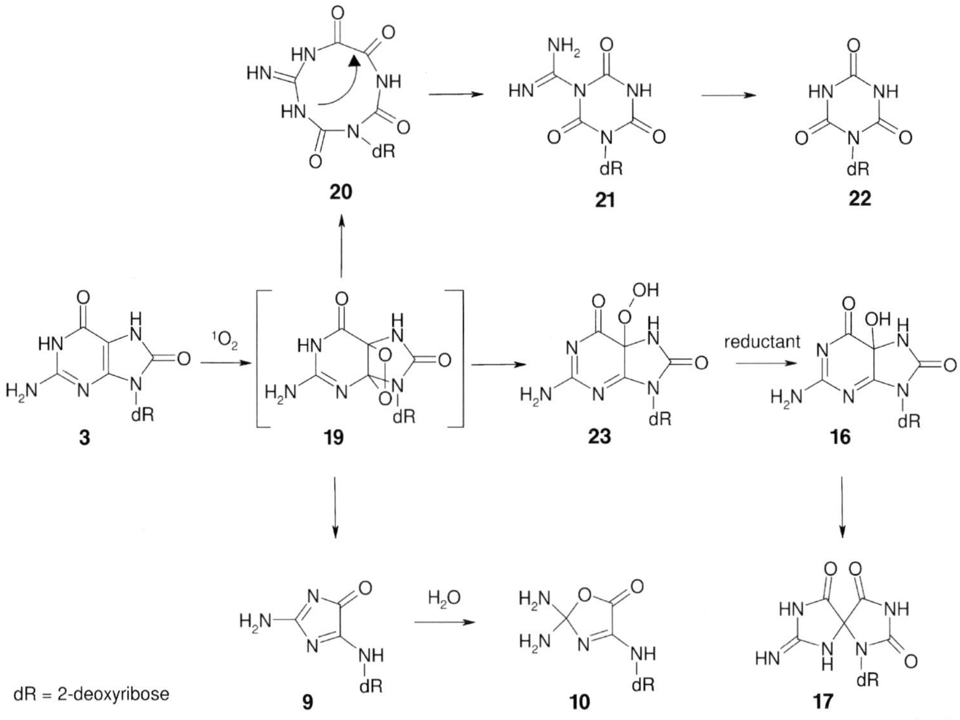

(2.5)

2.3.2.2 **Oligonucleotide**

The reaction of $^1O_2$ with an 8-oxo-7,8-dihydro-2′-deoxyguanosine residue (**3**), site-specifically inserted within a single-stranded oligonucleotide, was found to be more specific [Eq. (2.6)]. Thus, the predominant oxidation product in this system was identified as oxaluric acid (**26**) [59]. A likely mechanism for the formation of the latter ureido derivative involves an initial formation of the dioxetane **19** by $^1O_2$ addition across the 5,6-ethylenic bond [49] that, upon rearrangement, is converted into the unstable 5-hydroperoxide **23**. Cleavage of the 5,6-bond of **23** and subsequent decarboxylation gives rise to a dehydroguanidinohydantoin derivative **24**. Further decomposition of the latter nucleoside leads, by two successive hydrolytic steps, to the formation of oxaluric acid **26** through the parabanic acid (**25**) precursor [60]. Interestingly, the dehydroguanidohydantoin compound **24** that can be isolated was also found to be generated by two-electron oxidation of the guanine moiety of d(GpT) using the Mn-TMPyp/$KHSO_5$ oxidizing system [61].

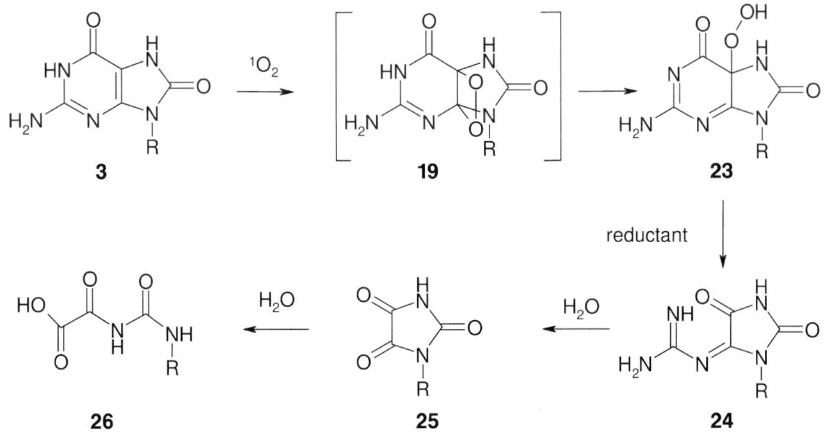

R = 2-deoxyribose (2.6)

2.4
Formation of 8-Oxo-7,8-Dihydro-2′-Deoxyguanosine in Cellular DNA

2.4.1
Methods of Measurement

The measurement of oxidized bases and nucleosides in cellular DNA may be used to gain insights into the nature and importance of chemical reactions that are mediated in cellular DNA by oxidative agents and processes. This could be achieved by singling out modified bases or nucleosides by HPLC associated with suitable detection techniques, or by using biochemical methods (alkaline elution technique, comet assay) in association with DNA repair enzymes that allow the detection of classes of modifications [9a, 43].

2.4.1.1 HPLC Methods (HPLC-ECD and HPLC-MS/MS)

Modified bases are usually separated – by using a chromatographic method – from the overwhelming normal DNA components after a suitable hydrolytic or enzymic digestion step. The detection of compounds of interest is made at the output of the column by employing a sensitive technique that should be able to single out one lesion per 10^6 to 10^7 nucleosides in a DNA sample size of at least 30 μg. Until recently, however, this approach has been hampered by the use of inappropriate methods that has led in most cases to overestimated values of the levels of DNA oxidized bases by factors that vary from one to three orders of magnitude. The origin of the main drawbacks associated with the use of the questionable gas chromatography-mass spectrometry (GC-MS) methods first introduced more than 20 years ago has now been identified [62]. Thus, the spurious oxidation of normal bases with an efficiency of about 0.1% has been shown to occur during the derivatization step that is required to make the samples volatile. This leads to an artifactual generation of oxidized purine and pyrimidine bases such as 8-oxoGua, 8-oxo-7,8-dihydroadenine and 5-(hydroxymethyl)uracil, thereby preventing any accurate measurement to be made.

A second matter of concern with the GC-MS assay, and also with assays requiring acid hydrolysis steps to release the bases, is the lack of stability of several modifications, including formamidopyrimidines derived from adenine and guanine [63] under hot formic acid treatment. A third source of artifacts – though usually of lower amplitude – is the occurrence of Fenton chemistry during DNA extraction and any subsequent work-up, including the digestion stage. It should be noted that a general consensus now exists on improved chromatographic protocols that have been evaluated within the European Standard Committee on Oxidative DNA Damage (ESCODD) network, to which 25 European laboratories have contributed. Today, recommended protocols are available that include suitable conditions of DNA extraction for which artifactual oxidation is minimized, followed by suitable HPLC analysis of the DNA digest [64]. The frequently used electrochemical detection technique (HPLC-ECD) was introduced more than 20 years ago [65] and is a robust method, but its application in oxidative detection mode is restricted to only a few electroactive DNA lesions, including 8-oxodGuo **3**, 8-oxodAdo and 5-hydroxysubstituted 2′-deoxyuridine and 2′-deoxycytidine compounds. In contrast, the most recently available tandem mass spectrometry (MS/MS) method [66] operating in electrospray ionization mode is more versatile and, on average, is more sensitive than HPLC-ECD, allowing the accurate measurement of up to 15 modifications in the cellular DNA of the approximately 70 model compounds identified to date. The detection of these oxidatively generated modifications is mostly achieved as nucleosides, with the notable exception of 4,6-diamino-5-formamidopyrimidine (FapyAde) and FapyGua bases 4, as the N-glycosidic bond of related nucleosides is labile at neutral pH.

2.4.1.2 **Enzymic Assays**

The comet assay or alkaline elution technique associated with base excision DNA repair enzymes (including bacterial Fpg and endo III with the comet assay) represents a suitable alternative for monitoring levels of base damage within the range of one to five lesions per 10^7 bases when only very small quantities (<10 μg) of DNA are available [67]. The radiation-induced levels of Fpg- sensitive sites that mostly comprised 8-oxoGua **3** and FapyGua **4** lesions were estimated to be 0.48 per 10^7 bases and per Gy, within the DNA of human monocytes exposed to γ-rays [68]. However, this number must be compared with that of strand breaks and alkali-labile sites (1.3 per 10^7 bases). Recently, by using an optimized version of the modified comet assay, the measurement of oxidative base damage to cellular DNA was performed for a dose of ionizing radiation as low as 0.2 Gy [69].

2.4.2
Indirect Effects of Ionizing Radiation (·OH Radical)

8-Oxo-7,8-dihydro-2′-deoxyguanosine(3)– an ubiquitous, oxidatively generated damage product of DNA that may be created by ·OH, one-electron oxidation, peroxynitrite, $^1O_2$ [9a, 43], or as a result of intrastrand addition with thymine 5(6)-hydroxy-6(5)-hydroperoxides [47] – has been found to be produced in cellular DNA upon exposure to gamma rays and high linear energy transfer (LET) heavy ions [70, 71]. This analysis was achieved after suitable enzymatic digestion of the extracted DNA and subsequent quantitative measurement by HPLC-MS/MS using the isotopic dilution method. FapyGua **4**, the related opened imidazole ring compound, was also found to be efficiently generated in the DNA of irradiated human cells, by using HPLC-MS/MS measurements. For this purpose, a dedicated protocol that takes into account the instability of the N-glycosidic bond of formamidopyrimidines derived from purine 2′-deoxyribonucleosides in order to obtain a quantitative release of the related free bases, was designed [73]. Interestingly, as for thymidine oxidation products, the radiation-induced formation yields of both 8-oxodGuo **3** and FapyGua **4** were found to decrease with the increase in LET of the incident photon or particle [70, 71]. This also is suggestive of the major implication of ·OH in the molecular effects of ionizing radiation on the guanine moiety of cellular DNA. Taking into consideration the available mechanistic information that has been gained from radical oxidation studies of model systems [43a], the following radiation-induced degradation pathway of the guanine DNA base in cells may be proposed. The addition of ·OH to the purine ring at C8 leads to the formation of a reducing 8-hydroxy-7,8-dihydro-7-yl (**2**) radical which, in the presence of oxidants such as O_2, gives rise to 8-oxo-7,8-dihydroguanine (**3**) [Eq. (2.1)]. A competitive reaction of the 7-yl radical **2** is a one-electron reduction that leads to the formation of FapyGua **4** through scission of the C8–N9 imidazole bond [11].

2.4.3
High-Intensity UV Laser Irradiation (One-Electron Oxidation)

The exposure of DNA to high-intensity UVC (266 nm) pulses has been shown to serve as a suitable means of generating radical cations by the bi-photonic ionization of purine and pyrimidine bases. The main one-electron oxidation product of DNA was found to be 8-oxodGuo **3** as the result of a hydration reaction and subsequent oxidation of the resulting 8-hydroxy-7,8-dihydroguanyl radical [18, 73]. In similar manner, 8-oxodGuo **3** was found to be generated in the DNA of TPH1 monocytes following nanosecond UVC laser irradiation, as inferred from the HPLC-MS/MS measurement of enzymatically released 2′-deoxyribonucleosides [71]. It is also likely that the predominant formation of 8-oxodGuo **3** within cellular DNA is accounted for by a charge transfer mechanism from initially generated pyrimidine and adenine radical cations to distant guanines that act as hole sinks.

2.4.4
UVA Photosensitization ($^1O_2$)

A suitably protected naphthalene endoperoxide [44] that can penetrate cells has been used to investigate the oxidation reactions of $^1O_2$ in nuclear DNA. Thus, the release of $^1O_2$ from the thermolabile endoperoxide precursor has shown to lead to the selective oxidation of DNA guanine base moieties by producing specifically 8-oxodGuo **3**, as monitored by HPLC-MS/MS [74]. Of equal relevance, it was established that the formation of 8-oxodGuo **3** in cellular DNA was due to singlet oxygen, and not to a putative oxidative stress. This finding was supported by labeling experiments involving a synthetically prepared $^{18}[^1O_2]$-endoperoxide of *N,N′*-di(2,3-dihydroxypropyl)-1,4-naphthalenedipropanamide (DHPNO$_2$) **27** [Eq. (2.7)]. The formation of 8-oxodGuo **3** in cellular DNA is likely to be accounted for by an initial Diels–Alder [4 + 2] cycloaddition of $^1O_2$ across the imidazole ring of the guanine moiety, leading to the generation of a pair of diastereomeric 4,8-endoperoxides 12, in agreement with the proposed mechanism that has been derived from model studies. It should be noted here that $^1O_2$ is not able to induce significant amounts of either direct DNA strand breaks or alkali-labile sites, as inferred by comet assay measurements [75]. Currently, there are growing lines of evidence showing that exposure to UVA irradiation (a major component of solar light) is able to generate 8-oxodGuo **3** in various mammalian and bacterial cells [76–85], as well as in human skin [86], and these are likely to differ in their content of endogenous photosensitizers at the origin of the observed photodynamic effects [7]. A detailed mechanistic study performed on human monocytes, has led to the conclusion that about 80% of the UVA-induced formation of 8-oxodGuo in cellular DNA was due to $^1O_2$ oxidation as the result of a likely type II photosensitization mechanism [68]. A Fenton-type radical mechanism that would involve the initial generation of a superoxide radical, followed by its spontaneous or enzymic dismutation into H_2O_2, are the most probable steps giving rise to the 20% remaining 8-oxodGuo **3** [68].

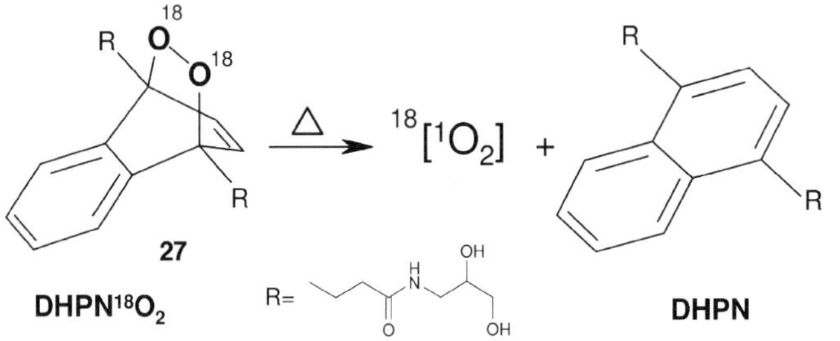

(2.7)

2.5
Synthesis of 8-OxodGuo and Insertion into Oligonucleotides

Today, several methods are available for the preparation of 8-oxodGuo **3** and its site-specific insertion into defined sequence DNA fragments that can be used as probes to assess DNA repair enzyme substrate specificity [43, 87], DNA polymerase replication, and mutagenic features [43, 88]. The preparation of a suitably protected 8-oxodGuo **3** building block can be achieved in several ways [89, 90]. A key step in the preparation of the synthon is the C8 hydroxylation; this involves in the first step the synthesis of 8-bromo-2′-deoxyguanosine (**28**) which is subsequently treated with sodium alkoxide (BnONa or MeONa) prior to hydrogenolysis or nucleophilic displacement [Eq. (2.8)] [90, 91]. The insertion of the resulting protected nucleotide into oligonucleotides was achieved by either solid-phase phosphotriester or phosphoramidite chemistry, such that several different protected monomers of **3** have been prepared [89, 90]. At this point it should be noted that several analogues of **3**, such as O^6-methyl and carbocyclic derivatives, are also available [91, 92].

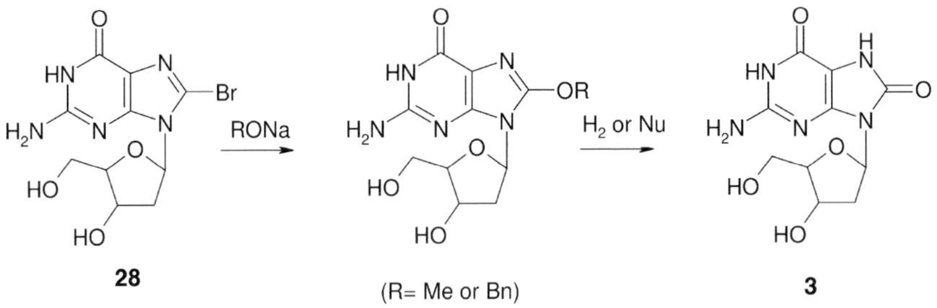

(2.8)

2.6 Conclusions

During the past decade, major progress has been achieved on the delineation of the mechanism of formation and measurement of oxidative by generated base damage in both isolated and cellular DNA. One-electron oxidation and ·OH radical-mediated degradation of 2′-deoxyguanosine (**1**) gives rise, among other oxidized nucleosides, to 8-oxo-7,8-dihydro-2′-deoxyguanosine (**3**), an ubiquitous DNA marker of oxidative reactions. The mechanism of singlet oxygen oxidation of the guanine bases in free nucleoside and isolated DNA was inferred from characterization of the final oxidation products, and of unstable intermediates. In this respect, it should be noted that one- and two-dimensional nuclear magnetic resonance measurements performed at low temperature on several guanosine derivatives in organic solvents by Foote and colleagues have constituted the key experiments. It may pointed out that insights into the chemical reactions of the ·OH radical, one-electron oxidants, and $^1O_2$ to cellular DNA, which lead to the predominant formation of 8-oxodGuo **3**, were gained from specific and accurate HPLC-MS/MS measurements. The use of a clean source of isotopically labeled $^1O_2$ has also provided relevant information on the mechanism of formation in the cell of 8-oxodGuo, which appears to be a key mutagenic lesion of UVA irradiation. This, in turn, clearly highlights the suitability of using 8-oxodGuo **3** as a relevant and ubiquitous cellular marker of oxidative stress.

Acknowledgments

The authors thank the Brazilian research funding institutions FAPESP (Fundação de Amparo à Pesquisa do Estado de São Paulo), CNPq (Conselho Nacional para o Desenvolvimento Científico e Tecnológico), CNPq-Insituto do Milênio Redoxoma, PRONEX/FINEP (Programa de Apoio aos Núcleos de Excelência) and Fundo Bunka de Pesquisa Banco Sumitomo Mitsui and the John Simon Guggenheim Memorial Foundation (fellowship for P.D.M.). In addition, J.C. acknowledges the support of the EU through the Training and Mobility Project "CLUSTOXDNA" (MRTN-CT2003).

References

1 (a) Sies, H., (ed.), *Oxidative Stress, Oxidants and Antioxidants*, Academic Press, New York, **1991**; (b) Lindahl, T. *Nature* **1993**, *362*, 709; (c) Ames, B. N., Gold, L. S., Willett, W. C., *Proc. Natl. Acad. Sci. USA*, **1995**, *92*, 5258.

2 (a) Marnett, L. J., Burcham, P. C., *Chem. Res. Toxicol.* **1993**, *6*, 771; (b) Loft, S., Poulsen, H. E., *J. Mol. Med.* **1996**, *74*, 297; (c) Floyd, R. A., West, M., Hensley, K., *Exp. Gerontol.* **2001**, *36*, 619.

3 (a) Babior, B. M., *IUBMB Life* **2000**, *50*, 267; (b) Marnett, L. J., Riggins, J. N., West, J. D., *J. Clin. Invest.* **2003**, *111*, 583.

4 Canova, S., Degan, P., Peters, L. D., Livingstone, D. R., Voltan, R., Venier, P., *Mutat. Res.* **1998**, *399*, 17.

5 (a) Koppenol, W. H., Liebman, J. F., *J. Phys. Chem.* **1984**, *88*, 99; (b) Rahhal, S., Ritcher, H. W., *J. Am. Chem. Soc.* **1988**, *110*, 3126; (c) Koppenol, W. H., *Free Radic. Biol. Med.* **1993**, *15*, 645; (d) Wardman, P., Candeias, L. P., *Radiat. Res.* **1996**, *145*, 523; (e) Lloyd, R. V., Hanna, P. M., Mason, R. P., *Free Radic. Biol. Med.* **1997**, *22*, 885.

6 von Sonntag, C., *The Chemical Basis of Radiation Biology*, Taylor and Francis, London, **1987**.

7 Cadet, J., Vigny, P., *Bioorganic Photochemistry*, Vol. 1 (Ed. H. Morrison), Chap. 1, Wiley-Interscience, New-York, **1990**, pp. 1–272.

8 (a) Kanofsky, J. R., Wright, J., Miles-Richardson, G. E., Tauber, A. I., *J. Clin. Invest.* **1984**, 1489; (b) Steinbeck, M. J., Khan, A. U., Kanofsky, J. R., *J. Biol. Chem.* **1992**, *267*, 13425; (c) Tatsuzawa, H., Maruyama, T., Hori, K., Sano, Y., Nakano, M., *Biochem. Biophys. Res. Commun.* **1999**, *262*, 647; (d) Kiryu, C., Makiuchi, M., Miyazaki, J., Fujinaga, T., Kakinuma, K., *FEBS Lett.* **1999**, *443*, 154.

9 (a) Cadet, J., Berger, M., Douki, T., Ravanat, J.-L., *Rev. Physiol. Biochem. Pharmacol.* **1997**, *131*, 1; (b) Cadet, J., Delatour, T., Douki, T., Gasparutto, D., Pouget, J.-P., Ravanat, J.-L., Sauvaigo, S., *Mutat. Res.* **1999**, *424*, 9; (c) Cadet, J. Di Mascio, P. *Patai Series: The Chemistry of Functional groups, The Chemistry of Peroxides*, Vol. 2 (Ed. Z. Rappoport) Wiley, New York, **2005**, pp. 915–1000.

10 (a) Ravanat, J.-L., Martinez, G. R., Medeiros, M. H. G., Di Mascio, P., Cadet, J., *Tetrahedron*, **2006**, *62*, 10709; (b) Cadet, J., Ravanat, J.-L., Martinez, G. R., Medeiros, M. H. G., Di Mascio, P., *Photochem. Photobiol.*, **2006**, *82*, 1219.

11 Stenken, S., *Chem. Rev.* **1989**, *89*, 503.

12 Candeias, L. P., Steenken, S., *Chem. Eur. J.* **2000**, *6*, 475.

13 (a) Cadet, J., Berger, M., Buchko, G. W., Joshi, P., Raoul, S., Ravanat, J.-L., *J. Am. Chem. Soc.* **1994**, *116*, 7403; (b) Raoul, S., Berger, M., Buchko, G. W., Joshi, P. C., Morin, B., Weinfeld, M., Cadet, J., *J. Chem. Soc., Perkin Trans.* **1996**, *2*, 371.

14 (a) Gasparutto, D., Ravanat, J.-L., Gérot, O., Cadet, J., *J. Am. Chem. Soc.* **1998**, *120*, 10283; (b) Kino, K., Saito, I., Sugiyama, H., *J. Am. Chem. Soc.* **1998**, *120*, 7373.

15 Shafirovich, V., Cadet, J., Gasparutto, D., Dourandin, A., Geacintov, N. E., *Chem. Res. Toxicol.* **2001**, *14*, 233.

16 Misiaszek, R., Crean, C., Joffe, A., Geacintov, N. E., Shafirovich, V., *J. Biol. Chem.* **2004**, *31*, 32106.

17 Vialas, C., Pratviel, G., Claparols, C., Meunier, B., *J. Am. Chem. Soc.* **1998**, *120*, 11548.

18 (a) Kasai, H., Yamaizumi, Z., Berger, M., Cadet, J., *J. Am. Chem. Soc.* **1992**, *114*, 9692; (b) Cullis, P. M., Malone, M. E., Jones, G. D. D., Merson-Davies, L. A., *J. Am. Chem. Soc.* **1996**, *118*, 2775.

19 (a) Douki, T., Angelov, D., Cadet, J., *J. Am. Chem. Soc.* **2001**, *123*, 11360; (b) O'Neill, P., Parker, A. W., Plumb, M. A., Siebbeles, L. D. A., *J. Phys. Chem. B* **2001**, *105*, 5283.

20 Ito, K., Kawanishi, S., *Biochemistry* **1997**, *36*, 1774.

21 Saito, I., Nakamura, T., Nakatani, K., *J. Am. Chem. Soc.* **2000**, *122*, 3001.

22 Luxford, C., Dean, R. T., Davies, M. J., *Chem. Res. Toxicol.* **2000**, *13*, 665.

23 Shafirovich, V., Dourandin, A., Huang, W., Geacintov, N. E., *J. Biol. Chem.* **2001**, *276*, 24621.

24 Milligan, J. R., Aguilera, J. A., Nguyen, J. V., Ward, J. F., *Int. J. Radiat. Biol.* **2001**, *77*, 281.

25 (a) Steenken, S., Jovanovic, S. V., *J. Am. Chem. Soc.* **1997**, *119*, 617; (b) Kang, P., Foote, C. S., *J. Am. Chem. Soc.* **2002**, *124*, 4865.

26 Ravanat, J.-L., Saint-Pierre, C., Di Mascio, P., Martinez, G. R., Medeiros, M. H. G., Cadet, J., *Helv. Chim. Acta* **2001**, *84*, 3702.

27 Ravanat, J.-L., Berger, M., Benard, F., Langlois, R., Ouellet, R., van Lier, J. E., Cadet, J., *Photochem. Photobiol.* **1992**, *55*, 809.

28 Ravanat, J.-L., Cadet, J., *Chem. Res. Toxicol.* **1995**, *8*, 379.

29 Buchko, G. W., Cadet, J., Berger, M., Ravanat, J.-L., *Nucleic Acids Res.* **1992**, *20*, 4847.

30 Niles, J. C., Wishnok, J. S., Tannenbaum, S. R., *Org. Lett.* **2001**, *3*, 963.

31 Luo, W., Muller, J. G., Rachlin, E. M., Burrows, C. J., *Org. Lett.* **2000**, *2*, 613.

32 Luo, W., Muller, J. G., Burrows, C. J., *Org. Lett.* **2000**, *3*, 2801.

33 Luo, W., Muller, J. G., Rachlin, E. M., Burrows, C. J., *Chem. Res. Toxicol.* **2001**, *14*, 927.

34 Adam, W., Arnold, M. A., Grune, M., Nau, W. M., Pischel, U., Saha-Möller, C. R., *Org. Lett.* **2002**, *4*, 537.

35 Martinez, G. R., Medeiros, M. H. G., Ravanat, J.-L., Cadet, J., Di Mascio, P., *Biol. Chem.* **2002**, *383*, 607.

36 Ye, Y., Muller, J. G., Luo, W., Mayne, C. L., Shalopp, A. J., Jones, R. A., Burrows, C. J., *J. Am. Chem. Soc.* **2003**, *125*, 13926.

37 McCallum, J. E. B., Kuniyoshi, C. Y., Foote, C. S., *J. Am. Chem. Soc.* **2004**, *126*, 16777.

38 Lia, L., Shafirovich, V., Shapiro, R., Geacintov, N. E., Broyde, S., *Biochemistry* **2005**, *44*, 6043.

39 Lia, L., Shafirovich, V., Shapiro, R., Geacintov, N. E., Broyde, S., *Biochemistry* **2005**, *44*, 13342.

40 Durandin, A., Jia, L., Crean, C., Kolbanovskiy, A., Ding, S., Shafirovich, V., Broyde, S., Geacintov, N. E., *Chem. Res. Toxicol.* **2006**, *19*, 908.

41 Karwowski, B., Dupeyrat, F., Bardet, M., Ravanat, J.-L., Krajewski, P., Cadet, J., *Chem. Res. Toxicol.* **2006**, *19*, 1357.

42 Cadet, J., Douki, T., Pouget, J.-P., Ravanat, J.-L., *Methods Enzymol.* **2000**, *319*, 143.

43 (a) Cadet, J., Douki, T., Gasparutto, D., Ravanat, J.-L., *Mutat. Res.* **2003**, *531*, 5; (b) Douki, T., Ravanat, J.-L., Angelov, D., Wagner, J. R., Cadet, J., *Top. Curr. Chem.* **2004**, *236*, 1.

44 Martinez, G. R., Ravanat, J.-L., Medeiros, M. H. G., Cadet, J., Di Mascio, P., *J. Am. Chem. Soc.* **2000**, *122*, 10212.

45 Boiteux, S., Gajewski, E., Laval, J., Dizdaroglu, M., *Biochemistry* **1992**, *31*, 106.

46 (a) Bourdat, A. G., Gasparutto, D., Cadet, J., *Nucleic Acids Res.* **1999**, *27*, 1015; (b) Bourdat, A. G., Douki, T., Frelon, S.,

Gasparutto, D., Cadet, J., *J. Am. Chem. Soc.* **2000**, *122*, 4549.

47 Douki, T., Rivière, J., Cadet, J., *Chem. Res. Toxicol.* **2002**, *15*, 445.

48 Gentil, A., Le Page, F., Cadet, J., Sarasin, A., *Mutat. Res.* **2000**, *452*, 51.

49 Sheu, C. Foote, C. S., *J. Am. Chem. Soc.* **1995**, *117*, 474.

50 (a) Prat, F., Houk, K. N., Foote, C. S., *J. Am. Chem. Soc.* **1998**, *120*, 845; (b) Bernstein, R., Prat, F., Foote, C. S., *J. Am. Chem. Soc.* **1999**, *121*, 464.

51 (a) Duarte, V., Muller, J. G., Burrows, C. J., *Nucleic Acids Res.* **1999**, *27*, 496; (b) Luo, W., Muller, J. G., Rachlin, E. M., Burrows, C. J., *Chem. Res. Toxicol.* **2001**, *14*, 927.

52 Sugden, K. D., Campo, C. K., Martin, B. D., *Chem. Res. Toxicol.* **2001**, *14*, 1315.

53 Adam, W., Kurz, A., Saha-Möller, C. R., *Chem. Res. Toxicol.* **2000**, *13*, 1199.

54 Ravanat, J.-L., Saint-Pierre, C., Cadet, J., *J. Am. Chem. Soc.* **2003**, *125*, 2030.

55 Doddridge, A. Z., Cullis, P. M., Jones, G. D. D., Malone, M. E., *J. Am. Chem. Soc.* **1998**, *120*, 10998.

56 (a) Hickerson, R. P., Chepanoske, C. L., Williams, S. D., David, S., Burrows, C. J., *J. Am. Chem. Soc.* **1999**, *121*, 9901; (b) Johansen, M. E., Muller, J. G., Xu, X., Burrows, C. J., *Biochemistry* **2005**, *44*, 5660.

57 Perrier, S., Gasparutto, D., Cadet, J., Favier, A., Ravanat, J.-L., *J. Am. Chem. Soc.* **2006**, *128*, 5703.

58 Raoul, S., Cadet, J., *J. Am. Chem. Soc.* **1996**, *118*, 1982.

59 Duarte, V., Gasparutto, D., Yamaguchi, L. F., Ravanat, J.-L., Martinez, G. R., Medeiros, M. H. G., Di Mascio, P., Cadet, J., *J. Am. Chem. Soc.* **2000**, *122*, 12622.

60 Buchko, G. W., Cadet, J., *Photochem. Photobiol.* **2006**, *82*, 191.

61 Chworos, A., Coppel, Y., Dubey, I., Pratviel, G., Meunier, B., *J. Am. Chem. Soc.* **2001**, *123*, 5867.

62 Cadet, J., Douki, T., Ravanat, J.-L., *Environ. Health Perspect.* **1997**, *105*, 1034.

63 Douki, T., Martini, R., Ravanat, J.-L., Turesky, R. J., Cadet, J., *Carcinogenesis* **1997**, *18*, 2385.

64 (a) Ravanat, J.-L., Douki, T., Duez, P., Gremaud, E., Herbert, K., Hofer, T., Lassere, L., Saint-Pierre, C., Favier, A., Cadet, J., *Carcinogenesis* **2002**, *23*, 1911; (b) Collins, A. R., Cadet, J., Möller, L., Poulsen, H., Viña, J., *Arch. Biochem. Biophys.* **2004**, *423*, 57; (c) Gedik, C. M., Collins, A. R., European Standards Committee on Oxidative DNA Damage (ESCODD) . *FASEB J.* **2005**, *19*, 82.

65 (a) Floyd, R. A., Watson, H., Wong, P. K., Altmiller, D. H., Rickard, R. C., *Free Rad. Res. Commun.* **1986**, *1*, 163; (b) Kasai, H., *Mutat. Res.* **1997**, *387*, 147.

66 Cadet, J., Douki, T., Badouard, C., Favier, A., Ravanat, J.-L., *Oxidative Damage to Nucleic Acids* (Eds M. Evans M. Cooke), Landes Biosciences, Georgetown, TX, **2007**, pp. 1–13.

67 (a) Collins, A. R., Dusinska, M., Gedik, C. M., Stétina, R., *Environ. Heath Perspect.* **1996**, *104* (suppl. 3), 465; (b) Epe, B., Hegler, J., *Methods Enzymol.* **1994**, *234*, 122.

68 Pouget, J.-P., Douki, T., Richard, M.-J., Cadet, J., *Chem. Res. Toxicol.* **2000**, *13*, 541.

69 Sauvaigo, S., Petec-Calin, C., Caillat, S., Odin, F., Cadet, J., *Anal. Biochem.* **2002**, *30*, 107.

70 Pouget, J.-P., Frelon, S., Ravanat, J.-L., Testard, I., Odin, F., Cadet, J., *Radiat. Res.* **2002**, *157*, 589.

71 Douki, T., Ravanat, J.-L., Pouget, J.-P., Testard, I., Cadet, J., *Int. J. Radiat. Biol.* **2006**, *82*, 119.

72 Frelon, S., Douki, T., Ravanat, J.-L., Pouget, J.-P., Tornabene, C., Cadet, J., *Chem. Res. Toxicol.* **2000**, *13*, 1002.

73 Douki, T., Ravanat, J.-L., Angelov, D., Wagner, J. R., Cadet, J., *Top. Curr. Chem.* **2004**, *236*, 1.

74 Ravanat, J.-L., Di Mascio, P., Martinez, G. R., Medeiros, M. H. G., Cadet, J., *J. Biol. Chem.* **2000**, *275*, 40601.

75 Ravanat, J.-L., Sauvaigo, S., Caillat, S., Martinez, G. R., Medeiros, M. H. G., Di Mascio, P., Favier, A., Cadet, J., *Biol. Chem.* **2004**, *385*, 17.

76 Fisher-Nilsen, A., Loft, S., Jensen, K. G., *Carcinogenesis* **1993**, *14*, 2431.

77 Rosen, J. E., Prahalad, A. K., Williams, G. M., *Photochem. Photobiol.* **1996**, *64*, 117.

78 Zhang, X., Rosenstein, B. S., Wang, Y., Lebwohl, M., Mitchell, D. L., Wei, H., *Photochem. Photobiol.* **1997**, *65*, 119.

79 Kvam, E., Tyrrell, R. M., *Carcinogenesis* **1997**, *18*, 2379.

80 Douki, T., Perdiz, D., Grof, P., Kuluncsics, Z., Moustacchi, E., Cadet, J., Sage, E., *Photochem. Photobiol.* **1999**, *70*, 184.

81 Duez, P., Hanocq, M., Dubois, J., *Carcinogenesis* **2001**, *22*, 771.

82 Courdavault, S., Baudoin, C., Charveron, M., Favier, A., Cadet, J., Douki, T., *Mutat. Res.* **2004**, *556*, 135.

83 Besaratinia, A., Synold, T. W., Chen, H.-H., Chang, C., Xi, B., Riggs, A. D., Pfeifer, G. P., *Proc. Natl. Acad. Sci. USA* **2005**, *102*, 10058.

84 Kozmin, S., Slezak, G., Reynaud-Angelin, A., Elie, C., de Rycke, Y., Boiteux, S., Sage, E., *Proc. Natl. Acad. Sci. USA* **2005**, *102*, 13538.

85 Cadet, J., Sage, E., Douki, T., *Mutat. Res.* **2005**, *571*, 3.

86 Mouret, S., Baudoin, C., Charveron, M., Favier, A., Cadet, J., Douki, T., *Proc. Natl. Acad. Sci. USA* **2006**, *103*, 13765-.

87 (a) Pascucci, B., Maga, G., Hubscher, U., Bjoras, M., Seeberg, E., Hickson, I. D., Villani, G., Giordano, C., Cellar, L., Dogliotti, E., *Nucleic Acids Res.* **2002**, *30*, 2124; (b) Barnes, D. E., Lindhal, T., *Annu. Rev. Genet.* **2004**, *38*, 445; (c) Hazra, T. K., Das, A., Das, S., Choudhury, S., Kow, Y. W., Roy, R., *DNA Repair, (Amst)* **2007**, *6*, 470–480.

88 Evans, M. D., Dizdaroglu, M., Cooke, M. S., *Mutat. Res.* **2004**, *567*, 1.

89 Bodepudi, V., Shibutani, S., Johnson, F., *Chem. Res. Toxicol* **1992**, *5*, 608.

90 Butenandt, J., Burgdorf, L. T., Carell, T., *Synthesis* **1999**, 1085.

91 Varaprasad, C. V., Bulychev, N., Grollman, A. P., Johnson, F., *Tetrahedron Lett.* **1996**, *37*, 9.

92 Johnson, F., Dorman, G., Rieger, R. A., Marumoto, R., Iden, C. R., Bonala, R., *Chem. Res. Toxicol.* **1998**, *11*, 193.

3
Modified DNA Bases: Probing Base-Pair Recognition by Polymerases

Eric T. Kool

3.1
Introduction

3.1.1
The Importance of Understanding DNA Polymerases

Since DNA polymerase I was first characterized by Kornberg in 1956 [1] there has been a rich history of the study of DNA polymerase enzymes. Many tens – if not hundreds – of these enzymes have been identified and examined experimentally [2–7]. Indeed, some single examples of these proteins (e.g., HIV reverse transcriptase; HIV-RT) have been addressed with thousands of published reports. Today, polymerases from all three kingdoms of life are recognized and are even available commercially. In addition, many biologically different and functionally distinct classes of polymerases have now been identified; for example, there are more than 15 different DNA polymerases active in human cells [8].

Despite this long period of progress, the study of DNA polymerases has remained a vital field. The reasons for this are many-fold. The first reason is the continuing importance of polymerases in medicine; for example, polymerase inhibitors have traditionally been one of the most important classes of anticancer agents [9], and compounds targeted towards HIV-RT were the first to be developed as successful anti-AIDS drugs [10]. The second reason is the central role of polymerases in biotechnology; for example, many different thermostable polymerases and reverse transcriptases are now used for various amplification methodologies and biological manipulations. The third reason is the ongoing discovery of polymerases having distinct biological roles in the cell, which leads to wide variations in structure, sequence, and mechanism. A fourth reason – and one which is related to the third reason – is the recent interest in DNA repair mechanisms and the roles that specialized polymerases play in them [11].

Modified Nucleosides: in Biochemistry, Biotechnology and Medicine. Edited by Piet Herdewijn

ISBN: 978-3-527-31820-9

3.1.2
The Utility of Modified Nucleobases in Probing Mechanisms

Much information about DNA polymerases has been learned from studies of their structures alone and with DNA in the active site, and from biochemical experiments with matched and mismatched pairs of the natural bases. However, non-natural bases can test mechanistic questions that are difficult or impossible to address using natural DNA alone. Some of these questions are addressed to basic physico-chemical interactions and mechanisms that allow these enzymes to function, while others are aimed at designing new forms of DNA and base pairs that can be replicated. For example, on the basic science side, we can ask how Watson–Crick hydrogen bonds play a role in DNA replication fidelity, and how the steric size and shape of nucleotides affect the outcome [12, 13]. As all natural nucleobases have hydrogen bonding groups, and do not vary widely in their architecture, designed molecules might be more useful than their natural counterparts in answering such questions.

The design of modified bases is not only an academic exercise aimed at basic understanding: one long-term goal of nucleic acids chemists in general is to see whether humans can design new base pairs (and even whole new forms of DNA) that can be replicated [14–16]. Thus, the synthesis and study of modified bases can reveal how broadly the structures can vary within the polymerase active site, outlining the future design parameters for man-made DNA-like systems.

3.1.3
The Scope of this Chapter

This chapter will outline the current state of research on the development and study of modified bases with polymerase enzymes. For reasons of space limitation, details of conjugates of natural DNA bases will not be addressed. Nonetheless, these conjugates are very important biotechnologically, and fluorescently conjugated bases are widely used in the detection and sequencing of DNA and RNA [17]. For example, 5-substitution of alkyl, alkenyl, and alkynyl tethers on the base uracil is a common label-carrying substitution for thymine. Indeed, the abundant literature on such conjugates alone is sufficient for a review. However, such bases do not generally alter the mechanisms by which DNA bases recognize each other, or by which polymerases recognize pairs. In the present discussion, attention is focused primarily on bases that are more radically altered, either by changing the Watson–Crick recognition face, or by changing the ring systems altogether.

3.2
Basic Principles and Methods in Replication

3.2.1
The Chemistry of Polymerases

Before delving into base modification it is worthwhile first briefly to review polymerase biochemistry [18]. Most DNA polymerases are template-dependent, which means that their substrate is double-stranded DNA (a template–primer duplex) having a recessed 3′ end (the primer terminus). The locally reactive group is the free hydroxyl on the 3′ nucleoside of the primer, which acts as a nucleophile to displace the leaving group pyrophosphate on a nucleoside triphosphate (dNTP) monomer. The polymerase catalyzes this reaction by binding the dNTP together with the DNA such that the new nucleobase is correctly positioned in a pair opposite the template base immediately downstream from the primer end. Also helping the chemistry is binding of the phosphate by a Mg^{2+} (occasionally Mn^{2+}) ion bound near the active site. Base-pair synthesis requires multiple steps, including opening of the polymerase to allow access of the new nucleotide, binding of dNTP, closing, making the new bond, and translocating to the next position. Detailed kinetic treatments of this process are now available for a few enzymes [19, 20].

Two other polymerase activities should be mentioned at this point as they also impact modified bases and pairs. The first activity is *base-pair extension*, which is simply the action of adding the next base beyond a base pair already in place. It is formally the same as the original pair synthesis, but from the standpoint of a modified pair it is different, since the synthesis of a modified pair might be different from the synthesis of the next pair which, in most cases, is a natural pair. This forces the modified pair one position further down into the enzyme, making non-natural DNA–protein contacts at new positions. It is not uncommon for a modified pair to be synthesized with high efficiency, but then to be extended with low efficiency (see below); of course, efficient ongoing DNA replication requires that a pair be made and extended all the way out of the enzyme (four or five pairs downstream).

The final polymerase function worth mentioning is that of 3′-exonuclease (3′-exo) activity. Some polymerases – but not all – have this activity in a site near the polymerase site. This mechanism serves to increase replication fidelity overall, because mismatched primer bases at the end of the DNA are shuttled to the exonuclease site (where they are removed) more often than correctly matched bases [3]. When studying modified bases and pairs with polymerases it is important to know whether there is a 3′-exo activity present and how the modified base(s) affect this activity.

3.2.2
Different Classes of Polymerases

It is now recognized that there are many different classes of polymerases, both in terms of differences in sequence and structure, and differences in function [2–7]. It must be stressed, therefore, that conclusions drawn from the mechanistic studies of one enzyme may not always hold for others, especially for phylogenetically different

Table 3.1 Classes of DNA polymerases, with examples from each family shown.

DNA family	Polymerase
A	pol I, T7, Taq
B	pol II, RB69, T4, pol alpha
C	pol III
X	pol beta, TdT
Y	Dpo4, pol eta, pol kappa, pol iota
RT	HIV-RT, AMV-RT

RT, reverse transcriptase.

classes of enzymes. Indeed, widely different responses have been observed of different enzymes to modified DNA bases and base pairs (see below). Families of enzymes, along with specific examples of polymerases from each class, are listed in Table 3.1. The families include A, B, C, X, Y, and reverse transcriptases (RT). Not only are these classes different in their sequences, but they are also distinct in their behavior. For example, the A-family enzymes are replicative polymerases and typically operate with high fidelity, making mismatch errors (using natural bases) only once in ~10 000 base pairs. In contrast, the Y-family enzymes are classified as repair enzymes, and are often error-prone, making errors (in some cases) once in every ten to 100 pairs [4].

3.2.3
Methods Used in Polymerase Studies

In the studies described below, reference will be made to a number of methods that are used to study DNA polymerases. Of course, structural studies are important, serving to reveal the geometries and locations of enzyme side chains and DNA, and the responses made when the DNA changes, such as with mismatched pairs. Single-crystal X-ray structures have been solved for several DNA polymerases, and this certainly adds mechanistic insight. However, such studies do not replace functional studies, which address interactions at the transition state, and address kinetics and thermodynamics rather than structure alone.

Functional studies can be carried out either qualitatively or quantitatively, and with varying levels of precision. For example, a simple qualitative gel-based (PAGE) assay can be conducted using a labeled primer. PAGE analysis can show a product yield (as when a single nucleotide is being added) compared with the amount of starting material remaining at a given time and set of conditions (dNTP concentration, polymerase and DNA concentration, etc.). For more quantitative information, it is possible to measure yields as a function of varied [dNTP] or varied time, and use Michaelis–Menten methods in the steady state to determine efficiency (commonly reported as k_{cat}/K_M or V_{max}/K_M) [21]. This yields numerical information about the transition state in the rate-limiting step. For a greater level of detail, stopped-flow methods can be used to obtain pre-steady-state kinetic information [20], which allows one to separate binding effects (as K_M) and chemistry rate effects (k_{cat}) with confidence, and to analyze individually the multiple steps of the reaction.

3.3 Alternative Hydrogen-Bonding Schemes

3.3.1 Thioguanine-Pyridone

The first serious attempts made to modify bases and study them with polymerases involved altering the hydrogen bonding groups. During the late 1980s, two laboratories began working simultaneously on new pair designs. One very interesting case was that of Rappaport, who designed a base pair between 6-thioguanine and 5-methyl-2-pyrimidinone (Figure 3.1) [23]. The design concept was to make a pair that could function in the presence of the natural base pairs; that is, to have an "orthogonal" pair having bases that did not pair well (during replication) with A, C, G, or T. Thioguanine, with sulfur at the 6-position, is (at least in principle) too large to pair well with cytosine, and might have a steric clash between the 6-S and the 4-amino group of C. The partner, pyrimidinone, lacks the amino group of C and thus it was thought that it might not pair as well opposite G as one H-bonding interaction was absent.

The early data (using gel-based assays) showed that the pair was indeed replicated by the Klenow enzyme lacking exonuclease (exo-) [24]. Substantial cross-pairing of cytosine with Gs and of guanosine with Th was seen, however. More recently, Rappaport carried out steady-state kinetics studies with T7 DNA polymerase (including exonuclease activity), using three pairs: adenine/thymine, hypoxanthine/cytosine (H/C) and 6-thiopurine/5-methyl-2-pyrimidinone (HS/TH) (Figure 3.1). Of the non-natural pairs, H/C was efficiently synthesized, while HS/TH was moderately well synthesized [25]. Although further studies are needed, this last six-base set (three different pairings) might be suitable for applications that require general replication and amplification.

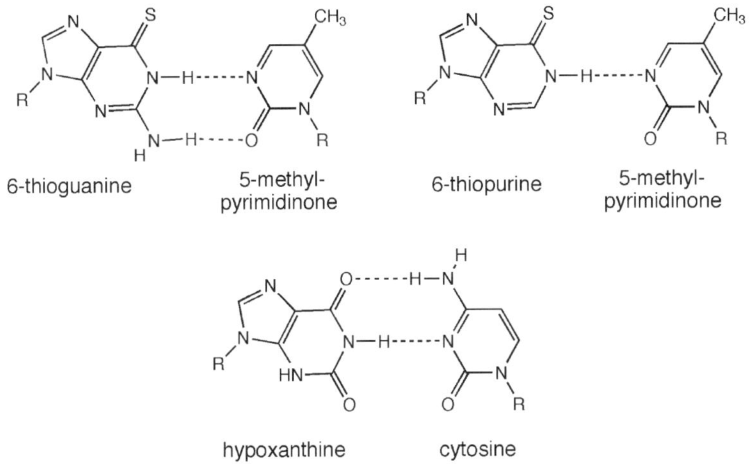

Figure 3.1 Three modified base pairs studied by Rappaport [22–24]. R = deoxyribose.

3.3.2
Benner Hydrogen-Bonding Variants

Also during the late 1980s, the Benner laboratory conceived the generalized concept of rearranging hydrogen-bonding groups on the bases, adopting the purine–pyrimidine scaffold as the constant factor. An important report in 1990 [25] drew structures of several possible pairs (Figure 3.2). It was recognized that, given three H-bonding donor (D) or acceptor (A) positions, there are eight possible combinations (DDD, ADD, DAD, DDA, AAD, ADA, DAA, AAA) on a given pyrimidine or purine framework, and thus eight different pairing schemes. This followed the early observation (by A. Rich) that isocytosine might pair with isoguanine in a way analogous to the canonical pairs [26].

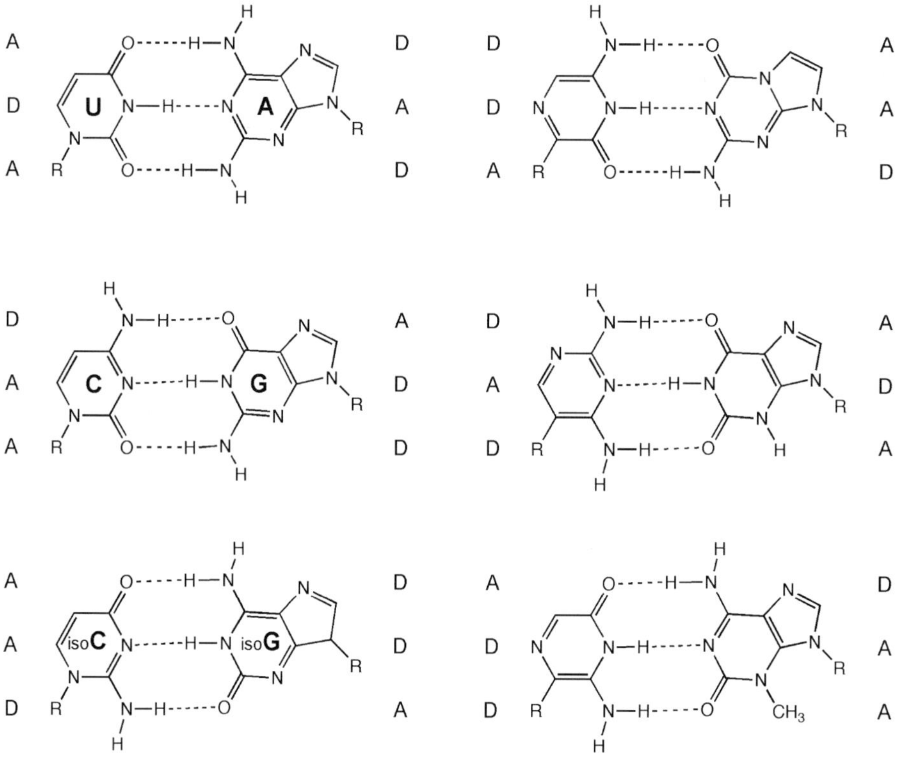

Figure 3.2 Six different modified pairs drawn by Benner in a 1990 report [25]. The structures were meant to represent some of the possible arrangements of H-bond donors (D) and acceptors (A) in Watson–Crick-like pairs. A number of examples have since been tested by synthesis, and isoC–isoG (shown) has been studied in numerous experiments. R = deoxyribose.

Over the subsequent years the Benner group [27–34] and other laboratories [35, 36] have explored the synthesis and study of some of these pair types using gel-based qualitative studies. The first – and among the most successful – of these was the isoC–isoG pair. Early on it was shown that a DNA polymerase was able to construct such a pair [25], and in a landmark subsequent experiment it was shown that the isoC–isoG pair could be used in *in-vitro* translation experiments to encode a non-natural amino acid [34]. This was the first successful effort to "expand the genetic code" with a non-natural base pair.

Subsequent studies showed that not only DNA polymerases could function with this pair. Dervan showed that T7 RNA polymerase could transcribe isoG, yielding RNAs with isoC [35]. This was demonstrated as a method for introducing a single label at a specific position in an enzymatically synthesized (transcribed) RNA.

Despite the many reports published on the pairing and polymerase properties of the isoC–isoG pair, there are apparently no existing quantitative studies measuring the kinetic effectiveness of that pair. The qualitative experiments suggest that its efficiency approaches that of a natural A–T or G–C pair with some common polymerases [32]. However, it is also clear that in the presence of mixtures of natural nucleotides also containing disoCTP, polymerases using the template base isoG incorporate either isoC or T reasonably well [33, 35]. This may be because isoG commonly exists in equilibrium with the tautomer 2-hydroxyadenine, which forms normal hydrogen-bonded geometry with thymine [33]. Thus, there is a practical problem with the specificity of the isoC–isoG pair when it is used in replication experiments.

Despite this limitation, the isoC–isoG pair is important as a precedent for many studies to follow. Moreover, the selective pairing (in DNA hybridization) of the two bases led to the early adoption of that pair in a commonly used diagnostic assay for detection of HIV and other viral nucleic acids [37].

The Benner laboratory has also studied other hydrogen-bonding arrangements of the classes shown in Figure 3.2. With polymerase enzymes it appears that (of the pairs tested so far) none has been as successful as the isoC–isoG pair. For example, the "Z–P" pair appears to suffer from mispairing between G and Z, possibly because of tautomerization of Z, leading it to resemble C [30, 38]. One interesting way to avoid tautomerization mispairing is to use steric effects to avoid it; for example, Benner recently used 2-thioT as a replacement for T, which allows it to pair with A but not 2-hydroxyA. In that report, three base pairs (A–2thioT, G–C, isoG–isoC) were used successfully and simultaneously in PCR amplifications [33].

On the whole, the strategy of rearranging hydrogen-bonding groups appears to be a potentially successful one in the search for new base pair substrates for polymerases, especially in cases where mispairing by tautomerization can be avoided. To date, there is a lack of kinetic data that would help in comparing such pairs with other designs (see below), but the recent studies using the isoC–isoG pair in amplifications [33] and the promising activity of the Rappaport three-pair system [25] suggests that efficiency and selectivity are approaching the levels needed for practical applications.

3.4
Non-Polar Nucleoside Isosteres

3.4.1
The Concept of Nucleobase Isosteres

In 1994, Schweitzer and Kool introduced the concept of non-polar isosteres of the natural DNA bases [39], proposing four structures that were the closest possible structural mimics of the natural bases, but which lack NH and O hydrogen-bonding groups. The structures started with benzene as a framework for pyrimidines, and indoles (and later, benzimidazoles) [40] as the architecture for purines. Exocyclic carbonyl groups were replaced with C−F groups, and N−H groups with C−H (Figure 3.3). The intended purpose of these molecules was to use them as probes of the importance of hydrogen bonding to base pairing in general. Two of these compounds, difluorotoluene (F, a thymine mimic) and 4-methylindole (an adenine mimic), were prepared and studied early on in several investigations [40–44].

3.4.2
Synthesis, Structure, and Physical Properties

The synthesis of indole and benzimidazole deoxynucleosides is generally simple, and relies on the displacement of a 1-chloro group from an ester-protected deoxyribose derivative. However, the synthesis of the pyrimidine isosteres was significantly more difficult, as they were C-glycosides. At the time, organocadmium- or organozinc-mediated chemistry was adopted to react with the same chlorosugar as was used in the N-nucleosides [44]. The disadvantage of this approach was poor yields and predominance of the undesired α-isomer. Later studies performed in many laboratories have

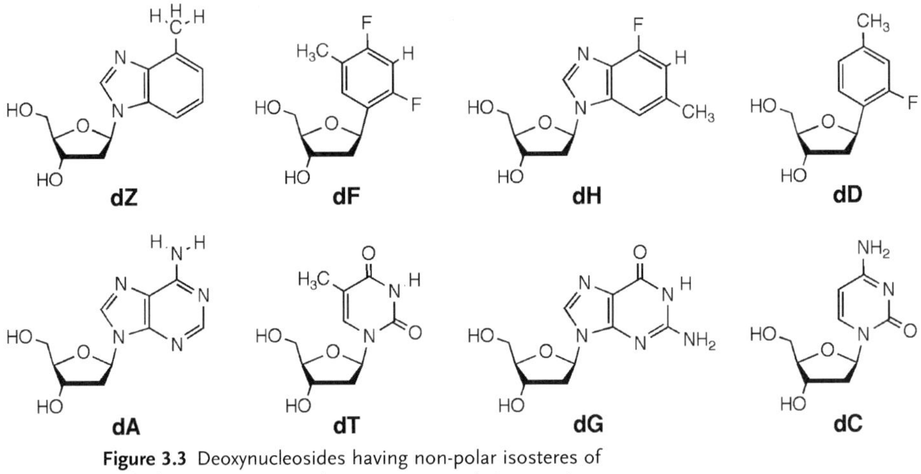

Figure 3.3 Deoxynucleosides having non-polar isosteres of nucleobases (top) [39] with comparison to their natural counterparts.

improved the synthetic approaches somewhat [45, 46], although the development of more efficient methods would still be welcome. Nevertheless, the difluorotoluene deoxyriboside and the 4-methylindole deoxyriboside are now commercially available as suitably protected phosphoramidite derivatives, and can thus be incorporated into oligonucleotides by virtually any DNA synthesis laboratory.

Kool and coworkers described the analysis of the structure of deoxyadenosine and thymidine non-polar isosteres using both solution-phase NMR and single-crystal X-ray analysis [40, 47]. Both compounds adopted conformations very close to those of their natural congeners. Because of the electronegativity of fluorine, the question of whether difluorotoluene would form hydrogen bonds with adenine in low-polarity solvents (where H-bonding strength is magnified relative to water) was examined [41]. However, no evidence for hydrogen bonding was found for the fluorine nucleobase. This was also confirmed by measurements showing very low polarity of dF (by water–octanol partitioning) [48], by poor pairing of F with adenine in DNAs [49] (see below), and recently, by X-ray analysis of F positioned opposite A in an RNA helix [50]. A number of *ab-initio* and semi-empirical theoretical studies of F in DNA have concluded that hydrogen bonding is either weak (in the gas phase) or non-existent (in water) [51–53]. It is generally true of all non-polar isosteres studied to date, that they are strongly hydrophobic and pair with little or no selectivity opposite natural bases (see below). Nevertheless, when situated in duplex DNA, they are commonly found by structural studies to be stacked within the helix; two NMR structures of DNA involving F or 4-methylbenzimidazole (Z, another adenine mimic) show the non-polar bases stacked within the DNA opposite their natural partners [54, 55].

3.4.3
Base-Pairing Properties

Although this chapter focuses on the properties of modified DNA bases with polymerase enzymes, it is useful to know the properties of such compounds in DNA in the absence of these enzymes. This gives a baseline of properties of the DNA and modifications alone, thus helping to clarify what constraints the enzymes add. In base-pairing experiments with synthetic DNAs, it has been found that the non-polar nature of the above nucleobase isosteres dominates their pairing properties. For example, when F is paired opposite an adenine in a 12-base-pair duplex, the melting temperature (T_m) of the helix drops by about 12 °C, and the free energy of helix formation becomes ca. 4 kcal mol^{-1} less favorable [49]. Importantly, while thymine at the same position as F shows a strong preference for adenine (as measured by T_m or free energy), F shows very little pairing preference among the natural bases. The interpretation of this result is that hydrogen bonds must be important for the stabilization of DNA base pairs at the center of a DNA duplex. Of course, a significant part of the observed destabilization is the cost of desolvation of the polar base (adenine in the F–A pair). Indeed, when both thymine and adenine were replaced simultaneously with non-polar isosteres, the T_m and free energy recovered substantially [56]. However, the F–Z pair was still not as stable as a natural T–A pair in that context.

Virtually all of the non-polar nucleoside isosteres show very similar behavior when paired in DNA. They all destabilize the helix, and show little or no selectivity for pairing for any of the natural bases. In addition, they all pair more favorably with other non-polar partners, although little selectivity among them is seen. Overall, the results suggest a substantially positive influence of Watson–Crick hydrogen bonds on selectivity of DNA pairing (in DNA alone), and they point out the large cost of desolvation when natural bases have their polar edges removed from solvent and placed against a non-polar surface [56].

Interestingly – and importantly– non-polar nucleoside isosteres behave quite differently at the ends of a DNA helix. This is significant in the context of polymerase enzymes because DNA synthesis by polymerases occurs at the end of DNA, not the center. At the helix end, an F–A pair is actually more stabilizing than a natural T–A pair, and similar effects are seen for other non-polar isosteres [56]. Selectivity remains low, although (again, interestingly) even natural bases lose much of their pairing selectivity when they are present in the helix terminal pair. The reasons for much greater stability of isostere-base pairs at the end of the duplex are twofold. First, non-polar isosteres stack more strongly than their natural counterparts do against the adjacent bases [48], and stacking at the helix end is an important stabilizing factor for DNA duplexes. In addition, the polar partner (e.g., adenine paired opposite F) has access to solvent at the end base pair, and so the cost of desolvation is much less, perhaps as small as zero.

3.4.4
Polymerase Behavior

The first experiments to test polymerase activity with non-polar nucleoside isosteres were reported in 1997, and focused on the difluorotoluene isostere of thymine [47, 57]. The initial investigations tested the well-studied enzyme DNA Pol I (Klenow fragment) from *Escherichia coli*, an A family replicative polymerase. In order to avoid possible confusion from simultaneous synthesis and editing of the pair, Moran *et al.* used a commercially available exonuclease-free mutant (Kf exo-). Using a primer–template duplex with one dF residue immediately downstream of the primer end, these authors added single natural nucleotides (in separate experiments) and quantitated products of addition of the single nucleotides catalyzed by this enzyme. Steady-state kinetics were also measured for the process.

Surprisingly, the results showed that the Kf enzyme inserted adenine opposite this thymine shape mimic with high efficiency (within fivefold of the natural base pair activity; see Figure 3.4) [57]. Also surprising was the selectivity of the process: adenine was added more than 1000 times more efficiently than any other nucleotide. The difluorotoluene-containing dTTP analogue (dFTP) was also prepared and studied with natural nucleotides in the template [47]. Results showed that dFTP was inserted with reasonably high efficiency (about 40-fold below wild-type) opposite adenine, and that the selectivity of insertion opposite A (as compared with opposite T, C, G) was the same as the natural pair, at over 1000-fold.

These experiments suggested that the DNA Pol I enzyme could synthesize base pairs efficiently and with high fidelity, without the need for Watson–Crick hydrogen

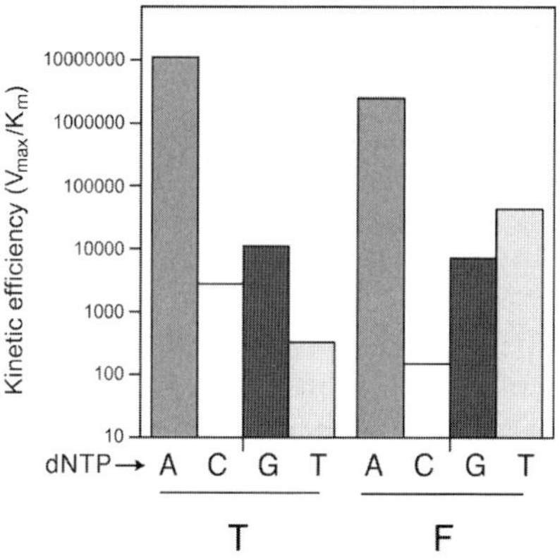

Figure 3.4 Steady-state kinetics data for synthesis of modified base pairs by the Klenow (Kf) enzyme [58]. Data are shown for insertion of natural nucleotides (dATP, dCTP, dGTP, dTTP) opposite thymine (T) or difluorotoluene (F). The V_{max}/K_m data show enzymatic efficiency on a log scale. Note the similar efficiencies for insertion of dATP opposite thymine or its non-polar isostere (F).

bonds. This concept has since been proven in multiple laboratories using over a dozen different non-hydrogen-bonded pairs [15, 16, 58]. After these initial experiments, Moran *et al.* hypothesized that selective DNA replication might gain its specificity from steric effects more than from hydrogen-bonding effects. Later experiments using compounds with varied size and shape tested this hypothesis directly (see below).

After the initial experiments with the dF analogue of thymidine, Guckian *et al.* also studied the dZ compound – the non-polar isostere of deoxyadenosine with the same enzyme [59]. The results showed that this compound was selectively replicated opposite T, as one might expect based on the analogue's adenine-like shape. However, this analogue was a considerably poorer enzyme substrate than was difluorotoluene. Two possible explanations for this low activity were considered. The first suggestion was that thymine (its natural partner) is more strongly solvated than adenine (which was used as a partner for the earlier dF compound). Thus, pairing thymine with Z might invoke a greater energetic cost at the transition state for base pair synthesis. A second hypothesis was that Z is a flawed shape mimic of adenine, since it has a proton on its pairing edge that adenine lacks [40]. This explanation would require that the enzyme be sensitive to a very small structural difference, which later experiments confirmed (see below).

3.4.5
Other Classes of Polymerases

Subsequent to the early studies of non-polar isosteres with A-family enzymes, Kool and coworkers carried out a survey of the successful difluorotoluene (F) mimic of thymidine with several other enzymes. The early data suggested that the B family enzyme RB69 could use this compound as an efficient substrate (E.T. Kool and

W. Konigsberg, unpublished data), and reverse transcriptases (in their DNA synthesis mode) also worked efficiently with analogue dF [60]. However, with pol beta (X family) and pol alpha (B family), the dF analogue was a very poor substrate. Although it is not yet clear why this is the case, it is clear that different enzymes have very different polymerization properties.

More recently, Kool and collaborators examined non-polar isosteres with Y-family enzymes. To date, Dpo4, pol kappa, and pol eta have been examined [61–63]. Interestingly, dF is a very poor substrate for these repair enzymes, with efficiencies commonly 1000-fold below that of natural thymine. The current hypothesis for explaining this poor activity is that these repair enzymes actually require Watson–Crick hydrogen bonding for activity, in stark contrast to pol A family replicative enzymes. It is possible that these sterically "loose" enzymes need the hydrogen bonding to properly align the nucleotide for bond formation in the active site [61].

3.4.6 Summary of Watson–Crick H-Bonding Effects in Polymerase Active Sites

The observation of quantitatively high efficiency and fidelity with non-polar nucleoside isosteres has paved the way for several laboratories to pursue the design of whole modified base pairs that function effectively with polymerases (see below). Published examples from several groups have made it clear that individual modified base pairs can be synthesized enzymatically, without any need for Watson–Crick hydrogen bonds [15, 16, 56, 58]. However, when one base is polar and the other is non-polar, the cost of desolvation can be significant; this issue can be avoided if both bases are non-polar [59].

Three exceptions to the above generalization should be noted. First, it appears that Watson–Crick hydrogen bonds are still helpful in stabilizing and selectively pairing bases in a DNA duplex [56, 57]. Thus, it is not yet clear whether many non-H-bonding pairs (or several consecutive pairs) can be made enzymatically in one DNA, as the duplex may have a tendency to fall apart. Second, it is clear that some enzymes (Y-family polymerases in particular) may have an entirely different H-bonding requirement than family A enzymes [60, 61]. Third, although Watson–Crick hydrogen bonds may not be important for many common replicative enzymes, other hydrogen bonds (e.g., minor groove interactions) may be important for the extension of base pairs [64] (see below).

3.5 Non-Polar Steric Probes

3.5.1 Isomers of Hydrocarbons Illustrate Hydrophobic "Packing" Effects

Once it became clear that base pairs could be enzymatically synthesized without polar groups on the bases, a number of research groups began to generate a range of

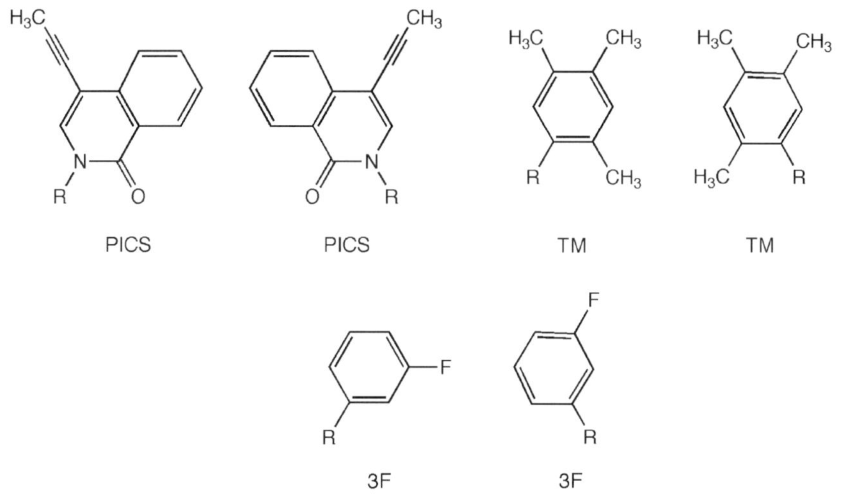

Figure 3.5 Examples of non-natural base pairs reported by Romesberg and Schultz [66–76]. The pairs shown are all relatively efficiently synthesized by Kf polymerase. R = deoxyribose.

structures based on simple hydrocarbon frameworks. If the earlier experiments with non-polar nucleoside isosteres made it difficult to explain the fidelity of DNA synthesis from Watson–Crick hydrogen bonding alone, it was clear that steric effects were the leading candidate for a more satisfying hypothesis. To test such a hypothesis, a variety of DNA bases having varied size and shape were needed.

One of the most productive groups in the study of modified bases with polymerases has been that of Romesberg and Schultz, who have reported dozens of non-polar base designs (Figure 3.5) [15, 65–75]. Two recent series of substituted benzenes prepared in those laboratories are especially useful in examining the effects of sterics; the authors consider hydrophobic contacts to be important, and thus varied sterics can be interpreted as altering the "hydrophobic packing" in a given modified base pair. One series used methyl-substituted benzenes [69, 72], while the other used fluorine-substituted benzenes [70]. The chief focus of the studies was to find non-polar bases that paired well with each other, rather than studying pairing with natural bases.

Experiments with the methyl-substituted series using Kf exo-enzyme showed that there were significant effects for mono-, di-, and tri-methylbenzenes depending on the substitution position. For example, 3-substitution was disruptive for pairing, presumably by a steric clash, while 2,4-disubstitution was favorable. The most efficiently synthesized pairs were the self-pairs involving two 2,4-dimethylbenzenes (D-D) or two 2,4,5-trimethylbenzenes (TM-TM), which were handled by the Klenow polymerase with efficiency approaching that of a natural pair. Those two particular isomers were reported previously by Ren *et al.* [43] and Chaudhuri *et al.* [44], but studies with polymerases were first published by Romesberg and coworkers.

The Hunziker [76], Romesberg [70], and Kool [77–79] laboratories have all studied polyfluorinated benzenes as bases in DNA, and the latter two have studied them with polymerases [70, 79]. In studies with a six-membered fluorine-substituted benzene series, Henry *et al.* observed that 3-fluorobenzene paired well with another copy of itself, presumably because one base was able to rotate 180 ° to allow a fluorine to move out of the way of the other. Another observation was that 2-fluorobenzene paired almost as well opposite adenine as did 2,4-difluorobenzene.

3.5.2
Systematic Size Variants

Many of the studies of modified nucleosides had examined different ring systems and a variety of substitution patterns spread over these ring systems in almost all directions. Studies with polymerases have shown wide variations in activity, but often provided little basic insight into the physical factors that contribute towards making substrates efficient or inefficient. Since a steric-based hypothesis had been developed to explain the selection of bases for pairing during replication, there was a need to test this hypothesis directly in a systematic way.

Using the previous difluorotoluene skeleton as a starting point, Kim and Kool prepared a series of non-polar thymine base analogues, all having shapes similar to that of thymine but with varying size [80]. This was achieved by varying the exocyclic substituents at the 2,4 positions (Figure 3.6) with a series of atoms of varied size (H, F, Cl, Br, I). This gave a variation across the series of 1.0 Å, thus allowing size effects to be tested over a carefully controlled range of perturbation. This series was prepared as nucleosides and nucleotides, and it was shown that they all are similarly hydrophobic

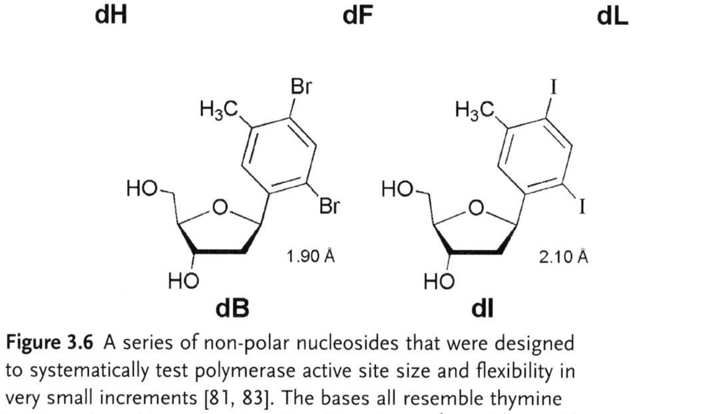

Figure 3.6 A series of non-polar nucleosides that were designed to systematically test polymerase active site size and flexibility in very small increments [81, 83]. The bases all resemble thymine in shape, but with varying size. Bond lengths (in Å) at the variable 2,4 positions are listed.

and pair similarly poorly in DNA [83], as was originally found for difluorotoluene nucleoside (dF) itself.

Importantly, studies with DNA polymerases showed that replicative enzymes could be extremely sensitive to such small steric changes. The Kf enzyme, for example, showed a two to four orders-of-magnitude change in efficiency across the series [82]. The smallest analogue (dH) was handled with poor efficiency, as was the largest (dI). In marked contrast to this, the dichloro compound (dL) was highly efficient (in fact, equal to natural thymidine) as a substrate. Similar results were also found for T7 DNA polymerase, another high-fidelity enzyme. The results showed that polymerases can gate nucleotide choice by steric effects very precisely, with even 0.5 Å changes causing large kinetic consequences. The data also suggested two ways in which polymerases can exert their steric preferences: (1) by presenting a specific size of active site pocket; and (2) by regulating the tightness or flexibility of that pocket. A systematic series of substrates such as those of Kim *et al.* can be used to probe both mechanisms. Indeed, the same series was used to investigate steric effects in replication in living *E. coli* cells (see below).

3.5.3
Systematic Shape Variants

Steric effects are not only dependent on the sizes of molecules and their substituents, but also on shape. It is the atom-level location and orientation of groups that leads to steric clashes and the presence of voids when two molecules interact, not just the overall size of the molecule, that matters. To test this in the DNA context, Sintim and Kool constructed a series of mono- and dihalotoluene C-nucleosides (Figure 3.7), and evaluated their pairing and replication properties [83]. These compounds were based on the original difluorotoluene [39] and dichlorotoluene [80] mimics of thymine, but with varied locations and sizes of substituents.

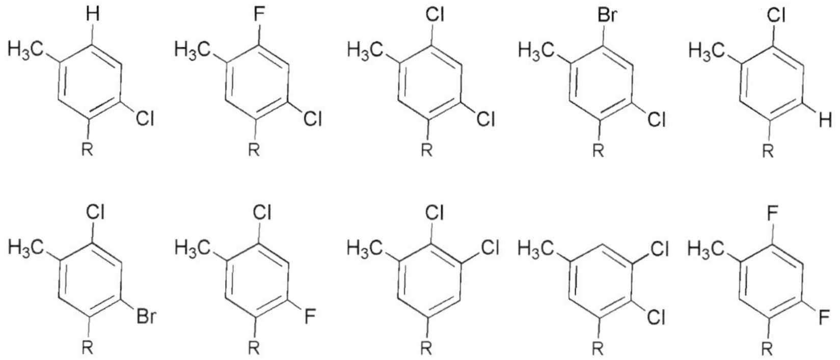

Figure 3.7 A set of designed non-polar nucleobases having systematically varied shape. These were used to assess the positional steric effects in the Kf polymerase, and activities were found to vary by a factor of >3000 among the compounds [84]. R = deoxyribose.

The first experiments with this series were carried out with DNA alone, pairing these compounds opposite natural bases. The results showed that most had no selectivity [85], while a few had a small selectivity for adenine, which might suggest a small shape preference enforced by the DNA backbone.

However, experiments with DNA Pol I gave striking results, with large changes in efficiency being observed for different shape variants. For example, a comparison of closely related compounds (such as three different dichlorotoluenes) showed that only the compound shaped like thymine (2,4-dichlorotoluene) was accepted as a good polymerase substrate opposite adenine. In addition, Sintim and Kool varied the substituent sizes at 2 and 4 individually, and found that the 2-bromo, 4-chloro compound was most efficient in replication opposite adenine. In fact, the compound was shown to be replicated with higher fidelity than natural thymine itself [83].

Overall, the results suggested strongly that not just size, but shape and size, matter strongly in DNA replication, and that the large majority of the selectivity comes not from the DNA alone, but rather from enforcement by the polymerase.

3.6 Minor Groove Hydrogen Bonds in Polymerases

3.6.1 Base Analogues Testing Minor Groove Interactions

A number of early structures of DNA polymerases bound to DNA inspired the general observation that polymerases make numerous hydrogen-bonded contacts between donor side chains and the minor groove of the DNA. Simple examination of the four natural bases shows that all four have potential H-bond acceptors (C=O in C and T; pyridine-type =N– groups in the purines) at similar positions. Thus, the natural pairs have two acceptors on either side of the groove, and many of these are involved in hydrogen bonds in the published structures of polymerases with DNA. This series of side-chain donors in the enzymes was termed the "minor groove binding track" [84], and was assumed to be important to replication. However, this assumption required more testing in order to be validated.

Morales and Kool carried out an early test of this suggestion by synthesizing a non-polar nucleoside isostere of dA which lacked Watson–Crick hydrogen bonding ability (like the earlier analogue Z) but which had the minor groove nitrogen (analogous to that of adenine) engineered in [85]. This new analogue (dQ; Figure 3.8) was shown to be accepted as a substrate for Kf in nucleotide incorporation, just as Z is, showing that minor groove interactions are not important on initial pair synthesis with this enzyme. However, extension beyond a Z (adding a natural nucleotide next) was quite inefficient. When Q was used instead, however, the efficiency was better by almost 300-fold. This pinpoints a single H-bond to an arginine in the minor groove as energetically quite important to DNA synthesis. An H-bond acceptor on the other (template) side of the groove had little effect, and neither did acceptors at positions further down the DNA. Thus, a single minor groove H-bond was more energetically

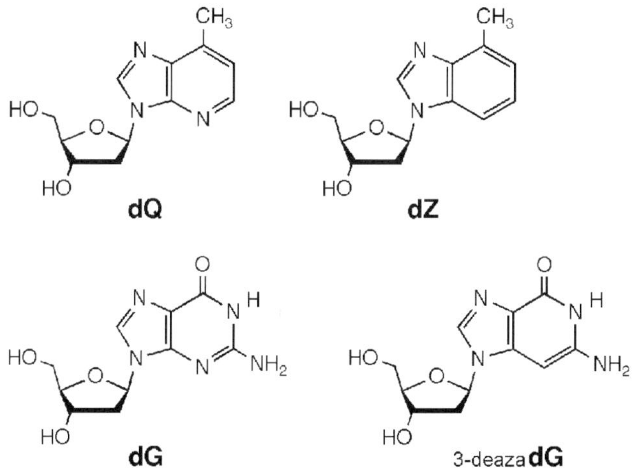

Figure 3.8 Modified nucleosides that have been used to study minor groove hydrogen bonding effects in DNA replication [86, 87]. dQ and dZ differ by only one nitrogen; the same is true of dG and 3-deazaG.

important than Watson–Crick hydrogen bonds [85]. This finding was consistent with the earlier revelations of Spratt, who used 3-deazaguanosine (Figure 3.8) as a probe [86]. Morales and Kool subsequently carried out studies with other enzymes, and found substantial differences in minor groove H-bonding requirements from case to case [87]. However, thus far it appears that most, if not all, A-family enzymes require at least one minor groove hydrogen bond for efficient extension. Subsequent studies with other analogues by Romesberg and coworkers [67] and Hirao *et al.* [16] have generally supported these findings.

3.7 Other Non-Polar Bases and Pairs

3.7.1 The Quest for New Base Pairs

Since the early studies of Rappaport and Benner, there has been a general desire among members of the nucleic acids research community to design new base pairs that could function alongside the natural ones. There are several motivations for this goal: first, additional base pairs might allow chemists to expand the genetic code, thus offering many more possible three-letter codons for amino acids. In principle, this might make it possible to incorporate multiple non-natural amino acids into proteins using ribosomal synthesis. A second motivation is that new base pairs might have biotechnological uses, as demonstrated by the early use of the isoC–isoG pair in HIV diagnosis [37]. A third reason is that new base pairs can be useful as tools in basic science; a good example of this is the use of isoC–isoG in adding labels to RNA molecules [35]. Finally, there is the long-term goal of some chemists to develop synthetic systems that act like functioning biological systems (a goal of "synthetic biology"); these systems offer both basic science interest as well as biotechnological advances.

The discovery that polymerases could be tolerant of large changes to DNA bases, such as completely rearranging the hydrogen-bonding groups, or eliminating hydrogen bonding altogether, has subsequently led to a rush of new molecular designs for base pairs. The principles used in these new designs included varying shapes, complementarity of shapes, base-stacking ability, hydrophobicity, and hydrogen-bonding ability. Minor groove H-bond acceptors were also taken into account in a number of cases.

3.7.2
A Broad Variety of Heterocycles and Hydrocarbons

The groups of Kool [56, 79, 82, 83], Romesberg [65–75] and Hirao [16, 88–93] have, to date, explored a wide range of DNA base structures as substrates for polymerase enzymes, including hydrocarbons of various sizes and shapes and also a number of heterocycles. A representative set of the different classes of compound skeletons that have been studied to date is shown in Figure 3.9. Although, in this chapter, space limitations preclude a review of all the results obtained with the many dozens of compounds and many hundreds of base pairs studied thus far, some generalizations from the data are worthy of mention. First, when pairing non-H-bonding compounds with each other, the most successful pairs in replication have been those that are small enough to fit well opposite one another on the "footprint" of a Watson–Crick purine–pyrimidine pair. The exceptions to this have been a few examples of larger non-polar compounds (e.g., the PICS–PICS pair; see Figure 3.5) that are expected to stack very strongly; these are likely to be accommodated as intercalated structures,

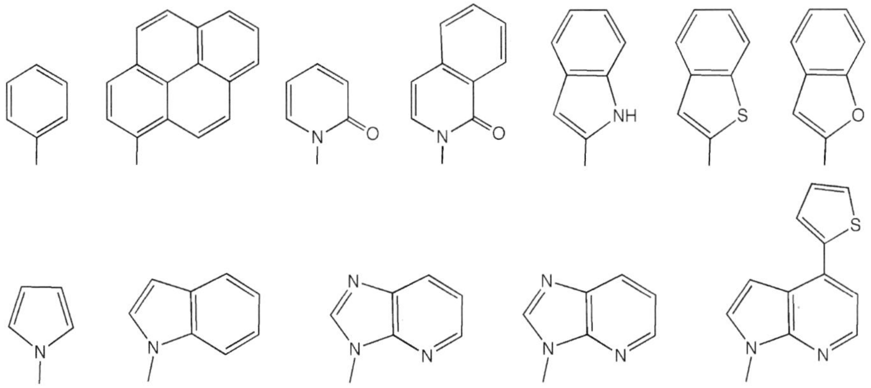

Figure 3.9 Illustration showing the varied carbon/heterocycle frameworks that have been synthesized as modified nucleobases and studied with polymerase enzymes during the past 10 years. See text for specific examples.

but are then subsequently extended very poorly. Second, vary large nucleobases (e.g., pyrene) can be quite efficient with polymerases, as long as the pairing partner is small enough to make room; in the case of pyrene the partner is an abasic sugar where the "partner" is formally hydrogen. Third, the non-polar examples tend to pair better with each other than with natural bases, apparently because of the cost of desolvating the polar natural structures. The one exception here is adenine, which is the least strongly solvated natural base. Finally, for most successful extension of DNA beyond a designed base pair, there appears to be a general advantage to the inclusion of a minor groove H-bond acceptor group.

3.7.3 Benzimidazoles Continue the Debate on Steric Effects

Despite a growing literature showing strong responses of DNA polymerases to changing steric sizes and shapes of nucleobases, two recent studies (by Berdis [94] and by Engels and Kuchta [95]) that dispute the importance of sterics should be mentioned at this point. Both studies used 5- and 6-substituted indoles and benzimidazoles as substrates for polymerases, and both found that the substituents could be varied at these positions without having much effect on polymerase activity. For example, various 5- and 6-substituted benzimidazoles all showed selectivity for pairing with guanine, despite the fact that their sizes and shapes changed. In response, it was argued [83] that this was because all of the compounds are too large to pair opposite any natural bases in the standard *anti* conformation, and as a result the analogues flipped to the *syn* conformation in the polymerase active site, which gives them all cytosine-like shapes and sizes. However, further studies to investigate this point would be helpful in clarifying the structural issue.

3.7.4 New Pairs of Hirao and Yokoyama

Recently, the Hirao laboratory has shown great success in new designed pairs (Figure 3.10) that can operate successfully in the presence of the four natural bases [16, 88–93]. This group's early studies focused on compounds that formed hydrogen bonds, and they evaluated their transcription to RNA by T7 RNA polymerase. They identified thienyl purines that could be paired well (by polymerases) opposite 2-pyridone compounds [88]. However, subsequent to that they studied fully non-H-bonded pairs, and identified carbaldehyde-substituted pyrroles that could pair well by DNA polymerases opposite the previously known azabenzimidazole of Morales and Kool [91]. Most recently, Hirao and colleagues extended this by pairing thienylazabenzimidazoles opposite aldehydopyrroles, leading to high efficiency both in DNA polymerase activity and in transcription [92, 93]. This last example is one of the most successful new base pairs that appears to function well and is orthogonal to the natural base pairs. The Hirao group has demonstrated the practical utility of this pair by using it to incorporate labels at specific sites in otherwise naturally substituted transcribed RNAs [92].

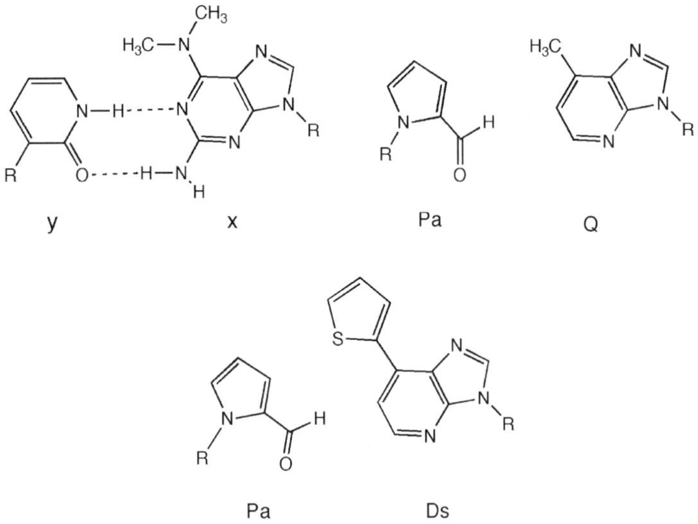

Figure 3.10 Examples of modified pairs studied recently by Hirao and Yokoyama [89–94]. Most have been studied both with RNA and DNA polymerases. R = deoxyribose or ribose.

3.8 Replication of Designed Bases in Living Cells

3.8.1 Effects of Hydrogen Bonding in *E. coli*

Studies of purified recombinant DNA polymerases functioning *in vitro* with pure synthetic DNA templates and nucleotides are far removed from the complexity that occurs in a living system. For example, in *E. coli* there are five different polymerases that are involved in the accurate transmission of genetic information from a mother cell to a daughter cell [96]. Two of these polymerases are chiefly replicative enzymes involved in the synthesis of the leading and lagging strands of the bacterial genome, while three are classified as repair enzymes that rescue damaged or mispaired sites in the genomic DNA. The timing of the entire replication and repair system is intricate, and localization and regulation of the varying enzymatic activities is crucial for cell survival. In addition, the cellular DNA itself has a structure which is more complex than that of simple synthetic DNAs, and both the DNA and the polymerases interact with proteins not found in a simple cellular model. Thus, the degree of relevance of *in-vitro* biochemical tests to actual living systems is often unclear.

It is for these reasons that the Kool laboratory worked with collaborators at MIT to begin studying the replication of non-polar nucleoside isosteres directly in live *E. coli* cells [97]. The first studies were directed to the question of whether base pair hydrogen bonding is necessary for efficient and accurate replication in the cell.

This was tested by inserting single isosteres dF or dQ into synthetic oligonucleotides, and ligating these at a single predetermined site in the M13 bacteriophage. These modified phage were then used to infect bacterial cells. The counting of daughter phage after an incubation period gives a measure of bypass efficiency (i.e., polymerase insertion of a nucleotide opposite the modified base, followed by bypass beyond the site). Recovery of the phage and sequence analysis provides quantitative data on which nucleotides replaced the modified base in the daughter phage [98].

Results showed that phage containing either dF or dQ were successfully replicated [97]. In the case of dQ, the phage showed 7–10% bypass, indicating moderate efficiency of overall bypass. For dF, however, the bypass was 25–30%, indicating moderate to high efficiency overall. More important was the analysis of which nucleotides replaced the modified nucleotides in the daughter phage. In the case of the adenine analogue dQ, the chief replacement was A, showing that T was inserted opposite Q most often in the cell. The fidelity was low, however, with G replacing Q for 20% of the time. Interestingly, with the better isostere dF, T replaced the analogue for 95% of the time, showing substantially higher fidelity. Overall, the data showed that hydrogen bonding is helpful – but not crucial – to base pair synthesis in the cell. Comparison with the earlier *in-vitro* data (with Pol I) [85] suggests that the minor groove hydrogen bond might be the important factor missing with analogue F.

3.8.2
Effects of Nucleobase Size in *E. coli*

As described above, a systematic series of thymine analogues of varying size had been used *in vitro* with a number of polymerases to test the effect of sterics on base-pair synthesis, and the data had shown that Pol I from *E. coli* was extremely sensitive to these size changes. Once again, it was of great interest to determine whether these effects would be relevant to a living system. Thus, Delaney *et al.* constructed a series of modified M13 phage containing these variably sized thymine analogues [82], and used them to infect *E. coli* cells, as described above. The results showed that replication bypass was indeed dependent on size, with the middle-sized compounds (F, L) being most efficiently bypassed, and analogues smaller or larger being less efficient in replication (Figure 3.11). An analysis of the fidelity of this replication showed that it also varied strongly with size. The highest fidelity (95% correct replacement) occurred with the optimum-sized compound (L), and the smallest and largest analogues showed very low fidelity. Hence, the results overall mirrored the earlier studies with DNA Pol I, showing the same range of size preferences and fidelity with size [82].

The data strongly suggest that replicative polymerases in *E. coli* are highly responsive to nucleobase size, even over a very small size range, and this is consistent with the steric hypothesis of DNA replication. Another important outcome of these experiments is that they represent the first example of a highly modified DNA base functioning correctly in a living system.

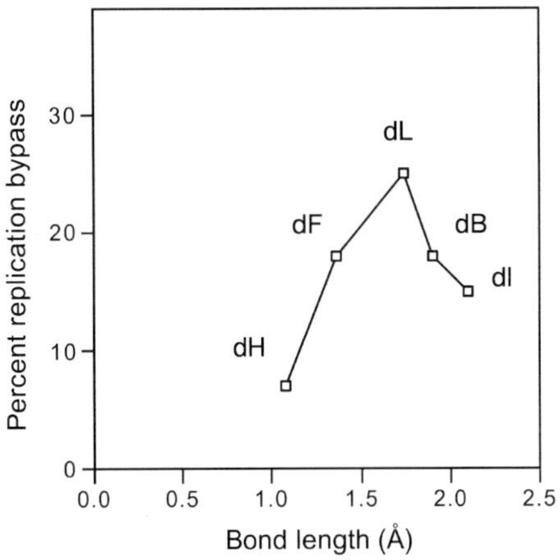

Figure 3.11 Plot showing replication of non-natural DNA base analogs in live *E. coli* cells as a function of their size (Angstroms). The variably sized thymidine analogues from Figure 3.6 were placed in one site in a phage genome and used to infect *E. coli* in separate experiments [83]. Quantitation of the daughter phage allowed an analysis of how well the replication machinery processes and continues beyond the non-natural nucleobases. For comparison, a natural nucleobase (guanine) is bypassed 100% relative to these, while an abasic site is bypassed at a level of only 4%.

3.9 Conclusions and Future Prospects

3.9.1 What We Know About Replication . . .

Studies conducted over the past decade and using modified DNA bases have revealed much new information concerning DNA polymerases and DNA replication in general. First, it is clear that many replicative DNA polymerases can operate quite efficiently to synthesize a base pair without hydrogen bonds. Second, it appears that steric effects are likely the best hypothesis at present for explaining the variations in efficiencies of various pairs of natural or non-natural bases, and even very small differences in sterics can cause large kinetic effects. Third, it is also clear that different classes of enzymes can behave quite differently with modified bases; for example, Y-family (repair) enzymes function very poorly with non-polar base analogues, and may in fact require Watson–Crick hydrogen bonding for high activity. Fourth, a few modified base pairs are currently successful enough as polymerase substrates to be biotechnologically useful, even without further modification, although it seems that more pairs would be welcomed for various applications. Finally, it is now clear that some non-hydrogen-bonding bases can be replicated with high efficiency and fidelity in live bacterial cells, thus opening the door for many future studies.

3.9.2
... And What Remains Unknown

The early experiments in which modified bases are tested with a variety of polymerases showed clearly that each new enzyme (especially in a different family) will require separate analysis of what is required (chemically speaking) in a base pair substrate. Thus, it is not known how polymerases (other than three or four commonly studied ones) function at the molecular level of detail. On the more biological level, the roles and activities of many cellular polymerases remain unclear. Moreover, nucleotides within the cells go through many metabolic steps aside from being inserted into DNA, and it is unclear how these other biological systems respond to such chemical changes. Thus, if non-natural nucleobases are to function in living, biologically relevant systems, many questions remain to be answered.

3.9.3
Future Directions

Although it is always difficult to predict the future of science, three trends will almost certainly be pursued during the coming years. First, there will be many biotechnological uses developed for modified DNA bases, for applications in biomedicine (therapeutic, diagnostic) and also as tools for basic science. Second, given that there are many existing polymerases and innumerable mutants that one could also make, there will no doubt be a broader range of base-pair substrates developed for operating with a broader range of enzymes. Finally, it is almost certain that modified bases will take bolder steps into biological pathways, and will be increasingly found to function – or engineered to function – in bacterial and human cells [99].

Acknowledgments

The author thanks the U.S. National Institutes of Health (grant GM072705) for support of these studies with modified nucleoside analogues, and also the many coworkers in the Kool laboratory who have contributed to these investigations and whose names are to be found within the references.

References

1 Kornberg, A., *Biochim. Biophys. Acta.* **1989**, *1000*, 53–56.

2 Kornberg, A., DNA replication. *Biochim. Biophys. Acta.* **1988**, *951*, 235–239.

3 Echols, H., Goodman, M. F., *Annu. Rev. Biochem.* **1991**, *60*, 477–511.

4 Kunkel, T. A., Bebenek, K., *Annu. Rev. Biochem.* **2000**, *69*, 497–529.

5 Kool, E. T., *Annu. Rev. Biochem.* **2002**, *71*, 191–219.

6 Prakash, S., Johnson, R. E., Prakash, L., *Annu. Rev. Biochem.* **2005**, *74*, 317–353.

7 Hogg, M., Wallace, S. S., Doublie, S., *Curr. Opin. Struct. Biol.* **2005**, *15*, 86–93.

8 Bebenek, K., Kunkel, T. A., Functions of DNA polymerases, *DNA Repair and*

Replication, Elsevier Press, San Diego, **2004**.
9 Bentle, M. S., Bey, E. A., Dong, Y., Reinicke, K. E., Boothman, D. A., *J. Mol. Histol.* **2006**, *37*, 203–218.
10 Castro, H. C., Loureiro, N. I., Pujol-Luz, M., Souza, A. M., Albuquerque, M. G., Santos, D. O., Cabral, L. M., Frugulhetti, I. C., Rodrigues, C. R., *Curr. Med. Chem.* **2006**, *13*, 313–324.
11 Lehmann, A. R., *Exp. Cell Res.* **2006**, *312*, 2673–2676.
12 Kool, E. T., *Biopolymers (Nucleic Acid Sciences)* **1998**, *48*, 3–17.
13 Kool, E. T., *Annu. Rev. Biophys. Biomol. Struct.* **2001**, *30*, 1–22.
14 Benner, S. A., *Acc. Chem. Res.* **2004**, *37*, 784–797.
15 Henry, A. A., Romesberg, F. E., *Curr. Opin. Chem. Biol.* **2003**, *7*, 727–733.
16 Hirao, I., *Curr. Opin. Chem. Biol.* **2006**, *10*, 622–627.
17 Berti, L., Xie, J., Medintz, I. L., Glazer, A. N., Mathies, R. A., *Anal. Biochem.* **2001**, *292*, 188–197.
18 Berg, J. M., Tymoczko, J. L., Stryer, L., *Biochemistry*, 5th edn. W. H. Freeman, New York, **2002**, p. 751.
19 Kuchta, R. D., Benkovic, P., Benkovic, S. J., *Biochemistry* **1988**, *27*, 6716–6725.
20 Patel, S. S., Wong, I., Johnson, K. A., *Biochemistry* **1991**, *30*, 511–525.
21 Creighton, S., Goodman, M. F., *J. Biol. Chem.* **1995**, *270*, 4759–4774.
22 Rappaport, H. P., *Nucleic Acids Res.* **1988**, *16*, 7253–7267.
23 Rappaport, H. P., *Biochemistry* **1993**, *32*, 3047–3057.
24 Rappaport, H. P., *Biochem. J.* **2004**, *381*, 709–717.
25 Piccirilli, J. A., Krauch, T., Moroney, S. E., Benner, S. A., *Nature* **1990**, *343*, 33–37.
26 Rich, A., *Horizons in Biochemistry* (Eds A. Kasha and B. Pullman), Academic Press, New York, **1963**.
27 Geyer, C. R., Battersby, T. R., Benner, S. A., *Structure* **2003**, *11*, 1485–1498.
28 Switzer, C. Y., Moroney, S. E., Benner, S. A., *Biochemistry* **1993**, *32*, 10489–10496.
29 Lutz, M. J., Held, H. A., Hottiger, M., Hubscher, U., Benner, S. A., *Nucleic Acids Res.* **1996**, *24*, 1308–1313.
30 Yang, Z., Hutter, D., Sheng, P., Sismour, A. M., Benner, S. A., *Nucleic Acids Res.* **2006**, *34*, 6095–6101.
31 Johnson, S. C., Marshall, D. J., Harms, G., Miller, C. M., Sherrill, C. B., Beaty, E. L., Lederer, S. A., Roesch, E. B., Madsen, G., Hoffman, G. L., Laessig, R. H., Kopish, G. J., Baker, M. W., Benner, S. A., Farrell, P. M., Prudent, J. R., *Clin. Chem.* **2004**, *50*, 2019–2027.
32 Lutz, M. J., Horlacher, J., Benner, S. A., *Bioorg. Med. Chem. Lett.* **1998**, *8*, 499–504.
33 Sismour, A. M., Benner, S. A., *Nucleic Acids Res.* **2005**, *33*, 5640–5646.
34 Bain, J. D., Switzer, C., Chamberlin, A. R., Benner, S. A., *Nature* **1992**, *356*, 537–539.
35 Tor, Y., Dervan, P. B., *J. Am. Chem. Soc.* **1993**, *115*, 4461–4467.
36 Johnson, S. C., Sherrill, C. B., Marshall, D. J., Moser, M. J., Prudent, J. R., *Nucleic Acids Res.* **2004**, *32*, 1937–1941.
37 Collins, M. L., Irvine, B., Tyner, D., Fine, E., Zayati, C., Chang, C., Horn, T., Ahle, D., Detmer, J., Shen, L. P., Kolberg, J., Bushnell, S., Urdea, M. S., Ho, D. D., *Nucleic Acids Res.* **1997**, *25*, 2979–2984.
38 Seela, F., Amberg, S., Melenewski, A., Rosemeyer, H., *Helv. Chim. Acta* **2001**, *84*, 1996–2014.
39 Schweitzer, B. A., Kool, E. T., *J. Org. Chem.* **1994**, *59*, 7238–7242.
40 Guckian, K. M., Morales, J. C., Kool, E. T., *J. Org. Chem.* **1998**, *63*, 9652–9656.
41 Schweitzer, B. A., Kool, E. T., *J. Am. Chem. Soc.* **1995**, *117*, 1863–1872.
42 Chaudhuri, N. C., Kool, E. T., *Tetrahedron Lett.* **1995**, 1795–1798.
43 Ren, X.-F., Schweitzer, B. A., Sheils, C. J., Kool, E. T., *Angew. Chem. Int. Ed. Engl.* **1996**, *35*, 743–746.
44 Chaudhuri, N. C., Ren, R. X.-F., Kool, E. T., *Synlett* **1997**, 341–347.
45 Wichai, U., Woski, S. A., *Org. Lett.* **1999**, *1*, 1173–1175.

46 Singh, I., Seitz, O., *Org. Lett.* **2006**, *8*, 4319–4322.

47 Guckian, K. M., Kool, E. T., *Angew. Chem. Int. Ed.* **1998**, *36*, 2825–2828.

48 Guckian, K. M., Schweitzer, B. A., Ren, R. X.-F., Sheils, C. J., Tahmassebi, D. C., Kool, E. T., *J. Am. Chem. Soc.* **2000**, *122*, 2213–2222.

49 Moran, S., Ren, R. X.-F., Kool, E. T., *Proc. Natl. Acad. Sci. USA* **1997**, *94*, 10506–10511.

50 Xia, J., Noronha, A., Toudjarska, I., Li, F., Akinc, A., Braich, R., Frank-Kamenetsky, M., Rajeev, K. G., Egli, M., Manoharan, M., *ACS Chem. Biol.* **2006**, *1*, 176–183.

51 Barsky, D., Colvin, M., Kool, E. T., *J. Biomol. Struct. Dyn.* **1999**, *16*, 1119–1134.

52 Meyer, R., Suhnel, J., *J. Biomol. Struct. Dyn.* **1997**, *15*, 619–624.

53 Wang, X., Houk, K. N., *Chem. Commun.* **1998**, *23*, 2631–2632.

54 Guckian, K. M., Krugh, T. R., Kool, E. T., *J. Am. Chem. Soc.* **2000**, *122*, 6841–6847.

55 Guckian, K. M., Krugh, T. R., Kool, E. T., *Nat. Struct. Biol.* **1998**, *5*, 954–959.

56 Kool, E. T., Morales, J. C., Guckian, K. M., *Angew. Chem. Int. Ed.* **2000**, *39*, 990–1009.

57 Moran, S., Ren, R. X.-F., Rumney, S., Kool, E. T., *J. Am. Chem. Soc.* **1997**, *119*, 2056–2057.

58 Matray, T. J., Kool, E. T., *Nature* **1999**, *399*, 704–708.

59 Morales, J. C., Kool, E. T., *Nat. Struct. Biol.* **1998**, *5*, 950–954.

60 Morales, J. C., Kool, E. T., *J. Am. Chem. Soc.* **2000**, *122*, 1001–1007.

61 Mizukami, S., Kim, T. W., Helquist, S. A., Kool, E. T., *Biochemistry* **2006**, *45*, 2772–2778.

62 Wolfle, W. T., Washington, M. T., Kool, E. T., Spratt, T. E., Helquist, S. A., Prakash, L., Prakash, S., *Mol. Cell. Biol.* **2005**, *25*, 7137–7143.

63 Washington, M. T., Helquist, S. A., Kool, E. T., Prakash, L., Prakash, S., *Mol. Cell. Biol.* **2003**, *23*, 5107–5112.

64 Morales, J. C., Kool, E. T., *J. Am. Chem. Soc.* **1999**, *121*, 2723–2724.

65 Berger, M., Wu, Y., Ogawa, A. K., McMinn, D. L., Schultz, P. G., Romesberg, F. E., *Nucleic Acids Res.* **2000**, *28*, 2911–2914.

66 Yu, C., Henry, A. A., Romesberg, F. E., Schultz, P. G., *Angew. Chem. Int. Ed. Engl.* **2002**, *41*, 841–3844.

67 Matsuda, S., Henry, A. A., Schultz, P. G., Romesberg, F. E., *J. Am. Chem. Soc.* **2003**, *125*, 6134–6139.

68 Henry, A. A., Olsen, A. G., Matsuda, S., Yu, C., Geierstanger, B. H., Romesberg, F. E., *J. Am. Chem. Soc.* **2004**, *126*, 6923–6931.

69 Matsuda, S., Romesberg, F. E., *J. Am. Chem. Soc.* **2004**, *126*, 14419–14427.

70 Henry, A. A., Olsen, A. G., Matsuda, S., Yu, C., Geierstanger, B. H., Romesberg, F. E., *J. Am. Chem. Soc.* **2004**, *126*, 6923–6931.

71 Hwang, G. T., Romesberg, F. E., *Nucleic Acids Res.* **2006**, *34*, 2037–2045.

72 Matsuda, S., Henry, A. A., Romesberg, F. E., *J. Am. Chem. Soc.* **2006**, *128*, 6369–6375.

73 Leconte, A. M., Matsuda, S., Hwang, G. T., Romesberg, F. E., *Angew. Chem. Int. Ed. Engl.* **2006**, *45*, 4326–4329.

74 Leconte, A. M., Matsuda, S., Romesberg, F. E., *J. Am. Chem. Soc.* **2006**, *128*, 6780–6781.

75 Kim, Y., Leconte, A. M., Hari, Y., Romesberg, F. E., *Angew. Chem. Int. Ed. Engl.* **2006**, *45*, 7809–7812.

76 Mathis, G., Hunziker, R., *Angew. Chem. Int. Ed.* **2002**, *41*, 3203–3205.

77 Lai, J. S., Qu, J., Kool, E. T., *Angew. Chem. Int. Ed.* **2003**, *42*, 5973–5977.

78 Lai, J. S., Qu, J., Kool, E. T., *J. Am. Chem. Soc.* **2004**, *126*, 3040–3041.

79 Lai, J. S., Kool, E. T., *Chemistry* **2005**, *11*, 2966–2971.

80 Kim, T., Kool, E. T., *Org. Lett.* **2004**, *6*, 3949–3952.

81 Kim, T., Kool, E. T., *J. Org. Chem.* **2005**, *70*, 2048–2053.

82 Kim, T. W., Delaney, J. C., Essigmann, J. M., Kool, E. T., *Proc. Natl. Acad. Sci. USA* **2005**, *102*, 15803–15808.

83 Sintim, H. O., Kool, E. T., *Angew. Chem. Int. Ed.* **2006**, *45*, 1974–1979.

84 Doublié S., Tabor, S., Long, A. M., Richardson, C. C., Ellenberger, T., *Nature* **1998**, *391*, 251–258.

85 Morales, J. C., Kool, E. T., *J. Am. Chem. Soc.* **1999**, *121*, 2723–2724.

86 Spratt, T. E., *Biochemistry* **1997**, *36*, 13292–13297.

87 Morales, J. C., Kool, E. T., *Biochemistry* **2000**, *39*, 12979–12988.

88 Fujiwara, T., Kimoto, M., Sugiyama, H., Hirao, I., Yokoyama, S., *Bioorg. Med. Chem. Lett.* **2001**, *11*, 2221–2223.

89 Hirao, I., Harada, Y., Kimoto, M., Mitsui, T., Fujiwara, T., Yokoyama, S., *J. Am. Chem. Soc.* **2004**, *126*, 13298–13305.

90 Endo, M., Mitsui, T., Okuni, T., Kimoto, M., Hirao, I., Yokoyama, S., *Bioorg. Med. Chem. Lett.* **2004**, *14*, 2593–2596.

91 Mitsui, T., Kitamura, A., Kimoto, M., To, T., Sato, A., Hirao, I., Yokoyama, S., *J. Am. Chem. Soc.* **2005**, *127*, 8652–8658.

92 Moriyama, K., Kimoto, M., Mitsui, T., Yokoyama, S., Hirao, I., *Nucleic Acids Res.* **2005**, *33*, e129.

93 Hirao, I., Kimoto, M., Mitsui, T., Fujiwara, T., Kawai, R., Sato, A., Harada, Y., Yokoyama, S., *Nat. Methods* **2006**, *3*, 729–735.

94 Zhang, X., Lee, I., Berdis, A., *J. Biochem.* **2005**, *44*, 13101–13110.

95 Kincaid, K., Beckman, J., Zivkovic, A., Halcomb, R. L., Engels, J. W., Kuchta, R. D., *Nucleic Acids Res.* **2005**, *33*, 2620–2628.

96 Johnson, A., O'Donnell, M., *Annu. Rev. Biochem.* **2005**, *74*, 283–315.

97 Delaney, J. C., Henderson, P. T., Helquist, S. A., Morales, J. C., Essigmann, J. M., Kool, E. T., *Proc. Natl. Acad. Sci. USA* **2003**, *100*, 4469–4473.

98 Delaney, J. C., Essigmann, J. M., *Methods Enzymol.* **2006**, *408*, 1–15.

99 Kool, E. T., Waters, M. L., *Nature Chem. Biol.* **2007**, *3*, 70–73.

4
2′-Deoxyribose-Modified Nucleoside Triphosphates and their Recognition by DNA Polymerases

Karl-Heinz Jung and Andreas Marx

4.1
Introduction

The entire blueprint of life is stored in the DNA of every living species. Generally, DNA is built from the four nucleotide building blocks that contain one of the four nucleobases, a phosphate that bridges the nucleotides via phosphodiester bonds, and a 2′-deoxyribose unit. The interconnected building blocks result in DNA strands that pair to form the well-known double helix via Watson–Crick pairing [1]. DNA is central to the evolution of life, and the transmission of genetic information from the parental DNA strand to the offspring is crucial for the survival of any living species [2–8]. This process must be sufficiently error-free (i.e., transmission according to the Watson–Crick pairing) to ensure the integrity of the genome, yet also sufficiently error-prone in order to spur evolution and foster the preservation of a species. Clearly, this process must be well balanced, and along these lines several intriguing aspects related to DNA have long inspired and challenged scientists of many disciplines. The mechanisms of the above-depicted information transfer from the ancestor to the offspring have been a matter of interest since the first proposal of the DNA structure, and even until now they are not fully understood. Following another aspect of DNA function, scientists have investigated the causative of the present-day nucleic acids, namely DNA and RNA [9, 10]. Insights into the chemical etiology of DNA and RNA are crucial for our understanding of the evolution, for example, of DNA as a generic genetic material.

In nature, all DNA synthesis is catalyzed by DNA polymerases [11]. These enzymes catalyze proceeding DNA synthesis in a template-directed manner by recognizing the template to correctly insert the complementary nucleotide. DNA polymerase selectivity is crucial for the survival of any living species. These enzymes are presented with a pool of four structurally similar dNTPs, and of course NTPs, from which the sole correct (i.e., Watson–Crick base-paired dNTP) substrate must be selected for incorporation into the growing DNA strand. The mechanisms by which these remarkable enzymes achieve this tremendous task have been a matter of interest and intensive

Modified Nucleosides: in Biochemistry, Biotechnology and Medicine. Edited by Piet Herdewijn

ISBN: 978-3-527-31820-9

discussion since the discovery of the first DNA polymerase, *Escherichia coli* DNA polymerase I, by Kornberg about half a century ago [12, 13]. While certain DNA polymerases show low error rates (as low as only one error within one million synthesized nucleotide linkages), other enzymes may exhibit high error rates of up to one error within one to ten synthesized nucleotide linkages [2, 4]. On the other hand, studies with nucleotide analogues that were synthesized in order to investigate the chemical etiology of DNA, in their interplay with DNA polymerases, should shed light on the ability of chemical alternatives to DNA to carry out its task of information storage and transfer. Recently, a wealth of valuable new insights along these lines concerning DNA polymerase selectivity mechanisms and substrate spectra was gained through the application of carefully designed synthetic nucleotides in functional enzyme studies [14–16]. This chapter reviews recent examples of dNTPs that comprise altered 2′-deoxyribose moieties, and highlights the merit of employment of the chemical probes in order to gain insight into complex biological processes.

4.2 Modified Nucleotides as Alternative Building Blocks to Natural Nucleic Acids

4.2.1 Introduction

Nature has evolved many carbohydrate entities imposing the question of why RNA and DNA have exclusive recourse to pentoses, namely ribose and 2′-deoxyribose. In order to be a suitable candidate for an ancestor of present-day nucleic acids, several criteria must be met. Among these, the capacities to form informational nucleobase pairing, and to be suitable substrates for information transfer, are of prime importance. The question “Why pentoses?” has attracted many scientists in their systematic search for alternative nucleobase pairing systems. Within this chapter, those nucleosides are detailed which have been evaluated in this context towards their action on DNA polymerases and harbor carbohydrate systems that are either downsized (fewer carbon atoms per repeating mononucleotide unit) or expanded (additional carbon atoms per repeating mononucleotide unit) compared to the pentoses ribose and 2′-deoxyribose used in RNA and DNA in nature.

4.2.2 Nucleotides with Downsized Residues: α-L-Threose-Derived Nucleotides

4.2.2.1 Introduction

In the search for potential nucleic acid alternatives, Eschenmoser and colleagues pursued systematic experimental studies of RNA and DNA analogues [9, 10]. Interestingly, these authors identified 3′-2′-α-L-threose oligonucleotides (TNA) (Figure 4.1) to undergo informational base pairing with each other, as well as cross-hybridization with RNA and DNA [17].

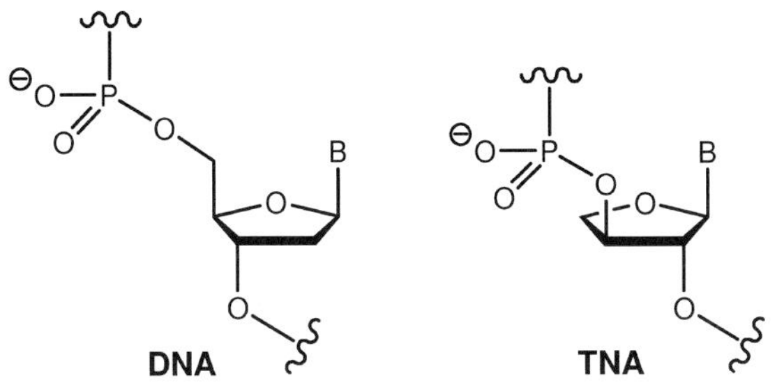

Figure 4.1 DNA and TNA.

The latter point is particularly interesting as TNA can be regarded as a simplified nucleic acid with a backbone that consists of five covalent bonds per repeating mononucleotide unit rather than the six-bond backbone of DNA and RNA. These properties of TNA suggest that this polymer might be a candidate for an evolutionary progenitor of the proposed "RNA world".

Recently, the respective dNTP analogues have been prepared and their interplay with DNA polymerases studied.

4.2.2.2 **Synthesis**

For the synthesis of α-L-threofuranosyl nucleosides, L-ascorbic acid is suitable as a starting material because it contains carbons C-4 and C-5 with the desired configuration (Scheme 4.1).

By oxidative degradation of L-ascorbic acid with hydrogen peroxide and calcium carbonate, the calcium salt of L-threonic acid **1** can be prepared very easily in large quantities [18, 19]. A one-pot lactonization [20] by treatment with an ion-exchange resin (H^+ form), followed by heating with *p*-toluenesulfonic acid (pTsOH) in acetonitrile and then reaction with benzoyl chloride (BzCl) led to the benzoyl-protected L-threonolactone derivative **2** [21]. Reduction with diisobutylaluminumhydride (DIBAL) and acylation with acetic anhydride or with benzoyl chloride gave the L-threofuranosyl acetate **3a** or benzoate **3b** [20], both of which are suitable as glycosyl donors for the nucleoside formation according to Vorbrüggen's method [22, 23]. Reaction of the acetate **3a** with benzoyl-protected adenine and bis(trimethylsilyl) acetamide (BSA) under tin tetrachloride catalysis gave stereoselectively the α-configured L-threofuranosyl nucleoside **4** $\mathbf{A^{Bz}}$; due to the neighboring group participating effect of the 2-*O*-acetyl group, the 1,2-trans-configured glycoside is formed exclusively. Accordingly, by reaction of the benzoate **3b** with thymine, uracil, benzoyl-protected cytosine, and diphenylcarbamoyl-protected (PAC) guanine under trimethylsilyl triftate (TMSTf) catalysis the α-L-threofuranosyl nucleosides **4** $\mathbf{T,U,C^{Bz},G^{PAC}}$ were prepared in good yields. Debenzoylation of the threose moiety with retention of the protecting groups at the nucleobase was performed with ammonia, triethylamine, or 0.1–0.2 M sodium hydroxide, leading to the nucleosides **5** $\mathbf{A^{Bz},T,U,C^{Bz},G^{PAC}}$. Attempts towards regioselective tritylation with dimethoxytrityl chloride under various conditions gave the 3′-O-protected compounds **6** $\mathbf{A^{Bz},T,U,C^{Bz},G^{PAC}}$ and the 2′-O-protected compounds **7** $\mathbf{A^{Bz},T,U,C^{Bz},G^{PAC}}$ in most cases with low

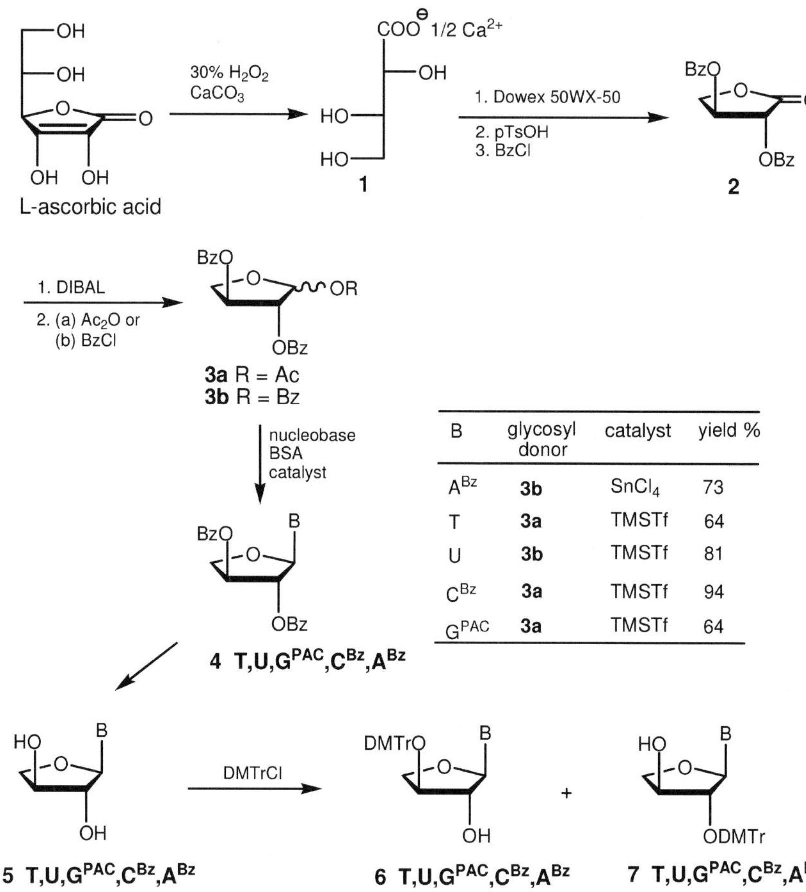

B	glycosyl donor	catalyst	yield %
A^{Bz}	**3b**	$SnCl_4$	73
T	**3a**	TMSTf	64
U	**3b**	TMSTf	81
C^{Bz}	**3a**	TMSTf	94
G^{PAC}	**3a**	TMSTf	64

Scheme 4.1 Synthesis of α-L-threofuranosyl nucleosides.

selectivity; nevertheless, after separation by column chromatography they were used as starting materials for the synthesis of phosphoramidite or triphosphate derivatives.

The synthesis of α-L-threofuranosyl nucleoside triphosphates was performed applying Eckstein's method (Scheme 4.2) [24, 25]. The 2′-O-protected nucleosides **7 T,G** which are obtained as major products in the mono-tritylation reaction, were applied to the reaction with 2-chloro-4*H*-1,3,2 benzodioxaphosphorin-4-one followed by bis(tri-*n*-butylammonium) pyrophosphate, and oxidation with iodine. Without isolation or purification, the resulting dimethoxytrityl-protected nucleoside triphosphate derivatives **10 T,G** were detritylated with aqueous acetic acid leading to the α-L-threofuranosyl nucleoside triphosphates **12 T,G.** For synthesis of the corresponding cytidine derivatives, an alternative strategy must be applied, because the 2′-O-protected cytidine derivative **7 C^{Bz}** was only available in negligible amounts. Thus, the 3′-O-protected cytidine derivative **6 C^{Bz}**, which is the major product in the

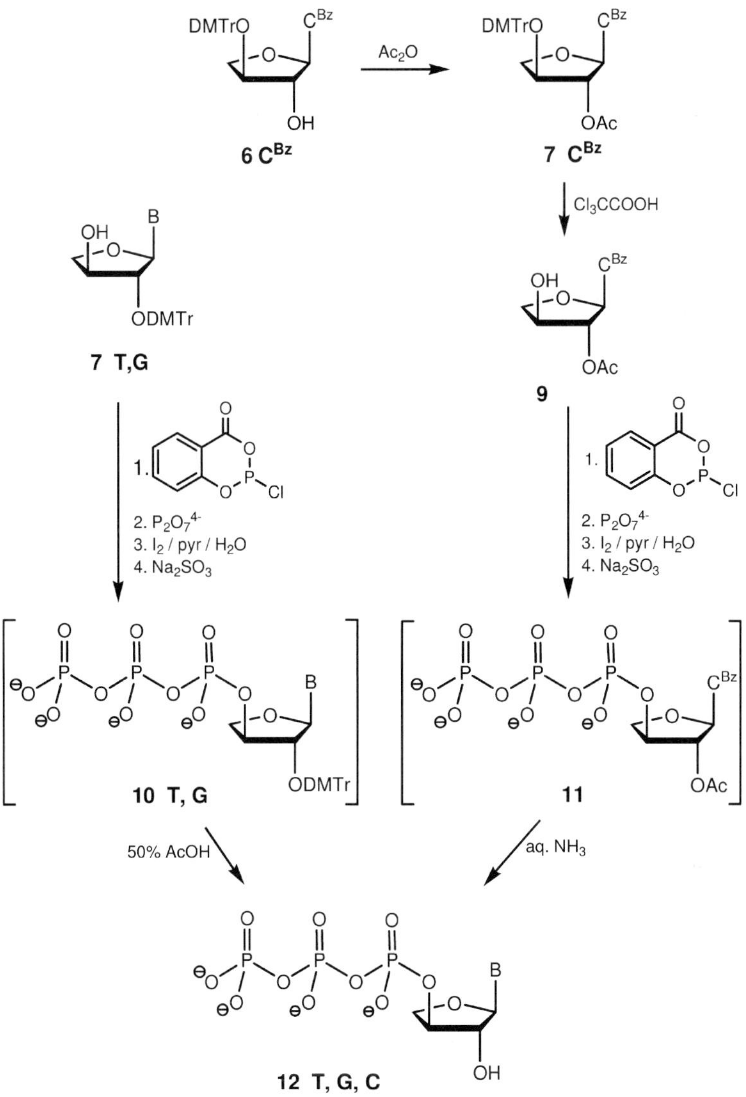

Scheme 4.2 Synthesis of α-L-threofuranosyl nucleoside triphosphates.

mono-tritylation reaction, was chosen as starting material. Acetylation with acetic anhydride leading to compound **8** and detritylation with trichloroacetic acid led to the 2′-*O*-acetyl-protected cytidine derivative **9**. Reaction with 2-chloro-4*H*-1,3,2 benzodioxaphosphorin-4-one, followed by bis(tri-*n*-butylammonium) pyrophosphate, and oxidation with iodine led to the acetyl-protected nucleoside triphosphate derivative **11**. Again, without isolation or purification, it was deacetylated with aqueous ammonia leading to the α-L-threofuranosyl nucleoside triphosphate **12 C**.

4.2.2.3 DNA Polymerase Studies

Recently, several remarkable DNA polymerase studies employing α-L-threofuranosyl nucleoside triphosphates (tNTPs) and TNA have been reported. In a highly interesting set of experiments, Szostak *et al.* investigated the proficiency of several DNA polymerases to bypass stretches of TNA with natural nucleotides. These authors found that, despite significant differences in the sugar–phosphate backbone, certain DNA polymerases are able to copy limited stretches of a TNA template [26]. The authors suggested that, due to the high activity of wild-type DNA polymerases, it might be possible to evolve a TNA-directed DNA polymerase with improved reactivity. In subsequent studies, Szostak *et al.* [27] and Herdewijn *et al.* [28] investigated the recognition of tNTPs by DNA and RNA polymerases. These authors found that some DNA polymerases are capable of incorporating α-L-threose nucleotides into a growing polymer, albeit the enzymatic synthesis of TNA is blocked after incorporation of a limited number of α-L-threose moieties. However, in subsequent experiments the A485L variant of a DNA polymerase from *Thermococcus* species 9°N-7 was identified to be more proficient in dealing with tNTPs and TNA [29–31]. This enzyme is commercially available as Therminator DNA polymerase, and has been shown to accept a wide range of modified nucleotides [32–35]. The enzyme also accepts tNTPs and is able to synthesize longer stretches of TNA. It was found that the fidelity of Therminator DNA polymerase to catalyze TNA synthesis is sufficient to allow *in-vitro* selection with TNA libraries to obtain TNA-derived binding and catalytically active molecules.

4.2.3 Nucleotides with Downsized Residues: Glycerol-Derived Nucleotides

4.2.3.1 Introduction

Another structurally further simplified nucleic acid was reported to be glycol nucleic acid (GNA) (Figure 4.2).

Meggers and colleagues reported that this three-carbon propylene glycol phosphodiester backbone-containing nucleic acid is able to form informative pairing with each other according to the Watson–Crick rule [36]. However, no cross-pairing was observed with DNA, whereas pairing of GNA could be detected with RNA.

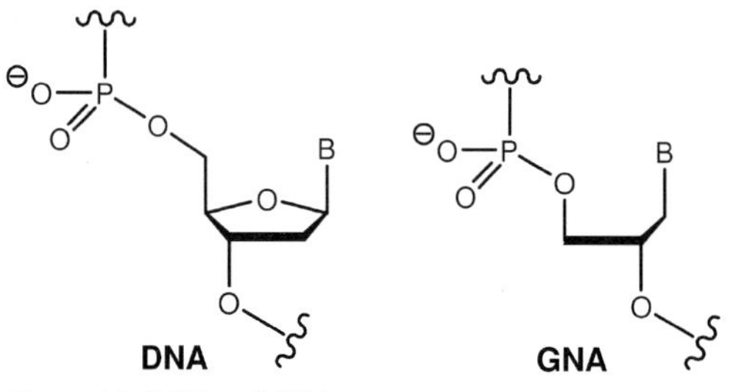

Figure 4.2 DNA and GNA.

4.2.3.2 Synthesis

For the synthesis of (*S*)-glycerol nucleosides, the commercially available (*R*)-glycidol **13** was chosen as starting material for the condensation with nucleobases under ring opening (Scheme 4.3). The stereochemistry at the secondary carbon is retained during the reaction.

The condensation of the nucleobases with (*R*)-glycidol (**13**) or the dimethoxytrityl-protected derivative **14** with sodium hydride as a base was reported in most cases without the need of protecting groups for the nucleobase [36, 37]. A modification was

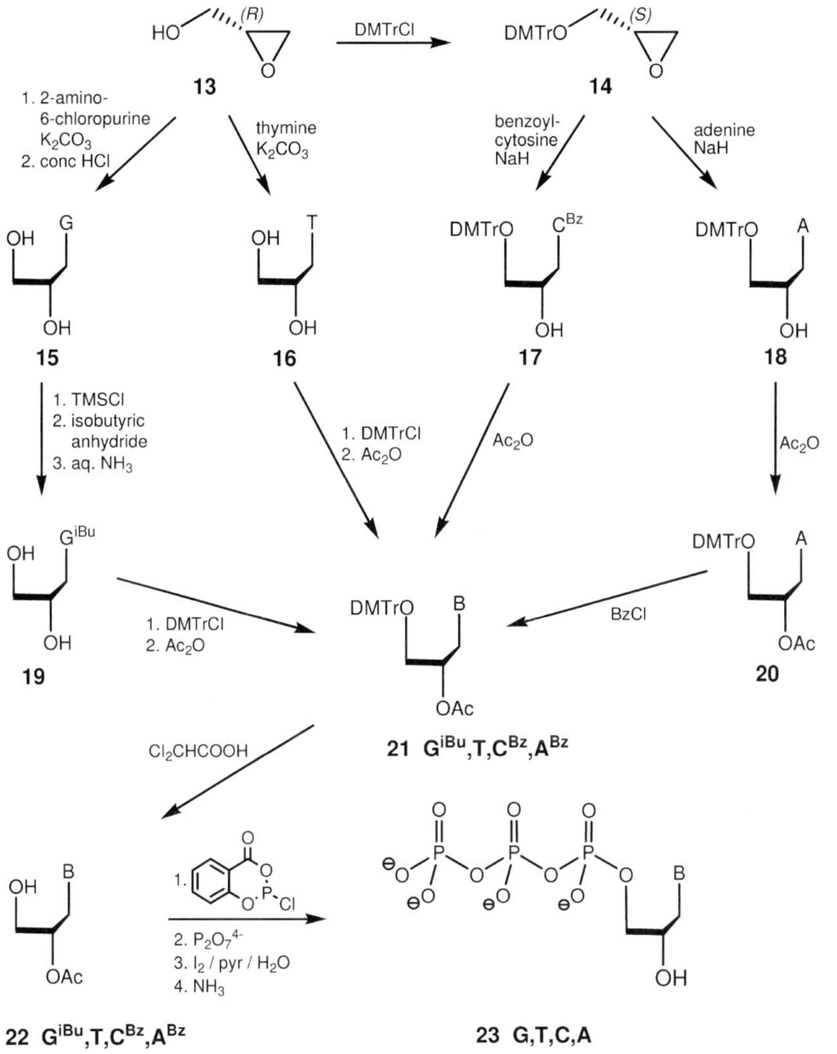

Scheme 4.3 Synthesis of (*S*)-glycerol nucleosides and conversion into the corresponding triphosphates.

reported replacing in some cases sodium hydride by potassium carbonate [38]. The reaction of (*R*)-glycidol (**13**) with 2-amino-6-chloropurine and potassium carbonate, followed by hydrolysis of the chloro group with aqueous hydrochloric acid, gave the guanosine analogous compound **15**. Accordingly, from thymine was prepared the thymidine analogue **16**. The dimethoxytrityl-protected (*S*)-glycidol derivative **14** which can be obtained easily from (*R*)-glycidol (**13**) by reaction with 4,4′-dimethoxytrityl chloride (DMTrCl) was used for the reaction with benzoyl-protected cytosine and sodium hydride to give the cytidine-analogous compound **17**. Accordingly, from unprotected adenine was prepared the adenosine analogue **16**. Prior to the conversion into the triphosphate derivatives by applying Eckstein's method [24], the introduction of protecting groups at the nucleobase and the glycerol moiety is required. Reaction of the guanosine analogue **15** with isobutyric anhydride leading to the isobutyryl derivative **19** followed by reaction with 4,4′-dimethoxytrityl chloride and finally with acetic anhydride, gave the fully protected guanosine analogue **21** $\mathbf{G^{iBu}}$. Accordingly, from compound **16** was prepared the thymidine analogue **21 T**. The partially protected cytidine analogue **17** was converted with acetic anhydride into the fully protected cytidine analogue **21** $\mathbf{C^{Bz}}$. From the adenosine analogue **18**, selective acetylation of the 2′-OH group with acetic anhydride under mild conditions, leading to compound **20** and subsequent benzoylation of the nucleobase with benzoyl chloride, gave the protected adenosine analogue **21** $\mathbf{A^{Bz}}$. After cleavage of the dimethoxytrityl protecting groups with dichloroacetic acid, the glycerol nucleoside analogues **22** $\mathbf{G^{iBu},T,C^{Bz},A^{Bz}}$ were converted into the corresponding glycerol nucleoside triphosphates **23 G,T,C,A** by reaction with 2-chloro-4*H*-1,3,2 benzodioxaphosphorin-4-one followed by bis (tri-*n*-butylammonium) pyrophosphate, oxidation by iodine, and deprotection with ammonia [38].

4.2.3.3 **DNA Polymerase Studies**

Very few results have been reported concerning the action of glycol-derived nucleoside triphosphates (gNTPs). McLaughlin et al. reported that among several DNA polymerases tested, the Therminator DNA polymerase is most proficient in extending an RNA or DNA primer by incorporation of one gNTP, regardless of whether RNA or DNA templates are used [38]. However, any further extension is significantly hampered and no extension from a glycol-derived nucleotide could be observed. The identified DNA polymerases that at least accept gNTPs for incorporation in an RNA- or DNA-composed primer template complex might be suited for a starting point of directed evolution to generate DNA polymerases.

4.2.4 **Nucleotides with Expanded Sugar Residues: 1,5-Anhydrohexitol Nucleotides**

4.2.4.1 **Introduction**

Previously, 1,5-anhydrohexitol nucleotides were developed wherein the 1,5-anhydrohexitol scaffold represents a sugar ring-size expanded structure that results from insertion of a methylene group between the 2′-deoxyribose oxygen atom and the anomeric center (C1′) (Figure 4.3).

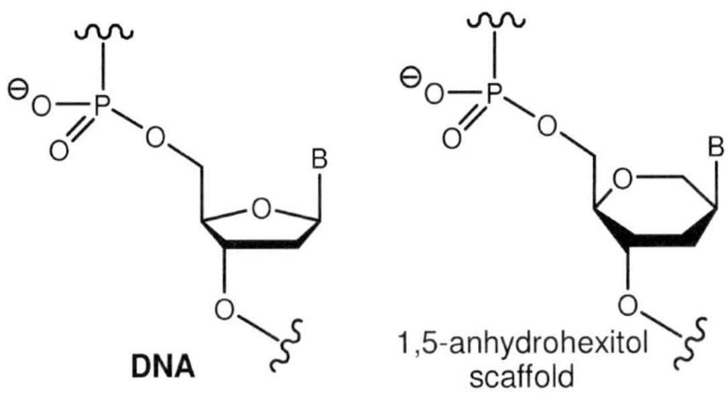

Figure 4.3 DNA and the 1,5-anhydrohexitol scaffold.

The 1,5-anhydrohexitol ring conformation is similar to the north conformation found in ribonucleotides, and contrasts that preferentially adopted in 2′-deoxyribose moieties. Thus, despite the absence of a 1,5-anhydrohexitol-derived scaffold, nucleoside triphosphates (hNTPs) do not bear a 2′-hydroxyl group and they are reminiscent of a size-expanded RNA building block in terms of their preferentially adopted conformation [39].

4.2.4.2 **Synthesis**

Synthesis of the 1,5-anhydrohexitol nucleosides starts from D-glucose, which is converted to acetobromoglucose (**24**) applying well-known one-pot procedures (Scheme 4.4) [40, 41]. Reductive dehalogenation with tributyltin hydride [42] and deacetylation with sodium methoxide led to the 1,5-anhydroglucitol **25** [43]. Benzylidenation with benzaldehyde and selective protection of the 2-OH group with dibutyltin oxide and toluoyl chloride gave the 3-OH unprotected compound **26**. Deoxygenation of the 3-OH group was performed by reaction with thiophosgene and 2,4-dichlorophenol, followed by reduction with tributyltin hydride and azoisobutyronitrile (AIBN); subsequent cleavage of the toluyl protecting group with sodium methoxide yielded compound **27**. The hydroxyl compound **27** can be used directly

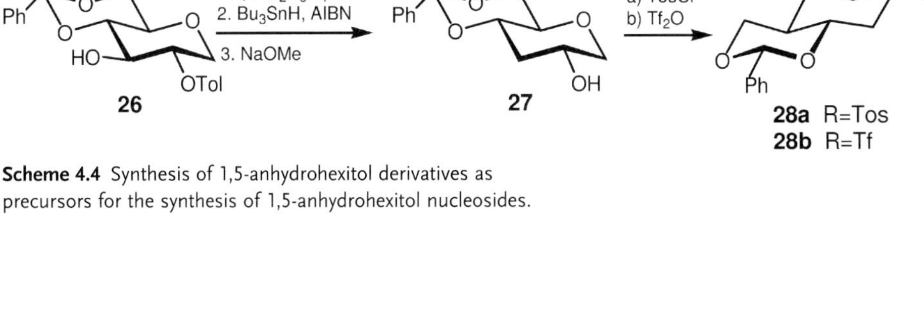

Scheme 4.4 Synthesis of 1,5-anhydrohexitol derivatives as precursors for the synthesis of 1,5-anhydrohexitol nucleosides.

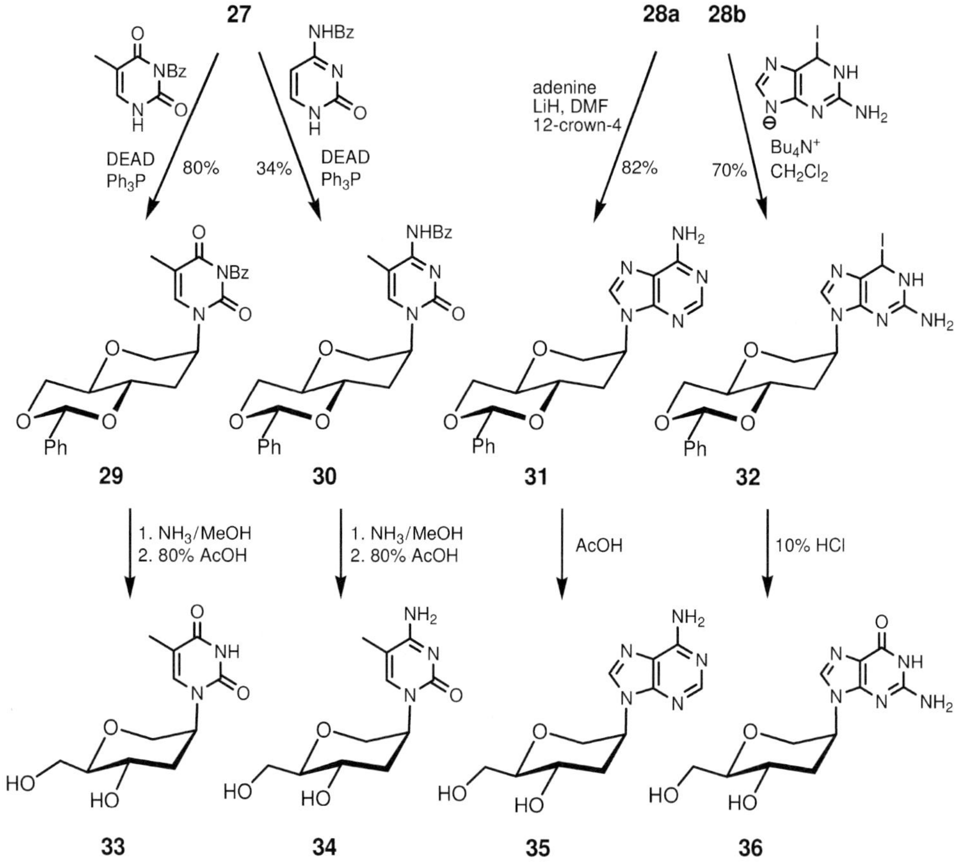

Scheme 4.5 Synthesis of 1,5-anhydrohexitol nucleosides.

for attachment of the nucleobase or converted into the tosyl or triflate derivatives **28a** or **28b** by reaction with *p*-toluenesulfonyl chloride or trifluoromethanesulfonic anhydride.

Attachment of the nucleobases to the 1,5-anhydrohexitol derivative **27** was performed by applying Mitsunobu conditions (Scheme 4.5). Reaction with N^3-benzoylthymidine in the presence of diethyl azodicarboxylate (DEAD) and triphenylphosphine led, under inversion of the configuration, to the thymidine analogue **29** [44] in high yield (80%). Reaction of the hydroxyl compound **27** with N^4-benzoylcytosine under the same conditions gave the cytidine analogue **30** in only low yield (34%). Other methods gave even lower yields; alkylation of cytosine derivatives leads generally to high amounts of byproducts by O^2-alkylation. The nucleoside analogues with purines as nucleobase were best prepared by nucleophilic substitution reactions. Reaction of the tosylate **28a** with adenine and lithium hydride gave exclusively the adenosine analogue **31** under inversion of the configuration. The triflate **28b** was used for the reaction with the readily soluble tetrabutylammonium salt of 2-amino-6-iodopurine in dichloromethane to give exclusively the desired

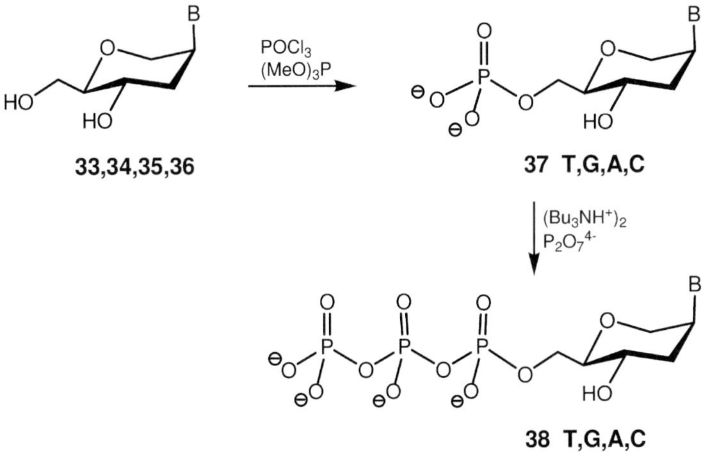

Scheme 4.6 Synthesis of 1,5-anhydrohexitol nucleoside triphosphates.

N-7 isomer **32** in good yield (70%). Debenzylidenation of compounds **29**, **30**, and **31** with acetic acid after cleavage of the protecting groups at the nucleobase, if required, led to the unprotected 1,5-anhydrohexitol nucleosides **33**, **34**, and **35**. For compound **32**, 10% hydrochloric acid was used to hydrolyze concomitantly the iodo group leading to the guanosine analogue **36**.

The unprotected 1,5-anhydrohexitol nucleosides **33**, **34**, **35**, and **36** were converted into the corresponding triphosphates applying a two-step synthesis (Scheme 4.6). The monophosphates **37 T,G,A,C** were prepared with phosphoryl chloride and trimethyl phosphate according to the method of Yoshikawa *et al.* [45] and converted according to Moffat and Khorana [46] with bis (tri-*n*-butylammonium) pyrophosphate into the 1,5-anhydrohexitol nucleoside triphosphates **38 T,G,A,C** [47].

4.2.4.3 Functional DNA Polymerase Studies

Herdewijn *et al.* investigated several DNA polymerases in their interplay with 1,5-anhydrohexitol nucleoside 5′-O-triphosphates (hNTPs) [39, 47]. These authors showed that all DNA polymerases tested were able to insert the artificial hNTPs, albeit extension from the inserted anhydrohexitol was hampered in some way. Interestingly, in comparison with the action of α-L-threose building blocks on the tested DNA polymerases (e.g., Vent DNA polymerase), Herdewijn *et al.* found that anhydrohexitol nucleotides are more easily accepted by DNA polymerases than the threose counterparts.

4.2.5 Nucleotides with Expanded Sugar Residues: Cyclohexenyl Nucleotides

4.2.5.1 Introduction

The cyclohexenyl nucleic acid (CeNA) is a DNA mimic in which the 2′-deoxyribose moiety is replaced by a six-membered cyclohexene ring which was developed by

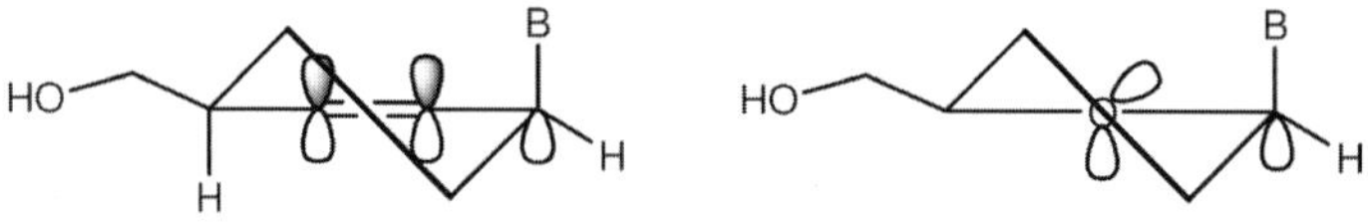

Figure 4.4 Stereoelectronic effects of cyclohexenyl nucleosides in comparison to the natural nucleosides [48, 49].

Herdewijn and colleagues [48, 49]. The cyclohexene moiety is a good mimic of the natural ribose because of its electronic similarity (Figure 4.4). In the $^3H_2$ conformation the pseudoaxial bond to the nucleobase is stabilized by the π–σ^* interaction (allylic effect), accordingly to the anomeric effect (n–σ^* interaction) in the N-conformation of the natural nucleosides.

Hybridization with both DNA and RNA has been observed, resulting in an increase in hybridization stability when cyclohexenyl nucleotides were incorporated into DNA strands. CeNA can be viewed as an expansion of the natural nucleic acid structure by substituting the 2′-deoxyribose oxygen atom with an ethylene group. These ring-size expanded analogues were tested as substrates for various DNA polymerases.

4.2.5.2 **Synthesis**

Synthesis of the cyclohexenyl moiety starts from the commercially available *R*-(−)-carvone (Scheme 4.7). Epoxidation with hydrogen peroxide led, both regio- and stereoselectively, to the epoxide **39**, as previously reported [50]. Stereoselective reduction [51] with lithium *sec*-butylborohydride (L-selectride) (diastereomeric excess <90%) and subsequent protection with *tert*-butyldimethylsilyl chloride (TBSCl) yielded compound **40** [52]. Oxidative cleavage of the double bond with osmium tetroxide and Baeyer–Villiger reaction of the resulting ketone **41** with *m*-chloroperbenzoic acid (MCPBA) led to the acetate **42**, with retention of the configuration. Deacetylation with potassium carbonate in methanol, and benzylation with benzyl bromide, yielded compound **43**. Regioselective ring opening of the epoxide **43** with lithium tetramethylpiperidide (LiTMP) and diethylaluminum chloride led to compound **44** containing an exocylic double bond [53]. Hydroboration with 9-borabicyclo [3.3.1]nonane (9-BBN) and work-up with hydrogen peroxide performed with high stereoselectivity (79 : 21); after regioselective protection of the primary hydroxyl group with TBSCl, compound **45** was obtained. Mesylation with methanesulfonyl chloride (MsCl), debenzylation with palladium on carbon and ammonium formate as hydrogen donor, and finally oxidation with manganese dioxide gave directly the enone derivative **47** via the ketone intermediate **46**, which was not isolated. Reduction of the enone **47** with sodium borohydride in the presence of cerium chloride led exclusively to the stereoisomer **48** which has the desired configuration and is a suitable intermediate for the attachment of the nucleobase.

Attachment of the nucleobases to the cyclohexenyl alcohol **48** was performed by applying Mitsunobu conditions (Scheme 4.8). Reaction with 2-amino-6-chloropurine in the presence of DEAD and triphenylphosphine was performed with inversion of the configuration; subsequent hydrolysis of the chloro substituent with

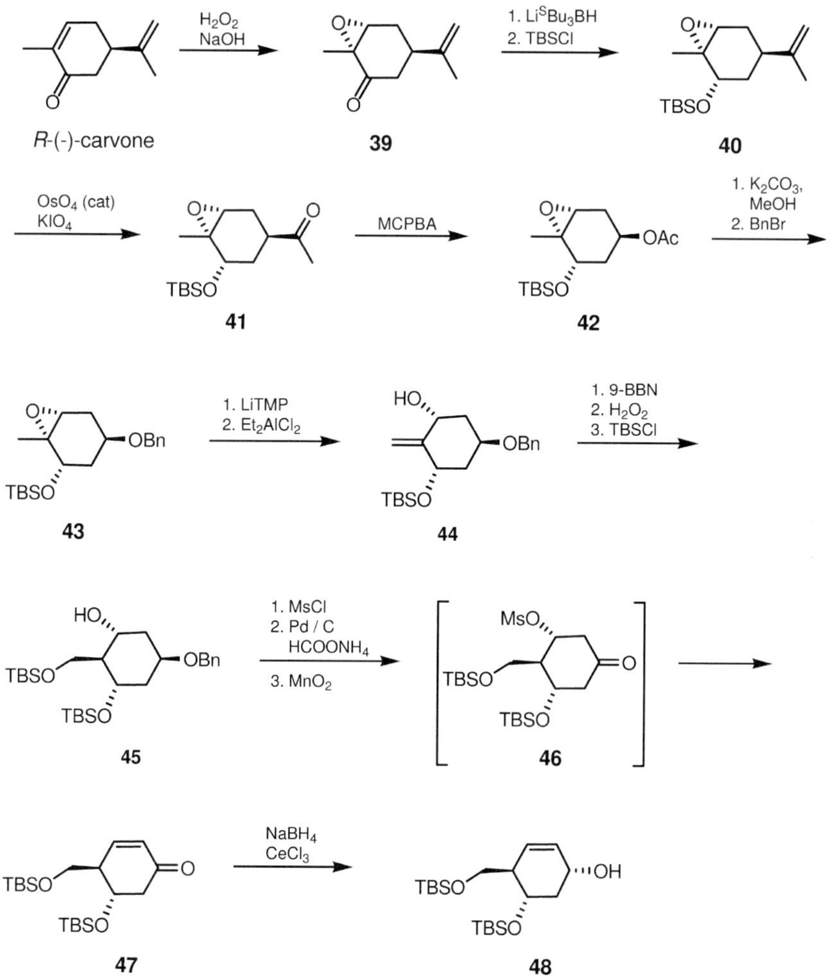

Scheme 4.7 Preparation of cyclohexenyl derivatives as precursors for the synthesis of cyclohexenyl nucleosides.

trifluoroacetic acid concomitant with cleavage of the *tert*-butylsilyl protecting groups led to the cyclohexenyl guanosine-analogue **49** [54]. Reaction of **48** with adenine under the same conditions gave the desired N_9-isomer (66%) together with 17% of the N_7-isomer; cleavage of the *tert*-butylsilyl protecting groups with trifluoroacetic acid was found to be better than with TBAF because the resulting cyclohexenyl adenosine-analogue **50** did not contain tetrabutylammonium salts [53]. By applying the one-pot synthesis of Ludwig [55] with phosphoryl chloride and trimethyl phosphate, followed by reaction with bis(tri-*n*-butylammonium) pyrophosphate, the unprotected cyclohexenyl nucleosides **49** and **50** were transformed into the corresponding guanosine and adenine analogous triphosphates **51 G,A** [56].

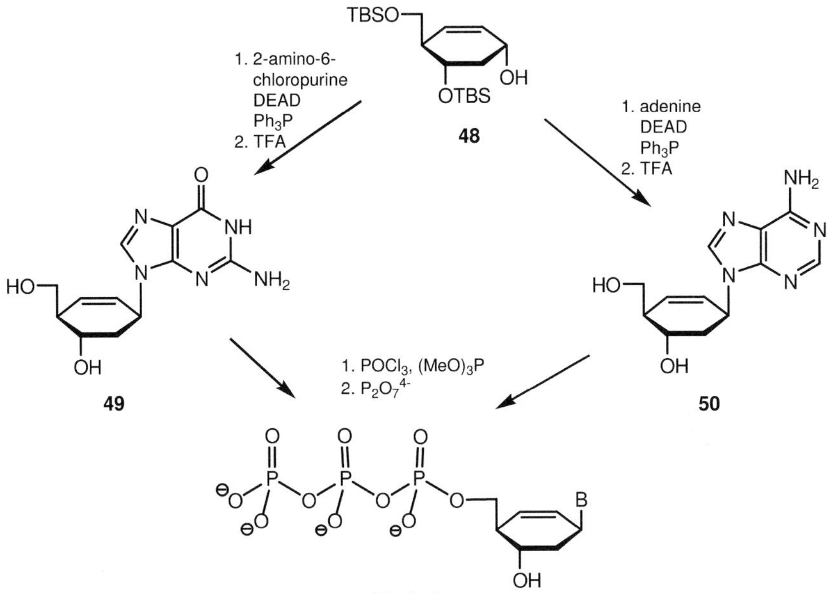

Scheme 4.8 Synthesis of cyclohexenyl nucleosides and conversion into the corresponding triphosphates.

4.2.5.3 **DNA Polymerase Studies**

Herdewijn and colleagues tested several DNA polymerases from different DNA polymerase sequence families for their ability to use cycleohexenyl-derived nucleoside triphosphates (CeNTPs) as substrates for DNA-templated CeNA synthesis [56]. All of the tested DNA polymerases were able to use CeNTPs as substrates for incorporation. *Thermococcus litoralis* (Vent) DNA polymerase and HIV-1 reverse transcriptase were shown to be capable of synthesizing even consecutive cyclohexenyl moieties. However (and this holds true for the above-mentioned analogues), this synthesis – especially when extending from a modified building block – shows a significantly lower efficiency compared to the natural substrates. It should be noted that superior results were obtained with Vent DNA polymerase. This enzyme derives from the same DNA polymerase sequence family as the above-mentioned Therminator DNA polymerase, and has shown the most promising results with tNTPs and gNTPs, namely family B. Thus, enzymes from this family may benefit from further improvement by directed enzyme evolution.

4.3 DNA Polymerase Selectivity: 4′-C-Modified Nucleotides

4.3.1 Introduction

The DNA polymerases involved in DNA replication processes exhibit low error rates of about only one error within 10^5 to 10^6 synthesized nucleotide linkages [2–8]. Hence, the question is, what are the mechanistic properties that enable a DNA polymerase to catalyze nucleotide incorporation with a selectivity far greater than that dictated by the thermodynamic differences between base pairs in free solution?

At first glance, the formation of distinct hydrogen-bonding patterns between the nucleobases of the coding template strand and the incoming 2′-deoxynucleoside-5′-*O*-triphosphate (dNTP) might be responsible for the selective information transfer. Nevertheless, as has been concluded by Goodman and colleagues on the basis of thermal denaturation studies of matched and mismatched DNA complexes, these interactions alone are not sufficient to explain the extent of selectivity commonly observed for enzymatic DNA synthesis [6, 7, 57]. Thus, several additional factors have been discussed as being involved in correct nucleotide recognition. Among these are the exclusion of water from the enzyme's active site, base stacking, solvation, minor groove scanning, and steric constraints within the nucleotide binding pocket. The contribution of each of these features to net-DNA polymerase selectivity remains to be disentangled. Recently, Kool described a functional strategy to evaluate the participation of hydrogen bonding in DNA replication selectivity mechanisms [2, 6]. Kool developed nucleotide analogues in which the polar natural DNA nucleobases are replaced by non-polar aromatic molecules, which closely mimic the shape and size of the natural nucleobases but have at least a significantly diminished ability to form stable hydrogen bonds. These non-polar nucleotide isosteres were applied as functional probes to elucidate the impact of hydrogen bonding on DNA polymerase selectivity. It was found that the non-polar isosteres were processed with remarkably high selectivity and efficiency by several DNA polymerases. Based on these results, it was concluded that hydrogen bonding is not required to achieve high incorporation efficiencies, and that significant levels of selectivity can be achieved without hydrogen bonds. A close fitting of Watson–Crick geometry and satisfaction of specific minor groove interactions are among the most important factors in DNA replication. These results led to the advent of the steric model for DNA polymerase selectivity. In brief, DNA polymerases form tight active sites composed from the nucleotide binding pocket that are believed to be fixed in size and shape in order to accurately process canonical nucleobase pairs. Consensus nucleotides are processed efficiently, while the non-consensus counterparts are excluded due to steric constraints imposed by the enzyme.

In order to probe steric constraints imposed in the active site to the nucleotide substrates by DNA polymerases, chemically modified nucleotides with graduate increasing steric demand were developed by Marx and colleagues [58]. The modifications were chosen in a way that they interfered with as few as possible further alterations of the analogue's properties such as hydrogen bonding capability or stacking. Thus, through subsequent functional studies of DNA polymerases in their interplay with the steric probes, some insight might be possible into the enzyme's selectivity mechanisms.

4.3.2
Design and Synthesis

Structural and functional studies of DNA polymerases suggest that the sugar moiety of the incoming dNTP is fully embedded in the nucleotide binding pocket [58]. Furthermore, DNA polymerases have to distinguish faithfully between 2′-deoxyriboses present in dNTPs and ribose moieties present in NTPs. Thus, the enzyme–sugar interactions are an integral part of the substrate recognition process. Clearly, there is no direct selectivity readout of the enzyme through the sugar moiety as all four nucleotides have the same sugar. Nevertheless, non-canonical nucleobase interactions should result in altered sugar conformations that might be edited by the DNA polymerase within the tight nucleotide binding pocket. Thus, the sugar residues might be indirectly involved in selectivity mechanisms.

To sense steric constraints in DNA polymerases within the nucleotide binding pocket acting on the sugar moiety of an incoming nucleoside triphosphate, Marx and colleagues developed a functional strategy based on 4′-*C*-alkylated thymidines (Figure 4.5) [58].

The steric probes were designed by substituting the 4′-C-hydrogen atom of thymidine-5′-O-triphosphate (TTP) with alkyl groups that continually increase in their steric bulk [59]. Alkyl substituents at the 4′-*C*-position were chosen since interference of the modification with hydrogen bonding, nucleobase pairing and stacking should be kept at minimum.

The synthesis of 4′-*C*-alkylated thymidines was performed via the alcohol **56**, which is easily accessible in high quantities as described recently (Scheme 4.9).

For the attachment of a 4′-*C*-branch to nucleoside derivatives, the most promising method, of applying an aldol reaction, was initially reported for a ribonucleoside [60], and later also applied to 2′-deoxythymidine [61]. The required aldehyde intermediate **53** was prepared from 2′-deoxythymidine by tritylation with dimethoxytrityl chloride (DMTrCl), silylation with TBSCl, and detritylation with

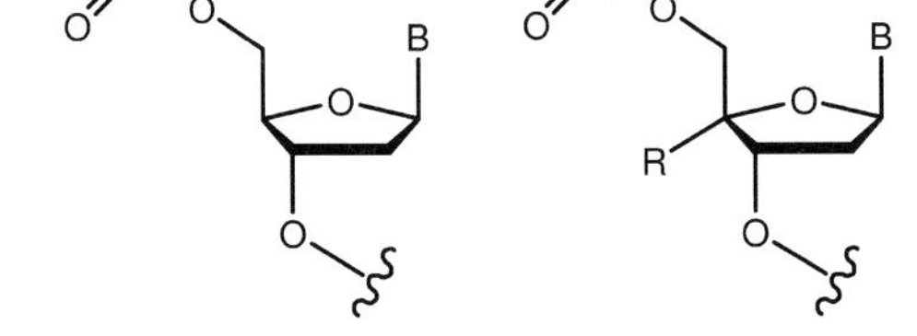

Figure 4.5 DNA and 4′-C-alkylated DNA; R = alkyl groups.

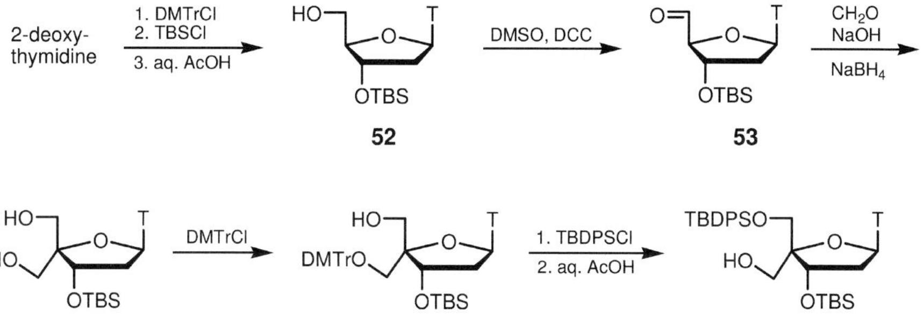

Scheme 4.9 Synthesis of a 4′-C-hydroxymethyl-modified 2′-deoxythymidine derivative.

acetic acid via the 5′-O-unprotected derivative **52**. Moffatt oxidation with dimethylsulfoxide (DMSO), dicyclohexylcarbodiimide (DCC), and pyridinium trifluoroacetate (Py-TFA) led to the aldehyde derivative **53**. A crossed aldol reaction with formaldehyde, followed by Canizzaro reduction or reduction with sodium borohydride according to an improved procedure [62], gave the 4′-*C*-hydroxymethyl-modified nucleoside **54**. The 5′-*O*-hydroxy group cannot be directly protected as the 4′-*C*-hydroxymethyl group was found to be more reactive [63]. Thus, the dihydroxy compound **54** was selectively tritylated with 4,4′-dimethoxytrityl chloride (DMTrCl) leading to compound **55** and, after protection of the 5′-*O*-hydroxy group with *tert*-butyldiphenylsilyl chloride (TBDPSCl), again desilylated with acetic acid to yield the 4′-*C*-hydroxymethyl derivative **56** [64, 65] which is a suitable intermediate for further 4′-*C*-modifications.

The hydrophobic 4′-*C*-modifications were introduced by conversion of the alcohol **56** into iodide **57** by treatment with iodine, triphenylphosphine, and imidazole (Scheme 4.10) [59, 66]. Reduction by hydrogenation with palladium on carbon in the presence of triethylamine and subsequent cleavage of the silylethers with tetra-*n*-butylammonium fluoride (TBAF) yielded 4′-*C*-methyl thymidine **58a**. 4′-*C*-ethyl modifications were introduced by Moffat oxidation of the alcohol **56** to the aldehyde **59** and Wittig reaction leading to the vinyl derivative **60**. Desilylation with tetra-*n*-butylammonium fluoride (TBAF) and subsequent hydrogenation of the double bond with palladium on carbon led to the 4′-*C*-ethyl thymidine **58b**. Nucleosides **58a**, **b** could be converted into the desired triphosphates **61a**, **b** ($\mathbf{T^RTP}$) by treatment with phosphoryl chloride and bis(tri-*n*-butylammonium) pyrophosphate in a one-pot multi-step reaction sequence.

4.3.3
DNA Polymerase Studies

If the steric model of DNA replication selectivity holds true, and enzyme interactions with the sugar moiety are involved in selectivity processes, the modifications would be expected to decrease the tolerance for geometrically altered conformations of nascent nucleotide pairs and result in an increase in nucleotide insertion

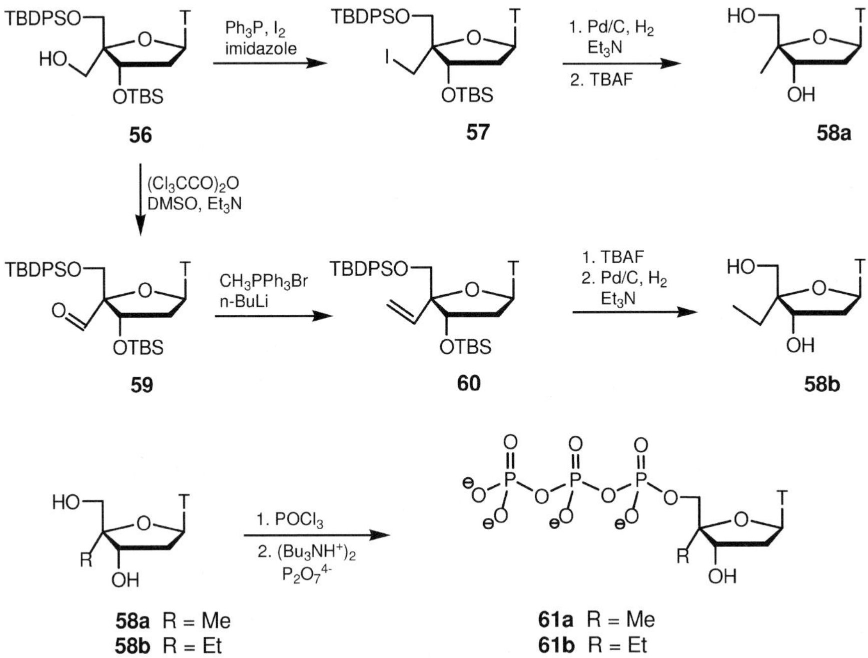

Scheme 4.10 Synthesis of 4′-*C*-alkylated nucleoside-5′-*O*-triphosphates.

selectivity. Indeed, it was found that increasing the bulk of nucleoside triphosphates through employment of the probes **T^{Me}TP** and **T^{Et}TP** led to a marked increase in nucleotide insertion selectivity catalyzed by a 3′ → 5′-exonuclease-deficient mutant of the Klenow fragment (KF^{-}) of *Escherichia coli* DNA polymerase I [59, 67]. Based on steady-state kinetic analysis, the enzyme is capable of inserting **T^{Me}TP** and **T^{Et}TP** with relative high efficiency opposite a canonical template base. On the other hand, misinsertion opposite non-canonical bases is approximately 100-fold less efficient compared to the natural substrate **T^{H}TP**. Thus, these results support the model that steric constraints are at least one crucial determinant of DNA polymerase selectivity.

Recent studies of several DNA polymerases show that the selectivity for Watson–Crick base-pair formation may vary by several orders of magnitude among different enzymes. The origin of this varying error propensity is not well understood, but it is assumed that DNA polymerases form nucleotide binding pockets that differ in properties such as shape and tightness [2]. Thus, high-fidelity DNA polymerases are believed to form tight and rigid substrate binding pockets that tolerate less geometric deviation. On the other hand, low-fidelity enzymes exhibit solvent-accessible and less flexible binding pockets, leading to a broader tolerance for aberrant geometries and thus, to a decreased fidelity.

Along these lines the efficacy of the steric probes **T^{R}TP** on KF^{-} with that on human immunodeficiency virus type-1 reverse transcriptase (HIV-1 RT) – an enzyme known for its error propensity – was investigated [68, 69]. Initially, it was reckoned that an error-prone DNA polymerase (such as HIV-1 RT) would process the bulkier thymidines more efficiently than the more selective enzyme KF^-. However, while investigating HIV-1 RT for “correct” insertion of the different $\mathbf{T^RTP}$, little difference between HIV-1 RT and KF^- was found. However, by analyzing misinsertion, the two enzymes are seen to behave differently. While 4′-*C*-methylation has little effect on the selectivity of HIV-1 RT, significant effects are observed for KF^-. Thus, based on the above-mentioned concept of active site tightness, these results suggest that both enzymes most significantly differ when promoting misinsertion rather than insertion opposite canonical template bases. This might be the result of differential active site conformations causing different steric constraints while promoting “incorrect” in comparison to “correct” nucleotide insertion.

In summary, these studies provide the first experimental evidence that variations of steric constraints within the nucleotide binding pocket of at least two DNA polymerases cause differences in nucleotide incorporation selectivity. However, for further insight into DNA polymerase selectivity mechanisms additional probes are required in future. For instance, the selectivity of nucleotide incorporation is dependent on the nature of the base pair. Thus, the dA–T base pair is formed with more different selectivity and efficiency than the dG–dC base pair. The origin of this phenomenon is not well understood, although steric factors were discussed. Steric probes such as those described above, that bear the four natural nucleobases, might be well-suited to gain mechanistic insight into these complex processes.

4.4 Concluding Remarks

As depicted above, numerous new insights in the function and substrate spectrum of DNA polymerases were gained through the employment of 2′-deoxyribose-modified substrates. It was found that several natural DNA polymerases accept even severely altered nucleotide structures. However, the nucleotide surrogates are processed with significantly diminished efficiency. The mutant from *Thermococcus* species 9°N-7 DNA polymerase, termed Therminator DNA polymerase, was found to exhibit significant proficiency in processing a variety of modified substrates, including those with modified sugar moieties. Thus, this enzyme variant might be a suitable starting point for directed enzyme evolution to generate new enzymes that process the depicted nucleotides with higher efficiency.

Through application of nucleotides with augmented size, new – and in part unexpected – insights into the selectivity of DNA polymerases have been gained. However, hitherto base-pair formation was investigated almost exclusively by the use of thymidine analogues. Further insights should be obtained by the development of approaches that use analogues of all four nucleotides.

References

1 Watson, J. D., Crick, F. H., *Nature* **1953**, *171*, 737–738.

2 Kool, E. T., *Annu. Rev. Biochem.* **2002**, *71*, 191–219.

3 Patel, P. H., Loeb, L. A., *Nat. Struct. Biol.* **2001**, *8*, 656–659.

4 Kunkel, T. A., *J. Biol. Chem.* **2004**, *279*, 16895–16898.

5 Kunkel, T. A., Bebenek, K., *Annu. Rev. Biochem.* **2000**, *69*, 497–529.

6 Kool, E. T., Morales, J. C., Guckian, K. M., *Angew. Chem. Int. Ed.* **2000**, *39*, 990–1009.

7 Goodman, M. F., *Proc. Natl. Acad. Sci. USA* **1997**, *94*, 10493–10495.

8 Echols, H., Goodman, M. F., *Annu. Rev. Biochem.* **1991**, *60*, 477–511.

9 Herdewijn, P., *Angew. Chem. Int. Ed.* **2001**, *40*, 2249–2251.

10 Eschenmoser, A., *Orig. Life Evol. Biosph.* **2004**, *34*, 277–306.

11 Kornberg, A., Baker, T. A., *DNA Replication*, W.H. Freeman & Co., New York, USA, **1992**.

12 Bessman, M. J., Kornberg, A., Lehman, I. R., Simms, E. S., *Biochim. Biophys. Acta* **1956**, *21*, 197–198.

13 Lehman, I. R., Bessman, M. J., Simms, E. S., Kornberg, A., *J. Biol. Chem.* **1958**, *233*, 163–170.

14 Marx, A., Detmer, I., Gaster, J., Summerer, D., *Synthesis* **2004**, 1–14.

15 Verma, S., Eckstein, F., *Annu. Rev. Biochem.* **1998**, *67*, 99–134.

16 Jung, K.-H., Marx, A., *Cell. Mol. Life Sci.* **2005**, *62*, 2080–2091.

17 Schöning, K.-U., Scholz, P., Guntha, S., Wu, X., Krishnamurthy, R., Eschenmoser, A., *Science* **2000**, *290*, 1347–1351.

18 Isbell, H. S., Frush, H. L., *Carbohydr. Res.* **1979**, *72*, 301–304.

19 Wie, C. C., Bernardo, S. D., Tengi, J. P., Borgese, J., Weigele, M., *J. Org. Chem.* **1985**, *50*, 3462–3467.

20 Schöning, K.-U., Scholz, P., Wu, X., Guntha, S., Delgado, G., Krishnamurthy, R., Eschenmoser, A., *Helv. Chim. Acta* **2002**, *85*, 4111–4153.

21 Weidenhagen, R., Wegner, H., *Chem. Ber.* **1939**, *72*, 2010–2020.

22 Niedballa, U., Vorbrüggen, H., *Angew. Chem. Int. Ed. Engl.* **1970**, *9*, 461.

23 Niedballa, U., Vorbrüggen, H., *J. Org. Chem.* **1974**, *39*, 3654–3660.

24 Ludwig, J., Eckstein, F., *J. Org. Chem.* **1989**, *54*, 631–635.

25 Zou, K., Horhota, A., Yu, B., Szostak, J. W., McLaughlin, L. W., *Org. Lett.* **2005**, *7*, 1485–1487.

26 Chaput, J. C., Ichida, J. K., Szostak, J. W., *J. Am. Chem. Soc.* **2003**, *125*, 856–857.

27 Chaput, J. C., Szostak, J. W., *J. Am. Chem. Soc.* **2003**, *125*, 9274–9275.

28 Kempeneers, V., Vastmans, K., Rozenski, J., Herdewijn, P., *Nucleic Acids Res.* **2003**, *31*, 6221–6226.

29 Ichida, J. K., Zou, K., Horhota, A., Yu, B., McLaughlin, L. W., Szostak, J. W., *J. Am. Chem. Soc.* **2005**, *127*, 2802–2803.

30 Horhota, A., Zou, K., Ichida, J. K., Yu, B., McLaughlin, L. W., Szostak, J. W., Chaput, J. C., *J. Am. Chem. Soc.* **2005**, *127*, 7427–7434.

31 Ichida, J. K., Horhota, A., Zou, K., McLaughlin, L. W., Szostak, J. W., *Nucleic Acids Res.* **2005**, *33*, 5219–5225.

32 Gardner, A. F., Joyce, C. M., Jack, W. E., *J. Biol. Chem.* **2004**, *279*, 11834–11842.

33 Gardner, A. F., Jack, W. E., *Nucleic Acids Res.* **2002**, *30*, 605–613.

34 Gardner, A. F., Jack, W. E., *Nucleic Acids Res.* **1999**, *27*, 2545–2553.

35 Renders, M., Emmerechts, G., Rozenski, J., Krecmerová, M., Holý, A., Herdewijn, P., *Angew. Chem. Int. Ed.* **2007**, *46*, 2501–2504.

36 Zhang, L., Peritz, A. E., Meggers, E., *J. Am. Chem. Soc.* **2005**, *127*, 4174–4175.

37 Zhang, L., Peritz, A. E., Carroll, P. J., Meggers, E., *Synthesis* **2006**, 645–653.

38 Horhota, A. T., Szostak, J. W., McLaughlin, L. W., *Org. Lett.* **2006**, *8*, 5345–5347.

39 Vastmans, K., Froeyen, M., Kerremanns, L., Pochet, S., Herdewijn, P., *Nucleic Acids Res.* **2001**, *29*, 3154–3163.

40 Lemieux, R. U., *Methods of Carbohydrate Chemistry*, Vol. II, (Eds R. L. Whistler M. L. Wolfrom), Academic Press, London, **1963**, pp. 221–222.

41 Kartha, K. P., Jennings, K. J. A., *J. Carbohydr. Chem.* **1990**, *9*, 777–781.

42 Kocienski, P., Pant, C. A., *Carbohydr. Res.* **1982**, *110*, 330–332.

43 Verheggen, I., Van Aerschot, A., Toppet, S., Snoeck, R., Janssen, G., Balzarini, J., De Clercq, E., Herdewijn, P., *J. Med. Chem.* **1993**, *36*, 2033–2040.

44 De Bouvere, B., Kerremans, L., Rozenski, J., Janssen, G., Van Aerschot, A., Claes, P., Busson, R., Herdewijn, P., *Liebigs Ann./Recueil* **1997**, 1453–1461.

45 Yoshikawa, M., Kato, T., Takenishi, T., *Tetrahedron Lett.* **1967**, *50*, 5065–5068.

46 Moffatt, J. G., Khorana, H. G., *J. Am. Chem. Soc.* **1961**, *83*, 649–658.

47 Vastmans, K., Pochet, S., Peys, A., Kerremans, L., Van Aerschot, A., Hendrix, C., Marlière, P., Herdewijn, P., *Biochemistry* **2000**, *39*, 12757–12765.

48 Wang, J., Verbeure, B., Luyten, I., Lescrinier, E., Froeyen, M., Hendrix, C., Rosemeyer, H., Seela, F., van Aerschot, A., Herdewijn, P., *J. Am. Chem. Soc.* **2000**, *122*, 8595–8602.

49 Wang, J., Froeyen, M., Hendrix, C., Andrei, G., Snoeck, R., De Clercq, E., Herdewijn, P., *J. Med. Chem.* **2000**, *43*, 736–745.

50 Rupe, H., Refardt, M., *Helv. Chim. Acta* **1942**, *25*, 836–859.

51 Nishiyama, S., Ikeda, Y., Yoshida, S., Yamamura, S., *Tetrahedron Lett.* **1989**, *30*, 105–108.

52 Wang, J., Busson, R., Blaton, N., Rozenski, J., Herdewijn, P., *J. Org. Chem.* **1998**, *63*, 3051–3058.

53 Wang, J., Herdewijn, P., *J. Org. Chem.* **1999**, *64*, 7820–7827.

54 Wang, J., Froeyen, M., Hendrix, C., Andrei, C., Snoeck, R., Lescrinier, E., De Clercq, E., Herdewijn, P., *Nucleosides Nucleotides* **2001**, *20*, 727–730.

55 Ludwig, J., *Acta Biochim. Biophys. Acad. Sci. Hung.* **1981**, *16*, 131–133.

56 Kempeneers, V., Renders, M., Froeyen, M., Herdewijn, P., *Nucleic Acids Res.* **2005**, *33*, 3828–3836.

57 Petruska, J., Goodman, M. F., Boosalis, M. S., Sowers, L. C., Cheong, C., Tinoco, I., Jr., *Proc. Natl. Acad. Sci. USA* **1988**, *85*, 6252–6256.

58 Marx, A., Summerer, D., *Synlett* **2004**, 217–224.

59 Summerer, D., Marx, A., *Angew. Chem. Int. Ed.* **2001**, *40*, 3693–3695.

60 Jones, G. H., Taniguchi, M., Tegg, D., Moffatt, J. G., *J. Org. Chem.* **1979**, *44*, 1309–1317.

61 Yang, C. O., Wu, H. Y., Fraser-Smith, E. B., Walker, M., *Tetrahedron Lett.* **1992**, *33*, 37–40.

62 Thrane, H., Fensholdt, J., Regner, M., Wengel, J., *Tetrahedron* **1995**, *51*, 10389–10402.

63 Yang, C. O., Kurz, W., Eugui, E. M., McRoberts, M. J., Verheyden, J. P. H., Kurz, L. J., Walker, K. A. M., *Tetrahedron Lett.* **1992**, *33*, 41–44.

64 Giese, B., Imwinkelried, P., Petretta, M., *Synlett* **1994**, 1003–1004.

65 Marx, A., Erdmann, P., Senn, M., Körner, S., Jungo, T., Petretta, M., Imwinkelried, P., Dussy, A., Kulicke, K. J., Macko, L., Zehnder, M., Giese, B., *Helv. Chim. Acta* **1996**, *79*, 1980–1994.

66 Detmer, I., Summerer, D., Marx, A., *Eur. J. Org. Chem.* **2003**, 1837–1846.

67 Strerath, M., Summerer, D., Marx, A., *ChemBioChem* **2002**, *3*, 578–580.

68 Strerath, M., Cramer, J., Restle, T., Marx, A., *J. Am. Chem. Soc.* **2002**, *124*, 11230–11231.

69 Cramer, J., Strerath, M., Marx, A., Restle, T., *J. Biol. Chem.* **2002**, *277*, 43593–43598.

5
Pyrimidine Dimers: UV-Induced DNA Damage

Shigenori Iwai

5.1
Introduction

As DNA is an organic molecule, like other laboratory chemicals, it is subjected to various chemical reactions in cells. The alteration of the chemical structure of DNA by such reactions is referred to as *DNA damage*, and this can be divided into two classes, termed "endogenous" and "environmental" [1].

Endogenous DNA damage includes: (1) hydrolysis of the glycosidic bonds, especially those of the purine bases, and the amino group of cytosine; (2) oxidation by reactive oxygen species, such as the hydroxyl radical, to form 7,8-dihydro-8-oxoguanine, thymine glycol, cytosine hydrate, etc.; (3) adduct formation with lipid peroxidation products, such as malondialdehyde; and (4) non-enzymatic methylation by *S*-adenosylmethionine.

Environmental DNA damage is caused by ionizing radiation, which results in direct and indirect chemical reactions, including oxidation and degradation of the bases and cleavage of the sugar–phosphate backbone, ultraviolet (UV) radiation, and various chemical agents, some of which are known as carcinogens. The UV radiation is divided into three parts, depending on wavelength: UV-A (320–400 nm); UV-B (290–320 nm); and UV-C (shorter than 290 nm) [2]. Although UV-B and UV-C are absorbed by the stratospheric ozone layer (the latter with very short wavelengths is also absorbed by the oxygen in the atmosphere), all of these wavelength regions can damage the base moieties of DNA [3].

In this chapter, the two major types of products formed by UV irradiation, namely cyclobutane pyrimidine dimers (CPDs) and pyrimidine(6–4)pyrimidone photoproducts [(6–4) photoproducts], are explained from both chemical and biological viewpoints. The products formed by the above-mentioned reactions, which are referred to as DNA lesions, induce mutations of DNA sequences that lead to carcinogenesis and cell death. Although living organisms possess repair systems to cope with DNA damage, exposure to UV radiation can cause skin cancer, the incidence of which is currently increasing at a high rate among fair-skinned populations [2]. Therefore, it is

Modified Nucleosides: in Biochemistry, Biotechnology and Medicine. Edited by Piet Herdewijn

ISBN: 978-3-527-31820-9

important to review the chemical and biochemical studies of UV-induced DNA lesions, which span a history of about 50 years.

5.2
Formation of Pyrimidine Dimers

5.2.1
Cyclobutane Pyrimidine Dimers

When DNA is exposed to UV radiation, a cyclobutane ring is formed between two adjacent pyrimidine bases by a [2 + 2] cycloaddition that converts two C5–C6 π-bonds into two σ-bonds, as shown in Figure 5.1 (structures 2–4) and Figure 5.2 (structures **9** and **10**). The cyclobutane formation of thymine was first described in 1960 [4], and in the following year, the *syn* structure was proposed for the UV lesion

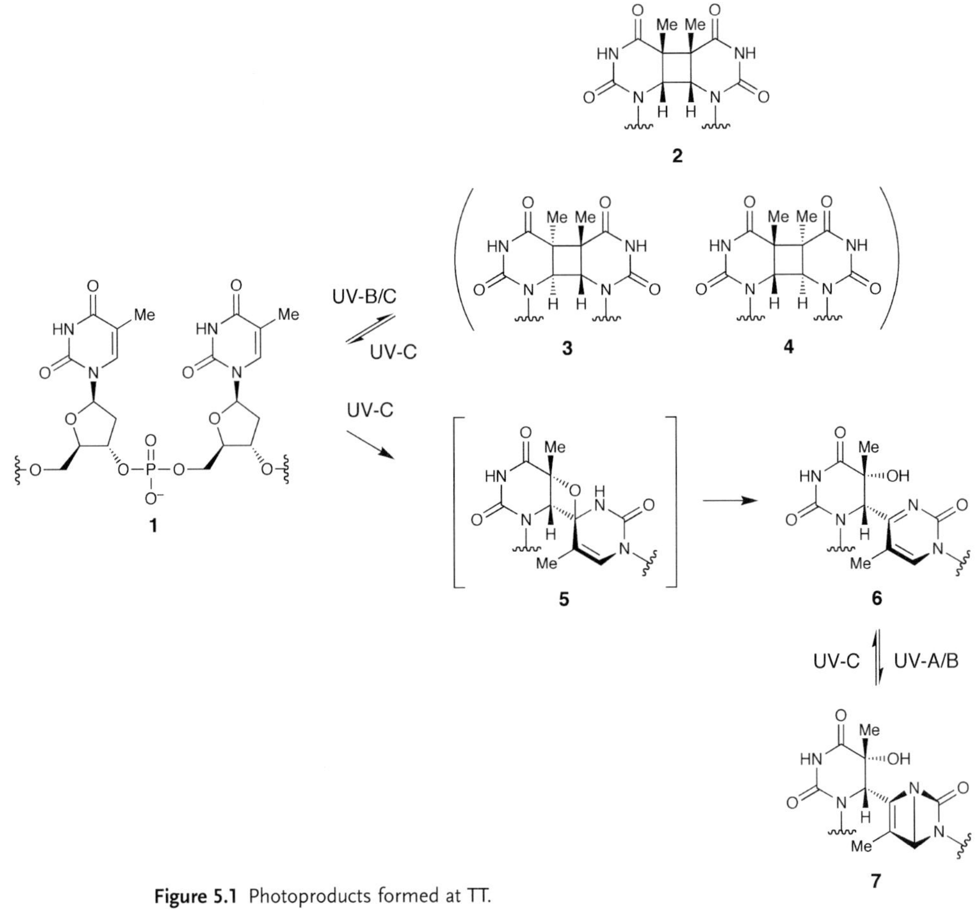

Figure 5.1 Photoproducts formed at TT.

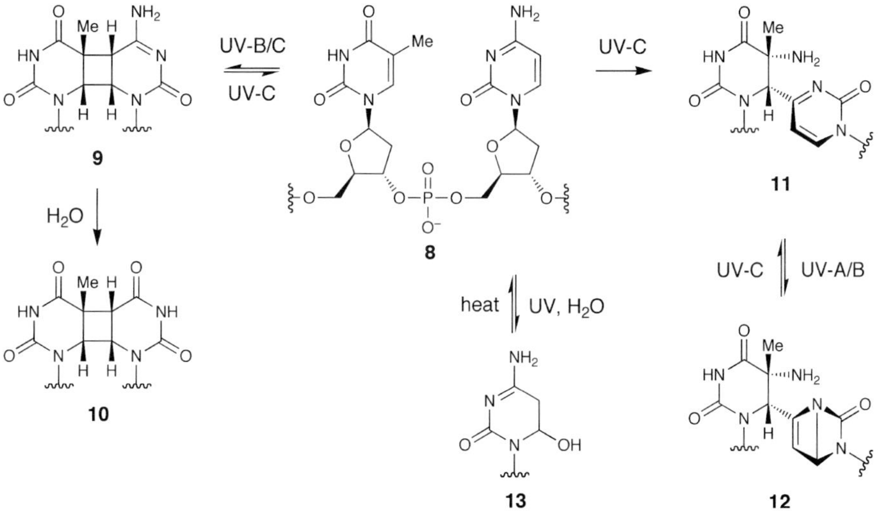

Figure 5.2 Photoproducts formed at TC.

in DNA [5, 6]. It was subsequently demonstrated, by a chemical reaction [7] and by nuclear magnetic resonance (NMR) [8] and infra-red (IR) [9] spectroscopy, that the major thymine–thymine dimer formed in DNA has the *cis–syn* configuration (**2**). This structure is quite reasonable, because the nucleobases are stacked on the adjacent base with the torsion angles of the glycosidic bonds in the *anti* conformation in double-stranded DNA. When denatured DNA is irradiated with UV, *trans–syn* thymine dimers (**3** and **4**) are also formed [10, 11]. In this case, the *trans–syn*-I thymine dimer (**3**), which is formed between the 5′ thymidine in the *syn* conformation and the 3′ thymidine in the *anti* conformation, is the major product [12], and the yield of the *trans–syn*-II product (**4**) is extremely low [13]. Formation of the thymine dimer reaches a plateau at a high UV dose, because it is a reversible process [14–16] that depends on the wavelength. At longer wavelengths (UV-B), the dimer is formed in a relatively efficient manner, but when DNA is irradiated with short-wavelength UV (UV-C), the photoequilibrium is inclined to the reversion of the thymine dimer to the original thymine bases, as shown in Figure 5.1. With irradiation at 254 and 280 nm, the percentages of the thymine dimer at saturation, with respect to thymine, are reportedly 3.8% and 7.6%, respectively [16].

Although the thymine dimer has been studied most intensively, the CPD is formed at all four of the pyrimidine–pyrimidine sites, namely TT, TC, CT, and CC. Mitchell *et al.* [17] investigated the sequence specificity of CPD formation by analyzing T4 endonuclease V (see Section 5.5.1)-digested fragments of irradiated DNA at nucleotide resolution, and the CPD formation was detected in the following order of preference: TT $\gg$ TC $\approx$ CT $>$ CC. Although this order was the same between the UV-C and UV-B irradiation, enhanced and reduced CPD formation was observed at the cytosine-containing sites and the TT site, respectively, in the case of the UV-B irradiation. The effect of the flanking sequence was also revealed in this study. A pyrimidine base on the 5′ side of the dipyrimidine site made a CPD hot spot, whereas a 5′ guanine prevented the CPD formation [17]. The quantum yields of the CPD formation were calculated to be 17.5×10^{-3} and 16.0×10^{-3} for TT and 9.1×10^{-3} and 7.2×10^{-3} for TC [16, 18], and the preferential CPD formation of TT over CT and CC has been investigated by the quantum-chemical method, using density functional theory (DFT) techniques [19]. The CPD is also formed at dipyrimidine sites containing 5-methylcytosine [20, 21]. The methylation alters the absorption spectrum of cytosine, and while a calculation using the DFT techniques suggested that the cytosine methylation did not contribute to the increase in the CPD yield [22], Pfeifer and coworkers [23–25] reported that 5-methylcytosine is a preferred target for CPD formation upon irradiation with natural sunlight. Another group reported the effects of the sugar conformation on the photoproduct formation [26]. When the two sugar moieties in thymidylyl(3′–5′)thymidine were changed to 2′-*O*-methylribose, the population of the C3′-*endo* conformer of the 5′ and 3′ sugars increased from 30% to 75% and from 37% to 66%, respectively, and the yield of the CPD after irradiation at 254 nm was 1.5-fold higher than that obtained for the normal deoxyribose counterpart.

While the pyrimidine bases are excited to a singlet state by UV-B or UV-C radiation [27], there is another mechanism for CPD formation. DNA does not absorb UV-A, but the triplet excited states of carbonyl compounds, such as acetone and acetophenone, generated by UV-A irradiation can produce the CPD exclusively, by the triplet energy transfer [28]. Endogenous UV-A photosensitizers exist in cells [29], and CPD formation by UV-A irradiation was demonstrated in Chinese hamster ovary cells [30, 31] and human fibroblasts and keratinocytes [32].

Gale *et al.* [33] determined the distribution of CPDs in nucleosomes by digesting the core DNA, which was prepared from UV-irradiated chromatin, with the 3′→5′ exonuclease activity of T4 DNA polymerase, and found that the CPD formation showed a 10.3 base periodicity. The sites of the maximum CPD yields in core DNA were found at the positions where the phosphate backbone was farthest from the histone surface. In contrast, the distribution of the CPD in linker DNA is nearly uniform [34]. This pattern is quite different from the distribution of the (6–4) photoproduct, as described in Section 5.2.2.

Among the four nucleobases in DNA, cytosine is the most susceptible to hydrolytic deamination, and when UV radiation causes cytosine to form a CPD with an adjacent pyrimidine base, the reaction rate of this deamination (**9** → **10** in Figure 5.2) is greatly accelerated, because the stabilization of the amino group by the aromatic resonance is lost due to the saturation of the 5,6-double bond. The reported deamination rate constants observed for the *cis–syn* CPDs in the form of dinucleoside monophosphates, d(TpC) and d(CpT), at 25 °C (or room temperature) are 2.5×10^{-5} and $2.8 \times 10^{-5}\,s^{-1}$, respectively [35, 36]. In DNA, various rate constants have been determined, with reported values of 3.9×10^{-5} and $1.2–1.8 \times 10^{-6}\,s^{-1}$ at 37 °C *in vitro*, and $1.5 \times 10^{-4}\,s^{-1}$ at 42 °C in *Escherichia coli* cells [37–39]. The 5′ methylation of cytosine in the CPD increases the stability of the amino group [21], and its rate constant was reported to be $10^{-5}\,min^{-1}$ ($=1.7 \times 10^{-7}\,s^{-1}$) [40]. Nevertheless, the deamination of the CPD containing 5-methylcytosine is an important component of mutagenesis [41].

5.2.2
The (6–4) Photoproducts and their Dewar Valence Isomers

Johns *et al.* [42] found that UV irradiation of thymidylyl(3′–5′)thymidine, d(TpT), resulted in the formation of a product (TpT^4) that differed from the CPD. This product had an absorption maximum at 325 nm, and was converted to a second product (TpT^3) by irradiation at 313 nm. TpT^3 could be reconverted to TpT^4 by irradiation at 240 nm. TpT^4 was produced irreversibly from d(TpT), and this compound fluoresced at 405 nm when excited at 325 nm [43]. The structure of TpT^4, in which a covalent bond is formed between C6 of the 5′ thymine and C4 of the 3′ thymine (structure **6** in Figure 5.1), was determined by Varghese and Wang [44, 45], and hence, this lesion was designated as the (6–4) photoproduct. In the formation of the (6–4) photoproduct, the first reaction is the [2 + 2] cycloaddition between the C5–C6 double bond and the carbonyl group, which produces an oxetane intermediate (**5**), as shown in Figure 5.1 [44]. This intermediate is stable at temperatures below −80 °C, but is converted to the (6–4) photoproduct at higher temperatures [46]. A similar structure (**11** in Figure 5.2) was determined for the photoproduct formed at the TC site [47].

UV irradiation of DNA produces alkaline-labile sites [48], and the product formed at these sites was identified as the (6–4) photoproduct [49]. We recently revealed that the first reaction in the alkali degradation of the (6–4) photoproduct is hydrolysis of the N3–C4 covalent bond of the 5′ base moiety [50]. Although the (6–4) photoproduct can be formed at all four of the dipyrimidine sites [49], its frequency is higher at the TC and CC sites [48, 51, 52], and 5-methylcytosine inhibits the formation of this type of lesion [52]. The distribution of the (6–4) photoproducts is random in the nucleosome core domain [53], and this type of lesion is formed preferentially in the linker DNA [54].

The chemical structure of TpT^3, produced by irradiation of the (6–4) photoproduct at 313 nm [42], was determined by Taylor and Cohrs [55]. These authors found that the 2-pyrimidone ring in the (6–4) photoproduct was photoisomerized to a Dewar-type structure (**7**), and therefore, this lesion has been termed the Dewar valence isomer or the Dewar photoproduct. This type of isomerization also occurs in the (6–4) photoproduct formed at the TC site (**12**) [56]. Both the (6–4) photoproduct and its Dewar valence isomer are formed simultaneously when DNA is exposed to sunlight, and because the Dewar photoproduct is also alkaline labile [57], these two photoproducts cannot be distinguished in the assays using piperidine treatment [48, 52]. The three types of photoproduct – namely the CPD, the (6–4) photoproduct, and the Dewar valence isomer – can be quantified separately using monoclonal antibodies [58] or by an HPLC–tandem mass spectrometry method [59]. When DNA was irradiated with simulated sunlight, the relative amounts of CPD, (6–4) photoproduct, and Dewar valence isomer formation were determined to be 1 : 0.18 : 0.06, respectively [60].

5.2.3
Other UV Lesions

Bose *et al.* [61] found that a photoproduct was formed between thymine and adenine at a TA site (**14** in Figure 5.3), and proposed a *cis–syn* cyclobutane dimer structure [62].

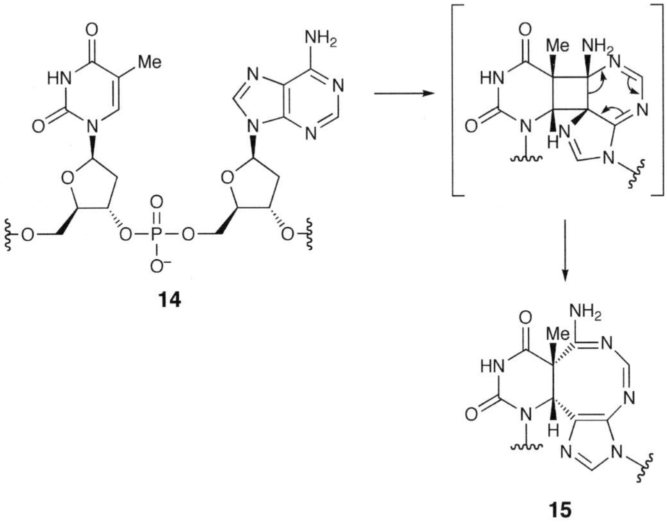

Figure 5.3 A photoproduct formed at TA.

Another structure with a *trans–syn*-I-type linkage was derived from an NMR study [63]. However, Taylor and coworkers [64] re-evaluated its structure by $^{1}H-^{13}C$ heteronuclear NMR experiments, and proposed a new structure containing an eight-membered ring (**15**), which was produced by a ring-expansion reaction of the cyclobutane intermediate, as shown in Figure 5.3. Recently, the reaction mechanism for the formation of this photoproduct was analyzed by quantum-chemical calculations [65].

In bacterial spores, the so-called "spore photoproduct", which differs from the major UV lesions, is formed between two thymine bases [66]. The chemical structure of this photoproduct was determined to be 5-thyminyl-5,6-dihydrothymine [67], which has two diastereomers (**16** and **17** in Figure 5.4). A dinucleoside monophosphate

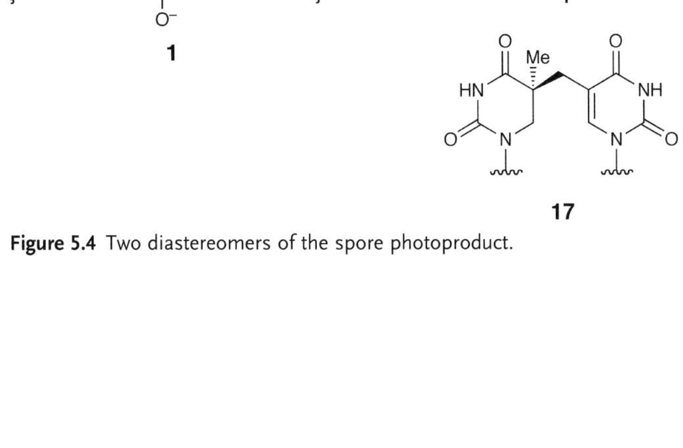

Figure 5.4 Two diastereomers of the spore photoproduct.

bearing the spore photoproduct [68, 69] and its analogue without the phosphodiester linkage [70, 71] have been synthesized.

UV irradiation of DNA causes photohydration of cytosine, and the product, cytosine hydrate, is a mixture of 6*R*- and 6*S*-hydroxy-5,6-dihydrocytosine (**13** in Figure 5.2) [72], which can be easily dehydrated back to cytosine. By irradiation with 100 kJ m^{-2} of 254-nm light at pH 8.0, 2.2% of the cytosine residues were reportedly converted to cytosine hydrate, which decayed with a half-life of 25 h at 37 °C [73], whereas the photohydration was not observed for thymine or 5-methylcytosine [74]. Cytosine hydrate undergoes accelerated deamination in the same way as the cytosine-containing CPD [75].

5.3
Chemical Synthesis of Oligonucleotides Containing Pyrimidine Dimers

5.3.1
Oligonucleotides Containing a CPD

In the past, DNA fragments treated with UV or γ radiation, alkylating or crosslinking agents, and other carcinogens were used as damaged DNA for biochemical studies. However, methods for the chemical synthesis of oligonucleotides containing a particular lesion in a defined sequence have been developed, as reviewed recently [76]. In principle, oligonucleotides are synthesized on a DNA synthesizer by linking nucleotides one by one in the 3′ → 5′ direction, and properly-protected nucleoside 3′-phosphoramidites are used as building blocks for chain elongation. Therefore, it is possible to incorporate a lesion into synthetic oligonucleotides if a phosphoramidite building block bearing a damaged base can be prepared. Taylor *et al.* [77] successfully synthesized a building block of the *cis–syn* CPD formed at the TT site, and incorporated it into oligonucleotides. As two neighboring bases are linked in the case of the UV lesions, these authors prepared protected thymidylyl(3′–5′)thymidine first, as shown in Figure 5.5. After removal of the 4,4′-dimethoxytrityl (DMT) group, which would interfere with the photoreaction, the dinucleoside monophosphate was irradiated with Pyrex-filtered light in the presence of acetophenone, as a triplet sensitizer. This reaction yielded four products, namely two diastereomers of the phosphotriester bearing the *cis–syn* and *trans–syn* CPDs. After chromatographic separation of these products, one of the *cis–syn* isomers was subjected to a three-step process to afford the building block (**18**) for the incorporation of the *cis–syn* CPD. The *trans–syn* isomer was similarly converted to a phosphoramidite building block (**19**) in their subsequent study [78]. Since the CPD is stable under both the acidic and alkaline conditions used for oligonucleotide synthesis, this type of UV lesion can be incorporated into oligonucleotides relatively easily. The same group recently reported the synthesis of an oligonucleotide containing a [3-^{15}N]-labeled *cis–syn* CPD [79].

In the study of T4 endonuclease V (see Section 5.5.1), we modified the synthetic method for the *cis–syn* CPD-containing oligonucleotides [80]. Although the *tert*-butyldimethylsilyl (TBDMS) group was used for protection of the 3′-hydroxyl

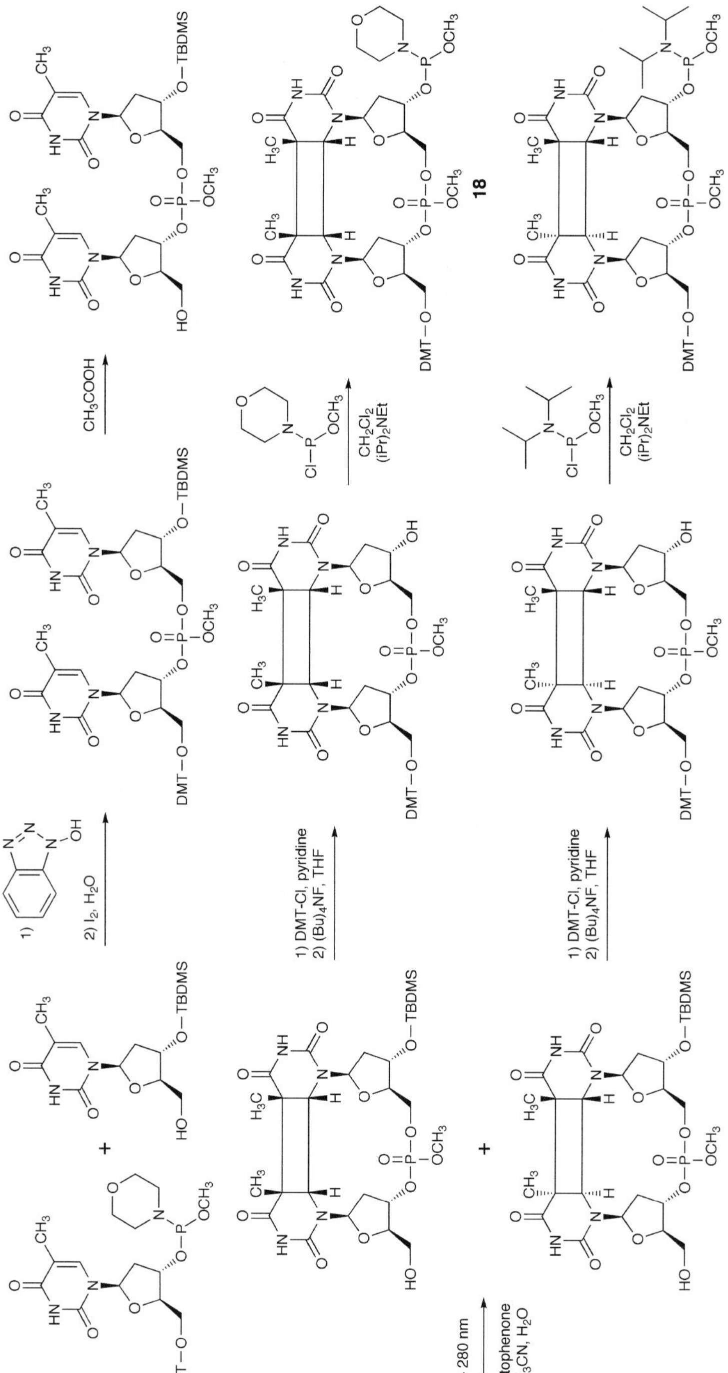

Figure 5.5 Synthesis of phosphoramidite building blocks of the *cis–syn* and *trans–syn* cyclobutane pyrimidine dimers.

function in the original study by Taylor's group [77], we used the levulinyl group shown in Figure 5.6 for this purpose, as this can be removed easily, without any side reaction, by treatment with hydrazine monohydrate in a buffered solution. Another improvement was that the 2-cyanoethyl group was used for protection of the phosphate, and the building block (**20**) was obtained, as shown in Figure 5.6. Clivio and coworkers [81] also reported a building block for incorporation of the *cis–syn* CPD (**23** in Figure 5.7), using the levulinyl group for protection of the 5′-hydroxyl function. In this case, removal and reattachment of this protecting group are not required before and after the photoreaction, respectively, but the procedure on the DNA synthesizer must be changed.

In order to study enzyme binding, the internucleotide linkage at the *cis–syn* CPD was modified. Thus, we incorporated a phosphorodithioate linkage at the CPD site, using **24** as a building block [80], while Carell and coworkers [82] changed the phosphodiester linkage into formacetal, which has the advantage of being non-chiral. Oligonucleotides containing a thymine–uracil-type CPD, which is a deamination product of the CPD formed at the thymine–cytosine site, were also synthesized using **26** in Figure 5.7 [83].

5.3.2
Oligonucleotides Containing the (6–4) or Dewar Photoproduct

As the (6–4) photoproduct is labile in alkali, it was believed that the chemical synthesis of oligonucleotides containing this lesion was impossible, as described in a review article [84]. However, a rather reckless attempt was made at the synthesis which turned out to be successful [85]. The starting material was the same as that used for the synthesis of the CPD building block, as shown in Figure 5.6, but instead of the triplet excitation, the dinucleoside monophosphate was subjected to irradiation with 254 nm light from germicidal lamps. Although the yield was low, two products with a UV absorption maximum at 326 nm, which were diastereomers caused by the chirality of the phosphorus atom, were obtained by this reaction. These products were purified on a reversed-phase column, and after the structures were confirmed by NMR spectroscopy, they were converted to the phosphoramidite building blocks (**21**), as shown in Figure 5.6. In this study, it was found that protection of the tertiary hydroxyl function at the base moiety was not necessary. The problem in the oligonucleotide synthesis using **21** was that the (6–4) photoproduct would be degraded during the deprotection step using aqueous ammonia at a high temperature. To avoid this, nucleoside 3′-phosphoramidites bearing the easily-removable *tert*-butylphenoxyacetyl group for protection of the exocyclic amino functions of the normal bases were used, and the oligonucleotides were deprotected by ammonia treatment at room temperature for 2 h. HPLC analysis of the crude products revealed that byproducts with longer retention times were formed during the chain assembly. It was assumed that the formation of these byproducts would be attributed to the coupling of phosphoramidites with the N3 imino function of the 5′ component of the (6–4) photoproduct. Moreover, it was found that the byproduct formation could be prevented by using benzimidazolium triflate as an alternative activator of the phosphoramidites [86].

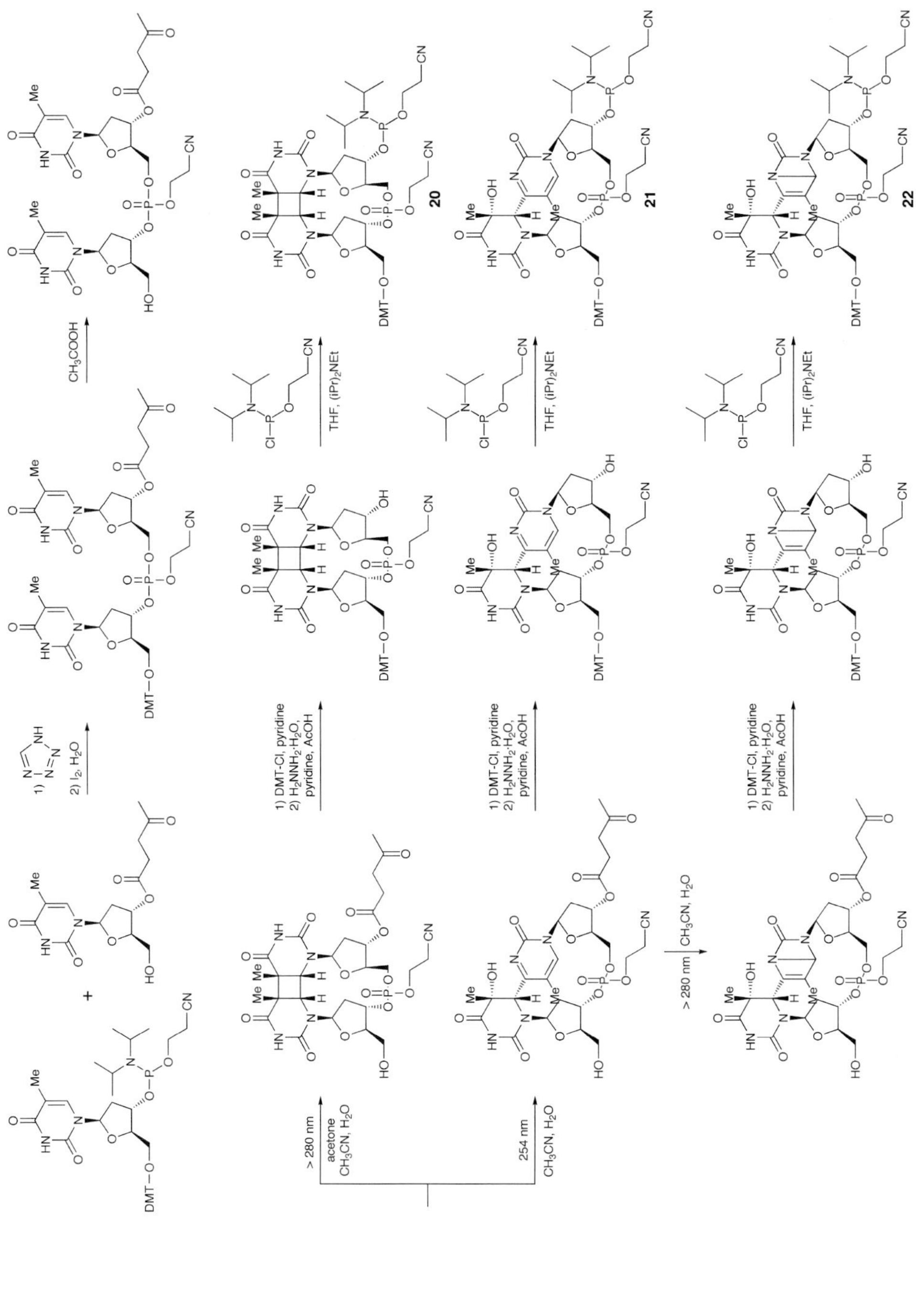

Figure 5.6 Synthesis of phosphoramidite building blocks of the *cis–syn* cyclobutane pyrimidine dimer, the (6–4) photoproduct, and the Dewar valence isomer.

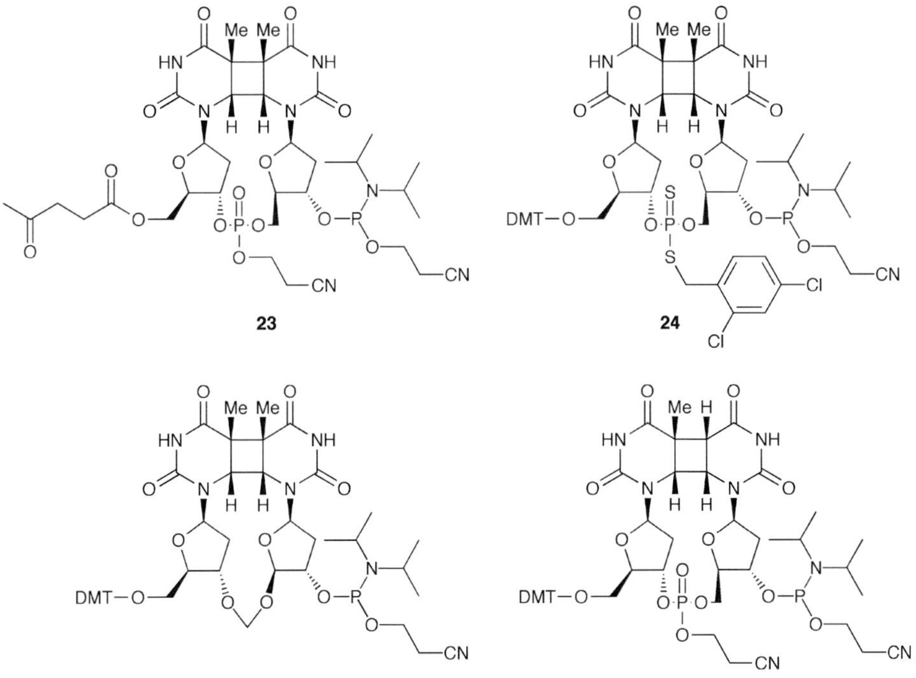

23 24 25 26

Figure 5.7 Other types of building blocks of the *cis–syn* cyclobutane pyrimidine dimers.

Recently, we synthesized a building block of the Dewar photoproduct [87]. The dinucleoside monophosphate bearing the (6–4) photoproduct, which was purified after UV irradiation, was isomerized to the Dewar photoproduct by a second irradiation with Pyrex-filtered light, and converted to the phosphoramidite building block (**22**), as shown in Figure 5.6. Before this method was reported, DNA fragments containing the Dewar photoproduct had been prepared by the two-step irradiation of very short oligonucleotides, followed by ligation with other oligonucleotides, but separation of the Dewar photoproduct-containing oligomer from that containing the (6–4) photoproduct was extremely difficult, even in the case of a hexamer [88]. Compound **22** enabled the preparation of "pure" Dewar photoproduct-containing oligonucleotides without contamination by the (6–4) photoproduct.

We also synthesized a building block of the (6–4) photoproduct formed at the TC sequence, which is the major site for this lesion, and incorporated it into oligonucleotides [89]. As shown in Figure 5.8, the amino group of cytosine, as well as the 5′-hydroxyl function of thymidine, was protected by the DMT group, and this protecting group was removed before UV irradiation. Although the yield of the (6–4) photoproduct was very low, due to the formation of cytosine hydrate during UV irradiation, one of the diastereomers of the phosphotriester was converted to the phosphoramidite (**27**). When this compound was used for the oligonucleotide synthesis, it was found that acylation of the amino function of the 5′ component

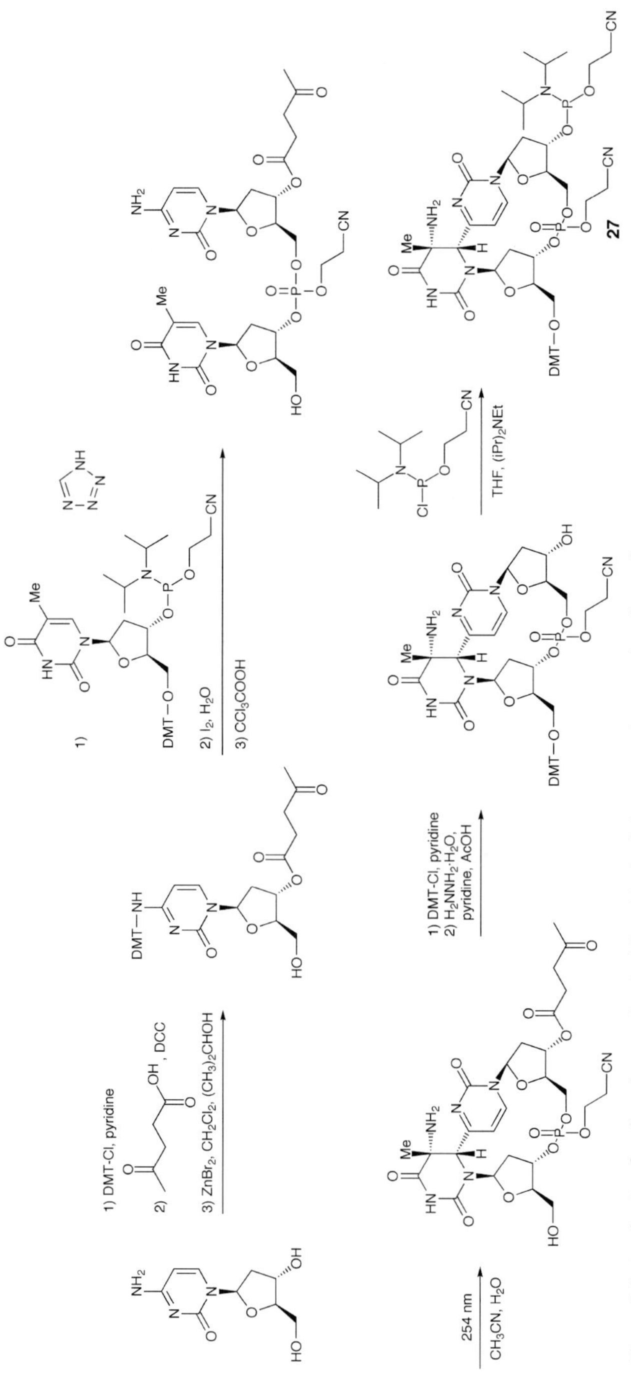

Figure 5.8 Synthesis of a phosphoramidite building block of the (6–4) photoproduct formed at TC.

occurred at the capping step. Oligonucleotides were synthesized by omitting the capping reaction after the incorporation of the (6–4) photoproduct.

As part of a study on the mechanism of (6–4) photoproduct formation, Clivio *et al.* [90] reported the results of the irradiation of thymidylyl(3′–5′)4-thiothymidine at 360 nm. These authors found that this reaction yielded a mixture of three compounds: a thietane intermediate; a thio analogue of the (6–4) photoproduct; and its Dewar valence isomer, as shown in Figure 5.9A; they also determined that a methyl disulfide was formed when the (6–4) photoproduct analogue was treated with methyl methanethiosulfonate. The same group [91] and Liu and Taylor [92] found that these sulfur-containing products could be efficiently photoreversed to the parent dinucleoside monophosphate by irradiation at 254 nm. This type of photoreaction occurred not only in the dinucleoside monophosphate but also in oligonucleotides [92, 93], and a phosphoramidite building block to incorporate the methyl disulfide analogue of the (6–4) photoproduct (see Figure 5.9B) was prepared by Clivio's group [94].

5.4 Structure and Mutagenesis of Pyrimidine Dimer-Containing DNA

5.4.1 Tertiary Structures of Pyrimidine Dimer-Containing Duplexes

DNA lesions generally alter the base-pair formation and the helix structure, at least locally. The tertiary structures of duplexes containing the *cis–syn* CPD have been analyzed mainly by NMR spectroscopy [95–99], and only small distortions from the canonical B-form conformation were reported in all of these studies. There are two possible puckering conformations of the cyclobutane ring, and while the two bases of the CPD were twisted in a left-handed helical form in dinucleoside monophosphates [100–102], a right-handed helical form was found for the CPD in duplexes [97–99]. In the recently reported structures [98, 99], which were determined using nuclear Overhauser effect (NOE) distances and restrained molecular dynamics, no significant bending of the helix was detected at the CPD site, although a subtle distortion was observed at the 3′ side of the CPD. The structures of DNA duplexes containing TA [103], GA [104], and GG [104] opposite the CPD formed at TT were also determined by NMR spectroscopy.

The crystal structure of a CPD-containing duplex is quite different from the solution structure [105]. One major difference is that a large helix bend (about 30 °) toward the major groove was found in the crystal structure, but this bend may have been caused by crystal packing forces, as observed for A-tract DNA duplexes [106]. Additionally, several groups reported structural studies by molecular dynamics (MD) simulations. Essential dynamics derived from MD trajectories of DNA duplexes revealed that the phosphodiester linkage between the two deoxyadenosines opposite the CPD had considerably greater mobility [107], and Monte Carlo simulations presented a theory of CPD-induced, prolonged large-amplitude oscillations of the two strands [108].

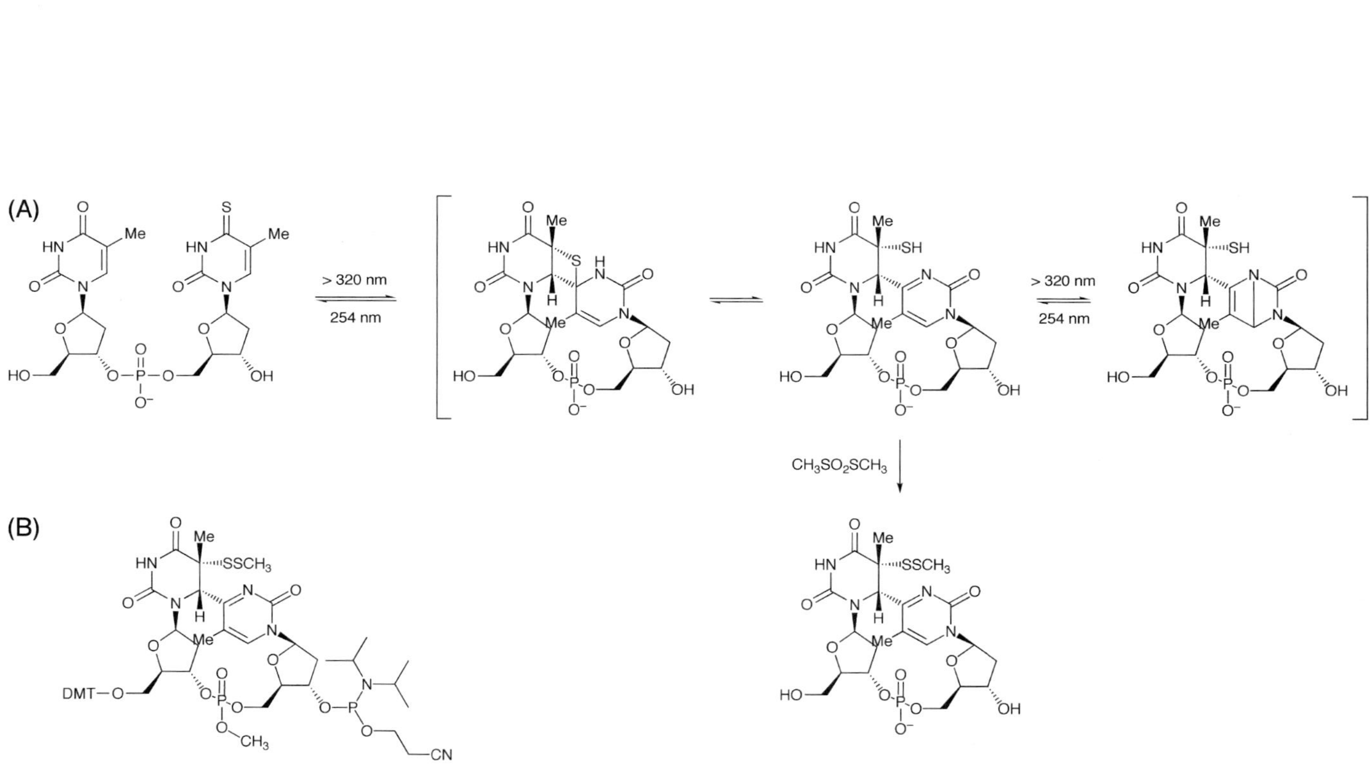

Figure 5.9 Formation of a thio analogue of the (6–4) photoproduct (A) and its building block for oligonucleotide synthesis (B).

Solution structures of duplexes containing the (6–4) photoproduct or its Dewar valence isomer were reported by Choi and coworkers [109–113]. The duplexes they used contained the (6–4) photoproduct of TT or its Dewar isomer opposite AA or GA, and a large helix bend (from 21 ° to 44 °) at the lesion site was reported for all of them. As there is a limitation in the determination of global DNA structures using only short-range NOE interactions, we analyzed the tertiary structure of a duplex containing the (6–4) photoproduct by measuring the fluorescence resonance energy transfer (FRET) [114]. A (6–4) photoproduct-containing duplex bearing fluorescein and tetramethylrhodamine was prepared using synthetic oligonucleotides, and its FRET efficiency was compared with that of the same type of duplex without the photoproduct. The fluorescence spectra and the decay curves did not change upon photoproduct formation, whereas a similar duplex containing a cisplatin adduct, which had been proved to be bent by several methods, showed a larger FRET efficiency. Although an unwinding at the lesion site was revealed in this study, the results indicated that helix bending was not induced by the photoproduct formation, and this observation agreed with the significantly small bend angle (5 °) derived from an unrestrained MD analysis [115].

5.4.2
Base-Pair Formation by Pyrimidine Dimers

Hydrogen-bonding interactions between the CPD formed at TT and the opposite adenine bases were investigated by NMR. Two earlier studies [96, 97] reported that the hydrogen bonds were formed in the same way as those in the A T base pair, but the duplexes were destabilized by 1.72 and 1.41 kcal mol^{-1} at 37 °C, respectively, upon CPD formation. A UV melting study also revealed destabilization of the duplex by 1.5 kcal mol^{-1} at 37 °C [116]. The detailed structure of the base pairs was determined by NMR spectroscopy [99], as shown in Figure 5.10A–C. Although the functional groups for hydrogen bond formation with adenine are intact in the CPD, the two thymine rings are not parallel to each other after formation of the cyclobutane ring. Therefore, the geometry of the functional groups deviates from the optimal alignment for normal hydrogen bonding.

In the case of the (6–4) photoproduct, hydrogen-bonding interactions were present only at the 5′ component of this lesion [109]. The structure determined by MD using NMR restraints revealed that hydrogen bonds were formed between the N3 of the 5′ component of the photoproduct and the N1 of the opposite adenine, between the O^4 of the 5′ component and the N^6 of the opposite adenine, and between the 5-OH of the 5′ pyrimidine and the N1 of the 5′ flanking adenine [110], as shown in Figure 5.10D and E. The lack of hydrogen-bonding interactions at the 3′ component of the (6–4) photoproduct caused remarkable destabilization of the duplex (6.1 kcal mol^{-1} at 37 °C) [116].

Conversion of the (6–4) photoproduct to its Dewar valence isomer drastically changes the hydrogen-bonding interactions. In the refined structure, hydrogen bonds were found only between the N3 atoms of both components of the Dewar photoproduct and the exocyclic amino group of the 3′ side adenine in the complementary

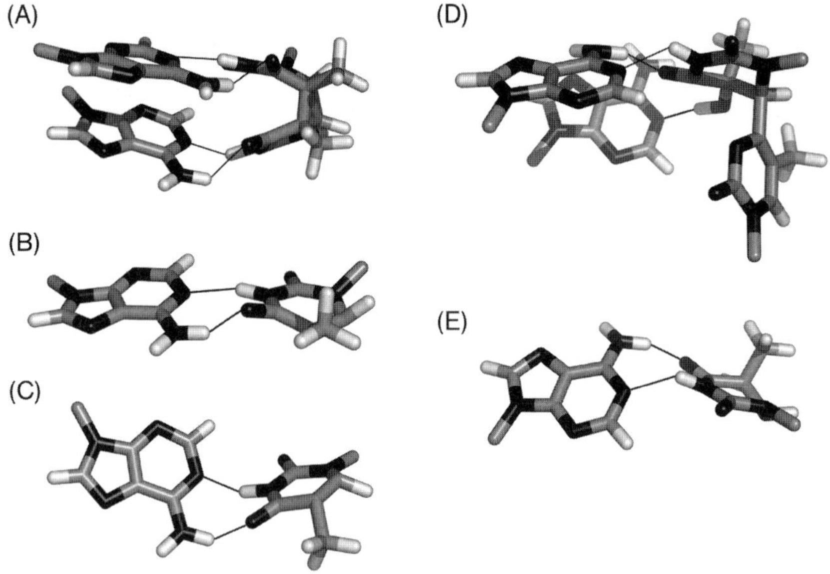

Figure 5.10 Base-pair formation found in the solution structures of duplexes containing the *cis–syn* cyclobutane pyrimidine dimer (A, viewed from the major groove side) and the (6–4) photoproduct (D, viewed from the minor groove side). The 5′ (B and E) and 3′ (C) components of each photodimer were extracted, and the hydrogen bonds are indicated by thin lines.

strand [112]. When the opposite bases were changed to AG, the base-pairing pattern was rearranged. Three hydrogen bonds were formed between the N3 of the 5′ component of the Dewar photoproduct and the N1 of adenine, between the O^4 of the 5′ component and the N1 of guanine, and between the O^2 of the 3′ component and the amino function of guanine [113]. However, there is a strange point in these NMR structures. When Taylor *et al.* [117] determined the chemical structure of the Dewar photoproduct in a dinucleoside monophosphate form, the configuration of the C6 of the 3′ component was *R*, whereas this carbon had an *S* configuration in the solution structures of the duplexes [112, 113]. This means that the bending of the 3′ base ring is reversed, and it is not a conformational change. In the case of the (6–4) photoproduct, the direction of the 3′ pyrimidone ring was the same between the dinucleoside monophosphate [118, 119] and the duplexes [109, 110].

5.4.3
Mutations Induced by Pyrimidine Dimers

Mutations caused by the photodimers have been analyzed by transforming cells, such as repair-deficient, SOS-induced *Escherichia coli*, with a vector containing each lesion at a single site, followed by sequencing of the replicated DNA [120–129]. The SOS-induction was required because the plaque-forming efficiencies of the phage vectors were very low without the induction, due to the replication block by the lesion. As shown in Table 5.1, the *cis–syn* CPD formed at the TT sequence was not very mutagenic. The mutation frequency was less than 10%, and the major type of mutation was a T → A transversion at the 3′ component of the CPD [120–124]. In yeast [124] and monkey [124, 125] cells, the mutation frequency was much lower. In our study [125], only the plasmids containing a mutation at the TT site were selected by digestion with a restriction endonuclease, whereas the results of the other studies were obtained by analyzing all of the DNA sequences.

Horsfall *et al.* [126] reported the mutagenic properties of the CPD formed at TC, but the aforementioned deamination made interpretation of the experimental results difficult. These authors concluded that the 3′ C → T transition could be attributed to accurate replication past a TU dimer formed by the deamination, because the same result was obtained by using a vector treated with DNA photolyase (see Section 5.5.2). The incorporation of A opposite a uracil-type CPD, which results in a C → T mutation, was demonstrated in *E. coli* [130].

In the case of the *trans–syn*-I CPD, a deletion at the lesion site and a T → A transversion were mainly observed in SOS-induced *E. coli* (see Table 5.1), and a deletion (4% of the analyzed sequences) was the only mutation detected for the SOS-uninduced cells [127]. It is surprising that the lesion-containing templates were replicated with high accuracy, because the glycosidic bond of the 5′ base of the *trans–syn*-I CPD is fixed at *syn*, which means that a base-pairing interaction is not possible. This accuracy was much higher than those observed for abasic site-containing templates [121], and it was proposed that the incorporation of dATP opposite the 5′ component of the *trans–syn*-I CPD might be attributed to a van der Waals interaction between the methyl group and the H2 of adenine [122].

A remarkable feature of the mutations induced by the (6–4) photoproduct is a T → C transition at the 3′ component, as shown in Table 5.2. LeClerc *et al.* [128] reported that 85% of the sequences obtained after replication of the (6–4) photoproduct-carrying vector contained this type of mutation. They also found that isomerization to the Dewar photoproduct significantly lowered the error frequency and specificity. Another group reported a similar tendency [122]. In order to elucidate the mechanism of the (6–4) photoproduct-induced mutagenesis, the thermodynamic parameters were determined by measuring the melting curves of template–primer-type duplexes containing UV lesions [131]. When guanine was located opposite the 3′ component of the (6–4) photoproduct, the duplex was more stable than those containing the other bases at this position. The same result was obtained for the (6–4) photoproduct formed at the TC sequence, but such a preference was lost upon conversion to the Dewar isomer. These results suggested that the 3′ component, 2-pyrimidone (Po), base-paired with guanine by forming hydrogen bonds between the N3 of Po and the N1 of G and between the O^2 of Po and the amino group of G. The incorporation of dGTP results in the T → C transition. When the primer strand was elongated by one nucleotide, adenine destabilized the duplex, although this base should form a base pair with the 5′ component of the (6–4) photoproduct. It should be noted that these results, using duplexes with the target base pair at the end [131], are different from those using duplexes containing a lesion at the center [116].

Table 5.1 Mutation spectra of the cyclobutane pyrimidine dimers.

Lesion	Cell	TT	AT	CT	GT	TA	TC	TG	Double	Deletion	Other	Total	Ref.
cis–syn TT	SOS-induced *E. coli*	347	0	3	0	15	8	0	0	0	0	373	[120]
		2443	0	6	0	130	28	0	0	0	1	2608	[121]
		58	0	0	0	0	0	0	0	0	0	58	[122]
		131	0	1	0	1	0	1	0	0	0	134	[122]
		188	0	0	0	18	2	0	0	0	0	208	[123]
		667	0	0	0	49	6	0	0	0	2	724	[124]
	S. cerevisiae	686	1	0	0	0	0	0	1	0	2	690	[124]
	monkey COS7	120	0	0	0	1	2	0	0	0	0	123	[123]
	(lagging)	ND[a]	2	4	2	1	6	7	1	2	6	31[a),b)]	[125]
	(leading)	ND[a]	3	2	2	1	4	2	3	3	4	24[a),b)]	[125]
cis–syn TC	SOS-induced *E. coli*	19[c]	0	0	0	0	84	0	0	0	0	103	[126]
trans–syn-I TT	SOS-induced *E. coli*	536	16	4	0	0	1	0	0	33	15	605	[127]
		82	1	0	0	0	0	0	0	0	0	83	[122]
		93	6	2	0	0	0	1	0	0	0	102	[122]
		457	16	4	0	0	12	1	1	42	2	535	[124]
	S. cerevisiae	100	41	0	0	0	1	1	0	19	4	166	[124]

a) Only the numbers of mutants were counted.
b) The mutation frequencies were 0.68% (the lagging strand) and 0.20% (the leading strand).
c) This mutation was experimentally demonstrated to be caused by accurate replication past a TU dimer formed by deamination.

Table 5.2 Mutation spectra of the (6–4) and Dewar photoproducts.

Lesion	Cell	TT TC[a]	AT AC[a]	CT CC[a]	GT GC[a]	TA	TC TT[a]	TG	Double	Deletion	Other	Total	Ref.
(6–4) TT	SOS-induced *E. coli*	40	0	1	0	0	18	0	1	0	0	60	[122]
		20	2	0	0	1	31	0	0	0	0	54	[122]
		14	0	0	0	0	35	1	1	0	0	51	[122]
		16	0	2	2	0	158	1	5	1	0	185	[128]
		29	0	3	0	0	10	0	0	0	3	45	[123]
		3	0	1	0	0	28	0	0	27	0	59	[123]
	monkey COS7	41	0	0	1	0	2	0	0	0	60[b]	104	[123]
	(lagging)	ND[c]	0	1	0	5	17	3	0	0	0	26[c],[d]	[125]
	(leading)	ND[c]	0	1	1	4	10	0	0	0	4	20[c],[d]	[125]
(6–4) TC[a]	SOS-induced *E. coli*	135	10	0	0	3	57	0	0	1	0	206	[129]
Dewar TT	SOS-induced *E. coli*	48	3	2	0	3	3	4	3	0	0	66	[122]
		28	4	0	0	2	3	3	2	0	0	42	[122]
		27	2	0	0	0	1	1	1	0	0	32	[122]
		17	5	3	1	3	23	1	2	0	0	55	[128]
		49	4	7	1	4	11	2	7	0	0	85	[128]
Dewar TC[a]		50	35	4	9	10	83	3	39	0	0	233	[129]

a) The original sequence was TC.
b) Most of these mutants carried a G → T mutation at the 5′ flanking site with the wild-type TT sequence.
c) Only the numbers of mutants were counted.
d) The mutation frequencies were 4.7% (the lagging strand) and 2.3% (the leading strand).

5.5 Repair of Pyrimidine Dimers in Cells

5.5.1 T4 Endonuclease V

Since the UV-induced pyrimidine dimers cause replication errors (as described in Section 5.4), they must be removed or repaired in order to maintain the genetic integrity. Endonuclease V from bacteriophage T4 is an enzyme that initiates the removal of *cis–syn* CPDs in DNA. An enhanced UV resistance of bacteriophage T4 was initially reported in 1947 [132]. An endonuclease activity specific for UV-irradiated DNA, encoded by the *denV* gene of bacteriophage T4, was subsequently purified from T4-infected *E. coli* by two groups [133, 134], and the sequence analysis of the *denV* gene revealed that T4 endonuclease V is a basic protein consisting of 138 amino acids with a molecular weight of 16 078 [135, 136]. This enzyme hydrolyzes the glycosidic bond of the 5′ component of the CPD [137, 138], and subsequently cleaves the sugar–phosphate backbone at the resultant abasic site by a β-elimination mechanism [139, 140] (see Figure 5.11). Lloyd and coworkers [141, 142] found that an intermediate with a covalent bond between the C1′ of the abasic site and the N-terminal amino group of the enzyme was formed in the first reaction, and this

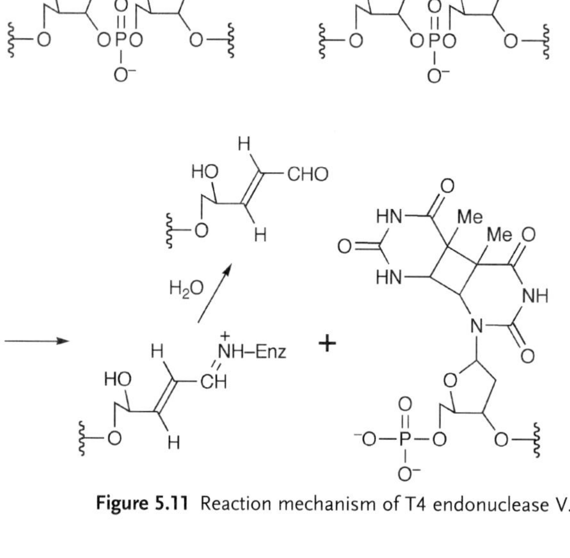

Figure 5.11 Reaction mechanism of T4 endonuclease V.

intermediate could be trapped under $NaBH_4$-reducing conditions. It was shown that the attachment of a fluorine atom to the 2′ position of the 5′ component of the CPD stabilized the covalent intermediate, probably by inhibiting the ring opening of the sugar moiety [143]. The formation of this Schiff base-type intermediate was generally shown for other enzymes [144], and the crystal structures of covalent complexes of several enzymes with DNA, obtained by reducing the intermediate, have been solved [145, 146]. A study of the stereochemical course of the second reaction revealed that it proceeds by a *syn* β-elimination involving the abstraction of the 2′-pro-*S* proton and the formation of a *trans* α,β-unsaturated product [147].

Morikawa and coworkers solved the crystal structures of T4 endonuclease V [148, 149] and its complex with a CPD-containing DNA [150]. The present author participated in the latter study, and a mutant (E23Q), which could bind the substrate but lacked the glycosylase activity [151], was used for crystallization of the enzyme–DNA complex. Enzyme binding occurred in the minor groove at the CPD site, as predicted in the study using modified duplexes [152], and the helix was kinked at an angle of 60°, as shown in Figure 5.12A. At the CPD site, the phosphates formed an interaction network with the amino acid side chains, but the damaged base was not recognized directly. Interestingly, the adenine base opposite the 5′ component of the CPD was completely flipped out of the helix (see Figure 5.12B), and was accommodated in a cavity on the enzyme surface without forming any hydrogen bonds. The empty space generated in the duplex by the base flipping was occupied by several

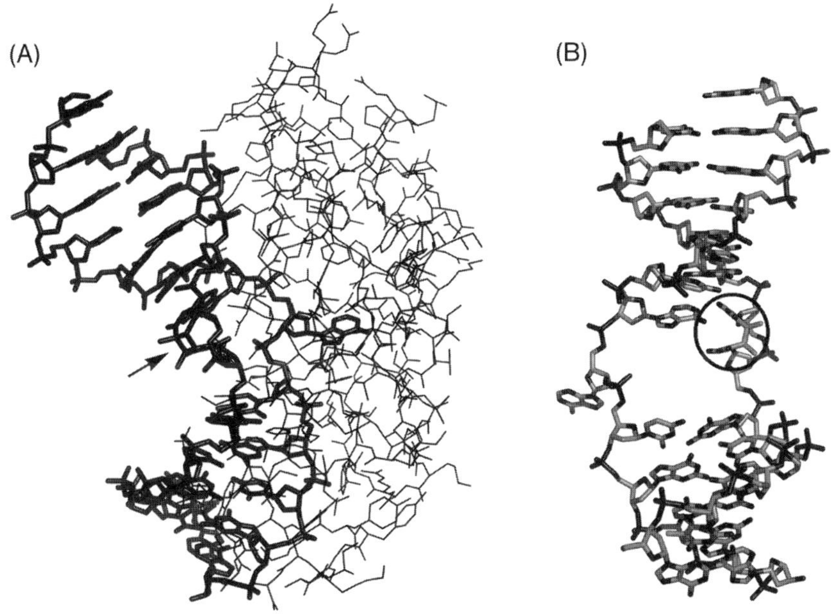

Figure 5.12 Crystal structure of a T4 endonuclease V–DNA complex (A) and the DNA structure in the complex viewed from the major groove side (B). The cyclobutane pyrimidine dimer is indicated by an arrow (A) and by a circle (B).

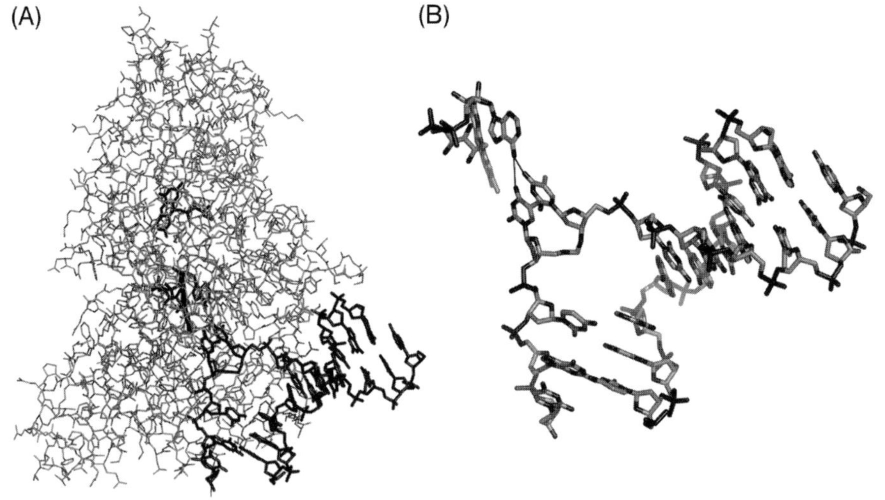

Figure 5.14 (A) Crystal structure of a CPD photolyase–DNA complex. DNA, FAD, and HDF are shown with thick lines. (B) Structure of DNA with FAD in the complex with CPD photolyase. The hydrogen bonds between the thymine carbonyl groups in the substrate and the adenine amino group in the FAD cofactor are shown with thin lines.

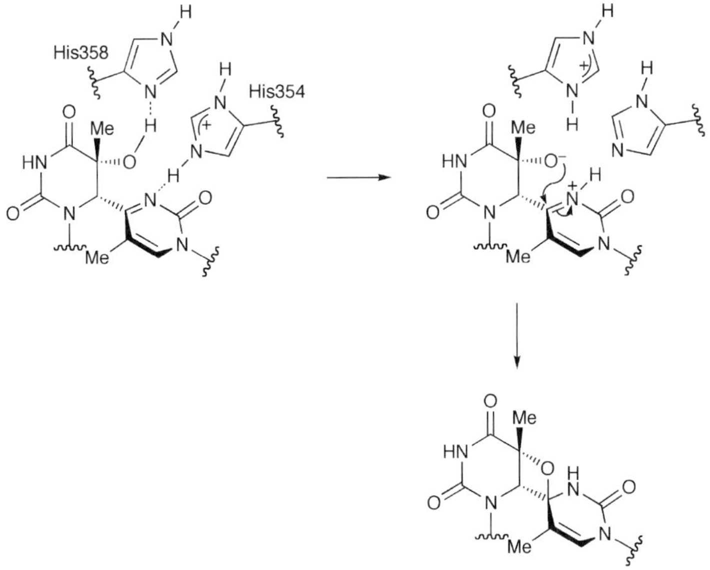

Figure 5.15 Proposed (6–4) photolyase-catalyzed formation of the oxetane intermediate.

length of 27 to 29 nucleotides is excised [183], and the resultant single-stranded region is filled by DNA polymerase and DNA ligase. The (6–4) photoproduct is repaired more rapidly than the CPD [184]. NER is classified into two subpathways: global genome repair (GGR); and transcription-coupled repair (TCR). While the stalling of RNA polymerase at a lesion site initiates TCR [185], GGR requires the recognition of lesions in DNA. The mechanism of NER has been investigated in relation to a human hereditary disease, xeroderma pigmentosum (XP), which is caused by a disorder in the NER pathway [186]. In the case of GGR, the primary recognition factor is the XPE protein, which is inactivated by a mutation in XP group E cells, and is usually called the UV-damaged DNA-binding (UV-DDB) protein. This protein is a heterodimer consisting of the p127 and p48 subunits, which are designated as DDB1 and DDB2, respectively, and it has greater affinity for the (6–4) photoproduct than for the CPD and several other lesions [187–189]. It was found that the DNA helix is bent at an angle of about 55° upon binding of the UV-DDB protein [188]. After the UV-DDB protein binds to damaged DNA with high affinity, the associated ubiquitin ligase E3 ubiquitylates both the UV-DDB and XPC proteins, and this ubiquitylation transfers the damaged DNA from the UV-DDB protein to the XPC–HR23B complex [190], which initiates GGR [191]. As the binding of the XPC–HR23B complex to the CPD-containing duplex cannot be detected in *in-vitro* experiments [192], it is thought that the UV-DDB protein recruits this complex to the CPD site [193, 194]. After binding of the XPC–HR23B complex, TFIIH is recruited, and its helicase subunits – the XPB and XPD proteins – unwind the duplex. The replication protein A and the XPA and XPG proteins then form a complex, replacing the XPC–HR23B heterodimer. Finally, the ERCC1–XPF complex enters, and the XPG and XPF proteins hydrolyze the phosphodiester linkages on the 3′ and 5′ sides of the lesion, respectively [195].

The NMR solution structures of a truncated XPA protein containing the DNA-binding domain [196, 197] and the crystal structures of the DDB1 subunit in complex with a viral protein and with ubiquitin ligase [198, 199] have been determined, but no structures of complexes of NER-related proteins with lesion-containing DNA have been reported.

5.5.4 UV Damage Endonuclease (UVDE)

An alternative DNA excision repair pathway has been found in *Schizosaccharomyces pombe* [200], *Neurospora crassa* [201], and *Bacillus subtilis* [202]. The enzyme, UVDE (also called Uve1p), recognizes both the *cis–syn* CPD and the (6–4) photoproduct [202, 203]. This enzyme similarly recognizes the *trans–syn*-I and *trans–syn*-II CPDs and the Dewar photoproduct [204]. Although the efficiency is lower than that observed for the UV lesions, an abasic site, 5,6-dihydrouracil, and a cisplatin adduct are also recognized [204, 205]. After binding to these lesion sites, UVDE catalyzes ATP-independent hydrolysis of the phosphodiester linkage immediately 5′ to the lesion, to generate a 3′-OH and a 5′-phosphate. Three more enzymes – flap endonuclease, DNA polymerase, and DNA ligase – are required for this pathway, and two repair mechanisms have been proposed [206, 207]. In one of these, a 5′ → 3′ exonuclease activity removes several nucleotides including the lesion, after which the single-stranded region is filled by DNA polymerase and DNA ligase. Alternatively, the nicked strand is extended by DNA polymerase, followed by cleavage of the junction in the flap structure and ligation of the nick.

5.6 Bypass of Pyrimidine Dimers by DNA Polymerases

Replication by DNA polymerases is usually blocked by UV-induced pyrimidine dimers [208]. In 1999, Prakash and coworkers [209] reported that the *RAD30* gene of *Saccharomyces cerevisiae* encodes a DNA polymerase that can efficiently replicate past the *cis–syn* CPD formed at TT, incorporating two adenines. Since this enzyme was the seventh eukaryotic DNA polymerase, they named it DNA polymerase η. In the same year, Hanaoka and coworkers [210] found that human XP variant (XPV) cells lack a DNA polymerase activity that replicates CPD-containing DNA, and sequence analysis of the protein purified from HeLa cells revealed that this enzyme is human DNA polymerase η [211]. The human enzyme not only replicates past the *cis–syn* CPD formed at TT efficiently and accurately, but also catalyzes translesion synthesis past several other lesions [212]. It preferentially incorporates adenine and guanine opposite an apurinic/apyrimidinic site, and cytosine opposite acetylaminofluorene or a cisplatin adduct of guanine. This enzyme also incorporates other bases opposite these lesions, but it can continue chain elongation only when the correct base is incorporated. Accurate syntheses past thymine–uracil [83] and 5-methylcytosine–thymine [213] CPDs were subsequently demonstrated, using the human and yeast enzymes, respectively. Since DNA polymerase η has low fidelity and low processivity when it replicates undamaged DNA [214, 215], polymerase switching should occur during the CPD bypass. Kunkel and coworkers [216] reported that both the processivity and fidelity of human DNA polymerase η are higher at the CPD site than those observed for the undamaged control, and proposed a model of polymerase switching.

Among other DNA polymerases capable of translesion synthesis, DNA polymerases ζ [217] and ι [218] can bypass the *cis–syn* CPD, although the efficiency is low. At the 3′ component of the CPD, DNA polymerase ι produces a G T mispair, which induces a T → C mutation, as well as the correct A T pair [218]. This enzyme bypasses a thymine–uracil CPD more efficiently than the thymine–thymine CPD, incorporating guanine opposite uracil at a high frequency [219]. This type of bypass may decrease the number of mutations caused by the deamination of cytosine-containing CPDs (see Section 5.2.1). DNA polymerases κ [220, 221] and Q [222] cannot bypass either type of UV lesion. DNA polymerase μ correctly incorporates two adenines opposite the CPD, but the DNA synthesis is aborted after a few nucleotides are incorporated beyond the lesion [223]. DNA polymerase β, which is required in the base excision repair pathway, is also able to catalyze translesion synthesis past the *cis–syn* CPD, although this enzyme causes cytosine misincorporation and error-prone extension [224]. Mutants of yeast DNA polymerase α [225] and *Thermus aquaticus* DNA polymerase [226] that can bypass the CPD have been isolated.

Human DNA polymerase η cannot bypass the (6–4) photoproduct. This enzyme incorporates all four of the bases opposite the 3′ component of the (6–4) photoproduct [212], but among them, guanine is preferred [227]. Although DNA synthesis by DNA polymerase η stops at this step, Johnson *et al.* [227] reported that DNA polymerase ζ can extend the strand, incorporating the correct adenine base opposite the 5′ component of the (6–4) photoproduct. In their model, the efficient bypass of the (6–4) photoproduct is achieved by the combined action of DNA polymerases η and ζ, and this type of DNA synthesis induces a T → C mutation. Johnson *et al.* [228] also reported the same type of bypass of the (6–4) photoproduct by the sequential action of DNA polymerases ι and ζ. Another group [229] reported that DNA polymerase ζ bypassed the (6–4) photoproduct by

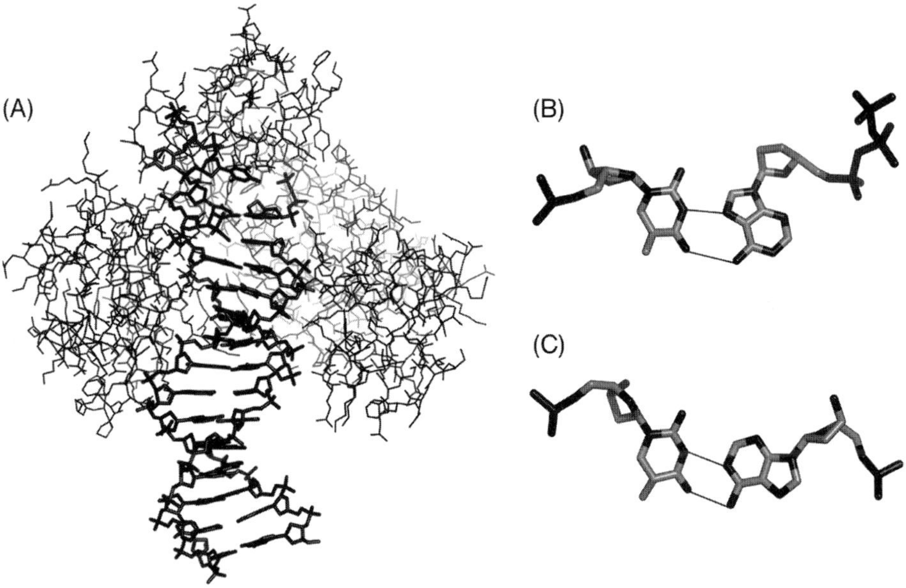

Figure 5.16 (A) Crystal structure of a Dpo4–DNA–ddATP complex. DNA and ddATP are shown with thick lines. (B) The Hoogsteen-type base pair formed between the 5′ component of the CPD and the adenine base of ddATP. (C) The Watson–Crick-type base pair formed at the 3′ component of the CPD.

itself, causing the same mutation. Other DNA polymerases that reportedly catalyze DNA synthesis past the (6–4) photoproduct include *E. coli* DNA polymerase V [230], *Drosophila* DNA polymerase η [231], and mammalian DNA polymerase β [224].

In the structural biology of translesion synthesis, the crystal structures of yeast DNA polymerase η [232], *Sulfolobus solfataricus* DNA polymerase IV (Dpo4), which is an archaeal homologue of DNA polymerase η, in complex with DNA containing the *cis–syn* CPD [233], an abasic site [234], and a benzo[*a*]pyrene adduct [235], and bacteriophage T7 DNA polymerase in complex with a CPD-containing duplex [236], have been solved. In the case of the Dpo4 complex with a CPD-containing duplex (Figure 5.16A), the 3′ end of the extending strand was adenine opposite the 3′ component of the CPD, and 2′,3′-dideoxyadenosine 5′-triphosphate (ddATP) was used as an incoming nucleotide opposite the 5′ component [233]. In this structure, the 5′ thymine formed a Hoogsteen base pair with ddATP in a *syn* conformation, whereas the 3′ base formed a Watson–Crick base pair with adenine, as shown in Figure 5.16B and C. However, two groups independently demonstrated, in experiments using modified nucleotides, that eukaryotic DNA polymerase η incorporates dATP opposite the 5′ thymine of the CPD via Watson–Crick base pairing, and not by Hoogsteen base pairing [237, 238].

Acknowledgement

The author thanks Junpei Yamamoto (Osaka University) for preparing the structures in the figures.

References

1 Friedberg, E .C., Walker, G. C., Siede, W., Wood, R. D., Schultz, R. A., Ellenberger, T., *DNA Repair and Mutagenesis*, 2nd edn. ASM Press, USA, **2005**.

2 Godar, D. E., *Photochem. Photobiol.* **2005**, *81*, 736–749.

3 Cadet, J., Courdavault, S., Ravanat, J.-L., Douki, T., *Pure Appl. Chem.* **2005**, *77*, 947–961.

4 Beukers, R., Berends, W., *Biochim. Biophys. Acta* **1960**, *41*, 550–551.

5 Wacker, A., Dellweg, H., Lodemann, E., *Angew. Chem.* **1961**, *73*, 64–65.

6 Beukers, R., Berends, W., *Biochim. Biophys. Acta* **1961**, *49*, 181–189.

7 Blackburn, G. M., Davies, R. J. H., *Biochem. Biophys. Res. Commun.* **1966**, *22*, 704–706.

8 Varghese, A. J., Wang, S. Y., *Nature* **1967**, *213*, 909–910.

9 Weinblum, D., *Biochem. Biophys. Res. Commun.* **1967**, *27*, 384–390.

10 Ben-Hur, E., Ben-Ishai, R., *Biochim. Biophys. Acta* **1968**, *166*, 9–15.

11 Douki, T., *J. Photochem. Photobiol. B* **2006**, *82*, 45–52.

12 Kemmink, J., Boelens, R., Kaptein, R., *Eur. Biophys. J.* **1987**, *14*, 293–299.

13 Kao, J. L.-F., Nadji, S., Taylor, J.-S., *Chem. Res. Toxicol.* **1993**, *6*, 561–567.

14 Wacker, A., *Prog. Nucleic Acid Res.* **1963**, *1*, 369–399.

15 Herbert, M. A., LeBlanc, J. C., Weinblum, D., Johns, H. E., *Photochem. Photobiol.* **1969**, *9*, 33–43.

16 Garcés, F., Dávila, C. A., *Photochem. Photobiol.* **1982**, *35*, 9–16.

17 Mitchell, D. L., Jen, J., Cleaver, J. E., *Nucleic Acids Res.* **1992**, *20*, 225–229.

18 Lemaire, D. G. E., Ruzsicska, B. P., *Photochem. Photobiol.* **1993**, *57*, 755–769.

19 Durbeej, B., Eriksson, L. A., *Photochem. Photobiol.* **2003**, *78*, 159–167.

20 Mitchell, D. L., *Photochem. Photobiol.* **2000**, *71*, 162–165.

21 Celewicz, L., Mayer, M., Shetlar, M. D., *Photochem. Photobiol.* **2005**, *81*, 404–418.

22 Li, X., Eriksson, L. A., *Chem. Phys. Lett.* **2005**, *401*, 99–103.

23 Tommasi, S., Denissenko, M. F., Pfeifer, G. P., *Cancer Res.* **1997**, *57*, 4727–4730.

24 You, Y.-H., Li, C., Pfeifer, G. P., *J. Mol. Biol.* **1999**, *293*, 493–503.

25 You, Y.-H., Pfeifer, G. P., *J. Mol. Biol.* **2001**, *305*, 389–399.

26 Ostrowski, T., Maurizot, J.-C., Adeline, M.-T., Fourrey, J.-L., Clivio, P., *J. Org. Chem.* **2003**, *68*, 6502–6510.

27 Durbeej, B., Eriksson, L. A., *J. Photochem. Photobiol. A* **2002**, *152*, 95–101.

28 Zhang, R. B., Eriksson, L. A., *J. Phys. Chem. B* **2006**, *110*, 7556–7562.

29 Wondrak, G. T., Jacobson, M. K., Jacobson, E. L., *Photochem. Photobiol. Sci.* **2006**, *5*, 215–237.

30 Rochette, P. J., Therrien, J.-P., Drouin, R., Perdiz, D., Bastien, N., Drobetsky, E. A., Sage, E., *Nucleic Acids Res.* **2003**, *31*, 2786–2794.

31 Douki, T., Reynaud-Angelin, A., Cadet, J., Sage, E., *Biochemistry* **2003**, *42*, 9221–9226.

32 Courdavault, S., Baudouin, C., Charveron, M., Favier, A., Cadet, J., Douki, T., *Mutation Res.* **2004**, *556*, 135–142.

33 Gale, J. M., Nissen, K. A., Smerdon, M. J., *Proc. Natl. Acad. Sci. USA* **1987**, *84*, 6644–6648.

34 Pehrson, J. R., *J. Biol. Chem.* **1995**, *270*, 22440–22444.

35 Douki, T., Cadet, J., *J. Photochem. Photobiol. B* **1992**, *15*, 199–213.

36 Lemaire, D. G. E., Ruzsicska, B. P., *Biochemistry* **1993**, *32*, 2525–2533.

37 Barak, Y., Cohen-Fix, O., Livneh, Z., *J. Biol. Chem.* **1995**, *270*, 24174–24179.

38 Peng, W., Shaw, B. R., *Biochemistry* **1996**, *35*, 10172–10181.

39 Burger, A., Fix, D., Liu, H., Hays, J., Bockrath, R., *Mutation Res.* **2003**, *522*, 145–156.

40 Douki, T., Cadet, J., *Biochemistry* **1994**, *33*, 11942–11950.

41 Lee, D.-H., Pfeifer, G. P., *J. Biol. Chem.* **2003**, *278*, 10314–10321.

42 Johns, H. E., Pearson, M. L., LeBlanc, J. C., Helleiner, C. W., *J. Mol. Biol.* **1964**, *9*, 503–524.

43 Pearson, M. L., Ottensmeyer, F. P., Johns, H. E., *Photochem. Photobiol.* **1965**, *4*, 739–747.

44 Varghese, A. J., Wang, S. Y., *Science* **1968**, *160*, 186–187.

45 Karle, I. L., Wang, S. Y., Varghese, A. J., *Science* **1969**, *164*, 183–184.

46 Rahn, R. O., Hosszu, J. L., *Photochem. Photobiol.* **1969**, *10*, 131–137.

47 Wang, S. Y., Varghese, A. J., *Biochem. Biophys. Res. Commun.* **1967**, *29*, 543–549.

48 Lippke, J. A., Gordon, L. K., Brash, D. E., Haseltine, W. A., *Proc. Natl. Acad. Sci. USA* **1981**, *78*, 3388–3392.

49 Franklin, W. A., Lo, K. M., Haseltine, W. A., *J. Biol. Chem.* **1982**, *257*, 13535–13543.

50 Higurashi, M., Ohtsuki, T., Inase, A., Kusumoto, R., Masutani, C., Hanaoka, F., Iwai, S., *J. Biol. Chem.* **2003**, *278*, 51968–51973.

51 Brash, D.E., Haseltine, W. A., *Nature* **1982**, *298*, 189–192.

52 Pfeifer, G. P., Drouin, R., Riggs, A. D., Holmquist, G. P., *Proc. Natl. Acad. Sci. USA* **1991**, *88*, 1374–1378.

53 Gale, J. M., Smerdon, M. J., *Photochem. Photobiol.* **1990**, *51*, 411–417.

54 Mitchell, D. L., Nguyen, T. D., Cleaver, J. E., *J. Biol. Chem.* **1990**, *265*, 5353–5356.

55 Taylor, J.-S., Cohrs, M. P., *J. Am. Chem. Soc.* **1987**, *109*, 2834–2835.

56 Taylor, J.-S., Lu, H.-F., Kotyk, J. J., *Photochem. Photobiol.* **1990**, *51*, 161–167.

57 Kan, L.-S., Voituriez, L., Cadet, J., *J. Photochem. Photobiol. B* **1992**, *12*, 339–357.

58 Clingen, P. H., Arlett, C. F., Rosa, L., Mori, T., Nikaido, O., Green, M. H. L., *Cancer Res.* **1995**, *55*, 2245–2248.

59 Douki, T., Cadet, J., *Biochemistry* **2001**, *40*, 2495–2501.

60 Perdiz, D., Gróf, P., Mezzina, M., Nikaido, O., Moustacchi, E., Sage, E., *J. Biol. Chem.* **2000**, *275*, 26732–26742.

61 Bose, S. N., Davies, R. J. H., Sethi, S. K., McCloskey, J. A., *Science* **1983**, *220*, 723–725.

62 Bose, S. N., Kumar, S., Davies, R. J. H., Sethi, S. K., McCloskey, J. A., *Nucleic Acids Res.* **1984**, *12*, 7929–7947.

63 Koning, T. M. G., Davies, R. J. H., Kaptein, R., *Nucleic Acids Res.* **1990**, *18*, 277–284.

64 Zhao, X., Nadji, S., Kao, J. L.-F., Taylor, J.-S., *Nucleic Acids Res.* **1996**, *24*, 1554–1560.

65 Colón, L., Crespo-Hernández, C. E., Oyola, R., García, C., Arce, R., *J. Phys. Chem. B* **2006**, *110*, 15589–15596.

66 Donnellan, J. E., Setlow, R. B., *Science* **1965**, *149*, 308–310.

67 Varghese, A. J., *Biochem. Biophys. Res. Commun.* **1970**, *38*, 484–490.

68 Kim, S. J., Lester, C., Begley, T. P., *J. Org. Chem.* **1995**, *60*, 6256–6257.

69 Chandor, A., Berteau, O., Douki, T., Gasparutto, D., Sanakis, Y., Ollagnier-de-Choudens, S., Atta, M., Fontecave, M., *J. Biol. Chem.* **2006**, *281*, 26922–26931.

70 Friedel, M. G., Berteau, O., Pieck, J. C., Atta, M., Ollagnier-de-Choudens, S., Fontecave, M., Carell, T., *Chem. Com.*, **2006**, 445–447.

71 Friedel, M. G., Pieck, J. C., Klages, J., Dauth, C., Kessler, H., Carell, T., *Chem. Eur. J.* **2006**, *12*, 6081–6094.

72 Liu, F.-T., Yang, N. C., *Biochemistry* **1978**, *17*, 4877–4885.

73 Boorstein, R. J., Hilbert, T. P., Cunningham, R. P., Teebor, G. W., *Biochemistry* **1990**, *29*, 10455–10460.

74 Zuo, S., Boorstein, R. J., Cunningham, R. P., Teebor, G. W., *Biochemistry* **1995**, *34*, 11582–11590.

75 O'Donnell, R. E., Boorstein, R. J., Cunningham, R. P., Teebor, G. W., *Biochemistry* **1994**, *33*, 9875–9880.

76 Iwai, S., *Nucleosides Nucleotides Nucleic Acids* **2006**, *25*, 561–582.

77 Taylor, J.-S., Brockie, I. R., O'Day, C. L., *J. Am. Chem. Soc.* **1987**, *109*, 6735–6742.

78 Taylor, J.-S., Brockie, I. R., *Nucleic Acids Res.* **1988**, *16*, 5123–5136.

79 Bdour, H. M., Kao, J. L.-F., Taylor, J.-S., *J. Org. Chem.* **2006**, *71*, 1640–1646.

80 Murata, T., Iwai, S., Ohtsuka, E., *Nucleic Acids Res.* **1990**, *18*, 7279–7286.

81 Mayo, J. U. O., Thomas, M., Saintomé, C., Clivio, P., *Tetrahedron* **2003**, *59*, 7377–7383.

82 Butenandt, J., Eker, A. P. M., Carell, T., *Chem. Eur. J.* **1998**, *4*, 642–654.

83 Takasawa, K., Masutani, C., Hanaoka, F., Iwai, S., *Nucleic Acids Res.* **2004**, *32*, 1738–1745.

84 Taylor, J.-S., *Pure Appl. Chem.* **1995**, *67*, 183–190.

85 Iwai, S., Shimizu, M., Kamiya, H., Ohtsuka, E., *J. Am. Chem. Soc.* **1996**, *118*, 7642–7643.

86 Iwai, S., Mizukoshi, T., Fujiwara, Y., Masutani, C., Hanaoka, F., Hayakawa, Y., *Nucleic Acids Res.* **1999**, *27*, 2299–2303.

87 Yamamoto, J., Hitomi, K., Todo, T., Iwai, S., *Nucleic Acids Res.* **2006**, *34*, 4406–4415.

88 Smith, C. A., Taylor, J.-S., *J. Biol. Chem.* **1993**, *268*, 11143–11151.

89 Mizukoshi, T., Hitomi, K., Todo, T., Iwai, S., *J. Am. Chem. Soc.* **1998**, *120*, 10634–10642.

90 Clivio, P., Fourrey, J.-L., Gasche, J., *J. Am. Chem. Soc.* **1991**, *113*, 5481–5483.

91 Clivio, P., Fourrey, J.-L., *Chem. Commun.*, **1996**, 2203–2204.

92 Liu, J., Taylor, J.-S., *J. Am. Chem. Soc.* **1996**, *118*, 3287–3288.

93 Warren, M. A., Murray, J. B., Connolly, B. A., *J. Mol. Biol.* **1998**, *279*, 89–100.

94 Matus, S. K. A., Fourrey, J.-L., Clivio, P., *Org. Biomol. Chem.* **2003**, *1*, 3316–3320.

95 Kemmink, J., Boelens, R., Koning, T. M. G., Kaptein, R., van der Marel, G. A., van Boom, J. H., *Eur. J. Biochem.* **1987**, *162*, 37–43.

96 Kemmink, J., Boelens, R., Koning, T., van der Marel, G. A., van Boom, J. H., Kaptein, R., *Nucleic Acids Res.* **1987**, *15*, 4645–4653.

97 Taylor, J.-S., Garrett, D. S., Brockie, I. R., Svoboda, D. L., Telser, J., *Biochemistry* **1990**, *29*, 8858–8866.

98 Kim, J.-K., Patel, D., Choi, B.-S., *Photochem. Photobiol.* **1995**, *62*, 44–50.

99 McAteer, K., Jing, Y., Kao, J., Taylor, J.-S., Kennedy, M. A., *J. Mol. Biol.* **1998**, *282*, 1013–1032.

100 Cadet, J., Voituriez, L., Hruska, F. E., Grand, A., *Biopolymers* **1985**, *24*, 897–903.

101 Hruska, F. E., Voituriez, L., Grand, A., Cadet, J., *Biopolymers* **1986**, *25*, 1399–1417.

102 Kan, L., Voituriez, L., Cadet, J., *Biochemistry* **1988**, *27*, 5796–5803.

103 Lee, J.-H., Choi, Y.-J., Choi, B.-S., *Nucleic Acids Res.* **2000**, *28*, 1794–1801.

104 Lee, J.-H., Park, C.-J., Shin, J.-S., Ikegami, T., Akutsu, H., Choi, B.-S., *Nucleic Acids Res.* **2004**, *32*, 2474–2481.

105 Park, H.-J., Zhang, K., Ren, Y., Nadji, S., Sinha, N., Taylor, J.-S., Kang, C.-H., *Proc. Natl. Acad. Sci. USA* **2002**, *99*, 15965–15970.

106 Dickerson, R. E., Goodsell, D. S., Neidle, S., *Proc. Natl. Acad. Sci. USA* **1994**, *91*, 3579–3583.

107 Yamaguchi, H., van Aalten, D. M. F., Pinak, M., Furukawa, A., Osman, R., *Nucleic Acids Res.* **1998**, *26*, 1939–1946.

108 Blagoev, K. B., Alexandrov, B. S., Goodwin, E. H., Bishop, A. R., *DNA Repair* **2006**, *5*, 863–867.

109 Kim, J.-K., Choi, B.-S., *Eur. J. Biochem.* **1995**, *228*, 849–854.

110 Lee, J.-H., Hwang, G.-S., Choi, B.-S., *Proc. Natl. Acad. Sci. USA* **1999**, *96*, 6632–6636.

111 Hwang, G.-S., Kim, J.-K., Choi, B.-S., *Eur. J. Biochem.* **1996**, *235*, 359–365.

112 Lee, J.-H., Hwang, G.-S., Kim, J.-K., Choi, B.-S., *FEBS Lett.* **1998**, *428*, 269–274.

113 Lee, J.-H., Bae, S.-H., Choi, B.-S., *Proc. Natl. Acad. Sci. USA* **2000**, *97*, 4591–4596.

114 Mizukoshi, T., Kodama, T. S., Fujiwara, Y., Furuno, T., Nakanishi, M., Iwai, S., *Nucleic Acids Res.* **2001**, *29*, 4948–4954.

115 Spector, T. I., Cheatham, T. E., Kollman, P. A., *J. Am. Chem. Soc.* **1997**, *119*, 7095–7104.

116 Jing, Y., Kao, J. F.-L., Taylor, J.-S., *Nucleic Acids Res.* **1998**, *26*, 3845–3853.

117 Taylor, J.-S., Garrett, D. S., Cohrs, M. P., *Biochemistry* **1988**, *27*, 7206–7215.

118 Rycyna, R. E., Alderfer, J. L., *Nucleic Acids Res.* **1985**, *13*, 5949–5963.

119 Taylor, J.-S., Garrett, D. S., Wang, M. J., *Biopolymers* **1988**, *27*, 1571–1593.

120 Banerjee, S. K., Christensen, R. B., Lawrence, C. W., LeClerc, J. E., *Proc. Natl. Acad. Sci. USA* **1988**, *85*, 8141–8145.

121 Lawrence, C. W., Banerjee, S. K., Borden, A., LeClerc, J. E., *Mol. Gen. Genet.* **1990**, *222*, 166–168.

122 Smith, C. A., Wang, M., Jiang, N., Che, L., Zhao, X., Taylor, J.-S., *Biochemistry* **1996**, *35*, 4146–4154.

123 Gentil, A., Le Page, F., Margot, A., Lawrence, C. W., Borden, A., Sarasin, A., *Nucleic Acids Res.* **1996**, *24*, 1837–1840.

124 Gibbs, P. E. M., Kilbey, B. J., Banerjee, S. K., Lawrence, C. W., *J. Bacteriol.* **1993**, *175*, 2607–2612.

125 Kamiya, H., Iwai, S., Kasai, H., *Nucleic Acids Res.* **1998**, *26*, 2611–2617.

126 Horsfall, M. J., Borden, A., Lawrence, C. W., *J. Bacteriol.* **1997**, *179*, 2835–2839.

127 Banerjee, S. K., Borden, A., Christensen, R. B., LeClerc, J. E., Lawrence, C. W., *J. Bacteriol.* **1990**, *172*, 2105–2112.

128 LeClerc, J. E., Borden, A., Lawrence, C. W., *Proc. Natl. Acad. Sci. USA* **1991**, *88*, 9685–9689.

129 Horsfall, M. J., Lawrence, C. W., *J. Mol. Biol.* **1994**, *235*, 465–471.

130 Jiang, N., Taylor, J.-S., *Biochemistry* **1993**, *32*, 472–481.

131 Fujiwara, Y., Iwai, S., *Biochemistry* **1997**, *36*, 1544–1550.

132 Luria, S. E., *Proc. Natl. Acad. Sci. USA* **1947**, *33*, 253–264.

133 Friedberg, E. C., King, J. J., *Biochem. Biophys. Res. Commun.* **1969**, *37*, 646–651.

134 Yasuda, S., Sekiguchi, M., *Proc. Natl. Acad. Sci. USA* **1970**, *67*, 1839–1845.

135 Radany, E. H., Naumovski, L., Love, J. D., Gutekunst, K. A., Hall, D. H., Friedberg, E. C., *J. Virol.* **1984**, *52*, 846–856.

136 Valerie, K., Henderson, E. E., de Riel, J. K., *Nucleic Acids Res.* **1984**, *12*, 8085–8096.

137 Radany, E. H., Friedberg, E. C., *Nature* **1980**, *286*, 182–185.

138 Gordon, L. K., Haseltine, W. A., *J. Biol. Chem.* **1980**, *255*, 12047–12050.

139 Kim, J., Linn, S., *Nucleic Acids Res.* **1988**, *16*, 1135–1141.

140 Manoharan, M., Mazumder, A., Ransom, S. C., Gerlt, J. A., Bolton, P. H., *J. Am. Chem. Soc.* **1988**, *110*, 2690–2691.

141 Schrock, R. D., Lloyd, R. S., *J. Biol. Chem.* **1991**, *266*, 17631–17639.

142 Dodson, M. L., Schrock, R. D., Lloyd, R. S., *Biochemistry* **1993**, *32*, 8284–8290.

143 Iwai, S., Maeda, M., Shirai, M., Shimada, Y., Osafune, T., Murata, T., Ohtsuka, E., *Biochemistry* **1995**, *34*, 4601–4609.

144 Sun, B., Latham, K. A., Dodson, M. L., Lloyd, R. S., *J. Biol. Chem.* **1995**, *270*, 19501–19508.

145 Zharkov, D. O., Golan, G., Gilboa, R., Fernandes, A. S., Gerchman, S. E., Kycia, J. H., Rieger, R. A., Grollman, A. P., Shoham, G., *EMBO J.* **2002**, *21*, 789–800.

146 Fromme, J. C., Verdine, G. L., *EMBO J.* **2003**, *22*, 3461–3471.

147 Mazumder, A., Gerlt, J. A., Rabow, L., Absalon, M. J., Stubbe, J., Bolton, P. H., *J. Am. Chem. Soc.* **1989**, *111*, 8029–8030.

148 Morikawa, K., Matsumoto, O., Tsujimoto, M., Katayanagi, K., Ariyoshi, M., Doi, T., Ikehara, M., Inaoka, T., Ohtsuka, E., *Science* **1992**, *256*, 523–526.

149 Morikawa, K., Ariyoshi, M., Vassylyev, D. G., Matsumoto, O., Katayanagi, K., Ohtsuka, E., *J. Mol. Biol.* **1995**, *249*, 360–375.

150 Vassylyev, D. G., Kashiwagi, T., Mikami, Y., Ariyoshi, M., Iwai, S., Ohtsuka, E., Morikawa, K., *Cell* **1995**, *83*, 773–782.

151 Doi, T., Recktenwald, A., Karaki, Y., Kikuchi, M., Morikawa, K., Ikehara, M., Inaoka, T., Hori, N., Ohtsuka, E., *Proc. Natl. Acad. Sci. USA* **1992**, *89*, 9420–9424.

152 Iwai, S., Maeda, M., Shimada, Y., Hori, N., Murata, T., Morioka, H., Ohtsuka, E., *Biochemistry* **1994**, *33*, 5581–5588.

153 Golan, G., Zharkov, D. O., Grollman, A. P., Dodson, M. L., McCullough, A. K., Lloyd, R. S., Shoham, G., *J. Mol. Biol.* **2006**, *362*, 241–258.

154 Rupert, C. S., *J. Gen. Physiol.* **1960**, *43*, 573–595.

155 Todo, T., Takemori, H., Ryo, H., Ihara, M., Matsunaga, T., Nikaido, O., Sato, K., Nomura, T., *Nature* **1993**, *361*, 371–374.

156 Kanai, S., Kikuno, R., Toh, H., Ryo, H., Todo, T., *J. Mol. Evol.* **1997**, *45*, 535–548.

157 Kim, S.-T., Malhotra, K., Smith, C. A., Taylor, J.-S., Sancar, A., *Biochemistry* **1993**, *32*, 7065–7068.

158 Payne, G., Heelis, P. F., Rohrs, B. R., Sancar, A., *Biochemistry* **1987**, *26*, 7121–7127.

159 Li, Y. F., Heelis, P. F., Sancar, A., *Biochemistry* **1991**, *30*, 6322–6329.

160 Byrdin, M., Eker, A. P. M., Vos, M. H., Brettel, K., *Proc. Natl. Acad. Sci. USA* **2003**, *100*, 8676–8681.

161 Saxena, C., Sancar, A., Zhong, D., *J. Phys. Chem. B* **2004**, *108*, 18026–18033.

162 Lukacs, A., Eker, A. P. M., Byrdin, M., Villette, S., Pan, J., Brettel, K., Vos, M. H., *J. Phys. Chem. B* **2006**, *110*, 15654–15658.

163 Weber, S., *Biochim. Biophys. Acta* **2005**, *1707*, 1–23.

164 Kao, Y.-T., Saxena, C., Wang, L., Sancar, A., Zhong, D., *Proc. Natl. Acad. Sci. USA* **2005**, *102*, 16128–16132.

165 Park, H.-W., Kim, S.-T., Sancar, A., Deisenhofer, J., *Science* **1995**, *268*, 1866–1872.

166 Tamada, T., Kitadokoro, K., Higuchi, Y., Inaka, K., Yasui, A., de Ruiter, P. E., Eker,

A. P. M., Miki, K., *Nature Struct. Biol.* **1997**, *4*, 887–891.
167 Komori, H., Masui, R., Kuramitsu, S., Yokoyama, S., Shibata, T., Inoue, Y., Miki, K., *Proc. Natl. Acad. Sci. USA* **2001**, *98*, 13560–13565.
168 Mees, A., Klar, T., Gnau, P., Hennecke, U., Eker, A. P. M., Carell, T., Essen, L.-O., *Science* **2004**, *306*, 1789–1793.
169 Vande Berg, B. J., Sancar, G. B., *J. Biol. Chem.* **1998**, *273*, 20276–20284.
170 Christine, K. S., MacFarlane, A. W., Yang, K., Stanley, R. J., *J. Biol. Chem.* **2002**, *277*, 38339–38344.
171 Torizawa, T., Ueda, T., Kuramitsu, S., Hitomi, K., Todo, T., Iwai, S., Morikawa, K., Shimada, I., *J. Biol. Chem.* **2004**, *279*, 32950–32956.
172 Kim, S.-T., Malhotra, K., Smith, C. A., Taylor, J.-S., Sancar, A., *J. Biol. Chem.* **1994**, *269*, 8535–8540.
173 Zhao, X., Liu, J., Hsu, D. S., Zhao, S., Taylor, J.-S., Sancar, A., *J. Biol. Chem.* **1997**, *272*, 32580–32590.
174 Hitomi, K., Kim, S.-T., Iwai, S., Harima, N., Otoshi, E., Ikenaga, M., Todo, T., *J. Biol. Chem.* **1997**, *272*, 32591–32598.
175 Li, J., Uchida, T., Todo, T., Kitagawa, T., *J. Biol. Chem.* **2006**, *281*, 25551–25559.
176 Heelis, P. F., Liu, S., *J. Am. Chem. Soc.* **1997**, *119*, 2936–2937.
177 Hitomi, K., Nakamura, H., Kim, S.-T., Mizukoshi, T., Ishikawa, T., Iwai, S., Todo, T., *J. Biol. Chem.* **2001**, *276*, 10103–10109.
178 Schleicher, E., Hitomi, K., Kay, C. W. M., Getzoff, E. D., Todo, T., Weber, S., *J. Biol. Chem.* **2007**, *282*, 4738–4747.
179 Todo, T., Ryo, H., Yamamoto, K., Toh, H., Inui, T., Ayaki, H., Nomura, T., Ikenaga, M., *Science* **1996**, *272*, 109–112.
180 Hsu, D. S., Zhao, X., Zhao, S., Kazantsev, A., Wang, R.-P., Todo, T., Wei, Y.-F., Sancar, A., *Biochemistry* **1996**, *35*, 13871–13877.
181 Todo, T., Tsuji, H., Otoshi, E., Hitomi, K., Kim, S.-T., Ikenaga, M., *Mutation Res.* **1997**, *384*, 195–204.
182 Cashmore, A. R., *Cell* **2003**, *114*, 537–543.
183 Huang, J.-C., Svoboda, D. L., Reardon, J. T., Sancar, A., *Proc. Natl. Acad. Sci. USA* **1992**, *89*, 3664–3668.
184 Szymkowski, D. E., Lawrence, C. W., Wood, R. D., *Proc. Natl. Acad. Sci. USA* **1993**, *90*, 9823–9827.
185 Mellon, I., *Mutation Res.* **2005**, *577*, 155–161.
186 Cleaver, J. E., *Nature Rev. Cancer* **2005**, *5*, 564–573.
187 Reardon, J. T., Nichols, A. F., Keeney, S., Smith, C. A., Taylor, J.-S., Linn, S., Sancar, A., *J. Biol. Chem.* **1993**, *268*, 21301–21308.
188 Fujiwara, Y., Masutani, C., Mizukoshi, T., Kondo, J., Hanaoka, F., Iwai, S., *J. Biol. Chem.* **1999**, *274*, 20027–20033.
189 Wittschieben, B. O., Iwai, S., Wood, R. D., *J. Biol. Chem.* **2005**, *280*, 39982–39989.
190 Sugasawa, K., Okuda, Y., Saijo, M., Nishi, R., Matsuda, N., Chu, G., Mori, T., Iwai, S., Tanaka, K., Tanaka, K., Hanaoka, F., *Cell* **2005**, *121*, 387–400.
191 Sugasawa, K., Ng, J. M. Y., Masutani, C., Iwai, S., van der Spek, P. J., Eker, A. P. M., Hanaoka, F., Bootsma, D., Hoeijmakers, J. H. J., *Mol. Cell* **1998**, *2*, 223–232.
192 Sugasawa, K., Okamoto, T., Shimizu, Y., Masutani, C., Iwai, S., Hanaoka, F., *Genes Dev.* **2001**, *15*, 507–521.
193 Hwang, B. J., Ford, J. M., Hanawalt, P. C., Chu, G., *Proc. Natl. Acad. Sci. USA* **1999**, *96*, 424–428.
194 Fitch, M. E., Nakajima, S., Yasui, A., Ford, J. M., *J. Biol. Chem.* **2003**, *278*, 46906–46910.
195 Gillet, L. C. J., Schärer, O. D., *Chem. Rev.* **2006**, *106*, 253–276.
196 Ikegami, T., Kuraoka, I., Saijo, M., Kodo, N., Kyogoku, Y., Morikawa, K., Tanaka, K., Shirakawa, M., *Nature Struct. Biol.* **1998**, *5*, 701–706.
197 Buchko, G. W., Isern, N. G., Spicer, L. D., Kennedy, M. A., *Mutation Res.* **2001**, *486*, 1–10.
198 Li, T., Chen, X., Garbutt, K. C., Zhou, P., Zheng, N., *Cell* **2006**, *124*, 105–117.
199 Angers, S., Li, T., Yi, X., MacCoss, M. J., Moon, R. T., Zheng, N., *Nature* **2006**, *443*, 590–593.

200 Freyer, G. A., Davey, S., Ferrer, J. V., Martin, A. M., Beach, D., Doetsch, P. W., *Mol. Cell. Biol.* **1995**, *15*, 4572–4577.
201 Yajima, H., Takao, M., Yasuhira, S., Zhao, J. H., Ishii, C., Inoue, H., Yasui, A., *EMBO J.* **1995**, *14*, 2393–2399.
202 Takao, M., Yonemasu, R., Yamamoto, K., Yasui, A., *Nucleic Acids Res.* **1996**, *24*, 1267–1271.
203 Bowman, K. K., Sidik, K., Smith, C. A., Taylor, J.-S., Doetsch, P. W., Freyer, G. A., *Nucleic Acids Res.* **1994**, *22*, 3026–3032.
204 Avery, A. M., Kaur, B., Taylor, J.-S., Mello, J. A., Essigmann, J. M., Doetsch, P. W., *Nucleic Acids Res.* **1999**, *27*, 2256–2264.
205 Kanno, S., Iwai, S., Takao, M., Yasui, A., *Nucleic Acids Res.* **1999**, *27*, 3096–3103.
206 Yoon, J.-H., Swiderski, P. M., Kaplan, B. E., Takao, M., Yasui, A., Shen, B., Pfeifer, G. P., *Biochemistry* **1999**, *38*, 4809–4817.
207 Alleva, J. L., Zuo, S., Hurwitz, J., Doetsch, P. W., *Biochemistry* **2000**, *39*, 2659–2666.
208 Smith, C. A., Baeten, J., Taylor, J.-S., *J. Biol. Chem.* **1998**, *273*, 21933–21940.
209 Johnson, R. E., Prakash, S., Prakash, L., *Science* **1999**, *283*, 1001–1004.
210 Masutani, C., Araki, M., Yamada, A., Kusumoto, R., Nogimori, T., Maekawa, T., Iwai, S., Hanaoka, F., *EMBO J.* **1999**, *18*, 3491–3501.
211 Masutani, C., Kusumoto, R., Yamada, A., Dohmae, N., Yokoi, M., Yuasa, M., Araki, M., Iwai, S., Takio, K., Hanaoka, F., *Nature* **1999**, *399*, 700–704.
212 Masutani, C., Kusumoto, R., Iwai, S., Hanaoka, F., *EMBO J.* **2000**, *19*, 3100–3109.
213 Vu, B., Cannistraro, V. J., Sun, L., Taylor, J.-S., *Biochemistry* **2006**, *45*, 9327–9335.
214 Washington, M. T., Johnson, R. E., Prakash, S., Prakash, L., *J. Biol. Chem.* **1999**, *274*, 36835–36838.
215 Matsuda, T., Bebenek, K., Masutani, C., Hanaoka, F., Kunkel, T. A., *Nature* **2000**, *404*, 1011–1013.
216 McCulloch, S. D., Kokoska, R. J., Masutani, C., Iwai, S., Hanaoka, F., Kunkel, T. A., *Nature* **2004**, *428*, 97–100.
217 Nelson, J. R., Lawrence, C. W., Hinkle, D. C., *Science* **1996**, *272*, 1646–1649.
218 Tissier, A., Frank, E. G., McDonald, J. P., Iwai, S., Hanaoka, F., Woodgate, R., *EMBO J.* **2000**, *19*, 5259–5266.
219 Vaisman, A., Takasawa, K., Iwai, S., Woodgate, R., *DNA Repair* **2006**, *5*, 210–218.
220 Ohashi, E., Ogi, T., Kusumoto, R., Iwai, S., Masutani, C., Hanaoka, F., Ohmori, H., *Genes Dev.* **2000**, *14*, 1589–1594.
221 Zhang, Y., Yuan, F., Wu, X., Wang, M., Rechkoblit, O., Taylor, J.-S., Geacintov, N. E., Wang, Z., *Nucleic Acids Res.* **2000**, *28*, 4138–4146.
222 Seki, M., Masutani, C., Yang, L. W., Schuffert, A., Iwai, S., Bahar, I., Wood, R. D., *EMBO J.* **2004**, *23*, 4484–4494.
223 Zhang, Y., Wu, X., Guo, D., Rechkoblit, O., Taylor, J.-S., Geacintov, N. E., Wang, Z., *J. Biol. Chem.* **2002**, *277*, 44582–44587.
224 Servant, L., Cazaux, C., Bieth, A., Iwai, S., Hanaoka, F., Hoffmann, J.-S., *J. Biol. Chem.* **2002**, *277*, 50046–50053.
225 Niimi, A., Limsirichaikul, S., Yoshida, S., Iwai, S., Masutani, C., Hanaoka, F., Kool, E. T., Nishiyama, Y., Suzuki, M., *Mol. Cell. Biol.* **2004**, *24*, 2734–2746.
226 Ghadessy, F. J., Ramsay, N., Boudsocq, F., Loakes, D., Brown, A., Iwai, S., Vaisman, A., Woodgate, R., Holliger, P., *Nature Biotechnol.* **2004**, *22*, 755–759.
227 Johnson, R. E., Haracska, L., Prakash, S., Prakash, L., *Mol. Cell. Biol.* **2001**, *21*, 3558–3563.
228 Johnson, R. E., Washington, M. T., Haracska, L., Prakash, S., Prakash, L., *Nature* **2000**, *406*, 1015–1019.
229 Guo, D., Wu, X., Rajpal, D. K., Taylor, J.-S., Wang, Z., *Nucleic Acids Res.* **2001**, *29*, 2875–2883.
230 Tang, M., Pham, P., Shen, X., Taylor, J.-S., O'Donnell, M., Woodgate, R., Goodman, M. F., *Nature* **2000**, *404*, 1014–1018.
231 Ishikawa, T., Uematsu, N., Mizukoshi, T., Iwai, S., Iwasaki, H., Masutani, C., Hanaoka, F., Ueda, R., Ohmori, H., Todo, T., *J. Biol. Chem.* **2001**, *276*, 15155–15163.

232 Trincao, J., Johnson, R. E., Escalante, C. R., Prakash, S., Prakash, L., Aggarwal, A. K., *Mol. Cell* **2001**, *8*, 417–426.

233 Ling, H., Boudsocq, F., Plosky, B. S., Woodgate, R., Yang, W., *Nature* **2003**, *424*, 1083–1087.

234 Ling, H., Boudsocq, F., Woodgate, R., Yang, W., *Mol. Cell* **2004**, *13*, 751–762.

235 Ling, H., Sayer, J. M., Plosky, B. S., Yagi, H., Boudsocq, F., Woodgate, R., Jerina, D. M., Yang, W., *Proc. Natl. Acad. Sci. USA* **2004**, *101*, 2265–2269.

236 Li, Y., Dutta, S., Doublié, S., Bdour, H. M., Taylor, J.-S., Ellenberger, T., *Nature Struct. Mol. Biol.* **2004**, *11*, 784–790.

237 Johnson, R. E., Prakash, L., Prakash, S., *Proc. Natl. Acad. Sci. USA* **2005**, *102*, 12359–12364.

238 Hwang, H., Taylor, J.-S., *Biochemistry* **2005**, *44*, 4850–4860.

6
Locked Nucleic Acids: Properties, Applications, and Perspectives

Poul Nielsen and Jesper Wengel

6.1
Introduction

A decade has passed since the introduction of locked nucleic acid (LNA) [1–3]. This chapter provides a review and a current status on the features and applications of LNA. Attention is focused on the structural chemistry of LNA and on its applications for therapeutics and probes. With regards to the latter point, the chapter is based on former comprehensive reviews on LNA applications [4–6] and, for that reason, is focused on results reported subsequently, with no intention of being comprehensive.

The field of nucleic acid chemistry has evolved considerably during the past few decades, leading from synthetic developments via molecular recognition into nucleic acid-based therapeutics, powerful diagnostic probes, and modern nanobiotechnology. From the structural perspective, the central issue has been the thorough understanding of the basis for molecular recognition and the selective formation of nucleic acid complexes, initially the classic Watson–Crick-type double helix. The central creative challenge for synthetic chemists has been to design simple chemical perturbations of the natural duplex in order to increase our knowledge within nucleic acid chemical biology, and to bring this knowledge into applied nucleic acids research [7–9].

The major motivation behind the design of nucleic acid analogues has, over the past two decades, been the antisense strategy aimed at the silencing of genes by strong binding towards mRNA, with or without associated RNA-cleavage. Two central problems have motivated the chemical effort: (1) the physiological instability of natural oligonucleotides; and (2) the relatively moderate affinity of short natural oligonucleotides for their RNA-targets [10, 11]. If a solution to these problems – as well as to the delicate question of cellular delivery – can be found, then oligonucleotides will present unlimited therapeutic perspectives. Among the more than 1000 chemical analogues tested for these properties, a large part seems to induce a high degree of resistance towards nucleolytic degradation, although this is dependent on the sequence composition, the number and positioning of the chemical modifications within the sequence, and on the degree of perturbation compared to the natural chemical structure. On the other hand, significantly increasing the RNA-affinity of an

Modified Nucleosides: in Biochemistry, Biotechnology and Medicine. Edited by Piet Herdewijn

ISBN: 978-3-527-31820-9

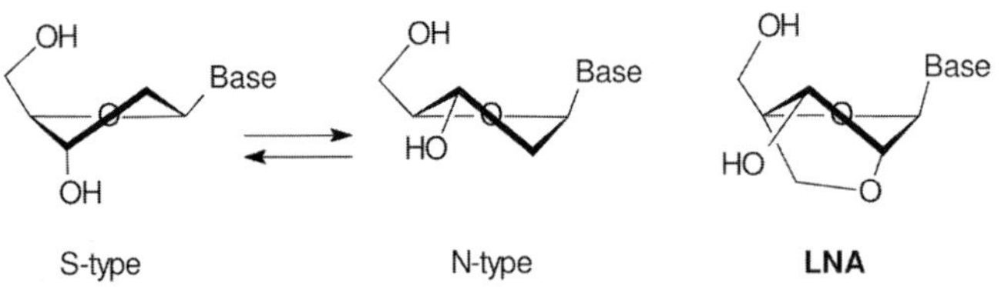

Figure 6.1 Nucleoside conformations and the structure and locked conformation of LNA.

oligonucleotide – in other words, strengthening the natural nucleic acid recognition – has been a much more difficult task, and the first-generation antisense oligonucleotides (phosphorthiate DNA oligomers) in which one of the non-bridging oxygen atoms of the natural phosphordiester linkage is replaced by a sulfur atom, the overall RNA-affinity is compromised.

Nucleic acid duplexes fall into two major conformational types, the A-type and the B-type, dictated by puckering of the single nucleotides, a C3′-endo (N-type) conformation in the A-type and a C2′-endo (S-type) conformation in the B-type (Figure 6.1) [12]. The A-type is adopted by RNA when double-stranded (ds)RNA duplex regions are found, whereas the dsDNA in the genome adopts a B-type; the latter can be locally transformed to the A-type during dehydration. The targeting of RNA is therefore closely related to the ability of an oligonucleotide to form an A-type duplex, and this can be effected by a conformational restriction in an N-type conformation. This has been shown also to be the key feature behind the second-generation antisense oligonucleotides containing various 2′-alkoxy groups, all of which lead to some degree of restriction into N-type conformation and a limited increase in the thermal stability of the duplex formed with complementary RNA (as is observed for fully modified 2′-O-alkyl oligonucleotides) [13]. One member of this family is the 2′-MOE (2′-methoxyethyl) modification first introduced by Martin in 1995 [14]. Other nucleic acid analogues which mimic the N-type conformation include hexitol nucleic acids (HNA), introduced by Herdewijn and coworkers [15], and the 3′-NH-phosphoramidates introduced by Gryaznov and coworkers [16].

The most powerful way to introduce a conformational restriction is, however, by covalently linking two different positions in the pentofuranose part of the nucleoside, thereby obtaining a bicyclic nucleoside [8, 9]. The first member of this family was introduced by Leumann and coworkers in 1993 [17], and was termed bicycloDNA, which was later developed by the introduction of tricycloDNA in 1997 [18]. The conformational restriction did not, however, lead to perfect N-type mimics in these cases. Both, an N-type and an S-type mimic using a bicyclo[3.1.0]hexane skeleton were introduced in 1994 [19, 20], leading later in the Marquez group to a number of significant results within nucleoside chemistry and biology [21, 22]. Subsequently, the present authors' group and the group of Imanishi independently introduced LNA (locked nucleic acid) in 1998 [1–3]. In LNA, the O2′ and C4′ atoms are linked by a methylene group, thereby introducing a conformational lock of the molecule into a near-perfect N-type conformation. Following the pseudorotational cycle introduced by Altona and Sundaralingham [23], the LNA-monomers adopt a ^{3}E envelope conformation (which means that the C3′ is positioned above the plane formed by the other four ring-atoms) (Figure 6.1) with a pseudorotational angle of $P = 17°$, as shown by nuclear magnetic resonance (NMR) and X-ray crystallography [2, 3, 24]. This is very near the average sugar puckering of 14° found in natural RNA [12]. The discovery of LNA was based on chemical experience with C4′-branched nucleosides, and with other bicyclic nucleosides mimicking S-type or intermediate (i.e., E-type) conformations [25]. Today, LNA is defined as oligonucleotides containing one or more of the 2′-*O*,4′-*C*-methylene-β-D-ribofuranosyl nucleosides called LNA monomers (Figure 6.1).

A major structural characteristic of LNA is its close resemblance to the natural nucleic acids. This leads to easy handling, as LNA sequences have similar physical properties (including water-solubility) to natural oligodeoxynucleotides (ODNs). Furthermore, LNAs are synthesized following the conventional phosphoramidite chemistry, allowing the automated synthesis of fully modified LNA-sequences as well as chimeras with DNA, RNA or other modifications. Currently, LNA oligonucleotides and LNA phosphoramidites are available commercially.

The most remarkable and conclusive feature of LNA is the unprecedented hybridization to complementary nucleic acids. Hence, the increase in thermal stability (T_m values) of duplexes formed between a short ODN and its complementary RNA is up to approximately 10 °C for each LNA-monomer introduced into the ODN (for DNA–LNA chimera containing one or a few LNA monomers). Although promising results were obtained with other nucleic acid analogues – for example, like the HNA mentioned above – none of the other conformationally restricted N-type nucleoside mimics, and no bicyclic analogues restricted in other conformations (S or E), has shown duplex stabilizations near the stabilizations obtained with LNA and its close analogues (see below). Another unique feature of LNA is the fact that it flexibly can be combined with DNA while still inducing the remarkable hybridization properties (see below). That the increase in thermal stability of duplexes formed with complementary single-stranded DNA is also up to approximately 10 °C is perhaps even more remarkable. This opens the perspective of applying LNA not only for therapeutically interesting RNA-targeting but also as powerful DNA-targeting probes.

6.2
LNA in High-Affinity Hybridization: Designing Sequences

The duplexes formed by LNA-sequences are, relative to their length, among the most thermally stable known. A plethora of LNA-oligonucleotides has been synthesized for various purposes, including basic hybridization studies, and some general rules for the scope in sequence design can now be given. As mentioned above, LNA can easily be made as chimeras with DNA by following either a mixmer or a gapmer approach. The hybridization behavior of some typical LNA constructs are described in the following section and listed in Table 6.1. The chosen sequences represent a broad spectrum of the numerous LNA sequences reported in the literature. Entries 1 and 3 show the original sequences tested by the present authors in their first publications

Table 6.1 Thermal stabilities of representative duplexes and triplexes formed by LNA sequences with matched complementary strands.[a)]

Entry	LNA sequence	Duplexes – T_m/°C Complementary RNA	Complementary DNA	Complementary LNA	Ref.
1	5′-GTGATATGC	50 (+7.3)	44 (+5.3)	63 (+11.3)	1, 27
2	5′-GTGATATGC RNA	63 (+8.3)	55 (+9.3)	74 (+15.7)	26, 27
3	5′-GTGATATG^{m}C	74 (+5.1)	64 (+4.0)	>93 (>+7)	2, 27
4	5′-GCGTTTTTTGCT	49 (+4.0)	49 (+2.0)	–	3
5	5′-GCGTTTTTTGCT	71 (+4.3)	58 (+1.8)	–	3
6	5′-CCATTGCTACC	–	50.6 (−0.3)	–	28
7	5′-CCATTGCTACC	–	54.4 (+3.5)	–	28
8	5′-CACGGCTC	–	47.4 (+3.4)	–	28
9	5′-CACGGCTC	–	51.2 (+7.2)	–	28
10	5′-TGCTCCTG	–	31.0 (+2.5)	–	29
11	5′-TGCTCCTG	–	41.0 (+4.1)	–	29
12	5′-ACUACCA 2′-OMe NA	47.9 (+6.9)	–	–	30
13	5′-TAAGCGGGTCGCTGC	–	86 (+2.2)	–	31
14	5′-AGGGTCGCTCGGTGT	–	78 (+0.9)	–	31
15	5′-CATGTCATGACGGTTAGG	70 (+4.0)	–	–	32
16	5′-CATGTCATGACGGTTAGG	85 (+2.7)	–	–	32

		Triplexes – T_m/°C Complementary dsDNA pH 6.3 or 6.6	pH 7.2	
17	5′-T^{m}CT^{m}CT^{m}CT^{m}CC^{m}CTTTT	55 (+4.5)	33 (~+4)	36
18	5′-TCTCTCTCCCTTTT	No transition	No transition	36
19	5′-TTTTCTTTTCCCCCCT	22 (+1.0)	No transition	37
20	5′-TTTTCTTTTCCCCCCT	35 (+4.3)	26 (~ +5.5)	37
21	5′-TTTTCTTTTCCCCCCT	43 (+5.0)	23 (~ +3.8)	37

a) Melting temperatures shown as published in the indicated references not taking different buffer conditions into account. In brackets are shown (as published in the indicated references) the change in melting temperature per LNA modification compared to the corresponding unmodified sequences. LNA monomers are shown in bold red. RNA and 2′-*O*-Me-RNA are shown in blue and green, respectively. ^{m}C = 5-methylcytosine.

of LNA [1, 2, 26, 27]. Increases in T_m per LNA incorporation of 4.0 to 7.3 °C were observed with complementary RNA and DNA. The relatively increase per LNA nucleotide was slightly lower for the fully modified LNA (entry 3) than for the LNA-DNA chimeras (entry 1). LNA can also be combined with RNA-monomers (entry 2), leading to an even more pronounced enhancement of duplex stability. LNA : LNA base pairing was examined, and showed that a fully modified LNA : LNA duplex was stable above the detection limit of 93 °C, even after lowering the Na^+ concentration. Entries 4–5 show the sequence initially tested by Imanishi and coworkers [3]. With complementary DNA the relative increase in T_m was somewhat lower, demonstrating some sequence dependency in LNA-mediated stabilization. Entries 6–9 show four out of 100 different LNA-sequences studied with their DNA-complements by McTigue *et al.* in order to develop a general affinity prediction set for LNA

sequences [28]. Entries 6 and 9 represent the second lowest versus the second highest stabilization obtained within the entire set of sequences, whereas entries 7 and 8 are representative intermediate results. Typical increases in T_m are +2 to +5 °C, but clearly much larger increases and even a small destabilization can be seen in rare cases. A recent study on LNA : DNA duplexes by Kaur *et al.* (entries 10 and 11) showed similar increases in thermal stability [29]. Entry 12 demonstrates that LNA can also be mixed in chimeras with other modifications, such as 2′-OMe nucleosides [30]. The increase in thermal stability in the present case is very high (+6.9 °C) compared to the corresponding 2′-OMe sequence, which is again +2.5 °C (in total) above the corresponding unmodified DNA sequence (not shown). Entries 13 to 16 demonstrate the difference between typical mixmer (entries 13 and 15) and typical gapmer sequences (entries 14 and 16) [31, 32]. In the latter two, stretches of seven or eight 2′-deoxynucleotides (DNA nucleotides) are framed with stretches of LNA. As expected, the relative increase in T_m for each LNA-incorporation is higher for mixmers than for gapmers.
In the following, some general observations on LNA hybridization are listed:

- The increase in duplex stability by introducing one or several LNA monomers is strong in all cases with complementary RNA, and in most cases with complementary DNA.
- The exact increase in duplex stability is dependent on the base sequence.
- The affinity-enhancing effect per monomer is generally more pronounced against complementary RNA than against complementary DNA (normally from +3 to +7 against RNA and from +2 to +5 against DNA).
- The relative increase observed against complementary DNA is more variable (from −1 to +9) than against complementary RNA (from +2 to +9).
- LNA monomers incorporated in the 3′- or 5′-termini induce a smaller increase in thermal stability than do centrally placed LNA monomers.
- The relative increase in thermal stability per LNA monomer is larger for a single or a few non-consecutive incorporations in DNA-sequences (mixmers) than for a stretch of LNA bases in either fully modified or LNA-DNA gapmer sequences.
- The relative increase in thermal stability per LNA monomer is larger in shorter sequences than in longer sequences.
- LNA monomers with pyrimidine bases tend to induce larger thermal stability increases than LNA monomers with purine bases [28].
- LNA monomers introduced into RNA, 2′-OMe and/or phosphorthioate sequences also strongly stabilize duplexes.
- LNA : LNA duplexes are inherently very stable, and this should be carefully considered when designing LNA-sequences for probes or therapeutic purposes.

McTigue *et al.* have used their large set of LNA : DNA duplexes to suggest some predictive rules for hybridization enthalpy and entropy for all 32 possible nearest neighbors in LNA-DNA : DNA hybridization [28]. In terms of $\Delta\Delta G^\circ$, the order of induced stabilization of the four LNA-monomers were found to be C > T > G ≫ A. The thermodynamic data suggest that LNA can stabilize the duplex by either preorganization or improved stacking in terms of $\Delta\Delta S^\circ$ or $\Delta\Delta H^\circ$, but not both simultaneously. Other studies have similarly suggested a contribution to stability from either enthalpy or entropy [3, 27, 33]. The results of a recent study by Kaur *et al.* suggest that the formation of LNA duplexes is associated with a lower hydration and a higher uptake of counter ions relative to unmodified duplexes [29]. A kinetic study indicated that the increase in duplex stability obtained with LNA is due to slower rates of dissociation, whereas association rates are unchanged [33]. Finally, it has been shown that LNA also recognizes other nucleic acid systems, and can even form very stable duplexes with peptide nucleic acids (PNAs) [34].

LNA monomers stabilize triplexes and are useful in the design of triplex-forming oligonucleotides (TFOs). In general, homopyrimidine ODNs can form stable triplexes with complementary dsDNA through Hoogsteen base pairing in the major groove of the DNA duplex, with the TFO being parallel to the homopurine strand of the duplex [35]. Thymine forms Hoogsteen hydrogen bonds to adenine, whereas cytosine needs protonation before hydrogen bonds to guanine are formed in the triplex mode. Therefore, the stability of triplexes increases at lower pH. When a mixmer approach is followed, LNA increases the thermal stability of triplexes to the same extent as of duplexes (Table 6.1). Thus, entries 17, 20, and 21 demonstrate that LNA monomers in every second or third position lead to large increases in triplex thermal stability [36, 37]. On the other hand, fully modified LNA (entry 18) cannot form stable triplexes at all, and a stretch of LNA in the 5′-end of a TFO leads only to a minor stabilization (entry 19). Much more stable triplexes are in all cases found at pH 6.3 or 6.6 compared to pH 7.2. A kinetic study has also in the case of triplexes suggested a lower rate of dissociation for an LNA TFO compared to the corresponding DNA TFO [38, 39]. Thermodynamic measurements indicate both enthalpic and entropic contributions of the LNA TFO, though somewhat contradictory results have been reported. Thus, isothermal titration experiments indicated a significantly less unfavorable entropic contribution to the triplex formation with an LNA TFO, but also some contribution from a favorable enthalpic contribution [38]. A recent study based on surface plasmon resonance experiments indicated similarly an entropic gain but also an enthalpic loss compared to unmodified TFOs [39]. Nevertheless, it seems that the favorable preorganization of the LNA strands for triplex formation, rather than increased stacking, is the key factor.

LNA has also been examined as a building block to stabilize quadruplexes that are structurally similar to those found in human telomere sequences. Single incorporations of LNA monomers into an intramolecular 5′-$(G_4T_4)_3T_3$ quadruplex led to small thermal destabilizations and in some cases to a change in the thermodynamically most preferred structure [40]. Thus, as LNA monomers have been reported to prefer *anti* glycosidic conformations, their incorporation into positions in which a *syn* conformation is preferred leads to a geometry change of the quadruplex. A fully modified LNA quadruplex (formed by 5′-TG_3T) was stabilized relative to the corresponding DNA quadruplex [41]. This stabilization was reported to by caused by an entropic effect, and kinetic experiments displayed slower dissociation and faster association for the LNA quadruplex [42]. In another case studying the quadruplex formed by 5′-GGTTGGTGTGGTTGG, a fully LNA-modified quadruplex was not formed at all, whereas LNA-modifications in all non-loop positions led to a destabilized quadruplex [43]. In a recent study of the quadruplexes formed by 5′-TGGGGT, 5′-TG$\mathbf{G}^{L}$G$\mathbf{G}^{L}$T and 5′-T$\mathbf{G}^{L}\mathbf{G}^{L}\mathbf{G}^{L}\mathbf{G}^{L}$T, both the LNA-modified quadruplexes were thermally stabilized and showed only local structural alterations compared to the unmodified quadruplex [44].

6.3 Structural Studies

As mentioned above, an LNA monomer mimics the N-type conformation of an RNA monomer and induces the formation of A-type duplexes, as confirmed by circular dichroism (CD)-spectroscopy and intensive NMR studies. In particular, the latter approach has provided important contributions towards an understanding of the high-affinity hybridization properties of LNA.

In a study of a 9-mer LNA-sequence (5′-C$\mathbf{T}^{L}$GA$\mathbf{T}^{L}$A$\mathbf{T}^{L}$GC), the population of the different unmodified 2′-deoxynucleotides (in N- and S-type conformations) was studied. A strong tendency to adopt N-type conformation was found [45], which indicates a strong conformational steering by the LNA nucleotides on their more flexible 2′-deoxynucleotide neighbors, especially in the 3′-direction. Full structural studies of the above and other LNA sequences in duplexes formed with complementary DNA have shown that the B-type character of a dsDNA duplex decreases by the incorporation of LNA monomers [46, 47]. Thus, the 2′-deoxynucleotides in an LNA-sequence are more or less forced into adapting much larger populations of N-type sugar puckering, whereas the nucleotides of the complementary DNA-sequence display only a slight increase in the population of N-type sugars. In line of this, a strong resemblance of a fully modified LNA : DNA duplex with an unmodified RNA : DNA duplex was found [48]. In other words, LNA is a true RNA-mimic. The conformational changes in an LNA-sequence have been associated with an increased stacking of the nucleobases [45–47].

Related studies of LNA : RNA duplexes demonstrated that a fully modified LNA : RNA duplex adopts an almost canonical A-type duplex structure with a very regular geometry [48]. However, it has been shown that only three LNA monomers in a 9-mer ODN sequence is enough to convert the entire LNA : RNA duplex into an A-form [49, 50]. In other words, the number of modifications reaches a saturation level with respect to structural changes. This is in good agreement with the hybridization experiments showing relatively larger increases in duplex stability with a few non-consecutive incorporations of LNA rather than with fully modified sequences.

Petersen *et al.* suggested that the vital conformational steering of the LNA-monomers on their 3′-neighboring 2′-deoxynucleotides might be due to the 2′-oxygen of LNA altering the charge distribution in the minor groove, thereby facilitating the conformational shift of the neighbor [50]. A recent study on the same LNA : RNA sequence focused on the internal dynamics of the duplex, as studied by ^{13}C NMR relaxation measurements. This study demonstrated a high degree of order in the duplex as compared to unmodified dsRNA or dsDNA duplexes. Thus, the hypothesis of a strong preorganization of the LNA-DNA mixmer strand for an A-type duplex formation is supported [51].

The X-ray crystal structure of an LNA-DNA duplex ([GCGTAT^{L}ACGC]$_2$) has been studied at 1.4 Å resolution [52]. All 2′-deoxynucleotides were found to adopt an N-type conformation similar to that seen for the two LNA-nucleotides. No backbone deviations around the LNA modification were detected, and the 2′-oxygens of the two LNA monomers were shown to form hydrogen bonds to water molecules.

The NMR studies, together with other structural investigations, indicate that many different factors influence the intrinsic power of LNA for inducing extremely thermally stable duplexes. Thus, both entropic and enthalpic contributions are of significance, though apparently only rarely simultaneously. The high order in an LNA : RNA duplex indicates that preorganization is indeed very important, but simultaneously, the same preorganization might induce a more efficient base-stacking. The associations of water molecules to the duplexes seem equally important, perhaps through the LNA 2′-oxygen and its influence on the conformational steering and preorganization of the LNA-strand.

A triplex structure formed by an LNA sequence (5′-TC^{L}TC^{L}TC^{L}TT) with its parallel dsDNA complement has also been investigated using NMR [53]. The dsDNA shows a geometry being an A/B-type intermediate in order to accommodate the LNA-strand in the major groove. The triplex has a regular geometry, which is the same at both pH 5.1 and pH 8.0, though being more thermally stable at the lower pH. The pyrimidines of the duplex and the LNA TFO are close in space, and a spiral-like hydrogen bonding pattern through the triplex, which seems to be an LNA-specific motif, was found. This corresponds well to the thermodynamic experiments indicating preorganization (favorable entropy) and unfavorable stacking (enthalpy) for the triplex formation [39]. The LNA C-monomers were found not to steer their thymidine neighbors into N-type conformation. This is consistent with the fact that fully modified LNA-sequences are not able to form stable triplexes (Table 6.1, entry 20).

The extreme affinity of LNA towards complementary nucleic acids is closely related to the locked N-type conformation of the monomers. Nevertheless, the study of close analogues of LNA and thermodynamic studies of these have confirmed that factors such as the puckering amplitude and hydration are very important for the structure and stability of LNA-containing duplexes (see below).

6.4 Analogues of LNA and their Structural Impact

In order to explore the scope of locked nucleic acids, a series of LNA-analogues has been introduced, all of which contain a 2′-4′-linkage and all being preorganized into an N-type conformation, but with a variation in the constitution of the 2′-4′-linkage (Figure 6.2). Thus, the oxymethylene linkage of LNA has been replaced by amino-methylene (amino-LNA) and thiomethylene linkages (thio-LNA) of the same length, and therefore also with a similar puckering amplitude of the ribose ring [54, 55]. Replacement with longer linkages has been accomplished given a CCO-linkage (ENA) [56, 57], a CCN-linkage (amino-ENA) [58], and a COC-linkage [59, 60]. Recently, a carbocyclic ring with a saturated CCC-linkage or an unsaturated C=CC-linkage has

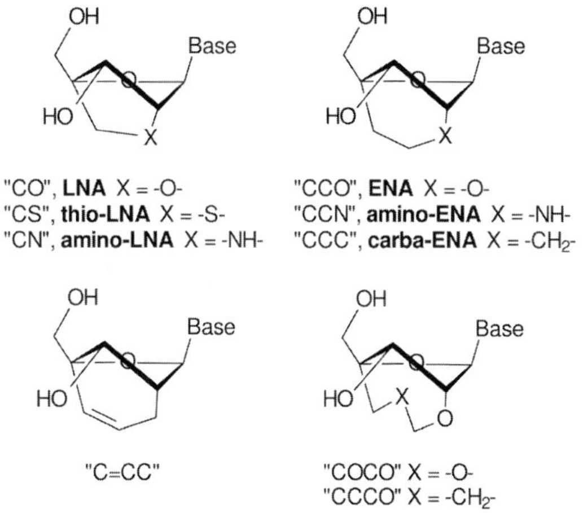

Figure 6.2 LNA-analogues with alternative 2′-4′-linkages.

been introduced [61], and even longer linkages (CCCO [57] and COCO [62]) have been studied. A review of the available NMR and X-ray data, combined with molecular modeling, demonstrates that the puckering amplitude is closely related to the number of atoms in the 2′-4′-linkage [61]. Thus, the puckering of LNA and analogues with two-atom linkages have amplitudes of approximately 57 °, three-atom bridges about 48 °, and four-atom bridges about 38 °. On the other hand, the pseudorotational angle P is in all cases found within a narrow N-type range of 12 to 20 °, with only the C=CC-linkage deviating with $P = 27$ °.

The ability to form stable duplexes with complementary DNA and RNA sequences has been studied for all of these analogues of LNA, but unfortunately in varying sequence contexts. All analogues – except for those with the longest 2′-4′-linkages – were found to induce increased affinity for complementary RNA, whereas more variable results were obtained with complementary DNA. Due to the different sequence contexts, a direct comparison of duplex stabilities in terms of melting temperatures is difficult. However, a general ranking with respect to RNA-affinity of sequences with few and non-consecutive incorporations of the LNA-type monomers in ODNs is suggested to occur as follows:

LNA thio-LNA	>	amino-LNA ENA	>	CCN COC CCC CC=C	>	COCO	>	CCCO
+3° to +9°		+3° to +6°		+2° to +5°		+1° to +2°		–0.5°

Although, clearly, increasing the linker length has a negative effect on the RNA-affinity, the presence of a 2′-heteroatom is also important.

A corresponding general ranking with respect to DNA-affinity of sequences with few and non-consecutive incorporations of the bicyclic monomers in ODNs could be as follows:

LNA thio-LNA amino-LNA	>	ENA	>	COC	>	CCN CCC CC=C COCO CCCO
+2° to +5°		+0.5° to +2°		–0.5° to +1°		–3° to –1°

The length of the linkage thus seems to have a more crucial effect when targeting a DNA complementary strand. The presence of a 2′-heteroatom also seems to be essential, as all analogues lacking this heteroatom (or having four-atom linkers) display a decreased affinity for single-stranded DNA.

It has been proven, by using NMR spectroscopy, that a key feature responsible for an LNA-DNA mixmer strand to form stabilized A-type duplexes is the ability of LNA-nucleotides to conformationally steer the neighboring 2′-deoxynucleotides into N-type conformation. Bases on the results of hybridization experiments and on CD investigations with several of the analogues, including LNA, ENA and the CCC and C=CC-linked bicyclic nucleosides, this steering is directly linked to the presence of a 2′-oxygen atom. In other words, the conformational steering is decreased by two factors: (1) a decreased puckering amplitude; and (2) an increased hydrophobicity. This result is in line with the NMR experiments which suggested that preorganization of the LNA sequences is associated with the steering of neighbors through a charge distribution in the minor groove from the 2′-oxygen [50]. The final analogue that would enlighten this even further, the carba-analogue of LNA ("CC" in terms of Figure 6.2), awaits synthetic realization.

In summary, LNA shows the most prominent affinity-enhancing effect among all of the above-mentioned LNA-type analogues with different 2′-4′-linkages. Nevertheless, based on the few studies reported to date, thio-LNA appears to display the same hybridization behavior as LNA [54]. Also, amino-LNA and ENA demonstrate comparable – though slightly lower – degrees of duplex stabilization. On the other hand, ENA has gained interest in TFOs as fully modified ENA sequences, in contrast to fully modified LNA sequences, form stable triplexes [63]. This has been connected with the smaller puckering amplitude of ENA and the slightly more flexible three-atom 2′-4′-linkage [53].

Surprising is the power in nucleic acid recognition disclosed for stereoisomers of LNA (Figure 6.3). These have shown that the strong N-type conformation and hydrophilic nature can be transformed to other systems [64], and most intriguing is that α-L-LNA demonstrates a very high affinity for both complementary DNA and RNA, and forms duplexes that are only slightly less stable than the corresponding LNA-modified duplexes [65]. Notably, α-L-LNA monomers are also affinity-enhancing when mixed with DNA. Also of interest is that TFO-forming properties have been

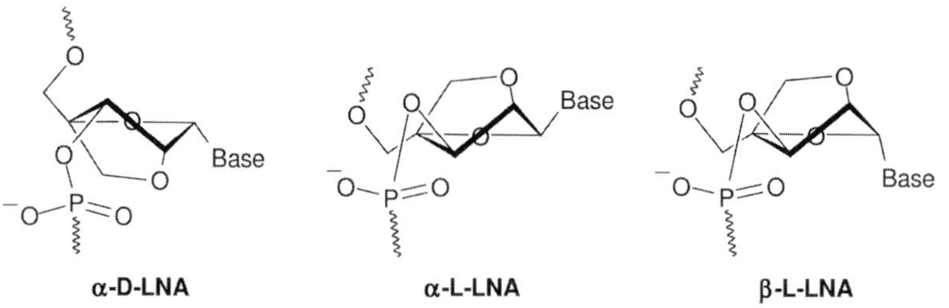

Figure 6.3 Stereoisomers of LNA.

identified for α-L-LNA [66]. Structural studies have revealed that, whereas LNA in structural terms is a true RNA-mimic, α-L-LNA can be considered as a DNA-mimic due to the B-type mimicking properties of α-L-LNA sequences [67]. The strong affinity for complementary RNA found for α-D-LNA forming a parallel duplex was more expected, although a structure which is seemingly different from other parallel duplexes has been indicated [68]. As a consequence of α-L-LNA being a DNA-mimic, β-L-LNA was investigated as an α-DNA mimic, and indeed, stable parallel duplexes with both complementary RNA and DNA were demonstrated in a mirror-image study [69].

6.5
LNA as Potential Therapeutics

Currently, LNA oligonucleotide sequences are commercially available, and have been investigated in a large number of biological studies. These have been focused on gene silencing experiments following different RNA-targeting approaches such as steric blocking antisense, RNase H-mediated RNA-cleavage, DNAzymes and siRNA studies. In addition, the cellular delivery, physiological stability and toxicity of LNA oligonucleotides have also been studied [4–6].

The issue of physiological stability mostly concerns the potential degradation of oligonucleotides by nucleases. A fully modified LNA sequence has been reported to be fully resistant towards the 3′-exonuclease SVPDE [70], whereas only minor protection against the same enzyme is obtained with one LNA monomer in the 3′-end or in the middle of a sequence. In general, LNA strands were found to be more resistant towards nucleolytic attack, and much more stable in serum than the corresponding DNA strands [71]. Kurreck *et al.* also found that end-blocked sequences – that is, LNA-DNA gapmers such as entry 16 in Table 6.1 – display a high stability in human serum compared to similar 2′-OMe modified sequences [32]. Another study showed that two terminal LNA monomers provided a significant protection against a Bal-31 exonucleolytic degradation [72]. Recently, LNA incorporations have been found significantly to improve not only thermal stability but also blood retention, plasma stability and tumor uptake of aptamers when compared to a range of other chemical modifications [73]. LNA oligonucleotides can be delivered into cells using standard cationic transfection agents [4, 71]. Recently, a rapid nuclear uptake of cholesterol–LNA conjugates using the membrane-damaging protein streptolysin-O has been proven, using fluorescence microscopy [74].

The central questions in the antisense field [10] include the necessity for RNase H recruitment in relation to the alternative steric block-based approach, and the potential of these two strategies versus the siRNA approach. Briefly, LNA has proved to be enabling (or at least useful) for all of these approaches due to its biostability and universal ability to increase RNA-affinity as both fully modified LNA and as chimeric constructs. It is clear, however, that LNA-nucleotides cannot induce the cleavage of target RNA by RNase H. The reason for this is the A-type duplex conformation induced by LNA nucleotides that is incompatible with the intermediate A/B-type duplex formed between natural DNA and RNA, and that is required by RNAse H for target RNA cleavage. As LNA directs its near 2′-deoxynucleotide neighbors into N-type conformations, and thereby induces a local A-type duplex conformation, not even an LNA-DNA mixmer approach (as in Table 6.1, entries 13 and 15) is sufficient [31, 32]. The RNase H-mediated cleavage of RNA induced by LNA sequences is therefore dependent on a gap of DNA nucleotides in these sequences. Thus, LNA-DNA-LNA gapmers (such as entries 14 and 16 in Table 6.1), with a gap size of at least five but preferably seven or more 2′-deoxynucleotides, can elicit the RNase H-mediated cleavage of a complementary RNA-sequence. Accordingly, a study by Kurreck *et al.* showed that a gap of six DNA nucleotides in an LNA sequence is necessary to give 65% RNase H activity, whereas a gap of seven DNA nucleotides allows complete RNase H degradation of the RNA-complement [32]. A study by Frieden *et al.* confirmed that a gap of seven or more DNA nucleotides is optimal [75].

As mentioned in Section 6.1, the antisense approach has been the central motivation behind the field of nucleic acid chemistry, and the emergence of LNA has opened a wide range of new opportunities within the field. As LNA antisense studies have been reviewed on several occasions [4–6], the present chapter is focused on the most recent results.

Rekasi *et al.* studied the suppression of serotonin *N*-acetyltransferase transcription and the subsequent melatonin secretion by LNA antisense oligonucleotides [76]. Both, an LNA-DNA-LNA gapmer designed for RNase-H activation and an LNA-DNA mixmer sequence designed for the steric block approach, were applied, and both were found to induce significant suppression of the transcript [76]. Lennox *et al.* studied a range of different chemically modified antisense oligonucleotides targeting a *Xenopus laevis* survivin gene. Among the oligonucleotides microinjected into *Xenopus* embryos [77], an LNA-DNA-LNA gapmer with the DNA part comprising phosphorthioate linkages was the most potent in reducing gene expression at 40 nM compared to oligonucleotides containing 2′-OMe modifications, phosphorthioates only, neutral methoxyethyl phosphoramidates or cationic *N,N*-dimethylethylenediamine phosphoramidates. Only at 400 nM was the latter modification as effective as LNA. As expected, the LNA sequence also demonstrated the highest thermal affinity for complementary RNA, and the best mismatch discrimination, the latter fact translating into the highest degree of specificity [77].

In a direct comparative study, an LNA gapmer antisense oligonucleotide downregulated two adoptosis inhibitors that are overexpressed in human tumors, more effectively than the isosequential 2′-MOE oligonucleotide [78]. Both sequences were fully phosphorthioate modified. Another study compared LNA directly to PNA with respect to blocking translation of the so-called internal ribosomal entry site of hepatitis C virus (HCV). Different regions of the site were targeted by an LNA-DNA-LNA gapmer, four different LNA-DNA mixmers, and a range of PNA sequences, all of which were introduced into HCV-infected cells by lipid-mediated transfection [79]. Both, the LNA and PNA approaches revealed inhibition of translation, with EC_{50} values of 50 to 150 nM.

A recent antisense study by the group of F. Baas compared four isosequential antisense oligonucleotides containing LNA with three LNA-analogues, namely thio-LNA, amino-LNA and α-L-LNA, respectively (see Figure 6.2) [80]. The four sequences were all-phosphorthioate-modified LNA-DNA-LNA gapmers. The thermal stability of the duplexes formed with complementary DNA was almost the same for all four oligonucleotides (amino-LNA 66.6 °C, LNA 69.4 °C, α-L-LNA, 69.5 °C, and thio-LNA 71.1 °C), and serum stability was also high in all cases. The four constructs were tested in cancer-cell cultures and in a mouse model. An efficient knockdown of H-Ras mRNA was observed for all four antisense oligonucleotides *in vitro* at concentrations below 5 nM, but the α-L-LNA sequence was found to be most potent. Both, α-L-LNA and LNA mediated efficient tumor-growth inhibition, and were also non-toxic in mice at the tested doses.

LNA oligonucleotides have proven useful as inhibitors of gene expression by other approaches than traditional single-stranded antisense. Crinelly *et al.* studied double-stranded oligonucleotides as decoy molecules for transcription factors [72, 81]. With LNA-modifications, resistance towards exo- and endonucleolytic degradation was obtained, but insertion of internal LNA monomers led to a decreased affinity, most likely because of the RNA mimicking structure induced by the LNA monomers. This problem was partly solved by introducing α-L-LNA for internal insertion.

LNA-sequences have been used to show that miRNA mediates an antiviral response in human cells [82]. Darfeuille *et al.* used short LNA-DNA chimeras as hairpin aptamers against HIV-I TAR RNA [83]. One of these was shown to bind in the same low nanomolar range as the original RNA aptamer, but also displayed an improved nuclease resistance. As mentioned above, high tumor uptake of other LNA aptamers has been found [73]. Andrieu-Soler *et al.* used an LNA-DNA-LNA gapmer approach for targeted mutagenesis with 25-mer ODNs with one or two LNA-monomers in each terminal to induce nuclease resistance. The LNA sequences were found to be superior to corresponding phosphorthioate sequences [84].

The specific cleavage of target RNA-strands has been achieved using catalytically active DNA molecules called DNAzymes. The so-called 10-23 motif was found by *in-vitro* selection [85] to be a specific DNAzyme, and the introduction of two LNA nucleotides into each of the two binding arms gave an LNAzyme with an enhanced RNA-cleavage [86]. Recently, the 8-17 motif DNAzyme has also been modified to give an 8-17 LNAzyme with improved RNA-cleaving properties [87]. The binding arm length and content of LNA or α-L-LNA monomers were varied. The fastest cleavage reaction was detected with two α-L-LNA nucleotides in each binding arm. The LNAzyme design was further explored by Fahmy and Khachigian [88], who showed a

>50% inhibition of serum-inducible smooth muscle cell proliferation under conditions where the DNAzyme control showed no inhibition. LNAzymes have been used for allele-specific inhibition – that is, targeting a single-nucleotide polymorphism (SNP) in RNA polymerase II [89]. However, the LNAzyme approach was less efficient than a corresponding RNase-H recruiting LNA antisense approach.

RNA interference (RNAi), in which double-stranded RNA induces the degradation of target RNA [90], has attracted much attention as a novel means of mediating potent gene silencing. A crucial step in the RNAi mechanism is the generation of short interfering RNAs (siRNAs; double-stranded RNAs that are about 19 to 22 nt long). siRNAs are themselves candidate molecules for the incorporation of modified nucleotides for improved biostability and RNA targeting [91]. LNA-modified siRNA constructs have been shown to be compatible with RNAi. The exact positioning and overall number of LNA monomers seem to be essential when optimizing LNA-containing siRNA constructs. A systematic study on LNA-modified siRNAs showed that LNA 3′-end modifications substantially enhance the serum half-life when evaluated relative to the unmodified siRNAs [92]. Most importantly, the results of this study also revealed that LNA-modified sense strands reduce sequence-related off-target effects. It was reported that LNA-modifications led to an improved efficacy at various RNA motifs relevant to targeting the SARS-CoV virus. In short, these results underline the promise of LNA-modified siRNA in relation to realizing the promises of RNAi for therapeutic applications [91, 92].

As shown above, LNA monomers as building blocks in TFOs significantly increase the affinity for dsDNA when mixmers of alternating LNA and DNA nucleotides are used [36, 37]. An LNA TFO has been found to produce an antagonistic inhibition of binding of NF-κB to the target dsDNA at neutral pH [36]. Recently, LNA TFOs have been found to inhibit transcription both *in vitro* and in cells [39].

In general, LNA sequences have been found to be devoid of toxicity at the dose levels tested. Not surprisingly, however, immune stimulation has been identified for LNA-phosphorthioate chimeras, as phosphorthioate sequences are known to mediate this effect. Potent immune stimulation has especially been obtained by phosphorthioate ODNs containing CpG dinucleotides [93]. The introduction of LNA monomers, either in the dinucleotide CpG moiety or in the flanking ends, reduced the level of immune modulation, which suggests that LNA might be useful for alleviating the detrimental immune response found with higher doses of some antisense oligonucleotides with phosphorthioate linkages. Recently, based on studies in mice, a higher level of liver toxicity was claimed for LNA-DNA-LNA phosphorothioate gapmers relative to the corresponding MOE-DNA-MOE phosphorthioate gapmers [94]. Nonetheless, this study may be considered as supportive for LNA antisense applications, as a very efficient gene silencing was achieved at the lower, non-toxic doses, with toxicities being base sequence-dependent. At this point, it would be naïve to draw wide-ranging conclusions based on studies in mice. Rather, with regards to possible therapeutic applications in humans, a comparative study should have been conducted in primates. Nonetheless, LNA-DNA-LNA gapmers are currently undergoing Phase I/II clinical trails, and have therefore successfully passed through preclinical toxicity studies [95].

6.6 LNA-Probes

The properties of LNA for targeting both RNA and DNA clearly show LNA to be an unprecedented tool for generating superior nucleic acid probes. Absolutely crucial in this context is the freedom in design due to the excellent cooperativity between LNA monomers and, for example, neighboring 2′-deoxynucleotides. In other words, the high-affinity hybridization of an oligonucleotide can be programmed by introducing LNA monomers at selected positions in a DNA (or RNA) sequence. This is an appealing option when designing arrays containing probes of approximately the same length. In this context, it should be noted that Exiqon A/S can now probe for the entire human genome using an array of only 96 short LNA-sequences [96]. As LNA-based probes have been used intensively for a wide range of purposes, comprehensive reviews have been published on the subject [4, 97]. Nonetheless, a brief update on recent results is provided in the following section.

LNA has been applied as a capture probe for the direct isolation of RNA-sequences containing polyA-tracts (from cell extracts). The poly-T probe, when using a 20-mer alternating thymidine/LNA-thymine probe, was 30- to 50-fold more efficient than an unmodified poly-T probe, and was also efficient in a low-salt buffer [98]. A similar approach (a 20-mer polyT containing seven LNA nucleotides) was used successfully to detect polyA-RNA accumulation within the nucleus/nucleolus of wild-type cells [99]. Other fluorescent LNA-based probes were used to detect RNAs in fixed yeast cells.

In a recent important study the *in-situ* detection of micro-RNAs in the zebrafish embryo was achieved using LNA-modified DNA-probes [100, 101]. Furthermore, previously unidentified expression of miRNAs in mouse embryos was detected [101]. The optimal RNA-detection was found at hybridization temperatures 20–25 °C below the melting temperature of the probe, and it was possible to use short probes (~14-mers) due to the high binding affinities induced by the LNA monomers. A microarray platform known as "miChip", and based on T_m-normalized LNA-modified probes, has been recently developed for the detection of miRNAs [102]. Efficient LNA-probes at the 8-mer length have also been developed [103]. RNA-transcripts, into which pseudocomplementary bases have been incorporated, have a smaller propensity to form secondary structures and therefore unrestricted accessibility to the short LNA probes, in contrast to the longer 15-mer RNA-probes.

A thorough study of mismatch discrimination by LNA probes based on melting temperatures was recently reported, and this has led to guidelines for the design of LNA probes [104]. For this, three LNA nucleotides with the mismatch site in the center should be placed centrally in short probes. The practical use of LNA-probes to detect SNPs has also been reported [6]. For example, Ugozzoli *et al.* reported the detection of SNPs in real-time PCR, and an improved specificity of LNA probes in assays for allelic discrimination [105]. Goldenberg *et al.* showed that LNA-probes could be used as hybridization/FRET probes for detection and quantification (by real-time PCR) [106]. Hummelshøj *et al.* used LNA as a tool for inhibiting the PCR amplification of contaminating DNA sequences, thereby enhancing the specificity of real-time PCR [107]. Recently, LNA-modified primers were used in developing assays for the two alleles of an SNP-specific PCR applied to identify *Chlamydia pneumoniae* genotypes [108].

LNA-based molecular beacons as recognition probes have also been reported [109]. Here, the main advantages of using LNA are high thermostability, enhanced selectivity and nuclease resistance. Hence, it seems that LNA-based molecular beacons may be used not only to detect SNPs but also to increase the sensitivity of PCR systems [110].

6.7
Concluding Remarks

LNA has, within a very short time, found widespread use in the development of therapeutics and the optimization of probes. The central feature is the high-affinity recognition of complementary nucleic acids, whether these are RNA-sequences, single-stranded DNA-sequences, or double-stranded DNA, while the high flexibility in sequence design is due to the valuable cooperativity between the LNA-monomers and the unmodified (2′-deoxy)nucleotides within the same oligonucleotide chimeric sequence. These properties allow for the design of relatively short LNAs with a power of recognition that can otherwise be obtained only with much longer sequences. This in turn increases the specificity of recognition, thereby making possible the creation of highly efficient probes for SNP detection. The same flexibility that allows both gapmer and mixmer designs is also pivotal in the development of therapeutic sequences. Thus, RNA-targeting sequences can be designed for RNase H-mediated cleavage of the target using a gapmer approach, or with even higher affinity and higher nucleolytic stability by using a mixmer approach. Thus, the combination of high affinity, high nuclease resistance and low toxicity, bode well for the future introduction of LNA-based therapeutics.

These properties are, in the main, shared by the close analogues of LNA mentioned above (especially thio-LNA, amino-LNA and ENA), and complemented by the interesting LNA-stereoisomer α-L-LNA. However, further studies are required in order to realize the scope of these analogues with regards to therapeutic use. A further benefit of the original LNA is its availability, which is based on a remarkably simple chemical synthesis that allows the preparation of LNA-monomers on a kilogram scale. Indeed, today LNA-sequences are available commercially and barely more expensive than RNA-sequences. Whilst the preparation of amino-LNA, thio-LNA and α-L-LNA is likewise straightforward on the small scale, the preparation of analogues with longer 2′-4′-linkages appears more complicated, and potential difficulties for up-scaling might be anticipated.

With regards to the future, one aspect of LNA not yet mentioned here is an introduction into applied nanobiosciences [111], as high-affinity recognition, coupled with flexibility and predictability in sequence design, will undoubtedly create new opportunities for nucleic acid-based nanoscale engineering. Moreover, amino-LNA may serve as an excellent branching point for decorating the LNA sequences with various entities, and several relevant studies have already been conducted [112, 113]. Elsewhere in the field of nanotechnology, LNA sequences have been used to block a polymerase extension reaction in the development of a universal surface DNA computer [114].

References

1 Singh, S. K., Nielsen, P., Koshkin, A. A., Wengel, J. *Chem. Commun.* **1998**, *4*, 455–456.
2 Koshkin, A. A., Singh, S. K., Nielsen, P., Rajwanshi, V. K., Kumar, R., Meldgaard, M., Olsen, C. E. Wengel, J. *Tetrahedron* **1998**, *54*, 3607–3630.
3 Obika, S., Nanbu, D., Hari, Y., Andoh, J., Morio, K., Doi, T. Imanishi, T. *Tetrahedron Lett.* **1998**, *39*, 5401–5404.
4 Petersen, M. Wengel, J. *TrendsBiotechnol.* **2003**, *21*, 74–81.
5 Jepsen, J. S. Wengel, J. *Curr.Opin. Drug Discovery Dev.* **2004**, *7*, 188–194.
6 Vester, B. Wengel, J. *Biochemistry* **2004**, *43*, 13233–13241.
7 Kool, E. T. *Chem. Rev.* **1997**, *97*, 1473–1487.
8 Herdewijn, P. *Biochim. Biophys. Acta* **1999**, *1489*, 167–179.
9 Leumann, C. J. *Bioorg. Med. Chem.* **2002**, *10*, 841–854.
10 Kurreck, J. *Eur. J. Biochem.* **2003**, *270*, 1628–1644.
11 Aboul-Fadl, T. *Curr. Med. Chem.* **2005**, *12*, 2193–2214.
12 Saenger, W. *Principles of Nucleic Acid Structure*, **1984**, Springer, New York.
13 Manoharan, M. *Biochim. Biophys. Acta* **1999**, *1489*, 117–130.
14 Martin, P. *Helv. Chim. Acta* **1995**, 486–504.
15 Van Aerschot, A., Verheggen, I., Hendrix, C. Herdewijn, P. *Angew. Chem. Int. Ed.* **1995**, *34*, 1338–1339.
16 Gryaznov, S. Chen, J.-K. *J. Am. Chem. Soc.* **1994**, *116*, 3143–3144.
17 Tarköy, M., Bolli, M., Schweizer, B. Leumann, C. *Helv. Chim. Acta* **1993**, *76*, 481–510.
18 Steffens, R. Leumann, C. J. *J.Am. Chem. Soc.* **1997**, *119*, 11548–11549.
19 Altmann, K.-H., Kesselring, R., Francotte, E. Rihs, G. *Tetrahedron Lett.* **1994**, *35*, 2331–2334.
20 Altmann, K.-H., Imwinkelried, R., Kesselring, R. Rihs, G. *Tetrahedron Lett.* **1994**, *35*, 7625–7628.
21 Marquez, V. E., Siddiqui, M. A., Ezzitouni, A., Russ, P., Wang, J., Wagner, R. W. Matteucci, M. D. *J. Med. Chem.* **1996**, *39*, 3739–3747.
22 Marquez, V. E., Hughes, S. H., Sei, S. Agbaria, R. *Antiviral Res.* **2006**, *71*, 268–275.
23 Altona, C. Sundaralingam, M. *J. Am. Chem. Soc.* **1972**, *94*, 8205–8212.
24 Obika, S., Nanbu, D., Hari, Y., Morio, K., In, Y., Ishida, T. Imanishi, T. *Tetrahedron Lett.* **1997**, *38*, 8735–8738.
25 Wengel, J. *Acc. Chem. Res.* **1999**, *32*, 301–310.
26 Singh, S. K. Wengel, J. *Chem.Commun.* **1998**, 1247–1248.
27 Koshkin, A. A., Nielsen, P., Meldgaard, M., Rajwanshi, V. K., Singh, S. K. Wengel, J. *J. Am. Chem. Soc.* **1998**, *120*, 13252–13253.
28 McTigue, P. M., Peterson, R. J. Kahn, J. D. *Biochemistry* **2004**, *43*, 5388–5405.
29 Kaur, H., Arora, A., Wengel, J. Maiti,S. *Biochemistry* **2006**, *45*, 7347–7355.
30 Kierzek, E., Ciesielska, A., Pasternak, K., Mathews, D. H., Turner, D. H. Kierzek, R. *Nucleic Acids Res.* **2005**, *16*, 5082–5093.
31 Braasch, D. A., Liu, Y. Corey, D. R. *Nucleic Acids Res.* **2002**, *30*, 5160–5167.
32 Kurreck, J., Wyszko, E., Gillen, C. Erdmann, V. A. *Nucleic Acids Res.* **2002**, *30*, 1911–1918.
33 Christensen, U., Jacobsen, N., Rajwanshi, V. K., Wengel, J. Koch, T. *Biochem. J.* **2001**, *354*, 481–484.
34 Ng, P.-S. Bergstrom, D. E. *Nano Lett.* **2005**, *5*, 107–111.
35 Buchini, S. Leumann, C. J. *Curr. Opin. Chem. Biol.* **2003**, *7*, 717–726.
36 Obika, S., Uneda, T., Sugimoto, T., Nanbu, D., Minami, T., Doi, T. Imanishi, T. *Bioorg. Med. Chem.* **2001**, *9*, 1001–1011.
37 Sun, B. W., Babu, B. R., Sørensen, M. D., Zakrzewska, K., Wengel, J. Sun, J. S. *Biochemistry* **2004**, *43*, 4160–4169.
38 Torigoe, H., Hari, Y., Sekiguchi, M., Obika, S. Imanishi, T. *J. Biol. Chem.* **2001**, *276*, 2354–2360.

39 Brunett, E., Alberti, P., Perrouault, L., Babu, R., Wengel, J. Giovannangeli, C. *J. Biol. Chem.* **2005**, *280*, 20076–20085.

40 Dominick, P. K. Jarstfer, M. B. *J. Am. Chem. Soc.* **2004**, *126*, 5050–5051.

41 Randazzo, A., Esposito, V., Ohlenschläger, O., Ramachandran, R. Mayol, L. *Nucleic Acids Res.* **2004**, *32*, 3083–3092.

42 Petraccone, L., Erra, E., Randazzo, A. Giancola, C. *Biopolymers* **2006**, *83*, 584–594.

43 Randazzo, A., Esposito, V., Ohlenschläger, O., Ramachandran, R., Virgilio, A. Mayol, L. *Nucleosides Nucleotides Nucleic Acids* **2005**, *24*, 795–800.

44 Nielsen, J. T., Arar, K. Petersen, M. *Nucleic Acids Res.* **2006**, *34*, 2006–2014.

45 Petersen, M., Nielsen, C. B., Nielsen, K. E., Jensen, G. A., Bondensgaard, K., Singh, S. K., Rajwanshi, V. K., Koshkin, A. A., Dahl, B. M., Wengel, J. Jacobsen, J. P. *J. Mol. Recognit.* **2000**, *13*, 44–53.

46 Nielsen, K. E., Singh, S. K., Wengel, J. Jacobsen, J. P. *Bioconjugate Chem.* **2000**, *11*, 228–238.

47 Jensen, G. A., Singh, S. K., Kumar, R., Wengel, J. Jacobsen, J. P. *J. Chem. Soc., Perkin Trans.* **2001**, *2*, 1224–1232.

48 Nielsen, K. E., Rasmussen, J., Kumar, R., Wengel, J., Jacobsen, J. P. Petersen, M. *Bioconjugate Chem.* **2004**, *15*, 449–457.

49 Bondensgaard, K., Petersen, M., Singh, S. K., Rajwanshi, V. K., Kumar, R., Wengel, J. Jacobsen, J. P. *Chem.Eur. J.* **2000**, *6*, 2687–2695.

50 Petersen, M., Bondensgaard, K., Wengel, J. Jacobsen, J. P. *J. Am. Chem. Soc.* **2002**, *124*, 5974–5982.

51 Nielsen, K. E. Spielmann, H. P. *J. Am. Chem. Soc.* **2005**, *127*, 15273–15282.

52 Egli, M., Minasov, G., Teplova, M., Kumar, R. Wengel, J. *Chem.Commun.* **2001**, 651–652.

53 Sørensen, J. J., Nielsen, J. T. Petersen, M. *Nucleic Acids Res.* **2004**, *20*, 1–8.

54 Kumar, R., Singh, S. K., Koshkin, A. A., Rajwanshi, V. K., Meldgaard, M. Wengel, J. *Biorg. Med. Chem. Lett.* **1998**, *8*, 2219–2222.

55 Singh, S. K., Kumar, R. Wengel, J. *J. Org. Chem.* **1998**, *63*, 10035–10039.

56 Morita, K., Hasegawa, C., Kaneko, M., Tsutsumi, S., Sone, J., Ishikawa, T., Imanishi, T. Koizumi, M. *Bioorg. Med. Chem. Lett.* **2002**, *12*, 73–76.

57 Morita, K., Takagi, M., Hasegawa, C., Kaneko, M., Tsutsumi, S., Sone, J., Ishikawa, T., Imanishi, T. Koizumi, M. *Bioorg. Med. Chem.* **2003**, *11*, 2211–2226.

58 Varghese, O. P., Barman, J., Pathmasiri, W., Plashkevych, O., Honcharenko, D. Chattopadhyaya, J. *J. Am. Chem. Soc.* **2006**, *128*, 15173–15187.

59 Wang, G., Girardet, J.-L. Gunic, E. *Tetrahedron* **1999**, *55*, 7707–7724.

60 Wang, G., Gunic, E., Girardet, J.-L. Stoisavljevic, V. *Bioorg. Med. Chem. Lett.* **1999**, *9*, 1147–1150.

61 Albæk, N., Petersen, M. Nielsen, P. *J. Org. Chem.* **2006**, *71*, 7731–7740.

62 Hari, Y., Obika, S., Ohnishi, R., Eguchi,K., Osaki, T., Ohishi, H. Imanishi, T. *Bioorg. Med. Chem.* **2006**, *14*, 1029–1038.

63 Koizumi, M., Morita, K., Daigo, M., Tsutsumi, S., Abe, K., Obika, S. Imanishi, T. *Nucleic Acids Res.* **2003**, *31*, 3267–3273.

64 Rajwanshi, V. K., Håkansson, A. E., Sørensen, M. D., Pitsch, S., Singh, S. K., Kumar, R., Nielsen, P. Wengel, J. *Angew. Chem. Int. Ed.* **2000**, *39*, 1656–1659.

65 Sørensen, M. D., Kvaernø L., Bryld, T., Håkansson, A. E., Verbeure, B., Gaubert, G., Herdewijn, P. Wengel, J. *J.Am. Chem. Soc.* **2004**, *124*, 2164–2176.

66 Kumar, N., Nielsen, K. E., Maiti, S. Petersen, M. *J. Am. Chem. Soc.* **2006**, *128*, 14–15.

67 Petersen, M., Håkansson, A. E., Wengel, J. Jacobsen, J. P. *J. Am. Chem. Soc.* **2001**, *123*, 7431–7432.

68 Nielsen, P., Christensen, N. K. Dalskov, J. K. *Chem. Eur. J.* **2002**, *8*, 712–722.

69 Christensen, N. K., Bryld, T., Sørensen, M. D., Arar, K., Wengel, J. Nielsen, P. *Chem. Commun.* **2004**, 282–283.

70 Frieden, M., Hansen, H. F. Koch, T. *Nucleosides Nucleotides Nucleic Acids* **2003**, *22*, 1041–1043.

71 Wahlestedt, C., Salmi, P., Good, L., Kela, J., Johnsson, T., Hökfelt, T., Broberger, C., Porreca, F., Lai, J., Ren, K., Ossipov, M., Koshkin, A., Jakobsen, N., Skouv, J., Oerum, H., Jacobsen, M. H. Wengel, J. *Proc. Natl Acad. Sci. USA* **2000**, *97,* 5633–5638.

72 Crinelli, R., Bianchi, M., Gentilini, L. Magnani, M. *Nucleic Acids Res.* **2002**, *30,* 2435–2443.

73 Schmidt, K. S., Borkowski, S., Kurreck, J., Stephens, A. W., Bald, R., Hecht, M., Friebe, M., Dinkelborg, L. Erdmann, V. A. *Nucleic Acids Res.* **2004**, *32,* 5757–5765.

74 Holasová S., Mojzisek, M., Buncek, M., Vokurková D., Radilová H., Safárová M., Cervinka, M. Haluza, R. *Mol.Cell. Biochem.* **2005**, *276,* 61–69.

75 Frieden, M., Christensen, S. M., Mikkelsen, N. D., Rosenbohm, C., Thrue, C. A., Westergaard, M., Hansen,H. F., Orum, H. Koch, T. *Nucleic Acids Res.* **2003**, *31,* 6365–6372.

76 Rekasi, Z., Horvath, R. A., Klausz, B., Nagy, E. Toller, G. L. *Mol. Cell. Endocrinol.* **2006**, *249,* 84–91.

77 Lennox, K. A., Sabel, J. L., Johnson, M. J., Moreira, B. G., Fletcher, C. A., Rose, S. D., Behlke, M. A., Laikhter, A. L., Walder, J. A. Dagle, J. M. *Oligonucleotides* **2006**, *16,* 26–42.

78 Simões-Wüst, A. P., Hopkins-Donaldson, S., Sigrist, B., Belyanskaya, L., Stahel, R. A. Zangemeister-Wittke, U. *Oligonucleotides* **2004**, *14,* 199–209.

79 Nulf, C. J. Corey, D. *Nucleic Acids Res.* **2004**, *32,* 3792–3798.

80 Fluiter, K., Frieden, M., Vreijling, J., Rosenbohm, C., De Wissel, M. B., Christensen, S. M., Koch, T., Ørum, H. Baas, F. *ChemBioChem* **2005**, *6,* 1104–1109.

81 Crinelli, R., Bianchi, M., Gentilini, L., Palma, L., Sørensen, M. D., Bryld, T., Babu, R. B., Arar, K., Wengel, J. Magnani, M. *Nucleic Acids Res.* **2004**, *32,* 1874–1885.

82 Lecelillier, C.-H., Dunoyer, P., Arar, K., Lehmann-Che, J., Eyquem, S., Himber, C., Saïb, A. Voinnet, O. *Science* **2005**, *308,* 557–560.

83 Darfeuille, F., Hansen, J. B., Orum, H., Di Primo, C. Toulmé, J.-J. *Nucleic Acids Res.* **2004**, *32,* 3101–3107.

84 Andrieu-Soler, C., Casas, M., Faussat, A.-M., Gandolphe, C., Doat, M., Tempé, D., Giovannangeli, C., Behar-Cohen, F. Concordet, J.-P. *Nucleic Acids Res.* **2005**, *33,* 3733–3742.

85 Santoro, S. W. Joyce, G. F. *Proc.Natl. Acad. Sci. USA* **1997**, *94,* 4262–4266.

86 Vester, B., Lundberg, L. B., Sørensen, M. D., Babu, B. R., Douthwaite, S. Wengel, J. *J. Am. Chem. Soc.* **2002**, *124,* 13682–13683.

87 Vester, B., Hansen, L. H., Lundberg, L. B., Babu, B. R., Sørensen, M. D., Wengel, J. Douthwaite, S. *BMC Mol. Biol.* **2006**, *7,* 19.

88 Fahmy, R. G. Khachigian, L. M. *Nucleic Acids Res.* **2004**, *32,* 2281–2285.

89 Fluiter, K., Frieden, M., Vreijling, J., Koch, T. Baas, F. *Oligonucleotides* **2005**, *15,* 246–254.

90 Elabshir, S. M., Harborth, J., Lendeckel, W., Yalcin, A., Weber, K. Tuschl, T. *Nature* **2001**, *411,* 494–498.

91 Braasch, D. A., Jensen, S., Liu, Y., Kaur, K., Arar, K., White, M. A. Corey, D. R. *Biochemistry* **2003**, *8,* 7967–7975.

92 Elmén, J., Thonberg, H., Ljungberg, K., Frieden, M., Westergard, M., Xu, Y., Wahren, B., Liang, Z., Ørum, H., Koch, T. Wahlestedt, C. *Nucleic Acids Res.* **2005**, *33,* 439–447.

93 Vollmer, J., Jepsen, J. S., Uhlmann, E., Schetter, C., Jurk, M., Wader, T., Wüllner, M. Krieg, A. M. *Oligonucleotides* **2004**, *14,* 23–31.

94 Swayze, E. E., Siwkowski, A. M., Wancewicz, E. V., Migawa, M. T., Wyrzykiewicz, T. K., Hung, G., Monia, B. P. Bennett, C. F. *Nucleic Acids Res.* **2007**, *35,* 687–700.

95 http://www.santaris.com

96 http://www.exiqon.com

97 Mouritzen, P., Toftgaard Nielsen, A., Pfundheller, H. M., Choleva, Y., Kongsbak, L. Møller, S. *ExpertRev. Mol. Diagn.* **2003**, *3,* 89–100.

98 Jacobsen, N., Nielsen, P. S., Jaffares, C. C., Eriksen, J., Ohlsson, H., Arctander, P.

Kauppinen, S. *Nucleic Acids Res.* **2004**, *32*, e64.

99 Thomsen, R., Nielsen, P. S. Jensen, T. H. *RNA* **2005**, *11*, 1745–1748.

100 Wienholds, E., Kloosterman, W. P., Miska, E., Alvarez-Saavedra, E., Berezikov, E., deBruijn, E., Horvitz, H. R., Kauppinen, S. Plasterk, R. H. A. *Science*, *205*, 309, 310–311.

101 Kloosterman, W. P., Wienholds, E., DeBruijn, E., Kauppinen, S. Plasterk, R. H. A. *Nature Methods* **2006**, *3*, 27–29.

102 Castoldi, M., Schmidt, S., Benes, V., Noerholm, M., Kulozik, A. E., Hentze, M. W. Muckenthaler, M. U. *RNA* **2006**, *12*, 913–920.

103 Gamper, H. B., JR., Arar, K., Gewirtz, A. Hou, Y.-M. *RNA* **2005**, *11*, 1441–1447.

104 You, Y., Moreira, B. G., Behlke, M. A. and Owczarzy, R. *Nucleic Acids Res.* **2006**, *34*, e60.

105 Ugozzoli, L. A., Latorra, D., Pucket, R., Arar, K. Hamby, K. *Anal. Biochem.* **2004**, *324*, 143–152.

106 Goldenberg, O., Landt, O., Swchumann, R. R., Göbel, U. B. Hamann, L. *BioTechniques* **2005**, *38*, 29–32.

107 Hummelshoj, L., Ryder, L. P., Madsen, H. O. Poulsen, L. K. *BioTechniques* **2005**, *38*, 605–610.

108 Rupp, J., Solbach, W. Gieffers, J. *Appl. Environ. Microbiol.* **2006**, *72*, 3785–3787.

109 Wang, L., Yang, C. J., Medley, C. D., Benner, S. A. Tan, W. *J. Am. Chem. Soc.* **2005**, *127*, 15664–15665.

110 Sidon, P., Heimann, P., Lambert, F., Dessars, B., Robin, V. El Housni, H. *Clin. Chem.* **2006**, *52*, 1436–1438.

111 Wengel, J. *Org. Biomol. Chem.* **2004**, *2*, 277–282.

112 Sørensen, M. D., Petersen, M. Wengel, J. *J. Chem. Commun.* **2003**, 2130–2131.

113 Hrdlicka, P. J., Babu, B. R., Sørensen, M. D., Harrit, N. Wengel, J. *J.Am. Chem. Soc.* **2005**, *127*, 13293–13299.

114 Su, X. Smith, L. M. *Nucleic Acids Res.* **2004**, *32*, 3115–3123.

7
Synthesis and Properties of Oligonucleotides Incorporating Modified Nucleobases Capable of Watson–Crick-Type Base-Pair Formation

Mitsuo Sekine, Akio Ohkubo, Itaru Okamoto, and Kohji Seio

7.1
Introduction

Most modified oligonucleotides have been developed as tools for gene regulation and detection in biological studies and medicinal applications [1–4]. The principle of the study is based essentially on the inherent hybridization property of nucleic acids where guanine and adenine bind to cytosine and thymine, respectively. As exemplified by the formation of various mismatched base pairs, DNA probes cannot always bind to target sequences according to the well-known principle of Watson–Crick base pair formation [5–11]. For example, thymine often forms a T–G mismatched base pair with guanine, as shown in Figure 7.1 [12–16]. Guanine is known to form various base pairs with adenine and guanine [17–24], while adenine can bind to cytosine in its protonated form under acidic conditions [25].

Natural living cells have sophisticated systems where, even if such a mismatched base pair is formed, repairing enzymes detect it and cut the part to regenerate the original normal base pair [26–29]. However, unfavorable detection of base sequences different from the target sequence occurs when natural nucleobases are used as components of DNA probes in DNA chips/microarrays [4, 5, 30–32]. In order to avoid this mismatched base-pair formation, the base recognition ability of the naturally occurring nucleobases should be improved by chemical modification. In ongoing studies on the development of new modified nucleobases, it was concluded that if protecting groups used for the base residues did not interfere with the formation of Watson–Crick-type base pairs upon hybridization with target DNA or RNA sequences, it would be unnecessary to remove the base-protecting groups at the final stage of oligonucleotide synthesis. Such a sophisticated design was very difficult to realize when simple modifications were employed. Therefore, an extensive search was conducted for a suitable structural change of the base aglycones, and strategy using "Protected DNA probes" composed of 2-thiothymine or 2-thiouracil [33–35], 6-acetyl-8-aza-7-deazaadenine [36], 4-*N*-acetylcytosine [37, 38], and 2-carbamoylguanine [39]. When controlled pore grass (CPG) was used as a

Modified Nucleosides: in Biochemistry, Biotechnology and Medicine. Edited by Piet Herdewijn

ISBN: 978-3-527-31820-9

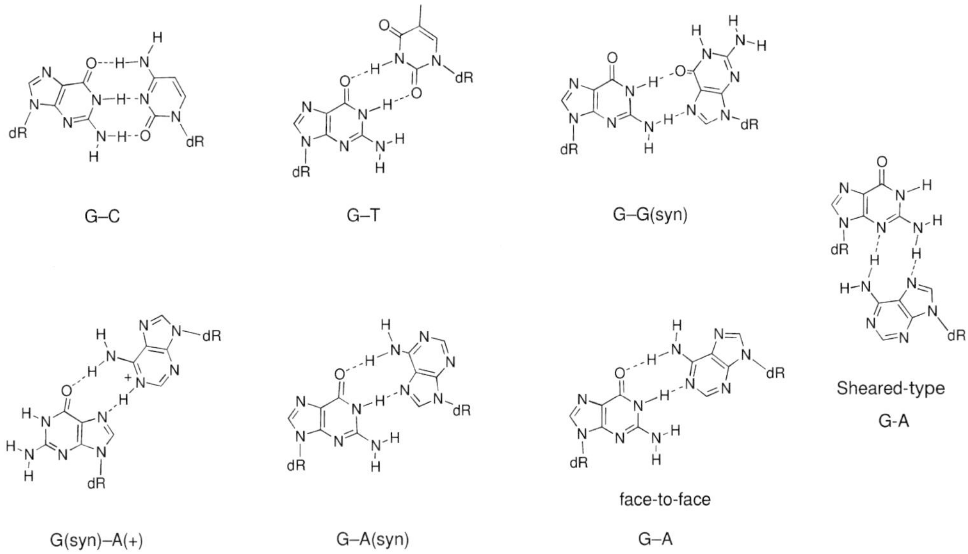

Figure 7.1 Various mismatched base pairs of guanine with other bases.

polymer support for the synthesis of such protected DNA probes, protected DNA probe–CPG conjugates could be used directly for gene detection [36]. In addition, modified base structures were designed to improve the base recognition ability. Consequently, more accurate gene detection could be performed using these new protected DNA probes (PDP strategy) (Figure 7.2) [36].

Herein is reported, in great detail, the entire background of the present authors' recent studies and the creation of new modified nucleobases that can discriminate the correct counterpart from mismatched nucleobases.

7.2
Natural, Enzyme-Assisted Sophisticated Devices for Maintaining Correct Base Recognition of Canonical Nucleobases

In the DNA replication of living cells, the fidelity of DNA polymerases can be maintained by DNA repairing enzymes that play an important role in repairing incorrect base pairs [26–29]. DNA-dependent RNA polymerases also exhibit precise transcription with the help of specific repairing enzymes. In translation, tRNA molecules bind to the cognate triplet codons on mRNA. In order to avoid any errors in the codon–anticodon interaction, first-letter nucleosides are often modified on their nucleobase moieties [40–46]. Thus far, known modifications have led us to examine the possibility that such modified nucleobases could be used in place of the canonical bases as they show excellent base recognition ability for the codon–anticodon interaction. For example, 4-*N*-acetylcytosine at the first letter of non-initiator $tRNA^{Met}$ is known to form a Watson–Crick-type base pair only with guanine of the third letter of an AUG codon on mRNAs [47, 48]. Located at the first letter of tRNAs, 2-thiouracil

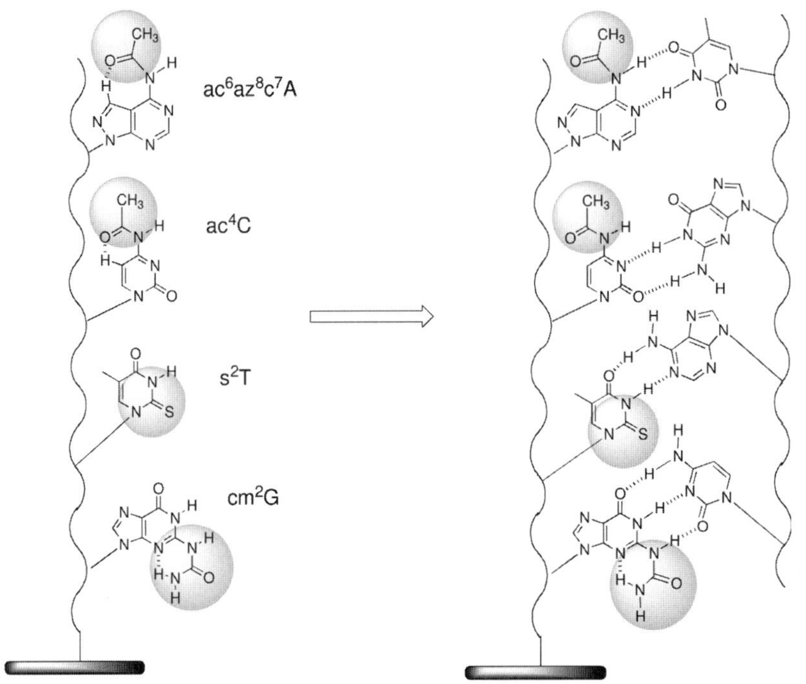

Figure 7.2 A protected DNA probe capable of the precise detection of target genes.

also binds to adenine on mRNAs [49, 50]. The reason for this is closely related to its predominance of C3′-*endo* conformation in sugar puckering over the C2′-*endo* conformation [51–55]. The former is more rigid than the latter, so that the conformational restriction leads to an allowance of base pairing only with adenine. These modifications, when observed in Nature, should be considered as clues to improve the inherent base recognition ability of the naturally occurring nucleobases.

7.3 Synthesis and Properties of Oligodeoxynucleotides Incorporating 4-*N*-Acylated Cytosine Derivatives

In an initial study, particular attention was paid to the accurate base pair formation of the 4-*N*-acetylcytosine base with guanine in the codon–anticodon recognition [47, 48]. Until then, however, no studies had been reported on the base recognition ability of this modified base at the level of oligonucleotides. In addition, no methods have been reported for the synthesis of oligonucleotides incorporating 4-*N*-acetylcytosine. Therefore, a method was first established for the synthesis of such modified oligonucleotides. The acetyl group has long been known to be used as a protecting group for the cytosine base in oligonucleotide synthesis, and to be removed by treatment with ammonia [56]. In the usual protocol of DNA synthesis, it is necessary to use ammonia to remove the base-protecting groups at the final stage. Thus, the

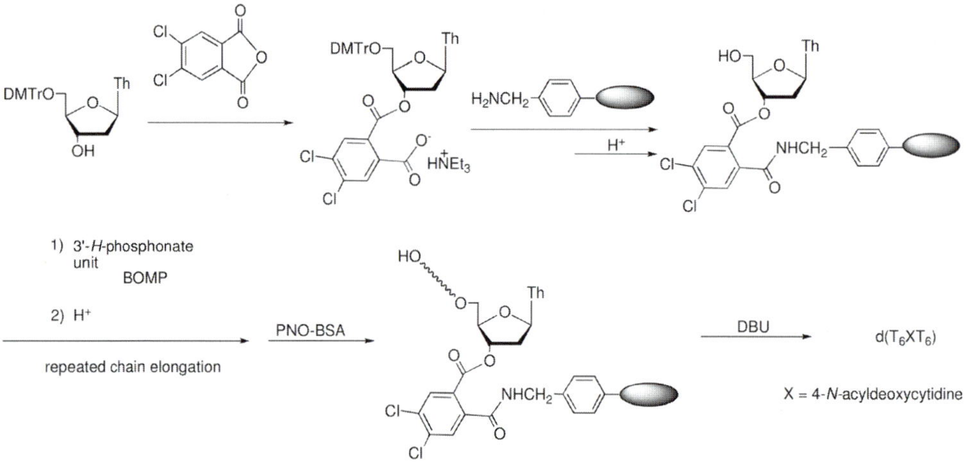

Figure 7.3 Synthesis of oligothymidylates containing 4-*N*-acylcytosine derivatives.

usual strategy could not be used for the synthesis of oligonucleotides containing this modified base. Hence, to obtain these bases, a new *H*-phosphonate method was used for the synthesis of oligonucleotides, without base protection (Figure 7.3) [57–59].

This strategy, along with the use of a 1,8-diazabicyclo[5.4.0]undec-7-ene (DBU)-labile linker and 3,4-dichlorophthaloyl, was employed to synthesize oligodeoxynucleotides containing one to three ac^{4}C bases [60, 61]. Within a 5-min period, DBU had simultaneously eliminated the cyanoethyl group used for the phosphate protection and the linker. It was noted that this modified base recognized guanine more strictly than the unmodified C, which showed a stronger base-pairing ability towards guanine.

The reason why the 4-*N*-acetylcytosine base is able to form a Watson–Crick-type base pair with guanine has been discussed in terms of its characteristic structure. This involves an intramolecular hydrogen bond between the carbonyl group and the 5-vinylic proton, as shown in Figure 7.4. This type of hydrogen bond seems to

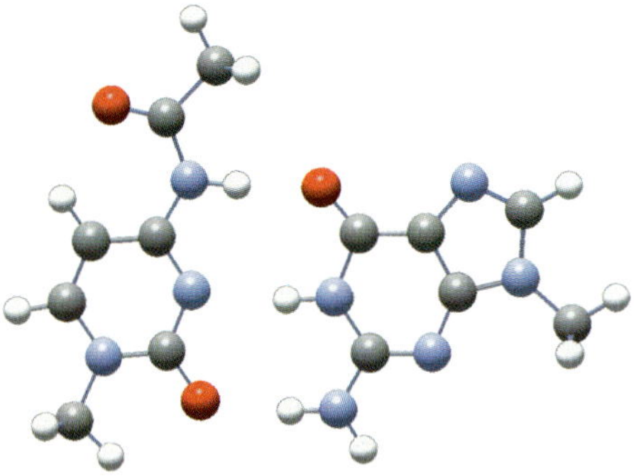

Figure 7.4 Structure of the base pair of 4-*N*-acetyl-1-methylcytosine with 9-methylguanine.

be unusual; however, the X-ray analysis data indicated that the distance between the carbonyl group and the 5-vinylic proton is within the range of hydrogen bonds [62, 63].

The ^{1}H NMR data of a series of 4-*N*-acylated cytosine derivatives also showed significantly lower magnetic field shifts of the vinyl proton, by approximately 1.5 ppm [61]. In addition, theoretical studies using *ab-initio* molecular orbital (MO) calculations implied that the most stable conformers of these modified cytosine derivatives have similar intramolecular hydrogen-bonding systems [61]; this was seen to be true for 4-*N*-alkoxycarbonyl cytosine derivatives [64]. Moreover, the 4-*N*-acetyl group affected the equilibrium between the C3′-*endo* and C2′-*endo* conformation of the ribose puckering when incorporated into cytidine. This effect was first reported by Kawai et al. [65, 66]. In 4-*N*-acetylcytidine, the C3′-*endo* conformer that is seen in RNA duplexes is predominant over the C2′-*endo* observed in DNA duplexes. The role of ac^4C at the wobble position of *Escherichia coli* $tRNA^{Met}$ may be to prevent errors that might arise during protein synthesis due to reading of the AUA isoleucine codon by $tRNA^{Met}$ [47]. Removal of the acetyl group from the wobble base allows $tRNA^{Met}$ to misread a non-methionine codon containing only A and U.

Although the 4-*N*-acetyl group on the cytosine base increased the thermodynamic stability of DNA duplexes, 4-*N*-acyl groups having longer alkyl substituents decreased such stability [61]. The introduction of aromatic acyl groups such as benzoyl, 2-picolyl, and 2-nicotinoyl into the 4-amino group significantly destabilized DNA duplexes [61]. The benzoyl group induces steric repulsion with the 5-vinyl proton so that the most stable conformer of 4-*N*-benzoyl-1-methylcytosine has a twisted torsion angle (6.1°) around O–C–N(4)–C(4). However, theoretical calculations suggested that the base pair formed between 4-*N*-benzoyl-1-methylcytosine and 9-methylguanine has a completely planar structure [61]. It is likely that disturbance of the hydration structure around the major group, due to the aliphatic benzoyl group and the energy loss due to the rotation of this twisted structure to the planar structure, were the main factors for the destabilization of the DNA duplexes. The hydration structure is of crucial importance to retain duplex structures, since the incorporation of less-hindered aroyl groups such as furoyl, thenoyl, and 2-picolyl into DNA also resulted in the destabilization of duplexes. In this regard, the (smallest) acetyl group is superior to the other aroyl groups.

7.4 Base-Recognition Ability of 4-*N*-Alkoxycarbonylcytosine Derivatives

Oligodeoxyribonucleotides incorporating 4-*N*-alkoxycarbonyldeoxycytidine derivatives were synthesized on polystyrene-type ArgoPore resins having a new benzyloxy(diisopropyl)silyl linker using $ZnBr_2$ as the detritylating agent (Figure 7.5) [64].

The first 3′-terminal deoxynucleoside derivative was attached to the resin by successive *in-situ* reactions of a 5′-*O*-DMTr-deoxynuculeoside derivative with diisopropylsilanediyl ditriflate and an ArgoPore resin containing hydroxyl groups. The modified oligodeoxynucleotides were easily released from the resin by treatment with

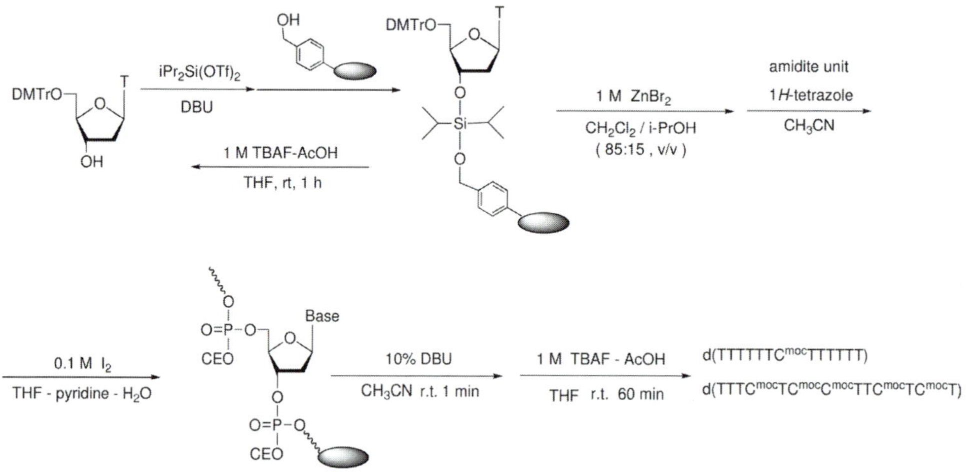

Figure 7.5 Solid-phase synthesis of oligodeoxynucleotides incorporating 4-*N*-methoxycarbonylcytosine bases.

tetrabutylammonium fluoride (TBAF). The incorporation of 4-*N*-alkoxycarbonyl-deoxycytidines into DNA strands resulted in a higher hybridization affinity for the complementary DNA strands than that of 4-*N*-acyldeoxycytidines with a similar size of substituents. The alkoxyacyl ($RO(CH_2)_nC(O)$-) groups tend to exhibit higher T_m values than the acyl (RC(O)-) groups. The 4-*N*-alkoxycarbonylcytosine base can form base pairs not only with guanine but also with adenine [64].

Based on the *ab-initio* calculations of the hydrogen bond energies of the possible base pairs formed between 4-*N*-methoxycarbonyl-1-methylcytosine and 9-methyladenine (Figure 7.6), it was concluded that the base pair involves two unique hydrogen bonds between the cytosyl 4-NH group and the adenyl N1 atom, and between the O atom of the ester group and the adenyl 6-NH group. This conclusion

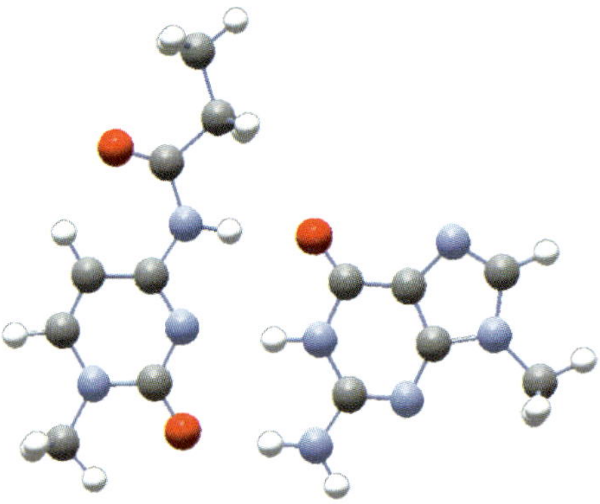

Figure 7.6 Structure of the base pair of 4-*N*-methoxycarbonyl-1-methylcytosine with 1-methylguanine.

was also supported by the NMR analysis of the base pairs of [15]*N*-labeled 4-*N*-alkoxycarbonyldeoxycytidines with deoxyadenosine derivatives.

7.5
Synthesis and Properties of Oligonucleotides Incorporating 4-*N*-Carbamoylcytosine Derivatives

Compared to the excellent base-recognition ability of the acetylcytosine base, 4-*N*-carbamoylcytosine derivatives exhibit their inherent base-pairing ability with canonical bases [67]. This property is deduced from an intramolecular hydrogen bond. This inner interaction prevents base-pair formation with bases at the opposite site. Moreover, 4-*N*-carbamoylcytosine (C^{cmy}) derivatives form a dimer using four intermolecular hydrogen bonds in non-polar solvents; however, in polar solvents they exist in a monomeric form having two intramolecular hydrogen bonds (Figure 7.7).

Regardless of these results, the simplest 4-*N*-carbamoylcytosine C^{cmy} that was incorporated into oligodeoxynucleotides could bind to guanine in a manner similar to that of the unmodified cytosine, with a slightly lower T_m upon hybridization with the

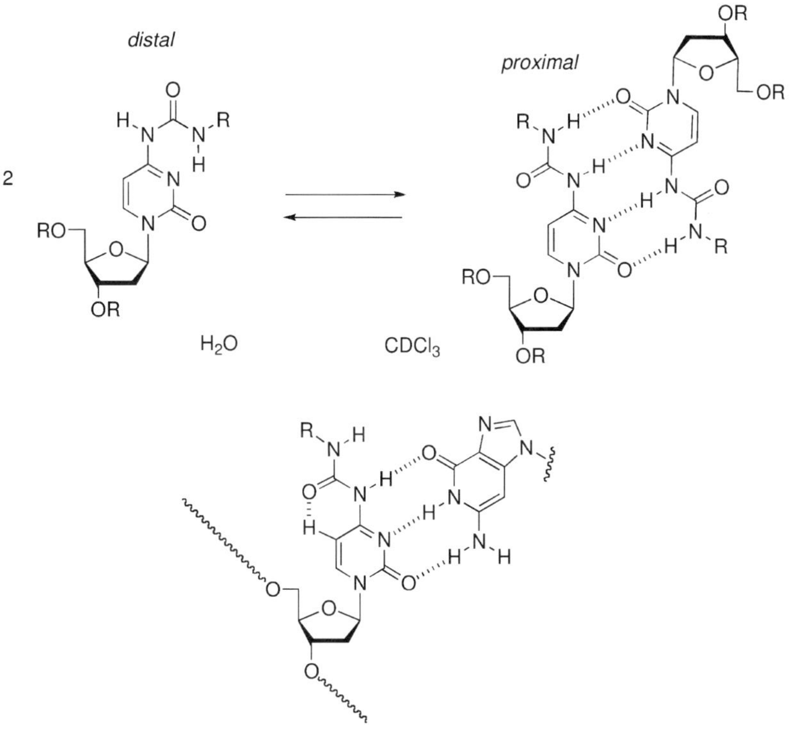

Figure 7.7 Equilibrium of *N*-carbamoylcytosine bases between the distal and proximal forms and its base pair with a guanine base in DNA duplexes.

complementary strands. Although the modified base binds to adenine more strongly than the unmodified cytosine, the binding between the unmodified base and adenine is as strong as that between the unmodified cytosine and thymine. Therefore, the carbamoyl group could be used as a protected probe base capable of forming a base pair with guanine, since ΔT_m between C^{cmy}-G and C^{cmy}-T is still sufficient (i.e., 13.7 °C), whereas it is 15.4 °C in the case of the unmodified cytosine. In DNA duplexes, the inner side is known to be more lipophilic than the outer side, and the intramolecular hydrogen bonds of C^{cmy} are dissociated so that the opposite guanine base can form a tighter Watson–Crick-type base pair composed of three hydrogen bonds. The *ab-initio* calculation suggested that the hydrogen bond energy of a base pair of 4-carbamoyl-1-methylcytosine and 9-methylguanine is $-27.4\,kcal\,mol^{-1}$, a value only slightly lower than that of 1-methylcytosine and 9-methylguanine ($-28.7\,kcal\,mol^{-1}$).

7.6
2-Thiouracil as an Improved Nucleobase in Place of Thymine

2-Thiouracil and its 5-substituted derivatives have been found as modified nucleobases located at the first letter of tRNAs [40, 49]. They are known to play an important role in the codon box recognizing only adenine (Figure 7.8).

2-Thiouracil derivatives also have an interesting property where the 2-thiocarbonyl group contributes to stabilization of duplexes due to its strong base-stacking effect toward upstream and downstream nucleobases [68–70]. This effect increased the T_m-values of modified oligonucleotides upon their incorporation. Kierzek *et al.* reported that 2′-*O*-methyl-2-thiouridine (s^2Um)-containing short probes proved to be useful for predicting the secondary structure of rRNA because of the accurate and stable hybridization abilities of s^2Um [71].

This accuracy is derived from the fact that a hydrogen bond between the sulfur atom of the thiocarbonyl group and the thyminyl N1 proton is considerably weakened, as reported previously [72, 73]. In fact, the replacement of uracil with 2-thiouracil in an RNA probe resulted in a marked improvement in base recognition when an RNA oligomer was used as its target molecule [35]. It was also reported that RNA duplexes containing s^2U or Um were thermodynamically more stable than unmodified natural RNA duplexes [74, 75]. Recently, more detailed studies were

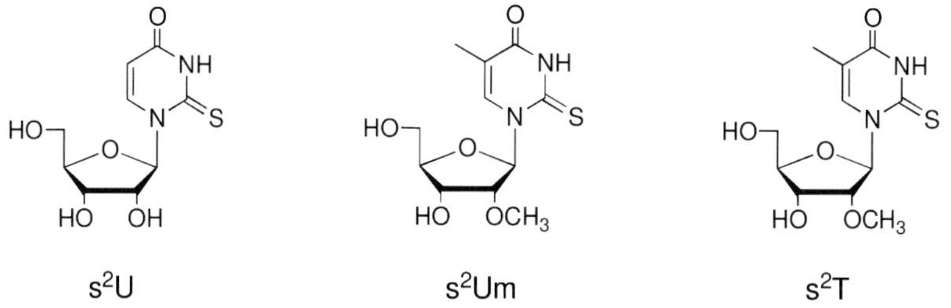

Figure 7.8 2-Thiouridine derivatives.

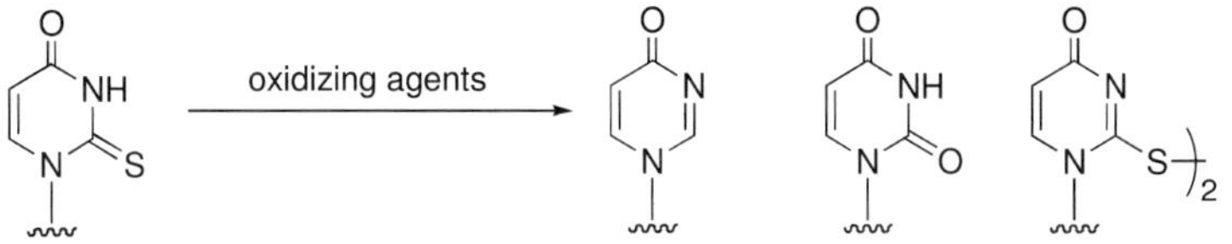

Figure 7.9 Reactions of 2-thiouracil derivatives with oxidizing agents.

conducted on the base-recognition ability of the s^2U or s^2T base [76]. When s^2Um was incorporated into RNA or DNA strands, the hybridization and base-discrimination abilities of the modified RNA or DNA oligomers for the complementary RNA strands were superior to those of the corresponding unmodified ones. On the other hand, their base-discrimination abilities for complementary DNA strands were almost the same as those of the unmodified ones.

From the viewpoint of synthesis, there was a serious problem at the oxidation step in the phosphoramidite chemistry-based solid-phase synthesis of oligonucleotides incorporating 2-thiouridine or 2-thiothymidine derivatives. It is known that these 2-thiolated species were reactive to oxidizing agents such as iodine [77], mCPBA [77], dimethyldioxirane [78], oxone [79], and oxaziridines [80]. Several studies have been conducted to improve the oxidation step which caused the side reactions, as shown in Figure 7.9.

Although these studies recommended the use of *t*BuOOH [81], serious side reactions were still observed when this reagent was used. Recently, it was reported that a 0.02 M solution of I_2 in pyridine/water (9 : 1, v/v) did not cause these side reactions and, under these conditions, the synthesis of oligoribonucleotides containing five s^2Um units in satisfactory yield was successfully achieved, without difficulty. In contrast, when a 0.1 M I_2 solution was employed, considerable amounts of byproducts were observed.

7.7 Modified Adenine Bases Capable of Recognizing the Thymine Base

The introduction of an acyl-type of base-protecting group into an unmodified adenine moiety leads to blocking of the site for the Watson–Crick base pair by electronic repulsion between the carbonyl oxygen of the acyl group and the adenyl 7-nitrogen (Figure 7.10). For example, 6-*N*-acetyladenine and 6-*N*-benzoyladenine could not

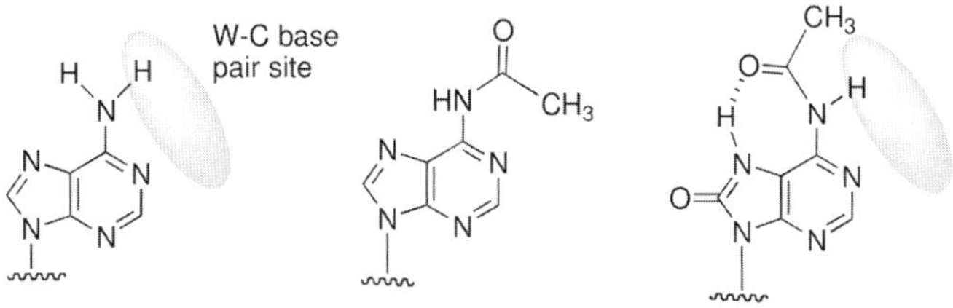

A ac⁶A ac⁶ox⁸A

Figure 7.10 Structures of modified adenine moieties.

bind to thymine, so that oligodeoxynucleotides incorporating these simple *N*-acylated adenine bases exhibited poor hybridization affinity for complementary DNA strands. Therefore, special designs are needed in this case.

Previously, phosmidosine, a naturally occurring nucleotide antibiotic which incorporates an 8-oxoadenine moiety, was synthesized [82]. In order to achieve this, an acetyl group was introduced into the *exo*-amino group. However, at the final stage, when attempts were made to remove this acetyl group from a protected phosmidosine derivative, the protecting group proved to be rather resistant to concentrated aqueous ammonia. This unexpected resistance seemed to be due to an intramolecular hydrogen bond between the acetyl oxygen and the 8-NH proton. Resistance of this acetyl group to ammonia and orientation of this acetyl group via the intramolecular hydrogen bond were favorable in the PDP strategy. Therefore, this modified base was selected in place of adenine. Details of the phosphoramidite building block required for the incorporation of 6-acetyl-8-oxoadenine (ac^6ox^8A) into oligodeoxynucleotides were reported earlier by Essigmann and coworkers [83],who used it for the synthesis of oligodeoxynucleotides containing 8-oxoadenine. Therefore, for present purposes it was unnecessary to develop a new building block.

Oligodeoxynucleotides incorporating ac^6ox^8A were synthesized using the new strategy [84, 85] for DNA synthesis, without base protection in the phosphoramidite approach (Figure 7.11).

The modified oligomers thus obtained were found to have a slightly lower hybridization affinity for the complementary DNA strand, but to have a sufficient base-discrimination ability at the unmodified adenine level [36].

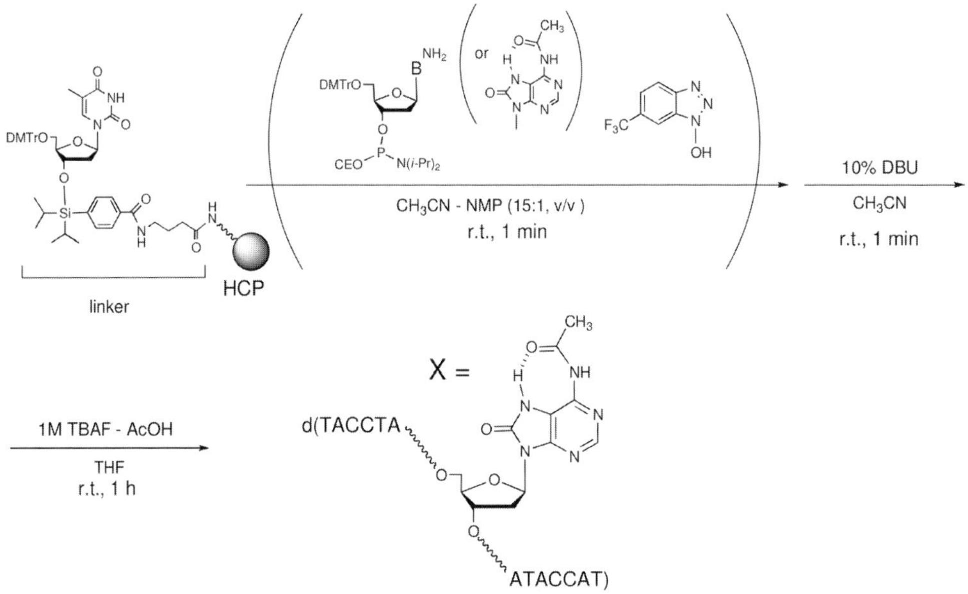

Figure 7.11 The activated phosphite method for the synthesis of oligodeoxynucleotides without base protection.

Tsukahara and Nagasawa reported an interesting approach called the "Probe-on-Carrier method" using DNA–CPG conjugates where CPG was used not only as a polymer support for the synthesis of DNA oligomer probes but also as beads carrying DNA probes which were attached to a slide glass plate [86]. Since it is well established that DNA oligomers may be synthesized on CPG with high purity, by using the phosphoramidite approach, it was possible to prepare DNA–CPG conjugates obtained by the ammonia-mediated selective removal of the phosphate and base-protecting groups from the completely protected oligomers elongated on CPG. This strategy is very advantageous, because a high sensitivity can be obtained as a considerable number of DNA probes can be mounted on the three-dimensional large surface of CPG that is a macroporous bead. This is an outstanding feature of the method, since DNA probes on the normally used glass plates can only be planted on the two-dimensional surface.

To demonstrate the utility of these modified bases as components of actual DNA probes based on the probe-on-carrier method, several CPG-linked oligodeoxynucleotides containing these modified bases were synthesized. The thymine and guanine bases were not protected, since Gryaznov and Letsinger initially reported that there was no need to protect the base residues of these bases as far as the phosphoramidite approach was concerned [87]. These modified oligodeoxynucleotides were rapidly synthesized by solid-phase synthesis with CPG as the polymer support. Treatments of the resin only with 1% trifluoroacetic acid and 1,8-diazabicyclo[5.4.0]undec-7-ene (DBU) for 1 min each are necessary to use PDP–CPG conjugates for detection of gene fragments (Figure 7.12).

This PDP–CPG conjugate was used for detection of single nucleotide polymorphisms (SNPs) as a new probe. The result of the accuracy of base-recognition ability is

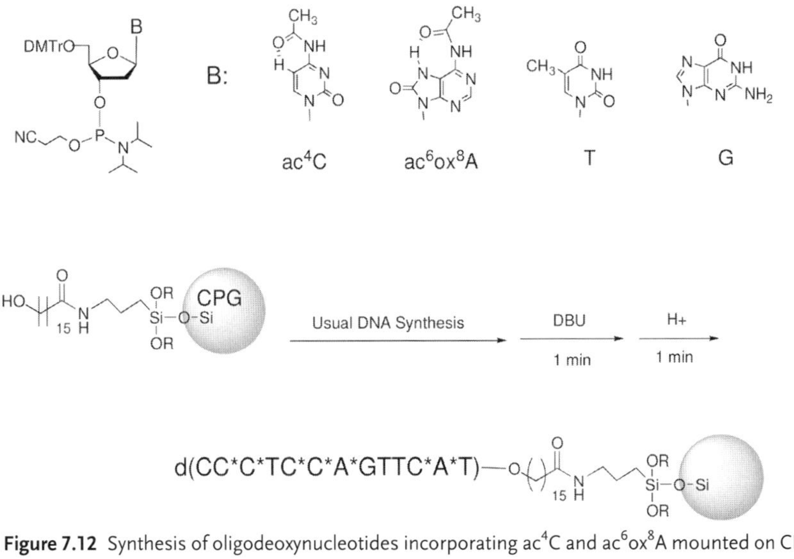

Figure 7.12 Synthesis of oligodeoxynucleotides incorporating ac^4C and ac^6ox^8A mounted on CPG.

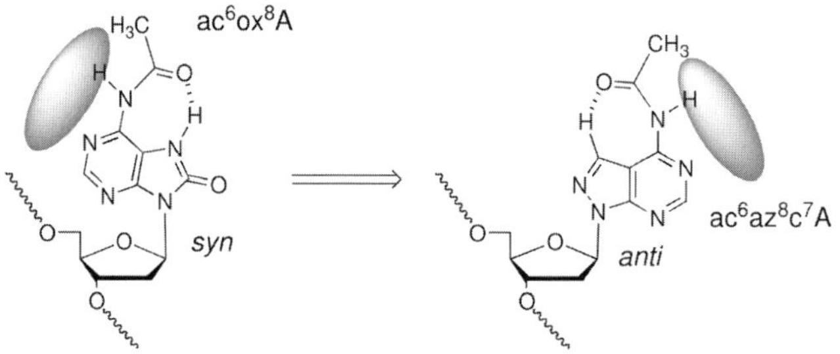

Figure 7.13 A new modified adenine moiety capable of Watson–Crick base-pair formation with *anti* orientation around the glycosyl bond.

almost the same as that of the unmodified DNA–CPG conjugate that was synthesized using the activated phosphite method.

8-Oxoadenine derivatives are known to be found in *syn*-conformation around the glycosyl bond because of the steric hindrance of the 8-oxo group [88–91]. Thereby, the Watson–Crick base-pair formation requires the initial rotation of the *syn* form to an *anti* form. When considering this unfavorable rotation, an adenine analogue was designed in which an acetyl group was used for protection of the 6-amino group of the analogue. The adenine analogue thus designed was 8-aza-7-deaza-6-acetyladenine ($az^8c^7ac^6A$) (Figure 7.13).

Previously, deacetylated species, such as 8-aza-7-deazaadenine (az^8c^7A), were extensively incorporated into oligodeoxynucleotides in a series of Seela's pioneering studies on the deaza effect of the nucleobases on hybridization of oligodeoxynucleotides [92, 93]. It was expected that the shift of the 7-nitrogen to position 8 would make it possible to create an intramolecular hydrogen bond system between the 6-acetyl oxygen and the 7-vinylic proton, as for the 5-vinyl proton of 4-*N*-acylated cytosine derivatives [61–63]. Since the 8-nitrogen atom shift is less-hindered than the oxo group of 8-oxoadenine, the most stable conformer around the glycosyl bond is an *anti*-form. The expectation was subsequently verified, since the oligodeoxynucleotides that incorporated this modified adenine analogue exhibited a higher T_m-value for hybridization with the complementary DNA strand, with somewhat improved base-recognition ability compared to that of the adenine base. The synthesis of oligodeoxynucleotides incorporating ac^4C and $az^8c^7ac^6A$ mounted on a flattened CPG disk is illustrated schematically in Figure 7.14.

The hybridization and base-recognition ability of 4-*N*-acetylcytosine and 8-aza-7-deaza-6-acetyladenine, when incorporated into DNA probes, was estimated. As a result, the combined use of these modified bases proved to be very effective for considerable improvement of these important factors.

In order to compare the PDP strategy with the normal DNA detection method using DNA probes covalently linked to glass plates, oligodeoxynucleotides incorporating ac^4C and $az^8c^7ac^6A$ were synthesized and isolated. In this synthesis, a Bu_4NF-labile linker was developed to release the base-protected oligomers [94, 95]. At the 5′-end, an amino group was attached to the oligonucleotide chain via a spacer. Consequently, these modified oligonucleotides were linked to active esters on a glass

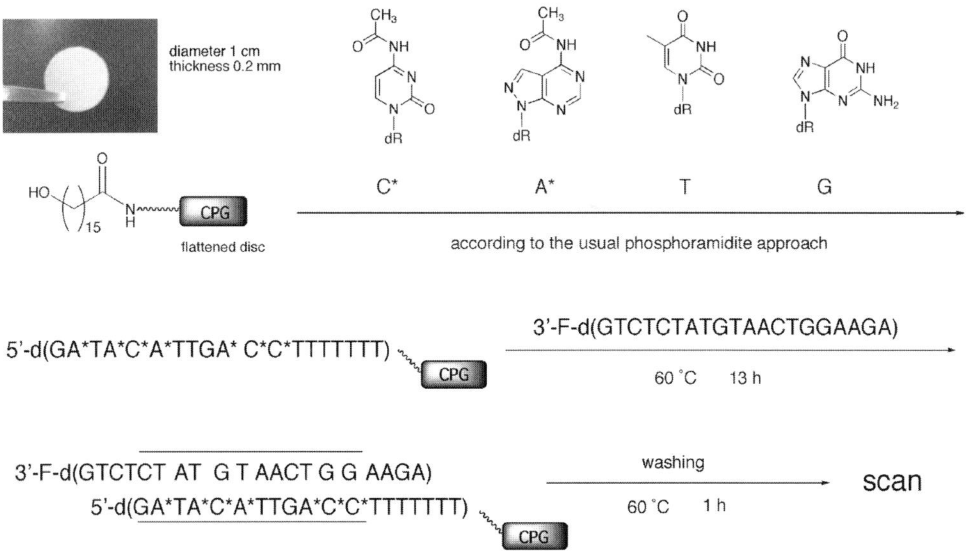

Figure 7.14 Preparation of a PDP–CPG conjugate containing ac^4C and $ac^6az^8c^7A$ on a flattened CPG disk using the normal phosphoramidite approach, and its application to single nucleotide polymorphism detection.

plate as described for the usual DNA chip protocol. As a result, and as expected, the PDP strategy proved to have greater sensitivity and base-recognition ability.

7.8 Design of Modified Guanine Bases Capable of Recognizing Cytosine

As mentioned above, it was possible to use the 2-*N*-unprotected deoxyguanosine 3′-phosphoramidite building block in the phosphoramidite approach. Because the solubility of this unit was very poor, it often precipitated during DNA synthesis and caused mechanical blockage problems. The average coupling yield of this building block was also lower than that of other completely protected deoxynucleoside 3′-phosphoaramidite units and, accordingly, the overall yields of oligodeoxynucleotides decreased. Another problem with this building unit was that its synthesis required the selective 5′-*O*-dimethoxytritylation of deoxyguanosine [96]. Unfortunately, this reaction was accompanied by the formation of a 2-*N*-dimethoxytritylated byproduct, which could not easily be separated. Hence, to overcome these problems, a range of alternative new guanine analogues was studied [97]. Various unusual base pairs formed between guanine and the other bases or itself are summarized as shown in Figure 7.1. Among these, the most competitive base pairs were those of G–T, G–G, and G–A (Sheared type). The others were found from more complex structures such as tRNA and rRNA, and thereby were rather exceptional in DNA duplexes. Thus, major interest was focused on the avoidance of G–T, G–G, G–A (Sheared type) base

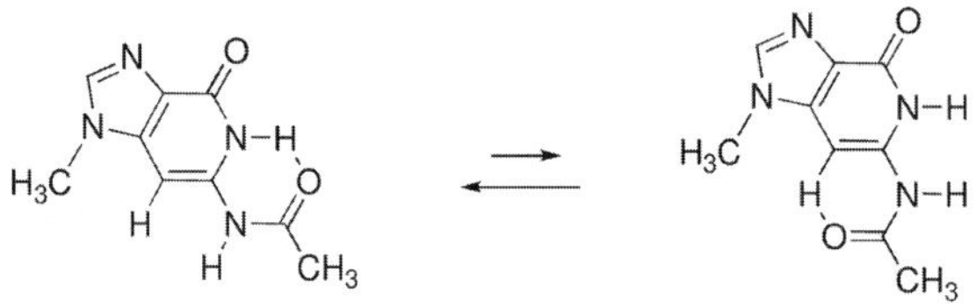

Figure 7.15 Two conformers of 2-*N*-acetyl-9-methylguanine.

pairs in the usual duplexes of DNA–DNA, DNA–RNA, and RNA–RNA. One strategy employed was to eliminate the nitrogen atom associated with the formation of such unusual base pairs from the guanine structure. In the case of the Sheared type G–A base pair, elimination of the 3-nitrogen enabled the creation of a guanine analogue to prevent this unusual base pair.

First, 2-*N*-acetyl-3-deazaguanine was designed as such a guanine analogue on the basis of *ab-initio* MO calculations [97]. The theoretical calculations of the ac^2c^3G–C and ac^2c^3G–U also suggested this result. The most stable conformer of 9-methyl-ac^2c^3G was calculated to be the "closed form" of the (*Z*)-*N*-acetyl rotamer (Figure 7.15). For the formation of a Watson–Crick base pair, this conformer must be converted by 180° rotation around the N(2)–C(ac) bond to the "open form", which is less stable by 2.3 kcal mol^{-1} than the "closed form". The hydrogen bond energy ($\Delta E = -28.5$ kcal mol^{-1}) of the Watson–Crick-type a^2c^3G(open)–C base pair was 4.0 kcal mol^{-1} greater than that (-24.5 kcal mol^{-1}) of the G–C pair. Therefore, when considering the energy cost of 2.3 kcal mol^{-1} to convert the closed form to the open form, this modified base pair is more stable by 1.7 kcal mol^{-1} than the canonical G–C base pair. On the other hand, the a^2c^3G(open)–U base pair was slightly more stable by 0.1 kcal mol^{-1} than a G–U base pair. The difference in energy between a^2c^3G(open)–C and a^2c^3G(open)–U is 1.6 kcal mol^{-1} greater than that between G–C and G–U.

Based on these theoretical studies, the base-discrimination ability of 2-*N*-acetyl-3-deazaguanine was examined. Oligonucleotides containing this base were proven to show a sufficient base-discrimination ability toward dC against dU. T_m experiments using RNA–RNA duplexes of 5′-CGGCXAGGAG-3′/3′-r(GCCGYUCCUC)-5′ (X = a^2c^3G, G) showed that the duplex (X = a^2c^3G, Y = C) had a thermostability (T_m = 70.1 °C) similar to that (T_m = 70.9 °C) of the duplex (X = G, Y = C); however, the T_m value (54.9 °C) of 5′-CGGCac^2c^3GAGGAG-3′/3′-r(GCCGUUCCUC)-5′ was considerably lower by 6.9 °C than that (61.8 °C) of 5′-CGGCGAGGAG-3′/3′-r(GCCGUUCCUC)-5′. Therefore, the modified guanine base ac^2c^3G proved to recognize distinctly the C base from U, compared to the unmodified G. A similar tendency was observed when the modified RNA oligomer was hybridized with its complementary DNA strands [97]. In addition, the incorporation of a^2c^3G into 5′-CGGCXAGGAG-3′/3′-r(GCCGAGCCUC)-5′ resulted in considerable destabilization of these duplexes containing a tandem GA mismatch.

In order to study the effect of the acetyl group of a^2c^3G on the hybridization and base-discrimination ability, modified oligoribonucleotides containing c^3G [98–103] were synthesized.

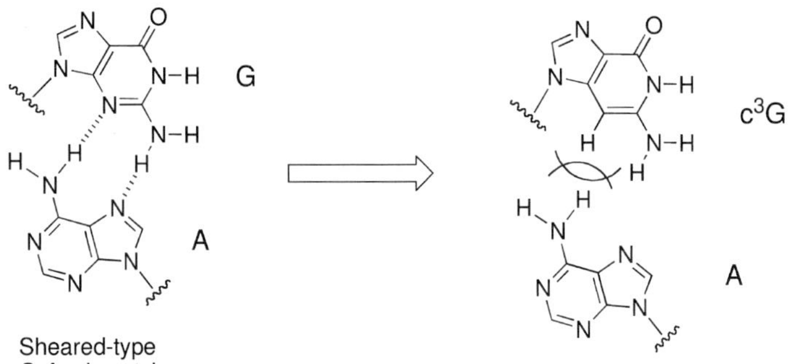

Figure 7.16 Prevention of Sheared-type G–A mismatch base pair using c^3G.

As expected, the elimination of the acetyl group resulted in a destabilization of the modified 2′-O-methyl-RNA–RNA and modified 2′-O-methyl-RNA–DNA duplexes. When c^3G was inserted in place of the 5th G in a duplex of 2′-O-Me-5′-r(CGGCGAG GAG-3′)/3′-GCCGAUCCUC-5′ that has tandem Sheared-type G/A mismatches at the central position, the modified 2′-O-methyl-RNA–RNA duplex exhibited more destabilization than when c^3G was inserted in place of the 5th G in a duplex of m(5′-CGGCGAGGAG-3′)/3′-GCCGAGCCUC-5′ that has tandem face-to-face G/A mismatches at the central position. Although the Sheared-type G/A mismatch requires the 3-nitrogen atom of G for hydrogen bonding, the face-to-face mismatch does not (Figure 7.16). The result obtained supported the significant effect of c^3G on the avoidance of a Sheared-type mismatch base pair.

Although a^2c^3G showed excellent base-recognition ability, multi-step reactions were required for the synthesis of the skeleton of its ribonucleoside or deoxynucleoside. Therefore, a search was conducted for a more convenient compound capable of accurate base recognition. Consequently, interest was focused on the intramolecular hydrogen bonding system as seen in a^2c^3G. If such an intramolecular hydrogen bond could be formed between the N3-nitrogen and a substituent on the 2-amino group of guanine, then the Watson–Click base pair site could be preserved. Therefore, 2-*N*-carbamoylguanine derivatives were designed as new guanine analogues capable of selectively forming base pairs with C, as shown in Figure 7.17 [104].

In this modified base, there is an equilibrium between open and closed forms such as a^2c^3G. The closed form can be stabilized by an intramolecular hydrogen bond between the carbonyl oxygen and the N1-nitrogen via a six-membered ring. On the other hand, the open form can be stabilized by an intramolecular hydrogen bond between an amide proton of the carbamoyl group and the N3-nitrogen. The *ab-initio* MO calculations suggested that the former was more stable than the latter by 1.56 kcal mol^{-1}. In connection with the present studies, Zimmerman and coworkers reported that 2-*N*-[*N*-(*n*-butyl)carbamoyl]guanine could form stable supramolecular complexes having four hydrogen bonds in open-type conformation [105, 106].

The thermostability of a duplex of 5′-(CGGC^{cm}GAGGAG)-3′/3′-d(GCCGCTCCTC)-5′ was slightly lower, by 0.9 °C, than that of the unmodified duplex. The base-recognition ability of this modified base was significantly higher than that of G, and therefore this

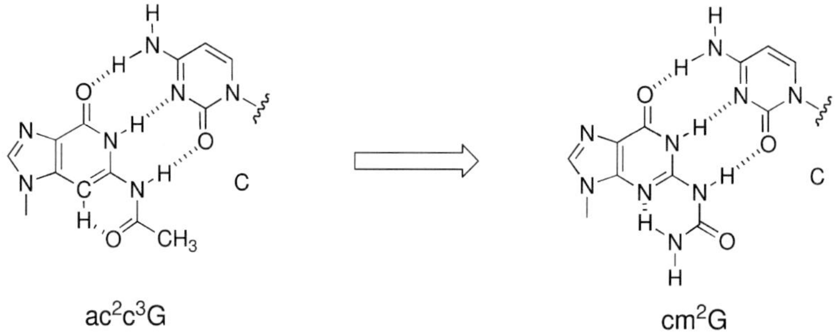

Figure 7.17 2-Carbamoylguanine capable of base pairing with cytosine and avoiding the formation of a Sheared-type mismatch base pair.

modified base would have been useful for gene detection. However, the synthesis of oligodeoxynucleotides incorporating ^{cm}G required a protecting group such as diphenylcarbamoyl for the 6-carbonyl oxygen. This limitation made it impossible to use acC and $ac^6az^8c^7A$ at the same time, since the DPC group must be deprotected by treatment with ammonia, which removes the acetyl group from acC and $ac^6az^8c^7A$.

7.9 Conclusions

During the past decade, many studies have been conducted with *N*-acylated oligonucleotides as new probe molecules for gene detection, as well as for gene therapy. These studies were conducted using several new strategies for the synthesis of DNA oligomers, without base protection, and using the phosphoramidite approach. Hence, Therefore, the potential use of protected DNA probes as tools for accurate gene detection has been disclosed. In the near future, more accurate systems will be required for the medical application of chemically oriented methods of gene detection. The strategy described in this chapter should provide a useful basis for improving the present approach to gene detection.

References

1 Manoharan, M., *Curr. Opin. Chem. Biol.* **2004**, *8*, 570–579.

2 Bumcrot, D., Manoharan, M., Koteliansky, V., Dinah, W. Y. S., *Nature Chem. Biol.* **2006**, *2*, 711–719.

3 Gewirtz, A. M. (Ed.), *Nucleic Acid Therapeutics in Cancer*, Hamana Press, **2004**.

4 Gao, X., Gulari, E., Zhou, X., *Biopolymers* **2004**, *73*, 579–596.

5 Pirrung, M. C., *Angew. Chem. Int. Ed.* **2002**, *41*, 1276–1289.

6 Sasaki, S., Nagatsugi, F., *Curr. Opin. Chem. Biol.* **2006**, *10*, 615–621.

7 SantaLucia, J., Allawi, H. T., Seneviratne, P. A., *Biochemistry* **1996**, *35*, 3555–3562.

8 Binder, H., Preibisch, S., Kirsten, T., *Langmuir* **2005**, *21*, 9287–9302.

9 Binder, H., Preibisch, S., *Biophys. J.* **2005**, *89*, 337–352.

10 Hunter, W. N., Brown, T., *Oxford Handbook of Nucleic Acid Structure* (ed. S. Neidle), Oxford University Press, **1999**, pp. 311–330.

11 Modrich, P., *Annu. Rev. Genet.* **1991**, *25*, 229–253.

12 Alvarez-Salgado, A., Desvaux, H., Boulard, Y., *Mag. Res. Chem.* **2006**, *44*, 1081–1089.

13 Isaacs, R. J., Rayerns, W. S., Spielmann, H. P., *J. Mol. Biol.* **2002**, *319*, 191–207.

14 Su, S., Gao, Y. G., Robinson, H., Liaw, Y. C., Edmondson, S. P., Shriver, J. W., Wang, A. H., *J. Mol. Biol.* **2000**, *303*, 395–403.

15 Otokiti, E. O., Sheardy, R. D., Richard, D., *Biochemistry* **1997**, *36*, 11419–11427.

16 Li, Y., Zon, G., Wilson, W. D., *Proc. Natl. Acad. Sci. USA* **1991**, *88*, 26–30.

17 Faibis, V., Cognet, J. A. H., Boulard, Y., Sowers, L. C., Fazakerley, G. V., *Biochemistry* **1996**, *35*, 14452–14462.

18 Fazakerley, G. V., Quignard, E., Woisard, A., Guschlbauer, W., van der Marel, G. A., van Boom, J. H., Jones, M., Radman, M., *EMBO J.* **2000**, *5*, 3697–3703.

19 Spackova, N., Berger, I., Sponer, J., *J. Am. Chem. Soc.* **2000**, *122*, 7564–7572.

20 Ke, S. H., Wartell, R. M., *Nucleic Acids Res.* **1996**, *24*, 707–712.

21 Maskos, K., Gunn, B. M., LeBlanc, D. A., Morden, K. M., *Biochemistry* **1993**, *32*, 3583–3595.

22 Brown, T., Leonard, G. A., Booth, E. D., Kneale, G., *J. Mol. Biol.* **1990**, *212*, 437–440.

23 Li, Y., Zon, G., Wilson, W. D., *Proc. Natl. Acad. Sci. USA* **1991**, *88*, 26–30.

24 Chou, S.-H., Zhu, L., Reid, B. R., *J. Mol. Biol.* **1997**, *267*, 1055–1067.

25 Boulard, Y., Cognet, J. A., Gabarro-Arpa, J., Le, B. M., Carbonnaux, C., Fazakerley, G. V., *J. Mol. Biol.* **1995**, *246*, 194–208.

26 Kunkel, T. A., Erie, D. A., *Annu. Rev. Biochem.* **2005**, *74*, 681–710.

27 Buermeyer, A. B., Deschenes, S. M., Baker, S. M., Liskay, R. M., *Annu. Rev. Genet.* **1999**, *33*, 533–564.

28 Kolodner, R. D., Marsischky, G. T., *Curr. Opin. Genet. Dev.* **1999**, *9*, 89–96.

29 Modrich, P., Lahue, R., *Annu. Rev. Biochem.* **1996**, *65*, 101–133.

30 Zhang, L., Miles, M. F., Aldape, K. D., *Nat. Biotech.* **2003**, *21*, 818–821.

31 Fidanza, J. A., McGall, G. H., *Nucleosides Nucleotides* **1999**, *18*, 1293–1295.

32 Forman, J. E., Walton, I. D., Stern, D., Rava, R. P., Trulson, M. O., *Molecular Modeling of Nucleic Acids* (eds. N. B. Leontis, J. SantaLucia), ACS Publishers, Washington, DC, **1997**, pp. 206–228.

33 Okamoto, I., Seio, K., Sekine, M., *Tetrahedron Lett.* **2006**, *47*, 583–585.

34 Okamoto, I., Shohda, K., Seio, K., Sekine, M., *J. Org. Chem.* **2003**, *68*, 9971–9982.

35 Shohda, K., Okamoto, I., Wada, T., Seio, K., Sekine, M., M. *Bioorg. Med. Chem. Lett.* **2000**, *10*, 1795–1798.

36 Ohkubo, A., Kasuya, R., Sakamoto, K., Miyata, K., Taguchi, H., Nagasawa, H., Tsukahara, T., Watanobe, T., Maki, Y., Seio, K., Sekine, M., (unpublished results).

37 Wada, T., Kobori, A., Kawahara, S., Sekine, M., *Tetrahedron Lett.* **1998**, *39*, 6907–6910.

38 Wada, T., Kobori, A., Kawahara, S., Sekine, M., *Eur. J. Org. Chem.* **2001**, *24*, 4583–4593.

39 Sasami, T., Seio, K., Ohkubo, A., Sekine, M., *Tetrahedron Lett.* **2006**, *48*, 5325–5329.

40 Agris, P. F., Vendeix, F. A. P., Graham, W. D., *J. Mol. Biol.* **2007**, *366*, 1–13.

41 Nishimura, S., Watanabe, K., *J. Biosci.* **2006**, *31*, 465–475.

42 Johansson, M. J. O., Bystrom, A. S., *Topics Curr. Genetics* **2005**, *12*, 87–120.

43 Steinberg, S., Misch, A., Sprinzl, M., *Nucleic Acids Res.* **1993**, *21*, 3011–3015.

44 Sprinzl, M., Hartmann, T., Weber, J., Blank, J., Zeidler, R., *Nucleic Acids Res. Sequence Suppl.* **1989**, *17*, r1–r67.

45 Sprinzl, M., Vassilenko, K. S., *Nucleic Acids Res.* **2005**, *33*, D139–D140.

46 Limbach, P. A., Crain, P. F., McCloskey, J. A., *Nucleic Acids Res.* **1994**, *22*, 2183–2196.

47 Stern, L., Schulman, L. H., *J. Biol. Chem.* **1978**, *253*, 6132–6139.

48 Stern, L., Schulman, L. H., *J. Biol Chem.* **1977**, *252*, 6403–6408.

49 Yokoyama, S., Nishimura, S., *tRNA. Structure, Biosynthesis, and Function* (eds D. Söll, U. L. RajBhandary), ASM Press, Washington, **1995**, pp. 207–224.

50 Sekiya, T., Takeishi, K., Ukita, T., *Biochim. Biophys. Acta* **1969**, *182*, 411–426.

51 Sierzputowska-Gracz, H., Sochaka, E., Malkiewicz, A., Kuo, K., Gehrke, C. W., Agris, P. F., *J. Am. Chem. Soc.* **1987**, *109*, 7171–7177.

52 Agris, P. F., Sierzputowska-Gracz, H., Smith, W., Malkiewicz, A., Sochacka, E., Nawrot, B., *J. Am. Chem. Soc.* **1992**, *114*, 2652–2656.

53 Sakamoto, K., Kawai, G., Watanabe, S., Niimi, T., Hayashi, N., Muto, Y., Watanabe, K., Satoh, T., Sekine, M., Yokoyama, S., *Biochemistry* **1996**, *35*, 6533–6538.

54 Kumar, R. K., Davis, D. R., *Nucleic Acids Res.* **1997**, *25*, 1272–1280.

55 Testa, S. M., Disney, M. D., Turner, D. H., Kierzek, R., *Biochemistry* **1999**, *38*, 16655–16662.

56 Chaix, C., Duplaa, A. M., Molko, D., Teoule, R., *Nucleic Acids Res.* **1989**, *17*, 7381–7393.

57 Wada, T., Mochizuki, A., Sato, Y., Sekine, M., *Tetrahedron Lett.* **1998**, *39*, 5593–5596.

58 Wada, T., Mochizuki, A., Sato, Y., Sekine, M., *Tetrahedron Lett.* **1998**, *39*, 7123–7126.

59 Wada, T., Sato, Y., Honda, F., Kawahara, S., Sekine, M., *J. Am. Chem. Soc.* **1997**, *119*, 12710–12721.

60 Wada, T., Kobori, A., Kawahara, S., Sekine, M., *Tetrahedron Lett.* **1998**, *39*, 6907–6910.

61 Wada, T., Kobori, A., Kawahara, S., Sekine, M., *Eur. J. Org. Chem.*, **2001**, 4583–4593.

62 Saenger, W., *Principles of Nucleic Acid Structure*, Springer-Verlag, New York, **1984**.

63 Parthasarathy, R., Ginell, S. L., De, N. C., Chheda, G. B., *Biochem. Biophys. Res. Commun.* **1978**, *83*, 657–663.

64 Kobori, A., Miyata, K., Ushioda, M., Seio, K., Sekine, M., *J. Org. Chem.* **2002**, *67*, 476–485.

65 Kawai, G., Hashizume, T., Yasuda, M., Miyazawa, T., McCloskey, J. A., Yokoyama, S., *Nucleosides Nucleotides* **1992**, *11*, 759–771.

66 Kawai, G., Ue, H., Yasuda, M., Sakamoto, K., Hashizume, T., McCloskey, J. A., Miyazawa, T., Yokoyama, S., *Nucleic Acids Symposium Ser.* **1991**, *25*, 49–50.

67 Miyata, K., Kobori, A., Tamamushi, R., Ohkubo, A., Taguchi, H., Seio, K., Sekine, M., *Eur. J. Org. Chem.*, **2006**, 3626–3637.

68 Smith, W. S., Sierzputowska-Gracz, H., Sochacka, E., Malkiewicz, A., Agris, P. F., *J. Am. Chem. Soc.* **1992**, *114*, 7989–7997.

69 Mazumdar, S. K., Saenger, W., *J. Mol. Biol.* **1974**, *85*, 213–229.

70 Kumar, R. K., Davis, D. R., *Nucleic Acids Res.* **1997**, *25*, 1272–1280.

71 Kierzek, E., Kierzek, R., Turner, D., Catrina, I., *Biochemistry* **2006**, *45*, 581–593.

72 Kawahara, S., Uchimaru, T., Sekine, M., *J. Mol. Struct. (Theochem)* **2000**, *530*, 109–117.

73 Kawahara, S., Wada, T., Kawauchi, S., Uchimaru, T., Sekine, M., *J. Phys. Chem. (A)* **1999**, *103*, 8516–8523.

74 Kumar, R. K., Davis, D. R., *Nucleic Acids Res.* **1997**, *25*, 1272–1280.

75 Kumar, R. K., Davis, D. R., *Nucleosides Nucleotides* **1997**, *16*, 1469–1472.

76 Okamoto, I., Seio, K., Sekine, M., *Nucleic Acids Res.* (unpublished results).

77 Muimelis, R. G., Nambiar, K. P., *Tetrahedron Lett.* **1993**, *34*, 3813–1816.

78 Crestini, C., Saladino, R., Nicoletti, R., *Tetrahedron Lett.* **1993**, *34*, 1631–1634.

79 Saladino, R., Mincione, E., Cearsini, C., Mezzetti, M., *Tetrahedron* **1996**, *52*, 6759–6780.

80 Sochacka, E., Fratczak, I., *Tetrahedron Lett.* **2004**, *45*, 6729–6731.

81 Kumar, R. K., Darrell, R. D., *Nucleic Acids Res.* **1997**, *25*, 1272–1280.

82 Moriguchi, T., Asai, N., Okada, K., Seio, K., Sasaki, T., Sekine, M., *J. Org. Chem.* **2002**, *67*, 3290–3300.

83 Wood, M. L., Esteve, A., Morningstar, M. L., Kuziemko, G. M., Essigmann, J. M., *Nucleic Acids Res.* **1992**, *20*, 6023–6032.

84 Ohkubo, A., Seio, K., Sekine, M., *Tetrahedron. Lett.* **2004**, *45*, 363–366.

85 Ohkubo, A., Seio, K., Sekine, M., *J. Am. Chem. Soc.* **2004**, *126*, 10884–10896.

86 Tsukahara, T., Nagasawa, H., *Sci. Tech. Adv. Mat.* **2004**, *5*, 359–362.

87 Gryaznov, S. M., Letsinger, R. L., *J. Am. Chem. Soc.* **1991**, *113*, 5876–5877.

88 Uesugi, S., Ikehara, I., *J. Am. Chem. Soc.* **1977**, *99*, 3250–3253.

89 Cysewski, P., Vidal-Madjar, C., Jordan, R., Olinski, R., *J. Mol. Struct. (THEOCHEM)* **1997**, *397*, 167–177.

90 Mariaggi, N., Teoule, R., *Bull. Soc. Chim. Fra.* **1976**, *9–10*, Pt. 2, 1595–1598.

91 Lipkind, G. M., Karpeiskii, M. I., *Mol. Biol.* **1982**, *16*, 712–719.

92 Seela, F., Kaiser, K., *Helv. Chim. Acta* **1988**, *71*, 1813–1823.

93 Seela, F., Kehne, A., A. *Biochemistry* **1985**, *24*, 7556–7561.

94 Kobori, A., Miyata, K., Ushioda, M., Seio, K., Sekine, M., *Chem. Lett.*, **2002**, 16–17.

95 Ohkubo, A., Kasuya, R., Aoki, K., Kobori, A., Taguchi, H., Seio, K., Sekine, M., *J. Org. Chem.* (unpublished results).

96 Kataoka, M., Hayakawa, Y., *J. Org. Chem.* **1999**, *64*, 6087–6089.

97 Seio, K., Sasami, T., Ohkubo, A., Ando, K., Sekine, M., *J. Am. Chem. Soc.* **2007**, *129*, 1026–1027.

98 Robins, R. K., Horner, J. K., Greco, C. V., Noell, C. Y., Beames, C. G. Jr., *J. Org. Chem.* **1963**, *28*, 3041–3046.

99 Cook, P. D., Rousseau, R. J., Mian, A. M., Dea, P., Meyer, R. B., Jr., Robins, R. K., *J. Am. Chem. Soc.* **1976**, *98*, 1492–1498.

100 Minakawa, N., Matsuda, A., *Tetrahedron* **1993**, *49*, 557–570.

101 Tanaka, H., Hirayama, M., Matsuda, A., Miyasaka, T., Ueda, T., *Chem. Lett.*, **1985**, 589–592.

102 Tanaka, H., Hirayama, M., Suzuki, M., Miyasaka, T., Matsuda, A., Ueda, T., *Tetrahedron* **1986**, *42*, 1971–1980.

103 Seela, F., Debelak, H., Andrew, L., Beigelman, L., *Helv. Chim. Acta* **2003**, *86*, 2726–2740.

104 Seio, K., Sasami, T., Tawarada, R., Sekine, M., *Nucleic Acids Res.* **2006**, *34*, 4324–4334.

105 Park, T., Zimmerman, S. C., Nakashima, S., *J. Am. Chem. Soc.* **2005**, *127*, 6520–6521.

106 Park, T., Todd, E. M., Nakashima, S., Zimmerman, S. C., *J. Am. Chem. Soc.* **2005**, *127*, 18133–18142.

8
The Properties of 4′-Thionucleosides

Masataka Yokoyama

8.1
Introduction

Thionucleosides are those nucleosides which have a sugar moiety wherein the ring oxygen is replaced by sulfur. After being stimulated initially by the synthesis of thio-D-xylose, 5-thio-D-ribose, and 5-thio-D-glucose, Reist and colleagues prepared the first so-called thionucleosides in 1964, with the aim of identifying their biological activities. Subsequently, Bobek *et al.* synthesized a series of thionucleosides and evaluated their activity against HIV. Today, many organic chemists have contributed to the synthesis of many types of thionucleoside, of which the reported forms may be classified into four groups: (1) 4′-thionucleosides; (2) isothionucleosides; (3) L-thionucleosides; and (4) thioxonucleosides (Figure 8.1). Although the first systematic review of the four thionucleoside groups was published in 2000 [1], Jeong and coworkers more recently reviewed the field of 4′-thionucleosides in greater detail [2]. The syntheses and biological activities of the four groups will be summarized in this chapter, which also includes a literature survey extending to May 2006.

8.2
Synthesis of 4′-Thionucleosides

Reist and colleagues [3] were initially interested in the synthesis of 4′-thio-D-ribose and its derivatives due to the widespread occurrence in biological systems of D-ribose, wherein the substitution of sulfur for oxygen in biologically important compounds has resulted in analogues of chemotherapeutic value. The synthesis commences with a cyclization of L-lyxose with 0.5% methanolic HCl to produce **1**, which was then allowed to react with 2,2-dimethoxy propane (DMP) to form **2**. Tosylation of **2** afforded **3** in 53% overall yield from L-lyxose. Reaction of **3** with potassium thiolbenzoate at 115 °C for 72 h gave **4**, which was then deacetonated with 66%

Modified Nucleosides: in Biochemistry, Biotechnology and Medicine. Edited by Piet Herdewijn

ISBN: 978-3-527-31820-9

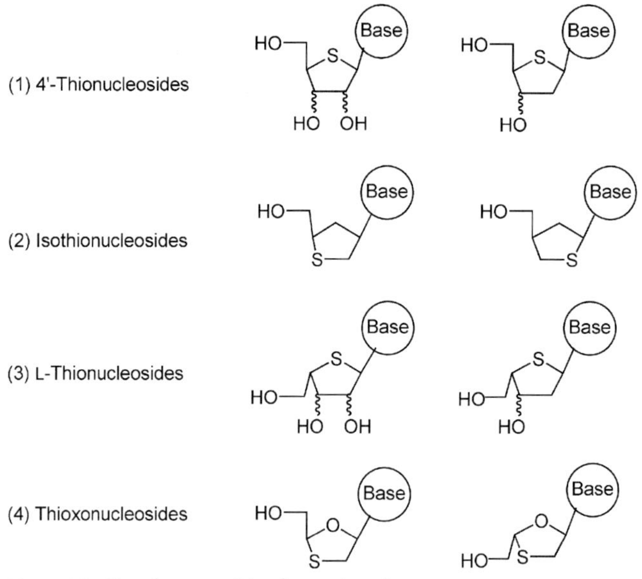

Figure 8.1 Classification of the thionucleosides.

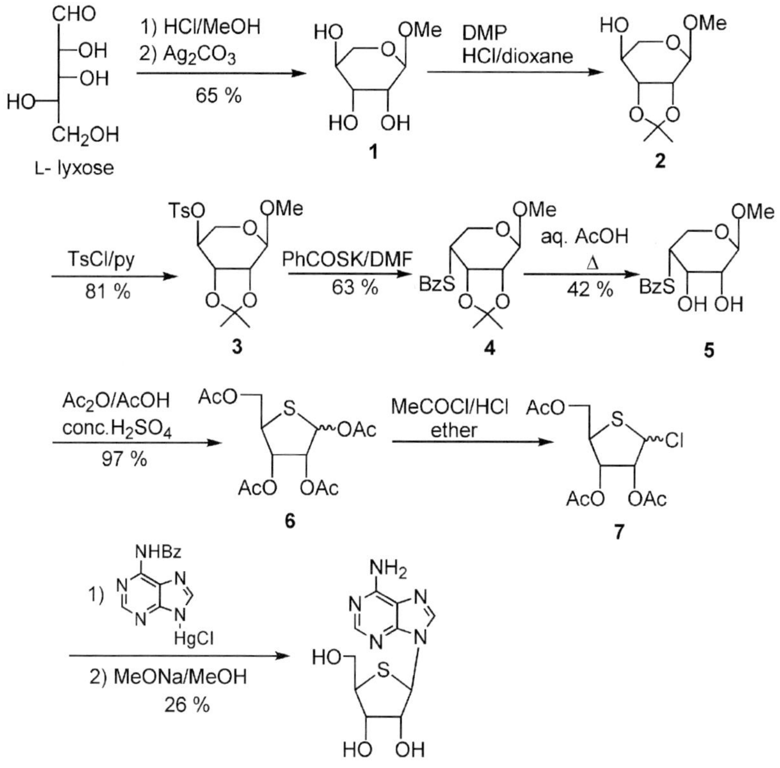

Scheme 8.1

8

aqueous acetic acid to **5**. The latter compound was then treated with acetic anhydride, acetic acid, and concentrated H_2SO_4 to form **6**, the reaction of which with acetyl chloride and anhydrous HCl in anhydrous diethyl ether at 0 °C for 4 days gave **7** in quantitative yield. Compound **7** was condensed with chloromercury-6-benzamidopurine and then deacylated to **8** (Scheme 8.1).

Next, **9** was allowed to react with potassium thiolbenzoate in dimethylformamide (DMF) to afford **10** in 75% yield. Acetolysis of **10** with acetic anhydride, acetic acid, and sulfuric acid gave **11**, the conversion of which to an adenine nucleoside **12** was accomplished by the chloro sugar condensation with chloromercury-6-benzamidopurine (as described for the synthesis of **8**). Acetonation of **12** gave **13**, which was converted to the mesylate **14** by methanesulfonylation (MsCl/pyridine). Deacetonation of **14** to **15** was followed by treatment with methanolic sodium methoxide to give the epoxide **16**, the nucleophilic cleavage of which with sodium acetate gave **17** (Scheme 8.2) [4].

Bobek and coworkers have reported the synthesis and biological activity of a number of adenosine analogues, modified both at the 6-position of heterocycle and in the 4-position of the carbohydrate [5]. A purine thionucleoside **19** was synthesized by condensation of **7** with the chloromercury derivative of 6-chloropurine, followed by

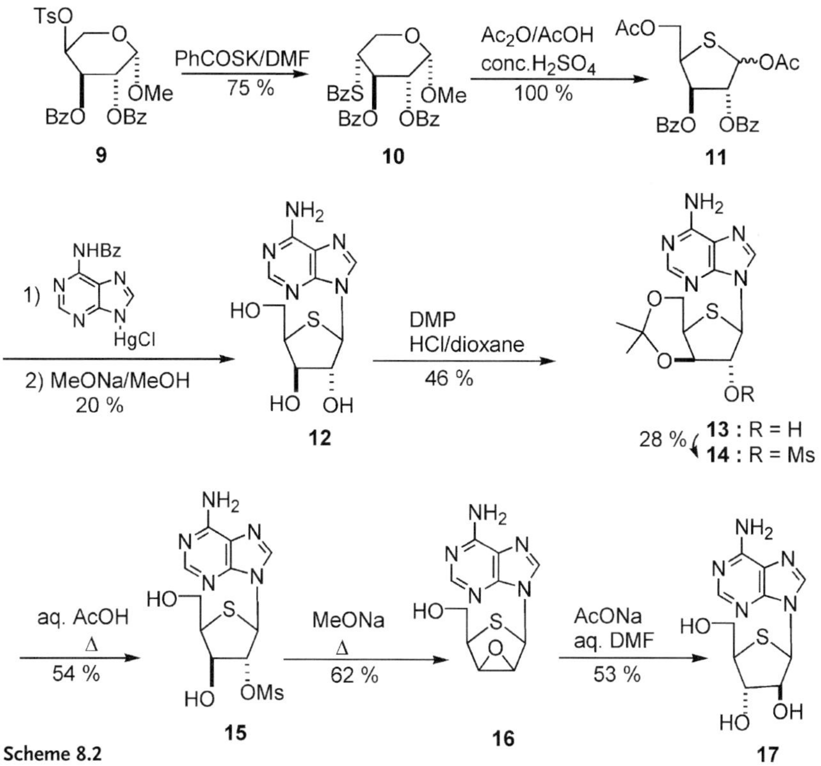

Scheme 8.2

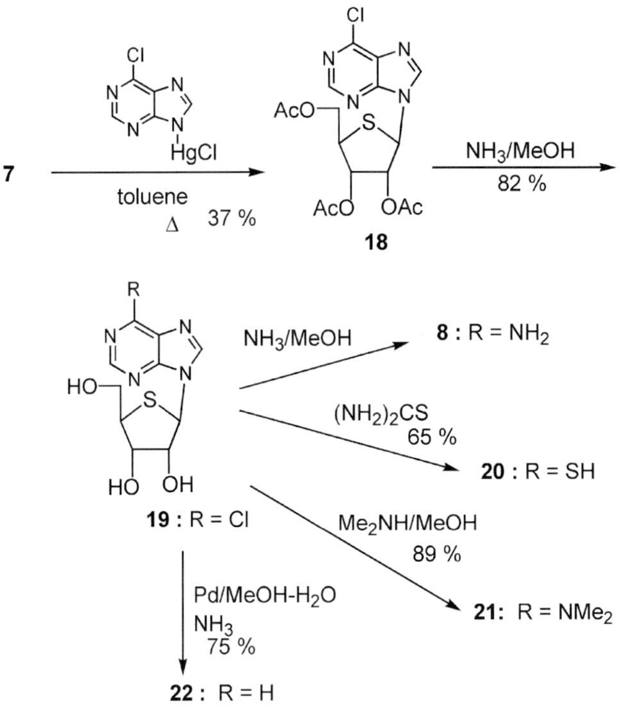

Scheme 8.3

ammonolysis. Nucleophilic substitution was used to replace the 6-chloro with NH_2, SH, Me_2N, or H to give **8**, **20**, **21**, or **22**, respectively (Scheme 8.3).

The antibiotic toyocamycin is an analogue of adenosine which contains a CN group. Although, in experimental systems this antibiotic showed marked anti-tumor activity, the severe localized toxicity which it caused in human patients limited its clinical value. In an attempt to decrease such toxicity, two structural modifications of the toyocamycin molecule were made. In one compound, the oxygen in the ring of the carbohydrate moiety was replaced by sulfur, while the other compound involved, in addition to this sulfur replacement, a substitution of the 4- and 6-positions of the heterocycle with Cl and NH_2 groups, respectively [6]. Thus, treatment of 2,3,5-tri-*O*-acetyl-4-thio-D-ribofuranosyl chloride **7** with the chloromercury derivative of 4-acetoamido-6-bromo-5-cyanopyrrolo-[2,3-d]pyrimidine gave **23**, which was deacetylated with methanolic NH_3 to **25**. Dehalogenation of **25** with H2-Pd in methanol gave 4′-thiotoyocamycin (**26**), while reaction with NH_3/MeOH at 115–120 °C produced **27**. The chloromercury salt of 6-bromo-4-chloro-5-cyanopyrolo[2,3-*d*]pyrimidine was condensed with **7** in anhydrous toluene to give **24** in 85% yield. Removal of the protection groups from **24** was accomplished with NH_3/MeOH at 5 °C to furnish **28** (Scheme 8.4).

Ritchie and Szarek subsequently synthesized 4′-thiocordycepin (**37**), the thioanalogue of the 3′-deoxyadenine nucleoside antibiotic, cordycepin [7]. This synthesis commenced from **29**, which was treated with silver nitrate and iodine to afford **30**.

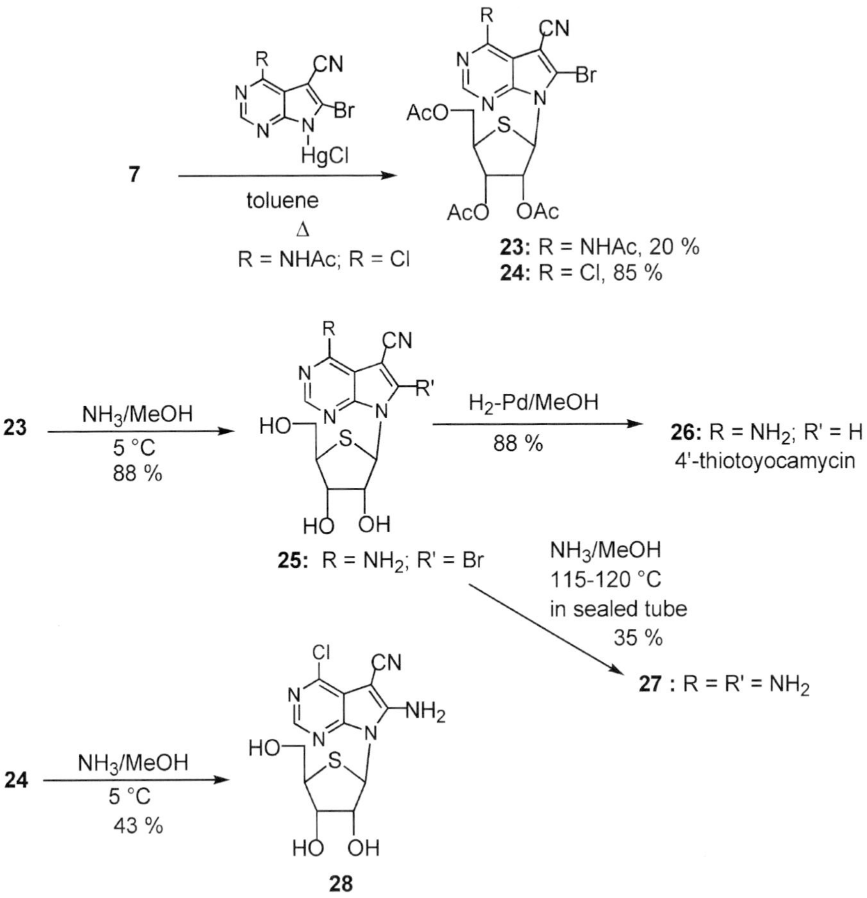

Scheme 8.4

Hydrogenolysis of **30** to **32** was carried out by a two-step reaction of de-esterification and deiodination. Further hydrogenation of **31** over Raney nickel afforded **32** in 55% overall yield from **29**. The usual tosylation of **32** gave **33**, which was then heated for 48 h at refluxing temperature in methanol containing toluene-a-thiol and sodium methoxide to produce **34** in almost quantitative yield. All of the benzoyl groups of **34** were removed with sodium in liquid ammonia, and the resulting free sugar was converted into **35** as 1 : 3 mixture of α- and β-anomers on treatment with benzoyl chloride/pyridine. Each anomer of **35** was separately treated with 6-benzamide-9-chloromercurypurine in the presence of titanium tetrachloride to give **36**, the debenzoylation of which with sodium methoxide afforded **37** (Scheme 8.5).

Jeong and coworkers synthesized 2-chloro-N^6-substituted-4′-thioadenosine-5′-uronamides as highly potent and selective agonists at the human A_3 adenosine receptor [8]. The synthesis commenced from **38**, which was protected with acetone to **39** in the usual manner. The ring-opening of **39** with lithium aluminum hydride (LAH) afforded **40**, which was mesylated to **41** and then recyclized with sodium

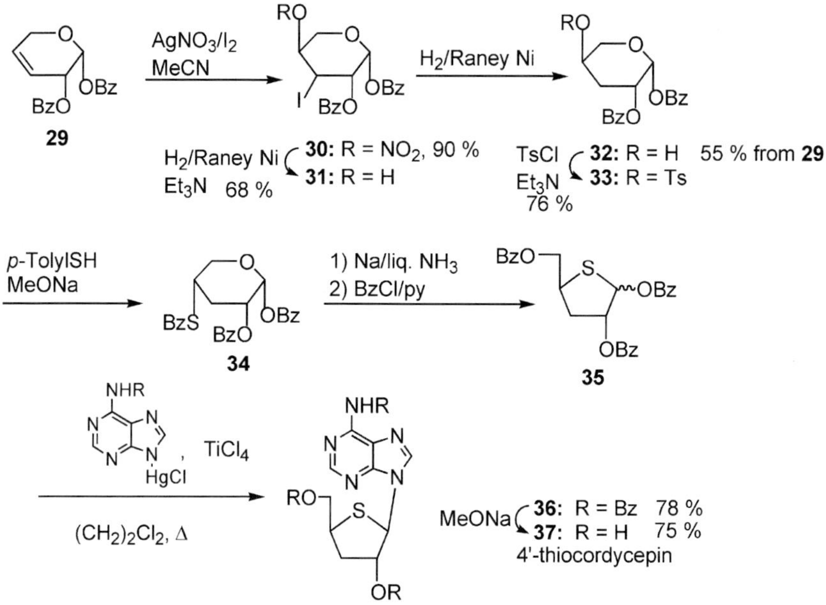

Scheme 8.5

sulfide to **42**. Deprotection of the 5,6-hydroxyl groups of **42** was then effected with 30% acetic acid to afford **43**. Oxidative cleavage with $Pb(OAc)_4$ and subsequent reduction with sodium borohydride of **43** led to the formation of **44**. Protection of the 5-hydroxyl group in **44** was achieved with benzoyl chloride/pyridine to yield **45**, the oxidation of which with *m*-chloroperbenzoic acid (MCPBA) afforded the sulfoxide **46**; the latter was converted to the glycosyl donor **47** by heating in acetic anhydride at 100 °C. Compound **47** was directly condensed with silylated 2,6-dichloropurine to give **48** (β-anomer) in 60% yield ($\alpha/\beta = 1/9$). The direct condensation of **46** with silylated 2,6-dichloropurine afforded **48** in 54% yield, with trace amounts of the α-anomer. Compound **48** was treated with various amines to give **49**, which was changed to **50** having another protective group [*tert*-butyldimethylsilyl (TBS)] in the usual manner. The oxidation of **50** with pyridinium dichromate (PDC) afforded the acid derivative **51**, the reaction of which with various amines using 1-ethyl-3-(3′-dimethylaminopropyl) carbodiimide (EDC) and 1-hydroxybenzotriazole (HOBt) afforded the desired uronamides **52** (Scheme 8.6).

1-β-D-Arabinofuranosylcytosine has well-known activity against both rodent and human neoplasms. Unfortunately, however, its clinical use produces megaloblastosis and chromosomal alteration in bone marrow, and consequently Whistler and coworkers determined to synthesize the sulfur analogue **56**, with the hope that it might be less toxic than the parent compound [9]. Hence, **6** and bis(trimethylsilyl)-*N*-acetylcytosine were condensed with stannic chloride as a catalyst to afford **53** together with its α-anomer in 3% yield. The acetyl groups were removed from **53** in NH_3/MeOH at 100 °C in a sealed tube to give **54**, and cyclization between the 2- and 2′-positions of **54** was conducted in the presence of phosphoryl chloride in DMF

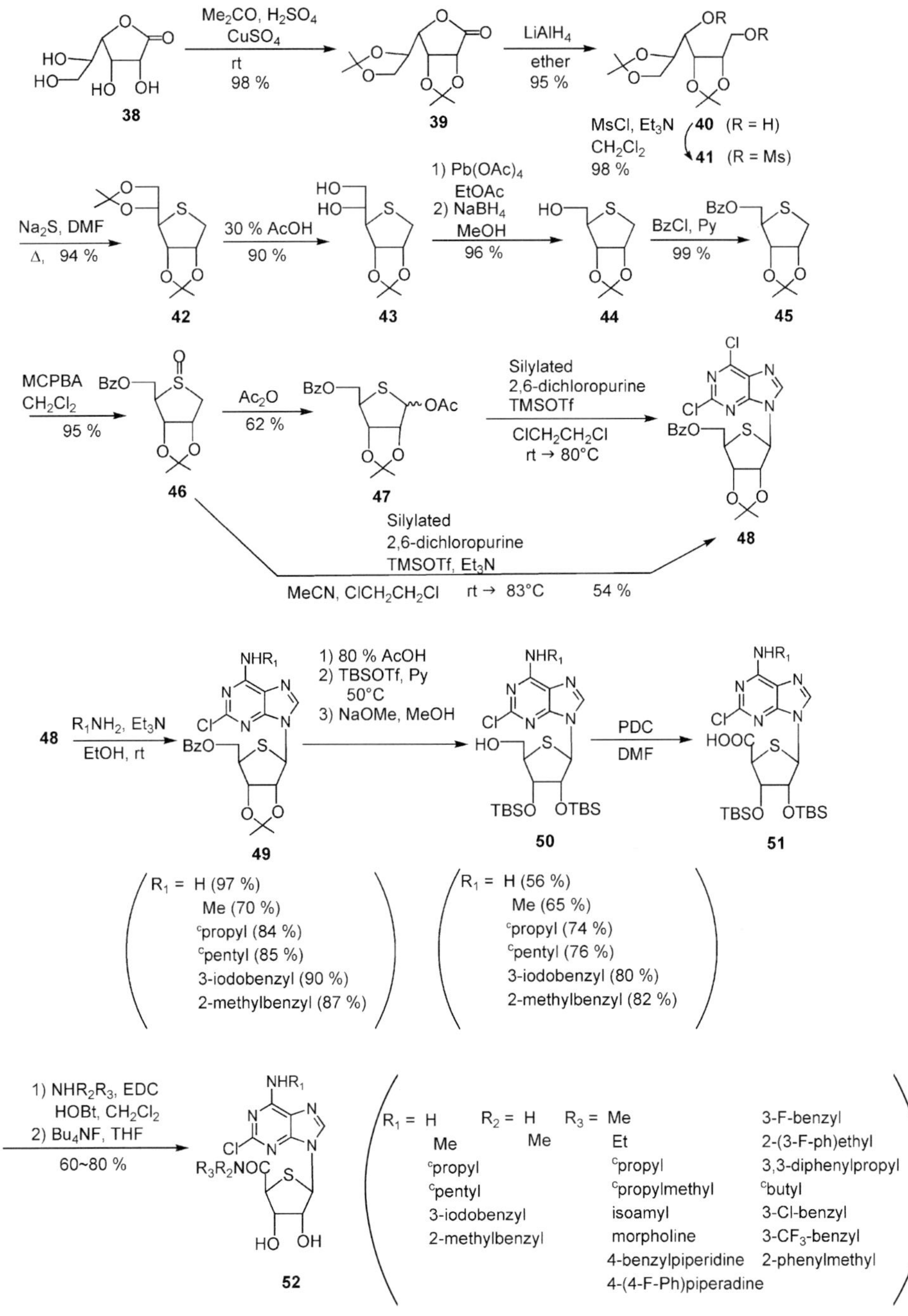

Scheme 8.6

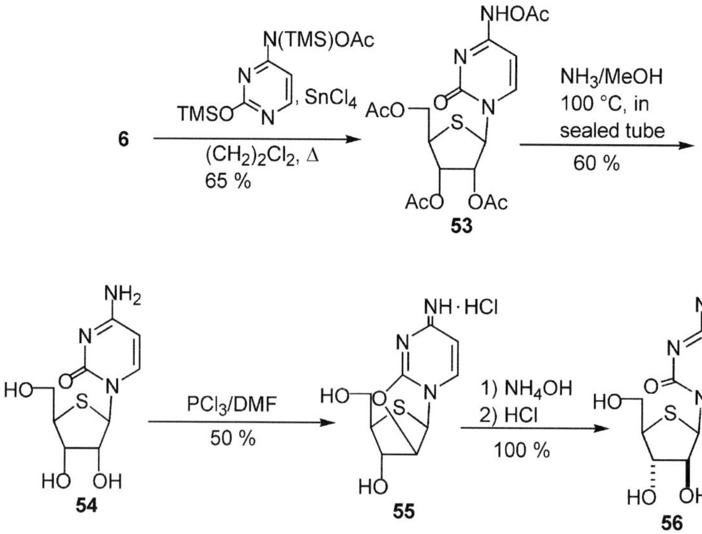

Scheme 8.7

at 25 °C; this resulted in the formation of **55**, the ring-opening of which by aqueous ammonia gave **56** in quantitative yield (Scheme 8.7).

Because of the pronounced biological activity of 5-fluorinated pyrimidines, and the fact that the 4′-thio derivatives of various purine nucleosides analogues have been shown to retain their growth inhibitory activity against cells resistant to the corresponding ribofuranosyl analogues, a number of 4′-thio-5-halogenopyrimidine nucleosides were synthesized [10]. Condensation between 2,4-bis(trimethylsilyl ethers) of

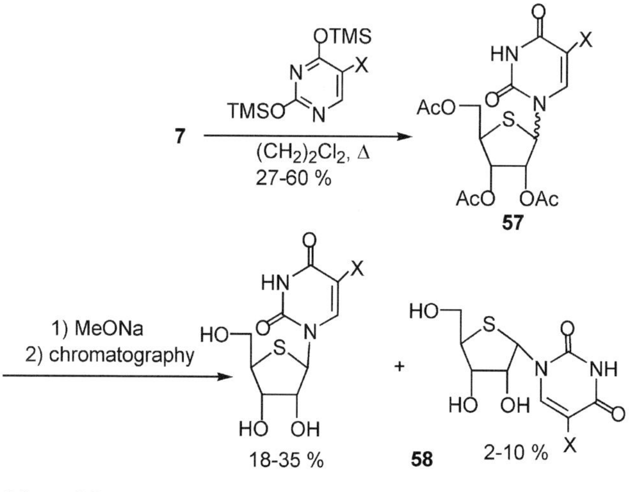

Scheme 8.8

X = F, Cl, Br, I

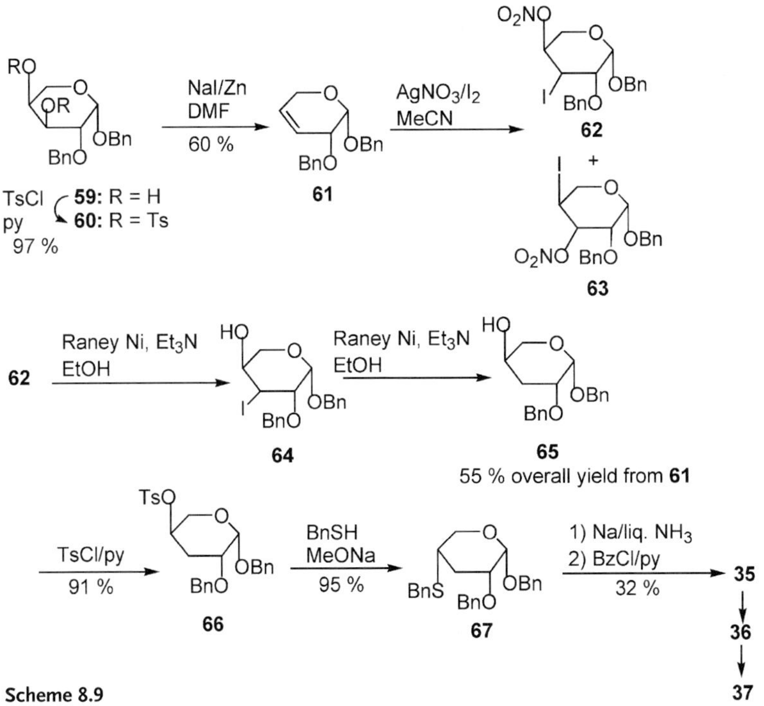

Scheme 8.9

5-halogenouracil and **7** was performed using a modified Hilbert–Johnson reaction. The obtained **57** was a mixture of α- and β-anomers, which were changed to the deprotected compound **58** with sodium methoxide. The α- and β-anomers of **58** could be separated by fractional crystallization from ethanol, and by chromatography on a dry column of silica gel using MeOH/$CHCl_3$ as an eluent (Scheme 8.8).

Szarek and coworkers also reported a better synthetic method for 4′-thiocordycepin (**37**) starting from benzyl-protected compound **61** in place of benzoyl-protected compound **29**, and using the same synthetic procedure as described in Scheme 8.5 (Scheme 8.9) [11].

For the study of biological activity, 1-(2-deoxy-4-thio-β-D-erythro-pentofuran osyl)-5-fluorouracil (**72**) was synthesized by the condensation of **70** with the trimethylsilyl derivative of 5-fluorouracil catalyzed by mercuric oxide/mercury bromide. The reaction provided a favorable yield of the desired β-D-anomer **72**, whereas the use of stannic chloride as catalyst gave primarily its α-anomer together with only a trace amount of **72** (Scheme 8.10) [12].

Some biologically active nucleosides such as 2′-deoxy-4′-thiocytidine (**79**), 2′-deoxy-4′-thiouridine (**76**), and 4′-thiothymidine (**77**) were synthesized by Secrist and colleagues [13]. The precursor **68** for all the three compounds was prepared from L-arabinose, by using the procedure reported by Bobek *et al.*, to convert a tolyl-protected **69** in the usual manner. Experimental modification of usual acetolysis conditions provided the acetyl sugar **73** as a 1 : 1 mixture of anomers which was stable

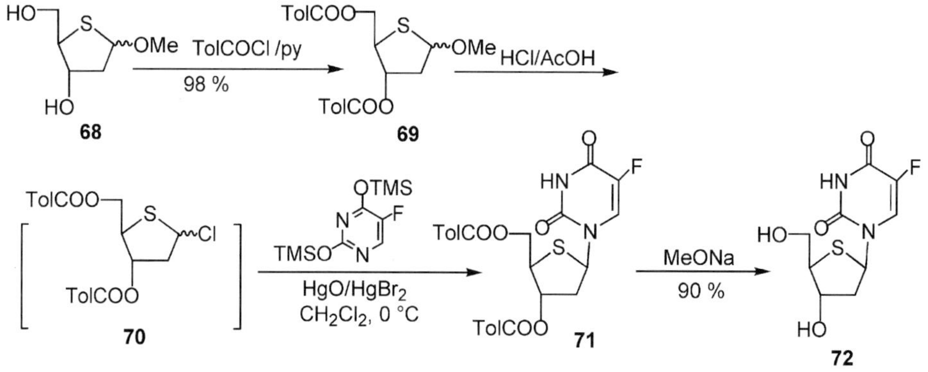

Scheme 8.10

for some time at room temperature. The trimethylsilyl triflate-catalyzed coupling of **73** with uracil, thymine, and cytosine afforded the corresponding nucleosides. Both, **74** and **75** could be separated as α- and β-anomers by fractional crystallization, whereas the cytosine analogue could not be separated as pure anomers. In order to obtain the pure cytosine analogue, **73** was treated with 1,2,4-triazole and *p*-chlorophenyl phosphorodichloridate in pyridine to give **78**, which was converted directly to **79** by sequential treatment with NH_4OH and MeONa (Scheme 8.11).

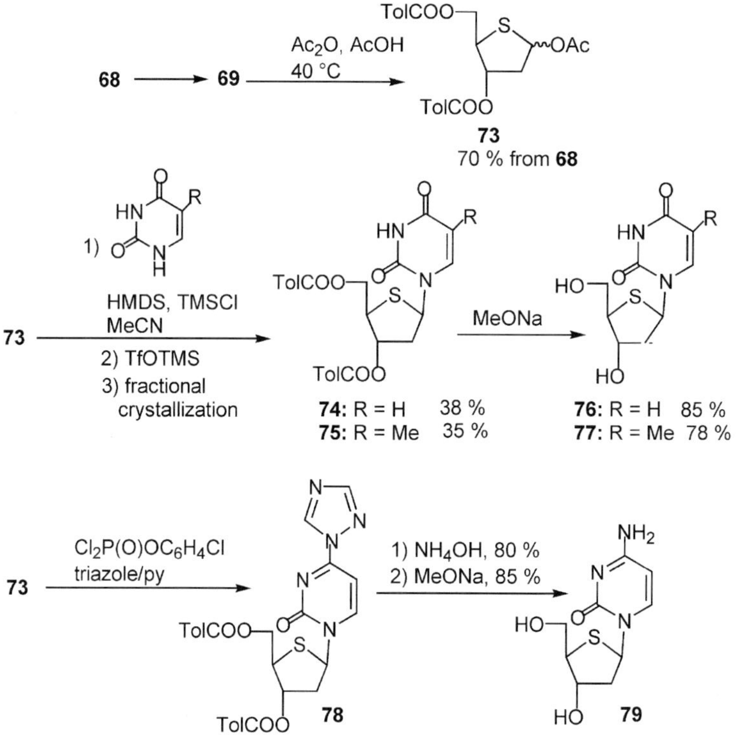

Scheme 8.11

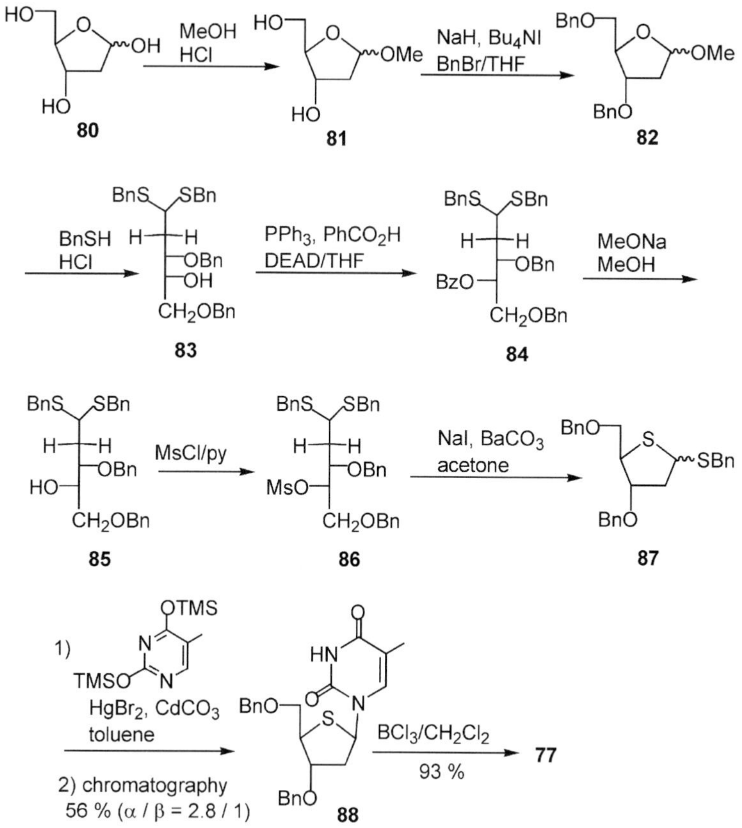

Scheme 8.12

A practical seven-step synthesis of benzyl 3,5-di-*O*-benzyl-2-deoxy-1,4-dithio-D-erythropentofuranoside **87** from 2-deoxy-D-ribose **80** was reported by Walker and coworkers, together with some 4′-thio-2′-deoxynucleosides which have potentially useful biological activity [14]. Initially, **80** was converted to the corresponding methyl riboside **81**, which was then protected with benzyl groups to **82**. Reaction of the latter with benzylmercaptan/HCl gave **83**. The stereostructure of the hydroxyl group in the 4-position of **83** was inverted by the Mitsunobu procedure to **85** via **84**. After being mesylated, **85** was treated with NaI and $BaCO_3$ to afford the desired thiosugar **87**. Next, the 4′-thio analogue of thymidine (**88**) was synthesized from **87** by the method of Horton and Marovs, as shown in Scheme 8.12 [15]. The obtained **88** consisted of a 2.8 : 1 mixture of the α- and β-anomers, which could be separated chromatographically or by fractional crystallization from MeOH. Deprotection of the separated anomers was achieved using borontrichloride to afford **77** in 93% yield.

Miller and coworkers synthesized the known 4′-thionucleosides (**94**) starting from the L-ascorbate-derived epoxide **89** [16]. Compound **89** was opened exclusively at the terminal carbon by a lithium salt of the formaldehyde dithioacetals **90** to give

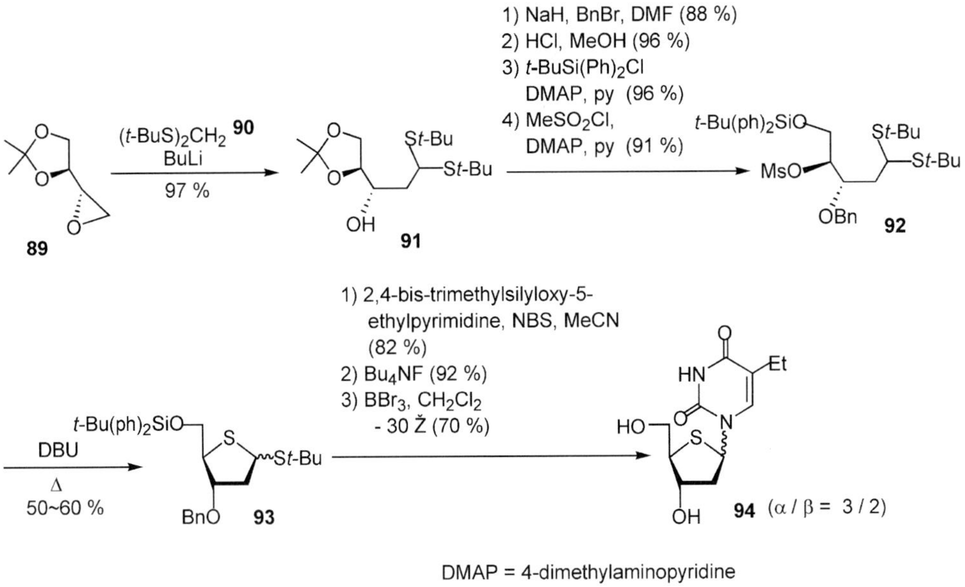

89 91 92 93 94 (α / β = 3 / 2)

Scheme 8.13

the hydroxydithioacetals **91**. The standard deprotection–protection procedures (benzylation of 3-OH; silylation of 5-OH: mesylation of 4-OH) of **91** afforded the mesylate **92**, which was cyclized by heating with 1,8-diazabicyclo[5.4.0]-7-undecene (DBU) to **93** as a mixture of α- and β-anomers. The desired compound **94** was obtained from **93** by coupling with nucleic base (82% yield), followed by standard deprotection procedures (Bu_4NF: 92%; BBr_3: 70% yield) (Scheme 8.13). The β-anomer of **94** has been known as the 4′- thionucleosides, and demonstrating powerful anti-herpes activity [17].

A series of 2′,3′-dideoxy-4′-thionucleoside analogues of purines and pyrimidines such as 4′-thioddI (**105**), 4′-thioddC, and 4′-thioAZT were synthesized and evaluated for their activity against HIV [18]. The synthesis was carried out as follows: the starting compound **96** was prepared via **95** in three steps from L-glutamic acid, using known methodology [19]. Compound **96** was opened with sodium hydroxide and converted to **97**, which was then transformed into 4(*R*)-iodo ester **98** by treatment with triphenylphosphine, imidazole, and iodine. Displacement of the iodo group in **98** by thioacetate in toluene occurred readily to produce **99**, which was then treated with diisobutyl aluminum hydride (DIBAH) to reductively deprotect the sulfur and reduce the methyl ester to an aldehyde, thereby giving rise to **100** via spontaneous cyclization. Sugar **101** was obtained by the acetylation of **100**, employing standard conditions. Subsequently, **101** was coupled to 6-chloropurine by a modified method of Niedballa and Vorbruggen [20] with diethylaluminum chloride as the catalyst to give **102** as a 1 : 1 α/β anomeric mixture in 60% yield. Coupling of **101** and 2,6-dichloropurine gave **103** in an approximate 2 : 3 α/β anomeric ratio in 60% yield. Deprotection of **102** with Bu_4NF afforded the anomers of **104**, which were tediously separated by preparative thin-layer chromatography (TLC) to give the pure β-anomer. Treatment of α,β-**104** with adenosine deaminase converted only the β-anomer to the dideoxyinosine analogue **105**. Ammonia treatment of α,β-**104** provided **106**, which was separated by ion-exchange chromatography to give a modest yield of β-**106**. Separation of α,β-**103** was straightforward, by using preparative TLC. Pure **103**β was treated with lithium azide to give the diazido nucleoside **109** in quantitative yield, and this was then reduced with LAH to afford **110** in 80% yield. Subsequent deprotection with tetrabutylammonium fluoride (Bu_4NF) of **110** afforded **111** in 54% yield after recrystallization. Amination at C-6 of β-**103** with ethanolic ammonia gave **107** in good yield, and deprotection of **107** was carried out in the usual manner to yield **108** in 70% yield (Scheme 8.14).

2′,3′-Dideoxy-3′-*C*-(hydroxymethyl)cytidine was reported to be a potent inhibitor of HIV-1 activity *in vitro*, and to represent a lead structure of a new type [21]. In order to investigate their structure–activity relationship, these thio analogues were synthesized as follows [22]. Benzyl dithiofuranoside **112** was condensed with silylated thymine in the presence of trimethylsilyl triflate and mercuric acetate in CH_2Cl_2 to produce an anomeric mixture of the corresponding protected nucleoside. Deblocking with methanolic ammonia and separation of the anomers by HPLC gave **113** and **114** in 35% and 24% yields, respectively. The conversion of **112** to **115** was achieved in 99% yield by treatment with mercuric acetate in acetic acid. Condensation of **115** with silylated cytosine, followed by deprotection, gave **116** and **117** in 36% and 28% yields, respectively. Trimethylsilyl triflate-promoted condensation of **115** with 6-chloropurine gave an anomeric mixture of the corresponding nucleosides, which were separated by HPLC after deblocking to give **118** and **119** in 33% and 28% yields, respectively (Scheme 8.15).

Walker and coworkers were interested in the relationship between the sugar conformation (C-2′-endo, C-3′-exo in **77**) and biological activity. Thus, the sulfone of 4′-thiothymidine **122** was prepared by the usual oxidation of **120** with MCPBA, followed by deprotection, and evaluated for anti-viral activity. Based on the results of X-ray analyses, the lack of biological activity was attributed to the glycosidic torsion angle encountered (Scheme 8.16) [23].

Rassu and colleagues synthesized anti-HIV-active 2′,3′-dideoxy-4′-thiocytidine (**129**) by using 2-(*tert*-butyldimethylsilyloxy)thiophene as a versatile carbon nucleophile and chiral glyceraldehyde acetonide **123** [24]. The diastereoselective addition of 2-(*tert*-butyldimethylsilyloxy)thiophene to **123** in the presence of $BF_3{\cdot}OEt_2$ in CH_2Cl_2 resulted in preferential formation of the 4*S*-adduct **124** (73%), together with trace amounts of the 4*R*-adduct **125**. This diastereoisomeric mixture was treated with successive procedures of reduction (H_2, Pd/C, AcONa), deblocking of acetonide (AcOH), oxidative cleavage of vicinal diol ($NaIO_4$), reduction of aldehyde ($NaBH_4$), protection of alcohol (TBDPSCl, imidazole), reduction of carbonyl group (LAH), and acetylation of alcohol (Ac_2O) to afford **128** in overall yield 50% based on the **124** and **125** mixture. The final coupling of **128** with cytosine was carried out in the usual manner to give an α/β (1 : 1) mixture of the desired nucleosides (**129**) in 65% yield. Following the separation of anomers by preparative TLC, **129** was obtained as a semicrystalline compound (Scheme 8.17).

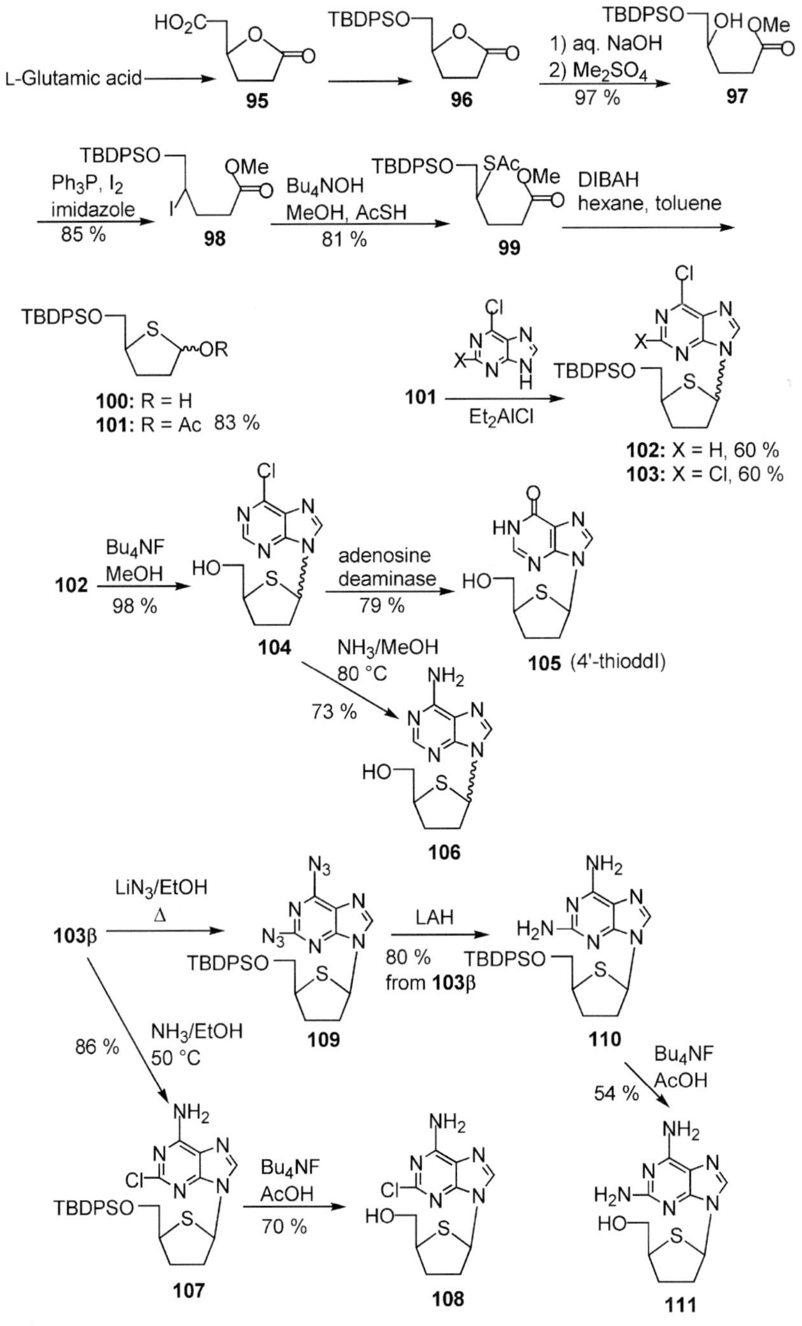
L-Glutamic acid
95
96
1) aq. NaOH
2) Me2SO4
97 %
97
Ph3P, I2
imidazole
85 %
98
Bu4NOH
MeOH, AcSH
81 %
99
DIBAH
hexane, toluene
100: R = H
101: R = Ac 83 %
101
Et2AlCl
102: X = H, 60 %
103: X = Cl, 60 %
102
Bu4NF
MeOH
98 %
104
adenosine
deaminase
79 %
105 (4'-thioddI)
NH3/MeOH
80 °C
73 %
106
103β
LiN3/EtOH
Δ
109
LAH
80 %
from 103β
110
Bu4NF
AcOH
54 %
111
NH3/EtOH
50 °C
86 %
107
Bu4NF
AcOH
70 %
108

Scheme 8.14

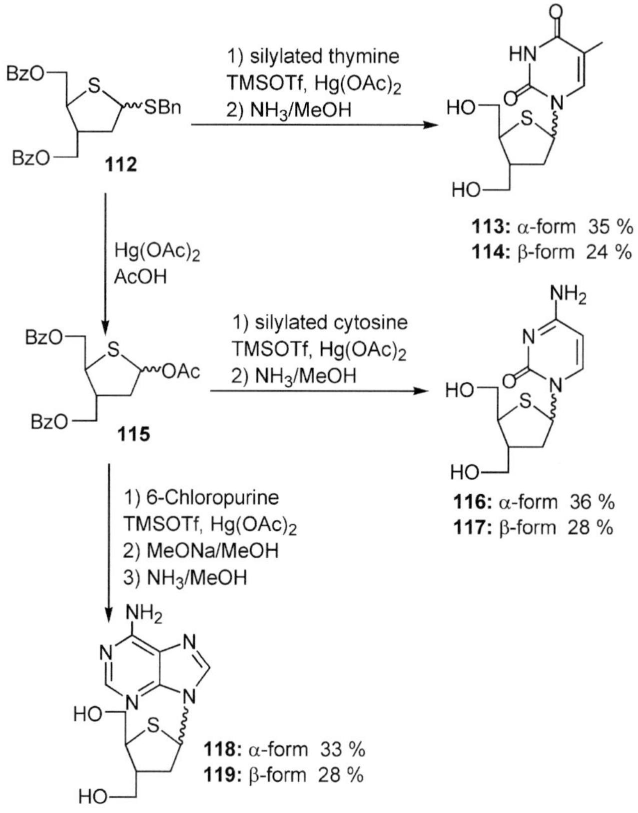

Scheme 8.15

An anti-HIV agent, 2′-deoxy-5-ethyl-4′-thiouridine **138** was synthesized starting from methyl 2-deoxy-3,4-*O*-thiocarbonyl-β-D-ribopyranoside **132**, which was prepared from 2-deoxy-D-ribose **130** via **131** [25]. A bromide ion-catalyzed O–S rearrangement of **132** produced the isomers **133** and **134**, which could be separated by

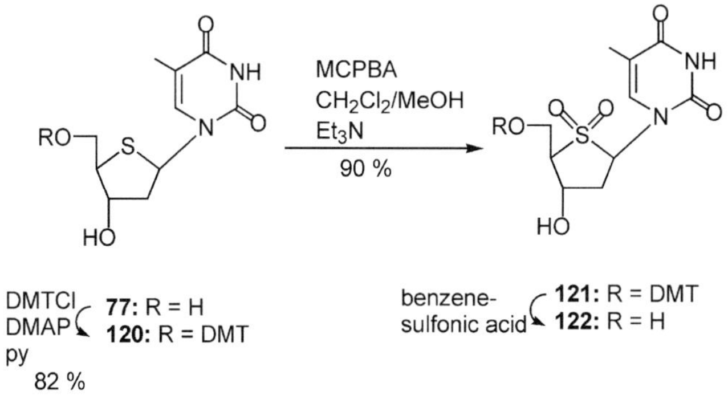

Scheme 8.16

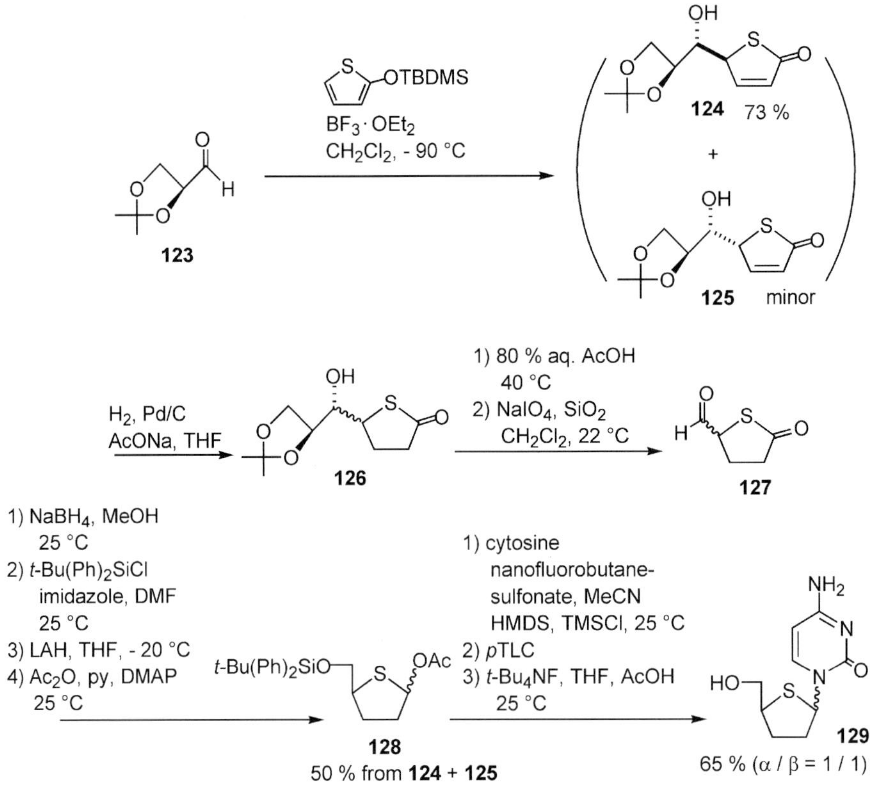

Scheme 8.17 HMDS = 1,1,1,3,3,3-hexamethyldisilazane

flash chromatography of the ammonolysis to products **135** and **136**, respectively. Reaction of **135** with the *in-situ* silylated 5-ethyluracil provided the furanose nucleoside **137** directly, presumably via ring opening, rearrangement, and subsequent elimination of MeOH. The desired thionucleoside **138** was obtained by treatment of **137** with sodium methoxide (Scheme 8.18).

A series of 2′-deoxy-4′-thioribonucleosides **139** was synthesized via trans-*N*-deoxyribosylase-catalyzed reaction of 2′-deoxy-4′-thiouridine **76** with a variety of purine bases by Van Draanen and coworkers. Compound **140** was obtained from **139a** (R = OMe) using adenosine deaminase (Scheme 8.19) [26]. This synthetic procedure is an improvement over methods used previously to prepare purine 4′-thionucleosides.

A series of 5-substituted 2′-deoxy-4′-thiopyrimidine nucleosides **143** was synthesized starting from **141** in the usual way (Scheme 8.20) [27], and evaluated as potential anti-viral agents.

The 4′-thioDMDC (4′-thio-2′-deoxy-2′-methylenecytidine) was synthesized with the hope of finding anti-HIV activity [28]. For this purpose, compound **144** was converted to **145** by the usual procedures shown in Scheme 8.21. Compound **145** was

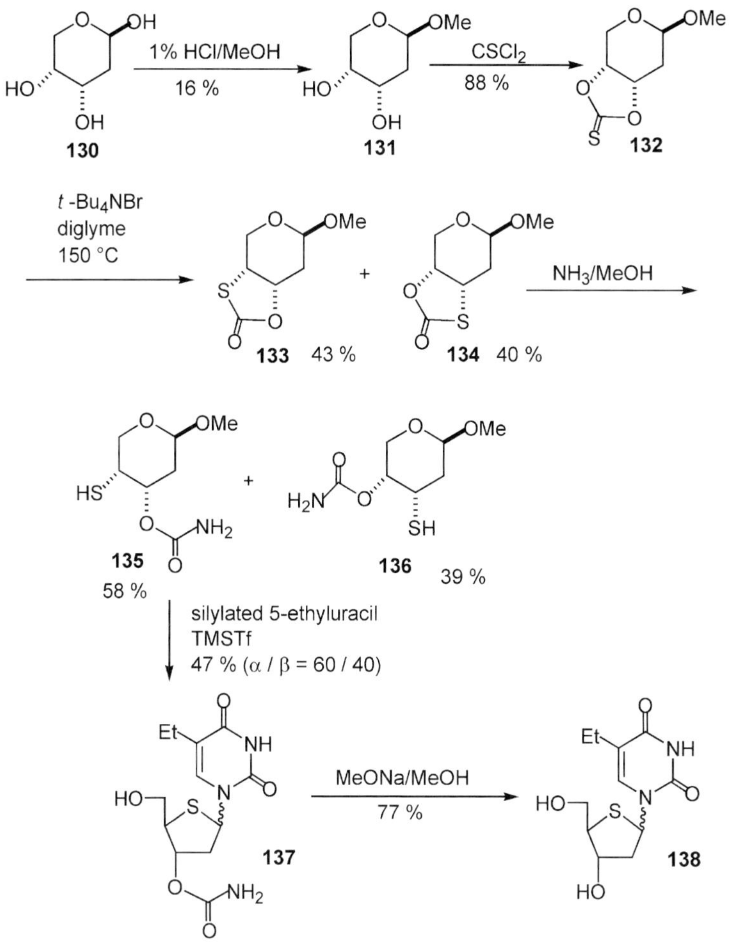

Scheme 8.18

then subjected to acidic methanolysis, producing an anomeric mixture of **146** in high yield. The anomers were easily separated by a silica gel column, and each of **146a** and **146b** was mesylated, producing α-**147** and β-**147** followed by treatment with sodium sulfide in DMF to give α-**148** and β-**148** in 78% and 73% yields, respectively. Acid hydrolysis and hydride reduction of α,β-**148** gave **149** in 90% yield. Selective protection of **149** with a TBDPS group afforded **150**, which was oxidized with DMSO/Ac_2O to the ketone **151**. A Wittig reaction of **151** gave **152** in 74% yield, based on **150**. The benzyl group in **152** was deprotected with BCl_3 to give **153** with over 90% efficiency. Compound **154**, obtained by MCPBA oxidation of **153**, was treated with silylated *N*-acetylcytosine and TMSOTf by application of the Pummerer reaction via C–N bond formation at the α-position of sulfoxide, producing the α,β-**155** in 74% yield (α : β = 2.5 : 1). Finally, the α,β-mixture of **155** was deprotected by tetrabutylammonium fluoride (TBAF), followed by aqueous ammonia in methanol treated with $NaBH(OAc)_3$ to afford **201** in excellent yield. Ring closure of the bis-mesylate **202** was effected upon exposure to anhydrous Na_2S in DMF at 100 °C. A concomitant loss of the benzoate ester occurred, furnishing **203** in good yield. Epimerization of the secondary OH was achieved by the Mitsunobu esterification, followed by deprotection to give **205**. Elaboration of **205** to the compound of interest was carried out via mesylate activation followed by S_N2 displacement by normal DNA bases, employing the method of Johnson. The thus-obtained **206** was converted to **209** by the usual transformation (Scheme 8.27).

L-2′,3′-Dideoxyisonucleosides such as **215** and **217** were synthesized starting from the readily available, optically active, C_2-symmetric bis-epoxide **210**. The addition of Na_2S to **210** in aqueous EtOH at 0 °C for several hours gave **211** in excellent yield. The subsequent silylation of **211** gave **212** in 57% overall yield from **210** [36]. The alcohol **212** was converted to the corresponding mesylate **213**, which was then allowed to react with the anion of adenosine to give **214** and desilylated to afford **215**. In a similar way, **217** was prepared from the reaction of **213** with uracil via **216** (Scheme 8.28).

8.4
Synthesis of L-Thionucleosides

4′-Thionucleoside analogues (**227**) were synthesized starting from D-xylofuranoside **218** via 1,4-dithio-L-arabinofuranoside **224** [37]. Compound **218** was methylated and benzylated according to standard procedures, to afford **220**, and this was allowed to react with phenylmethanethiol to yield **221** along with a small amount of **223**. Mesylation of **221** gave **222**, which was cyclized to **224**. Compound **226** was obtained from **224** under Seebach conditions; deprotection of **226** with BBr_3 then afforded **227** (Scheme 8.29).

1-*O*-Acetyl-2-deoxy-3,5-di-*O*-toluoyl-4-thio-D-erythropentfuranose **228** and 2-deoxy-1,3,5-tri-*O*-acetyl-4-thio-L-threopentofuranose **233** were coupled with 5-azacytosine by the usual method shown in Scheme 8.30. The obtained nucleosides were the α- and β-anomers of **230**, **232**, **235**, and **237** [38], the compounds of which were examined for biological activities.

The synthesis of L-thioarabinose derivatives and L-2′-deoxy-2,2′-disubstituted-4-thionucleosides such as **249** and **250** was reported by Jeong and coworkers as a novel type of thionucleoside [39]. First, benzylation of **238** to **239** was achieved in quantitative yield, after which removal of the acetonide moiety under acidic conditions followed by esterification of the 2-hydroxyl function, gave **240**. Ring-opening of **240** with $BF_3 \cdot Et_2O$ in the presence of BnSH in CH_2Cl_2 gave **241**, which was then recyclized to **242** via the corresponding methanesulfonate in the presence of Bu_4NI and $BaCO_3$. Compound **242** was converted to the alcohol **243** via the corresponding acetate by treatment with $Hg(OAc)_2/AcOH$ and removing the acetate subsequently with Et_3SiH and TMSOTf, followed by debenzoylation in 50% overall yield from **238**. Oxidation of **243** with $DMSO/Ac_2O$ afforded the corresponding ketone, which was

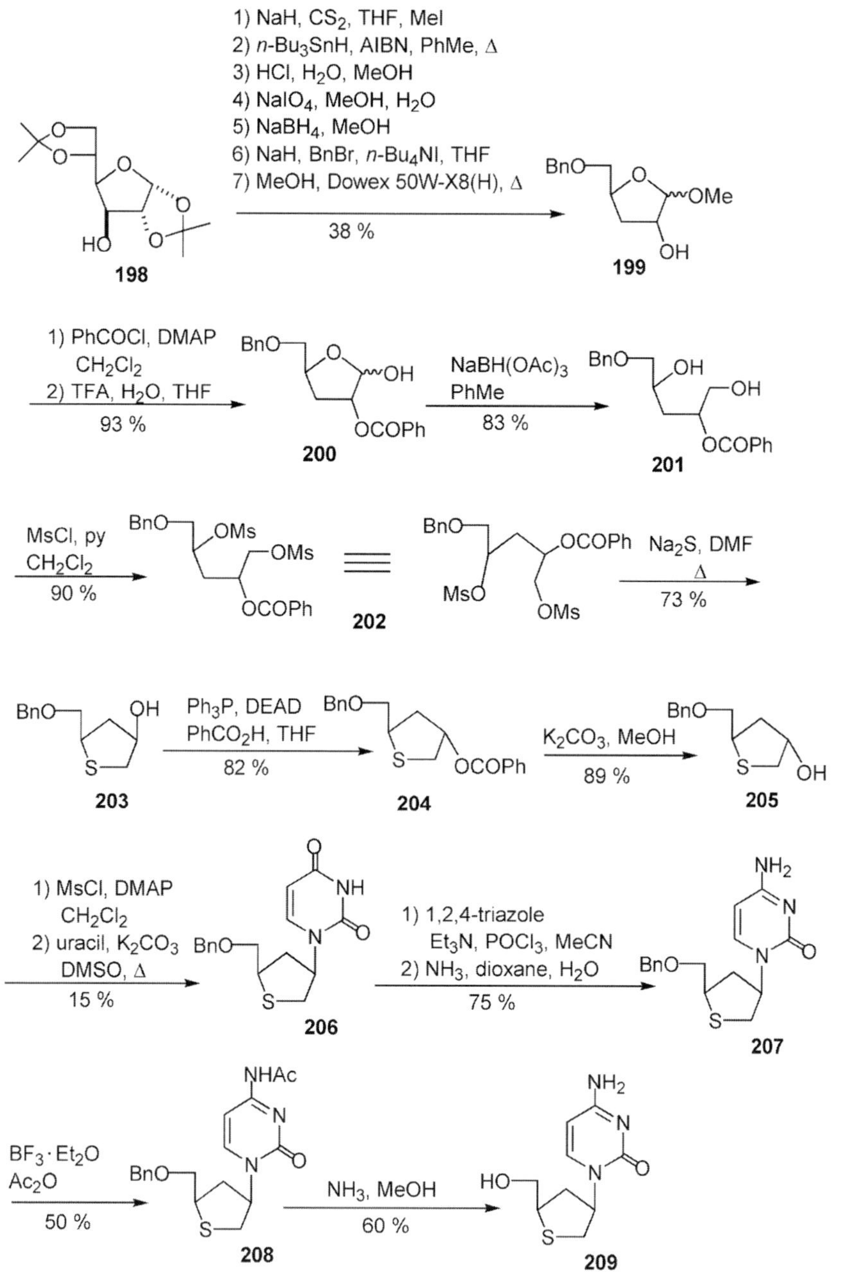
1) NaH, CS2, THF, MeI
2) n-Bu3SnH, AIBN, PhMe, Δ
3) HCl, H2O, MeOH
4) NaIO4, MeOH, H2O
5) NaBH4, MeOH
6) NaH, BnBr, n-Bu4NI, THF
7) MeOH, Dowex 50W-X8(H), Δ
38 %
198
199
1) PhCOCl, DMAP CH2Cl2
2) TFA, H2O, THF
93 %
200
NaBH(OAc)3 PhMe
83 %
201
MsCl, py CH2Cl2
90 %
202
Na2S, DMF Δ
73 %
203
Ph3P, DEAD PhCO2H, THF
82 %
204
K2CO3, MeOH
89 %
205
1) MsCl, DMAP CH2Cl2
2) uracil, K2CO3 DMSO, Δ
15 %
206
1) 1,2,4-triazole Et3N, POCl3, MeCN
2) NH3, dioxane, H2O
75 %
207
BF3·Et2O Ac2O
50 %
208
NH3, MeOH
60 %
209

Scheme 8.27

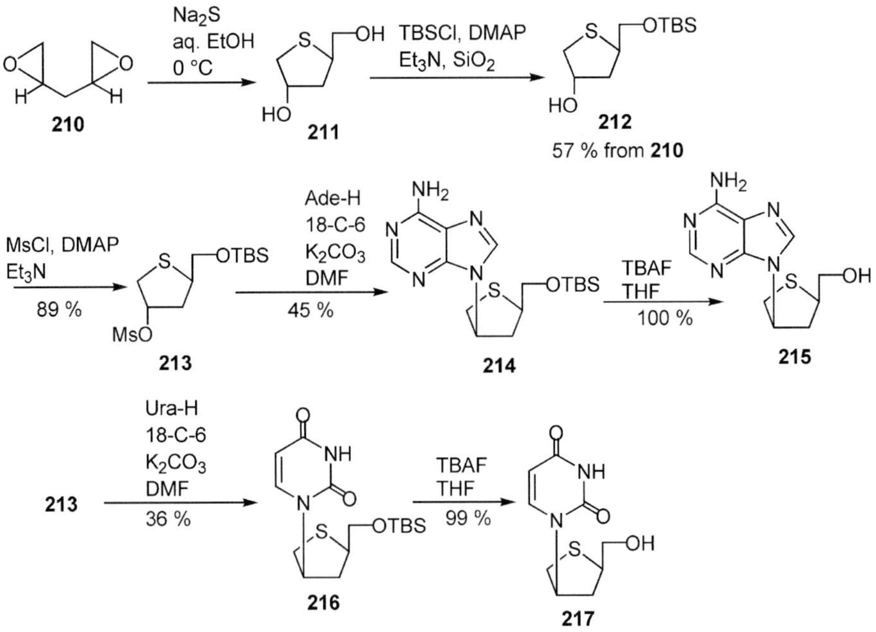

Scheme 8.28

converted to the difluoro derivative **244** after diethylaminosulfur trifluoride (DAST) treatment, or into the methylidene **245** by the Wittig reaction.

Next, **244** was oxidized to the sulfoxide with MCPBA, and this was condensed with persilylated *N*-benzoylcytosine and TMSOTf as Lewis acid catalyst to give **247α** (22%) and **247β** (18%). Treatment of **245** with BBr_3 in CH_2Cl_2, followed by benzoylation, gave **246**. Under the same conditions as used in the preparation of **247α,β**, the condensation reaction with **246** produced **248α** (29%) and **248β** (14%). Reaction of **247α** and **247β** with BBr_3, followed by hydrolysis with MeONa, gave **249α** (73%) and **249β** (85%), respectively. Likewise, the removal of all benzoyl groups in **248α** and **248β** with MeONa/MeOH gave **250α** (100%) and **250β** (100%) (Scheme 8.31).

L-Thionucleosides **254** were synthesized starting from a protected L-thioglycoside **251** by Miller and coworkers [40]. This was converted to iodonucleoside **252** with Icl, followed by the addition of silylated 5-ethylpyrimidine. Successive deiodination (Bu_3SnH, AIBN) and debenzylation (BBr_3) of **252** gave **254** in overall yield 58% from **252** (Scheme 8.32).

8.5 Synthesis of Thioxonucleosides

Chu and coworkers synthesized enantiomerically pure (+)-BHC (**259**) starting from D-mannose via 1,6-thioanhydro-D-mannose for the bioassay of anti-HIV activity [41]. The thioxo compound **258** was prepared in five steps from D-mannose: Selective

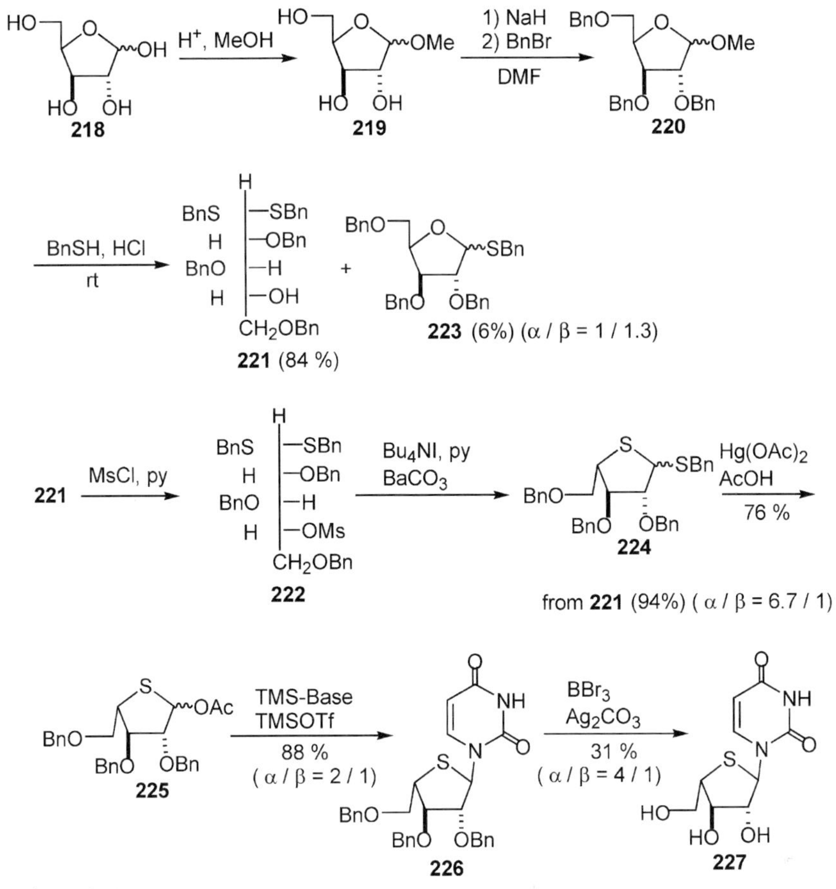

Scheme 8.29

tosylation of the primary hydroxyl group of D-mannose followed by acetylation gave **255**, which was converted to the bromo sugar **256** by treatment with HBr/AcOH. Compound **256** was treated with potassium *O*-ethylxanthate to give **257**, which was then converted to **258** by treatment with NH_4OH in MeOH. Protection of **258** as its isopropylidene, followed by benzoylation, gave **259**. The isopropylidene group was removed using 2% aqueous H_2SO_4 to afford **260**. Oxidative cleavage of **260** with $Pb(OAc)_4$ followed by reduction with $NaBH_4$ afforded **261** as an intermediate. Under the reaction conditions, **261** underwent a secondary to primary benzoyl migration to give **262**, the silylation of which, followed by removal of the benzoyl protection, gave **263**. Treatment of the latter with $Pb(OAc)_4$ gave the aldehyde which, without isolation, was further oxidized with sodium chlorite to afford **264** as a mixture of *endo*- and *exo*-sulfoxides. The carboxylic group in **264** was esterified with Me_2SO_4 and the sulfoxide was reduced with $BHCl_2$ to the sulfide **266**. Hydrolysis of **266** with LiOH in THF/H_2O (4 : 1) gave the acid which, without purification, was converted to **267** by treatment with $Pb(OAc)_4$/pyridine in AcOEt. Condensation of **267** with *N*-acetylcytosine gave

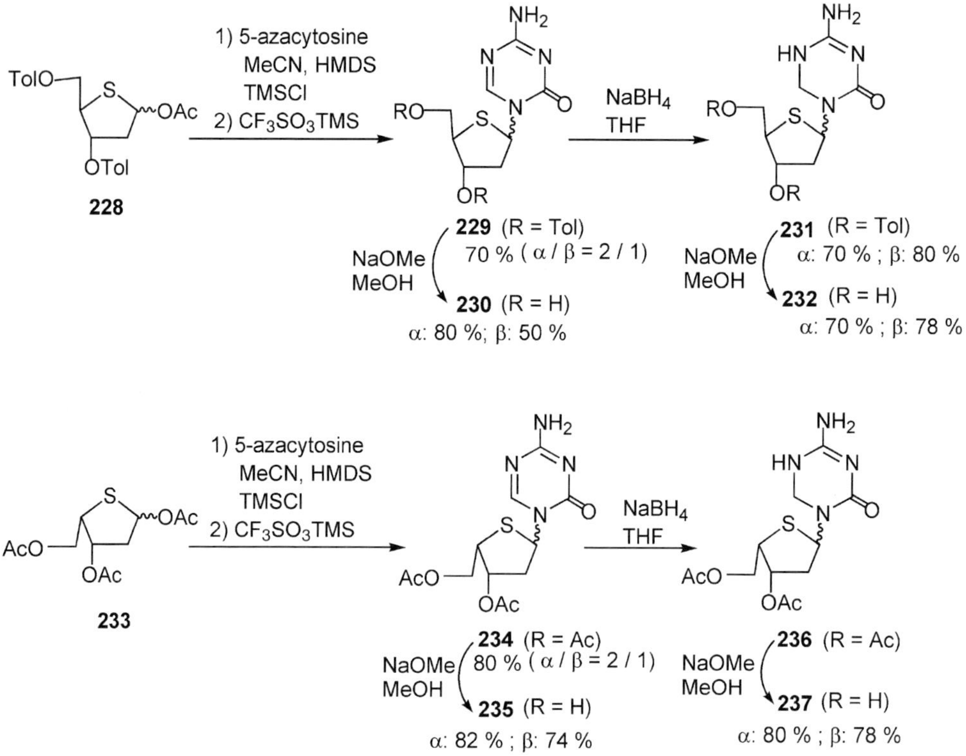

Scheme 8.30

268α and **268**β as a 1 : 2 mixture of α,β-anomers. Separation on a silica gel column, followed by deacetylation with NH_3 in MeOH and desilylation with *tert*-Bu_4NF (TBAF), gave **269** (Scheme 8.33).

Further, enantiomerically pure (2*R*,5*S*)-(−)-[2-(hydroxymethyl)oxathiolan-5-yl]cytosine (**280**) (known as 3TC) was synthesized as a potent anti-viral agent against HBV and HIV [42]. Selective 6-*O*-tosylation followed by acetylation of **270** gave **271**, which was treated with HBr to give a bromo sugar **272**. Reaction of **272** with potassium *O*-ethylxanthate gave 2,3,4-tri-*O*-acetyl-1,6-thionanhydro-L-gulose which, without isolation, was deacetylated to give **273**. Oxidation of the latter with $NaIO_4$ to cleave the 2,3-*cis* diol, followed by reduction and protection of the remaining primary alcohol and treatment with catalytic TsOH, gave **275**. Oxidative cleavage of **275** with $Pb(OAc)_4$, followed by further oxidation with PDC, furnished **276**. Oxidative decarboxylation of **276** gave **277** in 66% overall yield from **275**. Condensation of **277** with silylated N_4-acetylcytosine using TMSOTf as a Lewis acid catalyst gave an α,β-mixture (1 : 2) of **278** and **279**. Separation of these anomers, followed by deacetylation and desilylation, gave the final compounds **280** and **281** (Scheme 8.34).

Next, enantiomerically pure (+)-(2*S*,5*R*)-1-[2-(hydroxymethyl)-1,3-oxathiolan-5-yl] cytosine was synthesized from D-galactose [43]. The preparation of **282** was more straightforward and gave excellent yield, compared to that of 1,6-thioanhydro-

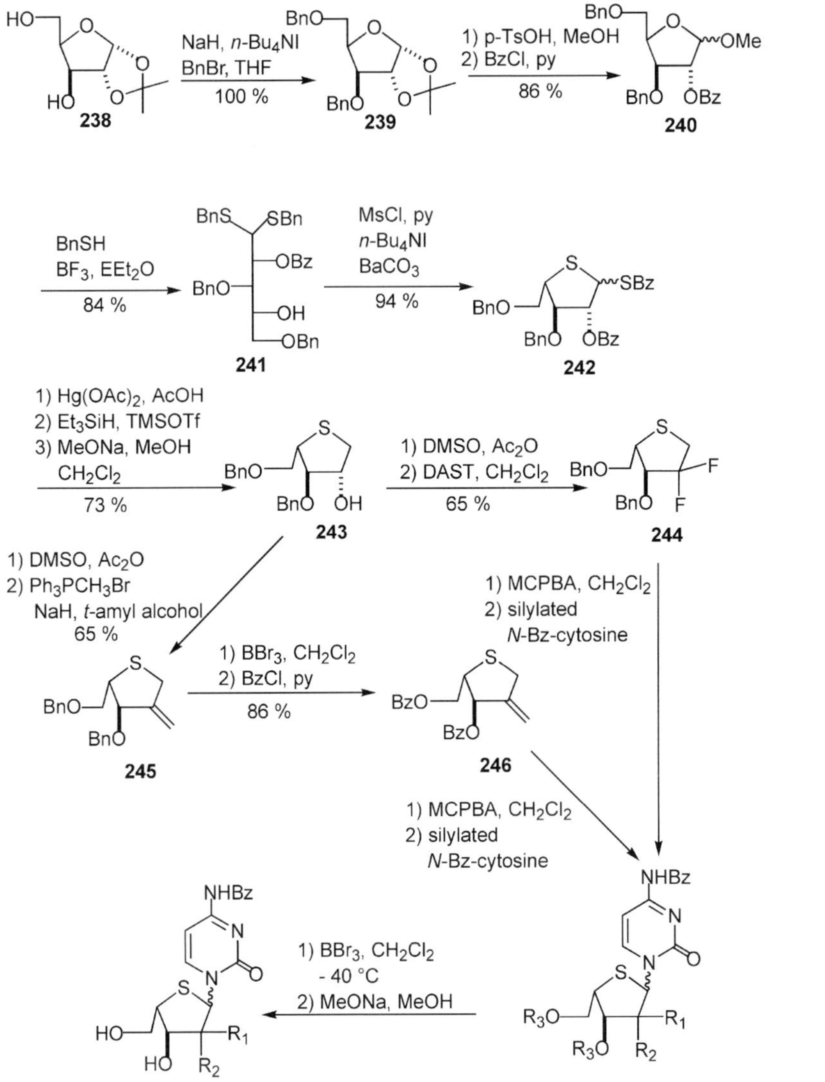

249: $R_1 = R_2 = F$, 73 %(α); 85 % (β)
250: $R_1 = R_2 = CH_2$, 100 %(α); 100 %(β)

247: $R_1 = R_2 = F$, $R_3 = Bn$, 22 %(α); 18 % (β)
248: $R_1 = R_2 = CH_2$, $R_3 = Bz$, 29 %(α); 14 % (β)

Scheme 8.31

D-mannose. The selective oxidative cleavage of 1,6-thioanhydro-D-galactose **282** by $NaIO_4$ to the corresponding aldehyde, and reduction with $NaBH_4$ followed by protection of resulting diol with 2,2-dimethoxypropane as isopropylidene derivative, gave **283**. The primary hydroxyl group of **283** was benzoylated to afford **284** and then converted to **285** by deprotection of isopropylidene group with 10% HCl, and oxidative cleavage of the resulting diol by $NaIO_4$ to the corresponding aldehyde followed by reduction with $NaBH_4$. Silyl protection of **285** followed by debenzoylation

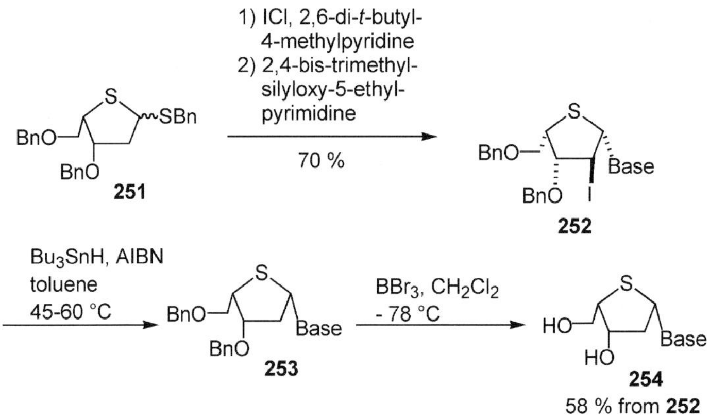

Scheme 8.32

with MeONa afforded **286**. Treatment of **286** with PDC gave **287** which, without further purification, was converted to **288** by $Pb(OAc)_4$. Condensation of **288** with silylated *N*-acetylcytosine in the presence of trimethylsilyl triflate gave a mixture of **289** (43%) and **290** (20%), which was purified using silica gel column chromatography. Deacetylation of **289** and **290** with NH_3 in MeOH, followed by desilylation with tetrabutylammonium fluoride, produced the desired **269** and **291**, respectively (Scheme 8.35).

A clinically significant (−)-2′-deoxy-3′-thiacytidine (3TC) **280** was synthesized in four steps from (+)-thiolactic acid **292** by Jones and coworkers [44]. Condensation of **292** with 2-benzoyloxyacetaldehyde occurred upon exposure to $BF_3 \cdot EOEt_2$ to give a 1 : 2 mixture of **293** and **294**. The less-polar acid **294** was treated with $Pb(OAc)_4$ to afford an anomeric mixture (2 : 1) of the *anti*- and *syn*-acetate **295**. When the mixture **295** was treated with silylated cytosine in the presence of iodotrimethylsilane (TMSI), **296** and **297** were obtained in a 1.3 : 1 ratio. A final deprotection of **296** was carried out by using a basic resin to give **280** (Scheme 8.36).

2′,3′-Dideoxy-3′-oxa-4′-thioribonucleoside **304** was synthesized starting from **298** by Jin and coworkers [45]. Treatment of **298** was effected with Et_3SiH and TMSOTf to give **299**, which was then reduced with $NaBH_4$ followed by silylation to give **300**. Compound **300** was oxidized with MCPBA to afford a diastereomeric mixture of **301**, which underwent the Pummerer rearrangement in acetic anhydride in the presence of Bu_4NOAc to give **302** as a mixture of *cis*- and *trans*-isomers. Condensation of **302** with silylated 6-chloropurine in the presence of TMSOTf gave **303** as a mixture of *cis*- and *trans*-isomers, which were then subjected to chromatographic separation. The pure *cis* compound belonging to **303** was desilylated followed by heating with NH_3 in MeOH to afford the desired **304** (Scheme 8.37).

An anti-viral agent, Lamivudine (**312**), was synthesized by Rayner and coworkers, by coupling of the sodium salt of **305** with bromoacetaldehyde diethylacetal to give **306** [46]. The latter compound was then converted to its sulfoxide, followed by the Pummerer rearrangement to afford **307**. The enzymatic reaction of **307** gave

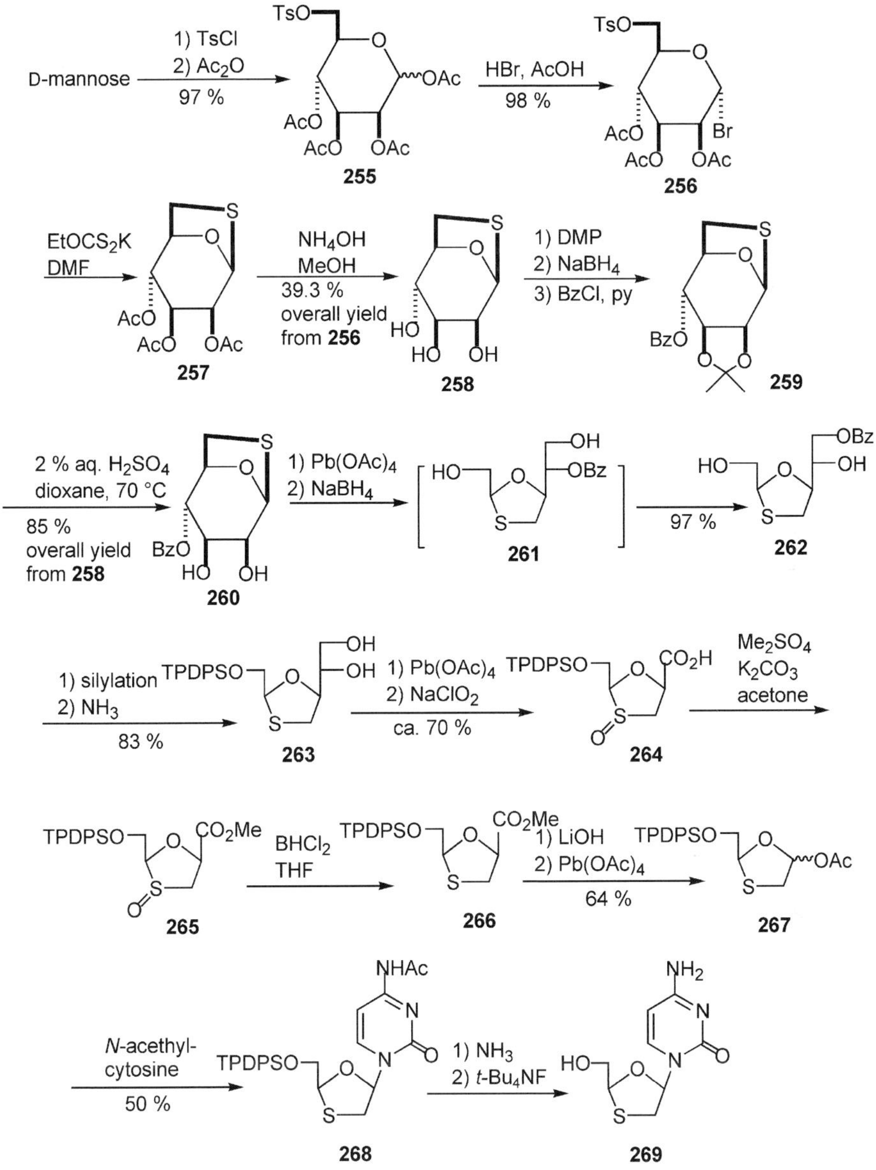

Scheme 8.33

(−)-**307**, which was converted to **309** via a deacetylated intermediate **308**. The ester moiety of **309** was reduced with $LiBH_4$ followed by benzoylation to afford **310**. The introduction of base gave the acetylated cytidine **311** as a 1 : 1 mixture of α- and β-anomers, which could be separated chromatographically. The individual anomers were deprotected to give **312α** and **312β** in excellent yield (Scheme 8.38).

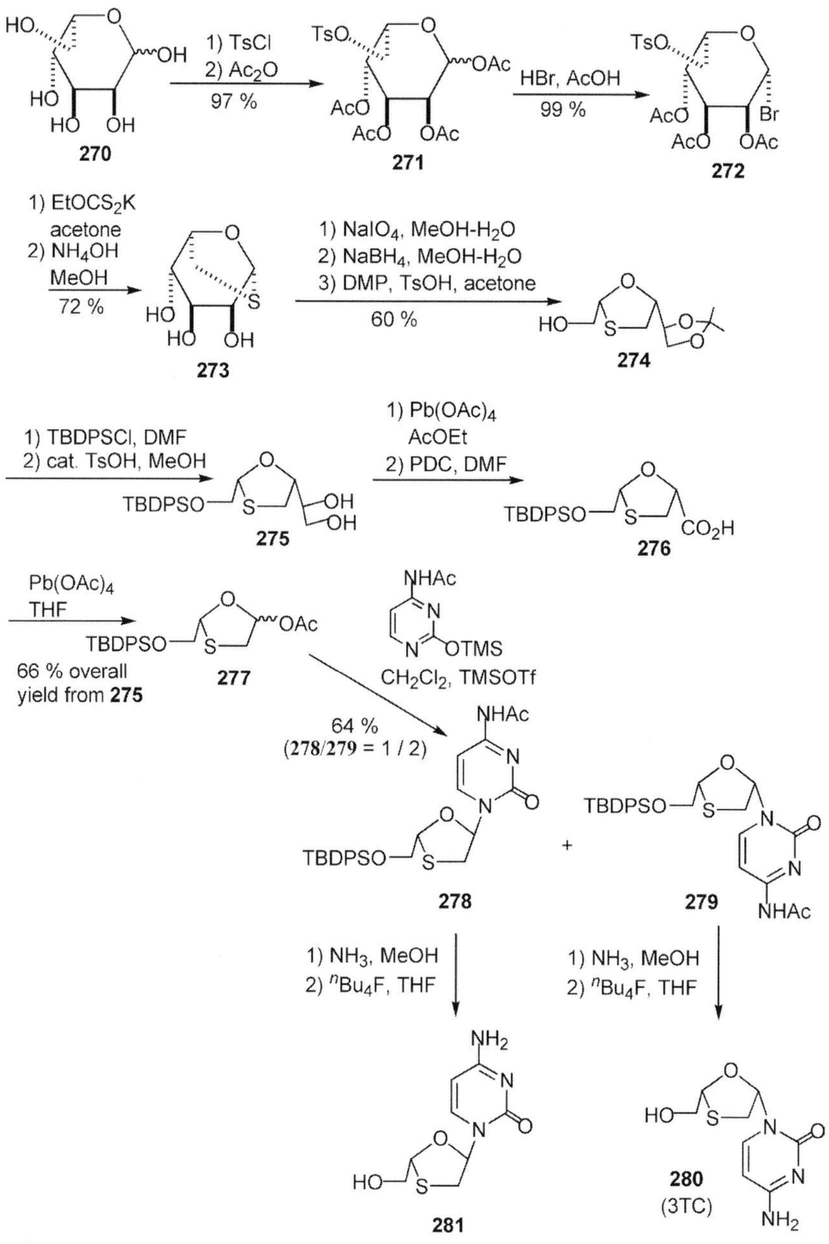

Scheme 8.34

1,3-Oxathiolane nucleoside **315** was synthesized as an agricultural chemical by Kubo and Yokoyama [47]. *R*-Glicidol **313** was protected by silylation and then converted to **314** with sodium hydrosulfide; the latter compound was then allowed to react with aryl aldehyde, followed by desilylation to afford **315** (Scheme 8.39).

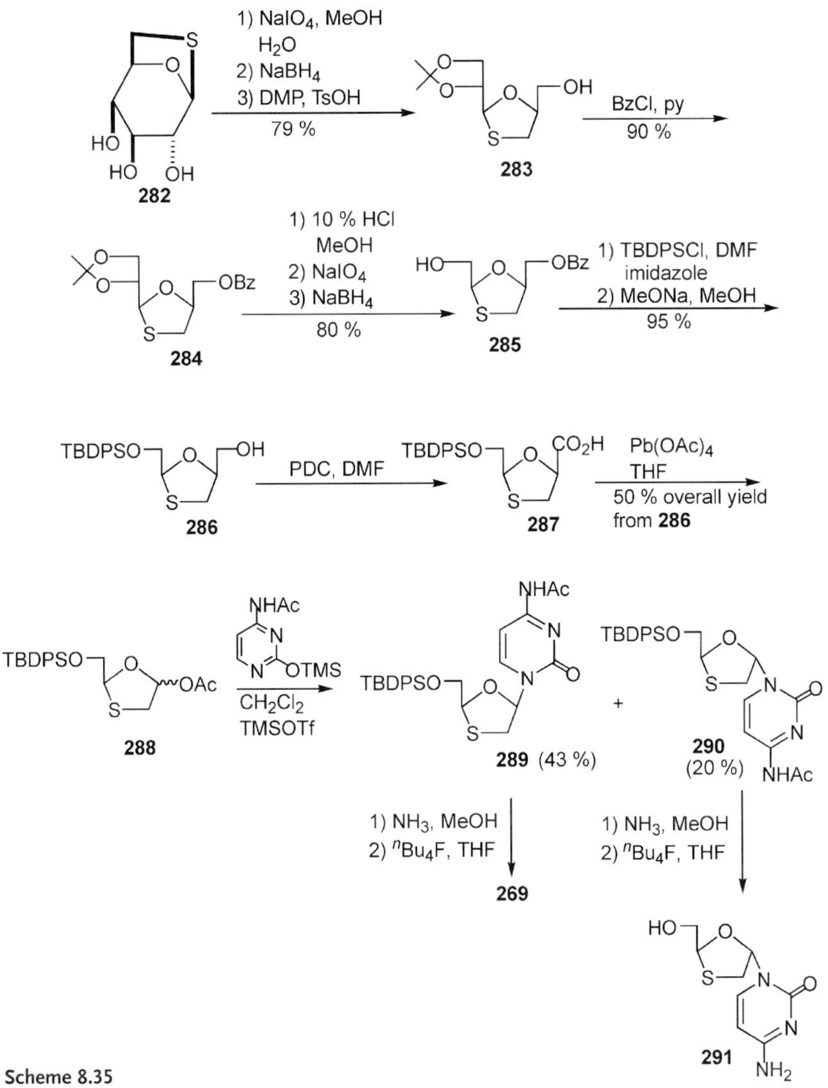

Scheme 8.35

8.6
Synthesis of Miscellaneous Thionucleosides

Those thionucleosides having 6-substituted purine as bases and 4-thio-D,L-erythrofuranose and 4-thio-D,L-threofuranose as sugars were synthesized for biological testing [48]. An absence of the 5′-phosphorylation and a decrease in solubility can be added to the effects arising from the introduction of a sulfur atom. The phenylboronate acetate **316** was condensed with 6-chloropurine in the presence of TsOH to give **317**, which was deboronated to **318**. The 6-chlorine atom was changed to

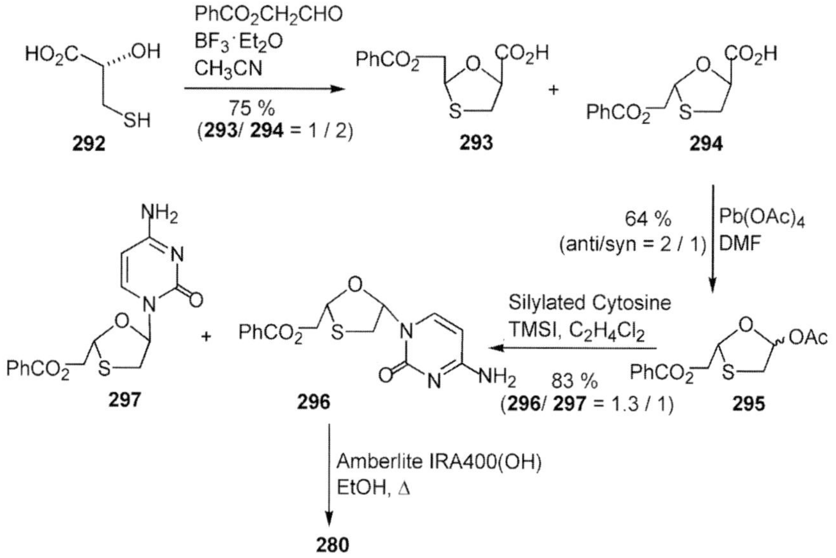

Scheme 8.36

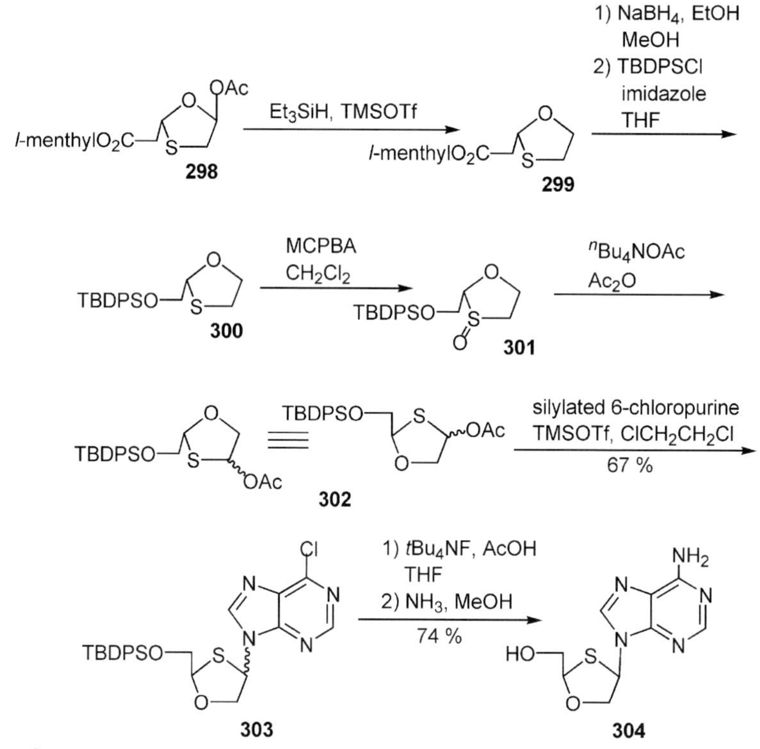

Scheme 8.37

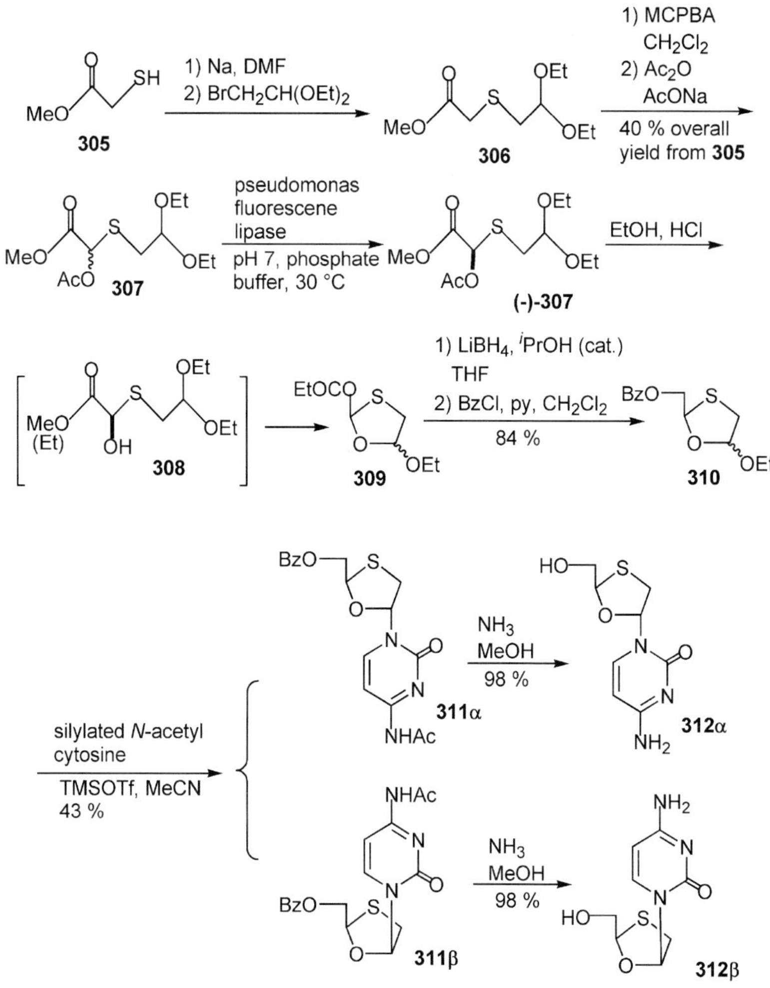

Scheme 8.38

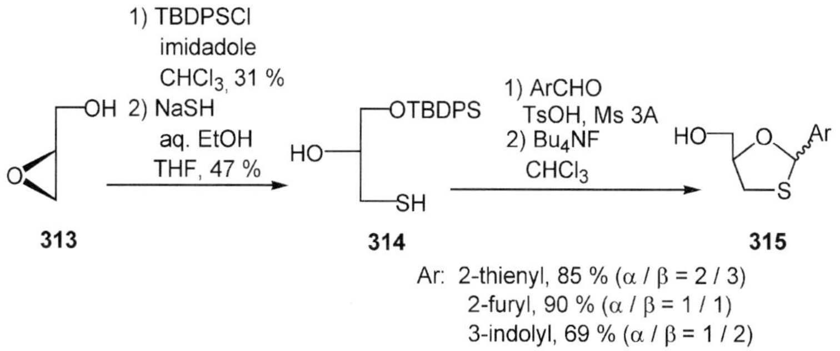

Scheme 8.39

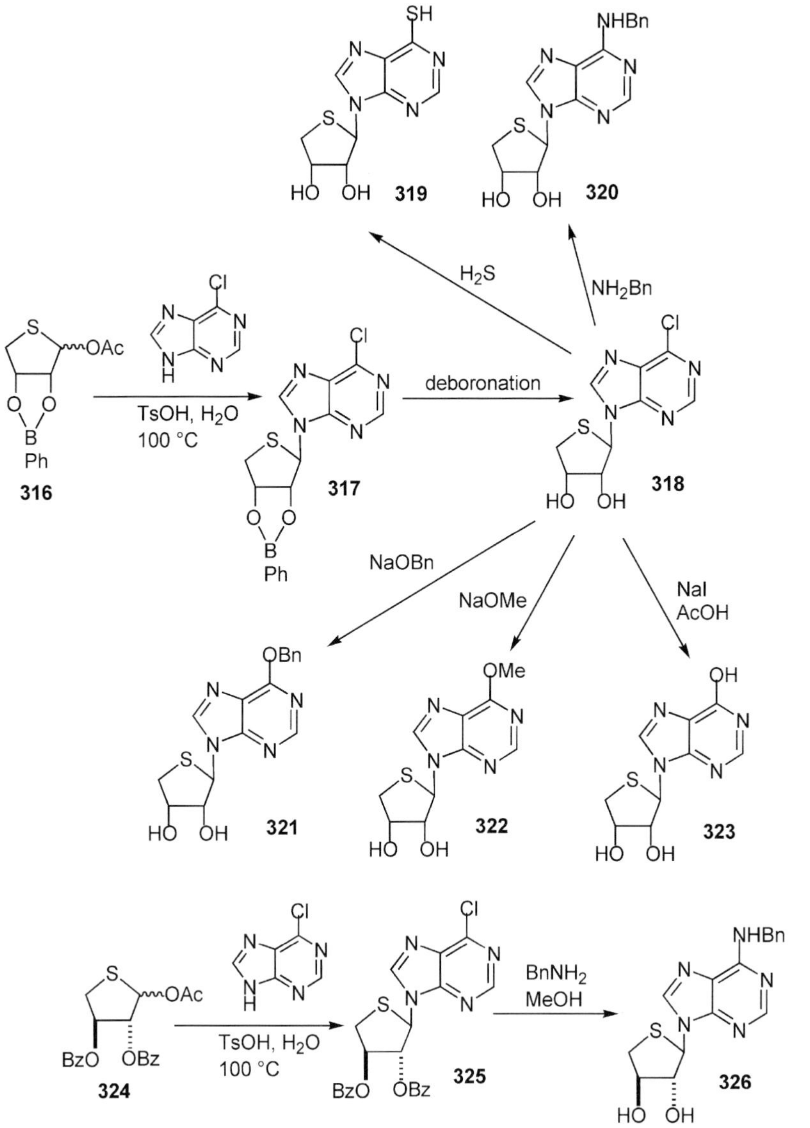

Scheme 8.40

various groups by the usual exchanging reactions (**319**∼**323**). Similarly, **324** was converted to **326** via **325** (Scheme 8.40).

Zard and coworkers prepared the difluorophosphonate analogues of thionucleosides **331** [49]. The xanthate **327** was made by treating ethyl bromoacetate with sodium *O*-neopentyl xanthate, the key radical addition to 1,1-difluoro-3-butenylphosphonate proceeding with reasonable efficiency (60% yield). Cleavage of the adduct xanthate **328** with ethylene diamine and exposure of the crude thiol to hot trifluor-

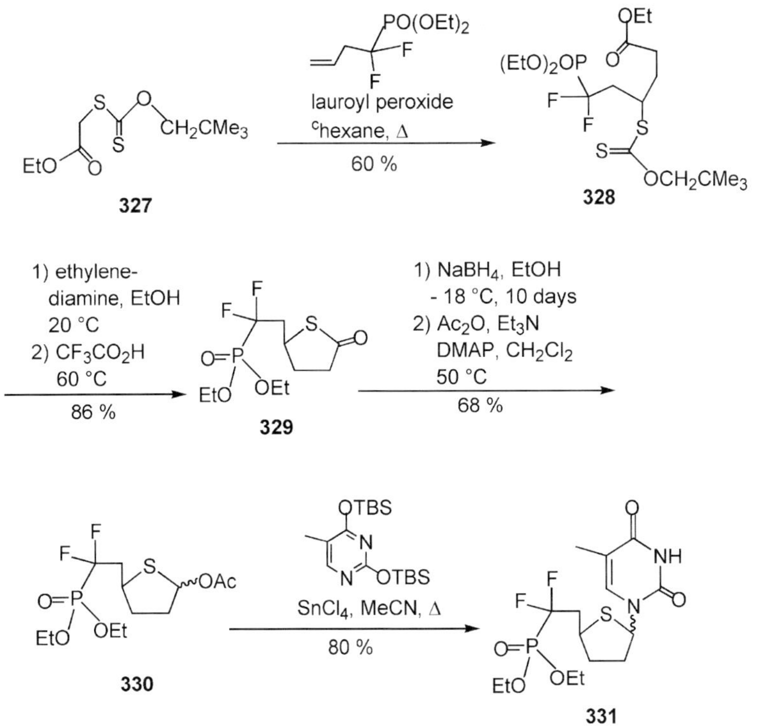

Scheme 8.41

oacetic acid provided **329**. Storage of **329** for 10 days at −18 °C with sodium borohydride in ethanol, followed by acetylation, furnished **330**. Finally, a Vorbruggen coupling of **330** with silylated thymine in the presence of tin(IV) chloride provided **331** as a 55 : 45 mixture of epimers (Scheme 8.41).

Matsuda and coworkers synthesized (*2R,3R,4S*)-1-[3,4-di-*O*-(2,4-dimethoxybenzoyl)thiolane-3,4-diol-2-yl]thymine by stereoselective coupling of thymine with sulfoxides derived from meso-thiolane-3,4-diol via the Pummerer reaction [50]. After protection of the hydroxyl groups of **332** with a benzoyl group, **333** was treated with OsO_4 and the resulting diol protected with an isopropylidene group, followed by deprotection of the benzoyl groups to give **334** in 84% yield. Reaction of **334** with methanesulfonyl chloride gave dimesylate, which was treated with sodium sulfide to give **335**. The isopropylidine group of **335** was removed by acidic conditions, and the introduction of a benzoyl group to the resulting hydroxyl groups afforded **337**. Oxidation of **337** with MCPBA gave **338** as diastereomeric mixture. When **338** was treated with thymine, the corresponding coupling compound **339** was obtained as a 1 : 6 mixture of α- and β-anomers (Scheme 8.42).

A new thymidine thietane nucleoside **347** was synthesized via the Pummerer rearrangement of the corresponding **345** in the presence of thymine, TMSOTf, Et_3N, and ZnI_2 as a key step [51]. Compound **341**, derived from D-isoascorbic acid **340** [52], was deacetalized followed by tritylation to **342**, which was reduced and then mesitylated to afford **343**. The latter compound was treated with Na_2S followed by

Table 8.4 Inhibition of growth of KB cell by 4′-thio-l-β-arabinofuranosylcytosine and derivatives.

Derivatives of cytosine	Concentration M required for 50% growth inhibition
2,2′-Anhydro-l-β-arabinosylcytosine hydrochloride	1.9×10^{-7}
1-β-D-Arabinosyl-cytosirie	1.6×10^{-7}
55	4.3×10^{-7}
56	4.2×10^{-7}

Table 8.5 Growth inhibition by 4′-thio-5-fluorouridine.

Compound	Concentration M required for 50% growth inhibition			
	Leukemia L-1210	Mammary carcinoma TA-3	*S. faecium*	*E. coli* B
58α (X=F)	4×10^{-7}	$>10^{-7}$	4×10^{-9}	6×10^{-7}
58β (X=F)	2×10^{-7}	3×10^{-8}	6×10^{-10}	4×10^{-5}
5-Fluorouridine	5×10^{-8}	–	6×10^{-11}	7×10^{-10}

Table 8.6 Cytotoxicity data.

Compound	IC_{50}(μM)		
	Leukemia L-1210	H-Ep-2	CCRF-CEM
79α,β	1.0	<0.20	3.5
79	1.2	<0.20	–
76	2.7	2.1	~4
77	0.12	0.14	0.67

α-4′-Thiothymidine (the α-anomer of **77**) appeared to have no significant activity, whereas β-4′-thiothymidine **77** was surprisingly active, particularly against human cytomegalovirus, although it was also highly toxic (Table 8.7) [15].

Purine nucleosides **106** and **111** showed slight activity against HIV replication in MT-2 and CEM cells, in comparison with azathionine (AZT) and 2′,3′-dideoxycytidine (DDC) (Table 8.8) [18].

Compounds **139** and **140** were tested in a number of anti-viral assays, including hepatitis B virus (HBV), human cytomegalovirus (HCMV), HSV-1 and -2, and varicella zoster virus (VZV) (Table 8.9) [24]. Compound **140** showed potent activity against both HBV and HCMV; however, significant leukemic cell cytotoxicity was observed with **140** and **139** (R = Cl, OMe). Although **139**(R = $OCH_2{}^{c}Pr$) was 10-fold less toxic to leukemic cells than was **140**, it was at least 10-fold less active against HBV and HCMV. Compound **139** (R = SMe) showed an even more drastic reduction in efficacy, with simple monoalkylamines such as *N*-propyl and *N*-allyl analogues being

Table 8.7 Anti-viral activities.

Compound	MIC_{50} (μM)					IC_{50}	
	HIV-1	HSV-1	HSV-2	VZV	CMV	VERO	MT-4
77	non	0.37	2.3	10	0.98	7.1	1
BVDU	–	0.03	12	0.02	>100	>500	–
acyclovir	–	1.3	1.8	14.5	60–100	>500	–

Table 8.8 Inhibition of HIV replication in MT-2 and CEM cells.

Compound	Cell line	IC_{50} (μg mL^{-1})	TC_{25} (μg mL^{-1})	SI	TAI
106	MT-2	80	>100	1.3	>17
111	CEM	37	97	2.6	21
AZT	MT-2	0.14	9.3	>160	>82
	CEM	<0.03	>10	>300	>93
DDC	MT-2	0.44	>9.8	>25	>52
	CEM	0.05	5.3	120	>63

Table 8.9 Anti-viral activity and cytotoxicity evaluation for **139** and **140**.

Compound (R of 139)	HBV		HCMV		Cytotoxicitj[a)]		
	Efficacy (IC_{50} μM)	Toxicity (CC_{50} μM)	Efficacy (IC50 μM)	Toxicity (CC_{50} μM)	IM-9	CEM	Molt 4
Cl	0.001	>0.2	0.1	2	42	10	3.8
OMe	0.0025	14	0.062	>**5**	5	8	2
$OCH_2{}^cPr$	0.035	>0.2	0.6	>20	61	41	13
SMe	0.45	>200	2	>20	88	78	73
NHPr	0.061	>0.2	2	>50	92	68	52
NHallyl	0.058	>0.2	1.5	>200	67	60	40
NH^iPr	0.3	>200	4	>100	90	80	77
NH^cPr	0.0072	77	0.2	**15**	78	68	11
N(Et)Me	0.19	>200	2	>20	85	69	77
$N(Me)^cPr$	0.3	101	2	>20	72	58	100
Piperidino	4	>200	6	>200	88	80	77
Pyrrolidino	0.85	>200	10	>200	99	71	112
140	0.002	13	0.06	2	4	9	14

a) (% of Control at 100 μM or CC_{50} μM).

10~50-fold less active than **140** against HBV and HCMV; an even greater decrease was apparent in terms of cytotoxicity. The bulkier isopropyl-amine analogue was inactive. This lack of activity cannot be accounted for by simple steric hindrance, because the cyclopropylamine compound was very potent against both HBV and HCMV. The disubstituted amines showed minimal anti-viral activity.

they do not synthesize these compounds [28]. Thus, the metabolism of AHL (**13**) is seen as a new target for antibiotics.

The plant hormone ethylene is obtained in two enzymatic reactions. In the first step, 1-aminocyclopropane-1-carboxylate (ACC, **14**) (Scheme 9.2, reaction c) is formed from AdoMet (**1**) by the pyridoxal-5′-phosphate-dependent ACC synthase (see Section 9.3.2 for a discussion of the enzymatic mechanism) [29–31]. The second step is catalyzed by ACC-oxidase, which converts ACC to ethylene [32–34].

Nicotianamine (**15**) (Scheme 9.2, reaction d) is a metal-chelating compound which is ubiquitous among higher plants [35, 36]. Besides metal transport within plants, nicotianamine serves as precursor for several phytosiderophores which are responsible for the uptake of iron ions needed for the biosynthesis of chlorophyll [37]. Nicotianamine (**15**) is biosynthesized from three AdoMet molecules; the fusion of three 3-amino-3-carboxypropyl groups, as well as azetidine ring formation, are catalyzed by nicotianamine synthase [38–40].

Polyamines are another group of biomolecules synthesized using AdoMet (**1**). In a first step, AdoMet (**1**) is decarboxylated by AdoMet decarboxylase, after which the activated aminopropyl group of the decarboxylated AdoMet (dcAdoMet) is transferred to putrescine, yielding spermidine (**16**), whilst a second transfer leads to spermine (**17**) (Scheme 9.2, reaction e). These polyamines are ubiquitous compounds that play a role as cationic mediators in cell proliferation and differentiation [41–43]. As higher levels of polyamines are found in cancer cells, their biosynthetic pathway represents an interesting target for potential antiproliferative drugs [44, 45].

9.2.3
AdoMet as an Adenosyl Group Donor

To date, very few naturally occurring organofluorine compounds have been recognized [46], and an enzyme which catalyzes the incorporation of fluoride ions was recently identified [47]. This enzyme, known as *fluorinase*, catalyzes the nucleophilic substitution of L-methionine in AdoMet (**1**) with a fluoride ion to yield 5′-desoxy-5′-fluoroadenosine (**18**) (Scheme 9.2, reaction f) [48, 49]. ^{18}F is the preferred radioisotope when conducting positron emission tomography (PET), and rapid, clean synthetic methods are required to obtain good radiochemical yields. Therefore, fluorinase is of major interest in this area, and an ^{18}F-labeled monosaccharide was produced with the help of additional enzymes [50]. Fluorinase can also utilize chloride ions for nucleophilic substitution, provided that the equilibrium is shifted towards the product 5′-chloro-5′-desoxyadenosine by removing the co-product L-methionine [51].

9.2.4
AdoMet as a Ribosyl Group Donor

AdoMet (**1**) may also serve as a source of ribosyl groups. The first evidence for this was obtained with the enzyme *S*-adenosylmethionine:tRNA ribosyltransferase-isomerase (QueA), which is involved in the penultimate step of queuosine biosynthesis

[52]. QueA catalyzes both the transfer of the ribosyl moiety of AdoMet (**1**) to 7-aminomethyl-7-deazaguanosine in tRNA and the isomerization to epoxyqueuosine (**19**) (Scheme 9.2, reaction g), which is reduced to queuosine in a subsequent coenzyme B_{12}-dependent step. During catalysis by QueA, adenine and L-methionine are released, a process for which a catalytic mechanism has been proposed [53, 54]. The initial step involves deprotonation at C5′ under sulfonium ylide formation and elimination of adenine, as seen in the decomposition pathway C of AdoMet (**1**) (compare Section 9.3.3). The vinyl sulfonium intermediate is the attacked at C4′ by the primary amino group of the 7-aminomethyl-7-deazaguanosine residue, and the newly formed sulfonium ylide at C5′ attacks the aldehyde at C1′. Finally, oxygen at C1′ closes next to the sulfonium center under epoxide formation and release of L-methionine.

9.2.5
AdoMet as a Radical Source

The reactions catalyzed by the "radical-SAM" superfamily enzymes use an iron-sulfur cluster and AdoMet (**1**) to catalyze diverse radical chemistries on a vast array of substrates. This superfamily was first identified in 2001 by using bioinformatic analysis, and is thought to contain several hundred members and to represent a major class of iron-sulfur proteins [55]. Although little is known about most members of this superfamily, several enzymes were investigated well before the superfamily was discovered, and have provided insights into the biological radical catalysis involving AdoMet (**1**). Early studies on pyruvate formate-lyase activating enzyme and lysine 2,3-aminomutase showed that AdoMet (**1**) can function as a radical source, and consequently the involvement of metal–adenosyl complexes was hypothesized [56–58]. However, more recent results – including X-ray structure analyses of coproporphyrinogen III oxidase and biotin synthase [59, 60] – have revealed that AdoMet (**1**) binds to free iron coordination sites of the iron-sulfur cluster via its amino and carboxylate groups, thus placing the sulfonium center in close proximity to the cluster. A one-electron transfer to AdoMet (**1**) leads to homolytic cleavage of the S–C5′ bond and generation of L-methionine and a 5'-deoxyadenosyl radical (**20**) (Scheme 9.2, reaction h). This radical then abstracts a hydrogen atom from various substrates, resulting in 5'-deoxyadenosine and free radicals which undergo further reactions [61]. This AdoMet-initiated radical chemistry is found in a wide variety of biological transformations involved in DNA precursor, vitamin, cofactor, antibiotic and herbicide biosynthesis, as well as in biodegradation and DNA repair [55]. For example, biotin synthase catalyzes the final step in the biosynthesis of biotin (**21**) (Scheme 9.2, reaction h), which is the insertion of sulfur between C6 and C9 of dethiobiotin [62].

9.2.6
AdoMet as an Amino Group Donor

Although most of the cofactor biochemistry of AdoMet (**1**) is related to the unique reactivity of the sulfonium center, there is one example where it functions as an amino group donor. 7,8-Diaminopelargonic acid aminotransferase, which is involved in an early stage of biotin biosynthesis, catalyzes the transamination of 7-keto-8-aminopelargonic acid using AdoMet (**1**) as an amino donor to yield 7,8-diaminopelargonic acid (**22**) (Scheme 9.2, reaction i) [63, 64]. This enzyme belongs to the large family of pyridoxal-5′-phosphate-dependent aminotransferases, and has recently attracted interest as a target for antibiotics directed against *Mycobacterium tuberculosis* [65].

9.2.7
AdoMet-Dependent Riboswitches

Riboswitches are complex folded regions in mRNAs which serve as highly specific receptors for small molecules. They are normally placed in non-coding regions, and function as genetic switches by adapting an alternative fold upon binding of their ligands. Riboswitches regulate gene expression in bacteria by controlling transcription elongation and initiation, depending on the concentration of key metabolites such as cofactors (thiamine pyrophosphate, coenzyme B_{12}, flavin mononucleotide), amino acids, or nucleobases [66–68].

Taking into account the importance of AdoMet-dependent reactions, it is not surprising to find a large number of AdoMet-recognizing riboswitches which are mostly referred to as "SAM-riboswitches" [69]. Riboswitches are thought to be good targets for the development of new antibiotics. The *metK* gene, which encodes AdoMet synthetase (compare Section 9.2.8), is predicted to be essential for the survival [70] and virulence [71] of *Staphylococcus aureus*. Its expression is regulated by an AdoMet-binding riboswitch located in the 5'-untranslated region [69, 72]. An AdoMet analogue that binds to this riboswitch could, in principle, inhibit *S. aureus* growth and/or virulence by repressing *metK* and thereby preventing AdoMet biosynthesis [73]. Therefore, the three-dimensional structure of a "SAM-riboswitch" in complex with AdoMet (**1**) [74] and the structural requirements for ligand binding are of special interest [75].

9.2.8
The Biosynthesis and Metabolism of AdoMet

The biosynthesis of AdoMet (**1**) is catalyzed by *S*-adenosylmethionine synthetase (also known as methionine adenosyltransferase, MAT) starting from L-methionine and ATP [1, 2]. In a first step, AdoMet synthetase catalyzes the nucleophilic attack of the L-methionine sulfur atom onto C5′ of ATP. The displaced triphosphate is then hydrolyzed by the enzyme to inorganic phosphate (P_i) and pyrophosphate (PP_i) in a second step. This anhydride hydrolysis provides the main driving force for the formation of the highly activated sulfonium compound (Scheme 9.3, reaction a). The formation and metabolism of AdoMet (**1**) are highly regulated because one equivalent of ATP is consumed during the synthesis, and an overproduction of AdoMet (**1**) would lead to a depletion of ATP in the cell. On the other hand, low levels of AdoMet (**1**) would affect a large number of crucial metabolic and regulatory processes, as discussed above. AdoMet synthetase is therefore strictly controlled by gene regulation and product inhibition, and all organisms possess recycling circuits for the byproducts of AdoMet-dependent reactions.

The first recycling pathway, termed the "methyl cycle" (Scheme 9.3, upper part), allows the conversion of AdoHcy (**2**) – the co-product formed by the reactions discussed in Section 9.2.1 – to L-homocysteine by AdoHcy hydrolase (Scheme 9.3, reaction b). L-Homocysteine is then remethylated to L-methionine by methionine synthase (also known as L-homocysteine MTase) using *N*5-methyltetrahydrofolate (Scheme 9.3, reaction c). There are two apparently unrelated families of methionine synthase. Those organisms that synthesize or transport vitamin B_{12} encode a cobalamin-dependent methionine synthase (MetH) [76, 77], whereas organisms that cannot produce vitamin B_{12} encode only the cobalamin-independent methionine synthase (MetE) [78]. *Escherichia coli* and many other species of bacteria express both enzymes, whereas mammals utilize only the cobalamin-dependent methionine synthase. In contrast, plants and yeasts utilize only the cobalamin-independent enzyme.

The second recycling pathway (Scheme 9.3, lower part) comprises a set of complex reactions that allow the direct synthesis of methionine from MTA (**3**), the co-product formed by the reactions discussed in Section 9.2.2. In this "MTA cycle" – which is not present in mammalian cells – the ribose moiety of MTA (**3**) gives rise to the four-carbon skeleton of L-methionine while conserving the methylthiol group [79].

9.3 The Chemistry and Biochemistry of Modified AdoMet

9.3.1 Synthetic Approaches to AdoMet Analogues

In general, AdoMet analogues can be obtained by either enzymatic or chemical synthesis. The enzymatic synthesis from L-methionine or ATP derivatives, using purified AdoMet synthetases, has the advantage that no protecting groups are needed and coupling leads directly to AdoMet analogues with the S-configuration at the sulfonium center. However, the enzymes from *E. coli*, yeast or rat liver have only a narrow substrate spectrum which greatly limits their general use as synthetic tools for the synthesis of AdoMet analogues [80, 81]. Recently, AdoMet synthetase from the archaeon *Methanococcus jannaschii* was found to accept a much wider range of nucleoside triphosphates, including ITP, GTP, CTP and UTP, and could be used to prepare novel AdoMet analogues [82].

Alternatively, many AdoMet analogues have been prepared using chemical syntheses which is focus on successive generation of the sulfonium center. Typically, AdoHcy analogues are synthesized, with the 5′-thioether function being alkylated in the last step to yield the less-stable 5′-sulfonium compounds (Scheme 9.4). In general, there are two synthetic routes towards the formation of the 5′-thioether bond in AdoHcy or side chain-modified analogues. In Scheme 9.4, route A starts from 2′,3′-*O*-isopropylidene-5′-*O*-tosyladenosine (**23**) [83, 84] or 5′-chloro-5′-deoxyadenosine (**24**) [85, 86] in which

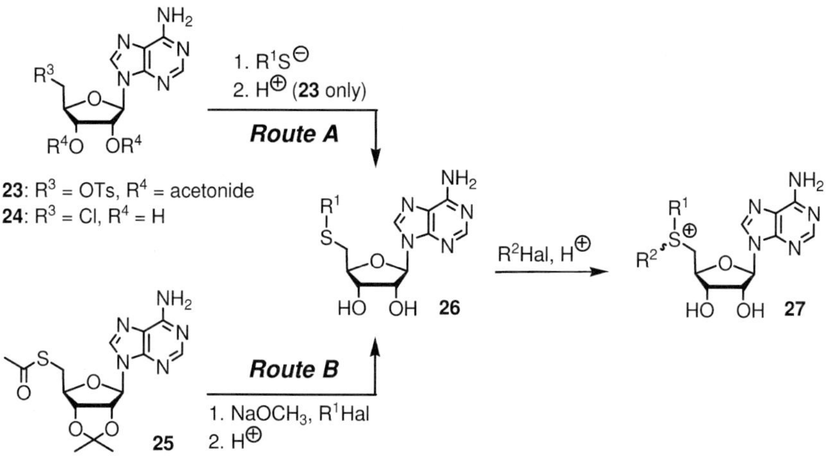

Scheme 9.4 General synthetic approaches towards AdoMet analogues.

the 5′ position is activated for nucleophilic substitution with a thiolate. Although many 5′-deoxyadenosine thioethers have been prepared, this route is particularly problematic with 5′-tosylates as starting material because 5′-deoxy-*N*3,5′-cycloadenosine can easily form by an intramolecular attack of the N3 nitrogen of the adenine ring, leading to mediocre yields [87]. This side reaction can be suppressed by N6-acylation [88] or by formation of the N1-oxide [89], which presumably leads to a decreased nucleophilicity of the N3 nitrogen but adds extra steps to the synthesis. The problem of cyclonucleoside formation is circumvented in route B of Scheme 9.4 by using a 5′-deoxy-5′-thioadenosine precursor [90] such as thioadenosine (**25**), which can be prepared directly from 2′,3′-*O*-isopropylideneadenosine under Mitsunobu conditions, with almost quantitative yield [91]. After cleavage of the thioacetate function under basic conditions, *in-situ* nucleophilic substitutions of alkyl halides yield the desired 5′-thioethers **26**. If oxygen is rigorously excluded from the reaction mixture, then nucleoside disulfide formation is efficiently suppressed and good yields can be obtained for the coupling reaction [92]. The AdoHcy analogues are deprotected and then treated with methyl iodid [93, 94] or other alkylating agents [95] under mildly acidic conditions to yield the 5′-sulfonium compounds **27**. No protecting groups are required in this step because the acidic conditions lead to protonation, and hence transient protection of all nucleophilic positions except for the sulfur atom. However, the AdoMet analogues **27** are obtained with minimal diastereoselectivity, and separation of the diastereoisomers at sulfur can often be achieved using reversed-phase high-performance liquid chromatography (HPLC) [96, 97].

9.3.2
Isotope-Labeled AdoMet

AdoMet-containing radioactive or stable isotopes have been widely used to study the function of AdoMet-dependent enzymes. Both, [*methyl*-^{3}H]AdoMet and [*methyl*-^{14}C] AdoMet with tritium or ^{14}C incorporated into the methyl group, are commercially available and commonly used to assay MTase activities. After separation of the product from the radioactive AdoMet, the radioactivity incorporated into the substrate is measured by scintillation counting. In the same way, [*carboxy*-^{14}C]AdoMet is used to monitor the activity of AdoMet decarboxylase by following the evolution of $^{14}CO_2$ [98].

In addition, [*methyl*-$^{2}H_1$, $^{3}H_1$]AdoMet (**28**) with a chiral methyl group has been prepared enzymatically from [*methyl*-$^{2}H_1$, $^{3}H_1$]methionine [99] and ATP with AdoMet synthetase [100], and was used to analyze the stereochemical course of transmethylation reactions (Scheme 9.5A). [*Methyl*-$^{2}H_1$, $^{3}H_1$]AdoMet was incubated with MTases (see Section 9.2.1) and their substrates, after which the methylated products were

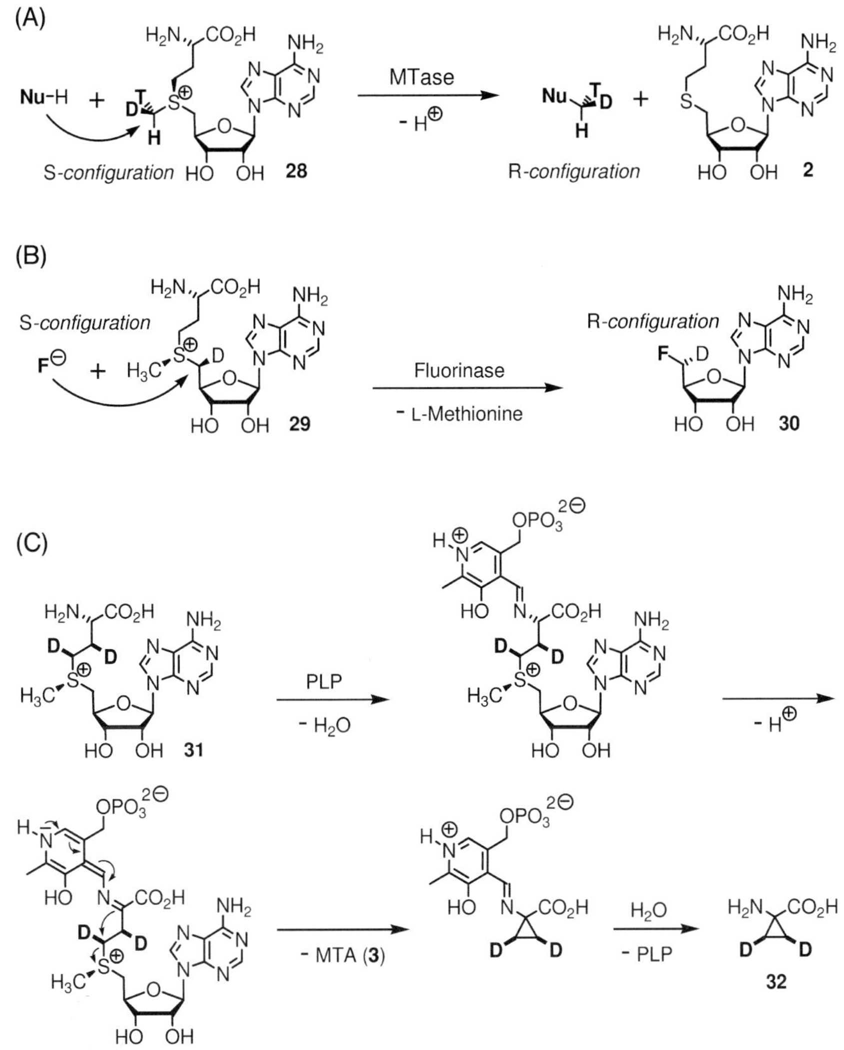

Scheme 9.5 Stereochemical analyses of the reactions catalyzed by (A) MTases, (B) fluorinase, and (C) ACC synthase.

isolated and converted to [*methyl*-$^{2}H_1$,$^{3}H_1$]acetic acid. The chirality of the methyl group was then determined by the enzymatic methods of Arigoni [101] and Cornforth [102]. Hence, it was found that the majority (if not all) of AdoMet-dependent MTases catalyze methyl group transfer with an inversion of configuration which is consistent with a single transfer step and an S_N2-type transition state [103].

Fluorinase catalyzes the nucleophilic attack of fluoride ions onto the 5′ position of AdoMet (**1**) (Section 9.2.3) and the stereochemistry of this reaction has been recently analyzed using (5′*S*)-[5′-^{2}H]-AdoMet (**29**) (Scheme 9.5B) [104]. *N*6-Benzoyl-2′,3′-*O*-isopropylideneadenosine-5′-aldehyde was reduced with $LiAlD_4$, the obtained deuterated *R*-configured alcohol (60% diastereomeric excess, d.e.) was converted into (5′*R*)-[5′-^{2}H]adenosine triphosphate, and the triphosphate transformed with inversion of configuration into (5′*S*)-[5′-^{2}H]AdoMet (**29**) using AdoMet synthetase. The deuterated 5′-deoxy-5′-fluoroadenosine (**30**) formed by fluorinase was analyzed using ^{2}H NMR in a chiral liquid-crystalline solvent, whereby it was concluded that fluorination occurs with an inversion of configuration at the 5′ position of AdoMet (**1**).

The mechanism and stereochemical course of the reaction catalyzed by ACC synthase (Section 9.2.2) has been studied using *syn*-[3,4-$^{2}H_2$]-AdoMet (**31**) (Scheme 9.5C) [105, 106]. ACC synthase uses pyridoxal-5′-phosphate (PLP) to activate Cα for nucleophilic attack onto the Cγ. In this case, MTA (**3**) serves as the leaving group and ACC is liberated after hydrolysis of the PLP-bound product [29] (Scheme 9.5C). For the stereochemical analysis, (*Z*)-[1,2-$^{2}H_2$]-ethene was chemically transformed into *syn*-[3,4-$^{2}H_2$]-methionine, and then into *syn*-[3,4-$^{2}H_2$]-AdoMet (**31**). After enzymatic transformation, the relative stereochemistry of the two deuterium atoms within the product was determined using ^{1}H NMR, whereupon it was found that the meso compound *cis*-[2,3-$^{2}H_2$]-ACC (**32**) with two enantiotopically related protons was formed. Thus, the intramolecular alkylation by ACC synthase proceeds with an inversion of configuration, and this reaction parallels the intermolecular substitutions catalyzed by MTases, fluorinase, and spermidine synthase [92, 107].

9.3.3
Selenium and Tellurium Analogues of AdoMet

The selenium analogue of AdoMet (AdoSeMet, **33**) was obtained from L-selenomethionine and ATP using AdoMet synthetases from different organisms [81, 108, 109]. In fact, L-selenomethionine serves as a better substrate for AdoMet synthetases than L-methionine. Recently, the tellurium analogue of AdoMet (AdoTeMet, **34**) was also synthesized enzymatically, and its chemical stability compared with that of AdoMet (**1**) and AdoSeMet (**33**) [110]. The three major pathways for the chemical decomposition of AdoMet (**10**) include: (i) epimerization at the sulfonium center, leading to the biological inactive *R*-diastereoisomer of AdoMet (**1**) [111]; (ii) an intramolecular nucleophilic attack of the carboxylate group onto Cγ, resulting in the formation of α-amino-γ-butyrolactone (**35**) and MTA (**3**) [112]; and (iii) a deprotonation at C5′ (sulfonium ylide formation) leading to the elimination of

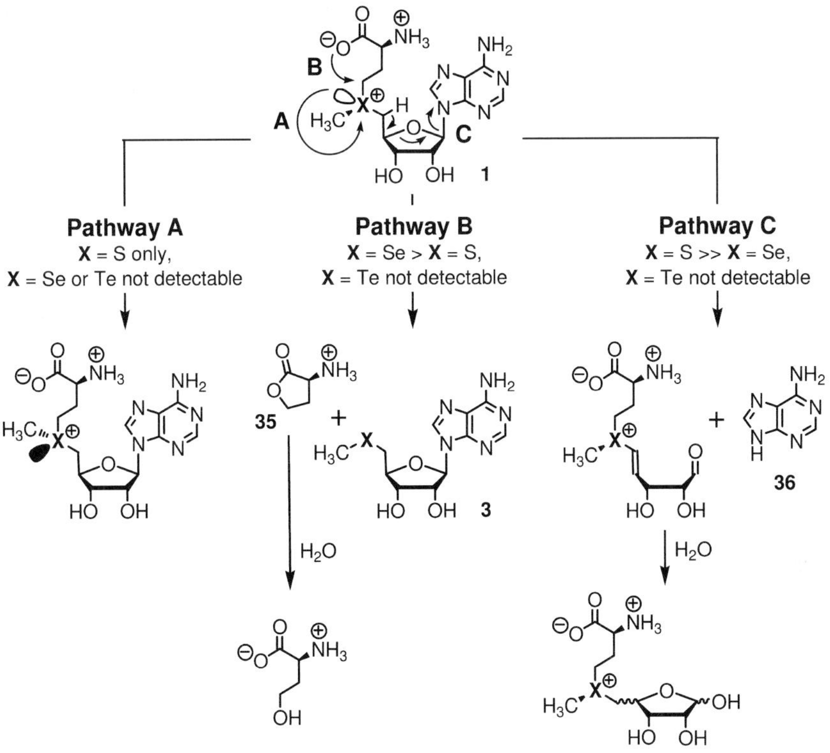

Scheme 9.6 Major chemical decomposition pathways of AdoMet (**1**, X = S) and AdoSeMet (**33**, X = Se). Note that no detectable decomposition via these pathways was observed for AdoTeMet (**34**, X = Te).

adenine (**36**) [113, 114] (Scheme 9.6) [115]. At pH values below 2, epimerization (Scheme 9.6, pathway A) is the predominant process, but as the pH is increased then intramolecular nucleophilic substitution (Scheme 9.6, pathway B) becomes more significant, and at neutral pH (or greater) decomposition via sulfonium ylide formation (Scheme 9.6, pathway C) is the major route. AdoSeMet (**34**) and AdoTeMet (**34**) do not epimerize at an appreciable rate, and decomposition via ylide formation is very slow for AdoSeMet (**33**) and not observed for AdoTeMet (**34**). In contrast, the rate of decomposition by nucleophilic substitution of AdoSeMet (**33**) is increased approximately 10-fold compared to AdoMet (**1**), whereas AdoTeMet (**34**) also does not decompose this way. This enhanced reactivity of AdoSeMet (**33**) implies that its methyl group is more reactive towards nucleophiles, which is consistent with an enhanced reactivity of trimethylselenonium hydroxide compared to trimethylsulfonium hydroxide [116]. These chemical differences between the three cofactors were used to study the enzymatic mechanism of CFA synthase acting on an isolated double bond [117]. The turnover number (TON) with AdoSeMet (**33**) is increased, whereas that with AdoTeMet (**34**) is decreased relative to AdoMet (**1**). These results imply that CFA synthase operates by direct methyl group transfer to the double bond, and not

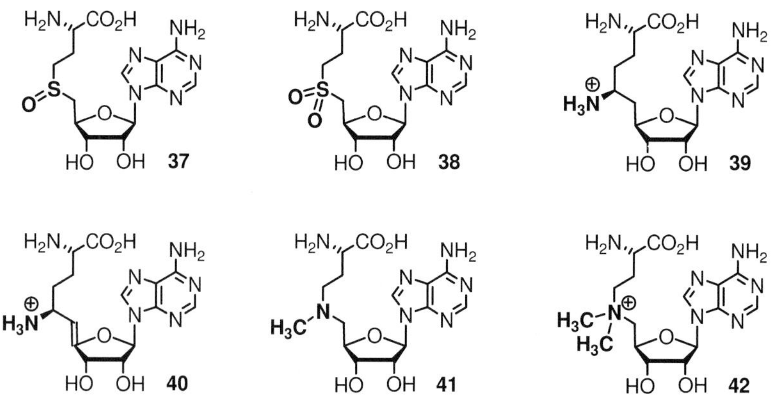

Figure 9.1 The structural formulae of AdoMet analogues.

by a ylide mechanism. However, in the case of α,β-unsaturated esters a conjugate addition by a sulfonium ylide derived from AdoMet (**1**) appears to be a more plausible mechanism [118].

9.3.4
Sulfoxide and Sulfone Analogues of AdoMet

The hydrogen peroxide oxidation of AdoHcy (**2**) leads to the sulfoxide analogue **37** (Figure 9.1), which can be regarded as an isostere of AdoMet (**1**) [119, 120]. The sulfoxide analogue **37** is typically obtained in an approximately 1 : 1 diastereomeric mixture at sulfur, and the diastereoisomers have been separated by using reversed-phase HPLC [121]. Stronger oxidation conditions lead to the sulfone analogue **38** (Figure 9.1) [120, 121]. Both AdoMet analogues show a moderate inhibitory effect on AdoMet-dependent enzymes such as MTases [122–124], ACC synthase [125], and CFA synthase [121], although in all cases AdoHcy (**2**) was the stronger inhibitor. Similarly, the inhibition of spermine synthase with sulfone analogue **38** and its decarboxylated analogue was comparable with that of the cofactor product MTA (**3**) [126]. Recently, the sulfone analogue **38** and AdoMet (**1**) were found to bind with comparable affinities to an AdoMet-binding riboswitch, whereas AdoHcy (**2**) was bound less well by about two orders of magnitude [75]. Such variation can be rationalized from the corresponding crystal structure in complex with AdoMet (**1**), where the methyl group is not directly recognized but two O2 atoms of two uridine residues point directly to the sulfur atom, making favorable electrostatic interactions [74]. Based on a correlation between binding constants and ^{1}H NMR chemical shifts of H5' and Hγ, it was concluded that the sulfone analogue **38** and AdoMet (**1**) have a similar positive charge density on sulfur, which explains their similar affinities to the AdoMet-binding riboswitch [75]. This charge recognition also explains why AdoHcy (**2**), though missing a positive charge at sulfur, has a much lower binding affinity. In this respect it is interesting to note that, for most MTases, AdoHcy (**2**) is a strong product inhibitor [120] and that the positive charge on sulfur of

AdoMet (**1**) is apparently less well-recognized by the enzymes. This is line with the catalytic role of the enzymes, because stabilization of the positive charge is expected to make AdoMet (**1**) less reactive.

9.3.5
Sinefungin

The natural product Sinefungin (**39**) may be regarded as a stable analogue of AdoMet (**1**) in which the sulfur atom is replaced by carbon, and the methyl group by a basic, protonated primary amino group (Figure 9.1). The compound was isolated from *Streptomyces griseolus* [127], and has been synthesized via several chemical routes [128–134]. Sinefungin (**39**) and even more its close analogue dehydrosinefungin (**40**, A9145C; Figure 9.1) are highly potent competitive inhibitors of MTases [135–138], CFA synthase [139] and ACC synthase [140], with the observed inhibition constants often being much lower than those with AdoHcy (**2**). In addition, Sinefungin (**39**) has antiviral [136, 137], antifungal [141] and antiparasitic activities [142–145], although due to a lack of specificity it also shows high *in-vivo* toxicity which greatly limits its clinical use [146].

9.3.6
Nitrogen Analogues of AdoMet

The replacement of sulfur by nitrogen also leads to stable AdoMet analogues. The chemical syntheses of AzaAdoMet (**41**) were started from 2′,3′-*O*-isopropylidene-5′-*O*-tosyladenosine (**23**), which was first converted with methylamine to protected 5′-deoxy-5′-methylaminoadenosine and then 5′-*N*-alkylated with iodides of amino acid side chain precursors. After deprotection, AzaAdoMet (**41**) was obtained as a diastereomeric mixture at the amino acid center [147, 148], and later as a single diastereoisomer [149, 150]. As expected, AzaAdoMet (**41**) does not serve as a methyl group donor for MTases [151]. Compared to AdoHcy (**2**), it is a week inhibitor for MTases [152, 153], but its structural binding mode to MTases is very similar to binding of the natural cofactor [153, 154]. Type III restriction endonucleases require the binding of AdoMet (**1**) for DNA cleavage, and AzaAdoMet (**41**) – but not AdoHcy (**2**) – can replace the natural cofactor to bring about a conformational change of the enzyme necessary for endonuclease activity [155]. The ternary amino group of AzaAdoMet (**41**) has an unusually low pK_a value of about 7.1, and thus acts as a charge-switchable analogue of AdoMet [149]. This property was utilized to probe the molecular mechanism of the *E. coli* methionine repressor MetJ, which does not undergo any significant structural change upon AdoMet binding, yet still binds its operator DNA sequence about 100-fold tighter in the presence of AdoMet (**1**). Only at a pH well below the pK_a did the addition of AzaAdoMet (**41**) to MetJ lead to tight binding of the repressor to its operator [156], a finding which is in agreement with a purely electrostatic DNA binding enhancement of MetJ upon AdoMet binding [157]. In addition, the methylated AzaAdoMet analogue **42**, with a positively charged quaternary amine, has been prepared [150] and investigated with the

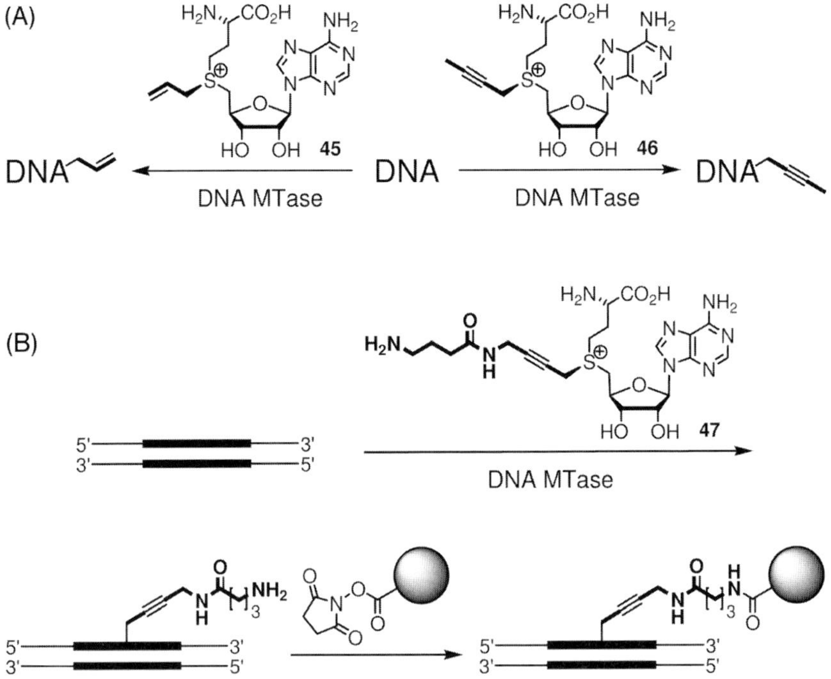

Scheme 9.8 (A) DNA MTase-catalyzed transfer of extended methyl group replacements from double-activated AdoMet analogues to DNA. (B) Transfer of a propargylic side chain with a primary amino group to specific DNA sequences (thick lines) for subsequent DNA labeling with activated esters of reporter groups (gray sphere).

to label selectively the primary amino groups on DNA with activated esters of fluorophores or biotin (Scheme 9.8) [174]. Targeted DNA labeling with these double-activated cofactors, as well as with the aziridine cofactors (see Section 9.3.7), offers interesting new applications in functional studies of DNA and DNA-modifying enzymes, molecular biology, medical diagnostics and DNA-based nanobiotechnology. In addition, it was suggested that these new classes of AdoMet analogues, in combination with RNA and protein MTases, might provide powerful tools for the targeted functionalization and labeling of RNA and proteins [175].

9.4
AdoMet as a Pharmaceutical

As discussed previously, AdoMet (**1**) is involved in a multitude of biological processes and is therefore a natural candidate for pharmaceutical use. For example, AdoMet (**1**) is important in the biosynthesis of various neurotransmitters in the brain, and low

AdoMet concentrations have been observed in the cerebrospinal fluid of depressed persons [176]. AdoMet (**1**) has also been prescribed as an antidepressant in Europe for decades, and is available as an over-the-counter dietary supplement in the USA. Several placebo-controlled clinical studies have suggested an efficacy of AdoMet (**1**) comparable to that of other antidepressants, and that it is well tolerated and relatively free of adverse side effects [177]. However, in patients with bipolar depression side effects such as anxiety, mania or hypomania were observed [178]. Overall, AdoMet (**1**) seems to offer the potential for effective antidepressant treatment of patients with Parkinson's disease [179], fibromyalgia [180] and HIV-positive individuals diagnosed with major depressive disorder [181]. Another pharmaceutical application of AdoMet (**1**) may be in the treatment of osteoarthritis and other degenerative joint diseases [182].

In humans, since about two-thirds of administered AdoMet (**1**) are metabolized in the liver, it has been investigated for the treatment of liver diseases, although the underlying molecular mechanisms remains mostly unclear and are likely to be complex. AdoMet (**1**) increases the survival of patients with alcoholic liver cirrhosis [183]. In fact, an impaired mitochondrial uptake of glutathione (GSH) has been postulated to be an important factor in alcoholic liver injury in rats, and AdoMet administration can restore GSH uptake [184].

AdoMet (**1**) can also prevent the development of liver tumors in rats [185], a finding which has been attributed to lower AdoMet levels in human hepatocellular carcinoma (HCC) leading to faster cell growth, whereas exogenous AdoMet treatment inhibits cell growth [186] and may induce apoptosis [187]. Interestingly, AdoMet administration led to anti-apoptotic events in healthy hepatocytes, but this effect was lacking in apoptosis-induced hepatoma cells [188]. These observations may make AdoMet (**1**) an attractive agent for both the chemoprevention and treatment of HCC.

9.5
Concluding Remarks

AdoMet (**1**) is involved in a myriad of biosynthetic transformations, and has interesting pharmaceutical properties. In addition, many AdoMet-dependent enzymes are important pharmacological targets. Unfortunately, close analogues of AdoMet with good inhibitory properties (e.g., Sinefungin (**39**) lack the specificity required for the development of drugs. Other inhibitors containing only structural elements of AdoMet and the second substrate (bisubstrate adduct inhibitors, (e.g., for catechol-*O*-MTase [189]), structural elements of the second substrate or even structurally unrelated compounds obtained from screening chemical libraries, are more suited. Nonetheless, AdoMet (**1**) analogues serve as good tools for mechanistic investigations of AdoMet-dependent enzymes, and represent promising molecules for the functionalization, labeling, and diversification of a large variety of biomolecules.

146 Zweygarth, E., Schillinger, D., Kaufmann, W., Roettcher, D., *Trop. Med. Parasitol.* **1986**, *37*, 255–257.

147 Davis, M., Dudman, N. P. B., White, H. F., *Aust. J. Chem.* **1983**, *36*, 1623–1627.

148 Minnick, A. A., Kenyon, G. L., *J. Org. Chem.* **1988**, *53*, 4952–4961.

149 Thompson, M. J., Mekhalfia, A., Jakeman, D. L., Phillips, S. E. V., Phillips, E., Porter, J., Blackburn, G. M., *Chem. Commun.* **1996**, *6*, 791–792.

150 Thompson, M. J., Mekhalfia, A., Hornby, D. P., Blackburn, G. M., *J. Org. Chem.* **1999**, *64*, 7467–7473.

151 Santi, D. V., Hardy, L. W., *Biochemistry* **1987**, *26*, 8599–8606.

152 Reich, N. O., Mashhoon, N., *J. Biol. Chem.* **1990**, *265*, 8966–8970.

153 Hausmann, S., Zheng, S. F. C., Schneller, S. W., Lima, C. D., Shuman, S., *J. Biol. Chem.* **2005**, *280*, 20404–20412.

154 Couture, J.-F., Hauk, G., Thompson, M. J., Blackburn, G. M., Trievel, R. C., *J. Biol. Chem.* **2006**, *281*, 19280–19287.

155 Bist, P., Sistla, S., Krishnamurthy, V., Acharya, A., Chandrakala, B., Rao, D. N., *J. Mol. Biol.* **2001**, *310*, 93–109.

156 Parsons, I. D., Persson, B., Mekhalfia, A., Blackburn, G. M., Stockley, P. G., *Nucleic Acids Res.* **1995**, *23*, 211–216.

157 Phillips, K., Phillips, S. E. V., *Structure* **1994**, *2*, 309–316.

158 Pignot, M., Siethoff, C., Linscheid, M., Weinhold, E., *Angew. Chem., Int. Ed.* **1998**, *37*, 2888–2891.

159 Petersen, S. G., Rajski, S. R., *J. Org. Chem.* **2005**, *70*, 5833–5839.

160 Weller, R. L., Rajski, S. R., *ChemBioChem* **2006**, *7*, 243–245.

161 Zhang, C., Weller, R. L., Thorson, J. S., Rajski, S. R., *J. Am. Chem. Soc.* **2006**, *128*, 2760–2761.

162 Comstock, L. R., Rajski, S. R., *J. Org. Chem.* **2004**, *69*, 1425–1428.

163 Comstock, L. R., Rajski, S. R., *J. Am. Chem. Soc.* **2005**, *127*, 14136–14137.

164 Comstock, L. R., Rajski, S. R., *Nucleic Acids Res.* **2005**, *33*, 1644–1652.

165 Weller, R. L., Rajski, S. R., *Org. Lett.* **2005**, *7*, 2141–2144.

166 Saxon, E., Bertozzi, C. R., *Science* **2000**, *287*, 2007–2010.

167 Rostovtsev, V. V., Green, L. G., Fokin, V. V., Sharpless, K. B., *Angew. Chem., Int. Ed.* **2002**, *41*, 2596–2599.

168 Tornoe, C. W., Christensen, C., Meldal, M., *J. Org. Chem.* **2002**, *67*, 3057–3064.

169 Pljevaljcic, G., Pignot, M., Weinhold, E., *J. Am. Chem. Soc.* **2003**, *125*, 3486–3492.

170 Pljevaljcic, G., Schmidt, F., Peschlow, A., Weinhold, E., *Methods in Molecular Biology 283: Bioconjugation Protocols* (ed. C. M. Niemeyer), Humana Press, NY, **2004**, pp. 145–161.

171 Pljevaljcic, G., Schmidt, F., Weinhold, E., *ChemBioChem* **2004**, *5*, 265–269.

172 Parks, L. W., *J. Biol. Chem.* **1958**, *232*, 169–176.

173 Schlenk, F., Dainko, J. L., *Biochim. Biophys. Acta* **1975**, *385*, 312–323.

174 Lukinavicius, G., Lapiene, V., Stasevskij, Z., Dalhoff, C., Weinhold, E., Klimasauskas, S., *J. Am. Chem. Soc.* **2007**, *129*, 2758–2759.

175 Klimasauskas, S., Weinhold, E., *Trends Biotechnol.* **2007**, *25*, 99–104.

176 Bottiglieri, T., Godfrey, P., Flynn, T., Carney, M. W., Toone, B. K., Reynolds, E. H., *J. Neurol. Neurosurg. Psychiatry* **1990**, *53*, 1096–1098.

177 Mischoulon, D., Fava, M., *Am. J. Clin. Nutr.* **2002**, *76*, 1158S–1161.

178 Spillmann, M., Fava, M., *CNS Drugs* **1996**, *6*, 416–425.

179 Di Rocco, A., Rogers, J., Brown, R., Werner, P., Bottiglieri, T., *Movement Disord.* **2000**, *15*, 1225–1229.

180 Bottiglieri, T., Ryland, K., *Acta Neurol. Scand. Suppl.* **1994**, *154*, 19–26.

181 Shippy, R. A., Mendez, D., Jones, K., Cergnul, I., Karpiak, S. E., *BMC Psychiatry* **2004**, *4*, 38.

182 Najm, W. I., Reinsch, S., Hoehler, F., Tobis, J. S., Harvey, P. W., *BMC Musculoskel. Dis.* **2004**, *5*, 6.

183 Mato, J. M., Camara, J., Fernandez de Paz, J., Caballeria, L., Coll, S., Caballero, A.,

Garcia-Buey, L., Beltran, J., Benita, V., Caballeria, J., Sola, R., Moreno-Otero, R., Barrao, F., Martin-Duce, A., Correa, J. A., Pares, A., Barrao, E., Garcia-Magaz, I., Puerta, J. L., Moreno, J., Boissard, G., Ortiz, P., Rodes, J., *J. Hepatol.* **1999**, *30*, 1081–1089.

184 Colell, A., Garcia-Ruiz, C., Morales, A., Ballesta, A., Ookhtens, M., Rodes, J., Kaplowitz, N., Fernandez-Checa, J. C., *Hepatology* **1997**, *26*, 699–708.

185 Pascale, R. M., Marras, V., Simile, M. M., Daino, L., Pinna, G., Bennati, S., Carta, M., Seddaiu, M. A., Massarelli, G., Feo, F., *Cancer Res.* **1992**, *52*, 4979–4986.

186 Lu, S. C., Mato, J. M., *Alcohol* **2005**, *35*, 227–234.

187 Yang, H., Sadda, M. R., Li, M., Zeng, Y., Chen, L., Bae, W., Ou, X., Runnegar, M. T., Mato, J. M., Lu, S. C., *Hepatology* **2004**, *40*, 221–231.

188 Ansorena, E., Berasain, C., Lopez Zabalza, M. J., Avila, M. A., Garcia-Trevijano, E. R., Iraburu, M., J. *Am. J. Physiol. Gastrointest. Liver. Physiol.* **2006**, *290*, G1186–G1193.

189 Lerner, C., Ruf, A., Gramlich, V., Masjost, B., Zürcher, G., Jakob-Roetne, R., Borroni, E., Diederich, F., *Angew. Chem., Int. Ed.* **2001**, *40*, 4040–4042.

Part II
Biotechnology

10
5-Substituted Nucleosides in Biochemistry and Biotechnology

Mohammad Ahmadian and Donald E. Bergstrom

10.1
Introduction

C-5-substituted pyrimidine nucleosides are an unusually significant class of compounds because of the role that they play as components of nucleotide-derived tools for molecular genetics. The revolution in molecular genetics has to a significant degree been driven by the development of techniques for the analysis of nucleic acids. The current level of activity to establish molecular level knowledge of human disease is astonishing. At the heart of this activity lie efforts to read and interpret nucleic acid sequences. For example, the human genome project required that more than three billion base pairs of sequence be determined. The molecular techniques used in the endeavor, as well as in many more recent sequencing projects, center around procedures involving modified oligonucleotides for use as primers and probes. It is often necessary to tag these primers and probes with fluorophores, haptens, or other signaling molecules. As a site for tethering molecular signaling modules, the C-5 position of the pyrimidine nucleosides is nearly ideal as even very large groups may be attached without interfering with DNA duplex formation. Moreover, as outlined in Section 10.3, polymerases are available that direct the incorporation of a wide variety of C-5 modified pyrimidine nucleosides in place of the natural substrates dCTP, dTTP, and UTP.

The chapter begins with a short section on the synthesis of C-5 pyrimidine nucleosides. Because this topic has been extensively reviewed in recent years, it has been decided to provide only a few highlights from recently published literature. There follows a more extensive discussion of the incorporation of C-5-substituted nucleosides into oligonucleotides via their 5′-triphosphates. Next, those studies which have focused on C-5 substituents designed to increase nucleic acid duplex stability are summarized; followed by details of the photochemical reactions of C-5.

Modified Nucleosides: in Biochemistry, Biotechnology and Medicine. Edited by Piet Herdewijn

ISBN: 978-3-527-31820-9

10.2
Synthesis

10.2.1
Organopalladium Coupling Reactions

In 2003, Agrofoglio and colleagues published a comprehensive review of organopalladium-mediated routes to nucleoside analogues [1]. Most investigators continue to use variations of the Heck, Sonogashira, Stille, and Suzuki reactions to couple 5-halocytosine and uracil nucleosides to alkenyl, alkynyl, carbonyl, and aryl moieties (Scheme 10.1). Although the scope of the reaction has been extended to linkers containing a variety of functional groups, the core reactions have not changed.

Of some interest here is the application of reagents, conditions and procedures that facilitate the coupling reactions under milder conditions. Shaughnessy and coworkers have found that unprotected 5-iodo-2′deoxyuridine as well as 8-bromopurine nucleosides can be coupled to aryl groups in high yield by the Suzuki reaction in aqueous solution using palladium acetate and the water-soluble phosphine, tris (3-sulfonatophenyl)phosphine [2]. Palladium-mediated reactions are often facilitated by microwave irradiation. For example, Petricci and workers found that microwave irradiation of unprotected 5-iodouridine and phenylacetylene gave a 70% yield of 5-(phenylethynyl)uridine in 5 min [3], though whether this reaction will be extendable to other alkynes in the Sonogashira coupling reaction is unknown. Carrying reactions out on a solid support often simplifies purification. For example, Aucagne *et al.* described the preparation of polystyrene-linked 5-iodo-2′-deoxyuridine and its application in Sonogashira and Stille coupling reactions [4]. Following the coupling

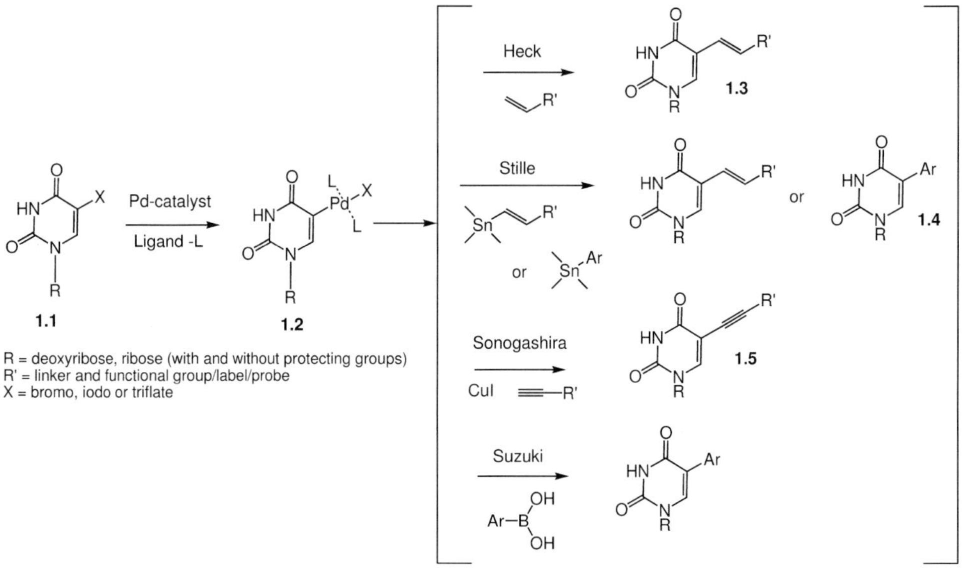

Scheme 10.1

reaction, deprotected C-5-modified 2′-deoxyuridine can be obtained in 40 to 80% yield. In comparison, the Heck and Stille reactions gave poor results. In 2003, Minakawa *et al.* incorporated 5-ethynyl-2′-deoxyuridine into an oligonucleotide on a solid support, and then investigated on column coupling with dimethylformamide (DMF)-soluble alkynes using CuCl-TMEDA [5]. These authors found that the scheme was useful for incorporating a fluorescent probe. Examples in which the organopalladium coupling reactions have been extended to novel linkers and labels will be referred to later in the chapter.

10.2.2
Strategies for Post-Oligonucleotide-Synthesis Modification through Pyrimidine C-5

Post-oligonucleotide-synthesis modification is useful if only one type of probe (at one or multiple sites) is required per sequence. This may be an efficient strategy if the aim is to create a library of modified oligonucleotides without having to synthesize the corresponding library of modified nucleoside phosphoramidites or triphosphates. However, there are stringent requirements for the successful use of this strategy. The modification reaction should provide a single product in high yield under mild conditions. Separation from unreacted oligonucleotide may not be an option in many cases. Ten examples of post synthesis modification are illustrated in Scheme 10.2.

Matsuda and coworkers first reported the use of 5-methoxycarbonyl-2′-deoxyuridine as a post-synthetic modification method during the mid-1990s [6, 7]. Phosphoramidite **2.1a** is incorporated into an oligonucleotide and deprotection carried out with a diamine rather than ammonia or methylamine. This results in a C-5 spacer terminated with an amino group (**2.2a** in Scheme 10.2), which is allowed to react with an active ester to link probes or other functional molecules (R^3 in **2.3a**). A later report described the use of the more reactive trifluoroethoxycarbonyl in place of the methoxycarbonyl [8]. In a similar strategy, Sawai and coworkers have described the application of a C-5 cyanomethoxycarbonylmethyl substituent for post-modification (**2.1b** to **2.2b**) [9, 10].

In order to place a series of propynylamino-linked substituents at C-5, Richert and coworkers synthesized 2′-deoxy-5′-*O*-(4,4′-dimethoxytrityl)-5-[*N*-(2-trimethylsilyl) ethoxycarbonyl-3-aminopropynyluridine-3′-*O*-(2-cyanoethyl-*N*,*N*′-diisopropylaminophosphoramidite (**2.1c**), which allows the protected aminopropynyl group to be positioned at any point within a sequence [11]. The Teoc protecting group can be selectively removed with TBAF in tetrahydrofuran (THF). This frees the amino group, which is then acylated by a carboxylic acid in the presence of HBTU, HOBt, and DIEA. Finally, the remaining protecting groups are removed from the oligonucleotide. In order to increase the diversity of groups that could be incorporated into oligonucleotide sequences, these authors also employed the Sonogashira reaction on 5-iodo-2′-deoxyuridine contained in oligonucleotide sequences (**2.2d**). The oligonucleotide is then deprotected by conventional means following the organopalladium coupling reaction. In a more recent variation on this approach, Richert and coworkers have described a 5-iodo-3′-amino-3′-deoxyuridine derivative that can be incorporated into oligonucleotides at the 5′-terminus and then coupled to alkynes by the Sonogashira

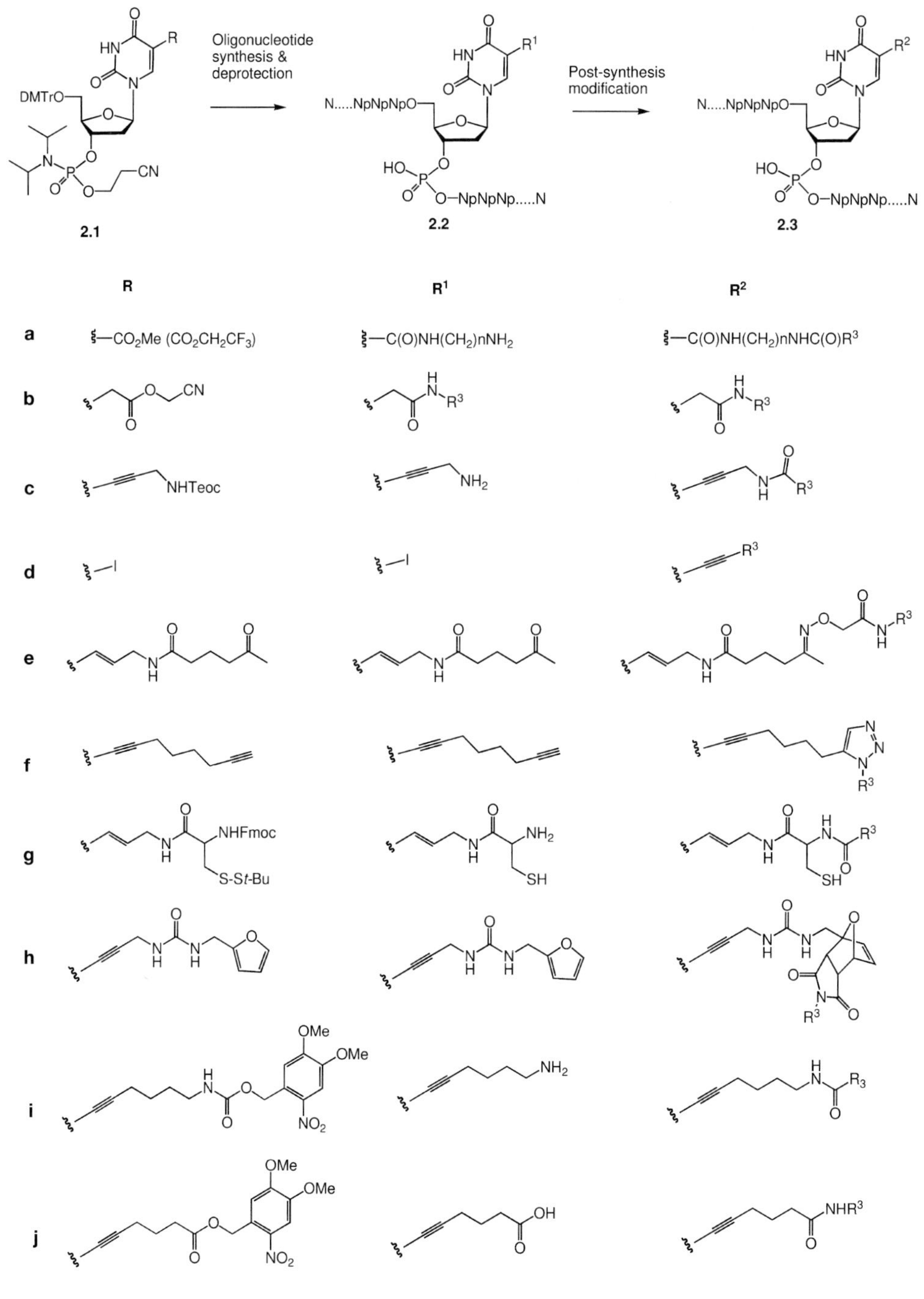

Oligonucleotide synthesis & deprotection
Post-synthesis modification
DMTrO
N.....NpNpNpO
O-NpNpNp.....N
2.1
2.2
2.3
R
R1
R2
a
b
c
d
e
f
g
h
i
j
—CO2Me (CO2CH2CF3)
—C(O)NH(CH2)nNH2
—C(O)NH(CH2)nNHC(O)R3
NHTeoc
NHFmoc
S-St-Bu

Scheme 10.2

reaction and acylated at the 3′-amino group [12]. The modified oligonucleotides constructed by this strategy were designed to explore chemical ligation.

The fifth example in Scheme 10.2 illustrates the preparation of a ketone-derivatized DNA (**2.2e**; Scheme 10.2) that reacts selectively with *O*-alkyl hydroxylamine to yield an oxime linkage connected to the oligonucleotide through a C-5 spacer (**2.3e**) [13]. Because of its reactivity orthogonality to functional groups found on biomolecules, the 3 + 2 cycloaddition reaction between terminal alkynes and organic azides has gained considerable popularity for modifications of biomolecules. Carell and coworkers found that this cycloaddition reaction can be applied to nucleic acids through a terminal alkyne linked to C-5 of 2′-deoxyuridine (transformation of **2.2f** to **2.3f**) [14]. These authors compared the alkyne shown in Scheme 10.2 to 5-ethynyl-2′-deoxyuridine, and determined that the spacer was necessary for optimal coupling yields. Also, it was critical to use tris(benzyltriazolylmethyl)amine as the copper ligand to prevent unwanted cleavage of the nucleic acid.

One reaction that has worked to advantage for protein–protein conjugation is native chemical ligation, in which a terminal cysteine on one component is allowed to react with a thioester on the second component. This reaction can be applied to a deoxyuridine derivative carrying a C-5 spacer terminated by a cysteine [15]. The phosphoramidite for this strategy contains an amino group that is protected by Fmoc and a thiol protected as a mixed *tert*-butyl disulfide (**2.1g**). For optimum specificity, a thioester is required for the conversion of **2.2g** to **2.3g**. Yet another cycloaddition reaction, the Diels–Alder reaction, allows highly specific conjugation. Reaction between a C-5 appended furan (**2.2h**) and a maleimide yielded adduct **2.3h**. Graham and coworkers used this method to attach a series of fluorophores to an oligonucleotide [16]. The final post-synthetic modification illustrated in Scheme 10.2 is the use of a photolabile protecting group to yield functional groups (amine in **2.2i** and carboxyl in **2.2j**) that can be modified specifically while the synthetic oligonucleotide is still on the solid support [17]. This is important because it allows high coupling yields to be achieved without added purification problems.

10.3 Incorporation of C-5-Substituted Pyrimidine Nucleotides into Nucleic Acids through Modified Nucleotide 5′-Triphosphates

10.3.1 The Early Studies

The development of technologies for the controlled introduction of probes into nucleic acid sequences has been extensively investigated for well over three decades. The seminal discovery occurred in 1973, when Ward and coworkers found that pyrimidine nucleotide triphosphates with relatively large substituents at C-5 could be utilized as substrates by many different DNA polymerases [18–20]. These authors prepared a series of C-5 mercurated UTP and dUTP derivatives, including those shown in Scheme 10.3. Other R groups were investigated and many were substrates,

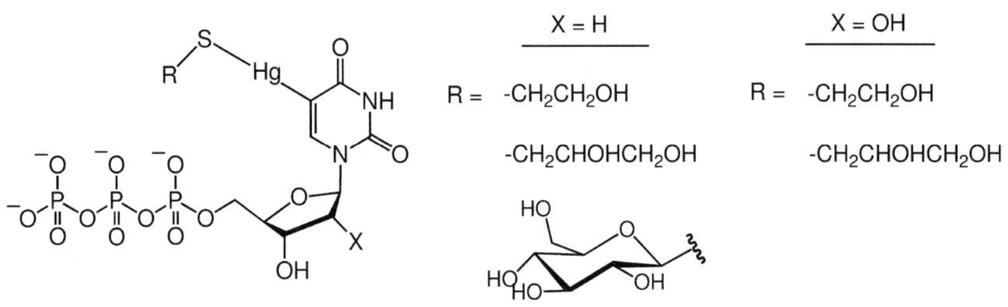

Scheme 10.3

but those shown in Scheme 10.3 were the most efficiently incorporated. The DNA polymerases that accepted these substrates included *Escherichia coli* Pol I, avian myeloblastosis virus (AMV), and calf-thymus terminal transferase. *E. coli* and T7 RNA polymerases accepted uridine analogues (Scheme 10.3) in templated RNA synthesis. Since C5 of cytidine and uridine within intact oligonucleotides can be mercurated directly, Hopman and coworkers developed a strategy for exploiting the transformation to introduce haptens for the antibody-mediated detection of probe sequences [21–24]. The group was in fact able to identify thiol-terminated linkers containing haptens that formed stable linkages to the mercurated nucleic acids. Unfortunately, however, this approach was never widely adapted, possibly because at the same time Ward and coworkers [25] found that the labile mercurated nucleic acids could be avoided entirely, by using stable C-5 alkenyl linkers introduced at the triphosphate stage by a palladium-mediated coupling reaction [26, 27]. The biotin-terminated dUTP analogue (Scheme 10.4) proved to be a substrate for *E. coli* Pol I, T4, murine α and β, human HeLa, and herpes simplex virus (HSV) DNA polymerases, while the UTP analogue was a substrate for *E. coli* and T7 DNA polymerases. In contrast to the dUTP analogues shown in Scheme 10.3, dUTP containing simple alkyl chains at C5 were not effective DNA polymerase substrates when extended beyond three carbons [28, 29]. Thus, it became apparent early on that the C-5 spacer had to include a short region near the pyrimidine that remained in the plane of the aromatic ring. An extensive body of research on alkynyl spacers in particular followed over the next decades, mainly because of the particular ease with which this group is introduced at C5 [30]. However, rather than provide an extensive review of these past investigations, the remainder of this section will focus on studies conducted over the past five years. References to previous investigations can found within this more recent body of work.

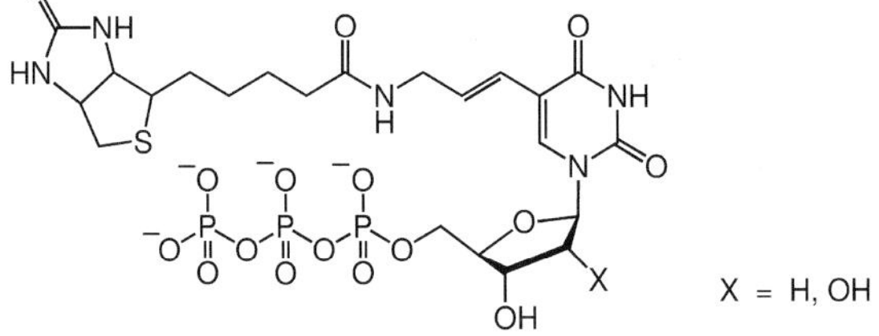

Scheme 10.4

10.3.2 Incorporation of Diverse Functionality into DNA

Over the past two decades, our knowledge of DNA polymerases – as well as the number of DNA polymerases (particularly those that are thermostable) available commercially – has expanded dramatically. Systematic studies with thermostable DNA polymerases have led to the development of a considerable variety of C-5 substituted analogues that can be incorporated into DNA polymerase. Some combinations of C-5 modifications and enzymes allow both incorporation of the modified nucleoside triphosphates, as well as the efficient replication of the resulting modified nucleic acid. In following up on the early findings of Ward, Sakthivel and Barbas prepared a novel series of acylated derivatives of 5-(3-amino-1-propenyl)-2′-deoxyuridine, and investigated the substrate properties of their 5′-triphosphates in a polymerase chain reaction (PCR) with the *Thermus aquaticus* DNA polymerase Taq, the *Thermococcus litoralis* DNA polymerase Vent, the *Pyrococcus furiosus* DNA polymerase Pfu, and the *Thermus thermophilius* DNA polymerase rTh [31]. In these studies, dTTP was replaced by a modified base in the replication of a 519-bp template of which 246 nt were thymidines. Amplification of the full-length product demonstrated that the modified nucleosides were both effective substrates, and that the replication of modified nucleic acid could be supported in subsequent rounds.

With these thermostable DNA polymerases, spacer structure proved to be critical. Those triphosphates that contained the $-CH{=}CH-CH_2-NHC(O)CH{=}CH-$ spacer were substrates for all four DNA polymerases (Scheme 10.5). Three analogues containing either $-CH{=}CH-CH_2-NHC(O)CH_2CH_2-$ or $-CH{=}CH-CH_2-NHC(O)-$phenyl$-$ (**5a–c**) were not substrates. Triphosphate **5i** was a suitable substrate for rTh DNA polymerase, but not the other polymerases. The variety of functional groups at the end of the spacer tolerated by the polymerases is noteworthy, and suggests the potential for creating nucleic acids with as much functional diversity as proteins.

Research groups at other laboratories have continued to expand the repertoire of acceptable functional groups. As illustrated in Scheme 10.6, Benner and coworkers were able to devise suitable syntheses of a series of C-5 linked protected thiols for incorporation of the thiol group into nucleic acids. For nucleosides **6.1a–d**, a detailed analysis was reported of their behavior with family A (*Taq*, *Tfl*, Hot*Tub*, and *Tth*) and B (*Pfu*, *Pwo*, Vent and Deep Vent) polymerases, examining first the steps of substrate insertion, multiple insertions, and template properties [32]. One of the nucleotides, **6.1c**, proved to be unsuitable for reasons of chemical instability. However, among the remaining three, **6.1d** proved to have the best combination of substrate and template properties for use in PCR with the preferred enzyme *Pwo*. In general, the family B polymerases more readily accept C-5 substituents than family A enzymes. A later study on a series of deoxycytidine analogues, including **6.2d**, again showed that a family B polymerase, Vent (exo-), accepted the triphosphate as a substrate whereas a family A polymerase, *Taq*, did not [33].

In a series of reports from 2001 onwards, Sawai and coworkers have described the substrate properties of deoxyuridine 5′-triphosphates containing substituents linked

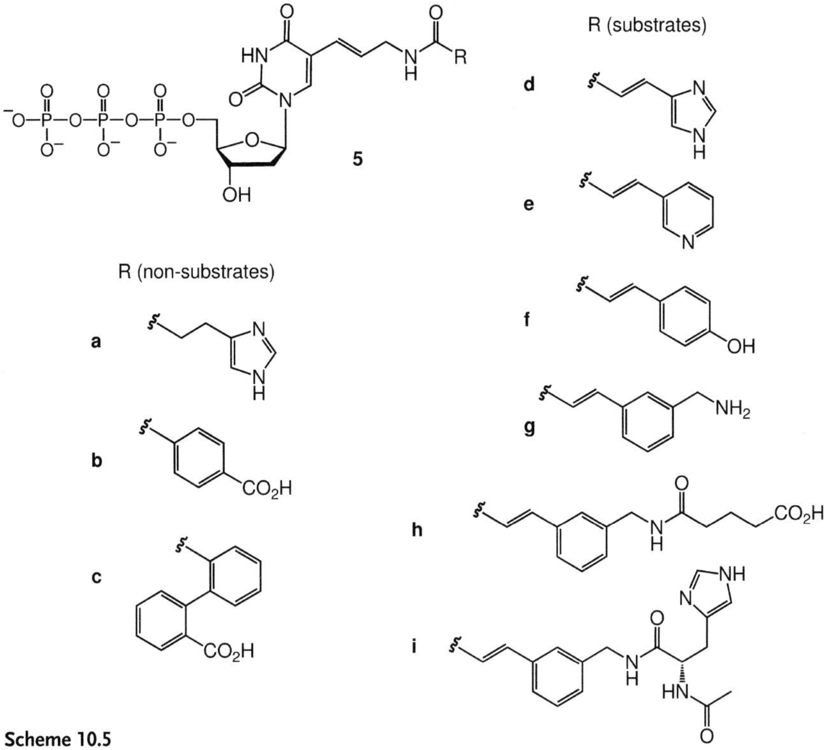

Scheme 10.5

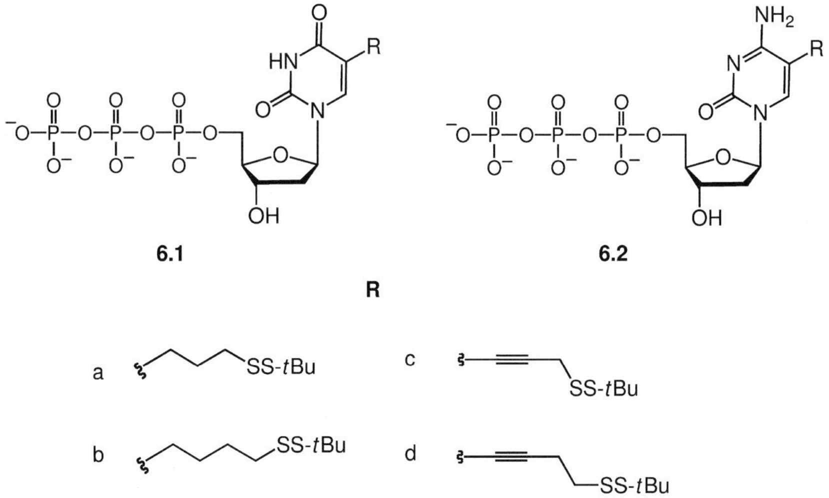

Scheme 10.6

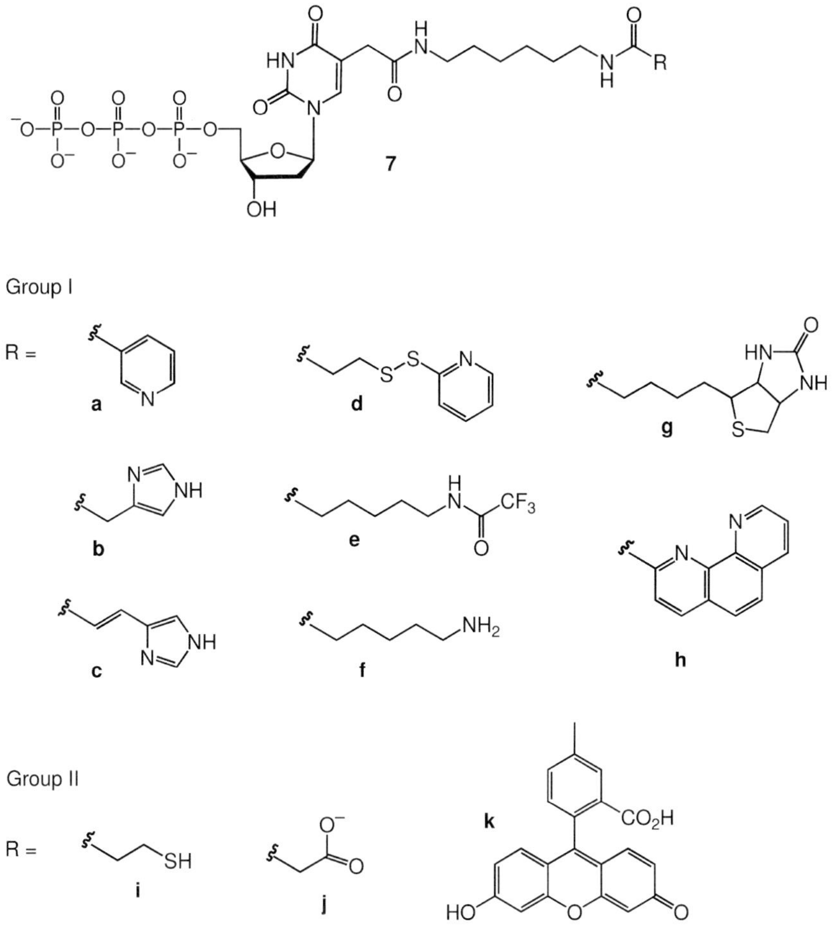

Scheme 10.7

through a C-5 carbonylmethylene [34–37]. These substituents are accommodated by the B-family thermophilic enzyme KOD Dash DNA polymerase from *Pyrococcus kodakaraensis*. The resulting modified oligonucleotides are accepted as templates and, as a result, PCR amplification takes place. However, even with a relatively long spacer the nature of the substituent at the end of the chain has a significant impact of acceptability as a substrate. As shown in Scheme 10.7, the Group I substituents were accepted as substrates, but the Group II were not [37]. In general, it appears that carboxylate is not tolerated.

Given that the results from most other studies suggest that a C-5 alkynyl or alkenyl linker is critical for enzyme recognition, Kuwahara *et al.* synthesized the pyrimidine nucleoside triphosphates shown in Scheme 10.8 and compared their substrate properties with the thermostable DNA polymerases, *Taq, Tth, Vent(exo-), KOD Dash,* and *KOD(exo-)*. The results of these experiments make it clear that there is no one ideal combination of substituent and enzyme [38].

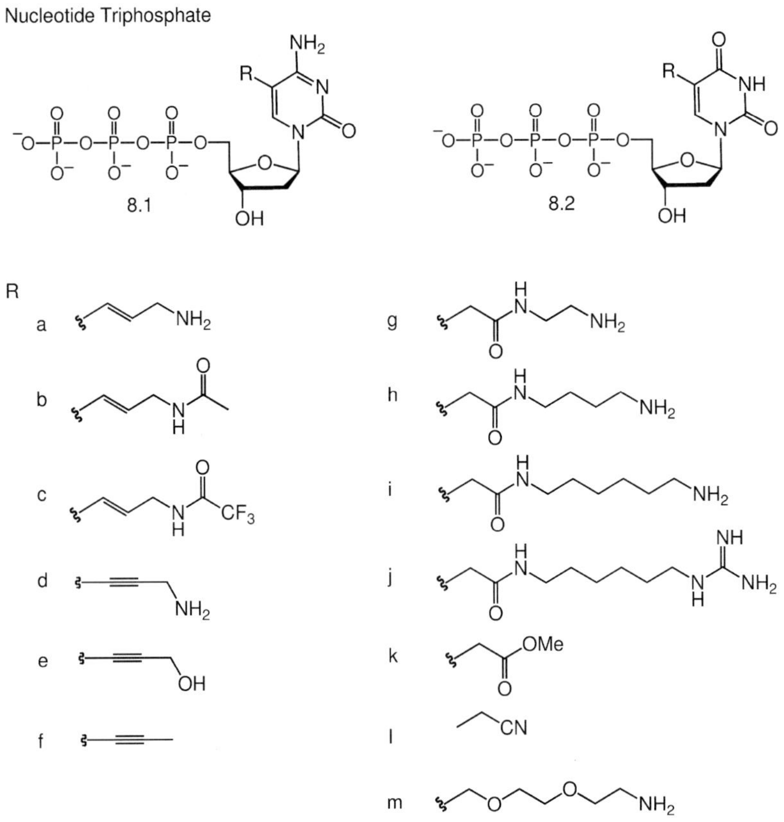

Scheme 10.8

In order to create oligonucleotides containing most of the natural amino acids linked through C-5, Kuwahara reported the synthesis the nucleoside triphosphates shown in Scheme 10.9. Full-length PCR products could be obtained from all using *KOD Dash* DNA polymerase under suitable conditions [39]. Matsui and coworkers have used *KOD Dash DNA* polymerase, and found that dUTP analogues substituted by disaccharides linked through a C-5 $-CH{=}CHC(O)NH(CH_2)_6-$ spacer can be incorporated and amplified by PCR [40].

Although this chapter is dedicated to C-5-substituted pyrimidine nucleosides, it is important to describe efforts to synthesize highly modified nucleic acids that contain both C-5-substituted pyrimidine nucleosides and C-5-substituted pyrrolo[2,3-*d*] pyrimidine nucleosides. The latter are purine analogues in which N7 is replace by a carbon atom, which is then available for linking spacers that in nucleic acid duplexes are positioned in the major groove similar to a C-5 spacer attached to a pyrimidine. Attempts to increase functional diversity has led to efforts to develop optimal strategies for the synthesis of oligonucleotides in which all of the natural nucleosides are replaced with C-5-substituted pyrimidine and pyrrolo[2,3-*d*]pyrimidine nucleosides. In a series of three reports, Famulok and coworkers have described the

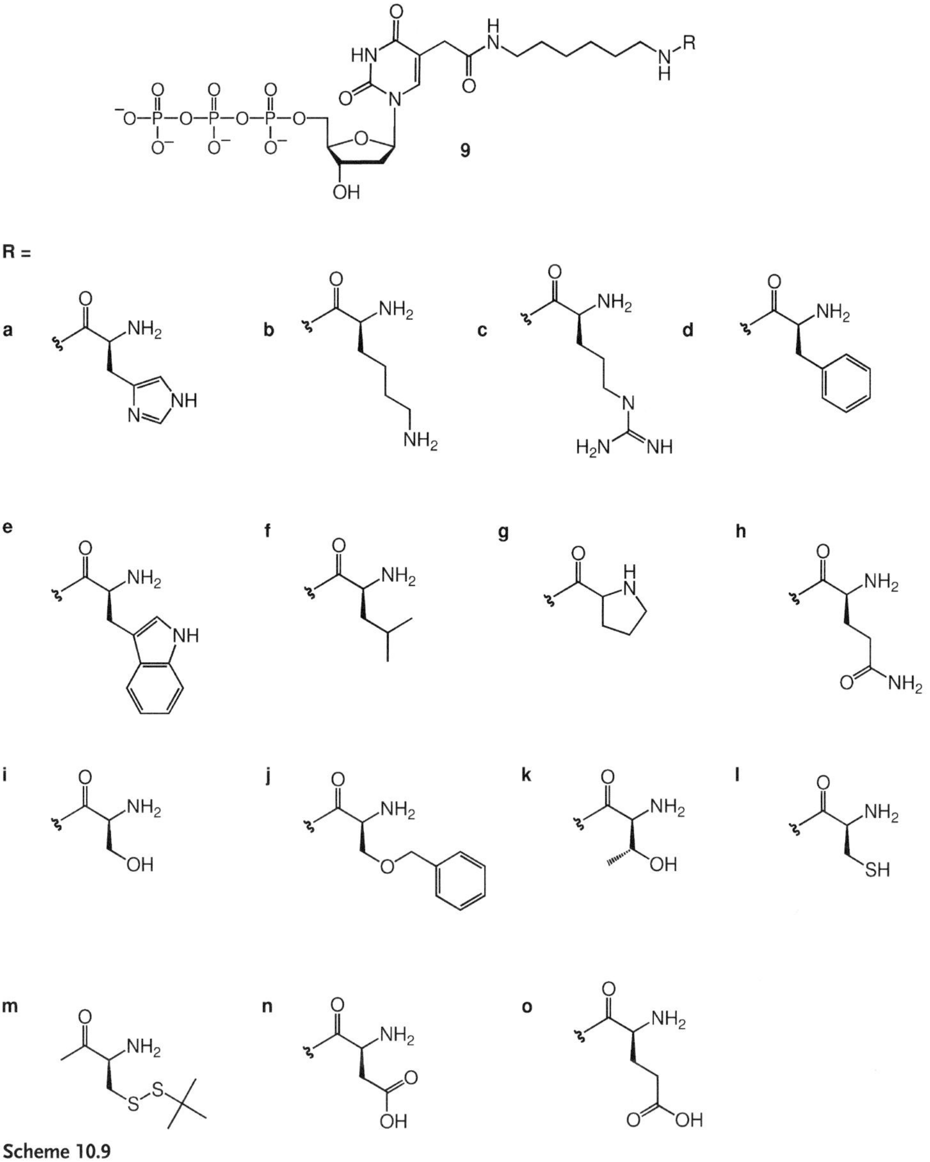

Scheme 10.9

enzymatic synthesis of oligonucleotides by Tth DNA polymerase in which all bases are modified [41–43]. The modified oligonucleotide can be copied without error to a sequence containing all natural nucleotides by Pwo DNA polymerase or replicated by Vent(exo-) DNA polymerase to create double-stranded DNA with every base modified. Scheme 10.10 shows a set of eight nucleotides, two each are analogues of dA, dC, dG, and dT. With this set of nucleoside triphosphates the group was able to prepare duplex DNA containing eight different modifications and with every base in the duplex substituted.

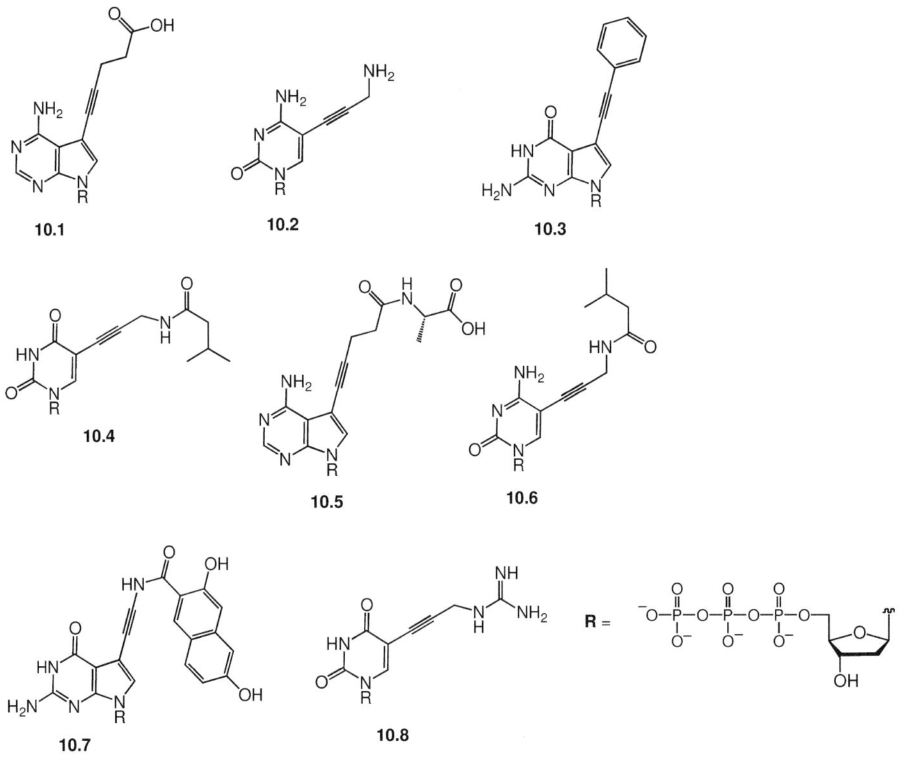

Scheme 10.10

10.3.3
T7 RNA Polymerase-Mediated Synthesis of Modified RNA

It is also possible to synthesize RNA using UTP derivatives substituted at C5. As mentioned above, Ward and coworkers were the first to accomplish this with UTP derivatives substituted by C-5 –HgSR. More recently, Eaton and coworkers prepared the UTP derivative shown in Scheme 10.11 [44, 45]. These proved to be substrates for T7 RNA polymerase in transcription of a 93 nt DNA template. Vaish *et al.* have reported transcription of a 100 nt DNA template using the two C5-substituted UTP derivatives shown in Scheme 10.12 [46]. Although transcription was less efficient than with UTP, full-length transcripts were obtained. Importantly, an RNA containing all modified U could be reverse-transcribed to natural DNA using SuperScript™ II RNase H^{-} reverse transcriptase.

10.3.4
Incorporation of C-5-Appended Fluorophores

The studies described above were dedicated to the development of the technology for incorporation of diverse functional groups through C-5 spacers. Given the need for

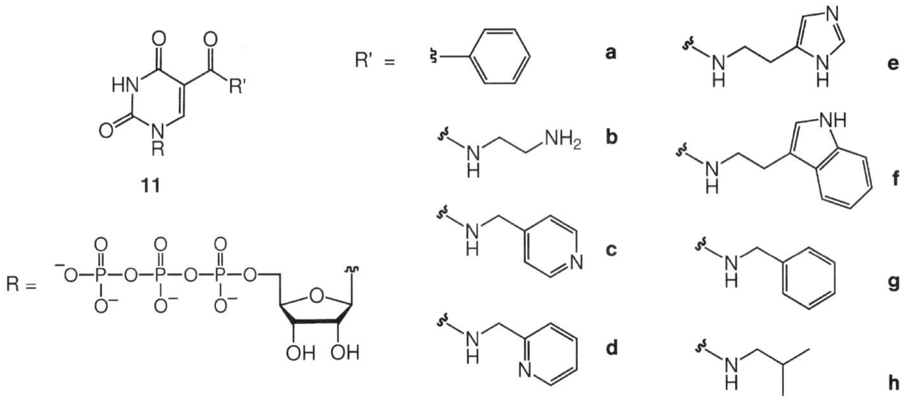

Scheme 10.11

fluorophore-derivatized oligonucleotides, however, several groups have focused on the incorporation of modified nucleoside triphosphates in which the C-5 spacer is linked to a fluorescent probe. Klenow fragment, a family A polymerase, can accept pyrimidine nucleosides containing rhodamine appended through C-5 (spacer structure unspecified). Brakmann and coworkers showed that complete pyrimidine replacement in long DNA sequences could be achieved, and that the modified DNA was a suitable template for incorporation of the same modified nucleotides in the complementary strand [47, 48]. A modified dsDNA containing a rhodamine-modified nucleotide at every base pair can be completely degraded by *E. coli* Exonuclease III [49]. *KOD Dash* DNA polymerase also appears useful for the incorporation fluorophores appended through the C-5 $-CH_2C(O)NH(CH_2)_6-$ spacer discussed above [50].

A number of these studies have sought to achieve complete modification using a combination of C-5-modified pyrimidine nucleosides and 7-deaza-modified purine (pyrrolo[2,3-*d*]). For example, Giller *et al.* [51] prepared a set of nucleoside analogues **13.1–13.4**, in which the fluorophores were attached by reaction of the amine (R = H) by acylation with the fluorophore NHS esters (Scheme 10.13). Thirty different probe (fluorophore or biotin) base combinations were investigated. With Vent exo-DNA polymerase and biotin-modified bases it was possible to synthesize a 300 bp DNA with complete replacement of the natural nucleotides [52]. Although none of the fluorophore-base combinations was effective with this length of sequence, it was

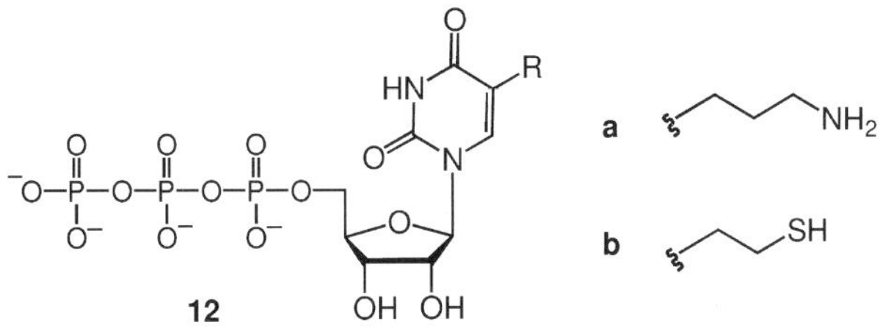

Scheme 10.12

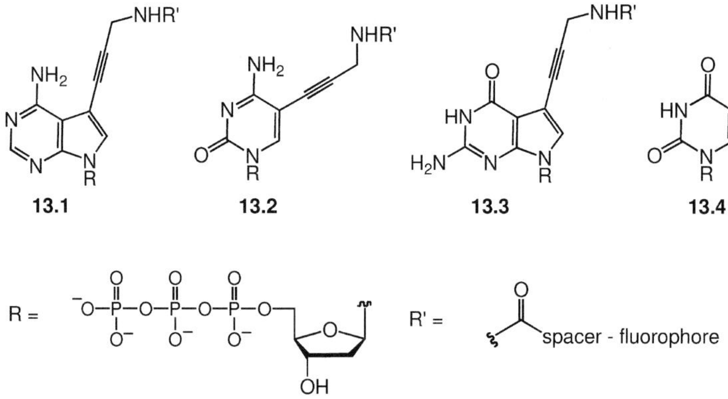

Scheme 10.13

possible to extend a sequence that required the insertion of 40 nucleotides with the combination GB3-dUTP, Evo30-dATP, Cy5-dCTP, and RhG-dGTP. The modified pyrimidine nucleotides were observed to be generally better substrates than the pyrrolo[2,3-*d*]pyrimidines.

A much more extensive screening of DNA polymerases has been reported by Anderson and colleagues [53], who examined *Taq*, Vent, Vent exo-, Deep Vent, Deep vent exo-, *Pfu*, AMV-RT, MLV-RT, T4, and Pol I-Klenow. As substrates, these authors investigated the commercially available dUTP derivatives, biotin-16-dUTP, 7-amino-4-methylcoumarin-3-acetic acid (AMCA)-6-dUTP, fluorescein-11-dUTP, tetramethylrhodamine (TMR)-6-dUTP, digoxigenin-11-dUTP, fluorescein-dATP, IR-770-dATP, Rhodamine Green™-X-dUTP, and Cy™5-dCTP. In order to assess incorporation into a long 1203-nt sequence, an assay was developed in which the template was linked to a solid support through a biotin to streptavidin association. Restriction sites were added in a way that would allow the release of fragments of varying lengths following the polymerase-mediated extension reaction. The DNA was labeled during the extension reaction by incorporating [α-^{32}P]dGTP, which allowed visualization on a gel following treatment with the restriction enzymes. This "extension cut assay" allowed rapid determination of the efficiency of many different combinations of polymerase and modified substrate. All of the enzymes listed above were able to incorporate C-5-substituted dUTP derivatives, with Taq and Vent exo- generally most efficient. The incorporation of N4-modified dCTP or C8 modified dATP was less efficient.

10.4
C-5 Substituents that Stabilize DNA Duplexes

A great deal of effort has been expended to develop C5 substituents that influence duplex annealing. The effects of C5 substituents on the hybridization properties of oligonucleotides were discussed in a comprehensive review by Luyten and Herewijn in 1998 [54]. Although, in this chapter more recent results are reviewed, some historical studies will also be discussed to provide the correct perspective. Among the reasons for developing C5-substituted pyrimidines that influence hybridization are included:

- enhancing overall duplex stability,
- increasing mismatch penalties,
- increasing affinity in base pairing to adenine to achieve a leveling effect.

The desire to enhance duplex stability stems from the presumed benefit in antisense therapy applications, in which lower K_m values could translate into a lower drug dosage. With regard to mismatch penalties, increasing the differential affinity between perfectly matched and mismatched sequences would be beneficial in both therapeutic applications and for sequence detection technologies. One can imagine achieving this either through the design of modified bases that have increased discrimination in pairing with their complement, or that influence local structure in a way that increases discrimination by neighboring bases. Finally, enhancing thymine affinity for adenine through C-5 modification provides an opportunity to tune duplex stability such that a set of duplexes differing in sequence and percentage GC content would all melt at the same temperature (leveling effect). This has obvious utility in nucleic acid array technology, where it would be beneficial to achieve signal levels that are directly proportional to the amount of each complementary sequence in solution, and independent of sequence.

C5 modifications provide many different types of opportunity for tuning affinity. As applied to thymine (uracil), these include designing substituents that:

- Increase the acidity of the N3 hydrogen through electron-withdrawing substituents directly linked to C5.
- Increase stacking interactions with neighboring bases.
- Interact electrostatically with negatively charged phosphodiester groups.
- Interact specifically with other bases upstream or downstream in a sequence.
- Interact cooperatively.

There may also be other more subtle effects, such as the ability of certain substituents to interact with water molecules that help to stabilize a duplex. Rather than attempt to catalogue and characterize all of the C5 substituents described in the literature over the past decade, a few examples will be taken at this point in order to discuss how they fit within the above design principles. One of the simplest and most-studied C-5 modifications is replacement of the thymidine methyl by a propynyl group. The original studies on oligonucleotides containing this date back to the early 1990s, and the report by Froehler and coworkers of the antisense activity of oligonucleotides containing both 5-(1-propynyl)uracil and 5-(1-propynyl)cytosine [55–57]. Both bases, when introduced at multiple positions in a sequence, increased the affinity of that sequence for hybridization to complementary RNA. Subsequent studies have shown that the C5 propynyl group stabilizes sequences not only through base stacking interactions with neighboring base pairs, but also through cooperative interactions between consecutive 5-(1-propynyl)dUs in a sequence [58, 59]. Also of importance is the discovery that the 5-(1-propynyl)dU (dU^P)

discriminates against a G·dU^P wobble pair by a factor of ~100 compared to the G·U wobble pair [59, 60]. A detailed structure study using nuclear magnetic resonance (NMR) showed that a DNA·RNA duplex adopted a modified A-type helix that maximizes attraction interactions and appears to enhance the spine of water hydration in the minor groove [61]. The stabilization was determined as being primarily enthalpy-driven.

As discussed in their 1998 review, Luyten and Herdewijn have compared the effects on duplex stability of a considerable number of other substituents [54]. Given the potential for other types of stabilizing interactions (see list above), in theory there is a sizable number of C5 structural variations that might be worth investigation. Many of these were illustrated in Section 10.2, and will be discussed further in the following paragraphs.

The most comprehensive survey of C5 substituents based on the C5 acetylenic linker was reported by Richert and coworkers in 2004 [11]. Using the Sonogashira coupling reaction as described in Section 10.2 above, these authors prepared the oligonucleotide 5′-CTTTTCU*TTCTT-3′, where U* was one of the modified deoxyuridines shown in Scheme 10.14. Because sequence context may be a significant factor here, the results reported by these investigators may not correlate directly with

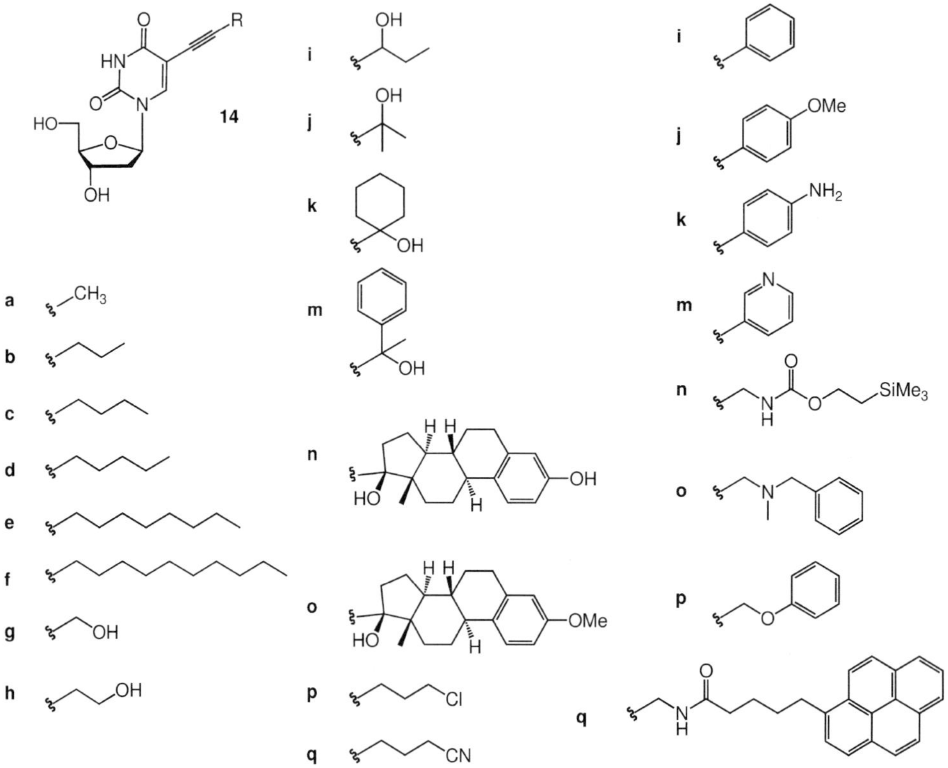

Scheme 10.14

other sequences; nevertheless, the observations are interesting. As discussed above, multiple propynyl-substituted pyrimidine nucleosides have considerable stabilizing effect when they stack with each other. A lone C-5 propynyl (**14a**) has only a very modest stabilizing effect compared to thymidine, whilst longer alkynyl substituents up through five carbons (**14b–d**) are slightly more stabilizing than propynyl. Most of the other substituents shown in Scheme 10.14 are slightly destabilizing. The most interesting exceptions are those side chains that contain hydroxyl groups, as all of these yield duplexes with higher T_m values. The 4-hydroxy-1-butynyl substituent is particularly stabilizing, giving a T_m value of 42.9 °C in 100 mM NaCl buffer compared to the propynyl with a T_m value of 39.4 °C and methyl (thymidine) with a T_m value of 39.0 °C. These investigators also prepared a sequence containing a pyrene linked as shown in **14q**. This proved to be the most stabilizing of all, providing a duplex that melted at 46.1 °C under the same conditions. The group also showed, with mismatched sequences, that the pyrene-substituted pyrimidine maintained high hybridization fidelity.

One of the most interesting classes of C-5 substituents are those containing protonated amino groups. These may be capable of both electrostatic attraction to the negatively charged phosphodiester backbone, as well as specific interactions with the other bases.

The history of amino-containing C5 side chains originated with the discovery of the hypermodified base 5-(4-aminobutylaminomethyl)uracil (deoxyribonucleoside, **15a**) in the DNA of bacteriophage ϕW-14 [62]. These authors determined that slightly greater than 50% of the thymidines in this bacteriophage are replaced by this modified base. The sharp melting profile suggested that the modified base occurred dispersed throughout the phage DNA. In addition, the DNA melted at a considerably higher temperature than would be expected for DNA of the same GC content containing no modified bases, and showed less salt concentration dependence. These results suggested that the terminal amino group, which would be protonated at physiological pH, could function as a salt bridge between the strands of the duplex. In following up on the discovery of Kropinski and colleagues, Takeda *et al.* synthesized a variety of oligodeoxyribonucleotides containing 5-(4-aminobutylaminomethyl)uracil in place of thymidine, and monitored the effect of this substitution on duplex melting [63]. The melting temperature showed significant sequence dependency. Of 16 possible nearest-neighbor sequence contexts, Takeda *et al.* examined five, TT*T, TT*C, CT*T, AT*A, and CT*A, where T* was the modified nucleoside 5-(4-aminobutylaminomethyl)-2′-deoxyuridine. Only one of these, CT*T, increased duplex stability in comparison to the natural sequence. Furthermore, in the sequence context TT*T the T_m values of a modified oligothymidine opposite poly A decreased in proportion to the number of modified nucleosides in the sequence. Given the substantial stabilization observed for the natural ϕW-14, it seems likely that the 5-(4-aminobutylaminomethyl) side chain occurs only in those sequence contexts that are stabilizing.

Following these early studies, considerable effort was expended to create similar amine-modified C5 chains, with the goal directed primarily towards optimizing stabilizing effects. Many of the C5-modified nucleosides studied during the period

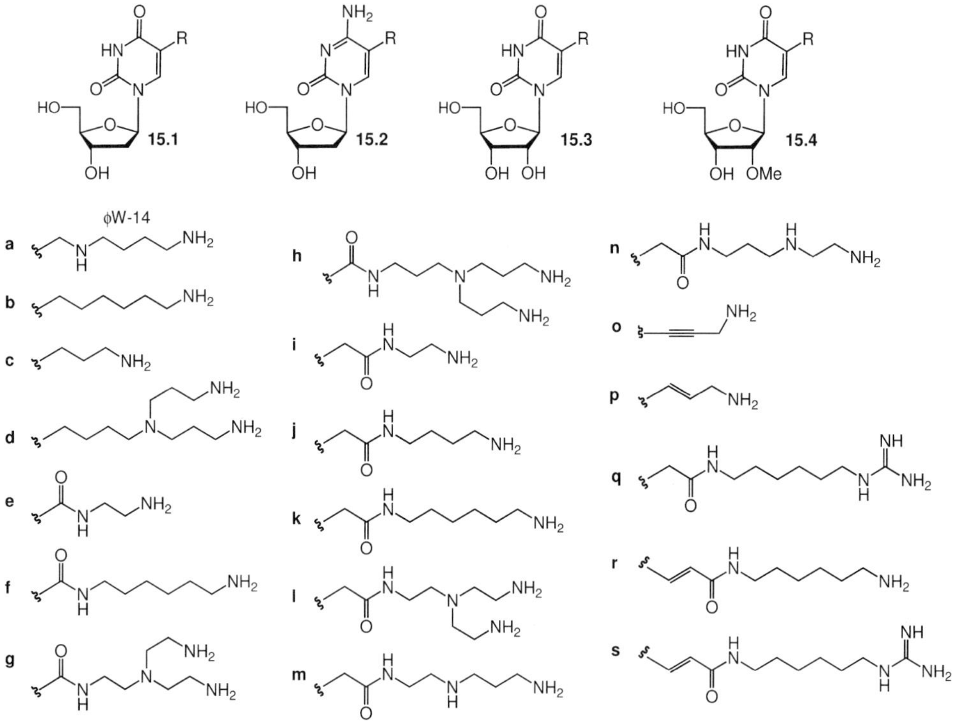

Scheme 10.15

are listed in Scheme 10.15. The earliest studies appended amino groups on saturated alkanes (**15b–d**) [64–66] which were not particularly stabilizing. However, the majority of amino-substituted side chains contain an amide functionality (**15e–n, r**) [10, 67–71], which in many cases does enhance stabilization, an effect that to some degree reflects the ability of the appended amide group to stack with neighboring bases. In compounds **15e–h** [8, 70], the carbonyl is directly conjugated to the ring, and an electron-withdrawing group would be expected to increase the acidity of the N-3 proton. Recent crystal structures of DNA·DNA and DNA·RNA containing **15f** show the side chain projecting from the major groove, thus placing the protonated amino group in close proximity to a phosphodiester group [72]. Several studies have been conducted to determine why the relatively small aminopropyl (**15c**) is not stabilizing [73–76]. Although the protonated amino can take the place of other cations in neutralizing the charged phosphodiester groups, it has been suggested that they negatively influence ordered water structure [76]. In contrast, the aminopropynyl group can enhance stacking interactions, thereby increasing duplex stability [77]. Overall, despite considerable effort to optimize amino-substituted C5 side chains, a simple substituent such as hydroxylpropynyl or hydroxybutynyl seems as effective as even the best amino-substituted C5 substituent.

In addition to amino substituents, several reports have been made of guanidinium-substituted C5 side chains (**15q, s**), which are of particular interest because of the facility with which guanidinium groups stimulate the cellular uptake of oligonucleotides [78, 79].

10.5 Photochemistry

Fujimoto and coworkers have recently described a strategy for ligating two nucleic acid strands through a 2 + 2 photocycloaddition reaction between a natural pyrimidine nucleic acid base (thymine, cytosine, or uracil) and a pyrimidine nucleoside containing either a C-5 acrylamide or cyanovinyl side chain (see Scheme 10.2) [80–82]. These followed earlier studies with 5-vinyl-2′-deoxyuridine and 5-carboxyvinyldeoxyuridine, which will not be described here [83]. The β-anomer of the 5-carbamoylvinyl-modified nucleoside, when placed at the 5′-end of a sequence, underwent photoinduced crosslinking with a natural base at the 3′-end of an adjacent sequence. Crosslinking in the opposite direction could be accomplished using an α-anomer of the C-5-modified nucleoside, which modeling indicated has the overlap necessary for the 2 + 2 cycloaddition reaction to occur without undue geometric distortion [84]. The investigators demonstrated efficient crosslinking both between oligonucleotides terminating with the C-5-modified nucleoside, as well as with oligonucleotides in which the modified nucleoside is embedded in the middle of the sequence (as shown in Scheme 10.16B). Even more significant, the photocrosslinking is reversible on irradiation at 312 nm. This opens up the potential for some interesting new applications, including the development of a photochemical DNA computer [85]. In yet another

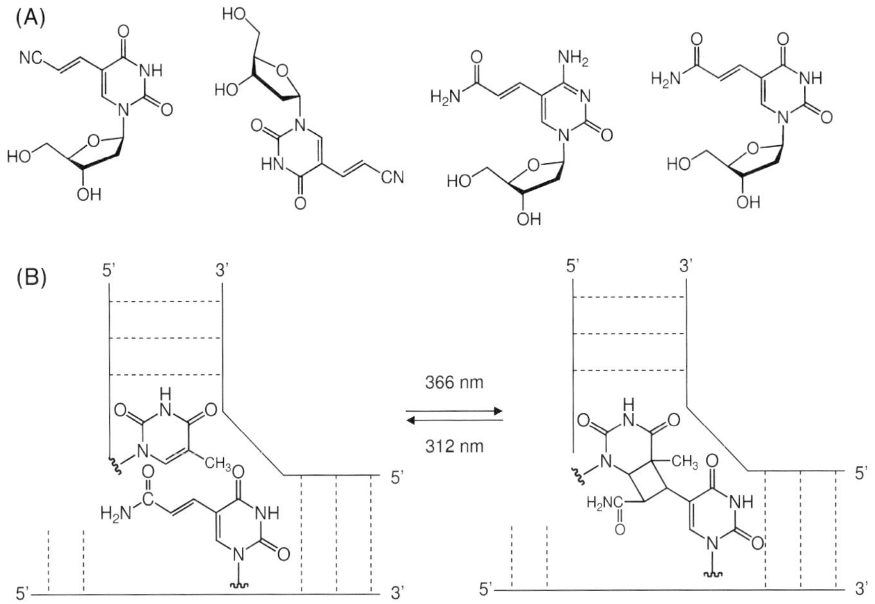

Scheme 10.16

7 Haginoya, N., Ono, A., *et al.* Nucleosides and nucleotides. 160. Synthesis of oligodeoxyribonucleotides containing 5-(N-aminoalkyl)carbamoyl-2′-deoxyuridine by a new postsynthetic modification method and their thermal stability and nuclease-resistance properties. *Bioconj. Chem.* **1997**, *8* (3), 271–280.

8 Ueno, Y. and Matsuda, A. Synthesis of oligonucleotides modified with polyamines and their properties as antisense and antigene molecules. *J. Synthetic Org. Chem. Japan* **2003**, *61* (9), 890–899.

9 Kohgo, S., Shinozuka, K., *et al.* Synthesis of a novel 2′-deoxyuridine derivative bearing a cyanomethoxy-carbonylmethyl group at C-5 position and its use for versatile post-synthetic functionalization of oligodeoxyribonucleotides. *Tetrahedron Lett.* **1998**, *39* (23), 4067–4070.

10 Shinozuka, K., Kohgo, S., *et al.* Multifunctionalization of oligodeoxynucleotide: a facile post-synthetic modification technique for the preparation of oligodeoxynucleotides with two different functional molecules. *Chem. Commun.* **2000**, 59–60.

11 Kottysch, T., Ahlborn, C., *et al.* Stabilizing or destabilizing oligodeoxynucleotide duplexes containing single 2′-deoxyuridine residues with 5-alkynyl substituents. *Chemistry – A European Journal* **2004**, *10* (16), 4017–4028.

12 Baumhof, P., Griesang, N., *et al.* Synthesis of oligonucleotides with 3′-terminal 5-(3-acylamidopropargyl)-3′-amino-2′,3′-dideoxyuridine residues and their reactivity in single-nucleotide steps of chemical replication. *J. Org. Chem.* **2006**, *71* (3), 1060–1067.

13 Dey, S. and Sheppard, T. L. Ketone-DNA: A versatile postsynthetic DNA decoration platform. *Org. Lett.* **2001**, *3* (25), 3983–3986.

14 Gierlich, J., Burley, G. A., *et al.* Click chemistry as a reliable method for the high–density postsynthetic functionalization of alkyne-modified DNA. *Org. Lett.* **2006**, *8* (17), 3639–3642.

15 Takeda, S., Tsukiji, S., *et al.* A cysteine-appended deoxyuridine for the postsynthetic DNA modification using native chemical ligation. *Tetrahedron Lett.* **2005**, *46* (13), 2235–2238.

16 Graham, D., Grondin, A., *et al.* Internal labeling of oligonucleotide probes by Diels-Alder cycloaddition. *Tetrahedron Lett.* **2002**, *43* (27), 4785–4788.

17 Kahl, J. D. and Greenberg, M. M. Introducing structural diversity in oligonucleotides via photolabile, convertible C5-substituted nucleotides. *J.Am. Chem. Soc.* **1999**, *121* (4), 597–604.

18 Dale, R. M. K., Livingston, D. C., Ward, D. C., and Martin, E. Synthesis and enzymatic polymerization of nucleotides containing mercury: potential tools for nucleicacid sequencing and structural-analysis. *Proc. Natl. Acad. Sci. USA* **1973**, *70* (8), 2238–2242.

19 Dale, R. M. K. and Ward, D. C. Mercurated polynucleotides – new probes for hybridization and selective polymer fractionation. *Biochemistry* **1975**, *14* (11), 2458–2469.

20 Dale, R. M. K., Martin, E., *et al.* Direct covalent mercuration of nucleotides and polynucleotides. *Biochemistry* **1975**, *14* (11), 2447–2457.

21 Hopman, A. H. N., Wiegant, J., *et al.* A nonradioactive in situ hybridization method based on mercurated nucleic-acid probes and sulfhydryl-hapten ligands. *Nucleic Acids Res.* **1986**, *14* (16), 6471–6488.

22 Hopman, A. H. N., Wiegant, J., *et al.* A new hybridocytochemical method based on mercurated nucleic-acid probes and sulfhydryl-hapten ligands. 1. Stability of the mercury-sulfhydryl bond and influence of the ligand structure on immunochemical detection of the hapten. *Histochemistry* **1986**, *84* (2), 169–178.

23 Hopman, A. H. N., Wiegant, J., *et al.* A new hybridocytochemical method based on mercurated nucleicacid probes and sulfhydryl-hapten ligands. 2. Effects of variations in ligand structure on the in situ detection of mercurated probes. *Histochemistry* **1986**, *84* (2), 179–185.

24 Hopman, A. H. N., Wiegant, J., *et al.* Mercurated nucleicacid probes, a new principle for nonradioactive in situ hybridization. *Exp. Cell Res.* **1987**, *169* (2), 357–368.

25 Langer, P. R., Waldrop, A. A., *et al.* Enzymatic-synthesis of biotin-labeled polynucleotides – novel nucleicacid affinity probes. *Proc. Natl. Acad. Sci. USA – Biol. Sci.* **1981**, *78* (11), 6633–6637.

26 Bergstrom, D. E. and Ruth, J. L. Synthesis of C-5 substituted pyrimidine nucleosides via organo-palladium intermediates. *J. Am. Chem. Soc.* **1976**, *98* (6), 1587–1589.

27 Bergstrom, D. E. and Ogawa, M. K. C-5 Substituted pyrimidine nucleosides. 2. Synthesis via olefin coupling to organopalladium intermediates derived from uridine and 2′-deoxyuridine. *J. Am. Chem. Soc.* **1978**, *100* (26), 8106–8112.

28 Sagi, J. T., Szabolcs, A., *et al.* Modified polynucleotides. 1. Investigation of enzymatic polymerization of 5-alkyl-dUTP-s. *Nucleic Acids Res.* **1977**, *4* (8), 2767–2777.

29 Sagi, J., Nowak, R., *et al.* Study of substrate-specificity of mammalian and bacterial-DNA polymerases with 5-alkyl-2′-deoxyuridine 5′-triphosphates. *Biochim. Biophys. Acta* **1980**, *606* (2), 196–201.

30 Robins, M. J. and Barr, P. J. Nucleicacid-related compounds. 31. Smooth and efficient palladium-copper catalyzed coupling of terminal alkynes with 5-iodouracil nucleosides. *Tetrahedron Lett.* **1981**, *22* (5), 421–424.

31 Sakthivel, K. and Barbas, C. F. Expanding the potential of DNA for binding and catalysis: highly functionalized dUTP derivatives that are substrates for thermostable DNA polymerases. *Angew. Chem. – Int. Ed.* **1998**, *37* (20), 2872–2875.

32 Held, H. A. and Benner, S. A. Challenging artificial genetic systems: thymidine analogs with 5-position sulfur functionality. *Nucleic Acids Res.* **2002**, *30* (17), 3857–3869.

33 Roychowdhury, A., Illangkoon, H., *et al.* 2′-Deoxycytidines carrying amino and thiol functionality: synthesis and incorporation by vent (exo(–)) polymerase. *Org. Lett.* **2004**, *6* (4), 489–492.

34 Sawai, H., Ozaki, A. N., *et al.* Expansion of structural and functional diversities of DNA using new 5-substituted deoxyuridine derivatives by PCR with superthermophilic KOD Dash DNA polymerase. *Chem. Commun.* **2001**, 2604–2605.

35 Sawai, H., Ozaki-Nakamura, A., *et al.* Synthesis of new modified DNAs by hyperthermophilic DNA polymerase: Substrate and template specificity of functionalized thymidine analogues bearing an sp3-hybridized carbon at the C5 alpha-position for several DNA polyme-rases. *Bioconj. Chem.* **2002**, *13* (2), 309–316.

36 Kuwahara, M., Takahata, Y., *et al.* Substrate properties of C5-substituted pyrimidine 2′-deoxynucleoside 5′-triphosphates for thermostable DNA polymerases during PCR. *Bioorg. Medicinal Chem. Lett.* **2003**, *13* (21), 3735–3738.

37 Ohbayashi, T., Kuwahara, M., *et al.* Expansion of repertoire of modified DNAs prepared by PCR using KOD dash DNA polymerase. *Org. Biomol. Chem.* **2005**, *3* (13), 2463–2468.

38 Kuwahara, M., Hanawa, K., *et al.* Direct PCR amplification of various modified DNAs having amino acids: Convenient preparation of DNA libraries with high-potential activities for in vitro selection. *Bioorg. Medicinal Chem.* **2006**, *14* (8), 2518–2526.

39 Kuwahara, M., Nagashima, J., *et al.* Systematic characterization of 2′-deoxynucleoside-5′-triphosphate analogs as substrates for DNA polymerases by polymerase chain reaction and kinetic studies on enzymatic production of modified DNA. *Nucleic Acids Res.* **2006**, *34* (19), 5383–5394.

40 Matsui, M., Nishiyama, Y., *et al.* Construction of saccharide-modified DNAs by DNA polymerase. *Bioorg. Medicinal Chem. Lett.* **2007**, *17* (2), 456–460.

41 Thum, O., Jager, S., *et al.* Functionalized DNA: A new replicable biopolymer. *Angew. Chem. - Int. Ed.* **2001**, *40* (21), 3990–3993.

42 Jager, S. and Famulok, M. Generation and enzymatic amplification of high-density functionalized DNA double strands. *Angew. Chem. - Int. Ed.* **2004**, *43* (25), 3337–3340.

43 Jager, S., Rasched, G., *et al.* A versatile toolbox for variable DNA functionalization at high density. *J. Am. Chem. Soc.* **2005**, *127* (43), 15071–15082.

44 Dewey, T. M., Mundt, A. A., *et al.* New uridine derivatives for systematic evolution of RNA ligands by exponential enrichment. *J. Am. Chem. Soc.* **1995**, *117* (32), 8474–8475.

45 Vaught, J. D., Dewey, T., *et al.* T7 RNA polymerase transcription with 5-position modified UTP derivatives. *J. Am. Chem. Soc.* **2004**, *126* (36), 11231–11237.

46 Vaish, N. K., Fraley, A. W., *et al.* Expanding the structural and functional diversity of RNA: analog uridine triphosphates as candidates for in vitro selection of nucleic acids. *Nucleic Acids Res.* **2000**, *28* (17), 3316–3322.

47 Brakmann, S. and Lobermann, S. High-density labeling of DNA: Preparation and characterization of the target material for single-molecule sequencing. *Angew. Chem. - Int. Ed.* **2001**, *40* (8), 1427–1429.

48 Brakmann, S. and Nieckchen, P. The large fragment of *Escherichia coli* DNA polymerase I can synthesize DNA exclusively from fluorescently labeled nucleotides. *Chembiochem* **2001**, *2* (10), 773–777.

49 Brakmann, S. and Lobermann, S. A further step towards single-molecule sequencing: *Escherichia coli* exonuclease III degrades DNA that is fluorescently labeled at each base pair. *Angew. Chem.-Int. Ed.* **2002**, *41* (17), 3215–3217.

50 Obayashi, T., Masud, M. M., *et al.* Enzymatic synthesis of labeled DNA by PCR using new fluorescent thymidine nucleotide analogue and superthermophilic KOD Dash DNA polymerase. *Bioorg. Medicinal Chem. Lett.* **2002**, *12* (8), 1167–1170.

51 Giller, G., Tasara, T., *et al.* Incorporation of reporter molecule-labeled nucleotides by DNA polymerases. I. Chemical synthesis of various reporter group-labeled 2′-deoxyribonucleoside-5′-triphosphates. *Nucleic Acids Res.* **2003**, *31* (10), 2630–2635.

52 Tasara, T., Angerer, B., *et al.* Incorporation of reporter molecule-labeled nucleotides by DNA polymerases. II. High-density labeling of natural DNA. *Nucleic Acids Res.* **2003**, *31* (10), 2636–2646.

53 Anderson, J. P., Angerer, B., *et al.* Incorporation of reporter-labeled nucleotides by DNA polymerases. *Biotechniques* **2005**, *38* (2), 257–264.

54 Luyten, I. and Herdewijn, P. Hybridization properties of base-modified oligonucleotides within the double and triple helix motif. *Eur. J. Medicinal Chem.* **1998**, *33* (7–8), 515–576.

55 Froehler, B. C., Wadwani, S., *et al.* Oligodeoxynucleotides containing C-5 propyne analogs of 2′-deoxyuridine and 2′-deoxycytidine. *Tetrahedron Lett.* **1992**, *33* (37), 5307–5310.

56 Wagner, R. W., Matteucci, M. D., *et al.* Antisense gene inhibition by oligonucleotides containing C-5 propyne pyrimidines. *Science* **1993**, *260* (5113), 1510–1513.

57 Wagner, R. W., Matteucci, M. D., *et al.* Potent and selective inhibition of gene expression by an antisense heptanucleo-tide. *Nature Biotechnol.* **1996**, *14*, 840–844.

58 Barnes, T. W. and Turner, D. H. Long-range cooperativity in molecular recognition of RNA by oligodeoxynucleotides, with multiple C5-(1-propynyl) pyrimidines. *J. Am. Chem. Soc.* **2001**, *123* (18), 4107–4118.

59 Barnes, T. W. and Turner, D. L. H. Long-range cooperativity due to C5-propynylation of oligopyrimidines enhances specific recognition by uridine of ribo-adenosine over ribo-guanosine. *J. Am. Chem. Soc.* **2001**, *123* (37), 9186–9187.

60 Barnes, T. W. and Turner, D. H. C5-(1-Propynyl)-2′-deoxy-pyrimidines enhance mismatch penalties of DNA: RNA duplex formation. *Biochemistry* **2001**, *40* (42), 12738–12745.

61 Gyi, J. I., Gao, D. Q., *et al.* The solution structure of a DNA center dot RNA duplex containing 5-propynyl U and C; comparison with 5-Me modifications. *Nucleic Acids Res.* **2003**, *31* (10), 2683–2693.

62 Kropinski, A. M. B., Bose, R. J., *et al.* 5-(4-Aminobutylamino)uracil, an unusual pyrimidine from the deoxyribonucleic acid of bacteriophage phiW-14. *Biochemistry* **1973**, *12* (1), 151–157.

63 Takeda, T., Ikeda, K., *et al.* Synthesis and properties of deoxyribonucleotides containing putrescinylthymine (Nucleosides and Nucleotides LXXVI). *Chem. Pharm. Bull.* **1987**, *35* (9), 3558–3567.

64 Hashimoto, H., Nelson, M. G., *et al.* Formation of chimeric duplexes between zwitterionic and natural DNA. *J. Org. Chem.* **1993**, *58* (16), 4194–4195.

65 Hashimoto, H., Nelson, M. G., *et al.* Zwitterionic DNA. *J. Am. Chem. Soc.* **1993**, *115* (16), 7128–7134.

66 Nara, H., Ono, A., *et al.* Nucleosides and nucleotides. 135. DNA duplex and triplex formation and resistance to nucleolytic degradation of oligodeoxynucleotides containing syn-norspermidine at the 5-position of 2′-deoxyuridine. *Bioconj. Chem.* **1995**, *6* (1), 54–61.

67 Ozaki, H., Nakamura, A., *et al.* Novel C5-substituted 2′-deoxyuridine derivatives bearing amino-linker arms – synthesis, incorporation into oligodeoxyribonucleotides, and their hybridization properties. *Bull. Chem. Soc. Japan* **1995**, *68* (7), 1981–1987.

68 Ueno, Y., Nakagawa, A., *et al.* Nucleosides and nucleotides. 165. Chemical ligation of oligodeoxynucleotides having a mercapto group at the 5-position of 2′-deoxyuridine via a disulfide bond. *Nucleosides Nucleotides* **1998**, *17* (1–3), 283–289.

69 Ozaki, H., Mine, M., *et al.* Effect of the terminal amino group of a linker arm and its length at the C5 position of a pyrimidine nucleoside on the thermal stability of DNA duplexes. *Bioorg. Chem.* **2001**, *29* (4), 187–197.

70 Ito, T., Ueno, Y., *et al.* Synthesis, thermal stability and resistance to enzymatic hydrolysis of the oligonucleotides containing 5-(N-aminohexyl)carbamoyl-2′-O-methyluridines. *Nucleic Acids Res.* **2003**, *31* (10), 2514–2523.

71 Ozaki, H., Mine, M., *et al.* Effect of imino group of a linker arm at the C5 position of a pyrimidine nucleoside on the thermal stabilities of DNA/DNA and DNA/RNA duplexes. *Nucleosides Nucleotides Nucleic Acids* **2004**, *23* (1–2), 339–346.

72 Juan, E. C. M., Kondo, J., *et al.* Crystal structures of DNA:DNA and DNA:RNA duplexes containing 5-(N-aminohexyl) carbamoyl-modified uracils reveal the basis for properties as antigene and antisense molecules. *Nucleic Acids Res.* **2007**, *35* (6), 1969–1977.

73 Gold, B. Effect of cationic charge localization on DNA structure. *Biopolymers* **2002**, *65* (3), 173–179.

74 Li, Z. J., Huang, L., *et al.* Structure of a tethered cationic 3-aminopropyl chain incorporated into an oligodeoxynucleotide: Evidence for 3′-orientation in the major groove accompanied by DNA bending. *J.Am. Chem. Soc.* **2002**, *124* (29), 8553–8560.

75 Moulaei, T., Maehigashi, T., *et al.* Structure of B-DNA with cations tethered in the major groove. *Biochemistry* **2005**, *44* (20), 7458–7468.

76 Shikiya, R., Li, J. S., *et al.* Incorporation of cationic chains in the Dickerson-Drew dodecamer: Correlation of energetics, structure, and ion and water binding. *Biochemistry* **2005**, *44* (37), 12582–12588.

77 Booth, J., Brown, T., *et al.* Determining the origin of the stabilization of DNA by 5-aminopropynylation of pyrimidines. *Biochemistry* **2005**, *44* (12), 4710–4719.

78 Ohmichi, T., Kuwahara, M., *et al.* Nucleic acid with guanidinium modification exhibits efficient cellular uptake. *Angew. Chem. – Int. Ed.* **2005**, *44* (41), 6682–6685.

79 Deglane, G., Abes, S., *et al.* Impact of the guanidinium group on hybridization and cellular uptake of cationic oligonucleotides. *Chembiochem* **2006**, *7* (4), 684–692.

80 Ogino, M., Yoshimura, Y., *et al.* Template-directed DNA photoligation via alpha-5-cyanovinyldeoxyuridine. *Org. Lett.* **2005**, *7* (14), 2853–2856.

81 Yoshimura, Y., Okamura, D., *et al.* Highly selective and sensitive template-directed photoligation of DNA via 5-carbamoylvinyl-2′-deoxycytidine. *Org.Lett.* **2006**, *8* (22), 5049–5051.

82 Yoshimura, Y., Noguchi, Y., *et al.* Highly sequence specific RNA terminal labeling by DNA photoligation. *Org. Biomol. Chem.* **2007**, *5* (1), 139–142.

83 Fujimoto, K., Ogawa, N., *et al.* Template directed photochemical synthesis of branched oligodeoxynucle-otides via 5-carboxyvinyldeoxyuridine. *Tetrahedron Lett.* **2000**, *41* (49), 9437–9440.

84 Ogasawara, S. and Fujimoto, K. A novel method to synthesize versatile multiple-branched DNA (MB-DNA) by reversible photochemical ligation. *Chembiochem* **2005**, *6* (10), 1756–1760.

85 Ogasawara, S. and Fujimoto, K. Solution of a SAT problem on a photochemical DNA computer. *Chem. Lett.* **2005**, *34* (3), 378–379.

86 Fujimoto, K., Matsuda, S., *et al.* Site-specific transition of cytosine to uracil via reversible DNA photoligation. *Chem. Commun.* **2006**, 3223–3225.

87 Fujimoto, K., Yoshimura, Y., *et al.* Photoinduced DNA end capping via N-3-methyl-5-cyanovinyl-2′-deoxyuridine. *Chem. Commun.* **2005**, 3177–3179.

88 Saito, I., Miyauchi, Y., *et al.* Template-directed photoreversible ligation of DNA via 7-carboxyvinyl-7-deaza-2′-deoxyadenosine. *Tetrahedron Lett.* **2005**, *46* (1), 97–99.

89 Kimoto, M., Endo, M., *et al.* Site-specific incorporation of a photo-crosslinking component into RNA by T7 transcription mediated by unnatural basepairs. *Chemistry Biology* **2004**, *11* (1), 47–55.

90 Zeng, Y., Cao, H. C., *et al.* Facile photocyclization chemistry of 5-phenylthio-2′-deoxyuridine in duplex DNA. *Org. Lett.* **2006**, *8* (12), 2527–2530.

11
Universal Base Analogues and their Applications to Biotechnology

Kathleen Too and David Loakes

11.1
Introduction

The concept of non-discriminatory base analogues has been known since Ohtsuka *et al.* described the use of hypoxanthine as an inert base at ambiguous codon positions [1, 2]. Hypoxanthine is still in use as an inert base because it forms hydrogen bonds with each of the native DNA or RNA bases; however, it does show significant discrimination in its base-pairing properties. Hypoxanthine forms the most stable base pair with cytosine, and when used in polymerase reactions it behaves almost exclusively as guanine [3]. As such, it does not conform to the idea of a universal base as discussed in this chapter, but is noted because it is one of the earliest analogues used. Although many reports have been made describing the use of hypoxanthine and its analogue xanthine as inert base analogues, these will not be discussed here, though some of their analogues are discussed later. In 1994, two new analogues were described, namely the 2′-deoxyribosyl-derivatives of 3-nitropyrrole (**1**) [4] and 5-nitroindole (**2**) [5]. These represent a new class of compound known as universal bases. As a general rule, this new class of compound is derived from aromatic, hydrophobic base analogues and, when incorporated into oligonucleotides, show little discrimination in their base-pairing properties as they have no hydrogen-bonding capability. The development of this class of compound, with emphasis on their biochemical applications, will be discussed in this chapter.

The general requirements for a universal base analogue are that it should: (1) pair with all the natural DNA (RNA) bases equally when opposed to them in oligonucleotides; (2) form a duplex which primes DNA (RNA) synthesis by a polymerase; (3) direct incorporation of the 5′-triphosphate derivatives of each of the natural DNA bases opposite it when copied by a polymerase; and (4) be substrates for polymerases as their 5′-triphosphate derivatives. To date, no universal base analogue meets all these requirements, almost certainly as a consequence of them having little or no hydrogen-bonding capability. Universal base analogues have particular use in hybridization applications, principally when employed in probes and primers, and

Modified Nucleosides: in Biochemistry, Biotechnology and Medicine. Edited by Piet Herdewijn

ISBN: 978-3-527-31820-9

this will be the focus of this chapter. The investigation of universal base analogues in polymerase assays and novel methods for improving their polymerase substrate efficiency will also be detailed.

11.2
General Methods of Synthesis

The most widely studied universal base analogues are 3-nitropyrrole (**1**) and 5-nitroindole (**2**), both of which are prepared by the stereospecific sodium salt glycosylation method described by Revankar [6], which produces only the β-anomer of these nucleosides. Thus, treatment of the nucleobase with sodium hydride followed by addition of the sugar, usually α-3,5-di-*O*-*p*-toluoyl-2-deoxyribofuranosyl chloride or 2′,3′-*O*-isopropylidene-5′-*O*-(^{t}butyldimethylsilyl) ribofuranosyl chloride, yields the deoxynucleoside or ribonucleoside respectively after deprotection (Scheme 11.1). An alternative method for the synthesis of ribonucleoside analogues uses the glycosylation procedure of Vorbrüggen, whereby the appropriate nucleobase is refluxed with *N,O*-bis(trimethylsilyl)acetamide and subsequently reacted with 1,2,3,5-tetra-*O*-acetyl-β-D-ribofuranose in the presence of the Lewis acid trimethylsilyl trifluoromethanesulfonate (Scheme 11.2A) [7].

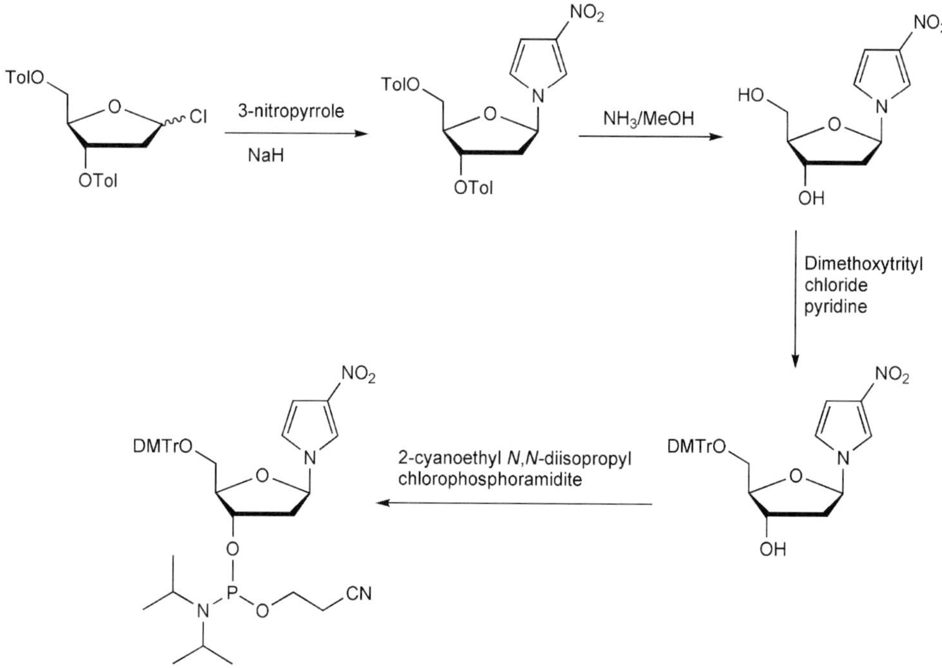

Scheme 11.1 Synthesis of 3-nitropyrrole via the stereospecific sodium salt glycosylation method.

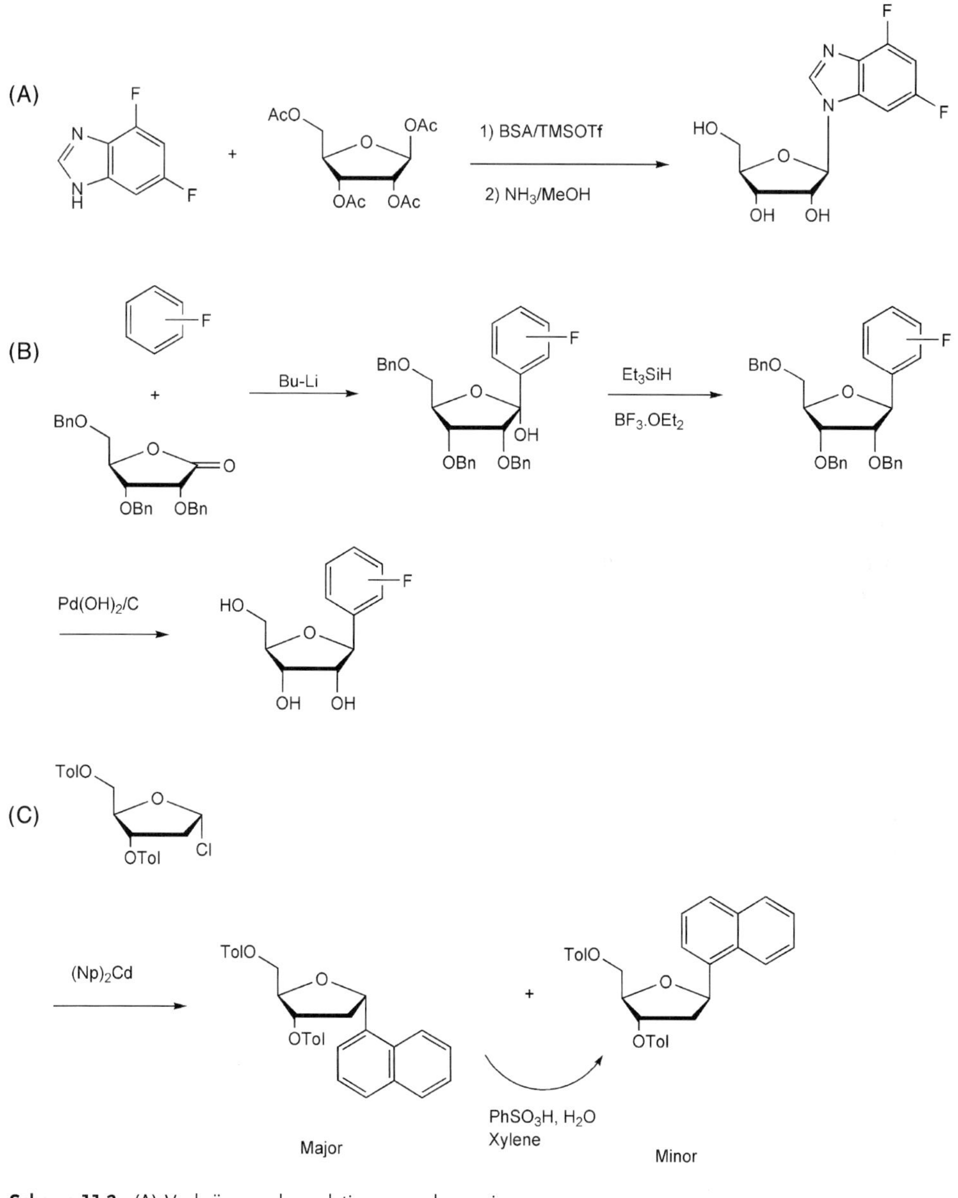

Scheme 11.2 (A) Vorbrüggen glycosylation procedure using *N*,*O*-bis(trimethylsilyl)acetamide (BSA) and trimethylsilyl trifluoromethane sulfonate (TMSOTf). (B) Synthesis of *C*-ribonucleoside using BuLi. (C) Synthesis of *C*-deoxynucleoside using diaryl cadmium reagent.

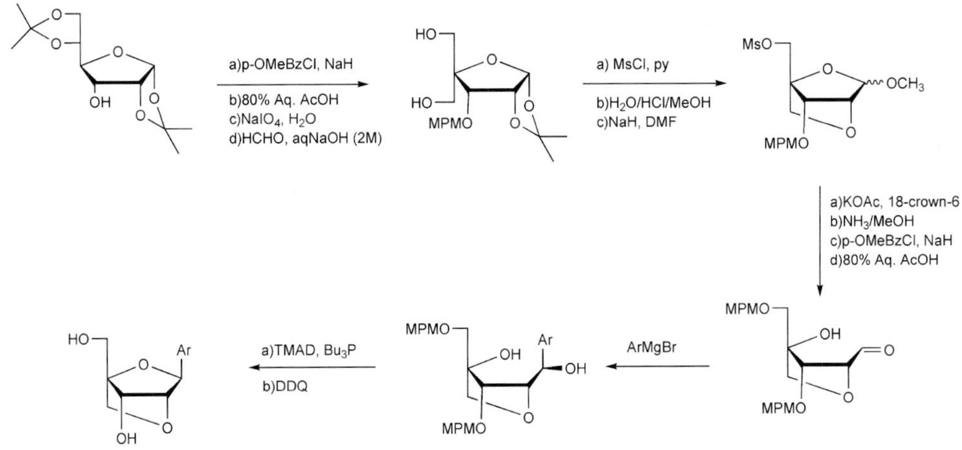

Scheme 11.3 Synthesis of aryl-β-C-LNA via Grignard chemistry.

Engels and coworkers have also described the synthesis of *C*-ribosyl analogues, prepared by the lithiation of a fluoroaryl bromide using BuLi in tetrahydrofuran (THF) at −78 °C, followed by the addition of 2,3,5-tri-*O*-benzyl-D-ribono-1,4-lactone to provide the corresponding lactol [7]. This is then dehydroxylated using triethylsilane and $BF_3{\cdot}OEt_2$, followed by deprotection with 20% $Pd(OH)_2/C$ in the presence of cyclohexene to yield the desired *C*-nucleoside (Scheme 11.2B). Alternatively, Kool has reported that α-*C*-deoxyribonucleosides containing the non-natural bases such as benzene, naphthalene, phenanthrene and pyrene, can be prepared using other organometallic reagents, for example the Grignard MgBr reagent, diaryl cadmium or diaryl zinc derivatives with a α-chlorosugar. The reaction of a dinaphthylcadmium species with 1′-α-chloro-3′,5′-di-*O*-toluoyl-2′-deoxyribose gives a 5 : 1 mixture of α- and β-anomers with 1′,2′-dideoxy-1′-(1-naphthyl)-α-D-ribofuranose as the major isomer (Scheme 11.2C). The latter was then epimerized to the desired β-anomer by refluxing in benzene sulfonic acid and xylene in the presence of a small amount of water [8, 9]. Babu *et al.* have used Grignard chemistry to synthesize a number of aryl-β-*C*-LNA monomers with universal base properties (Scheme 11.3) [10–13]. Pyrenyl LNA derivatives (**3**, **4**) retain their universal behavior both in DNA and 2′-OMe-RNA, which suggests that it is possible to increase the binding affinity of universal hybridization probes by using LNA analogues [11].

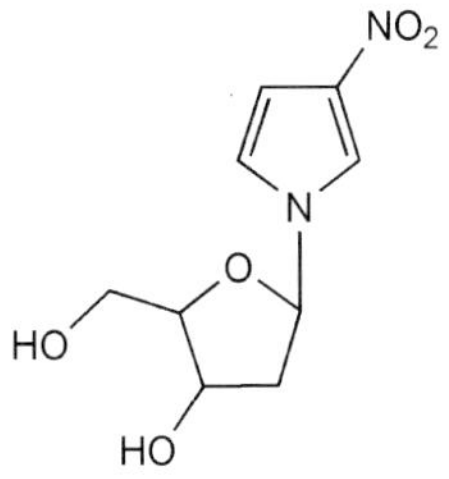

1

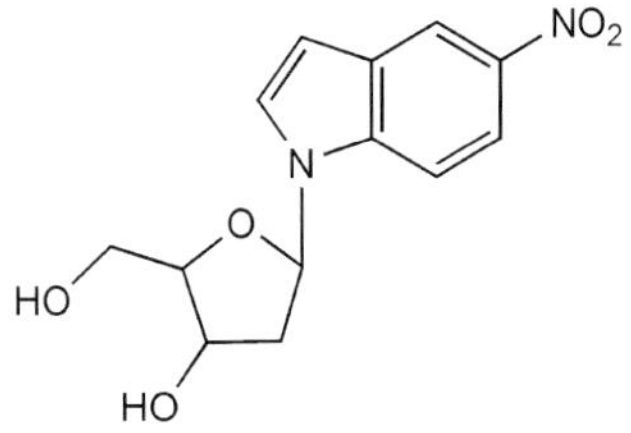

2

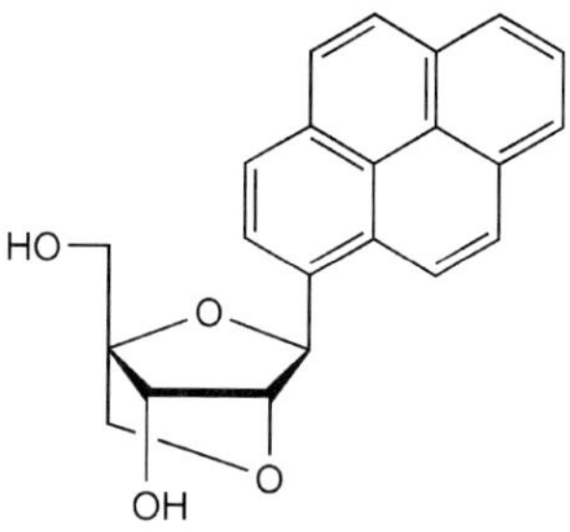

3

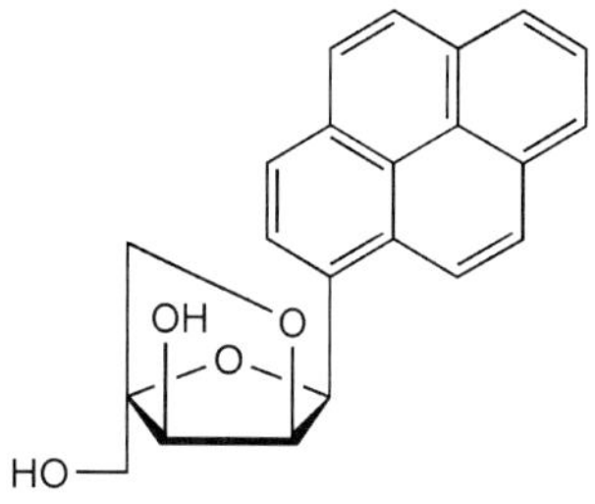

4

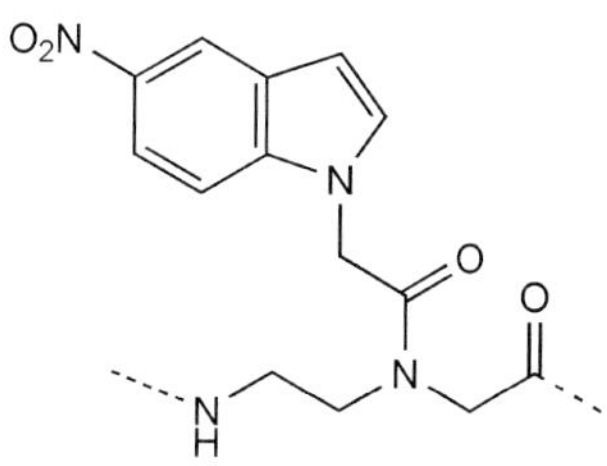

5

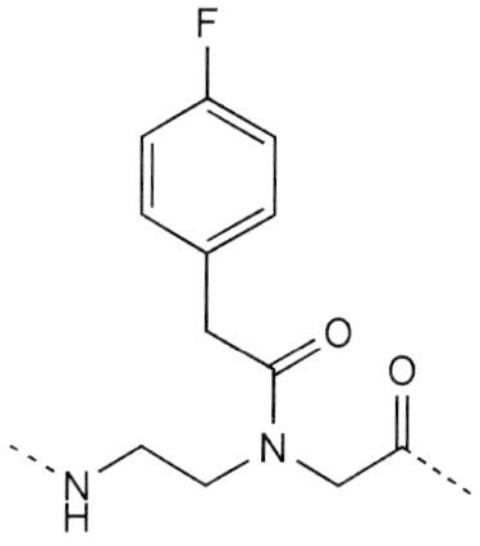

6

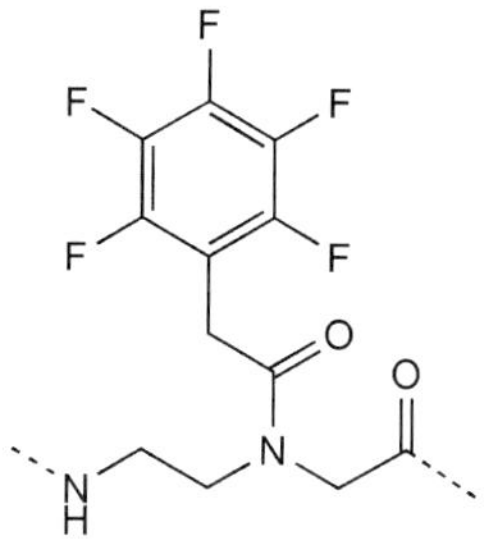

7

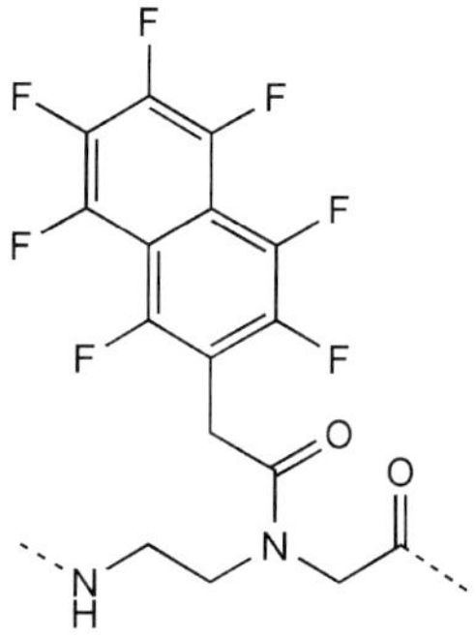

8

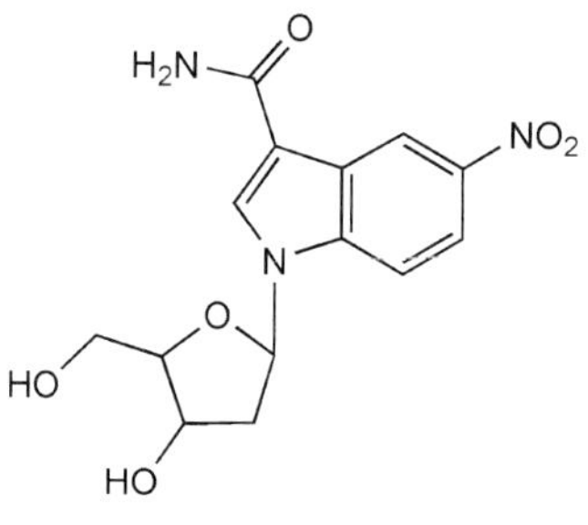

9

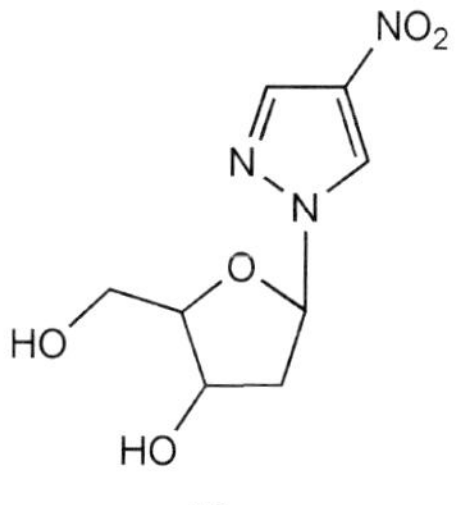

10

fluorescence intensity significantly increases upon the hybridization of matched duplexes. For example, when the thiazole orange is adjacent to a SNP site, the increase in fluorescence intensity is 26-fold compared to a mismatched sequence [66]. Such probes have applications in the field of genetic diagnostics as well as in the study of DNA–protein interactions. The thiazole orange PNA universal base has also been used to examine DNA-modifying enzymes, in particular DNA methyl transferases [67].

Seela *et al.* have investigated a series of benzotriazoles and 1,2,3-triazolo[4,5-d] pyrimidines for their base-pairing and fluorescent properties. A number of these analogues exhibited non-discriminate base-pairing properties, and were fluorescent, particularly at higher pH. In the case of the triazolopyrimidines, the fluorescent properties were dependent upon the glycosylation position, with the N2-regioisomer (**35**) exhibiting the best fluorescent profile for use in hybridization probes [39]. 7-Azaindole (**36**) is another analogue that may be used as a fluorescent universal base analogue in hybridization probes [68], and Kool and coworkers have described a number of non-discriminatory fluorescent base analogues [69, 70].

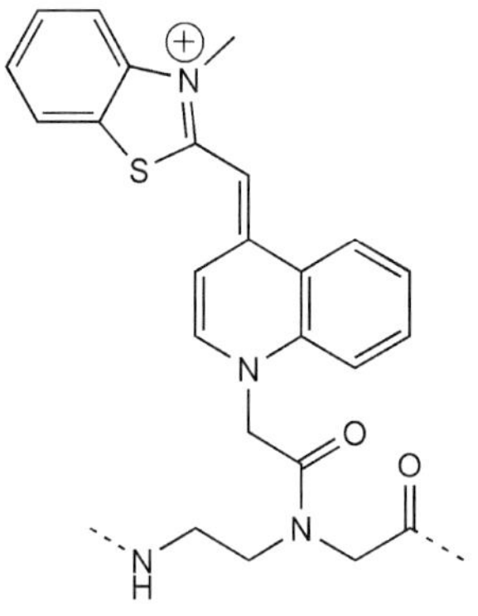

34

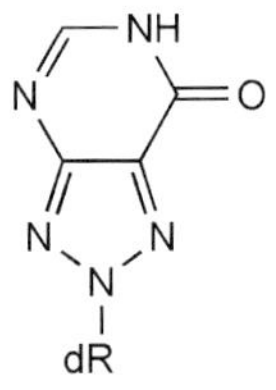

35

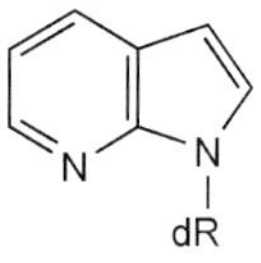

36

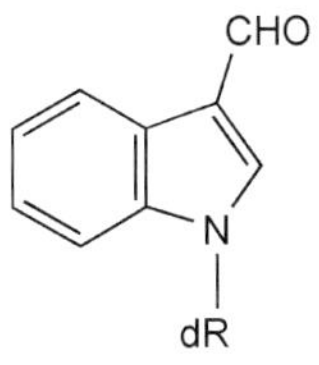

37

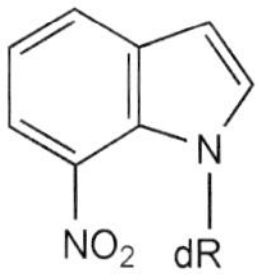

38

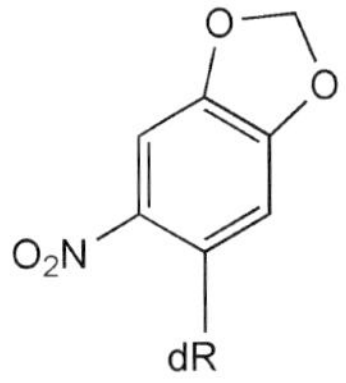

39

Universal bases have been used for post-synthetic modification. 3-Formylindole 2′-deoxyribonucleoside (**37**) retains the hybridization properties of a universal base, but as it bears an aldehyde group it allows for post-synthetic modifications of oligonucleotides containing it [71]. The treatment of DNA containing **37** with a variety of hydrazine derivatives, such as ferrocenecarbohydrazide, readily led to the hydrazone conjugate of **37**, allowing for site-specific oligonucleotide labeling.

The incorporation of 2′-aminoalkyl-substituted fluorobenzene and fluorobenzimidazole derivatives in the helix III stem of the hammerhead ribozyme resulted in enhanced self-cleavage reactions by a factor of up to 13, compared to the native ribozyme. It is suggested that the analogues alter the three-dimensional folding of the ribozyme, and that potential C−F•••H−C hydrogen bonding account for the improved geometry, and hence catalytic activity [72].

Universal base analogues have been shown to be substrates for ligase enzymes [73]. Luo *et al.* have shown that the presence of 3-nitropyrrole near the 3′-terminus and within the ligase active site leads to enhanced fidelity of ligation by *Thermus thermophilus* DNA ligase [63]. The ligase exhibits greatest fidelity when 3-nitropyrrole is incorporated at the third position from the discrimination site, and is more efficient in discriminating mismatches when 3-nitropyrrole is at the 3′-side of the mismatch rather than 5′. The incorporation of 5-nitroindole (**2**) in place of 3-nitropyrrole (**1**) did not lead to enhanced ligase fidelity. Burgner *et al.* also observed that improved allelic differentiation could be achieved using 3-nitropyrrole incorporated three nucleotides from a discrimination site, whilst 5-nitroindole had little effect, and was ineffective in detecting SNPs [62]. This is probably due to the fact that 3-nitropyrrole is more destabilizing in a duplex than 5-nitroindole, and the presence of a mismatch site close to 3-nitropyrrole in a duplex enhances this instability.

4-Methylindole (**16**), which has been used as a universal base analogue, was devised as a purine isostere. It has been incorporated into DNA to study charge transport, where the 4-methylindole radical cation may be detected by transient absorption and electroparamagnetic resonance (EPR) methods [74]. The mechanism and kinetics of charge hopping and A:T tunneling were studied using **16**.

The 7-nitroindole nucleoside (**38**) was introduced as a method for studying the 2′-deoxyribonolactone abasic site lesion [75], although it also behaves as a universal base analogue in its hybridization properties [76]. On photolysis, the 7-nitroindole nucleobase is removed as shown in Scheme 11.7, therefore behaving as a convertible nucleoside, the kinetics of which has been reported [77]. A further universal, photocleavable DNA base: nitropiperonyl deoxyribosyl-*C*-nucleoside (**39**) has been described by Pirrung *et al.* This was reported to have a slight preference for pairing with C, but was less discriminate with the remaining natural bases. Moreover, **39** was only destabilizing by an average of 7 °C compared to the natural base pair. Also, **39** could be used as a light-based DNA scissors as it can be readily cleaved by irradiation ($\lambda > 360$ nm, 20 min), followed by piperidine treatment [78]. In this case, photolysis of DNA containing **39** results in chain cleavage leading to excision of the nucleoside analogue generating an oligonucleotide-3′-phosphate.

Natural DNA can form a triple helix if the targeted strand is purine-rich where, in addition to Watson–Crick base pairing, Hoogsteen hydrogen bonds to a third strand

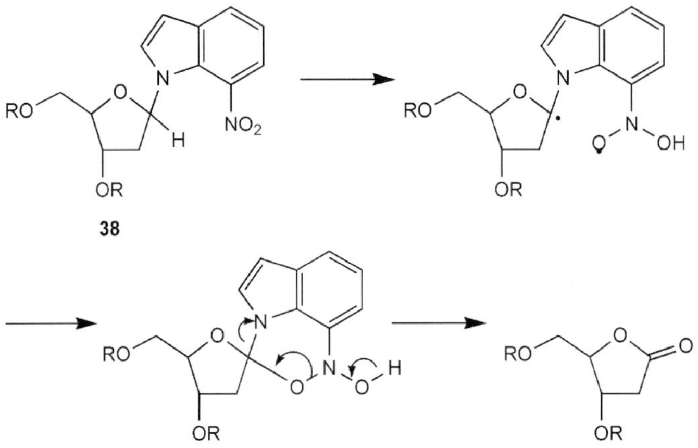

Scheme 11.7 Photolysis of 7-nitroindole (**38**) to produced the deoxyribonolactone abasic site.

oligonucleotide containing the matching pyrimidine sequence occur. Any Cs or Ts in the target strand of the duplex will bind only very weakly, as they contribute just one hydrogen bond leading to its destabilization. Moreover, the recognition of G requires the C in the probe strand to be protonated, so triplex formation will only work at low pH. In order to probe the stabilization of triplexes at neutral pH, Helene and coworkers examined a number of synthetic DNA base analogues as modifications in the third strand of a triplex [79]. These authors found that all the modified bases led to destabilized triplexes, and confirmed the report that 3-nitropyrrole (**1**) destabilized triplex formation [79–81]. Similar data were observed for the stability of acyclic nitroazole nucleosides in triplexes [82]. However, further studies showed that stable triplexes containing **1** in the third strand can be obtained in the presence of benzopyridoindole. In fact, **1** was found to discriminate GC from CG, and AT from TA [79] and, when incorporated in a polypurine strand, **1** had a marked stabilization opposite T compared with any other natural nucleotides [83]. This affinity for binding opposite thymidine was explained by molecular modeling, where it was demonstrated that the 3-nitropyrrole:A:T triad is isomorphous to A:A:T. A family of azole nucleosides was investigated in anti-parallel triplex formation. It was found that, when pyrazole, imidazole, 1,2,3-triazole and tetrazole were incorporated into oligodeoxyribonucleotides, each was capable of associating with C:G and T:A base pairs to form anti-parallel triplexes. The stability of the triplexes containing the azoles was considerably higher than that found with the natural nucleobases [84].

Hydrogen-bonding universal bases have also been used to stabilize triplex formation. A recent report showed that **29** causes destabilization at the center of a DNA duplex, but stabilization in a triplex. It was shown by molecular dynamic simulations that the amino group at position 8 is responsible for the electronic redistribution of the purine, favoring the Hoogsteen pairing over the Watson–Crick pairing in a triplex situation [43]. Parel and Leumann used quantitative DNase I footprint titration experiments to assay the triplex-forming properties of various aminopurines. These authors found that α-N^9-2-amino purine (**40**) showed no discrimination when associated with double-stranded DNA targets containing all four natural base pairs in an anti-parallel triplex formation. The order of base triplex stabilities was found to be **40**:A–T = G–C > T–A > C-G [85]. More interestingly, Rana and Ganesh described the first non-natural nucleoside base analogue 5-aminouracil (**41**) that recognizes all four bases A, T, C and G when placed in the central strand of a DNA triplex. In addition, **41** was found selectively to tolerate base-triples with the different natural bases and 2-aminopurine, depending on the parallel/anti-parallel orientation of the third strand [86].

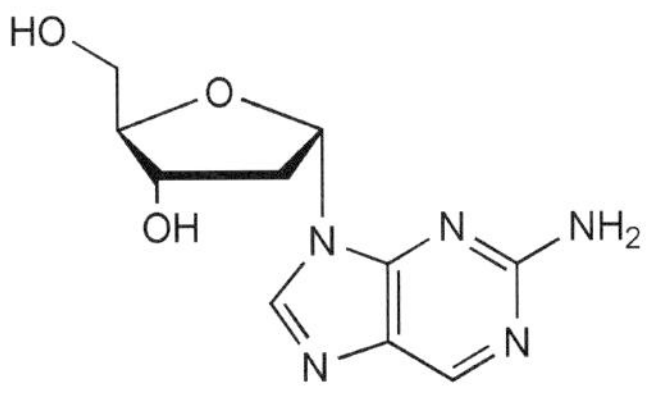

40

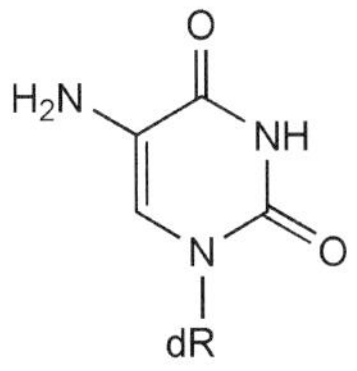

41

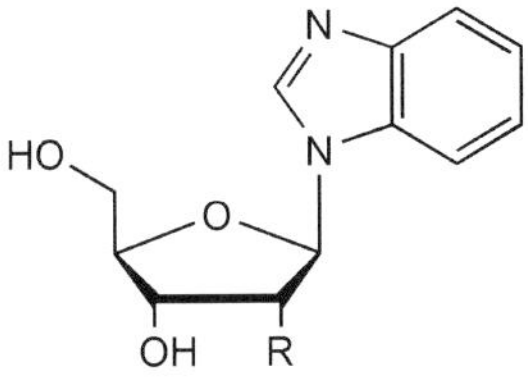

42

R=H
R=OH

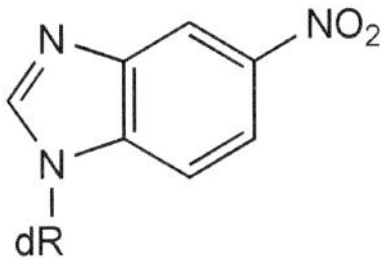

43

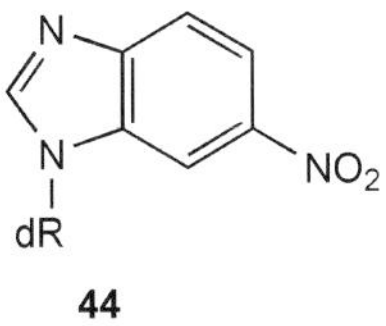

44

11.7
Triphosphate Derivatives

The 5′-triphosphate derivatives of a number of universal base analogues have been examined as substrates for various polymerases, where they have generally shown to be rather poor substrates. Indeed, hydroxylated phenyl *C*-nucleosides have been shown to be very inhibitory towards DNA polymerases [87]. Smith *et al.* showed that the 5′-triphosphates of 3-nitropyrrole (**1**) and 5-nitroindole (**2**) are incorporated opposite the native DNA bases without discrimination by Klenow fragment, but at low efficiency, and further extension is essentially inhibited [17]. In PCR reactions, as the concentration of either of these triphosphates is increased, the PCR reaction becomes increasingly inhibited. It is believed that **2** is a relatively good substrate for binding to polymerases, presumably through favorable stacking interactions, but is only poorly incorporated onto the growing DNA primer. Furthermore, once in the polymerase active site the dissociation from the polymerase is slow, leading to inhibition of the polymerase. Replication of template 5-nitroindole (**2**) and the photocleavable 7-nitroindole (**38**) by Klenow fragment showed a marked decrease in polymerization, with dAMP being preferentially incorporated opposite them [76]. By contrast, the 5′-triphosphate derivative of the ribonucleoside of 3-nitropyrrole is a substrate for the poliovirus 3Dpol, though it is preferentially incorporated opposite uridine and adenine [88].

The 5′-triphosphate derivatives of benzimidazole (**42**, R = H), 5- and 6-nitrobenzimidazole (**43** and **44**) and 5-nitroindole (**2**) have been examined as substrates for incorporation by DNA polymerase α and Klenow fragment. All four analogues were shown to be incorporated opposite the four canonical bases up to 4000 times more efficiently than an incorrect natural dNTP by DNA polymerase α, a rate approaching that of a correct dNTP [89]. Klenow fragment preferentially incorporated each analogue opposite template purines. Thus, DNA polymerase α was not able to discriminate between any of the four analogues, although it shows remarkable discrimination between correct and incorrect dNTPs, presumably due to the lack of hydrogen-bonding recognition. Interestingly, the arabino derivative of the benzimidazole (**45**) essentially inhibits DNA primase, the enzyme that synthesizes short RNA oligonucleotides as substrates for DNA polymerase α [90].

Berdis and coworkers have reported several instances of using universal base analogues to probe DNA polymerase recognition features. A comparison of the incorporation efficiencies of the triphosphate derivatives of 5-nitroindole and indole

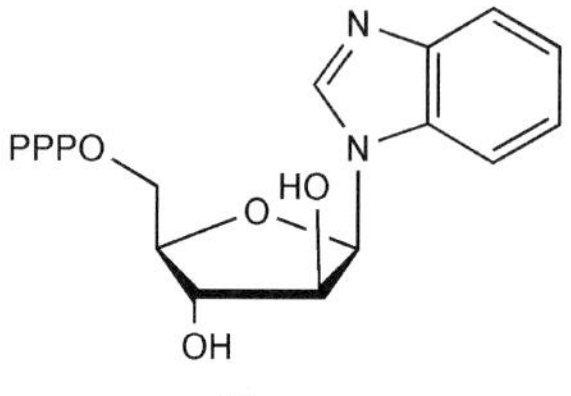

45

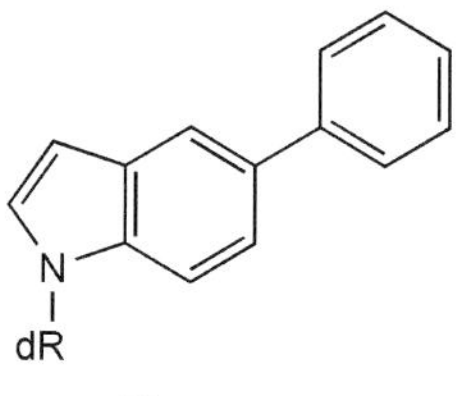

46

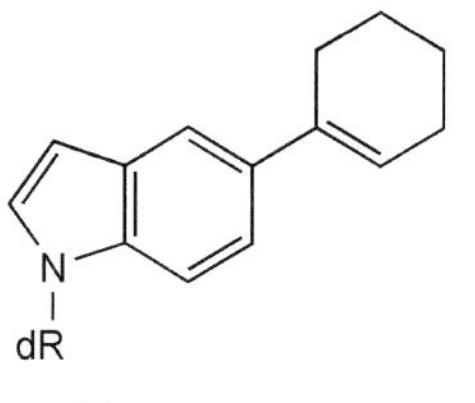

47

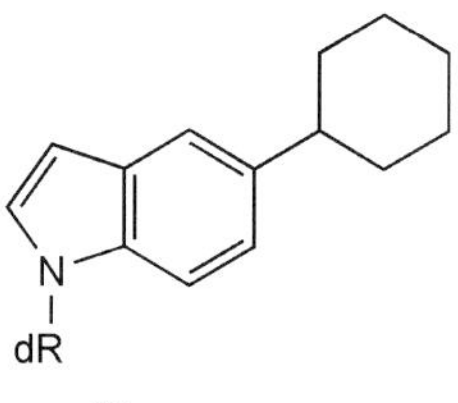

48

by T4 DNA polymerase opposite an abasic site showed marked differences. The indole is incorporated approximately 3600-fold less efficiently than 5-nitroindole (**2**) and, as both analogues have similar hydrophobicity, it is suggested that the difference is a result of the loss of base stacking by the nitro group, rather than being due to solvophobic effects [91]. The incorporation of a series of hydrophobic universal base analogues opposite an abasic site demonstrated that increasing the size of the aromatic base did not increase insertion efficiency [92]. The exception to this was 5-nitroindole (**2**), which was incorporated with an approximately 1000-fold greater efficiency compared to the incorporation of dAMP (the usual nucleotide inserted opposite an abasic site by DNA polymerases – the "A-rule"). This suggests that hydrogen-bonding groups are not a prerequisite, and can be accounted for by an increase in stacking interactions rather, again, than due to desolvation effects.

Further studies using 5-substituted indole analogues were carried out to probe for the contributions of shape complementarity and π-electron surface area during polymerization opposite an abasic site [93]. The efficiency of insertion of 5-phenylindole (**46**) was found to be about 1000-fold greater than for 5-fluoro- or 5-amino-indole derivatives, though about half the efficiency of incorporation of 5-nitroindole (**2**). The data suggest that the π-electron surface area plays a substantial role in the incorporation efficiency, rather than shape complementarity. Also, the incorporation efficiency of 5-cyclohexenyl-indole (**47**) is about 75-fold greater than for the cyclohexanyl derivative (**48**), providing further evidence that π-electron surface area plays a significant role in the polymerization process [94]. Despite the enhanced incorporation efficiencies of these analogues, they are still chain-terminating residues. In addition, **46** and **2** are excised much more slowly when incorporated opposite an abasic site than when incorporated opposite a natural DNA base [95]. As **46** is a fluorescent analogue, it has also been possible to carry out translesion synthesis by monitoring its fluorescence in real time [96].

As mentioned above, universal base analogues have proven to be at best only very poor polymerase substrates. This is true whether the universal base is in the template, or when using the 5′-triphosphate of the analogue. The one context where this proves to be different is the incorporation of a 5′-triphosphate of a universal base opposite a templated universal base – that is, in the formation of a self-pair. In the present authors' experience, the formation of a self-pair by thermophilic DNA polymerases approaches the rate of formation of a canonical base pair. Generally speaking, universal base analogues are blocking lesions to DNA polymerases. However, the scope of applications of universal bases would be significantly improved if they were recognized and used by polymerases. To this end, two groups in particular are endeavoring to evolve polymerases capable of accepting and replicating non-hydrogen-bonding universal base analogues, including universal bases.

The group of Romesberg uses phage display to evolve novel polymerases [97]. In this approach, the Stoffel fragment of *Taq* DNA polymerase and its DNA substrate primers are attached to phage particles via its minor phage coat protein pIII. This localization ensures intramolecular reactivity and ensures the association of genotype with phenotype. By using a library of proteins derived from the Stoffel fragment, novel polymerases may be assayed using biotinylated substrates to aid the recovery of active proteins. Successive rounds of screening under progressively rigorous selection conditions result in evolved polymerases with novel specificities. One polymerase isolated was shown capable of incorporating with reasonable efficiency the PICS analogue (**21**), although this polymerase requires further optimisation [98]. The same group also reported that, with a series of methyl-substituted phenyl *C*-nucleosides, the polymerases are acutely sensitive to the pattern of methyl group recognition, but more importantly that neither hydrogen bonding groups nor large aromatic surfaces are required for polymerase recognition [99].

A different strategy has been employed by the group of Holliger, known as compartmentalized self-replication (CSR). In this method, the polymerase library is transformed into *E. coli*, after which the cells containing the polymerase and its encoding gene are suspended in oil-in-water emulsions containing flanking primers and buffers, where they are segregated into aqueous compartments. Once the polymerase and its encoding gene are released from the cell, self-replication occurs – that is, the mutant polymerase amplifies its own encoding gene –but poorly active proteins fail to replicate. By using this method, the genotype and phenotype are associated by segregation within aqueous compartments [100].

Using a primer that has a distorting 3′-mismatch, a series of mutant polymerases were evolved that not only will extend a mismatch sequence, but also acquire a generic ability to bypass a number of otherwise blocking lesions [101]. As well as being able to bypass lesions, such as thymine dimers and abasic sites, they were also able to bypass the universal base 5-nitroindole (**2**). The polymerases also maintained catalytic turnover, processivity and fidelity. In more recent studies, the present authors' group are evolving polymerases specifically aimed at recognizing and replicating the universal base analogues 5-nitroindole (**2**) and 5-nitroindole-3-carboxamide (**9**) [18]. The main selection pressure applied is that the flanking primers contain the universal base at each 3′-end; hence, base recognition must occur in order to amplify the encoding gene. The resultant polymerases are able to copy universal base analogues as well as to incorporate the 5′-triphosphate without stalling. Whilst the efficiency of incorporation or replication are not as efficient as wild-type polymerase with canonical DNA, they are nevertheless one to two orders of magnitude more efficient than the wild-type. Interestingly, the newly evolved polymerases will synthesize a universal base self-pair with almost wild-type efficiency, but now the self-pair is an absolute termination site. Thus, with these new technologies emerging that enable the evolution of polymerases with designer specificity and activity, the scope of applications for universal base analogues – as well as many other aromatic moieties – will be greatly increased.

11.8 Therapeutic Applications

Recently, increasing evidence has emerged in support of an antiviral strategy known as "lethal mutagenesis". The hypothesis is based on the fact that RNA viruses exist as a pool of mutant species. However, the rate of mutation means that the viral population exists on the edge of viability, and that a slight increase in their mutation frequency might be enough to drive the viral population over its error threshold, resulting in a loss of viral viability (error catastrophe). Cameron and coworkers have been actively involved in the evaluation of universal bases as antiviral agents. Their idea is based on the fact that universal bases are indiscriminate towards the natural bases, and hence the presence of a universal base in an RNA strand could potentially lead to an increase in the mutation frequency during each round of viral replication. Whilst this antiviral strategy has been demonstrated for the degenerate base:ribavirin, 3-nitropyrrole (**1**) did not show any antiviral activity. Moreover, it was shown that the 5′-triphosphate of **1** was incorporated only opposite A or U by poliovirus RNA-dependent RNA polymerase (RdRP), but at a rate 100-fold slower than ribavirin triphosphate. It was argued that this slower incorporation rate might be due to the overly stable π-stacking interactions with the neighboring aromatic amino acid residues in the active site of the polymerase, preventing its efficient ejection after incorporation. In addition, it was concluded that hydrogen-bonding substituents are important for the RdRPs to incorporate ribonucleotides efficiently into RNA [88]. The same group have also studied various indole ribonucleosides as viral mutagens, though antiviral data were not reported [102].

Other universal bases with potential therapeutic uses include the previously discussed benzimidazole nucleotides (**42**, R = OH) as selective inhibitors of human DNA primase. DNA primase is a form of RNA polymerase which is activated by a DNA helicase to synthesize short RNA primers. During eukaryotic DNA replication, these short primers are elongated by DNA polymerase α in order to initiate the synthesis of all new DNA strands. Moore *et al.* have developed a nucleotide analogue that selectively inhibits the error-prone DNA primase with minimal effect on the high-fidelity DNA polymerase. After studying hydrophobic benzimidazoles lacking hydrogen-bonding properties, it was found that benzimidazole (**42**, R = OH), dichlorobenzimidazole and a 1 : 1 mixture of 5- and 6-nitrobenzimidazole (**43** and **44**) all inhibited human DNA primase activity, though polymerization was also significantly impaired. The addition of substituents on the benzimidazole ring failed to provide an increase in potency, but the most specific inhibitor was found to be the arabino derivative (**45**). The latter can be incorporated opposite all four natural bases by DNA Pol α with low efficiency, but proved to be a two- to four-fold better inhibitor compared to its ribofuranosyl analogue [90].

Universal base triphosphate derivatives are reported to have potential chemotherapeutic applications as selective inhibitors of promutagenic DNA synthesis. The triphosphate derivatives of 5-nitroindole (**2**) and 5-phenylindole (**46**) selectively inhibit replication when incorporated opposite an abasic site in a DNA duplex. The 5′-triphosphates of **2** and **46** have a high affinity for incorporation opposite the DNA lesion ($K_D \approx 10\,\mu M$) where they act as chain terminators with low IC_{50} values ($\sim 10\,\mu M$), and are resistant to enzymatic excision [95]. These results indicate that both 5-nitroindole derivatives **2** and **46** could be used to potentiate the cytotoxic effects of chemotherapeutic drugs such as temozolomide and cyclophosphamide, both of which are known to increase the rate of formation of abasic sites. As a combination therapy for cancer, they may inhibit the repair of lesions caused by DNA-damaging chemotherapeutic agents, and result in an increase in the potency of the anti-cancer drugs. Compounds **2** and **46** may also be used as potential chemopreventive agents;

the chain-terminating properties imply that, once incorporated, DNA polymerase cannot read across the primer, leading to a reduction in mutational errors caused by the inappropriate replication of unrepaired DNA lesions caused by DNA-damaging chemotherapeutic agents. Indeed, it is now recognized that the use of chemotherapeutic agents can lead to the development of secondary cancers caused by inadvertent mutagenesis [95].

References

1 Ohtsuka, E., Matsuki, S., Ikehara, M., Takahashi, Y., Matsubara, K., *J. Biol. Chem.* **1985**, *260*, 2605–2608.

2 Takahashi, Y., Kato, K., Hayashizaki, Y., Wakabayashi, T., Ohtsuka, E., Matsuki, S., Ikehara, M., Matsubara, K., *Proc. Natl. Acad. Sci. USA* **1985**, *82*, 1931–1935.

3 Liu, H., Nichols, R., *Biotechniques* **1994**, *16*, 24–26.

4 Nichols, R., Andrews, P. C., Zhang, P., Bergstrom, D. E., *Nature* **1994**, *369*, 492–493.

5 Loakes, D., Brown, D. M., *Nucleic Acids Res.* **1994**, *22*, 4039–4043.

6 Revankar, G. R., Robins, R. K., *Nucleosides & Nucleotides* **1989**, *8*, 709–724.

7 Parsch, J., Engels, J. W., *Helv. Chim. Acta* **2000**, *83*, 1791–1808.

8 Chaudhuri, N. C., Kool, E. T., *Tetrahedron Lett.* **1995**, *36*, 1795–1798.

9 Ren, R. X. -F., Schweitzer, B. A., Sheils, C. J., Kool, E. T., *Angew. Chem. Int. Ed. Engl.* **1996**, *35*, 743–746.

10 Babu, B. R., Prasad, A. K., Trikha, S., Thorup, N., Parmar, V. S., Wengel, J., *J. Chem. Soc. Perkin Trans. 1* **2002**, 2509–2519

11 Babu, B. R., Raunak, Sørensen, M. D., Hrdlicka, P. J., Trikha, S., Prasad, A. K., Parmar, V. S., Wengel, J., *Pure Appl. Chem.* **2005**, *77*, 319–326.

12 Babu, B. R., Wengel, J., *Chem. Commun.* **2001**, 2114–2115

13 Babu, B. R., Wengel, J., *Nucleosides, Nucleotides, Nucleic Acids* **2003**, *22*, 1317–1319.

14 Challa, H., Styers, M. L., Woski, S. A., *Org. Lett.* **1999**, *1*, 1639–1641.

15 Zhang, P., Egholm, M., Paul, N., Pingle, M., Bergstrom, D. E., *Methods* **2001**, *23*, 132–140.

16 Frey, K. A., Woski, S. A., *Chem. Commun.* **2002**, 2206–2207

17 Smith, C. L., Simmonds, A. C., Felix, I. R., Hamilton, A. L., Kumar, S., Nampalli, S., Loakes, D., Hill, F., Brown, D. M., *Nucleosides, Nucleotides Nucleic Acids* **1998**, *17*, 541–554.

18 Too, K., Brown, D. M., Holliger, P., Loakes, D., *Coll. Czech. Chem. Commun.* **2006**, *71*, 899–911.

19 Gallego, J., Loakes, D., *Nucleic Acids Res.* **2007**, *35*, 2904–2912.

20 Wheaton, C. A., Dobrowolski, S. L., Millen, A. L., Wetmore, S. D., *Chem. Phys. Lett.* **2006**, *428*, 157–166.

21 Millican, T. A., Mock, G. A., Chauncey, M. A., Patel, T. P., Eaton, M. A. W., Gunning, J., Cutbush, S. D., Neidle, S., Mann, J., *Nucleic Acids Res.* **1984**, *12*, 7435–7453.

22 Parsch, J., Engels, J. W., *J. Am. Chem. Soc.* **2002**, *124*, 5664–5672.

23 Petersheim, M., Turner, D. H., *Biochemistry* **1983**, *22*, 256–263.

24 Bergstrom, D. E., Zhang, P., Toma, P. H., Andrews, P. C., Nichols, R., *J. Am. Chem. Soc.* **1995**, *117*, 1201–1209.

25 Guckian, K. M., Schweitzer, B. A., Ren, R. X. -F., Sheils, C. J., Tahmassebi, D. C., Kool, E. T., *J. Am. Chem. Soc.* **2000**, *122*, 2213–2222.

26 Schweitzer, B. A., Kool, E. T., *J. Org. Chem.* **1994**, *59*, 7238–7242.

27 Schweitzer, B. A., Kool, E. T., *J. Am. Chem. Soc.* **1995**, *17*, 1863–1872.

28 Kool, E. T., Morales, J. C., Guckian, K. M., *Angew. Chem. Int. Ed. Engl.* **2000**, *39*, 990–1009.

29 Kool, E. T., *Cold Spring Harbor Symp. Quant. Biol.* **2000**, *65*, 93–102.

30 Matray, T. J., Kool, E. T., *J. Am. Chem. Soc.* **1998**, *120*, 6191–6192.

31 Parsch, J., Engels, J. W., *Nucleosides, Nucleotides Nucleic Acids* **2001**, *20*, 815–818.

32 Berger, M., Wu, Y., Ogawa, A. K., McMinn, D. L., Schultz, P. G., Romesberg, F. E., *Nucleic Acids Res.* **2000**, *28*, 2911–2914.

33 Wu, Y., Ogawa, A. K., Berger, M., McMinn, D. L., Schultz, P. G., Romesberg, F. E., *J. Am. Chem. Soc.* **2000**, *122*, 7621–7632.

34 Berger, M., Ogawa, A. K., McMinn, D. L., Wu, Y., Schultz, P. G., Romesberg, F. E., *Angew. Chem. Int. Ed.* **2000**, *39*, 2940–2942.

35 McMinn, D. L., Ogawa, A. K., Wu, Y., Liu, J., Schultz, P. G., Romesberg, F. E., *J. Am. Chem. Soc.* **1999**, *121*, 11585–11586.

36 Ogawa, A. K., Wu, Y., McMinn, D. L., Liu, J., Schultz, P. G., Romesberg, F. E., *J. Am. Chem. Soc.* **2000**, *122*, 3274–3287.

37 Vallone, P. M., Benight, A. S., *Nucleic Acids Res.* **1999**, *27*, 3589–3596.

38 Loakes, D., Hill, F., Brown, D. M., Salisbury, S. A., *J. Mol. Biol.* **1997**, *270*, 426–435.

39 Seela, F., Jawalekar, A., Münster, I., *Helv. Chim. Acta* **2005**, *88*, 751–765.

40 Seela, F., Mittelbach, C., *Nucleosides Nucleotides* **1999**, *18*, 425–441.

41 Seela, F., Chen, Y., *Nucleic Acids Res.* **1995**, *23*, 2499–2505.

42 Acedo, M., De Clerq, E., Eritja, R., *J. Org. Chem.* **1995**, *60*, 6262–6269.

43 Cubero, E., Guimil-Garcia, R., Luque, F. J., Eritja, R., Orozco, M., *Nucleic Acids Res.* **2001**, *29*, 2522–2534.

44 Seela, F., Debelak, H., *Nucleic Acids Res.* **2000**, *28*, 3224–3232.

45 Seela, F., Debelak, H., *Nucleosides, Nucleotides Nucleic Acids* **2001**, *20*, 577–585.

46 He, J. L., Seela, F., *Helv. Chim. Acta* **2002**, *85*, 1340–1354.

47 Cadena-Amaro, C., Delepierre, M., Pochet, S., *Bioorg. Med. Chem. Lett.* **2005**, *15*, 1069–1073.

48 Hill, F., Loakes, D., Brown, D. M., *Proc. Natl. Acad. Sci. USA* **1998**, *95*, 4258–4263.

49 Loakes, D., Brown, D. M., Linde, S., Hill, F., *Nucleic Acids Res.* **1995**, *23*, 2361–2366.

50 Smith, T. H., Latour, J., Leo, C., Muthini, S., Siebert, P., Nelson, P. S., *Clin. Chem.* **1995**, *41*, 10.

51 Yang, M., Hayashi, K., Hayashi, M., Fujii, J. T., Kurkinen, M., *J. Biol. Chem.* **1996**, *271*, 25548–25554.

52 Ball, S., Reeve, M. A., Robinson, P. S., Hill, F., Brown, D. M., Loakes, D., *Nucleic Acids Res.* **1998**, *26*, 5225–5227.

53 Loakes, D., Hill, F., Brown, D. M., Ball, S., Reeve, M. A., Robinson, P. S., *Nucleosides Nucleotides* **1999**, *18*, 2685–2695.

54 Lage, J. M., Leamon, J. H., Pejovic, T., Hamann, S., Lacey, M., Dillon, D., Segraves, R., Vossbrinck, B., González, A., Pinkel, D., Albertson, D. G., Costa, J., Lizardi, P. M., *Genome Res.* **2003**, *13*, 294–307.

55 Ganova-Raeva, L., Smith, A. W., Fields, H., Khudyakov, Y., *Virus Res.* **2004**, *102*, 207–213.

56 Smith, C. L., Simmonds, A. C., Hamilton, A. L., Martin, D. L., Lashford, A. G., Loakes, D., Hill, F., Brown, D. M., *Nucleosides, Nucleotides Nucleic Acids* **1998**, *17*, 555–564.

57 Zheng, D., Raskin, L., *Microb. Ecol.* **2000**, *39*, 246–262.

58 Missura, M., Buterin, T., Hindges, R., Hübscher, U., Kaspárková, J., Brabec, V., Naegeli, H., *EMBO J.* **2001**, *20*, 3554–3564.

59 Matlock, D. L., Heyduk, T., *Biochemistry* **2000**, *39*, 12274–12283.

60 Fotin, A. V., Drobyshev, A. L., Proudnikov, D. Y., Perov, A. N., Mirzabekov, A. D., *Nucleic Acids Res.* **1998**, *26*, 1515–1521.

61 Parinov, S., Barsky, V., Yershov, G., Kirillov, E., Timofeev, E., Belgovskiy, A.,

Mirzabekov, A. D., *Nucleic Acids Res.* **1996**, *24*, 2998–3004.

62 Burgner, D., D'Amato, M., Kwiatkowski, D. P., Loakes, D., *Nucleosides, Nucleotides Nucleic Acids* **2004**, *23*, 755–765.

63 Luo, J., Bergstrom, D. E., Barany, F., *Nucleic Acids Res.* **1996**, *24*, 3071–3078.

64 Frutos, A. G., Pal, S., Quesada, M., Lahiri, J., *J. Am. Chem. Soc.* **2002**, *124*, 2396–2397.

65 Köhler, O., Seitz, O., *Chem. Commun.* **2003**, 2938–2939

66 Köhler, O., Jarikote, D. V., Seitz, O., *ChemBioChem* **2005**, *6*, 69–77.

67 Köhler, O., Jarikote, D. V., Singh, I., Parmar, V. S., Weinhold, E., Seitz, O., *Pure Appl. Chem.* **2005**, *77*, 327–338.

68 Wang, K., Stringfellow, S., Dong, S., Jiao, Y., Yu, H., *Spectrochim. Acta A* **2002**, *58*, 2595–2260.

69 Ren, R. X. -F., Chaudhuri, N. C., Paris, P. L., Rumney, S., Kool, E. T., *J. Am. Chem. Soc.* **1996**, *118*, 7671–7678.

70 Strässler, C., Davis, N. E., Kool, E. T., *Helv. Chim. Acta* **1999**, *82*, 2160–2171.

71 Okamoto, A., Tainaka, K., Saito, I., *Tetrahedron Lett.* **2002**, *43*, 4581–4583.

72 Klöpffer, A. E., Engels, J. W., *ChemBioChem* **2004**, *5*, 707–716.

73 Loakes, D., Van Aerschot, A., Brown, D. M., Hill, F., *Nucleosides Nucleotides* **1996**, *15*, 1891–1904.

74 Pascaly, M., Yoo J., Barton, J. K., *J. Am. Chem. Soc.* **2002**, *124*, 9083–9092.

75 Kotera, M., Roupioz, Y., Defrancq, E., Bourdat, A.-G., Garcia, J., Coulombeau, C., Lhomme, J., *Chem. Eur. J.* **2000**, *6*, 4163–4169.

76 Crey-Desbiolles, C., Berthet, N., Kotera, M., Dumy, P., *Nucleic Acids Res.* **2005**, *33*, 1532–1543.

77 Roupioz, Y., Lhomme J., Kotera, M., *J. Am. Chem. Soc.* **2002**, *124*, 9129–9135.

78 Pirrung, M. C., Zhao, X., Harris, S. V., *J. Org. Chem.* **2001**, *66*, 2067–2071.

79 Kukreti, S., Sun, J.-S., Loakes, D., Brown, D. M., Nguyen, C. -H., Bisagni, E., Garestier, T., Helene, C., *Nucleic Acids Res.* **1998**, *26*, 2179–2183.

80 Amosova, O., George, J., Fresco, J. R., *Nucleic Acids Res.* **1997**, *25*, 1930–1934.

81 Kukreti, S., Sun, J.-S., Garestier, T., Helene, C., *Nucleic Acids Res.* **1997**, *25*, 4264–4270.

82 Walczak, K., Wamberg, M., Pedersen, E. B., *Helv. Chim. Acta* **2004**, *87*, 469–478.

83 Orson, F. M., Klysik, J., Bergstrom, D. E., Ward, B., Glass, G. A., Hua, P., Kinsey, B. M., *Nucleic Acids Res.* **1999**, *27*, 810–816.

84 Durland, R. H., Rao, T. S., Bodepuri, V., Seth, D. M., Jayaraman, K., Revankar, G. R., *Nucleic Acids Res.* **1995**, *23*, 647–653.

85 Parel, S. P., Leumann, C. J., *Nucleic Acids Res.* **2001**, *29*, 2260–2267.

86 Rana, V. S., Ganesh, K. N., *Nucleic Acids Res.* **2000**, *28*, 1162–1169.

87 Aketani, S., Tanaka, K., Yamamoto, K., Ishihama, A., Cao, H., Tengeiji, A., Hiraoka, S., Shiro, M., Shionoya, M., *J. Med. Chem.* **2002**, *45*, 5594–5603.

88 Harki, D. A., Graci, J. D., Korneeva, V. S., Ghosh, S. K. B., Hong, Z., Cameron, C. E., Peterson, B. R., *Biochemistry* **2002**, *41*, 9026–9033.

89 Chiaramonte, M., Moore, C. L., Kincaid, K., Kuchta, R. D., *Biochemistry* **2003**, *42*, 10472–10481.

90 Moore, C. L., Chiaramonte, M., Higgins, T., Kuchta, R. D., *Biochemistry* **2002**, *41*, 14066–14075.

91 Zhang, X., Lee, I., Berdis, A. J., *Org. Biomol. Chem.* **2004**, *2*, 1703–1711.

92 Reineks, E. Z., Berdis, A. J., *Biochemistry* **2004**, *43*, 393–404.

93 Zhang, X., Lee, I., Berdis, A. J., *Biochemistry* **2005**, *44*, 13101–13110.

94 Zhang, X., Lee, I., Zhou, X., Berdis, A. J., *J. Am. Chem. Soc.* **2006**, *128*, 143–149.

95 Zhang, X., Lee, I., Berdis, A. J., *Biochemistry* **2005**, *44*, 13111–13121.

96 Lee, I., Berdis, A. J., *ChemBioChem* **2006**, *7*, 1990–1997.

97 Xia, G., Chen, L., Sera, T., Fa, M., Schultz, P. G., Romesberg, F. E., *Proc. Natl. Acad. Sci. USA* **2002**, *99*, 6597–6602.

98 Leconte, A. M., Chen, L., Romesberg, F. E., *J. Am. Chem. Soc.* **2005**, *127*, 12470–12471.

99 Matsuda, S., Henry, A. A., Romesberg, F. E., *J. Am. Chem. Soc.* **2006**, *128*, 6369–6375.

100 Ghadessy, F. J., Ong, J. L., Holliger, P., *Proc. Natl. Acad. Sci. USA* **2001**, *98*, 4552–4557.

101 Ghadessy, F. J., Ramsay, N., Boudsocq, F., Loakes, D., Brown, A., Iwai, S., Vaisman, A., Woodgate, R., Holliger, P., *Nature Biotechnol.* **2004**, *22*, 755–759.

102 Harki, D. A., Graci, J. D., Morgan, R. L., Chain, W. J., Cameron, C. E., Peterson, B. R., *Abstracts Papers Am. Chem. Soc.* **2004**, *227*, 499-ORGN.

Part III
Medicinal Chemistry

12
The Properties of Locked Methanocarba Nucleosides in Biochemistry, Biotechnology, and Medicinal Chemistry

Victor E. Marquez

12.1
Introduction

The similar structural motifs in most nucleoside-based drugs, consisting of a furanose ring and an intact, or slightly modified DNA/RNA nucleobase, make them deceptively simple. On the other hand, understanding why the therapeutic range of these drugs varies from highly potent to totally inactive is complicated. Because the furanose ring can adopt two major antipodal conformations which affect the disposition of the nucleobase and the orientation of critical hydroxyl groups, it is important to construct and study conformationally locked nucleosides to learn about the nuances that Nature is able to detect in the conformation of these chameleon molecules to either utilize or reject them as substrates. In a similar manner, when nucleosides are incorporated into polymeric structures, such as DNA and RNA, the furanose moiety is able to respond to changes in environmental factors, such as degree of hydration, salt concentration, metal ion coordination, protein binding, and interactions with small molecules, giving rise to distinct helical arrangements where the constitutive 2′-deoxyribose or ribose moieties adopt one of the two major antipodal conformations. Here also, the construction of small DNA segments containing conformationally locked units contributes to our understanding of how the locally perturbed environment of a duplex affects biological function. Although there are potentially several ways of locking the conformation of the furanose moiety in a nucleoside, the use of the bicyclo[3.1.0]hexane platform described in this chapter is the only approach that simultaneously allows study of the two extreme versions of furanose conformations that exist in Nature. This chapter reviews the chemistry that provides these privileged scaffolds, and presents some specific biological examples where these locked analogues have contributed to the unraveling of some important biological puzzles.

Modified Nucleosides: in Biochemistry, Biotechnology and Medicine. Edited by Piet Herdewijn

ISBN: 978-3-527-31820-9

12.2 Structural Representation

12.2.1 The Bicyclo[3.1.0]hexane Template

The bicyclo[3.1.0]hexane scaffold was devised as a strategy to lock the embedded cyclopentane ring into a permanent envelope (E) conformation. If we consider the cyclopentane component of a bicyclo[3.1.0]hexane template as a surrogate of a five-membered furanose ring in a nucleoside, one can build on it at least two classes of conformationally restricted nucleosides that mimic a North (*N*) envelope (2′-*exo*, $_{2'}$E) conformation, or a South (*S*) envelope (3′-*exo*, $_{3'}$E) conformation as defined in the pseudorotational cycle [1, 2] (Figure 12.1).

Conformations in the pseudorotational cycle are defined by the phase angle of pseudorotation *P* (0–360 °), the value of which is a function of the five endocyclic torsion angles of the furanose (or surrogate cyclopentane ring). By convention, a phase angle $P = 0°$ corresponds to an absolute *N* conformation possessing a symmetrical twist form ($^{3'}T_{2'}$), whereas the *S* antipode ($_{3'}T^{2'}$) is represented by $P = 180°$. The value of *P* defines the type of sugar pucker (*N* or *S*), which together with the maximum out-of-plane pucker (ν_{max}) that corresponds to the radius of the circle, are the most important structural parameters that define the conformation of a

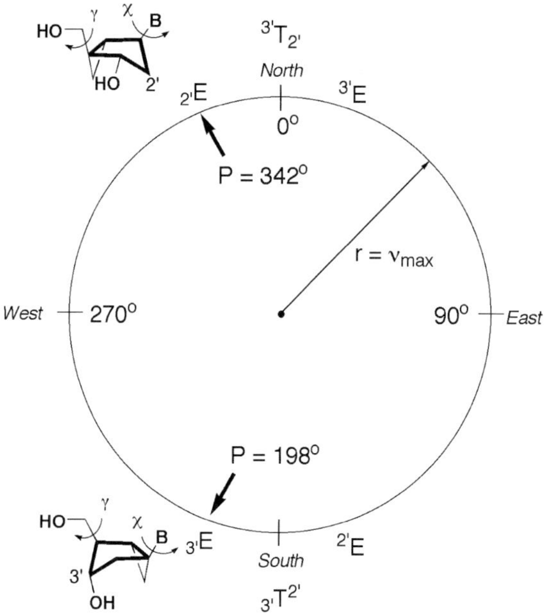

Figure 12.1 The pseudorotational cycle with the range of *N* and *S* conformations indicated. The radius of the circle is equal to the maximum out-of-plane pucker (ν_{max}). χ is the glycosyl torsion angle and γ is the CH_2OH torsion angle.

nucleoside. The value of two other parameters: (i) the glycosyl torsion angle χ, which determines the *syn* or *anti* disposition of the base relative to the sugar; and (ii) the torsion angle γ, which determines the orientation of the 5′-OH, complete defining the conformation of a nucleoside.

A close inspection of the available crystallographic data of individual nucleosides and nucleotides reveals that the puckering modes of the furanose ring tend to cluster into two antipodal domains around two envelope conformations ($^{3'}$E, *N* and $^{2'}$E, *S*), which occupy less than 10% of the total pseudorotational pathway. In solution, however, the sugar pucker fluctuates rapidly between these two conformational extremes. *N* conformations are defined to be in a range of *P* between 342° (−18°) and 18° [$_{2'}$E → $^{3'}$T$_2$, → $^{3'}$E (3′-*endo*)], whereas for the antipodal *S* conformation the range is between 162° and 198° [$^{2'}$E(2′-*endo*) → $_{3'}$T$^{2'}$ → $_{3'}$E] (Figure 12.1).

With bicyclo[3.1.0]hexane nucleosides the location of the fused cyclopropane ring determines whether the nucleoside is to be considered a locked *N* envelope ($_{2'}$E) conformer, or alternatively, a locked *S* envelope ($_{3'}$E) conformer (Figure 12.1). These two conformers are only ±18° away from the ideal *N* (0°) and *S* (180°) conformations.

12.2.2
Pseudoboat versus Pseudochair Conformations

A search for compounds containing an unrestricted bicyclo[3.1.0]hexane scaffold in the Cambridge Structural Database, revealed the pseudoboat conformation as the exclusive form for this system. Furthermore, electron diffraction microwave spectroscopy [3] and *ab-initio* calculations [4–9] confirmed the restricted presence of only the pseudoboat conformation. Specifically for the *N* and *S* thymidine analogues (compounds **17** and **75** in Schemes 12.2 and 12.11), quantum mechanical potential energy surfaces obtained after constraining a chosen target dihedral in 10° increments, and allowing the remaining degrees of freedom [except for χ (252°) and γ (51°)] to optimize to the default tolerance showed that the pseudoboat conformation is indeed more stable [10] (Figure 12.2).

Further confirmation that the pseudoboat conformation is maintained in solution came from key diagnostic coupling constants in the ^{1}H NMR spectra of all bicyclo [3.1.0]hexane carbocylic nucleosides where some dihedral angles approach 90° and the corresponding *J* coupling constants are zero or negligibly small [11–13] (Figure 12.3). These NMR results are in accord with the crystallographic data, and confirm unequivocally the exclusive pseudoboat pucker characteristic of this bicyclic system both in the solid state and in solution.

12.3
Synthesis of Locked Nucleosides

The inspiration to build bicyclo[3.1.0]hexane nucleosides was provided by Nature. During the early 1980s, a new family of carbocyclic nucleosides known generically as

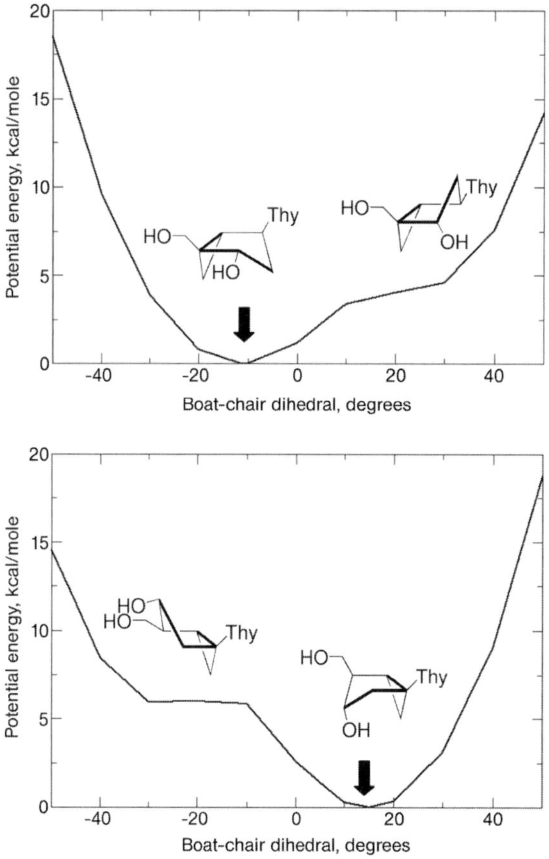

Figure 12.2 Potential energy surfaces representing pseudochair to pseudoboat transformations using Gaussian 03 at the MP2/6-31G* level for *N* (**17**, top) and *S* (**75**, bottom) bicyclo[3.1.0] hexane carbocyclic thymidines. The target dihedrals selected to monitor the change from pseudoboat to pseudochair are highlighted in bold. Stable conformers corresponding to the pseudoboat are above the arrows.

neplanocins were isolated from the culture filtrates of the mold *Ampuriella regularis* [14]. One of the minor metabolites of the most biologically important member of the family, neplanocin A, was the corresponding epoxide, neplanocin C [15]. The X-ray structure of neplanocin C revealed an oxobicyclo[3.1.0]hexane scaffold with the embedded cyclopentane ring clearly in the *N* conformation. Because the epoxide ring was considered potentially reactive [16], it was substituted for a fused cyclopropane ring to produce a similar *N* geometry. The generation of the *S* conformer simply required repositioning the cyclopropane ring relative to the base and the hydroxymethyl group as already explained.

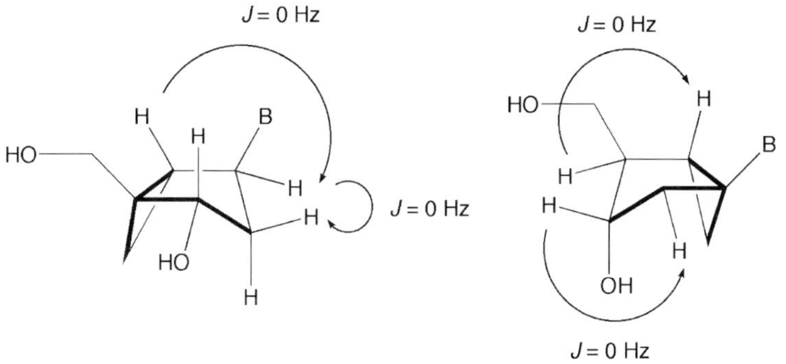

Figure 12.3 Some key vanishing coupling constants that confirm that the shape of the bicyclo[3.1.0]hexane scaffold in solution is pseudoboat. The values of these coupling constants do not change from 25 °C to 80 °C.

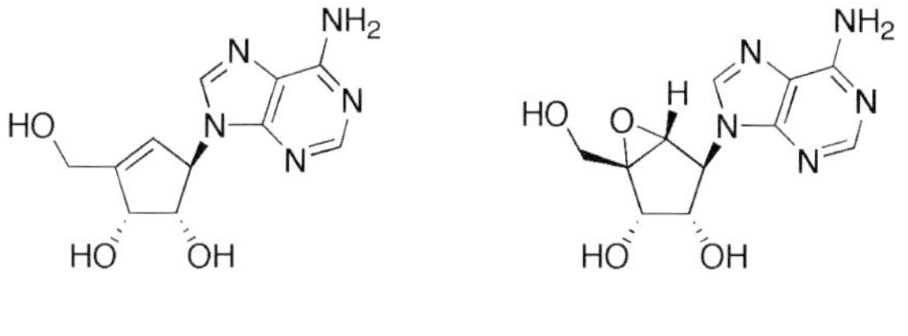

12.3.1
North (*N*) Conformer Mimics

12.3.1.1 **Dideoxyribonucleoside Analogues**

Because of the important role of dideoxynucleosides in inhibiting HIV reverse transcriptase (RT), the first nucleosides synthesized with this template were simple dideoxynucleoside analogues [11, 12]. Starting from (±)-3-[(benzyloxy)methyl]cyclopent-2-en-1-ol (**1**), a hydroxyl-directed cyclopropanation reaction via a samarium (II) carbenoid intermediate provided the desired pseudosugar (**2**) (Scheme 12.1). Following a convergent approach, all the nucleobases could be incorporated via a Mitsunobu coupling reaction. As expected, the Mitsunobu reaction worked better with purines than with pyrimidines, as the latter produced mixtures of *O*-alkylated products along with the desired *N*-alkylated nucleosides. The synthesis of the adenine analogue is exemplified. In this case, following the removal of the benzyl-protecting group, the identity of the D-like enantiomer [(+)-**5**] was achieved through the action of adenosine deaminase (ADA), which deaminated exclusively the D-enantiomer. Separation of the adenine enantiomers was performed by chiral HPLC. Enantioselective syntheses of other dideoxynucleosides were developed later from chiral precursors used in the syntheses of 2′-deoxynucleoside analogues (see Section 12.3.4).

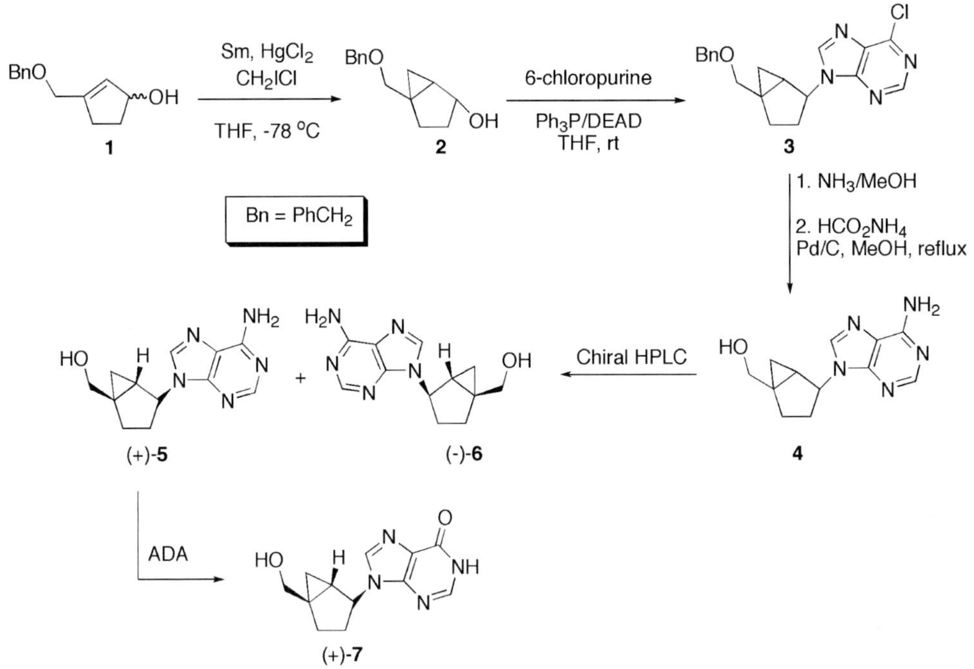

Scheme 12.1

12.3.1.2 2′-Deoxyribonucleoside Analogues

With the advent of the antisense field, there was a need to increase the thermodynamic stability of DNA/RNA heteroduplexes by chemical modification of the DNA strand. The desired modification involved the use of conformationally locked nucleosides to facilitate the preorganization of the annealed strands into an A-type conformation where all the sugars are puckered *N*. The first synthesis of a locked *N*, 2′-deoxynucleoside intended for this use was that of the thymidine analogue **17** reported by Altmann *et al.* [17] (Scheme 12.2). Starting from enantiomerically enriched cyclopentenol **8**, a hydroxyl-directed Simmons–Smith cyclopropanation provided the desired bicyclo[3.1.0]hexane **9** as a single diastereoisomer. Tosylation, followed by displacement of the tosyloxy group with azide and subsequent reduction of the azide moiety over Lindlar's catalyst, gave the partially protected aminotriol **10**. This compound was elaborated into the bicyclo ribo-thymidine analogue **12** by reaction with the acyl-isocyanate derived from β-methoxy-α-methylacrylic acid, and acid-catalyzed cyclization of the resulting acryloyl urea **11**, which also eliminated the acetonide group. Removal of the benzyl ether protection afforded the free ribo-thymidine **13**. Removal of the 2′-OH (nucleoside numbering) required the simultaneous protection of 3′- and 5′-hydroxyl groups as well as the NH of the thymine ring. These operations allowed the conversion of the 2′-OH to the thiocarbonate **15**, which underwent radical deoxygenation to the 2′-deoxy analogue **16** (84% *e.e.*). Optically pure **16** was obtained after preparative HPLC on Chiracel® OD, and removal of the TIPSi and BOM groups provided *N*-methanocarbathymidine (*N*-MCT, **17**).

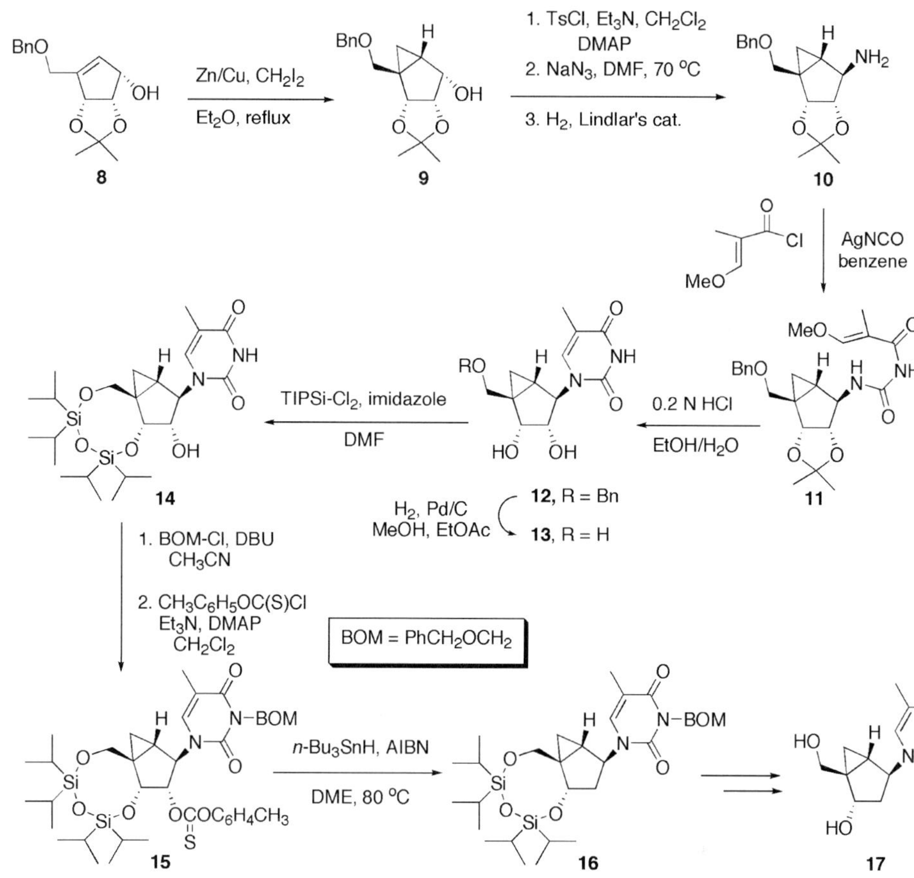

Scheme 12.2

Our group followed a similar method to obtain the corresponding adenosine analogue, but using instead a convergent approach that involved the coupling of **9** with adenine in the presence of K_2CO_3 and 18-crown-6 in DMF [18]. However, the lengthy process of removing the 2′-OH prompted us to seek other alternatives that would allow the preparation of 2′-deoxy analogues of all nucleobases from a common, universal starting material.

Starting with chiral cyclopentenol **8**, the idea was to perform the deoxygenation step prior to the elaboration of the nucleobase [19]. The key reaction was the regioselective cleavage of the contiguous *O*-isopropylidenetriol system in **8** with trimethylaluminum to give the corresponding carbocyclic 3-*t*-butoxy-1,5-glycol **18** (Scheme 12.3). Subsequently, reaction of **18** with *t*-butyldimethylsilyl chloride (TBDMSiCl), which occurred exclusively at the less-hindered allylic alcohol, gave **19**, thereby setting the stage for the ensuing radical deoxygenation from xanthate **20** to **21** under Barton's conditions. The unmasking of the silyl ether-protected secondary alcohol gave compound **22**, which directed the consequent cyclopropanation to give the pivotal intermediate **23**.

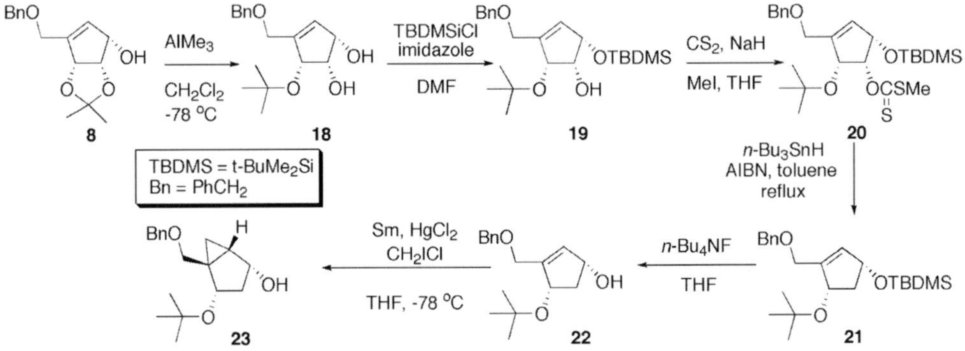

Scheme 12.3

Using this approach, the purine analogue **24** was synthesized via a convergent synthesis by coupling **23** with 6-chloropurine under Mitsunobu conditions. After ammonolysis and the simultaneous removal of the benzyl and *t*-butyl groups, the adenosine analogue **26** was obtained (Scheme 12.4). The versatility of **23** permitted access to 2′-deoxy-bicyclo[3.1.0]hexane nucleosides bearing the rest of the bases of the genetic code [20].

Because the synthesis of enantiomerically pure cyclopentenol **8** was difficult, efforts to eliminate the cumbersome deoxygenation step altogether were investigated. For this purpose, the readily accessible cyclopentenol synthon, (1*S*,2*R*)-2-[(benzyloxy)methyl]cyclopent-3-enol (**27*a***) was chosen (Scheme 12.5). This compound was initially developed by Biggadike *et al.* [21, 22] for the synthesis of plain 2′-deoxycarbocyclic nucleosides, and it was felt that this could be elaborated into appropriate synthons, such as **33** and **36**, for the syntheses of 2′-deoxy-bicyclo[3.1.0] hexane nucleosides via linear or convergent approaches, respectively. The reaction of either **27*b*** or **27*c*** with phenylselenyl chloride proceeded with complete stereochemical control to give exclusively *trans*-3-(phenylselenyl)-4-substituted cyclopentane intermediates **28*b*** and **28*c***, possibly due to the preferred pseudoequatorial disposition of both substituents in the cyclopentene ring, which biased formation of the episelenonium intermediate from the bottom face of a stable chair-like transition-state intermediate [23, 24] (Scheme 12.5). The regioselective opening of the episelenonium ion occurred by nucleophilic attack at the carbon atom farthest from the (benzyloxy)methyl group to give **28*b*** and **28*c***. An *in-situ* oxidation of the

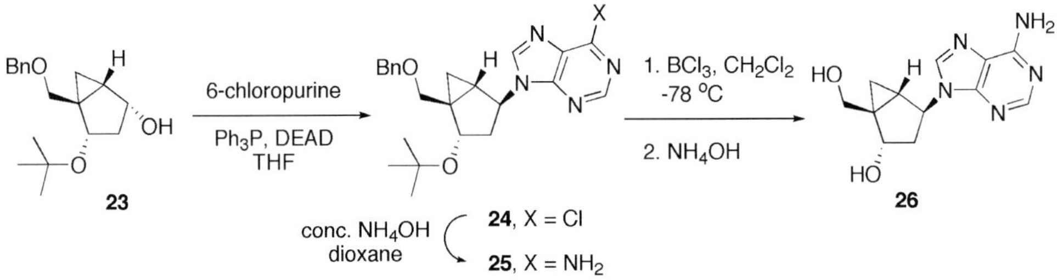

Scheme 12.4

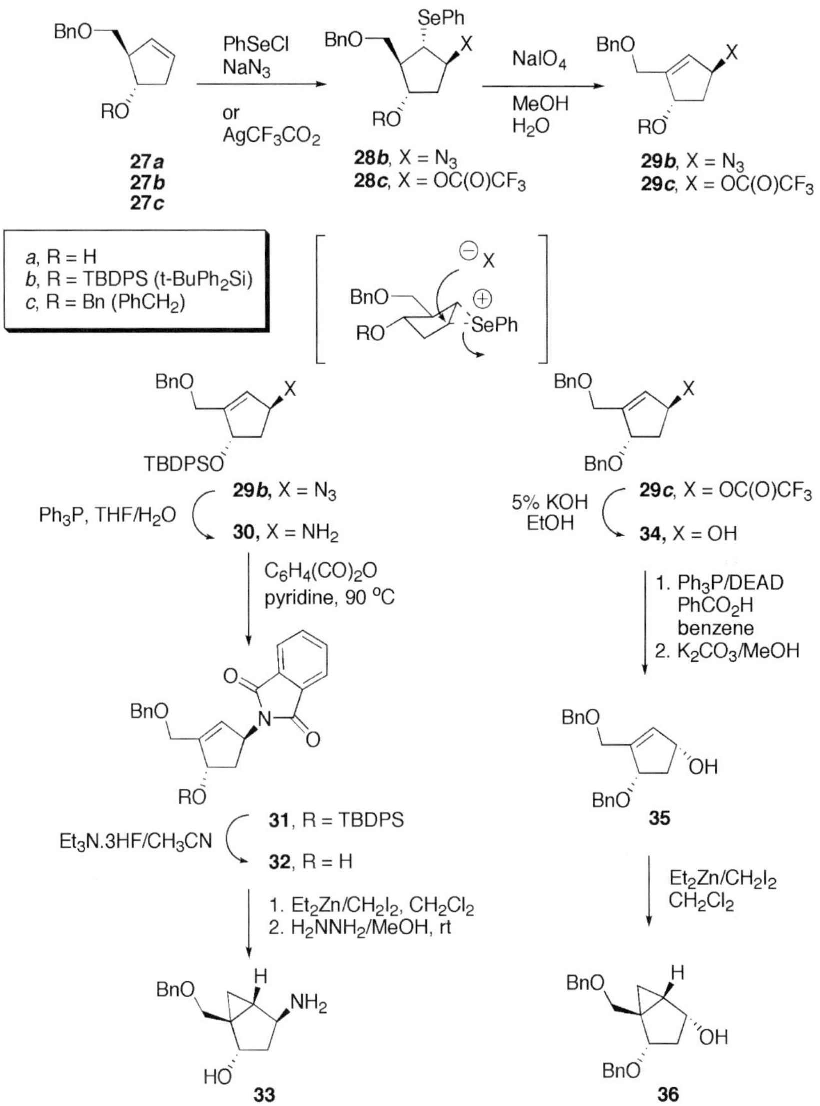

Scheme 12.5

phenylselenide group triggered the anticipated *syn* elimination to give almost exclusively the allylic products **29*b*** and **29*c***. Each of these synthons was elaborated accordingly to serve as nucleoside precursors via linear or convergent approaches, respectively. Compound **29*b*** was reduced to amine **30**, which required full protection as phthalimide **31**, to allow the hydroxyl group in **32** to direct the course of cyclopropanation in the ensuing Simmons–Smith reaction. In an identical system, an acyl-NH group took precedence over a hydroxyl group in directing the stereochemistry of the cyclopropanation [25]. Hydrazinolysis of the phthalimido group afforded amine **33**, which was linearly converted into *N*-MCT (**17**) in a similar fashion as depicted in Scheme 12.2.

For the convergent approach, hydrolysis of the trifluoroacetate ester of **29*c*** gave allylic alcohol **34**, the stereochemistry of which was inverted following a Mitsunobu esterification reaction. Hydrolysis of the inverted benzoate ester provided a new allylic alcohol **35** with the correct stereochemistry to direct the following cyclopropanation reaction to give intermediate **36**. This compound was utilized in the same manner as the related intermediate **23** (Scheme 12.4) for the synthesis of purines or pyrimidines.

In an effort to develop a synthesis of 2′-deoxy-bicyclo[3.1.0]hexane nucleosides from cheap starting materials, the decision was taken to combine an intramolecular olefin–ketocarbene cycloaddition from a simple linear precursor with an early stage lipase-catalyzed resolution to obtain the correct enantiomer [26]. The intermediate for the olefin–ketocarbene cycloaddition was prepared from ethyl acetoacetate (**37**) and acrolein in three steps (Scheme 12.6). The dianion of **37** that was generated from LDA reacted with acrolein at −78 °C, and the resulting alcohol was immediately protected as a silyl ether to give (±)-**38**. After diazo transfer with *p*-toluensulfonyl azide, the diazo compound (±)-**39** was obtained quantitatively. Intramolecular cyclopropanation of (±)-**39**, which presumably proceeds via a copper–carbenoid intermediate under thermolysis, generated a chromatographically separable mixture of (±)-**40** and (±)-**41**. That the major product (±)-**41** had the desired relative stereochemistry was unambiguously confirmed after converting it to a compound identical to one obtained from enantiomerically pure **27*a*** (Scheme 12.5). The preferred formation

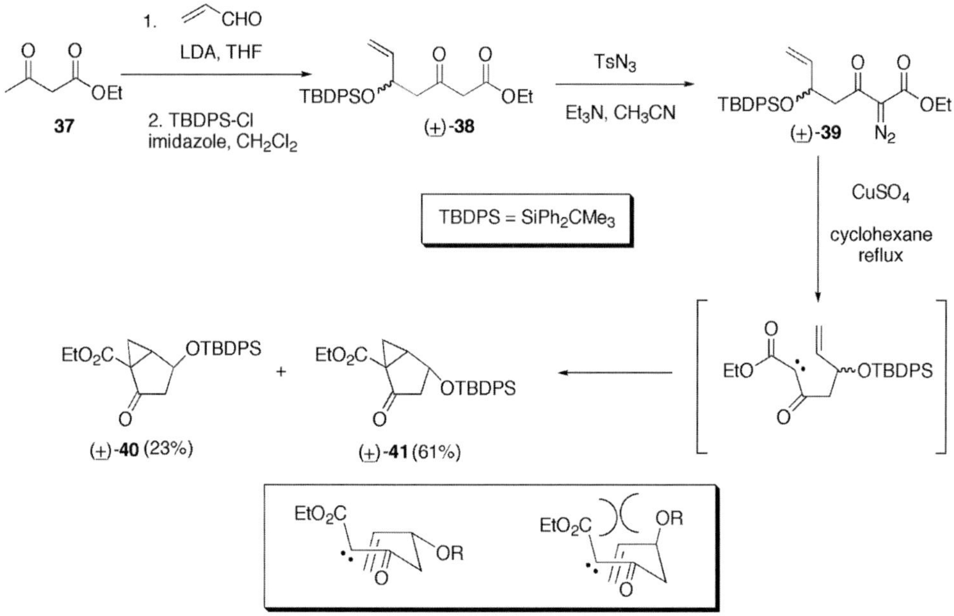

Scheme 12.6

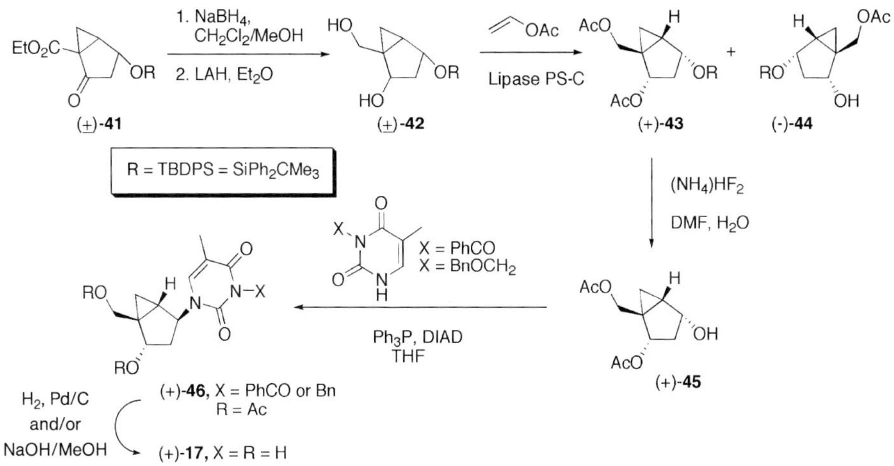

Scheme 12.7

of (±)-**41** can be envisaged by assuming that the transition state adopts a product-like, pseudoboat conformation where formation of the *trans*-intermediate is less sterically encumbered, as shown in the inset of Scheme 12.6.

The stepwise reduction of (±)-**41**, first with $NaBH_4$ to ensure hydride attack of the ketone carbonyl from the less-hindered convex face, followed by $LiAlH_4$ reduction of the ester to the primary alcohol, afforded (±)-**42** (Scheme 12.7). Then, commercially available lipase PS-C, in the presence of a large excess of vinyl acetate, was able to discriminate between the two enantiomers giving (+)-**43** as the diacetate and (−)-**44** as the monoacetate. Chiral HPLC showed that (−)-**44** was 96% *e.e.*, while (+)-**43** was contaminated with a byproduct which was removed by conventional chromatography after cleavage of the silyl group with $(NH_4)HF_2$ to give (+)-**45** with an estimated optical purity of 99% *e.e.* The source of the impurity was determined to arise from the reaction between the monoacetate (−)-**44** and excess acetaldehyde generated during the reaction. The identity of (+)-**45** as the enantiomer with the desired D-like configuration was confirmed by matching optical rotation values obtained with similar compounds derived from chiral intermediates **23** and **36** (Schemes 12.3 and 12.5). In the same manner as these intermediates, (+)-**45** was used in the syntheses of nucleoside analogues via convergent approaches, as exemplified by the synthesis of *N*-MCT (**17**; Scheme 12.7). Chiral resolution of adenine nucleosides obtained in a convergent manner via Mitsunobu coupling from the dibenzoate ester of racemate (±)-**42** was also achieved via the interaction with ADA after the removal of the silyl ether [27].

In a similar manner to that shown in Scheme 12.6, more elaborate bicyclo[3.1.0] hexane structures, such as racemic **51** and **52** (the β-D-enantiomer is shown for illustration), were obtained by functionalizing the terminal alkene by cross metathesis prior to the carbene-mediated intramolecular cyclopropanation [28] (Scheme 12.8). A modified approach to that described in Scheme 12.7 was employed

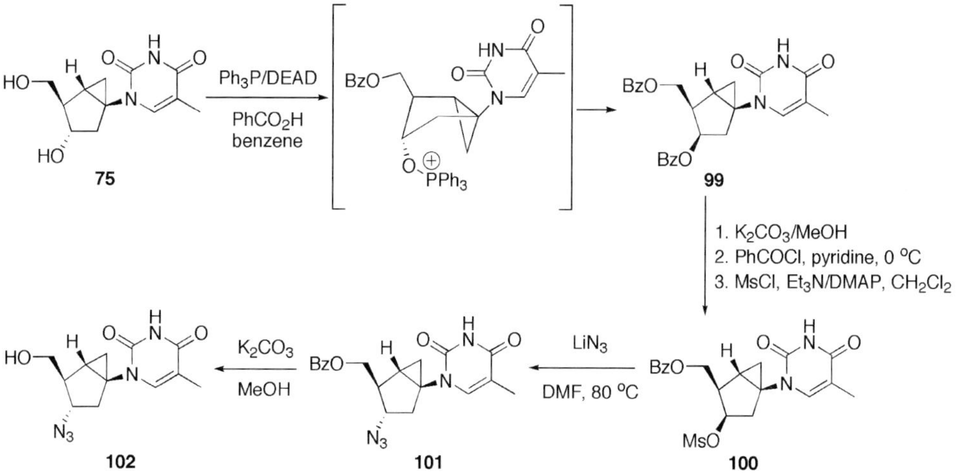

Scheme 12.14

the *N* template favors anhydride formation since for the *S* conformer, the C2 carbonyl oxygen is unable to reach the target electrophilic carbon.

12.3.4
Synthesis of Bicyclo[3.1.0]hexene Nucleosides

To impart a degree of planarity on the puckered bicyclo[3.1.0]hexane system, and to reduce ν_{max} in the pseudorotational cycle to values approximating those of important anti-HIV active nucleosides such as stavudine (D4T) and carbovir, the transformation of the bicyclo[3.1.0]hexane system into a bicyclo[3.1.0]hexene system was implemented [38, 39].

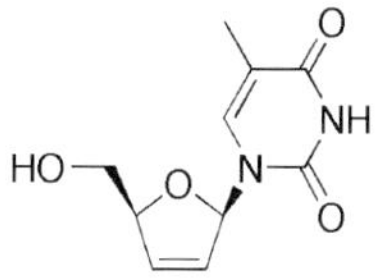

Stavudine (D4T)

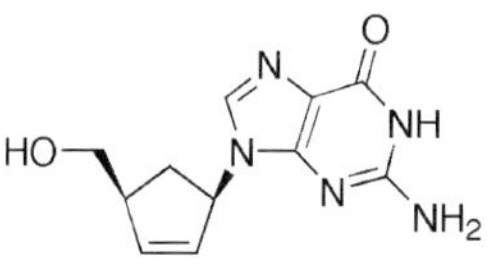

Carbovir

The chemistry to achieve these targets was also affected by the nature of the bicyclo [3.1.0]hexane template. Starting with compound **36** (see Scheme 12.5), compound **103** was prepared using a conventional deprotection/reprotection strategy (Scheme 12.15). This was followed by conversion of **103** to the mesylate ester **104**. Because the rigid bicyclo[3.1.0]hexane system fixes the position of the leaving OMs group and the two possible hydrogens available for elimination in an unfavorable gauche disposition, the attempted base-catalyzed elimination failed. With the intent to reorient the leaving group into a favorable antiperiplanar orientation for elimination, the synthesis of the iodo analogue was attempted because it is known

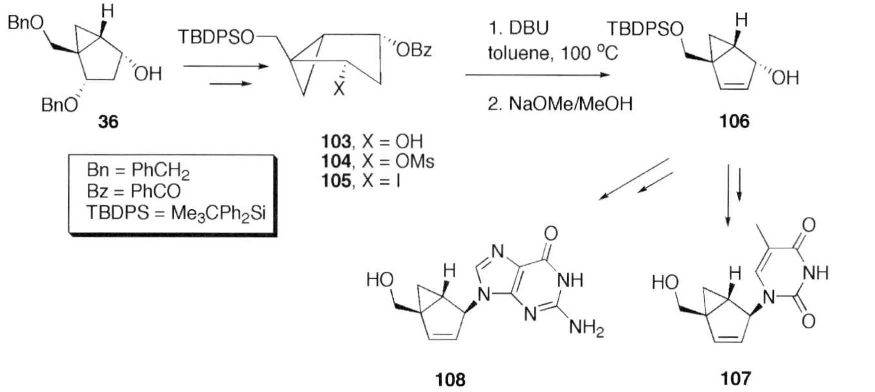

Scheme 12.15

that the iodination reaction (PPh_3, I_2, toluene, Δ) proceeds with inversion of configuration. Here again, the rigid bicyclo[3.1.0]hexane unexpectedly controlled the outcome of the reaction, giving instead **105** with retention of configuration. Because the bulky iodine in the inverted, axial configuration would not be stable, a second attack by iodide favored the formation of **105** having an equatorially disposed iodine resulting from a double inversion. Despite the fact that the geometry for the elimination was still not optimal, treatment of the compound with 1,9-diazabicyclo [5.4.0]undec-7-ene (DBU) at high temperature provided the desired alkene. After hydrolysis of the benzoate ester, the key bicyclic allylic alcohol **106** was obtained. This compound reacted with protected thymine, under similar Mitsunobu conditions as shown in Scheme 12.7, to give the D4Tanalogue **107**. The carbovir analogue (**108**) was also obtained by direct Mitsunobu coupling of **106** with *N*-(6-chloropurin-2-yl) acetamide following removal of the silyl ether and base-catalyzed conversion of the heterocyclic base to guanine.

Starting with compound **34** (Scheme 12.5), and using the same cyclopropanation strategy for the conversion of **35** to **36**, afforded compound **109** [40] (Scheme 12.16). In a similar fashion as described in Scheme 12.15, **109** was transformed into **110** which, under equivalent conditions (PPh_3, I_2, toluene, Δ), produced the iodo analogue **111** with inversion of configuration! In this case, an axially disposed group

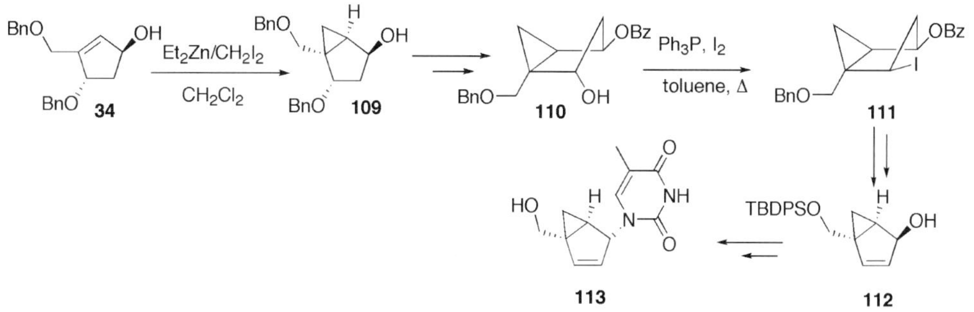

Scheme 12.16

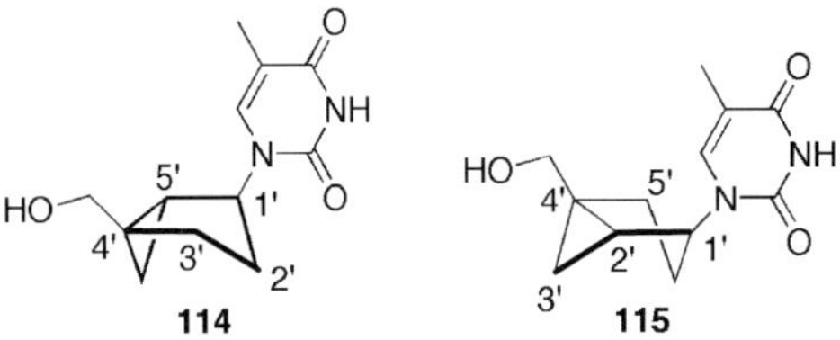

114

North ($_{2'}$E)-anti
D-isomer
[α] = +66.6°

115

South ($_{3'}$E)-anti
L-isomer
[α] = −67.0°

Scheme 12.17

had been converted into an equatorial iodo compound (**111**) that was stable. Treatment of **111** with DBU at high temperature provided alkene **112** which, under the same coupling conditions, afforded the enantiomer of the D4T analogue **107** (**113**).

Catalytic hydrogenation of **107** and **113** gave, respectively, two chiral deoxythymidine analogues (**114** and **115**) that represent compounds appearing to be locked at opposite extremes of the pseudorotational cycle [40] (Scheme 12.17).

12.3.5
Reshuffling of Groups on a Bicyclo[3.1.0]hexane Template

The bicyclo[3.1.0]hexane template can be considered to be a privileged platform upon which to generate diversity by relocating the position of the nucleobase, the hydroxyl groups, and the site of fusion of the cyclopropane ring. As an example of this strategy, the target compound **125** was designed by a two-step reshuffling depicted in Figure 12.4 [41, 42]. The reason for using this approach in drug discovery will be discussed in Section 4.1. The synthesis of **125** was achieved from plentiful D-ribose, which was converted into cyclopentenone **116** (Scheme 12.18) in six steps [42].

Using a modality of the Baylis–Hillman reaction, the treatment of cyclopentenone **116** with imidazole generated *in situ* a stabilized nucleophilic anion that reacted with formaldehyde. After elimination of the amine catalyst, the effective addition of a

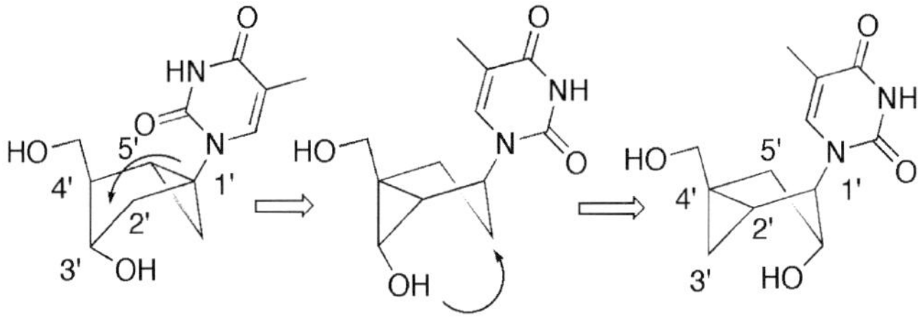

75 **124** **125**

Figure 12.4 Two-step structural "reshuffling" of bicyclo[3.1.0]hexane nucleoside **75** into **125**. The first step allows rotation of the C–N bond to the *anti* orientation, and the second step repositions the critical OH group to the opposite tip of the ring.

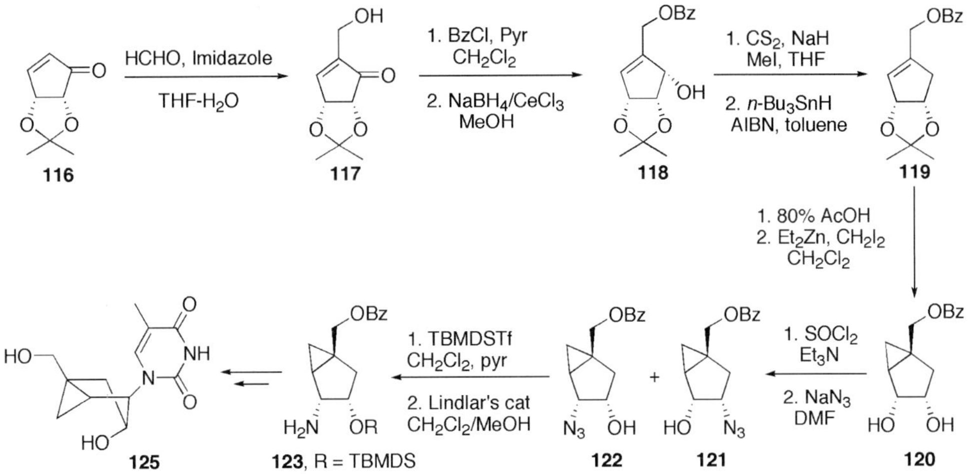

Scheme 12.18

hydroxymethyl group resulted in the formation of cyclopentenone **117**. Protection of the primary alcohol as a benzoyl ester proceeded smoothly, and stereospecific hydride reduction of the carbonyl group afforded exclusively the allylic alcohol **118**. Unfortunately, radical deoxygenation of **118** through the intermediate xanthate proceeded with racemization due to the energetic equivalence of the two possible radical intermediates. The synthesis from racemate **119** was continued after cleavage of the ketal ring and reaction of the corresponding diol under modified Simmons–Smith cyclopropanation conditions gave the critical bicylo[3.1.0]hexane skeleton **120** (relative stereochemistry shown) with the desired diastereoselectivity. Reaction of **120** with thionyl chloride gave almost quantitatively the corresponding cyclic sulfite that underwent nucleophilic attack with sodium azide to afford an easily separable 5 : 1 mixture of regioisomers in favor of the desired azido alcohol **122**. After protection of the free hydroxyl group as a silyl ether, reduction of the azide provided the requisite carbocyclic amine **123** in quantitative yield. Subsequently, the thymine ring was readily constructed from amine **123** following conventional, published methods. An X-ray structure of racemic **125** confirmed that all the desired stereochemical elements were in place: (i) the *anti* disposition of the thymine ring; and (ii) the location of the OH group at the other extreme of the concave face of the bicyclo [3.1.0]hexane template [42].

12.3.6
Bicyclo[3.1.0]hexane Pseudosugars as Surrogates of Abasic Nucleosides

Because of the important biological role of abasic sites in DNA, syntheses of conformationally locked abasic pseudosugars for incorporation into short oligodeoxynucleotides were implemented. These syntheses followed similar approaches to those described previously, and are shown succinctly in Scheme 12.19 [43, 44].

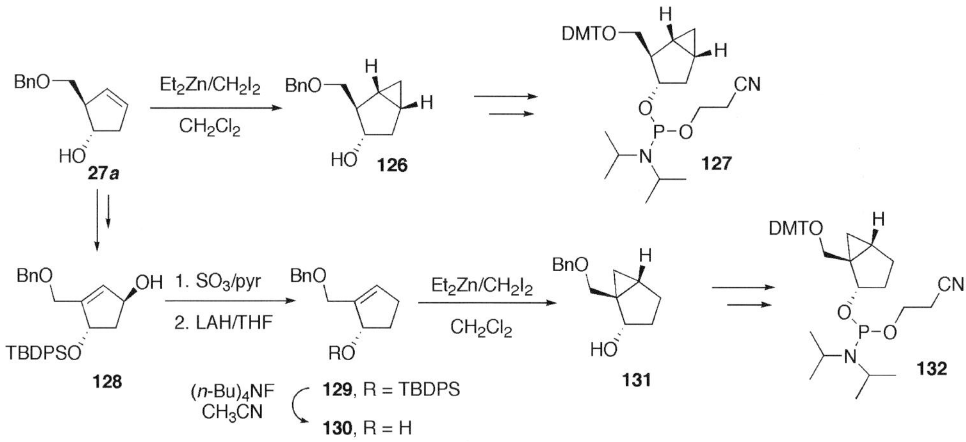

Scheme 12.19

From a common precursor (compound **27*a***; Scheme 12.5), homoallylic alcohol-directed cyclopropanation gave the desired *S* abasic template **126** that was readily converted to the 5′-*O*-DMT-phosphoramidite **127**. The conversion of **27*a*** to **128** was achieved using a similar approach to that described in Scheme 12.5. Deoxygenation of the allylic alcohol following reaction with a sulfur trioxide–pyridine complex and lithium aluminum hydride gave compound **129**. Freeing the allylic hydroxyl group after removal of the silyl ether and controlled cyclopropanation of **130** under Simmons–Smith conditions gave the *N* abasic template **131** that was also elaborated into the 5′-*O*-DMT-phosphoramidite **132**.

12.4 Synthesis of Oligodeoxynucleotides (ODNs) Containing Locked Nucleosides

The base sequence, the sugar pucker, and other base pair features control the degree of flexibility of the sugar–phosphate backbone in free DNA. When proteins bind to DNA they often distort its conformation, although the barrier to local deformability is more than compensated by the energy of the resulting protein–DNA complex [45]. Many of the structural adjustments made by the sugar–phosphate backbone in these complexes can be associated with a rearrangement of the furanose sugar pucker observed in the canonical A-type and B-type structures of DNA. In the A-form, sugar puckering is closely clustered around a 3′-*endo* (*N*) conformation, while the puckering distribution in the standard B-form is more diffuse around the C2′-*endo* (*S*) conformation. Because the locking of the conformation can be achieved with a rigid bicyclo [3.1.0]hexane pseudosugar, the embedded cyclopentane ring can be secured into either a 2′-*exo* ($_{2'}E$, *N*) or 3′-*exo* ($_{3'}E$, *S*), as shown previously in Figure 12.1. The introduction of locked 2′-*exo* (*N*) units into B-type DNA is expected to produce an A-like microenvironment commonly associated with the induction of a bend. On the other hand, the introduction of locked 3′-*exo* (*S*) units should resist bending. Both antipodal effects on conformation should restrict the degree of flexibility of the sugar–phosphate backbone, and are expected to influence the biophysical and biological properties of the duplex [46].

Normally, one would expect the synthesis of these modified ODNs to proceed as efficiently as with conventional ODNs. Unfortunately, during early attempts to make modified ODNs containing locked, 2′-*exo* (*N*) bicyclo[3.1.0]hexane units some serious difficulties were reported with the phosphoramidite approach [17]. These difficulties were not encountered with the hydrogen phosphonate protocol during the synthesis of a phosphorothioate 15-mer ODN containing 10 modified 2′-*exo* *N*-thymidine units [20]. Surprisingly, the same phosphoramidite chemistry when used with locked, 3′-*exo* (*S*) bicyclo[3.1.0]hexane units presented no such problems [34]. This contrasting chemical behavior between locked 2′-*exo* and 3′-*exo* units was later found to be related to a concept already discussed in Sections 12.3.3 and 12.3.4, whereby the rigid bicyclo[3.1.0]hexane scaffold controls the outcome of sharply different pathways [47]. During iodine oxidation of the internucleotide phosphite linkage to the corresponding phosphate triester, a strand cleavage occurred involving 2′-*exo* (*N*) nucleosides, which generated 5′-phosphate-containing oligonucleotide fragments on the resin. Under the reaction conditions, the ejection of the excellent leaving group formed during the reaction occurred either by direct nucleophilic attack (Scheme 12.20-c) or by an intramolecular Mitsunobu-type reaction leading to an anhydronucleoside intermediate (Scheme 12.20-b). These two reactions competed favorably against the expected oxidation path (Scheme 12.20-a). This problem was avoided by replacing iodine with *t*-butylhydroperoxide as oxidant, and with this small change several ODNs were successfully synthesized [47] (Table 12.1).

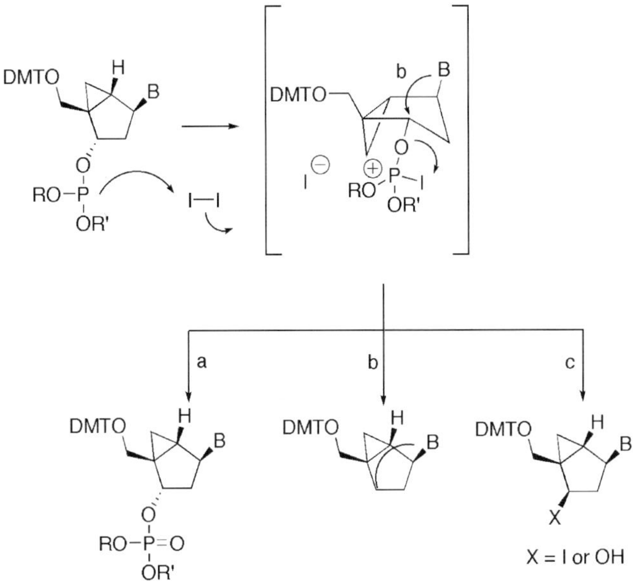

Scheme 12.20

Table 12.1 Sequence and modification of oligonucleotides containing locked carbocyclic pseudosugars.

Entry	Sequence (5′ → 3′)	Modification (X)	MW_{calc}	MW_{found}
ODN-1	CGCG**X**ATTCGCG	North methanocarba dA	3656.5	3656
ODN-2	CGCG**X**ATTCGCG	South methanocarba dA	3656.5	3658
ODN-3	CGCG**XX**TTCGCG	South methanocarba dA	3666.5	3668
ODN-4	CGCGAA**X**TCGCG	North methanocarba T	3656.5	3655.8
ODN-5	CGCGAAT**X**CGCG	North methanocarba T	3656.5	3655.7
ODN-6	CGCGAA**XX**CGCG	North methanocarba T	3666.5	3666
ODN-7	CGCGAA**X**TCGCG	South methanocarba T	3656.5	3658
ODN-8	CGCGAAT**X**CGCG	South methanocarba T	3656.5	3658
ODN-9	CGCGAA**XX**CGCG	South methanocarba T	3666.5	3668

12.4.1
The Dickerson–Drew (DD) Dodecamer

In order to gain a better understanding of how structural variations in the sugar moiety can contribute to both the local and global structural fold of ODNs, both *S*- and *N*-MCT analogues were inserted into a structurally well-characterized self-complementary B-DNA sequence (**ODNs 4–9**; Table 12.1). Since the furanose rings of the native DD dodecamer (CGCGAATTCGCG) are puckered almost exclusively in the standard B-type (*S*) conformation, it was anticipated that these substitutions would either stabilize or disrupt the local/global structure accordingly. Results from circular dichroism (CD), nuclear magnetic resonance (NMR) and differential scanning calorimetry (DSC) measurements have provided a descriptive picture of the effect of either modified nucleoside on duplex formation and stability [46]. The incorporation of *N*-locked analogues only minimally perturbed the DD system with regard to duplex formation and thermodynamics. However, it was shown that *N*-modified ODNs bend the central region more than the native DD [48–50]. Hence it may be surmised that, along with their ability to form stable helices, these analogues are "preorganized" to adjust the local structural environment (bend) in response to conformational restriction in the 2′-*exo* puckering mode.

Replacing the middle thymidine nucleotides with *S*-locked analogues led to a more complex equilibrium of species rather than the idealized assumption of a stable all-B-DNA duplex in a manner that was predicted *a priori* [46]. Interpreting the data regarding the effects of incorporating the *S*-locked analogues is more difficult, especially in light of the concentration dependence of the aforementioned equilibrium. However, the major species formed was indeed a duplex with a slightly higher thermodynamic stability compared to the native DD. NMR analysis showed that minor multiple species are present, but they are limited to solutions that are of low micromolar strand concentrations at salt concentrations of 100 μM. At higher salt concentrations, the lower temperature melting species seem to be replaced solely by the stable duplex. It may be surmised that the *S*-modified nucleotides actually do stabilize a B-like helix, and that the more hydrated environment associated with this type of helix requires a larger enthalpic change upon melting. Although the sequence of the DD is self-complimentary and, under certain conditions, has a tendency to form hairpin structures, there was no indication of these folds in any of the ODN species studied with the *S*-modified nucleotides [46].

12.5 Molecular Targets, Ligand Properties, and Binding Modes

Based on the principle that rigidity increases binding affinity because of lower entropy losses upon binding to the target proteins, many of these bicyclo[3.1.0] hexane nucleosides have been used to probe the conformational preferences of enzymes involved in the recognition of nucleosides, nucleotides, and nucleic acids. In addition to kinases and polymerases, there are other molecular targets for nucleosides and nucleoside analogues that bind to unphosphorylated forms, such as adenosine and cytidine deaminases, *S*-adenosylhomocysteine hydrolase, and families of adenosine receptors. Still, others bind to intermediate phosphorylated metabolites, principally at the monophosphate level, such as dCMP deaminase and thymidylate synthase (TS), and also at the diphosphate level, such as ribonucleotide reductase. Efficient binding to these molecular targets is also dependent on the nature of the nucleobase and the conformation of the phosphorylated or unphosphorylated sugar moiety.

12.5.1 Kinases and Polymerases

In the case of 2′-deoxyribonucleoside analogues, where the desired biological response is principally derived from the incorporation of the drug into DNA, the final outcome is determined by the ability of the molecule to interact effectively with several different classes of enzymes: (i) the three activating kinases; and (ii) the final target DNA polymerase(s). Therefore, structure–activity relationship (SAR) studies in this context are complicated by the intricacy of this process in which each of the steps represent a different drug–receptor interaction.

Because the sugar moiety behaves basically as a scaffold that supports the nucleobase, the dynamic changes in sugar conformation that take place simultaneously with correlated rotational movements about the torsion angles χ and γ [51] make the system behave like two "dancing partners" reacting to each other's motions in a predictable fashion. It is also expected that the architecture of the binding pocket will impose a specific conformational demand on the sugar ring (*N* or *S*) and the nucleobase (*anti* or *syn*) for optimal fit that should result in measurable differences in terms of the energy of binding and/or catalytic activity. Therefore, locked nucleoside analogues where these conformational freedoms are blocked represent excellent tools to study nucleoside and nucleotide-binding enzymes.

Table 12.2 Levels of *N*-MCT (**17**) and *S*-MCT (**75**) phosphates (pmoles per 10^6 cells) in uninfected and HSV-1-infected Vero cells. Cells were infected 2 h before treatment and the metabolites measured after 6 h.

	N-MCT		*S*-MCT	
Metabolite	**Uninfected cells**	**HSV-1-infected**	**Uninfected cells**	**HSV-1-infected**
MP	3.212 ± 0.152	76 ± 0.9	0.031 ± 0.034	72 ± 5.4
DP	0.032 ± 0.024	82 ± 6.3	0.021 ± 0.021	378 ± 29
TP	0.019 ± 0.012	197 ± 14.6	0.028 ± 0.011	490 ± 67

MP = monophosphate; DP = diphosphate; TP = triphosphate.

Because there is ample information about the crystal structure of HSV-tk, this enzyme was considered ideal to investigate the issue of conformation versus substrate specificity. Use of *N*-MCT (**17**) and *S*-MCT (**75**) to probe the enzymatic phosphorylations catalyzed by HSV-1 tk, demonstrated that – save for the first phosphorylation step – *S*-MCT was the better substrate [52] (Table 12.2). Indeed, for the first phosphorylation, the amount of *S*-MCT-MP produced was in fact comparable to that of *N*-MCT-MP, without any clear distinction by the enzyme between the *S* and *N* antipodes. It was hypothesized that the expected *S* conformational penchant of HSV-tk was being affected by other factors that countered this preference, such as a high *syn* ⇆ *anti* energy barrier that restricted free rotation of the C—N bond in *S*-MCT and forced the thymine ring to remain in the *syn* range, as shown both by crystallography and nuclear Overhauser (nOe) studies of **75** in solution [52]. Because crystal structures of *N*-MCT (**17**) and *S*-MCT (**75**) bound to HSV-tk show the base in the *anti* conformation [53, 54], *S*-MCT pays an entropic penalty incurred in the flipping of the thymine ring in **75** from *syn* to *anti* when the compound binds to HSV1-tk. Such an entropy penalty cancels the otherwise *S* preference of HSV1-tk, and the net phosphorylation for either compound (**17** and **75**) is virtually the same. This penalty is not observed for the subsequent phosphorylation steps because the presence of the 5′-*O*-monophosphate in *S*-MCT flips the ring to the *anti* conformation [52].

In seeking to alleviate this problem, it was decided to reposition the fused cyclopropane ring in **75** to the other end of the molecule, thereby generating a new compound (**124**) that still maintained an *S*-like conformation but allowed the thymine ring to sample the *anti* range, as shown earlier in Figure 12.4 [41, 42]. Because compound **124** was prone to undergo ring opening at room temperature via a retro-aldol reaction, the critical 3′-OH (nucleoside numbering) was relocated to the opposite end of the pseudoboat ring, where it could engage in H-bonding with the receptor in a manner akin to that of **75**. The synthesis of this rearranged compound was shown in Scheme 12.18. The course of phosphorylation of **125** relative to *S*-MCT (**75**) in HSV-1 infected Vero cells confirmed the success of this reshuffling strategy for all three important phosphorylation steps (Table 12.3) [42]. Because the first two key phosphorylations are catalyzed by HSV1-tk, the results obtained confirmed that this

Table 12.3 Phosphorylation levels of *S*-MCT (**75**) and (±)-**125** in HSV-1-infected Vero cells (pmoles per 10^6 cells). Cells were infected 2 h before treatment and the metabolites measured after 6 h.

	S-MCT		(±)-125	
Metabolite	**Uninfected cells**	**HSV-1-infected**	**Uninfected cells**	**HSV-1-infected**
MP	0.15 ± 0.06	54 ± 14	0.28 ± 0.11	252 ± 32
DP	0.30 ± 0.07	74 ± 17	0.08 ± 0.04	187 ± 15
TP	1.61 ± 0.24	153 ± 10	0.08 ± 0.63	342 ± 26

MP = monophosphate; DP = diphosphate; TP = triphosphate.

new compound had all the structural attributes for an effective and improved recognition by the viral enzyme. Also, because 5′-*O*-triphosphate levels were more elevated for the new compound, it was safe to infer that its molecular architecture was also favorably recognized by the cellular dinucleotide kinase. It is important to note that the levels of phosphorylated metabolites in Tables 12.2 and 12.3 for *S*-MCT cannot be compared because they represent different experiments, the outcomes of which depend on the level of infectivity of the virus. A similar experiment with a simpler version of compound **125** (compound **115**; Scheme 12.17), lacking the 3′-OH entirely, showed a more efficient monophosphorylation step than that observed for **75**; however, phosphorylation to the diphosphate level, also catalyzed by HSV1-tk, did not proceed at all because of the critical role of the OH group during the second phosphorylation [40]. These experiments demonstrated, conclusively, that HSV1-tk prefers to bind thymidine substrates in the *S-anti* conformation for two consecutive phosphorylation steps, and that the second phosphorylation step absolutely requires the presence of a 3′-OH or its equivalent.

Docking experiments showed that only one enantiomer of (±)-**125** closely matched the crystal structure of *S*-MCT (**75**) at the active site of HSV1-tk by forming a similar network of hydrogen bonds [42]. Despite the efficient formation of 5′-*O*-triphosphates from *S*-MCT and its rearranged isomer, both compounds were devoid of antiviral activity against HSV-1. This suggested that their inactivity was caused by failure of the cellular or viral DNA polymerases to incorporate the *S* triphosphates.

The examination of DNA extracts from tumor cells [MC38 (wild-type) and MC38/*HSVtk*] after a 24-h exposure to radiolabeled [methyl-^{3}H]-*N*-MCT and [methyl-^{3}H]-*S*-MCT showed that wild-type MC38 cells incorporated negligible amounts of both *N*- or *S*-MCT into DNA [52]. On the other hand, the *HSV-tk*-transduced cells incorporated significant quantities of *N*-MCT at a level even higher than for ganciclovir (GCV). The negligible amounts of *S*-MCT incorporated implies that viral or host DNA polymerase(s) may prefer to incorporate nucleotide triphosphate substrates with an *N* sugar pucker, effectively discriminating against the antipodal *S* conformer despite the presence of higher levels of the *S* triphosphate metabolite in the cell. These results are also consistent with our previous observation that *N*-methanocarba-AZT-triphosphate was more effective in inhibiting HIV RT than *S*-methanocarba-AZT-triphosphate [37]. Nucleosides are inherently flexible molecules and can, in principle, accommodate to the demands of both kinases and polymerases. On the other hand, the two probes, *S*-MCT (**75**) and *N*-MCT (**17**) each have been optimized to bind to either kinases or polymerases, respectively. The difference in biological activity stems from the fact that kinases appear to be more tolerant of *N*-MCT, allowing the formation of enough 5′-*O*-triphosphate to reach the ultimate polymerase target. These results also indicate that the polymerases are more stringent discriminators, and select almost exclusively in favor of the locked *N* triphosphate. As a result, *N*-MCT is a very effective antiviral against HSV1 and HSV2 infections, against Kaposi sarcoma-associated herpes virus (KSHV), and against orthopox virus infections [20, 55].

The contrasting conformational preference between kinases and polymerases is very intriguing, but can be explained on the basis of structural arguments. During phosphorylation, the 5′-position bearing the OH, or phosphate group, needs to be accessible. The orientation of the γ torsion angle with an *S* sugar pucker has more conformational freedom and the rotamers can sample the –*sc* and *ap* range. This is in sharp contrast to an *N* pucker, where the favored +*sc* conformation places the 5′-OH into a less-accessible disposition over the ribose. In addition, there is less interference between the 5′-OH and 3′-OH in the *S* conformation because these groups are approximately 7 Å apart, in contrast to a distance of 5.9 Å in the *N* conformation. These differences apparently do not result in an absolute discrimination between *N*-MCT and *S*-MCT as substrates for HSV-tk, since reasonable amounts of *N*-MCT-TP are still formed. In contrast, the polymerases appear to be very discriminating. Current models based on crystallographic analysis suggest that the intrinsic fidelity of a polymerase depends on its ability to constrain the nascent template–primer duplex in the A-form (*N*-sugars) providing a structural buffer to conformational variability that may contribute to polymerase fidelity by minimizing mismatches [56–58]. These observations help to explain why *N*-MCT and similar 5-substituted derivatives are the biologically active conformers [59].

From the above discussion, it appears that the decisive step to determine antiviral activity is the incorporation of *N*-MCT-TP by either the viral or the host DNA polymerase; however, the exact mechanism of action is not simple. *N*-MCT has an OH group on the pseudosugar ring in a position equivalent to the 3′-OH of a normal 2′-deoxyribose ring, which should allow DNA synthesis to continue.

Studies on the interaction of *N*-MCT-TP with HIV-1 RT showed that the compound operates as a delayed DNA chain terminator [60]. The ability of *N*-MCT-TP to be incorporated by HIV-1 RT and inhibit DNA synthesis was examined with a 5′-labeled 18-mer DNA primer annealed to a 43-mer template in the presence of 10 μM each dCTP, dGTP and dATP and 10 μM of (1) TTP, (2) dTTP, or (3) *N*-MCT-TP (Figure 12.5). When *N*-MCT-TP was the first nucleotide added to the primer strand, DNA synthesis was not blocked immediately; rather, it occurred after two nucleotides beyond the site where *N*-MCT-TP was incorporated. Additional experiments showed that the sequence of the template (RNA for example) and the number of consecutive *N*-MCT-TP units incorporated have a major role in determining where DNA synthesis will be blocked; in some cases, blocks were seen at positions corresponding

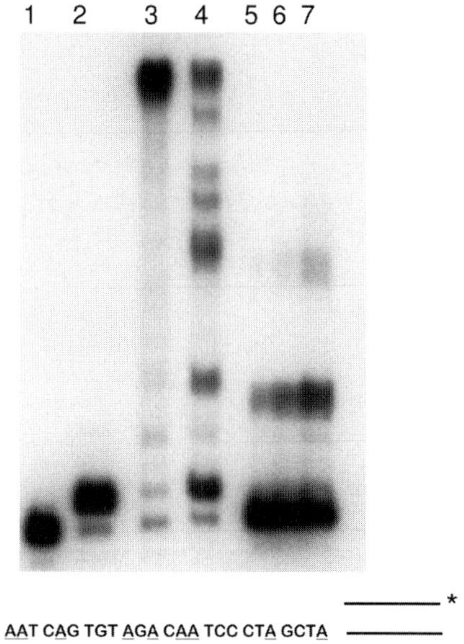

Figure 12.5 Inhibition of DNA synthesis. The template is shown with the sites of incorporation underscored and the primer was 5′-end-labeled (*). Lane 1 = reference lane, no RT; lane 2 = dTTP (termination of DNA synthesis); lane 3 = TTP (completion of DNA synthesis); lane 4 = 1 : 1 mixture of dTTP and TTP revealing the positions at which T analogues can be incorporated; lanes 5–7 = delayed chain termination by *N*-MCT-TP).

to five nucleotides beyond the site of incorporation of the first *N*-MCT-TP, or even at 17 nucleotides beyond the site of incorporation. A likely mechanism of the delayed chain termination that occurs after two to three nucleotides beyond the polymerase active site is that the added *N*-MCTanalogue makes an unfavorable steric contact with the amino acids in the thumb subdomain, thus blocking continued DNA synthesis. The delayed chain termination five nucleotides beyond might occur as the nucleic acid makes the transition from A form to B form DNA, and the 17 nucleotide block could involve contacts with the RNase H domain.

12.5.2 HIV Reverse Transcriptase

As discussed earlier, *S*-MCT (**75**) and *N*-MCT (**17**) were optimized to bind either the HSV-tk kinases or the viral/cellular polymerases, respectively. Despite the fact that *N*-MCT-TP was an effective inhibitor of HIV RT, the compound was inactive against HIV because phosphorylation by cellular kinases was ineffective. In order to determine the effect of *N*-MCT on viral replication, a human osteosarcoma (HOS) cell expressing HSV-tk was developed, after which the compound was able to block HIV-1 replication with an IC_{50} of 0.15 μM [60]. Decreased replication of HIV-1 was detected only at a 10-fold higher concentration of *S*-MCT. The fact that *N*-MCT-TP functions as a delayed chain terminator allows the drug to bypass the excision repair mechanism known to cause resistance to several antiretroviral agents, including AZT. When tested against three HIV RT variants – all of which are able to excise AZT-MP efficiently – *N*-MCT-MP was not efficiently excised, which suggested that the analogue is far enough from the 3′-end of the primer to make excision by the mutant RT inefficient. These data were also supported by a similar cell-based assay when HOS cells infected with excision-proficient RTs expressing the *HSV-tk* gene showed little or no resistance to *N*-MCT.

Although *N*-MCT is definitely a compound with great promise as a specific antiviral agent, the lack of recognition by cellular kinases makes it unsuitable to treat viral infections that depend on the latter enzymes for activation. Among the nucleoside analogues with significant clinical utility for the treatment of AIDS is staduvine (D4T) (see Section 3.4). One important structural characteristic of D4T is the presence of a double bond – a feature that renders the sugar ring near-planar and imparts a high degree of rigidity to the molecule. In order to impart a degree of planarity to the bicyclo[3.1.0]hexane template to a level comparable to that of D4T, a compound such as **107** (Scheme 12.15) was designed on a bicyclo[3.1.0]hexene template [38].

A comparison between the X-ray structure of *N*-MCD4T (**107**) and D4T revealed minor differences, and an RMS deviation of only 0.039 Å. In contrast, the pseudorotational parameters of these molecules are quite different, particularly with respect to the value of *P*. Although, both D4T and *N*-MCD4T are in the *N* hemisphere, they are 140° apart from each other, separated by an almost equal number of degrees (ca. 70°) from a perfect *N* ($P = 0°$) pucker toward the East and West, respectively. In terms of ring puckering, the ring of D4T is more planar ($\nu_{max} = 0.61°$), with a mean deviation from planarity of only 0.0025 Å. Relative to D4T, *N*-MCD4T is slightly more puckered ($\nu_{max} = 6.81°$) with a mean deviation from planarity of 0.025 Å. The reason for the structural similarity between D4T and *N*-MCD4T, despite their widely different *P*-values, is that as $\nu_{max} \rightarrow 0$ the relevance of *P* diminishes. With a ν_{max} radii of only 0.61 and 6.81, D4T and *N*-MCD4T are quite similar, making it likely that all the intervening enzymes would recognize them in a similar manner.

The relative anti-HIV activities of *N*-MCD4T and D4T in different cell types varied according to the specific cell type and assay conditions. In MT-2 cells the anti-HIV activity of *N*-MCD4T was only about threefold less than that of D4T, and in MT-4 it was about 10-fold less potent [38]. In order to identify the effect of ring pucker on the polymerase, the inhibition of HIV RT by D4T-TP and *N*-MCD4T-TP was determined. Under the same experimental conditions, the IC_{50} for D4T-TP was 0.08 μM, whereas that for *N*-MCD4T-TP was 0.65 μM, reflecting an approximately eight-fold difference in potency in favor of D4T-TP. This result appears to be consistent with all cell-based data, and the observed approximately 10-fold difference in ν_{max} between D4T and *N*-MCD4T may be responsible for the four- to 10-fold difference in potency seen in the various assays, including the polymerase assay with HIV RT.

The take-home lesson from this exercise is that, by flattening the ring, the new molecule is able to be recognized by both cellular kinases and the polymerase (HIV RT), indicating that a flat sugar moiety is a good compromise between the extreme *N* and *S* puckers that otherwise optimize the molecules toward one enzyme at the expense of the other.

12.5.3
DNA Methyltransferase

Several lines of evidence suggest that the control of gene expression in mammalian cells is related to the pattern of DNA methylation found in more than 70% of the CpG residues present in the genome. The methylation of DNA, which involves the covalent addition of a methyl group to the 5-position of the cytosine ring, is a post-replicative modification. The replication of a DNA that is methylated on both strands at CpG doublets produces two new hemimethylated helices. Subsequently, an enzyme with preferential affinity for hemimethylated sites (maintenance DNA methylase or DNMT1) preserves the methylation pattern with high fidelity from one cell generation to the next. Different protooncogenic pathways can up-regulate DNMT1 expression and, indeed, high levels of DNMT1 RNA have been observed in many cancer cells. The resulting hypermethylation at CpG islands in the promoter region of growth regulatory genes can silence their expression and provide a growth advantage to transformed cells. However, since methylation changes are reversible, the inhibition of aberrant DNMT1 activity could reactivate the expression of the growth regulatory gene, highlighting the importance of DNMT1 as a therapeutic target [61].

Initially, it was difficult to understand how DMNT1 and other DNA methylases acted on a target cytosine that is held at the solvent-accessible major groove surface by base pairing and stacking, and seemingly inaccessible to the concave active site pocket of the enzyme. The answer to this puzzle came with the resolution of the crystal structure of the bacterial methylase (M.*Hha*I) complexed with a synthetic DNA duplex chemically modified at position X with 5-fluorocytosine at the target position for methylation (Figure 12.6) [62].

In a process termed "base flipping", the enzyme simply rotates the target DNA base 180 ° (along an axis parallel to the DNA major axis) on its flanking phosphodiester bonds such that the base projects into the catalytic pocket. This strategy is used by other enzymes, such as those involved in DNA base excision repair, and helps explain the widespread use of this mechanism when an enzyme needs access to an individual base in double-stranded DNA or RNA substrates. Subsequent studies also demonstrated an enhanced binding affinity of M.*Hha*I to DNA when the target cytosine was replaced with an abasic furanose sugar at the target position [63, 64]. The new crystal structure revealed a common phosphodiester backbone conformation identical to

5'-C-C-A-T-G-M-G-C-T-G-A-C-A-3'

3'-G-G-T-A-C-G-X-G-A-C-T-G-T-5'

Figure 12.6 M. *Hha*I hemimethylated recognition sequence (box) showing the position of the target cytosine (X) or other modified nucleobases. The 5-methylcytosine in one strand directs methylation (arrow) to position X.

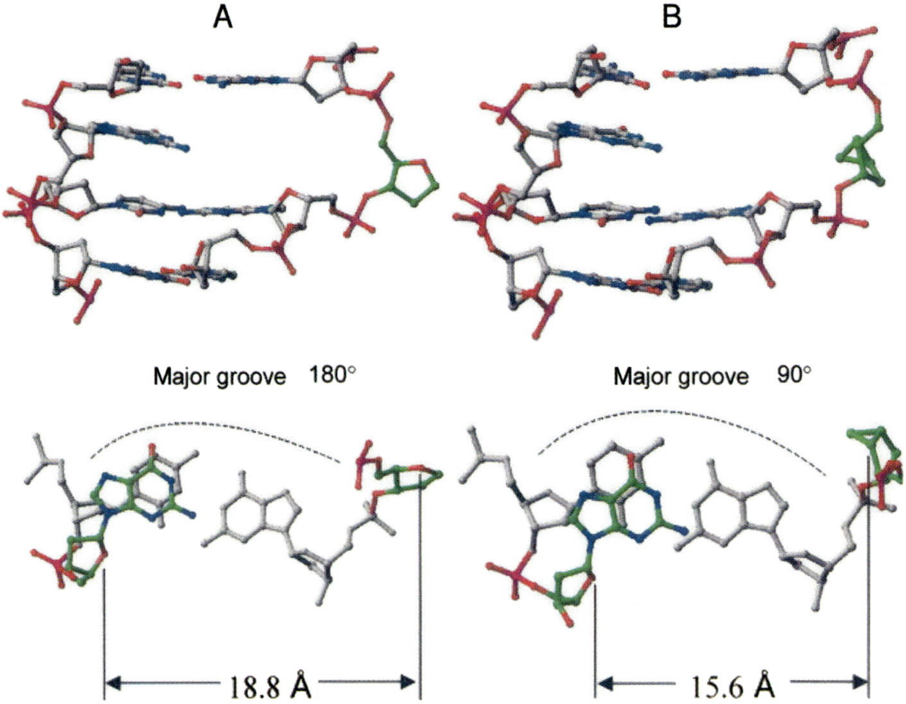

Figure 12.7 Close-up views of the 5′-GCGC-3′/5′-GXGC-3′ sequence (A: X = abasic furanose sugar; B: X = *S* abasic pseudosugar). The top panels show the view from the minor groove side, while the lower panels view along the helical axis. The O4*–O4* distances between Watson–Crick base-pair partners (A) and the corresponding O4*–C4* interstrand distance (B) are labeled.

that observed when the base was cytosine (or 5-fluorocytosine) and with the target, flipped-out sugar puckered in the *N* conformation (Figure 12.7A). When the abasic site in the same lower strand ODN (Figure 12.6) was replaced with *N* and *S* bicyclo [3.1.0]hexane pseudosugars, synthesized with phosphoramidites **127** and **132** (Scheme 12.19), there was an additional more than threefold enhancement in binding affinity when the abasic site was the constrained *S* conformer, but it was decreased by a similar amount when it was replaced with the antipodal *N* analogue [43, 44]. A number of crystal structures of M.*Hha*I/DNA complexes with either a flipped-out base (cytosine, adenine or uracil), or an abasic furanose sugar, revealed a common sugar–phosphate backbone conformation with the target sugar moiety in the *N* conformation [63, 64]. From these studies it was clear that the sugar–protein and phosphodiester backbone–protein interactions make significant contributions to rotation of the target base out of the DNA double helix by base-flipping enzymes. However, it seemed contradictory that the *N* abasic pseudosugar decreased the binding affinity when the fully flipped-out sugar appears always in the *N* conformation. This apparent discrepancy was resolved by examining the structure or the ternary complex of M.*Hha*I complexed with a 13-mer duplex where X was an abasic *S* bicyclo[3.1.0]hexane [65]. The *S* pseudosugar was trapped on the DNA major groove side by a 90° rotation corresponding to the mid-point along the flipping pathway, between the non-flipped native B-DNA (0° rotation) and the completely flipped-out state (180°) (Figure 12.7B). At the mid-point, the *S* pseudosugar appeared stabilized by an extensive network of interactions, primarily Van der Waals contacts, so that the complex does not move beyond this point as it would with a flexible furanose. The flexible furanose moiety disallows stabilization of the conformation at the mid-point and continues its trajectory to the fully flipped state with the change in sugar pucker observed in the crystal structures. In the case of the *N* pseudosugar, the conformational changes required for flipping are disallowed such that the same favorable interactions between the enzyme and DNA observed with the *S* pseudosugar never occur [65].

12.6
Concluding Remarks

The examples presented in this chapter describing the bicyclo[3.1.0]hexane strategy to lock the furanose pucker into the *N* and *S* conformational regions of the pseudorotational cycle provide compelling evidence for the argument that enzymes do indeed discriminate nucleoside and nucleotide substrates based on sugar pucker, and that the differences appear to be related to their inherent mechanisms of reaction. These differences extend beyond the single nucleoside(tide) units, and seem to apply to the recognition of oligomers that incorporate conformationally locked units. Albeit our efforts were primarily directed to the generation of biological probes, some of the compounds synthesized, such as *N*-MCT, are endowed with potent and selective activity against important viruses. Taken together, these results justify the efforts dedicated to the syntheses of these molecules, which additionally provide a treasure trove of interesting chemistry.

References

1 Altona, C., Sundaralingam, M., *J. Am. Chem. Soc.* **1972**, *94*, 8205–8212.
2 Saenger, W., *Principles in Nucleic Acid Structure*, Springer-Verlag: New York, Berlin, Heidelberg, **1984**.
3 Mastryukov, V. S., Osina, E. L., Vilkov, L. V., and Hilderbrandt, R. L., *J. Am. Chem. Soc.* **1977**, *99*, 6855–6861.
4 Mjöberg, P. J., Almlof, J., *Chem. Phys.* **1978**, *29*, 201–208.
5 Skancke, P. N., *Theochem.-J. Mol. Struct.* **1982**, *3*, 255–265.
6 Osawa, E., Szalontai, G., Tsurumoto, A., *J. Chem. Soc., Perkin Trans. II* **1983**, 1209–1216.
7 Siam, K., Ewbank, J. D., Schäfer, L., Van Alsenoy, C., *Theochem.-J. Mol. Struct.* **1987**, *35*, 121–128.
8 Okazaki, R., Niwa, J., Kato, S., *Bull. Chem. Soc. Jpn.* **1988**, *61*, 1619–1624.
9 Kang, P., Choo, J., Jeong, M., Kwon, Y., *J. Mol. Struct.* **2000**, *519*, 75–84.
10 Personal communication from Dr. Alexander D. MacKerell, Jr. Department of

Pharmaceutical Sciences, School of Pharmacy, University of Maryland, 20 Penn Street, Baltimore, MD 21201, USA.

11 Rodriguez, J. B., Marquez, V. E., Nicklaus, M. C., Barchi, J. J. Jr., *Tetrahedron Lett.* **1993**, *34*, 6233–6236.

12 Rodriguez, J. B., Marquez, V. E., Nicklaus, M. C., Mitsuya, H., Barchi, J. J. Jr., *J. Med. Chem.* **1994**, *37*, 3389–3399.

13 Ezzitouni, A., Barchi, J.J. Jr., Marquez, V. E., *J. Chem. Soc., Chem Commun.* **1995**, 1345–1346.

14 Yaginuma, S., Muto, N., Tsujino, M., Sudate, Y., Hayashi, M., Otani, M., *J. Antibiot.* **1981**, *34*, 359–366.

15 Kinoshita, K., Yaginuma, S., Hayashi, M., Nakatsu, K., *Nucleosides Nucleotides* **1985**, *4*, 661–668.

16 Comin, M. J., Rodriguez, J. B., Russ, P., Marquez, V. E., *Tetrahedron* **2003**, *59*, 295–301.

17 Altmann, K.-H., Kesselring, R., Francotte, E., Rihs, G., *Tetrahedron Lett.* **1994**, *35*, 2331–2334.

18 Jeong, L. S., Marquez, V. E., Yuan, C.-S., Borchardt, R. T., *Heterocycles* **1995**, *41*, 2651–2656.

19 Siddiqui, M. A., Ford, H. Jr., George, C., Marquez, V. E., *Nucleosides Nucleotides* **1996**, *15*, 235–250.

20 Marquez, V. E., Siddiqui, M. A., Ezzitouni, A., Russ, P., Wang, J., Wagner, R. W., Matteucci, M. D., *J. Med. Chem.* **1996**, *39*, 3739–3747.

21 Biggadike, K., Borthwick, A. D., Exall, A. M., Kirk, B. E., Roberts, S. M., Youds, P., Slawin, A. M. Z., Williams, D. J., *J. Chem. Soc., Chem. Commun.* **1987**, 255–256.

22 Biggadike, K., Borthwick, A. D., Evans, D., Exall, A. M., Kirk, B. E., Roberts, S. M., Stephenson, L., Youds, P., *J. Chem. Soc., Perkin Trans.* **1988**, *1*, 549–554.

23 Ezzitouni, A., Russ, P., Marquez, V. E., *J. Org. Chem.* **1997**, *62*, 4870–4873.

24 Marquez, V. E., Russ, P., Alonso, R., Siddiqui, M. A., Hernandez, S., George, C., Nicklaus, M. C., Dai, F., Ford, H. Jr., *Helv. Chim. Acta* **1999**, *82*, 2119–2129.

25 Russ, P., Ezzitouni, A., Marquez, V. E., *Tetrahedron Lett.* **1997**, *38*, 723–726.

26 Yoshimura, Y., Moon, H. R., Choi, Y., Marquez, V. E., *J. Org. Chem.* **2002**, *67*, 5938–5945.

27 Moon, H. R., Ford, H. Jr., Marquez, V. E., *Org. Lett.* **2000**, *2*, 3793–3796.

28 Comin, M. J., Parrish, D. A., Deschamps, J. R., Marquez, V. E., *Org. Lett.* **2006**, *8*, 705–708.

29 Joshi, B. V., Moon, H. R., Fettinger, J. C., Marquez, V. E., Jacobson, K. A., *J. Org. Chem.* **2005**, *70*, 439–447.

30 Jeong, L. S., Marquez, V. E., *Tetrahedron Lett.* **1996**, *37*, 2353–2356.

31 Jeong, L. S., Bae, M., Chun, M. W., Marquez, V. E., *Nucleosides Nucleotides* **1997**, *16*, 1059–1062.

32 Moon, H. R., Kim, H. O., Chun, M. W., Jeong, L. S., Marquez, V. E., *J. Org. Chem.* **1999**, *64*, 4733–4741.

33 Matot, I., Weiniger, C. F., Zeira, E., Galun, E., Joshi, B. V., Jacobson, K. A., *Critical Care* **2006**, *10*, R65, (http://ccforum.com/content/10/2/R65).

34 Altmann, K.-H., Imwinkelried, R., Kesselring, R., Rihs, G., *Tetrahedron Lett.* **1994**, *35*, 7625–7628.

35 Ezzitouni, A., Marquez, V. E., *J. Chem. Soc., Perkin Trans.* **1997**, *1*, 1073–1078.

36 Shin, J., Moon, H. R., George, C., Marquez, V. E., *J. Org. Chem.* **2000**, *65*, 2172–2178.

37 Marquez, V. E., Ezzitouni, A., Russ, P., Siddiqui, M. A., Ford, H. Jr., Feldman, R. J., Mitsuya, H. George, C., Barchi, J. J. Jr., *J. Am. Chem. Soc.* **1998**, *120*, 2780–2789.

38 Choi, Y. S., George, C., Comin, M. J., Barchi, J. J. Jr., Kim, H. S., Jacobson, K. A., Balzarini, J., Mitsuya, H., Boyer, P. L., Hughes, S. H., Marquez, V. E., *J. Med. Chem.* **2003**, *46*, 3292–3299.

39 Choi, Y. S., Sun, G., George, C., Nicklaus, M. C., Kelley, J. A., Marquez, V. E., *Nucleosides, Nucleotides Nucleic Acids* **2003**, *22*, 2077–2091.

40 Marquez, V. E., Choi, Y. S., Comin, M. J., Russ, P., George, C., Huleihel, M.,

Ben-Kasus, T., Agbaria, R., *J. Am. Chem. Soc.* **2005**, *127*, 15145–15150.

41 Marquez, V. E., Comin, M. J., *Nucleosides Nucleotides Nucleic Acids.* **2007**, *25*, 585–588.

42 Comin, M. J., Agbaria, R., Ben-Kasus, T., Hulaihel, M., Liao, C., Sun, G., Nicklaus, M. C., Deschamps, J. R., Parrish, D. A., Marquez, V. E., *J. Am. Chem. Soc.* **2007**, *129*, 6216–6222.

43 Wang, P., Brank, A. S., Banavali, N. K., Nicklaus, M. C., Marquez, V. E., Christman, J. K., MacKerell, A. D. Jr., *J. Am. Chem. Soc.* **2000**, *122*, 12422–12434.

44 Marquez, V. E., Wang, P., Nicklaus, M. C., Meier, M., Manoharan, M., Christman, J. K., Banavali, N. K., MacKerell, A. D. Jr., *Nucleosides, Nucleotides Nucleic Acids* **2001**, *20*, 451–459.

45 Parvin, D., McCormick, R. J. Sharp, P. A., Fisher, D. E., *Nature* **1995**, *373*, 724–727.

46 Maderia, M., Shenoy, S., Van, Q. N., Marquez, V. E., Barchi, J. J. Jr., *Nucleic Acids Res.* **2007**, *35*, 1978–1991.

47 Maier, A., Choi, Y. S., Gaus, H., Barchi, J. J. Jr., Marquez, V. E., Manoharan, M., *Nucleic Acids Res.* **2004**, *32*, 3642–3650.

48 Wu, Z., Maderia, M., Barchi, J.J. Jr., Marquez, V. E., Bax, A., *Proc. Natl. Acad. Sci. USA* **2005**, *102*, 24–28.

49 Maderia, M., Wu, J., Bax, A., Shenoy, S., O'Keffe, B., Marquez, V. E., Barchi, J. J. Jr., *Nucleosides, Nucleotides Nucleic Acids* **2005**, *24*, 687–690.

50 Macias, A. T., Banavali, N. K., Mackerell, A. D. Jr., *Biopolymers* **2006**, *85*, 438–449.

51 Lu, X. J., Olson, W. K., *Nucleic Acids Res.* **2003**, *31*, 5108–5121.

52 Marquez, V. E., Ben-Kasus, T., Barchi, J. J. Jr., Green, K. M., Nicklaus, M. C., Agbaria, R., *J. Am. Chem. Soc.* **2004**, *126*, 543–549.

53 Prota, A., Vogt, J., Pilger, B., Perozzo, R., Wurth, C., Marquez, V. E., Russ, P., Schulz, G. E., Folkers, G., Scapozza, L., *Biochemistry* **2000**, *39*, 9597–9603.

54 Schelling, P., Claus, M. T., Johner, R., Marquez, V. E., Schulz, G. E., Scapozza, L., *J. Biol. Chem.* **2004**, *279*, 32832–32838.

55 Marquez, V. E., Hughes, S. H., Sei, S., Agbaria, R., *Antiviral Res.* **2006**, *71*, 268–275, and references therein.

56 Timsit, Y., *J. Biomol. Struct. Dyn.* **2000**, 169–176, Special Issue S1.

57 Timsit, Y., *J. Mol. Biol.* **1999**, *293*, 835–853.

58 Mayer, C., Timsit, Y., *Cell. Mol. Biol.* **2001**, *47*, 815–822.

59 Russ, P., Schelling, P., Scapozza, L., Folkers, G., De Clercq, E., Marquez, V. E., *J. Med. Chem.* **2003**, *46*, 5045–5054.

60 Boyer, L., Julias, J. G., Marquez, V. E., Hughes, S. H., *J. Mol. Biol.* **2005**, *345*, 441–450.

61 Baylin, B., Ohm, J. E., *Nat. Rev. Cancer* **2006**, *6*, 107–116.

62 Klimasauskas, S., Kumar, S., Roberts, R. J., Cheng, X., *Cell* **1994**, *76*, 357–369.

63 Klimasauskas, S., Roberts, R. J., *Nucleic Acids Res.* **1995**, *23*, 1388–1395.

64 O'Gara, M., Horton, J. R., Roberts, R. J., Cheng, X., *Nat. Struct. Biol.* **1998**, *5*, 872–877.

65 Horton, J. R., Ratner, G., Banavali, N. K., Huang, N., Choi, Y., Maiers, M. A., Marquez, V. E., MacKerell, A. D., Jr., Cheng, X., *Nucleic Acids Res.* **2004**, *32*, 3877–3886.

13
Synthesis, Chemical Properties and Biological Activities of Cyclic Bis(3′–5′)diguanylic Acid (*c*-di-GMP) and its Analogues

Mamoru Hyodo and Yoshihiro Hayakawa

13.1
Introduction

Cyclic bis(3′–5′)diguanylic acid (*c*-di-GMP) is a naturally occurring super-small quantity of a compound which initially was found as an activator of a cellulose synthase complex in *Gluconabacter xylinus*, and first isolated from the bacterium *Acetobacter xylinum* by van Boom and coworkers in 1987 [1]. Later, it was disclosed that *c*-di-GMP exists in many bacteria, such as *Gluconabacter xylinus* [1], *Agrobacter tumefaciens* [2], *Caulobacter crescentus* [3], *Vibrio cholerae* [4], *Salmonella typhimurium* [5], *Rhodobacter spharoides* [6], *Synechocystis* sp. [6], *Borrelia burgdorferi* [6], *Pseudomonas aeruginosa* [7], *Yersinia pestis* [8], *Shewanella oneidensis* [9], *Xanthomonas campestris* [10], and *Pseudomonas fluorescens* [11]. It was also revealed that *c*-di-GMP is synthesized in bacteria cells by a family of enzymes known as diguanylate cyclases (DGC), which have a GGDEF motif, and that *c*-di-GMP is degraded by *c*-di-GMP phosphodiesterases (PDE) that each contain a conserved EAL domain [12]. Many proteins contain both of these domains, which suggests that they may be bifunctional enzymes with opposing activities, and that they may control the level of *c*-di-GMP. A number of proteins with DGC and PDE activities have been shown to regulate virulence-related traits of diverse pathogenic bacteria. Therefore, *c*-di-GMP is considered to control biofilm formation, cellulose production, motility, phenotype regulation, and virulence in bacteria. In other words, *c*-di-GMP is now considered a common second messenger in bacteria, controlling many biofunctions. In fact, many recent reports have suggested that *c*-di-GMP has a variety of important bioactivities and biofunctions.

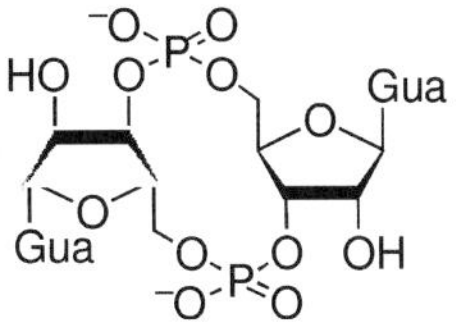

Modified Nucleosides: in Biochemistry, Biotechnology and Medicine. Edited by Piet Herdewijn

ISBN: 978-3-527-31820-9

Table 13.1 c-di-GMP-related biofunctions

Bacterium	Function	Author	Year
Acetobacter xylinum	Cellulose synthesis	Ross	1987
Agrobacterium tumefaciens	Cellulose synthesis	Amikam	1989
Pseudomonas aeruginosa	Twitching motility	Huang	2003
Vibrio cholerae	Swithching phenotype	Rashid	2003
Vibrio cholerae	Bioflim formation	Tischler	2004
Salmonella typhimurium	Sessility and motility	Simm	2004
Salmonella typhimurium	Cellulose synthesis	Garcia	2004
Salmonella typhimurium	Bioflim formation	Garcia	2004
Vibrio cholerae	Toxin production	Tischler	2005
Pseudomonas aeruginosa	Bioflim formation	Hickman	2005
Thermotoga maritima	Exopolysaccharide synthesis	Johnson	2005
Yersinia pestis	Bioflim formation	Bobrov	2005
Shewanella oneidensis	Biofilm formation	Thormann	2006
Vibrio cholerae	Polysaccharide synthesis	Lim	2007
Vibrio cholerae	Virulence factor production	Lim	2007
Pseudomonas fluorescens	Bioflim formation	Monds	2007

The representative bioactivities of c-di-GMP discovered to date are listed in Table 13.1 [1, 2, 4, 5, 9, 11, 13]. These include activation of cellulose synthase in the bacterium *Acetobacter xylinum*, acceleration of the DNA synthesis of cell division in Molt 4 cells, and elevation of expression of T-cell receptor CD4 expression in Jurkat cells [14]. Further, on the basis of studies conducted on the activity of a regulatory protein, the regulation of cell signaling (RocS) protein, in the switch between the smooth and rugose phenotypes of *V. cholerae* [15], it was proposed that c-di-GMP may have a function in regulating the biofilm formation of this bacterium. (Note: biofilm is a slimy polymeric substance, consisting mainly of various types of polysaccharide and water, which bacteria synthesize outside of their cells. This film strongly obstructs the approach of antibiotics to the cell and enhances bacterial adhesion to host cells. As a result, biofilm formation increases the antibiotic-resistant character and the infective power of bacteria [16].) A similar activity on biofilm formation was found in some bacteria, such as *E. coli*. Further, some recent investigations have suggested that c-di-GMP acts as a key compound in the process of biofilm biosynthesis of bacteria [17]. Thus, when microbes, moving in water, attach to a solid support (such as metal, plastic, rock or tissue) they start to form biofilm. Biofilm-covered microbes multiply to create a huge colony and communicate with each other, an event referred to as *quorum sensing*. In this process, c-di-GMP appears to be a key compound in the formation of biofilm and in quorum sensing [18]. These findings stimulated the present authors to carry out extensive investigations of the bioactivities of c-di-GMP, and particularly the activity of biofilm formation of various bacteria, because biofilm is closely connected with acquisition of the antibiotic-resistant character and enhancement of bacterial infective power. A further stimulation was to study the biological properties of artificial analogues of c-di-GMP with modified function or structure, because there are many cases where such modification allows enhancement or improvement of bioactivity of the parent naturally occurring compound. Thus, the authors' current research aim is to identify any previously unknown biological properties of *c*-di-GMP and its artificial analogues. Some of the results obtained to date are described in this chapter.

13.2 Synthesis of *c*-di-GMP and its Analogues

These research studies required the production of sufficient amounts (at least several tens of milligrams) of *c*-di-GMP and its analogues. On starting this research project, methods had been reported by van Boom for the synthesis of *c*-di-GMP and related compounds, which were classified as two types:

- An enzymatic method, synthesizing the target compound on the basis of cyclodimerization of 5′-GTP assisted by diguanylate cyclase [1].
- A chemical synthetic method, in which the formation of two internucleotide bonds of *c*-di-GMP was accomplished by the triester method [19].

Unfortunately, however, neither of these methods met the stated requirements, and consequently studies were commenced to develop new techniques for synthesizing the desired quantities of *c*-di-GMP and related compounds.

13.2.1 Synthesis of *c*-di-GMP

Initially, two methods were developed for the synthesis of *c*-di-GMP; these are illustrated in Scheme 13.1 (Method I) [20] and Scheme 13.2 (Method II) [21]. In the case of Method I, which is favored for large-scale synthesis, a few grams of *c*-di-GMP can be prepared at reasonable cost by one laboratory-scale synthesis. This noteworthy capability arises from the use of two novel strategies.

The first strategy is to use the di-*tert*-butylsilanediyl ribonuceloside-3′,5′-di-*O*-protection method [22] and selective removal of this protector from the 3′,5′-*O*-di-(*tert*-butylsilanediyl)-2′-*O*-(*tert*-butyldimethylsilyl)guanosine intermediate using HF·pyridine for the preparation of 2′-*O*-(*tert*-butyldimethylsilyl)guanosine derivative without 3′-*O*-protection. This strategy allowed 100% regioselective production of the desired 2′-*O*-silyl-protected derivative. Such perfect regioselective preparation of the 3′-*O*-free 2′-O-*tert*-butyldimethylsilyl-protected guanosine derivative could not be achieved by the method conventionally employed so far, namely the direct silylation of 2′,3′-di-*O*-unprotected substance [23].

The second strategy is to use an original internucleotide-linkage formation method using imidazolium perchlorate as a promoter [24] in the presence of molecular sieves 3A [25], in place of conventional 1*H*-tetrazole without molecular sieves, for condensation of the 2′-*O*-(*tert*-butyldimethylsilyl)guanosine 3′-phosphoramidite and 5′-*O*-free 2′-*O*-(*tert*-butyldimethylsilyl)guanosine 3′-phosphate to prepare the linear (3′–5′)-linked diguanylic acid intermediate. According to the conventional phosphoramidite

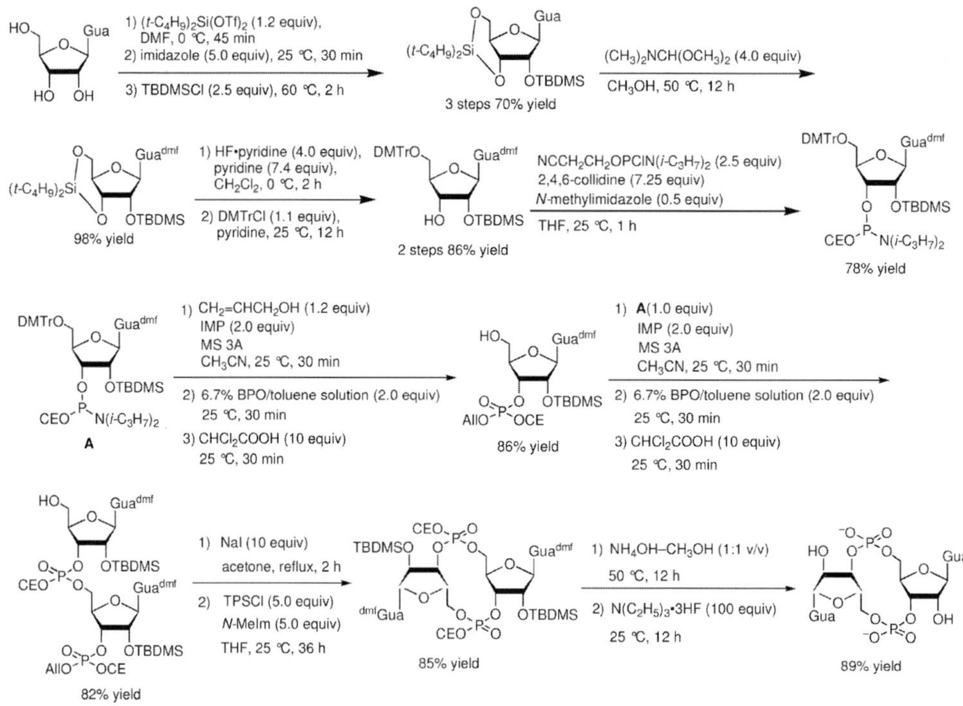

Scheme 13.1

method [26], use of an excess equivalent (2–4 equiv.) of the nucleoside phosphoramidite to the 5′-*O*-free nucleoside is necessary in order to efficiently achieve this condensation.

In contrast, our method allowed a high-yielding reaction by use of stoichiometric amounts of the nucleoside 3′-phosphoramidite and 5′-*O*-free nucleoside. This represents a major advantage in large-scale synthesis. On the other hand, Method II is useful for the synthesis of *c*-di-GMP analogues labile to acids and bases because, at the final stage of the synthesis, all protecting groups can be removed under neutral conditions. Thus, two *N*-allyloxycarbonyl (AOC) groups [27] in guanines and four *O*-allyl groups [28] in guanines and internucleotide linkage [29] could be removed simultaneously by treatment with a catalytic amount of $Pd_2[(C_6H_5CH{=}CH)_2CO]_3{\cdot}CHCl_3$ in the presence of triphenylphosphine and butylammonium hydrogen carbonate in tetrahydrofuran, and two *O*-*tert*-butyldimethylsilyl groups in riboses could be deblocked by exposure to $(C_2H_5)_3N{\cdot}3HF$ [30].

Following the reporting of Methods I and II, two further techniques for the synthesis of *c*-di-GMP were reported by the groups of Jones [31] and Giese [32]. A remarkable point in the Jones method is that cyclization of a linear diguanylic acid intermediate to a cyclic bis(3′–5′)diguanylic acid is performed via the *H*-phosphonate method [33]. Giese's method involves a unique strategy, in which first the cyclic bis (3′–5′)diribosyl diphosphate skeleton without nucleobases was constructed through the modified van Boom method (the triester method), after which two guanine bases

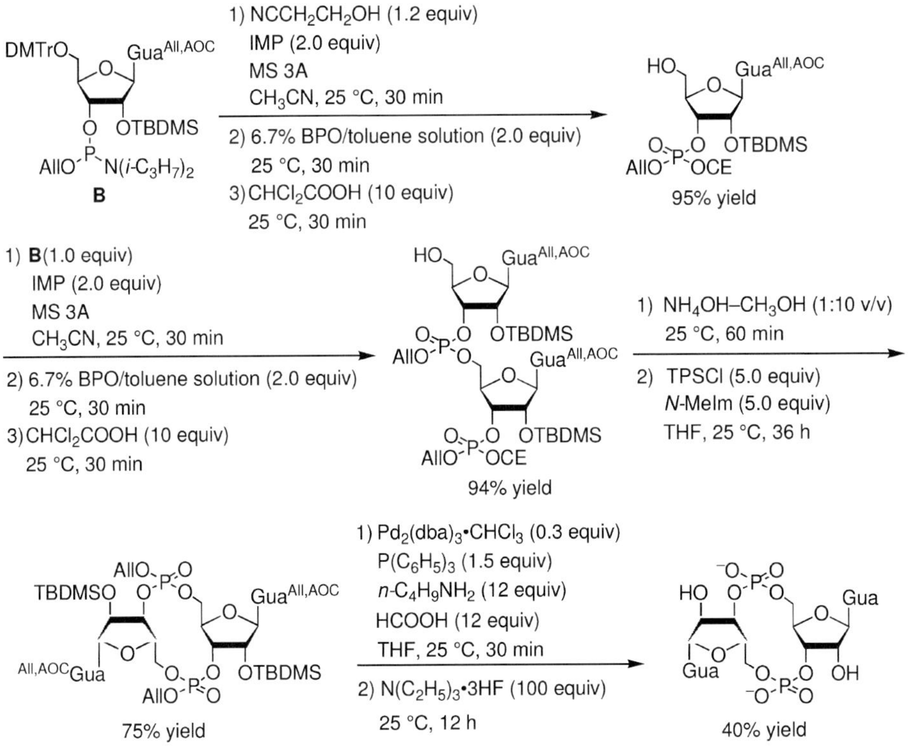

Scheme 13.2

are introduced to this cyclic diphosphate by means of a modified Vorbrüggen reaction to form the desired cyclic bis(3′–5′)diguanylic acid.

13.2.2
Synthesis of Artificial Analogues of *c*-di-GMP

The fundamental strategies used in Methods I and II could be applied to the synthesis of a variety of *c*-di-GMP artificial analogues with modified nucleobases, sugars, and backbones. In fact, to date the analogues shown in Figure 13.1 have been synthesized

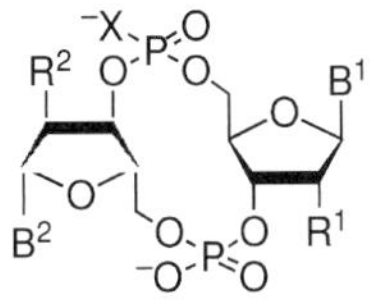

B^1 = Ade, B^2 = Gua, R^1 = OH, R^2 = OH, X = O
B^1 = Ino, B^2 = Gua, R^1 = OH, R^2 = OH, X = O
B^1 = Gua, B^2 = Gua, R^1 = H, R^2 = OH, X = O
B^1 = Gua, B^2 = Gua, R^1 = OH, R^2 = OH, X = S

Figure 13.1

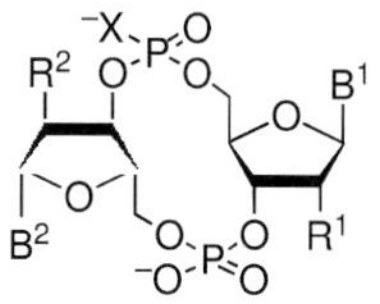

B^1 = Ade, B^2 = Ade, R^1 = OH, R^2 = OH, X = O
B^1 = Ade, B^2 = Gua, R^1 = OH, R^2 = OH, X = O
B^1 = Cyt, B^2 = Cyt, R^1 = OH, R^2 = OH, X = O
B^1 = Cyt, B^2 = Gua, R^1 = OH, R^2 = OH, X = O
B^1 = Ino, B^2 = Gua, R^1 = OH, R^2 = OH, X = O
B^1 = Ino, B^2 = Ino, R^1 = OH, R^2 = OH, X = O
B^1 = The, B^2 = The, R^1 = OH, R^2 = OH, X = O
B^1 = Thy, B^2 = Thy, R^1 = OH, R^2 = OH, X = O
B^1 = Ura, B^2 = Gua, R^1 = OH, R^2 = OH, X = O
B^1 = Ura, B^2 = Ura, R^1 = OH, R^2 = OH, X = O
B^1 = Xan, B^2 = Gua, R^1 = OH, R^2 = OH, X = O
B^1 = Xan, B^2 = Xan, R^1 = OH, R^2 = OH, X = O
B^1 = Gua, B^2 = Gua, R^1 = H, R^2 = OH, X = O
B^1 = Gua, B^2 = Gua, R^1 = H, R^2 = H, X = O
B^1 = Gua, B^2 = Gua, R^1 = OH, R^2 = OH, X = S

Figure 13.2

by means of strategies somewhat modified from those used in the two methods [34]. Further, the van Boom, Jones and Giese groups have prepared the analogues shown in Figure 13.2 according to their own methods.

13.3 Chemical Properties of *c*-di-GMP and its Analogues

13.3.1 Stability and Chemical Properties of *c*-di-GMP under Acidic, Basic, and Physiological Conditions

In order to investigate the biological activity of *c*-di-GMP and its analogues (cyclic diribonucleotides), it is important to know their stability and properties not only in water but also in aqueous solutions of an acid, a base, and in human serum. Thus, these characteristics were monitored under various conditions. First, *c*-di-GMP was seen to be quite stable in water; for example, when a solution of c-di-GMP in distilled water was left at 25 °C for one week and boiled for 1 h, no decomposition was observed in either case. *c*-di-GMP was also stable in diluted acid and base solutions; the cyclic nucleotide was left intact in a pH 3 HCl solution and in a pH 10 NaOH solution at 25 °C for 1 h. Likewise, *c*-di-GMP was seen to be stable in a human serum solution (Cambrex Bio Science Walkersville), and underwent no degradation at 37 °C for 24 h. Similar properties were observed for the *c*-di-GMP analogues prepared. The observed stability of the cyclic diribonucleotide is extremely high in comparison with that of the linear (3′–5′)-linked ribonucleotide, which is very labile to acid, base, and enzyme (ribonuclease). The high stability of the cyclic diribonucleotide is explained as follows. In the linear ribonucleotide, the 3′-phosphate function has sufficiently high conformational flexibility to approach the 2′-hydroxy group and to allow the attack of this hydroxy group. As a result, various decompositions, such as migration of the 3′-*O*-phosphoryl group to the 2′-hydroxy group to form the 2′-phosphate and cleavage of the $P-O^{5'}$ bond to give the 2′,3′-*O*-cyclic phosphate. Thus, in the linear (3′–5′)-linked ribonucleotide, neighboring participation of the 2′-hydroxy group easily occurs to cleave the 3′,5′-internucleotide linkage. In contrast, in cyclic diribonucleotides, the 3′-phosphate group has very low conformational flexibility and thus cannot approach the 2′-hydroxy group. Consequently, decomposition of the cyclic diribonucleotide does not take place.

13.3.2
Polymorphism of *c*-di-GMP in Aqueous Solutions

Extensive examination of this object revealed that *c*-di-GMP aggregates to assume varied forms (without skeletal degradation) in an aqueous solution containing a salt, such as sodium chloride or ammonium acetate, and the aggregate form is changed by the salt concentration. For example, ^{1}H NMR spectra indicated that *c*-di-GMP takes some aggregate forms in 150 mM NaCl solutions, and that the structure of the aggregate is changed by the salt concentration (see Figure 13.3 and compare the spectrum obtained in $NaCl/D_2O$ solution with that taken in DMSO-d_6, in which *c*-di-GMP seems to exist as a monomer).

High-performance liquid chromatography (HPLC) analysis provided more detail regarding this aggregation. According to analyses using 31 mM solutions of *c*-di-GMP in water containing various amounts of NaCl, *c*-di-GMP exists as a single aggregate in the 0.9% or higher concentration NaCl solution, and as a mixture of a number of aggregates in the lower concentration (<0.9%) NaCl solution (see Figure 13.4). It is interesting that these aggregates are all interconvertible. Thus, when NaCl was added to the less than 0.9% NaCl solution containing a number of aggregates to adjust the salt concentration to 0.9% or higher, all the aggregates become a single aggregate identical to that observed in the high-concentration solution. A similar polymorphism was observed in an ammonium acetate buffer and a phosphate buffer. In these solutions, *c*-di-GMP exists as a single compound in a 100 mM or higher salt concentration of solution, and as a mixture of many aggregates in a less than 100 mM salt concentration solution (see Figure 13.4). Although, in these examinations the structures of the aggregates were not determined, it is likely that the aggregates observed here may have dimeric, tetrameric and octameric structures, as proposed by the Jones group (see below).

Similar experiments and analyses were also conducted for some *c*-di-GMP analogues, such as *c*-GpAp, *c*-GpIp and *c*-dGpGp, to determine whether the polymorphism observed in *c*-di-GMP occurs generally in the cyclic bis(3′–5′)diribonucleic acids. The result showed that all these analogues exist as aggregates in water with or without a salt (e.g., NaCl), but they do not form polymorphisms when the salt

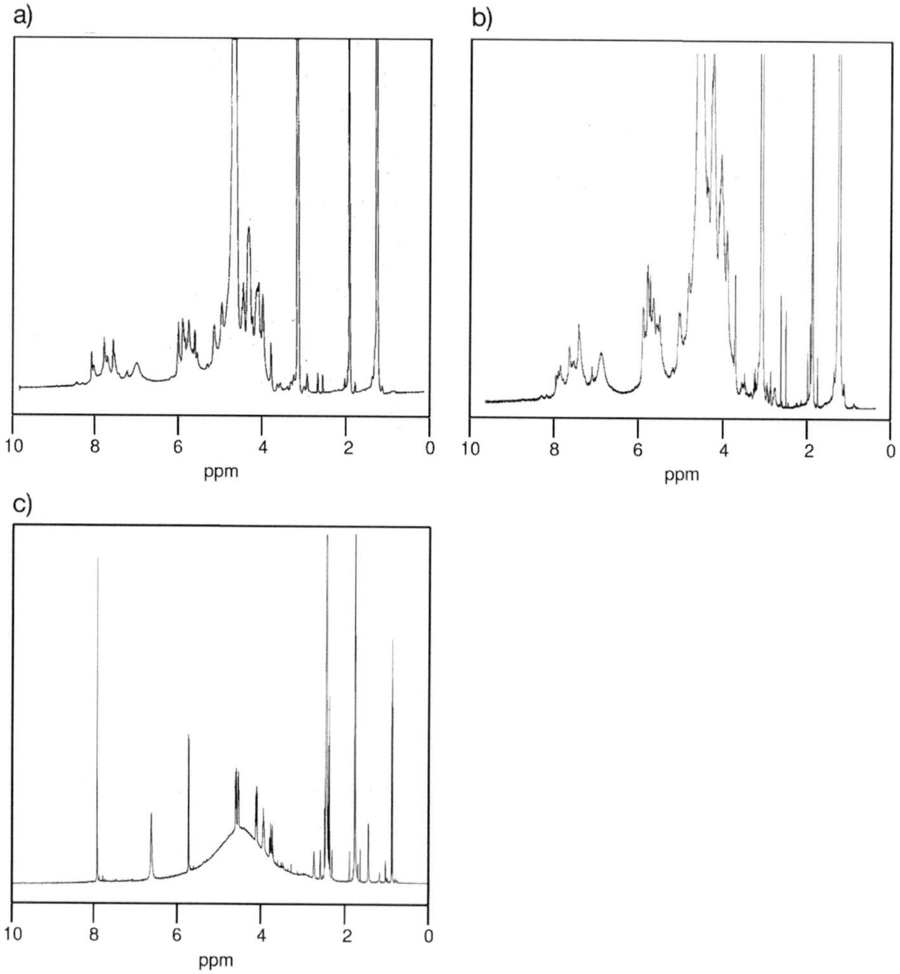

Figure 13.3

concentration is changed. Consequently, the polymorphism was shown to be a peculiar phenomenon occurring in *c*-di-GMP.

The polymorphism of *c*-di-GMP was also detected in aqueous solutions containing KCl or LiCl by Jones and coworkers [35]. who attempted to determine the structures of aggregates formed under various conditions. According to these investigations, *c*-di-GMP may exist as a mixture of a dimer, a tetramer and an octamer (see Figure 13.5), in a 100 mM KCl or LiCl solution, and the predominant species depended on the concentration of *c*-di-GMP in the solution. In a high-concentration *c*-di-GMP solution, the larger aggregates are predominant, whereas in a low-concentration *c*-di-GMP solution, the smaller aggregates are favored. Further, it was observed that the favored structure is also affected by the type of coexisting salt. In the same concentration (ca. 35 mM) of *c*-di-GMP solution, the tetramer or the octamer is a major aggregate in the LiCl or KCl solution, respectively.

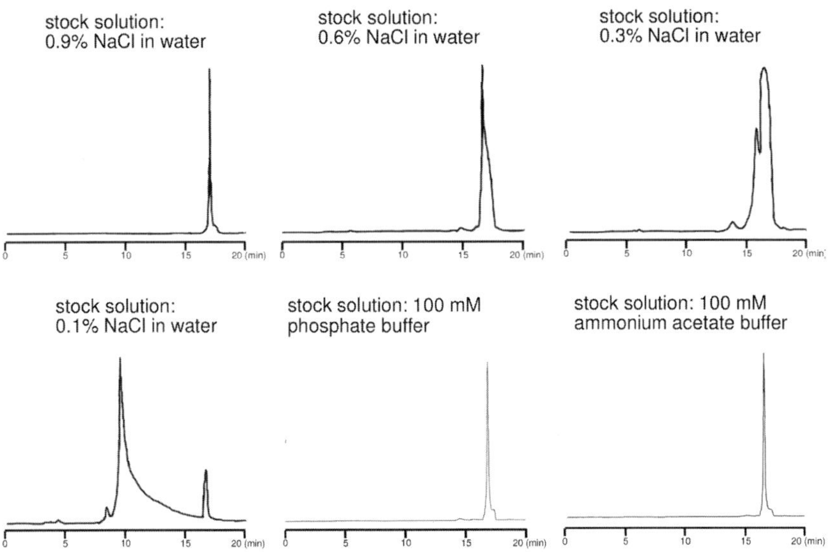

Conditions: COSMOSIL $5C_{18}$–AR–II column (4.6 x 200 mm); buffer A, stock solution; buffer B, 80% acetonitrile in water; gradient, 0–10 min A 100%, 10–60 min A:B = 100:0 to A:B = 40:60 in 50 min; detection 254 nm, flow rate, 1.0 mL/min, temperature 40 °C.

Figure 13.4

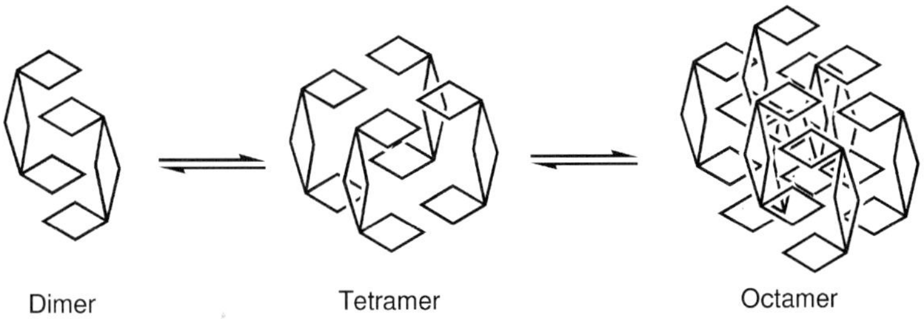

Figure 13.5

13.4
Bioactivities of *c*-di-GMP and its Analogues

Investigations were carried out (in collaboration with other groups) to identify and elucidate novel bioactivities and roles of *c*-di-GMP and related compounds, and the following important results have been obtained to date. The bioactivities and biological roles of *c*-di-GMP, and also those of *c*-dGpGp – an analogue of *c*-di-GMP – are described in the following section.

13.4.1
Activity of *c*-di-GMP on Biofilm Formation

As noted in Section 13.1, *c*-di-GMP is expected to have a variety of important bioactivities and biofunctions. Among these are included the regulation of bacterial biofilm, which today is attracting significant attention from bacteriologists and physicians. This interest is based on recent investigations having suggested that biofilm is one of the main factors to provide bacteria with antibiotic-resistance and strengthened infective powers. Thus, if biofilm formation were to be restrained, the bacteria might be prevented from acquiring these undesired characteristics [36]. It was for these reasons that we first examined the activity of *c*-di-GMP on biofilm formation of *Staphylococcus aureus*, in collaboration with Karaolis and coworkers.

13.4.1.1 Inhibition of Biofilm Formation and Prevention of Bacterial Infection of *S. aureus in vitro*

Initially, a series of *in-vitro* examinations was conducted whereby *S. aureus* was treated extracellularly with 20 μM, 200 μM and 400 μM *c*-di-GMP at 37 °C for 24 h, after which the degree of biofilm production in these samples was analyzed by measuring the absorbance of visible light at 570 μm (OD_{570}). The result showed that biofilm production was inhibited by approximately 50%, 65% and 85% in the samples treated with 20 μM, 200 μM and 400 μM of *c*-di-GMP, respectively (see Figure 13.6) [37]. A similar inhibitory effect was also observed in the highly adherent hyperbiofilm *S. aureus* strain 15981.

Next, we examined the inhibitory effect of *c*-di-GMP on the adherence of *S. aureus* to epithelial cells by using HeLa cells as the representative. The examination was carried out on bacteria with or without treatment with 200 mM *c*-di-GMP, and the degree of the effect was determined by counting the number of bacteria adhering to a cell (bacteria per cell). In the case of non-treatment with *c*-di-GMP, there were 12 bacteria per cell, whereas in the *c*-di-GMP-treated bacteria there were only four

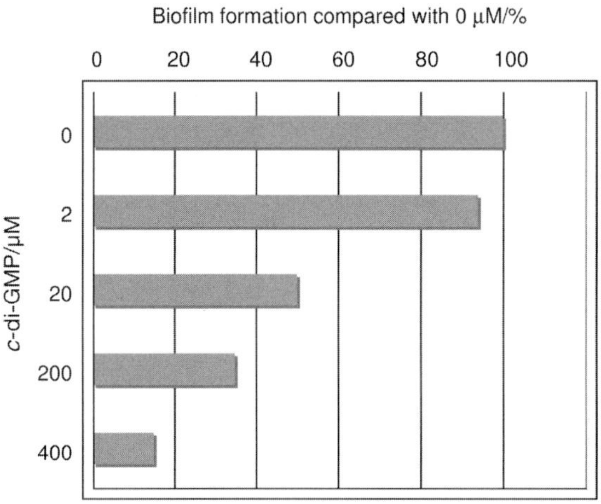

Figure 13.6

Table 13.2 *In-vivo* inhibition effect of *c*-di-GMP on *S. aureus* infection in a mouse mammary gland model.

c-di-GMP dose (nmol)	0	5	50	200
Number of glands used	14	14	12	15
Number of glands found non-infection	0	1	1	9

bacteria per cell. Thus, treatment with 200 mM *c*-di-GMP had reduced the *S. aureus* infection of cells by about 66%.

An inhibitory effect of *c*-di-GMP on the biofilm formation of *S. aureus* was also observed by conducting *in-vivo* tests in a mouse model of mammary gland infection. Among mice untreated with *c*-di-GMP, all were infected. In contrast, when mice were administered 200 nmol *c*-di-GMP at 0 h and again at 4 h post-infection, 60% of the mice were free of bacteria (see Table 13.2). Moreover, none of the mice treated with *c*-di-GMP showed any signs of toxicity, such as alterations in posture, breathing, piloerection or movement. These results indicated that *c*-di-GMP treatment attenuates virulence and prevents infection *in vivo* caused by biofilm-forming *S. aureus* [38].

13.4.1.2 **Activity as an Immunostimulatory Molecule**

Although *c*-di-GMP has no apparent inhibitory or bactericidal effect on *S. aureus in vitro*, when monitored *in vivo* it significantly reduced colonization of the mammary glands by a biofilm-forming *S. aureus* strain. These results imply that *c*-di-GMP might have biological activity in an *in-vivo* environment and in the host immune response. In other words, *c*-di-GMP might be expected to serve as not only as a microbial signaling molecule, but also as a novel immunostimulatory agent that could modulate the host immune response. The feasibility of this suggest was monitored in collaboration with Karaolis and coworkers [39].

First, intramammary treatment of mice with *c*-di-GMP at 12 h and 6 h before the *S. aureus* challenge provided a significant prophylactic effect, with a 1.5 and 3.8 log (ca. 10 000-fold) reduction in mean bacteria colony forming units (CFU) in tissues using 50- and 200-nmol doses, respectively, compared with the untreated control (Table 13.3).

These results suggested that *c*-di-GMP stimulates the host response and inhibits bacterial infection. Accordingly, to determine whether *c*-di-GMP does indeed act as an adjuvant on the host immune response, the following experiments were carried out.

First, two classes of mice were prepared: (1) those injected with only the recombinant C1fA antigen, which is a surface adhesion protein of *S. aureus*; and (2) those

Table 13.3 Numbers of colony-forming units (CFU) of *S. aureus* with or without *c*-di-GMP treatment.

c-di-GMP dose (nmol)	0	50	200
Colony forming unit (CFU)	$10^{7.3}$	$10^{5.8}$	$10^{3.5}$

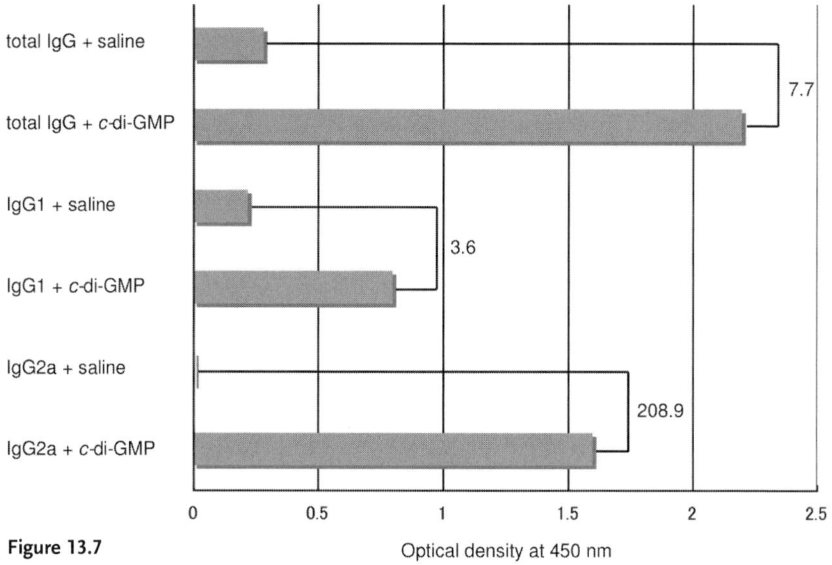

Figure 13.7

co-injected with *c*-di-GMP and the recombinant C1fA antigen. The extent of increase in immunoglobulin G (IgG) in each class was measured after 12 days, using an ELISA method. A remarkably higher increase of IgG was observed in mice co-injected with C1fA antigen and c-di-GMP, compared to mice injected with C1fA alone. Thus, in the *c*-di-GMP-injected sample, increases in total IgG, IgG1 and IgG2a were 7.7-, 3.6-, and 208.9-fold those in the sample without *c*-di-GMP injection, respectively (see Figure 13.7). Among these results, the remarkable increase in IgG2a observed in the sample in the presence of *c*-di-GMP may strongly indicate that the additional use of *c*-di-GMP to the recombinant C1fA antigen activates the Th1 pathway; that is, *c*-di-GMP improves antibody production. An activation of the immune response by *c*-di-GMP was also recently reported by Ebensen and colleagues [40].

Subsequently, investigations were made as to whether or not *c*-di-GMP causes the maturation of dendritic cells (DCs), as these cells have important functionality in the immune response to sense infection and respond approximately, both to induce T-cell immunity and to promote a Th1 immune response. Thus, it was proposed that the maturation of DCs with *c*-di-GMP might be critical for the initiation of immune responses. The examination was carried out using immature DCs either treated with 5 to 500 μM *c*-di-GMP for 24 h, or not treated. The results of the examination are shown in Figure 13.8, where the histogram compares the increased expression of CD83 in DCs treated with *c*-di-GMP to that of untreated cells (negative control) or treated with lipopolysaccharide (LPS; positive control). The increase in surface CD83 expression on cells was seen to depend on the *c*-di-GMP dose administered. Similar increases in expression were also observed in other maturation markers (CD80, CD86, CCR7 and MHC class II) on DCs treated with 200 μM *c*-di-GMP.

An investigation was also made as to whether or not cytokine and chemokine expression is altered in DCs treated with *c*-di-GMP, in order to determine the

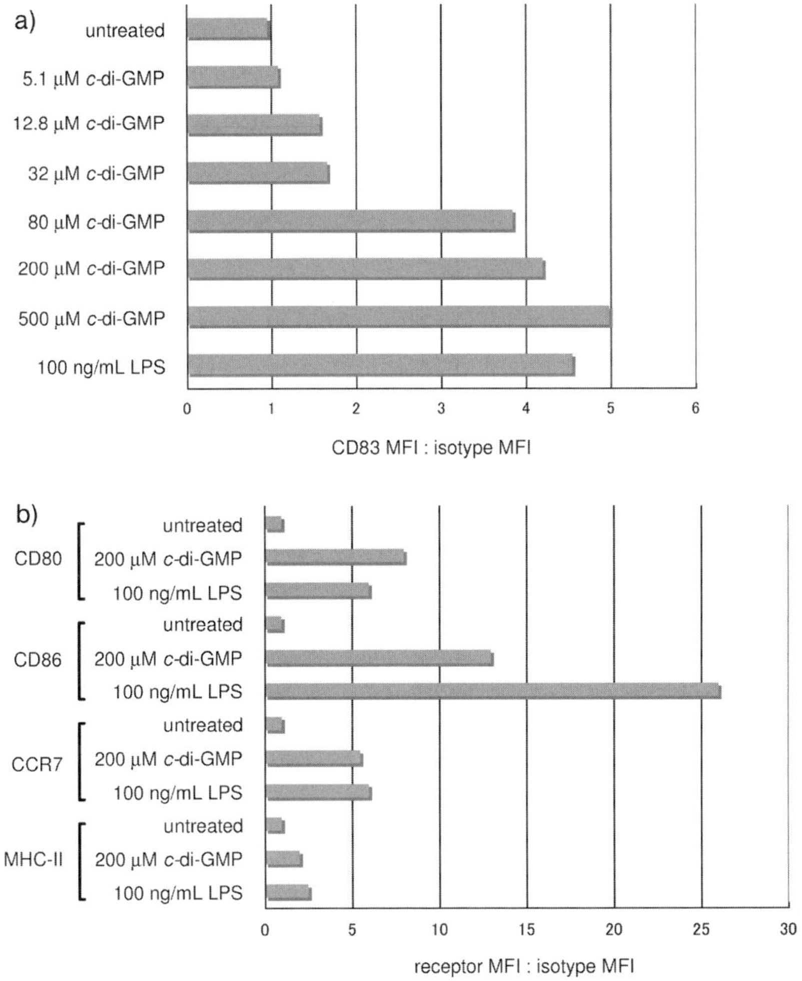

Figure 13.8

potential of *c*-di-GMP-treated DCs in activating and/or recruiting other effector cells to sites of infection. The results, which are displayed as histograms in Figure 13.9, indicated that treatment with *c*-di-GMP stimulates mRNA expression of several cytokines and chemokines to increase the production of the cytokines interferon (IFN)-α, interleukin (IL)-1β and tumor necrosis factor (TNF), all of which are considered Th1-related cytokines. These results were consistent with those obtained by the above-mentioned examinations of IgGs production.

As DCd may mature by the stimulation of T cells – which form part of the white blood cells and play a central role in cell-mediated immunity – the final point to be examined was whether or not *c*-di-GMP-treated DCs have the ability to stimulate T cells; this was assessed by measuring T-cell proliferation. The data in Figure 13.10 show that T-cell proliferation in samples treated with *c*-di-GMP was approximately

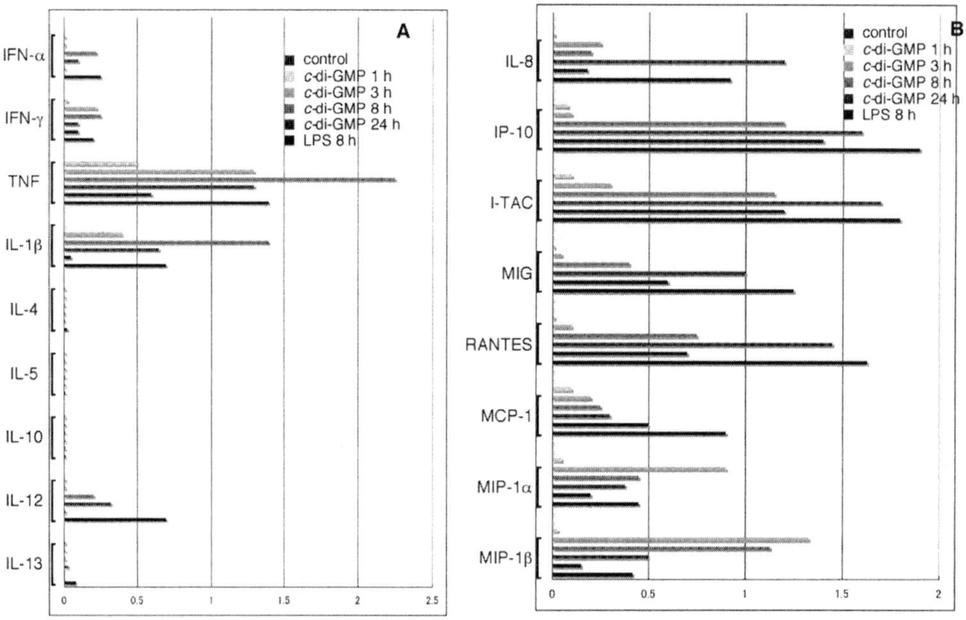

Figure 13.9

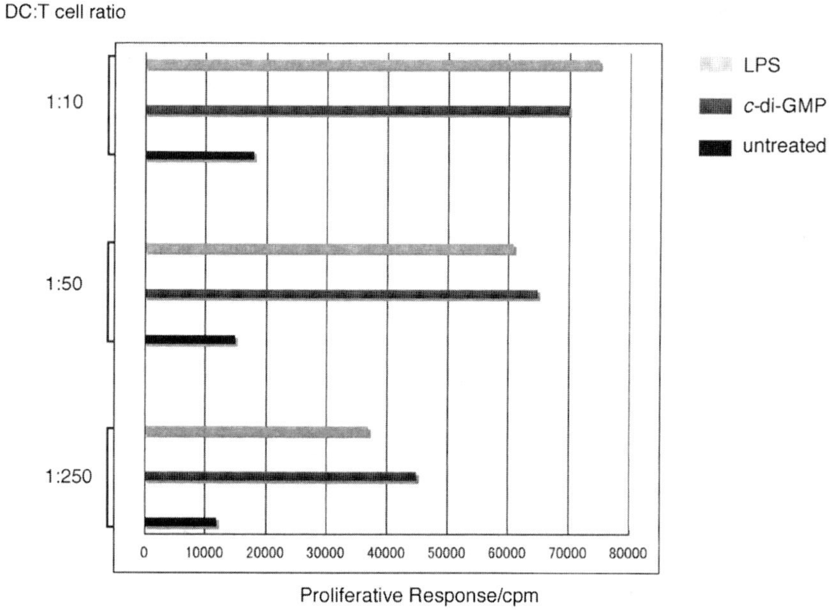

Figure 13.10

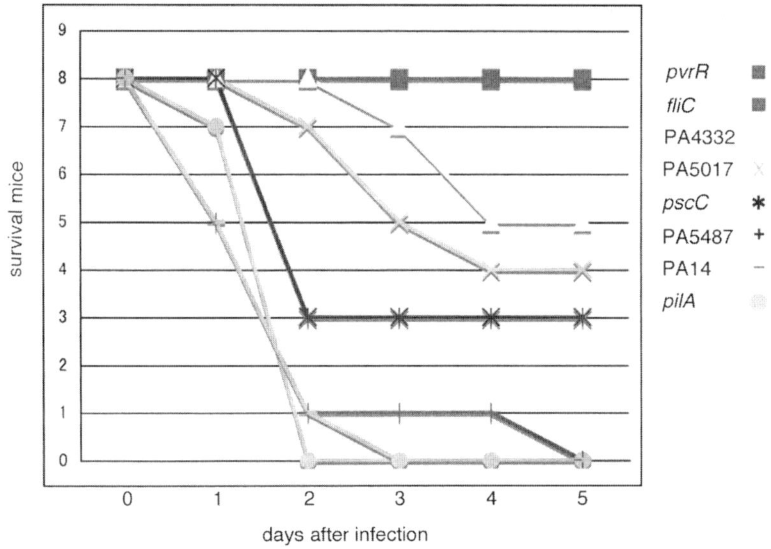

Figure 13.11

seven-fold higher than in untreated samples. This extent of proliferation caused by the addition of *c*-di-GMP was very similar to that seen in the above-described experiment with LPS as positive control.

13.4.1.3 **Activity on Biofilm Formation and Virulence Emergence of *P. aeruginosa***

A systematic analysis of phenotypes and cytotoxicities of DGC and PDE domains containing proteins in *P. aeruginosa* was carried out in collaboration with Lory and coworkers. The results showed that insertion into genes encoding DGC or DGC-PDE affects biofilm formation, whereas insertion into genes encoding only the PDE domain did not cause any biofilm-related phenotype. Thus, in order to assess whether virulence is influenced by mutations in genes encoding DGC and PDE, several mutants with biofilm or cytotoxicity phenotypes were tested in a murine thermal injury model. The test results indicated that, in this acute infection, the initiation of biofilm formation is apparently not important (*PA5017, PA5487*); however, hyperbiofilm formation may lead to a reduction in the virulence (*PA4332*) (Figure 13.11) [7].

13.4.2
Inhibition of Proliferation of Human Colon Cancer Cells with *c*-di-GMP

The group of Amikam found that *c*-di-GMP promotes cell cycle arrest in the lymphoblastoid CD4 Jurkat cell line. This implies that *c*-di-GMP inhibits the proliferation of cells, including cancer cells, and may therefore possess an anticancer effect. Hence, investigations were conducted to determine whether or not *c*-di-GMP has such an activity in human colon cancer cells.

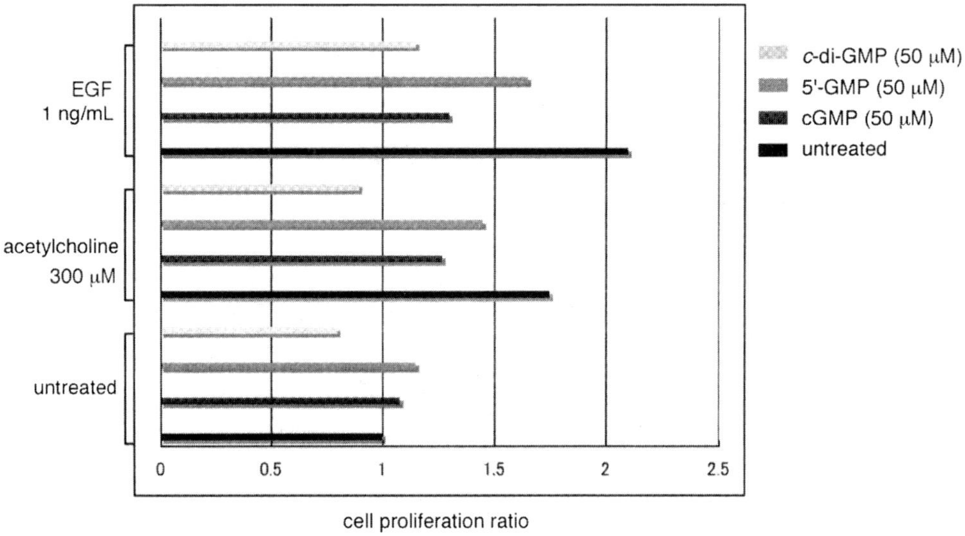

Figure 13.12

The potential therapeutic actions of *c*-di-GMP on basal and growth-factor-stimulated proliferation of cells derived from a moderately differentiated human cecal adenocarcinoma (H508 cells) were examined by incubating these cells in the absence or presence of growth factor (acetylcholine or epidermal growth factor; EGF), either alone or with *c*-di-GMP at 37 °C for 5 days. As a contrasting study, the same experiments were carried out using *c*-GMP or 5′-GMP in place of *c*-di-GMP. The results indicated that *c*-di-GMP reduced the proliferation of human colon cancer (HCC) cells more strongly than *c*-GMP and 5′-GMP, with or without growth factor (see Figure 13.12). Based on the assessment of the cytotoxicity of *c*-di-GMP towards human neuroblastoma cells, the EC_{50} of *c*-di-GMP was 350 μM). This result excluded the probability that the above-observed inhibitory effect of *c*-di-GMP on HCC cell proliferation was due to cell death following the addition of *c*-di-GMP. Hence, *c*-di-GMP is considered safe and non-cytotoxic at concentrations that inhibit cancer cell proliferation, and also functions on living HCC cells to inhibit cell proliferation [41].

13.4.3
Biological Activity of *c*-dGpGp

Current investigations include determination of the biological properties of those *c*-di-GMP analogues prepared and illustrated in Figure 13.1. Some preliminary results obtained, using *c*-dGpGp as a representative of the analogues, are presented in the following section.

Initially, the inhibitory effect of *c*-dGpGp on the biofilm formation of *V. parahaemolyticus* was studied, and the resulting activity compared with that of *c*-di-GMP. The examination was carried out using *V. parahaemolyticus* strain ATCC17802 with

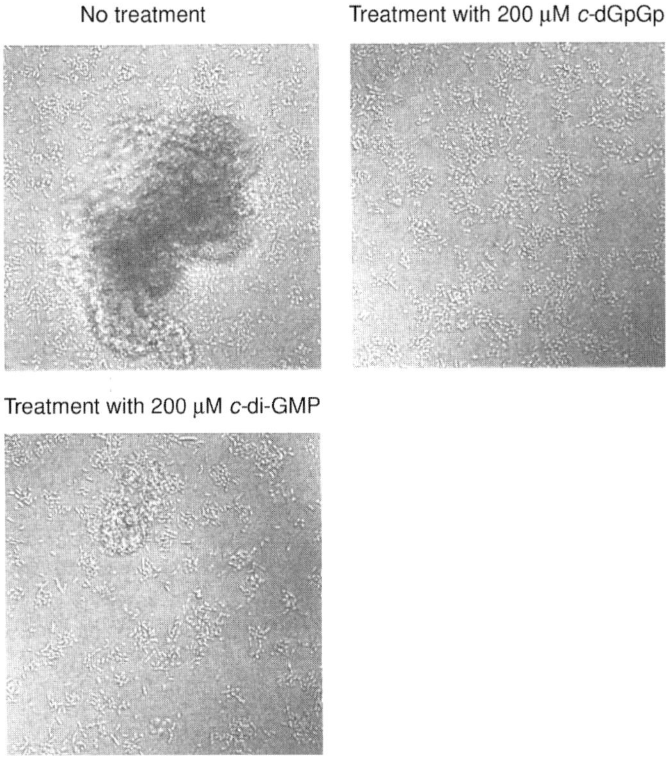

Figure 13.13

200 mM *c*-dGpGp and *c*-di-GMP, and by incubating these samples at 30 °C for 4 h on glass-based tissue. The results showed that both *c*-dGpGp and *c*-di-GMP inhibited biofilm formation by *V. parahaemolyticus*, and that the activity of *c*-dGpGp was higher than that of *c*-di-GMP (see Figure 13.13).

Subsequently, the effects of *c*-dGpGp and *c*-di-GMP on the motility of *V. parahaemolyticus* were examined, and the resulting data compared. The examinations were performed using *V. parahaemolyticus* ATCC17802 treated with 200 μM of *c*-di-GMP or *c*-dGpGp (controls were not treated), after which the swimming patterns of the bacteria were monitored. Compared to the motility of untreated bacteria, that of the bacteria treated with *c*-dGpGp was 10% repressed. In contrast, a 60% promotion in motility was observed in bacteria treated with *c*-di-GMP.

13.5 Conclusions

Recently, new methods have been developed for the synthesis of *c*-di-GMP and its various analogues, and these methods are capable of providing target compounds in large amounts, perhaps hundreds of milligrams to several grams. These syntheses

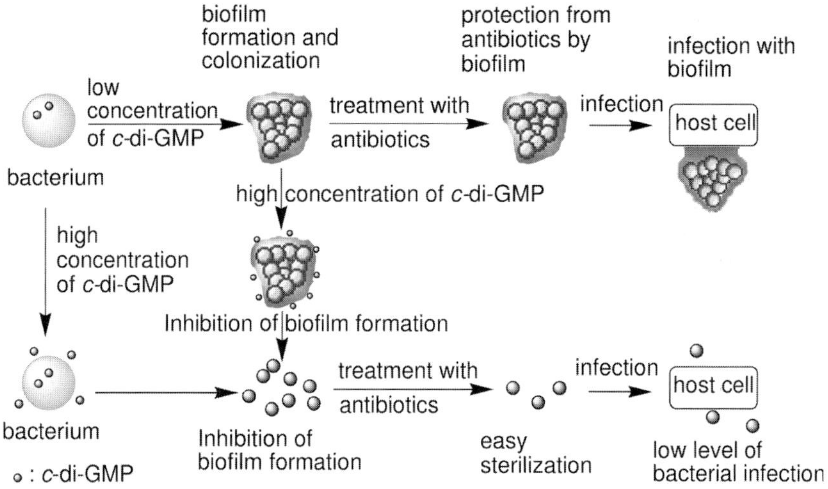

Figure 13.14

have allowed extensive investigations to be conducted on the biological activities of *c*-di-GMP and its analogues. As a result, it has been found that *c*-di-GMP appears to act as a second messenger in bacteria, and to show a variety of important bioactivities. Among such bioactivities, the most attractive is the regulation of biofilm formation and infection of host cells by various bacteria; that is, the inhibition of biofilm formation by *S. aureus*, *P. aeruginosa* and *E. coli*; reduction of infection by *S. aureus* to HeLa cells; reduction of the virulence of biofilm-forming *S. aureus* strains in a mouse model of mastitis infection *in vivo*; and the activation of immune response. The results obtained have suggested that *c*-di-GMP acts against biofilm formation and bacterial infection as shown in Figure 13.14, although this scheme is only speculation at this stage. Hence, further investigations are required to determine whether this speculation fits in generally with various microbes, or not. *c*-di-GMP has been shown to inhibit basal and growth factor-stimulated human colon cancer cell proliferation, while investigations with *c*-dGpGp – an artificial analogue of *c*-di-GMP – revealed that this artificial compound has different biological activities from those of *c*-di-GMP, despite the two compounds having very similar structures. Thus, whilst both *c*-di-GMP and *c*-dGpGp inhibit biofilm formation of *V. parahaemolyticus*, the strength of their inhibitory effects are dissimilar, with *c*-dGpGp having a stronger effect than *c*-di-GMP. Further, *c*-dGpGp shows a different effect from *c*-di-GMP on the motility of *V. parahaemolyticus*, whereby *c*-dGpGp promotes the motility of this bacterium while *c*-di-GMP depresses it.

In conclusion, the findings described in this chapter suggest that *c*-di-GMP and its analogues have great potential as antibacterial and immunotherapeutic agents, and consequently the importance of investigating their biological properties is raised. Clear evidence of this suggestion is seen in the drastic increase in recent years in the numbers of reports relating to *c*-di-GMP.

Acknowledgments

The authors are grateful to Dr. Rie Kawai, Ms. Reiko Nagata, Ms. Akiyoshi Hirata, Ms. Yumi Sato and Ms. Erina Mano for assisting with the synthesis of *c*-di-GMP and its analogues. They also thank Dr. David K. Karaolis of the University of Maryland, Dr. Stephen Lory of Harvard University, Dr. Jorge Membrillo-Hernández of Universidad Nacional Autonoma de Mexico, and Dr. Michio Ohta of Nagoya University for their collaboration in studies on the biological activities of *c*-di-GMP and related compounds.

References

1 Ross, P., Weinhouse, H., Aloni, Y., Michaeli, D., Weinberger-Ohana, P., Mayer, R., Braun, S., de Vroom, E., van der Marel, G. A., van Boom, J. H., Benziman, M., *Nature* **1987**, *325*, 279–281.

2 Amikam, D., Benziman M., *J. Bacteriol.* **1989**, *171*, 6649–6655.

3 Paul, R., Weiser, S., Amiot, N. C., Chan, C., Schirmer, T., Giese, B., Jenal, U., *Genes Dev.* **2004**, *18*, 715–727.

4 Tischler, A. D., Camilli A., *Mol. Microbiol.* **2004**, *53*, 857–869.

5 Simm, R., Morr, M., Kader, A., Nimtz, M., Römling, U., *Mol. Microbiol.* **2004**, *53*, 1123–1134.

6 Ryjenkov, D. A., Tarutina, M., Moskvin, O. V., Gomelsky, M., *J. Bacteriol.* **2005**, *187*, 1792–1798.

7 Kulesekara, H., Lee, V., Brencic, A., Liberati, N., Urbach, J., Miyata, S., Lee, D. G., Neely, A. N., Hyodo, M., Hayakawa, Y., Ausubel, F. M., Lory, S., *Proc. Natl. Acad. Sci. USA* **2006**, *103*, 2839–2844.

8 Simm, R., Fetherston, J. D., Kader, A., Römling, U., Perry, R. D., *J. Bacteriol.* **2005**, *187*, 6816–6823.

9 Thormann, K. M., Duttler, S., Saville, R. M., Hyodo, M., Shukla, S., Hayakawa, Y., Spormann, A. M., *J. Bacteriol.* **2006**, *188*, 2681–2691.

10 Ryan, R. P., Fouhy, Y., Lucey, J. F., Crossman, L. C., Spiro, S., He, Y. -W., Zhang, L. -H., Heeb, S., Camara, M., Williams, P., Dow, J. M., *Proc. Natl. Acad. Sci. USA* **2006**, *103*, 6712–6717.

11 Monds, R. D., Newell, P. D., Gross, R. H., O'Toole, G. A., *Mol. Microbiol.* **2007**, *63*, 656–679.

12 Tal, R., Wong, H. C., Calhoon, R., Gelfand, D., Fear, A. L., Volman, G., Mayer, R., Ross, P., Amikam, D., Weinhouse, H., Cohen, A., Sapir, S., Ohana, P., Benziman, M., *J. Bacteriol.* **1998**, *180*, 4416–4425.

13 (a) Huang, B., Whitchurch, C. B., Mattick, J. S., *J. Bacteriol.* **2003**, *185*, 7068–7076; (b) Rashid, M. H., Rajanna, C., Ali, A., Karaolis, D. K. R., *FEMS Microbiol. Lett.* **2003**, *227*, 113–119; (c) Garcia, B., Latasa, C., Solano, C., Portillo, F. G., Gamazo, C., Lasa, I., *Mol. Microbiol.* **2004**, *54*, 264–277; (d) Tischler, A. D., Camilli, A., *Infect. Immun.* **2005**, *73*, 5873–5882; (e) Hickman, J. W., Tifrea, D. F., Harwood, C. S., *Proc. Natl. Acad. Sci. USA* **2005**, *102*, 14422–1427; (f) Johnson, M. R., Montero, C. I., Conners, S. B., Shockley, K. R., Bridger, S. L., Kelly, R. M., *Mol. Microbiol.* **2005**, *53*, 664–674; (g) Lim, B., Beyhan, S., Yildiz, F. H., *J. Bacteriol.* 2007, **189**, 717–729.

14 (a) Amikam, D., Steinberger, O., Shkolnik, T., Ben-Ishai, Z., *Biochem. J.* **1995**, *311*, 921–927. (b) Steinberger, O., Lapidot, Z., Ben-Ishai, Z., Amikam, D., *FEBS Lett.* **1999**, *444*; 125–129.

15 Rashid, M. H., Rajanna, C., Ali, A., Karaolis, D. K. R., *FEMS Microbiol. Lett.* **2003**, *227*, 113–119.

16 For reviews about biofilm, see: (a) Gristina, A. G., *Science* **1987**, *237*, 1588–1595; (b) O'Toole, G., Kaplan, H. B., Kolter, R., *Annu. Rev. Microbiol.* **2000**, *54*, 49–79; (c) Lewis, K., *Antimicrob. Agents Chemother.* **2001**, *45*, 999–1007; (d) Stoodley, P., Sauer, K., Davies, D. G., Costerton, J. W., *Annu. Rev. Microbiol.* **2002**, *56*, 187–209; (e) Shirtliff, M. E., Mader, J. T., Camper, A. K., *Chem. Biol.* **2002**, *9*, 859–871; (f) Parsek, M. R., Singh, P. K., *Annu. Rev. Microbiol.* **2003**, *57*, 677–701; (g) Davies, D., *Nat. Rev. Drug Discov.* **2003**, *2*, 114–122; (h) Costerton, W. J., Wilson, M., *Biofilms* **2004**, *1*, 1–4; (i) Hall-Stoodley, L., Costerton, J. W., Stoodley, P., *Nat. Rev. Microbiol.* **2004**, *2*, 95–108.

17 Schachter, B., *Nat. Biotechnol.* **2003**, *21*, 361–365.

18 For reviews of *c*-di-GMP as a bacterial signaling molecule see: (a) Ross, P., Mayer, R., Benziman, M., *Microbiol. Rev.* **1991**, *55*, 35–58; (b) Jenal, U., *Curr. Opin. Microbiol.* **2004**, *7*, 185–191; (c) D'Argenio, D. A., Miller, S. I., *Microbiology* **2004**, *150*, 2497–2502; (d) Römling, U., Gomelsky, M., Galperin, M. Y., *Mol. Microbiol.* **2005**, *57*, 629–639; (e) Camilli, A., Bassler, B. L., *Science* **2006**, *311*, 1113–1116; (f) Jenal, U., Malone, J., *Annu. Rev. Genet.* 2006, *40*, 385–407; (g) Cotter, P. A., Stibitz, S., *Curr. Opin. Microbiol.* **2007**, *10*, 17–23.

19 Ross, P., Mayer, R., Weinhouse, H., Amikam, D., Huggirat, Y., Benziman, M., de Vroom, E., Fidder, A., de Paus, P., Sliedregt, L. A. J. M., van der Marel, G. A., van Boom, J. H., *J. Biol. Chem.* **1990**, *265*, 18933–18943.

20 Hayakawa, Y., Nagata, R., Hirata, A., Hyodo, M., Kawai, R., *Tetrahedron* **2003**, *59*, 6465–6471.

21 Hyodo, M., Hayakawa, Y., *Bull. Chem. Soc. Jpn.* **2004**, *77*, 2089–2093.

22 (a) Furusawa, K., Ueno, K., Katsura, T., *Chem. Lett.*, **1990**, 97–100; (b) Gundlach, C. W., Ryder, T. R., Glick, G. D., *Tetrahedron Lett.* **1997**, *38*, 4039–4042.

23 Heidenhain, S. B., Hayakawa, Y., *Nucleosides Nucleotides*, **1999**, *18*, 1771–1787.

24 Hayakawa, Y., Kawai, R., Hirata, A., Sugimoto, J., Kataoka, M., Sakakura, A., Hirose, M., Noyori, R., *J. Am. Chem. Soc.* **2001**, *123*, 8165–8176.

25 Hayakawa, Y., Hirata, A., Sugimoto, J., Kawai, R., Sakakura, A., Kataoka, M., *Tetrahedron* **2001**, *57*, 8823–8826.

26 Stec, W. J., Zon, G., *Tetrahedron Lett.* **1984**, *25*, 5279–5284.

27 (a) Hayakawa, Y., Kato, H., Uchiyama, M., Kajino, H., Noyori, R., *J. Org. Chem.* **1986**, *51*, 2400–2402; (b) Hayakawa, Y., Wakabayashi, S., Kato, H., Noyori, R., *J. Am. Chem. Soc.* **1990**, *112*, 1691–1696.

28 Hayakawa, Y., Hirose, M., Noyori, R., *J. Org. Chem.* **1993**, *58*, 5551–5555.

29 Hayakawa, Y., Uchiyama, M., Kato, H., Noyori, R., *Tetrahedron Lett.* **1985**, *26*, 6505–6508.

30 (a) Gasparutto, D., Livache, T., Bazin, H., Duplaa, A. -M., Guy, A., Khorlin, A., Molko, D., Roget, A., Téoule, R., *Nucleic Acids Res.* **1992**, *20*, 5159–5166; (b) Westman, E., Strömberg, R., *Nucleic Acids Res.* **1994**, *22*, 2430–2431; (c) Pirrung, M. C., Fallon, L., Lever, D. C., Shuey, S. W., *J. Org. Chem.* **1996**, *61*, 2129–2136.

31 Zhang, Z., Gaffney, B. L., Jones, R. A., *J. Am. Chem. Soc.* **2004**, *126*, 16700–16701.

32 Amiot, N., Heintz, K., Giese, B., *Synthesis*, **2006**, 4230–4236.

33 (a) Garegg, P. J., Lindh, I., Regberg, T., Stawinski, J., Strömberg, R., *Chem. Scr.* **1985**, *25*, 280–282; (b) Strömberg, R., Stawinski, J., *Current Protocols in Nucleic Acid Chemistry*, (Eds S. L. Beaucage, D. E. Bergstrom, P. Herdewijn, A. Matsuda), John Wiley & Sons, Inc. Hoboken, **2006**, pp.3.4.1–3.4.15.

34 Hyodo, M., Sato, Y., Hayakawa, Y., *Tetrahedron* **2006**, *62*, 3089–3094.

35 Zhang, Z., Kim, S., Gaffney, B. L., Jones, R. A., *J. Am. Chem. Soc.* **2006**, *128*, 7015–7024.

36 (a) O'Toole, G. A., Stewart, P. S., *Nat. Biotechnol.* **2005**, *23*, 1378–1379; (b) Rachid, S., Ohlsen, K., Witte, W., Hacker, J., Ziebuhr, W., *Antimicrob. Agents Chemother.* **2000**, *44*, 3357–3363; (c) Drenkard, E., Ausubel, F. M., *Nature* **2002**, *416*, 740–743; (d) Mah, T. -F., Pitts, B., Pellock, B., Walker, G. C., Stewart, P. S., O'Toole, G. A., *Nature* **2003**, *426*, 306–310; (e) Hoffman, L. R., D'Argenio, D. A., MacCoss, M. J., Zhang, Z., Jones, R. A., Miller, S. I., *Nature* **2005**, *436*, 1171–1175.

37 Karaolis, D. K. R., Rashid, M. H., Chythanya, R., Luo, W., Hyodo, M., Hayakawa, Y., *Antimicrob. Agents Chemother.* **2005**, *49*, 1029–1038.

38 Brouillette, E., Hyodo, M., Hayakawa, Y., Karaolis, D. K. R., Malouin, F., *Antimicrob. Agents Chemother.* **2005**, *49*, 3109–3113.

39 Karaolis, D. K. R., Means, T. K., Yang, D., Takahashi, M., Yoshimura, T., Muraille, E., Philpott, D., Schroeder, J. T., Hyodo, M., Hayakawa, Y., Talbot, B. G., Brouillette, E., Malouin, F., *J. Immunol.* **2007**, *178*, 2171–2181.

40 Ebensen, T., Schulze, K., Riese, P., Link, C., Morr, M., Guzman, C. A., *Vaccine* **2007**, *25*, 1464–1469.

41 Karaolis, D. K. R., Cheng, K., Lipsky, M., Elnabawi, A., Catalano, J., Hyodo, M., Hayakawa, Y., Raufman, J. -P., *Biochem. Biophys. Res. Commun.* **2005**, *329*, 40–45.

14
Siderophore Biosynthesis Inhibitors

Courtney C. Aldrich and Ravindranadh V. Somu

14.1
Introduction

The nucleoside antibiotic **1** (Sal-AMS) exhibited potent inhibition of *Mycobacterium tuberculosis* growth, the causative agent of tuberculosis [1, 2]. Additionally, **1** was shown to possess moderate activity against the Gram-negative pathogen *Yersinia pestis*, which is the etiological agent of the plague [2]. Nucleoside **1** was rationally designed to inhibit an adenylating enzyme involved in siderophore biosynthesis of these bacteria. Although, like almost all forms of life, microorganisms require iron for survival, this essential nutrient is highly sequestered in a mammalian host where the concentration of free iron is 10^{-18} M, which is far too low to sustain microbial growth. Bacteria and fungi have evolved a number of mechanisms to obtain this vital micronutrient, but the most common mechanism involves the synthesis, secretion, and reuptake of small molecule iron chelators which are known as siderophores [3] which have been demonstrated as essential components of the iron acquisition system (Figure 14.1). Thus, the inhibition of siderophore biosynthesis represents a logical strategy for the development of a new class of antibiotics, and Sal-AMS is the first confirmed inhibitor of siderophore biosynthesis *in vitro* [1, 2].

14.2
Synthesis, Physico-Chemical Properties, Metabolism, Mechanism of Action, and Biological Activity

14.2.1
Synthesis

Sal-AMS was synthesized by the sulfamoylation of commercially available 2′,3′-*O*-isopropylideneadenosine **2** (Scheme 14.1). Coupling to the *N*-hydroxysuccinimide ester of 2-*O*-benzylsalicylate **5** employing Cs_2CO_3 furnished acylsulfamate

Modified Nucleosides: in Biochemistry, Biotechnology and Medicine. Edited by Piet Herdewijn

ISBN: 978-3-527-31820-9

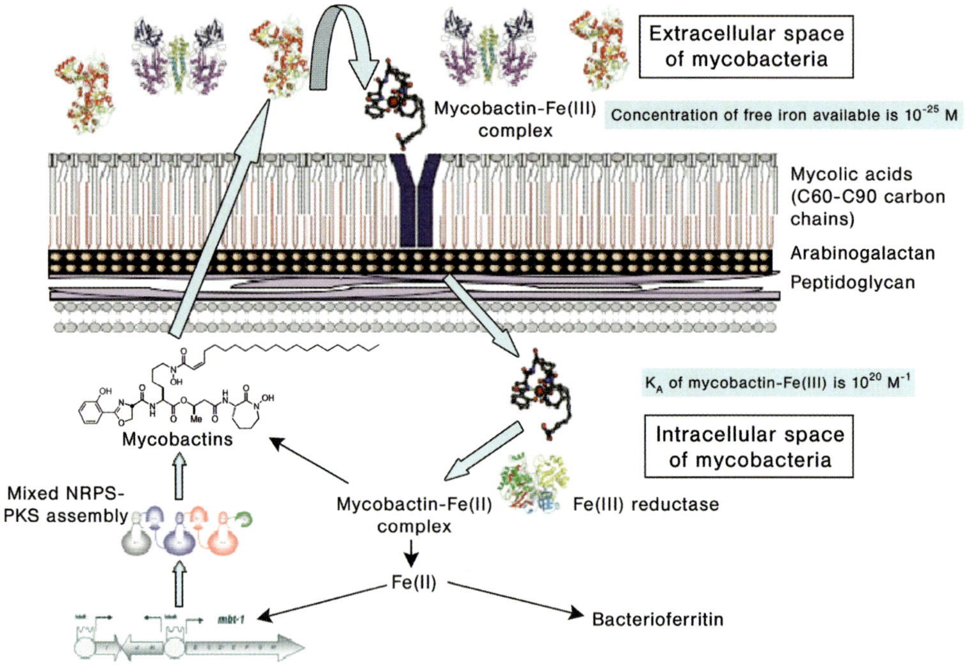

Figure 14.1 Iron acquisition in *Mycobacterium tuberculosis*. In pulmonary tuberculosis, the bacterium *Mycobacterium tuberculosis* lives in alveolar macrophages in the lung. Inside the macrophages, *M. tuberculosis* biosynthesize the siderophores known as the mycobactins, which are secreted across the mycobacterial cell envelope. The mycobactin siderophore abstracts iron from host proteins and the iron–siderophore complex is imported by a dedicated transport protein. In the cytosol, the iron is released, most likely by a reductase, which reduces mycobactin·Fe(III) to mycobactin·Fe(II). The reduced stability constant of the mycobactin·Fe(II) complex leads to the dissociation of Fe(II), which is subsequently stored in the bacterial iron storage protein bacterioferritin or utilized by one of the approximately 40 proteins that use iron as an obligate cofactor. Fe(II) also binds to the transcriptional repressor IdeR, which regulates the mycobactin operon.

derivate **6**. Sequential deprotection of the benzyl ether by catalytic hydrogenolysis and the isopropylidene ketal with aqueous trifluoroacetic acid (TFA) provided **1**. Although this compound was observed to be unstable during silica gel chromatography, it could be isolated by co-eluting with 1% triethylamine (TEA) to afford the TEA salt of **1** [1]. Ion-exchange of **1**·TEA to the sodium salt (**1**·Na) was readily achieved by passage through a Dowex-50WX2 column in the sodium form.

14.2.2
Physico-Chemical Properties

The acylsulfamate linkage of inhibitor **1** is extremely acidic with a calculated pK_a of 0.6, while the pK_a of **7** (see Scheme 14.1) was determined experimentally as 2.8 (calculated 3.4), demonstrating that the *ortho*-hydroxy function of the salicyl moiety

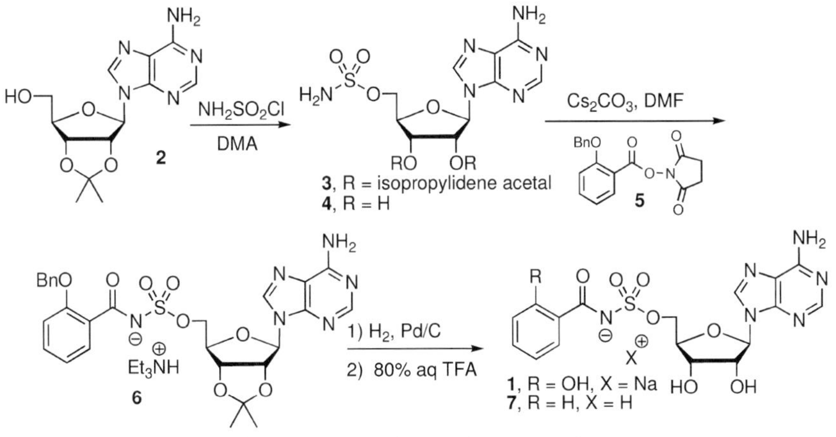

Scheme 14.1 Synthesis of 5′-*O*-[*N*-(Salicyl)sulfamoyl]adenosine (Sal-AMS, **1**).

modulates the acidity of the NH proton through resonance delocalization [4]. The acylsulfamate linkage is considerably more stable than the conjugate base, as the negative charge on the nitrogen atom deters nucleophilic attack on the carbonyl function. Significantly, it has been shown that **7**, which was isolated as the free acid, decomposed rapidly over several days at 25 °C in CD_3OD, but was unchanged after 30 days as the sodium salt. Sal-AMS is extremely polar, with a calculated logP (clogP) of −0.89; thus, analogues that increase the lipophilicity are expected to provide compounds with improved membrane permeability and corresponding enhanced biological activity and bioavailability.

14.2.3
Metabolism

A principle mechanism of metabolism of adenosine nucleoside derivatives is through oxidative deamination of the N^6-amino function catalyzed by adenosine deaminase, which is part of the purine salvage pathway. Another common enzyme responsible for the metabolism of purine nucleosides is purine nucleoside phosphorylase, which catalyzes the reversible phosphorolysis of ribonucleosides. Thus, the enzymatic stability of **1** toward adenosine deaminase and purine nucleoside phosphorylase was examined using established spectroscopic assays; however, **1** was not a substrate for either of these enzymes [5].

14.2.4
Toxicity

Preliminary investigations on the toxicity of **1** have been performed against murine P388 leukemia and Vero cells. Compound **1** did not display any toxicity up to the maximum concentration evaluated (200 μM) against both cell lines. 5′-*O*-(Sulfamoyl) adenosine **4** (Scheme 14.1) is one of the most cytotoxic compounds yet reported [6], and was used as a positive control in the cytotoxicity experiments; thus, the lack of toxicity of **1** is encouraging. Nevertheless, the potential for metabolism of **1** to release 5′-*O*-(sulfamoyl)adenosine is a legitimate concern for future development of this compound, and has motivated the synthesis of hydrolytically stable analogues that cannot release **4** [7].

14.2.5
Biochemical Target

Sal-AMS has been shown to be a potent reversible competitive nanomolar inhibitor of the adenylation enzymes MbtA, YbtE, and PchD which are involved in the first step of the biosynthesis of the siderophores mycobactin, yersiniabactin, and pyochelin, respectively (Table 14.1). The reported values do not measure the intrinsic potency, however, as the inhibition studies were conducted under supersaturating concentrations (50–150 × K_M) of the competitive substrates salicylic acid (SAL, 250 μM) and ATP (10 mM). Recent studies employing isothermal calorimety have revealed that the true dissociation constant of **1** toward MbtA is an astonishing 1.0×10^{-13} M, thereby demonstrating the exceptional potency of this bisubstrate inhibitor [8]. However, the apparent K_I values may reflect the true *in-vivo* potency as ATP and salicylate are abundant metabolites in many siderophore-producing microorganisms.

Table 14.1 A summary of the physical, biochemical, and biological properties of Sal-AMS (**1**).

Physical/biochemical properties		Biological properties		
K_I^{app}/IC_{50}		MIC_{99}	Iron-deficient[f]	Iron-rich[g]
MbtA	10.7 nM[b]/6.6 nM[a]	*M. tuberculosis*[a]	0.39 μM	1.56 μM
YbtE	14.7 nM[b]	*Y. pseudotuberculosis*[a]	20.0 μM	>400 μM
Pch	12.5 nM[b]	*Y. pestis*[b]	51.2 μM*[i]	>400 μM
		E. coli[d]	>200 μM	>200 μM
		P. aeruginosa[d]	>200 μM	>200 μM
cpKa[c]	0.6	CC_{50}[h]		
clogP[a]	–0.89	P388[a] >200μM		
		Vero[e] >200μM		

[a]Somu, R. V. *et al., J. Med. Chem.* (2006), *49*, 7623.
[b]Ferreras J. A. *et al., Nat Chem. Biol.* (2005), *1*, 29, Note the reported value is the IC_{50}.
[c]Somu, R. V. *et al., J. Med. Chem.* (2006), *49*, 37.
[d]Unpublished data; B. Beck, L. Celia, American Type Culture Collection, Manassas, Virginia.
[e]Unpublished data, A. Gupte, C. Aldrich, University of Minnesota, Minneapolis, Minnesota.
[f]Minimum inhibitory concentration (MIC) required to inhibit >99% growth under iron-deficient conditions.
[g]Minimum inhibitory concentration (MIC) required to inhibit >99% growth under iron-rich conditions.
[h]Cell cytotoxicity (CC) at which 50% cell-growth is inhibited.
[i]MIC_{50} value at which 50% inhibition of growth was observed.

14.2.6
Mechanism of Action

Compound **1** was evaluated for its ability to inhibit siderophore biosynthesis in *Mycobacterium tuberculosis* and *Yersinia pestis* by employing a radioassay with [7-^{14}C]-salicylic acid [9]. In the absence of inhibitor, siderophore production by these organisms was robust, whilst in the presence of 20 to 200 μM **1** complete inhibition of siderophore biosynthesis was observed [1, 2].

14.2.7
Biological Activity

Sal-AMS has been evaluated against several Gram-negative bacteria that require siderophore production for virulence, as shown in Table 14.1. Sal-AMS displayed impressive activity towards *M. tuberculosis* under iron-deficient conditions where the minimum inhibitory concentration (MIC_{99}) that inhibited growth >99% was 0.39 μM [1]. However, **1** also inhibited *M. tuberculosis* under iron-rich conditions, with an MIC_{99} value of 1.56 μM. Under iron-rich conditions, siderophore production is not essential, suggesting off-target binding and inhibition of alternate biochemical pathways. Nevertheless, the potent inhibition of *M. tuberculosis* is quite impressive, especially considering the imposing mycobacterial cell envelope, which provides intrinsic resistance to many antibiotics. By contrast, the first-line antitubercular agent isoniazid, which disrupts cell-wall biosynthesis, was equally effective under iron-rich and iron-deficient conditions. Sal-AMS was also effective against *Yersinia pseudotuberculosis* ATCC 6902, with an MIC_{99} of 20 μM under iron-deficient conditions and no inhibition under iron-rich conditions, a phenotype concordant with the designed mechanism of action [10]. Additionally, **1** exhibited moderate activity against *Y. pestis*, with an MIC_{50} of 51 μM and >400 μM under iron-deficient and iron-rich conditions, respectively [2]. Bioinformatic analysis of *M. tuberculosis* and *Y. pseudotuberculosis* suggests that the observed phenotypic difference between these organisms elicited by **1** may be due to the presence of over 30 functionally related fatty acid adenylating (FadD) enzymes found in *M. tuberculosis*, but not in *Y. pseudotuberculosis* [11]. Many of these FadD enzymes catalyze essential reactions in the biosynthesis of *M. tuberculosis* lipids, and may represent potential off-target proteins inhibited by **1**. An evaluation of Sal-AMS against *Escherichia coli* and *Pseudomonas aeruginosa* showed no bacteriostatic activity, revealing a rather limited spectrum of activity for this newly designed antibiotic [12]. The preparation of an antibiotic with a narrow therapeutic window against a particular pathogen such as *M. tuberculosis* is expected to be advantageous if the beneficial bacterial flora of the host are not affected. The lack of activity of **1** toward *E. coli* can be reconciled, however, as this Gram-negative organism has multiple redundant iron acquisition pathways, while the lack of activity toward *P. aeruginosa* may be attributed to drug-efflux pumps, which are known to confer this organism with resistance against many antibiotics.

14.3 Background of Siderophores: Molecular Target and Rationale for Inhibitor Design

Many of the most prominent pathogens rely on aryl-capped peptidic siderophores for iron acquisition. For example, *M. tuberculosis* (tuberculosis) produces the mycobactins [9], *Y. pestis* (plague) and *Klebsiella pneumoniae* (opportunistic infections) synthesize yersiniabactin [13, 14], *Bacillus anthracis* (anthrax) makes petrobactin [15], *E. coli* produces the enterobactins [16], and *Acinetobacter baumannii* (opportunistic infections) synthesizes the acinetobactins [17] (Figure 14.2). The biosynthesis of aryl-capped siderophores is performed by a class of multifunctional enzymes known as the non-ribosomal peptide synthetases (NRPSs), as they operate independently of the

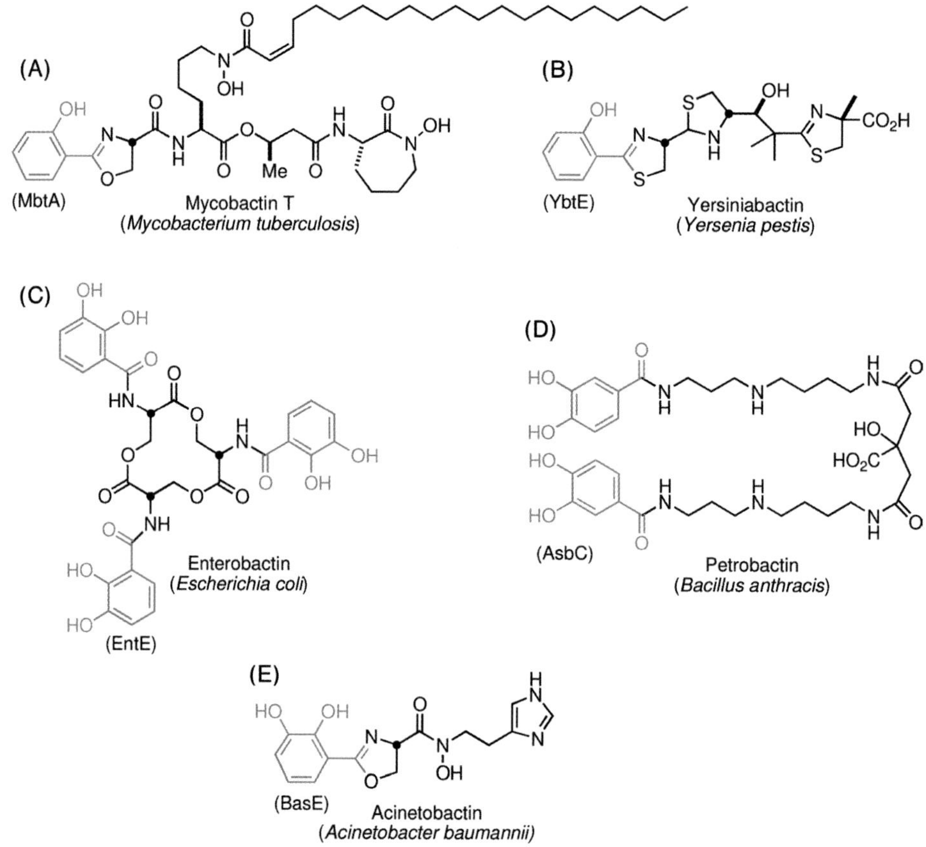

Figure 14.2 Representative siderophores of human pathogens. (A) *Mycobacterium tuberculosis* produces the mycobactins here represented by the lipid-soluble mycobactin-T. (B) *Yersinia pestis* and *Yersinia pseudotuberculosis* both require the yersiniabactins for iron acquisition. (C) *E. coli* produces the enterobactins. (D) *Bacillus anthracis* relies on petrobactin for iron acquisition *in vivo* (E) *Acinetobacter baumannii* synthesizes the acinetobactins.

mRNA templated ribosomal machinery and function analogously to the well-studied canonical type I polyketide synthases (PKSs) with their modular organization and use of a thiotemplate mechanism. At the core of an NRPS assembly line is a module, which is comprised of three essential domains: (i) the *Adenylation* domain that catalyzes the activation of an amino- or aryl acid to the corresponding AMP-ester; (ii) the *Thiolation* or *Carrier* domain on which the nascent peptides are covalently bound during synthesis as a thioester; and (iii) the *Condensation domain*, which catalyzes formation of the peptide bond between activated amino acids on adjacent carrier protein domains. The domains are usually embedded in multifunctional proteins and each module is responsible for one step of elongation.

The biosynthetic pathway of the mycobactins is shown in Figure 14.3, to illustrate a representative NRPS pathway. The mycobactins core scaffold is synthesized through the activity of six enzyme MbtA to MbtF that collectively comprises a five-module mixed NRPS–PKS assembly line. Biosynthesis is initiated by the stand-alone adenylation enzyme MbtA, which activates salicylic acid at the expense of ATP and loads this onto the N-terminal thiolation domain of MbtB where it undergoes sequential elongation by serine (catalyzed by MbtB), lysine (catalyzed by MbtE), two malonyl CoAs (to form the β-hydroxybutyrate residue; catalyzed by the PKS enzymes MbtC and MbtD), and another molecule of lysine (catalyzed by MbtF) [18]. Processing of this growing chain by the mixed NRPS–PKS assembly line leads to the fully elaborated chain intermediate that is released through lactamization by cyclization of the final lysine residue [19]. Additional tailoring modifications through sequential lipidation (catalyzed by MbtK) and *N*-hydroxylation (catalyzed by MbtG) of the lysine residues affords the mycobactins [20].

Adenylation domains/enzymes thus play a key role as they are responsible for selecting, activating, and catalyzing the loading of the appropriate amino- or aryl acid from the large pool of available building blocks (including the 20 proteinogenic amino acids) onto the downstream carrier domains, and do so with exquisite selectivity [21]. NRPS adenylation domains/enzymes are members of the adenylate-forming enzyme superfamily. These enzymes are between 500 and 700 residues in length, and contain a large N-terminal domain and a much smaller C-terminal domain [22]. The two-step reaction catalyzed by the adenylate-forming enzyme superfamily, as well as the functionally related aminoacyl tRNA synthetases that are responsible for the activation of amino acids and loading to the cognate tRNA molecule, is mechanistically identical and is illustrated in Figure 14.4 for the adenylating enzyme MbtA and it cognate thiolation domain involved in mycobactin biosynthesis in *M. tuberculosis*. In the first half-reaction, binding of both the substrate acid **8** and ATP is followed by nucleophilic attack of the substrate carboxylate on the α-phosphate of ATP to generate the acyladenylate **9** and the release of pyrophosphate (Figure 14.4a). In the second half-reaction, the enzyme binds the acceptor residue and transfers the acyladenylate **9** to a nucleophilic oxygen or sulfur atom of the acceptor, leading to product **10** (Figure 14.4b). The acceptor residue for aryl adenylating enzymes is the terminal sulfur atom of the phosphopantetheinyl cofactor of the carrier domain, whereas for aminoacyl tRNA synthetases this is the 2′ or 3′ alcohol from the ribose sugar of the terminal adenosine residue of the tRNA molecule [23, 24].

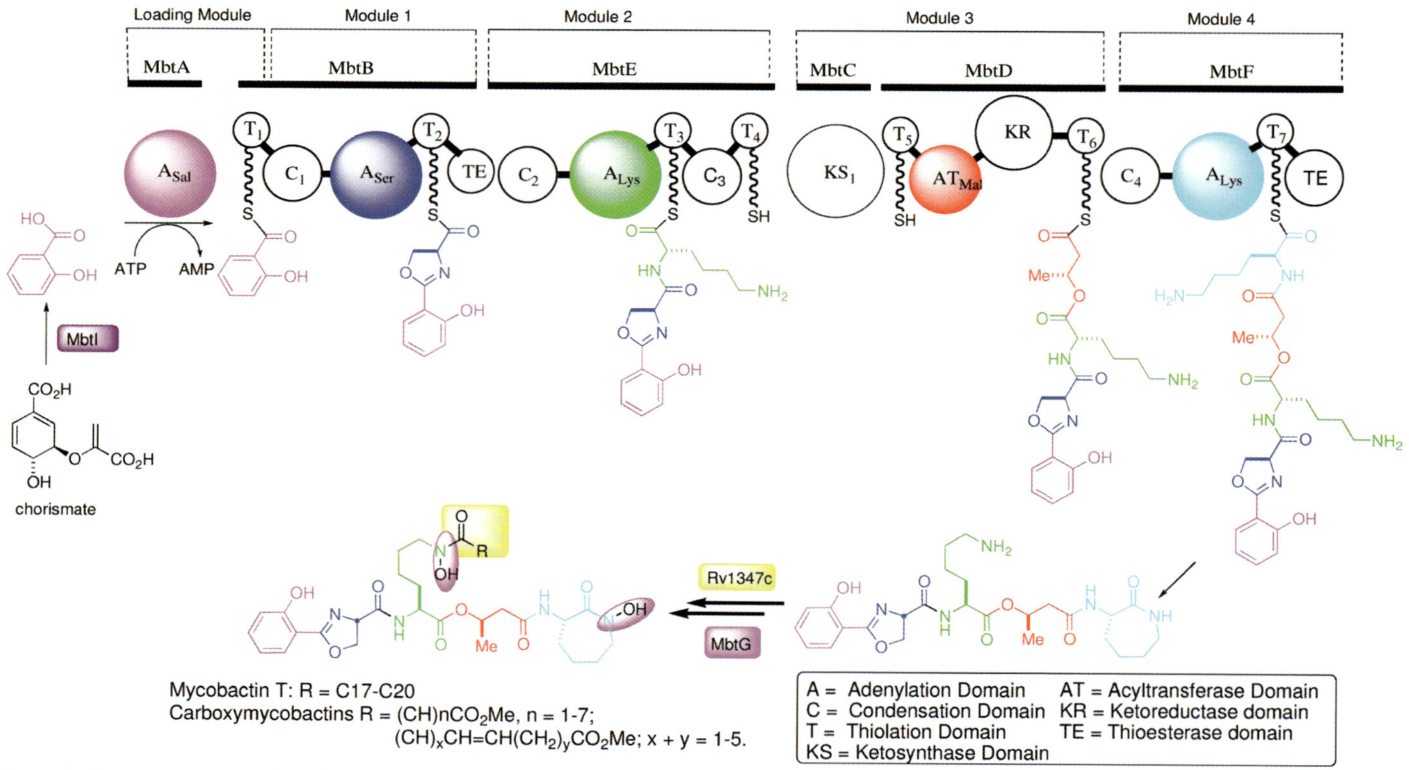

Figure 14.3 The mycobactin NRPS–PKS assembly line. The stereochemistry of the carboxymycobactins is assigned in analogy to the mycobactins. The individual domains are represented by circles, where the size of the circle is approximately proportional to the molecular weight of the individual protein domain, and the lines connecting the circles represent the interdomain linker regions. The curled lines represent the phosphopantetheinyl (ppan) cofactor of the carrier protein domain (T) on which the growing peptidic chain is covalently attached. The TE domain of MbtF was annotated as an epimerase; however, it has been defined as a thioesterase as it is expected that this domain will catalyze the macrolactamization and release of the full-length chain intermediate.

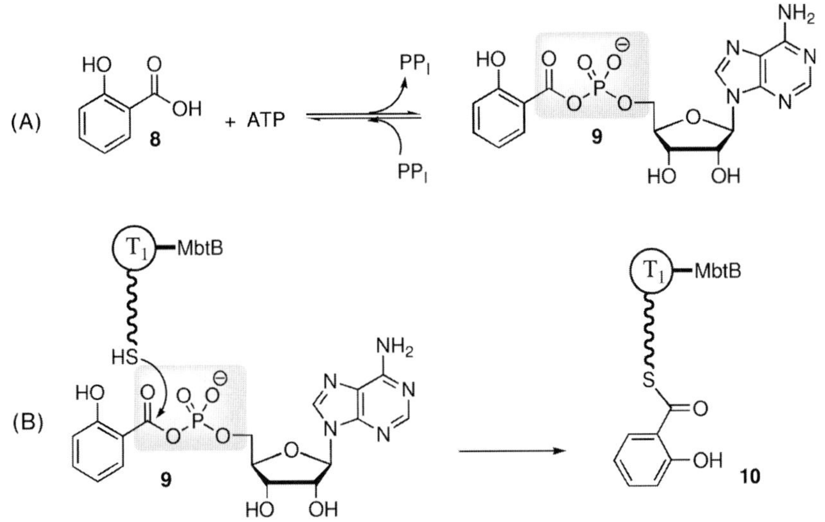

Figure 14.4 Mechanism of adenylate-forming enzymes and the functionally related aminoacyl-tRNA synthetases (aaRSs). (A) Adenylation half-reaction. (B) Acylation half-reaction.

The aminoacyl-tRNA synthetases (aaRSs) are responsible for loading each tRNA with their cognate amino acid, and have attracted considerable attention in recent years as antibacterial targets as they catalyze a fundamental and essential metabolic process [25, 26]. Aminoacyl tRNA synthetase inhibitors have already been developed and are used clinically (e.g., mupirocin, a natural product-derived topical antibiotic, which inhibits isoleucyl tRNA synthetase). A central strategy used to generate aaRS inhibitors, and which has been used to prepare aryl adenylation inhibitors, is based on the observation that the acyladenylate formed in the first half-reaction has been shown to bind tightly to the aaRSs with reported nanomolar dissociation constants, whereas the dissociation constants of the substrate amino acids are in the micromolar region [27]. The tight binding of the acyladenylate immediately suggests that a stabilized analogue of this intermediate would provide a potent enzyme inhibitor. This tight binding is essential to ensure that the acyladenylate is not lost to the bulk solvent through diffusion before this is channeled to the corresponding acceptor residue (tRNA-OH, T-domain–SH). Additionally, sequestering of the acyladenylate by the enzyme also prevents entry of water into the active site, thus preventing adventitious hydrolysis of the mixed phosphoric–carboxylic acid anhydride.

The acylsulfamate linkage, inspired from the natural product ascamycin **11** [28] has been extensively employed as a stable bioisostere of the labile acylphosphate linkage for the development of aaRS inhibitors (Figure 14.5) [29–36]. These inhibitors typically provide potent inhibition of the corresponding aaRSs. As an example, the L-prolyl analogue **12** displayed inhibition of the corresponding *E. coli* prolyl-aaRS with a $K_I = 4.3$ nM [37]. However, one of the difficulties of this approach has been the ability to obtain selective inhibition of bacterial enzymes, and hence this same analogue also inhibited the human prolyl-aaRS enzyme with even greater potency

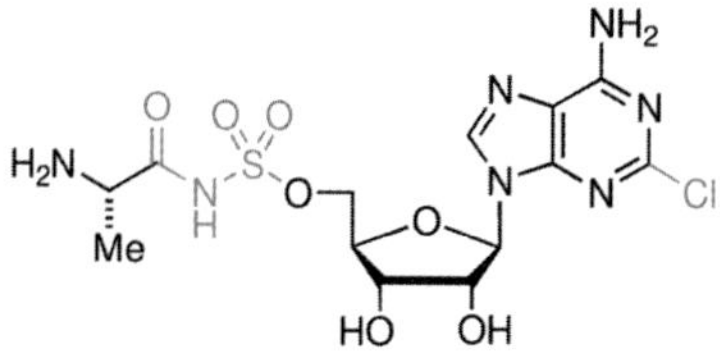

Ascamycin (**11**)

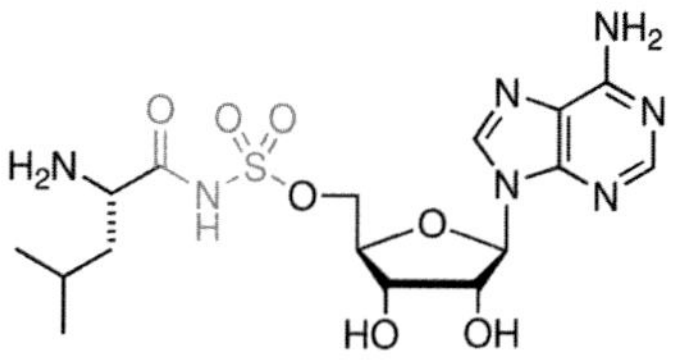

5'-*O*-[*N*-(L-Leucyl)sulfamoyl]adenosine (**13**)

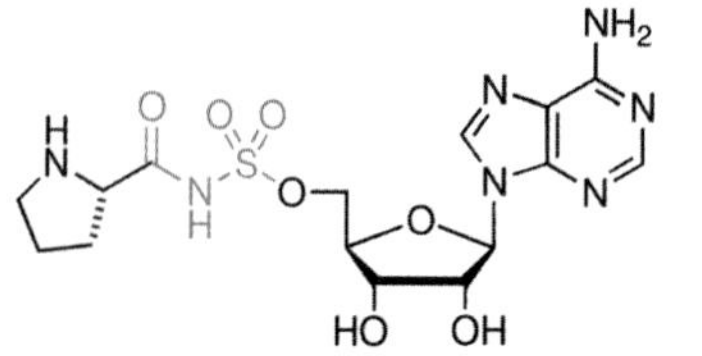

5'-*O*-[*N*-(L-Prolyl)sulfamoyl]adenosine (**12**)

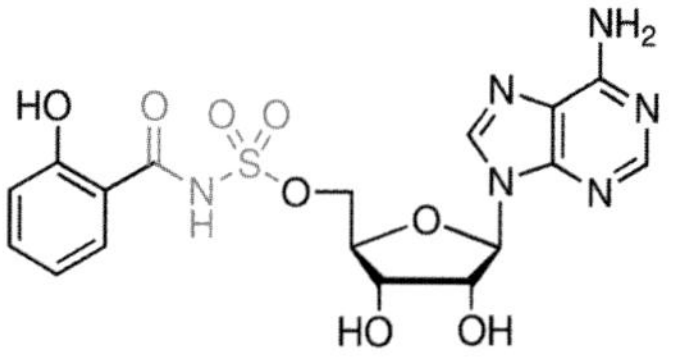

5'-*O*-[*N*-(Salicyl)sulfamoyl]adenosine (**1**)

Figure 14.5 Acyladenylate inhibitors of aminoacyl-tRNA synthetases and NRPS aryl adenylating enzymes.

($K_I = 0.6$ nM). NRPS adenylation enzymes are potentially much better targets than aaRSs as they catalyze the activation of many non-natural amino acids and aryl acids respectively for which there are no equivalent counterparts in human metabolism. Marahiel and coworkers demonstrated that inhibitors developed for the aaRSs could be used to inhibit adenylation domains from NRPS systems, despite the lack of either sequence or structural similarity. These authors showed that leucyl-AMS **13** was a potent inhibitor ($K_I = 8.4$ nM) of the leucyl adenylation domain involved in the biosynthesis of surfactin, a lipopeptide antibiotic. The observed inhibition constant was five orders of magnitude lower than the respective K_M value for leucine [38]. Application of this strategy to aryl acid adenylation domains involved in siderophore biosynthesis was subsequently successfully demonstrated by several research teams of microbiologists and chemists, including Quadri/Tan, Marahiel/Eustache, and Barry/Aldrich. Salicyl-AMS (**1**) is a nanomolar inhibitor of MbtA and YbtE, the adenylation enzymes that catalyze the first step of mycobactin- and yersiniabactin biosynthesis [1, 2, 39].

14.4 Ligand Properties/Binding Mode

Sal-AMS **1** was the first confirmed inhibitor of siderophore biosynthesis, and represents a prototype for a new class of antibiotics. As seen in Figure 14.6, the inhibitor scaffold can be conveniently disconnected into four modules: aryl, linker, sugar, and base. Structure–activity (SAR) studies of **1** have focused on the activity towards the adenylating enzyme MbtA responsible for mycobactin biosynthesis in *M. tuberculosis*, the causative agent of tuberculosis. Acyladenylate mimetics such as salicyl-AMS are considered bisubstrate inhibitors because they are expected to

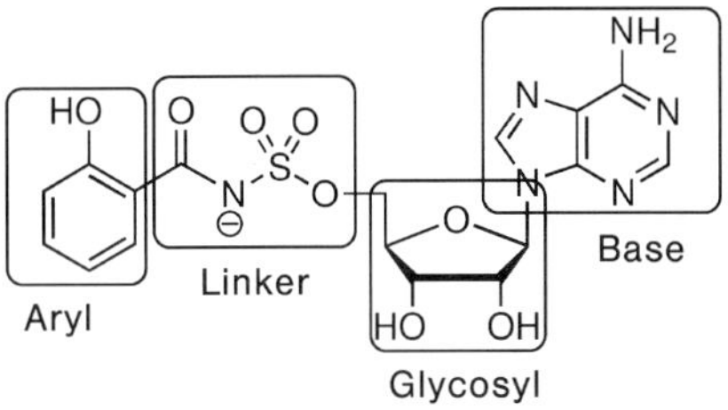

Figure 14.6 The modular scaffold of Sal-AMS (**1**).

interact with both substrate-binding pockets. Bisubstrate inhibitors that simply mimic the acyladenylate intermediate can be further classified as intermediate mimetics [40].

Although the three-dimensional structure of MbtA has not been determined, the sequence is highly homologous to the known co-crystal structure of DhbE, with an adenylated dihydroxybenzoate in the active site [41]. A sequence alignment shows 42% identity between the proteins overall, with significantly higher homology in the ligand binding site. Of the 21 residues contacting the adenylate ligand in DhbE (within 4 Å), 16 are identical in MbtA and the remaining five residues represent conservative changes (Y236F, S240C, A308S, V337L, T411S). The model of active compound **1** maintained all the ligand/protein hydrogen bonds observed for dihydroxybenzoate bound to DhbE, except for hydrogen bonds made by the benzoic acid meta-hydroxy, which is missing in **1** (Figure 14.7). In DhbE, this hydroxy hydrogen bonds to Ser240, which is a cysteine in MbtA and therefore a weaker hydrogen bonding partner. In addition, the nearby substitution V337L makes the presence of a meta hydroxy sterically unfavorable. Therefore, residues 240 and 337 probably contribute to the specificity differences of DhbE and MbtA.

14.4.1 Nature of the Linker

The most crucial consideration of the inhibitors is the linker, as this must be metabolically stabile as opposed to the natural acylphosphate moiety and appropriately position both the nucleoside and the aryl domains to obtain optimal interactions with their respective binding sites. Observations with related bisubstrate inhibitors developed for aminoacyl tRNA synthetases has shown that the linker region is very sensitive to modification [25, 26]. Attempts to either increase [35] or decrease [42] the length reduced potency dramatically. Thus, most analogues investigated have maintained the native linker spacing, but explored modifications to the both the molecular geometry and polarity of the linker pharmacophore. These modifications to the linker afforded intermediate mimetics and transition-state analogues of the adenylation and acylation half-reactions.

The acylsulfamide **14**, followed closely by the acylsulfamate **1**, provided the most potent compounds of all linkers examined, with apparent K_I values toward MbtA of 5.1 nM and 3.8 nM respectively (Table 14.2). The high activity of acylsulfamide **14** was unexpected as this function has not been well tolerated with the related aminoacyl

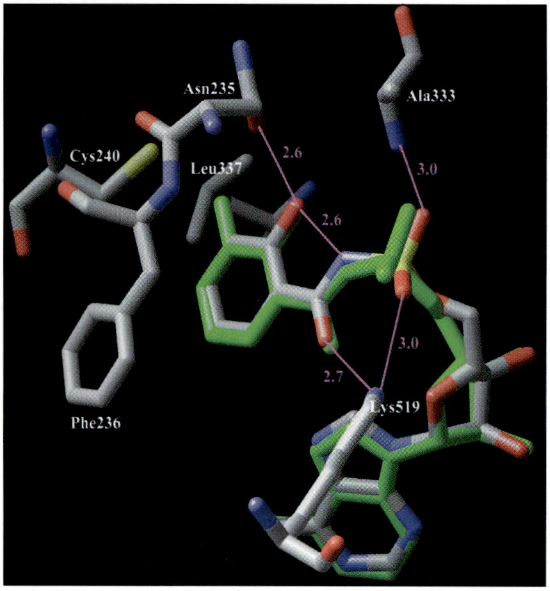

Figure 14.7 Model of **1** bound to MbtA. Legend: Compound **1** (colored by atom) was modeled into the DhbE active site after making several amino acid changes to reflect the sequence of MbtA. Dihydroxybenzoate from the DhbE X-ray structure is overlaid in green. Hydrogen bonds in the model of **1** are shown using purple lines with distances in Ångstroms. Only key residues affecting salicyl and linker binding are shown. For comparison with the DhbE structure, residues are numbered according to that sequence. Hydrogens removed for clarity.

tRNA synthetases, typically resulting in greater than a 1000-fold loss in binding affinity [32, 33]. Replacement of the central nitrogen atom of acylsulfamide **14** by a carbon atom in β-ketosulfonamide **16** resulted in a profound 870-fold loss in binding affinity (**16** versus **14**), while substitution with a CF_2 group in **17** led to more than a 20 000-fold loss of activity. The pK_a of the NH proton of a simple acylsulfamate linkage of **1** is acidic, with a calculated pK_a of 0.6 [10]. Consequently, the NH proton is fully ionized at physiological pH, and this negative charge may mimic the transition state of phosphoryl transfer during the adenylation half-reaction and may account for a large part of this 870-fold difference in activity. Docking studies of the parent ligand **1** into a homology model of MbtA, as well as quantum mechanic calculations free from the constraints of the active site, show that the salicyl moiety of **1** adopts a coplanar conformation that is stabilized by an internal hydrogen bond where the phenol acts as a hydrogen bond donor to the sulfamate nitrogen, which is deprotonated (Figure 14.8). The X-ray co-crystal structure of DhbE with adenylated 2,3-dihydroxybenzoic acid also shows a similar coplanar arrangement. By contrast, **16** and **17** deviated significantly from planarity.

Sulfamate derivative **15** (Table 14.2) that lacks the linkage carbonyl displayed a 80-fold loss in potency relative to **1** (Table 14.1, entry 1). This was consistent with modeling studies, which show that the carbonyl *syn*-lone pair of **1** H-bonds with Lys542 of MbtA. However, compound **15** is significantly less polar as the pK_a of the

Table 14.2 SAR of the linker domain.

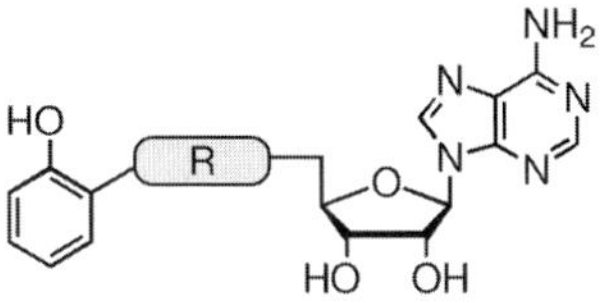

Compound	R=	K_i^{app}(nM)[a]	MIC_{99}(μM)[b] (iron-poor)	MIC_{99}(μM)[c] (iron-rich)
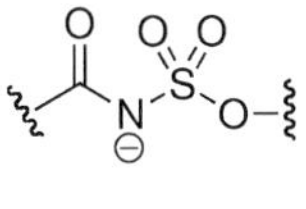	1	5.1 ± 1.1	0.29	1.56
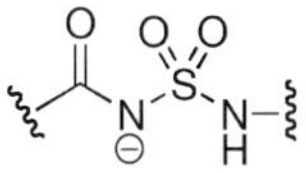	14	3.8 ± 0.6	0.19	0.39
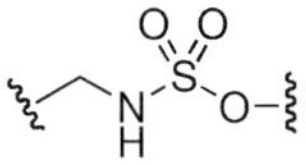	15	410 ± 60	>100	>100
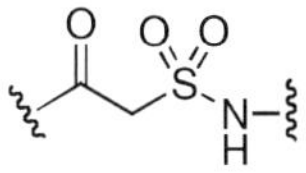	16	3300 ± 570	25	>100
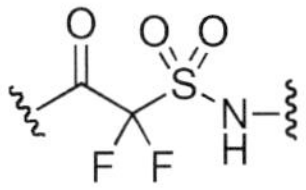	17	>100,000	>100	>100
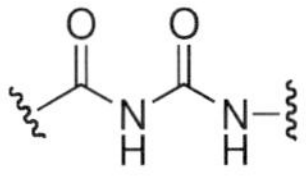	18	12.50 ± 50	200	>200
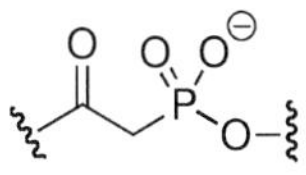	19	>100,000	>100	>100
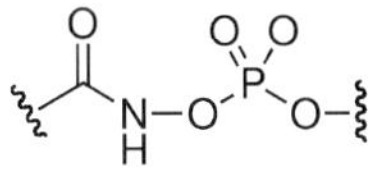	20	n.d.	n.d.	n.d.
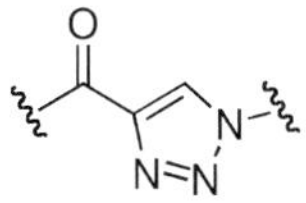	21	>100,000	>100	>100
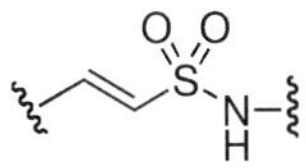	22[d]	$(143 \pm 9) \times 10^3$	100	>100

[a] ATP-PPi exchange assay.
[b] Grown in GAST media without Fe^{3+}.
[c] Grown in GAST media supplemented with 200 μM Fe^{3+}.

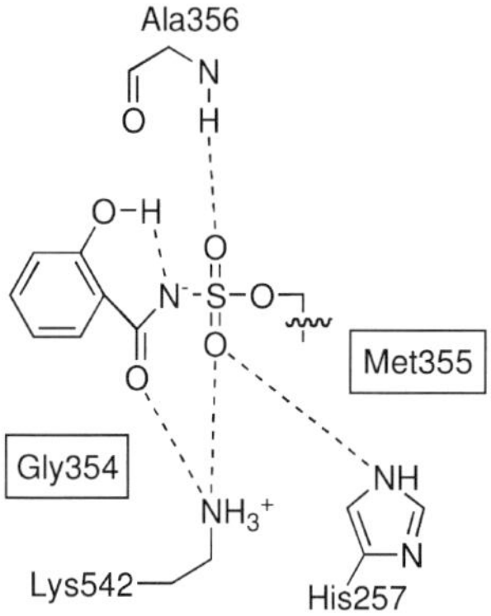

Figure 14.8 Putative binding interactions of the linker domain of **1** with a MbtA homology model.

sulfamate NH group is estimated as 7.5, in contrast to the ionized NH groups of the acylsulfamate linkage of **1**. Consequently, the modest decrease in binding potency of **15** was expected to be largely offset by its improved pharmacodynamic properties.

The complete loss of activity of compound **19** is likely due to its inability to adopt the required coplanar conformation as described for analogues **1** and **14**. The improved activity of β-ketosulfonamide **16** relative to **19** suggests that the sulfonyl group of **16** interacts more favorably than the phosphate function with MbtA. A β-ketophosphonate linkage was also found to be poor surrogate for the native acylphosphate linkage for a valine tRNA synthetase inhibitor [43]. Although MbtA and *E. coli* valyl tRNA synthetase are structurally unrelated, the result underscores how exquisitely sensitive the linker region is toward modification.

Acylurea analogue **18** explored the importance of the tetrahedral geometry of the sulfonyl moiety for activity and displayed modest activity (1.25 μM), which represents a 330-fold loss of binding affinity relative to acylsulfamide **14**. The acylurea contains an ionizable NH group in analogy to **1** and **14**, and thus may also represent a transition state analogue of the adenylation half-reaction. Acyltriazole **21** displayed no affinity towards MbtA, which was concordant with modeling studies that had shown that this was a poor fit for the active site, requiring out-of-plane bending of the ring substituents.

Callahan and coworkers recently synthesized compound **20** (Table 14.2), which contains the novel acylhydroxamoylphosphate linker that is approximately 2 Å longer than the native acylphosphate linkage [44]. These authors showed that **20** was a potent nanomolar reversible inhibitor of EntE, the adenylating enzyme involved in enterobactin biosynthesis, and is expected to be equally potent against MbtA as these enzymes share almost absolute conservation of their active site residues. Compound **20** can be considered a transition-state inhibitor of the adenylation half-reaction. Although the ionized phosphate group may hinder membrane permeability, the straightforward one-step synthesis and high activity of **8** are very encouraging.

Analogues **1** and **14** to **21** were designed as reversible adenylation inhibitors, whereas vinylsulfonamide **22** (Table 14.2) was designed as an irreversible inhibitor of the downstream thiolation domain [45]. Vinylsulfonamide **22** is an acyladenylate surrogate that contains a Michael acceptor at the precise position of the incoming nucleophilic thiol function of the phosphopantetheine cofactor arm of the downstream aryl carrier domain of MbtB during acyl group transfer (see Figure 14.4b). Roush and coworkers recently ranked the relative activities of Michael acceptors: enone > vinylsulfone > vinylsulfonate > enoate > vinylsulfonamide [46]. The vinylsulfonamide was chosen initially as this is the least reactive member in the series to minimize nonspecific thiol addition. Additionally, molecular modeling showed that the vinylsulfonamide adopted the required folded conformation of the bound acyladenylate. Compound **22** was found to be a relatively weak inhibitor of MbtA, with an apparent K_I of 143 μM. The approximately five orders of magnitude difference in activity between **22** and **14** can be attributed to the removal of the carbonyl group, which interacts with Lys519 of MbtA (~100-fold loss, compare **1** versus **15**) and the central nitrogen atom (~1000-fold loss, compare **14** versus **16**). However, the modest micromolar activity of **22** towards MbtA was deemed adequate to ensure binding to MbtA before channeling onto the N-terminal thiolation domain of MbtB. Using an *in-vitro* assay, Qiao and coworkers showed that compound **22** irreversibly modified the aryl carrier domain of MbtB, but only in the presence of the cognate adenylating enzyme MbtA. Thus, compound **22** represents a mechanistically distinct class of inhibitors and is a prototype for a new class of thiolation domain inhibitors [45].

Compounds **1** and **14** to **22** were evaluated against whole-cell *M. tuberculosis* H37Rv, and the minimum inhibitory concentrations (MIC_{99}) that inhibited >99% of growth are shown in Table 14.2. Overall, the *in-vitro* enzyme inhibition and whole-cell activity were well correlated. Analogues **1** and **14**, which contained the acylsulfamate and acylsulfamide linkages, displayed the highest activity with MIC_{99}s of 0.39 and 0.19 μM, respectively, under iron-limiting conditions. However, these compounds also exhibited activity under iron-rich conditions, which suggested that the inhibitors were operating by siderophore-independent mechanisms. The ratio of MIC_{99} (iron-poor)/MIC_{99} (iron-rich) is referred herein to as the *selectivity ratio*, and is a measure of the inhibitor specificity.

14.4.2
Importance of the Aryl Ring

In general, NRPS adenylation domains display a fairly strict substrate specificity; hence, Qiao and coworkers examined a series of conservative changes to the aryl subunit of the inhibitor [47]. In order to explore the significance of the *ortho*-hydroxy group, a systematic series of analogues was prepared bearing substitution at the 2-position. Several important trends emerged from this series. Deletion of the hydroxy afforded benzoyl analogue **24** and resulted in a concomitant 14-fold loss of binding affinity (Table 14.3). Substitution of the *ortho*-hydroxy group reduced potency in all cases, but by widely varying amounts: by six-fold for 2-fluoro **23**, 117-fold for 2-amino **25**, 1100-fold for 2-nitro **26**, and 2700-fold for 2-chloro **27**. Molecular mechanics simulations of **26** and **27** showed that the bulky nitro- and chloro- substituents disfavor the required coplanar conformation (as discussed in Section 14.4.1), and additionally are unable to form an internal hydrogen bond. The 117-fold loss of

Table 14.3 SAR of the aryl domain.

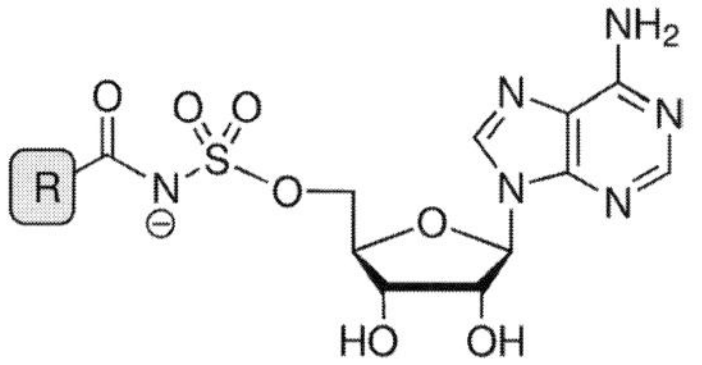

Compound	R =		K_i^{app}(nM)[a]	MIC_{99} (μM)[b] (iron-poor)	MIC_{99} (μM)[c] (iron-rich)
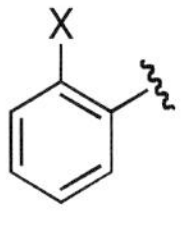	X =	**1** OH	5.1 ± 0.11	0.39	1.56
		23 F	38 ± 8	12.5	50
		24 H	92 ± 7	12.5	50
		25 NH_2	770 ± 120	>100	>100
		26 NO_2	7320 ± 900	>100	>100
		27 Cl	$(18.1 \pm 1.7) \times 10^3$	>100	>100
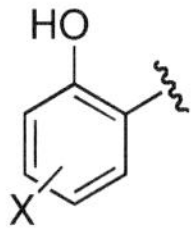	X =	**28** 3-Cl	61 ± 3	50	200
		29 4-Cl	12.0 ± 0.6	12.5	50
		30 5-Cl	20.0 ± 1.4	12.5	50
		31 6-F	7.3 ± 0.6	0.78	3.13
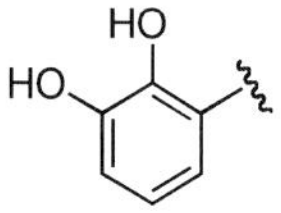		**32**	137 ± 12	>100	>100
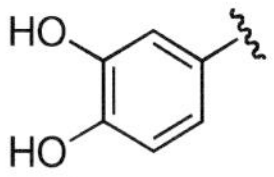		**33**	>100	>100	>100
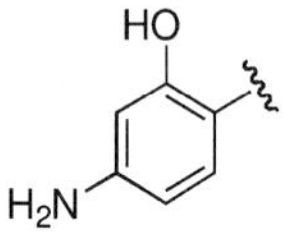		**34**	40 ± 4	1.56	25
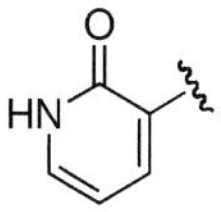		**35**	3700 ± 500	>100	>100
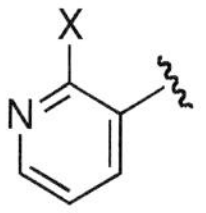	X =	**36** F	2900 ± 400	>100	>100
		37 Cl	175 ± 19	12.5	50
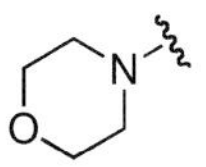		**38**	>100	>100	>100

Table 14.3 (*Continued*)

Compound	R =	K_I^{app}(nM)[a]	MIC_{99} (μM)[b] (iron-poor)	MIC_{99} (μM)[c] (iron-rich)
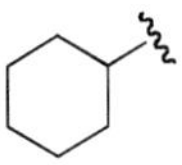	39	$(17.3 \pm 2.8) \times 10^3$	>100	>100
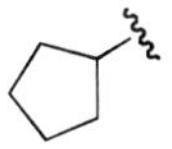	40	$(49.8 \pm 2.0) \times 10^3$	>100	>100
	41	>100	>100	>100
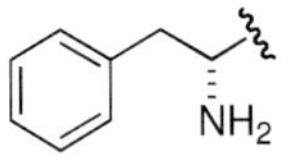	42	$(36.5 \pm 2.2) \times 10^3$	>100	>100

[a] ATP-PPi exchange assay.
[b] Grown in GAST media without Fe^{3+}.
[c] Grown in GAST media supplemented with 200 μM Fe^{3+}.

activity when the aryl hydroxy is replaced by an amino group suggests that the Asn258 side chain of MbtA must present its amino group to the inhibitor. Notably, the 2-fluoro analogue displayed the smallest destabilization of binding consistent with its ability to adopt the coplanar arrangement and importantly lacked the metabolically liable phenol function.

Analogues **28** to **30** were prepared to define the steric requirements of the salicyl binding pocket, and to identify potential sites for further modification. The homology model of MbtA reveals a shallow hydrophobic binding pocket for salicylic acid consisting of Phe259, Cys263, Leu360, Val352, Gly329, Gly330, and Gly334. The chloro group is isosteric with a methyl, and importantly can be modified by standard palladium-mediated coupling reactions. The SAR of these analogues showed that substitution of a chloro group at the 4-positions of the aryl ring was most tolerated and resulted in a modest two-fold decrease in potency relative to **1**, whereas substitution at the 3- and 5-positions reduced inhibitor potency by three- and nine-fold respectively (Table 14.3). Additionally, the 4-amino derivative **34** was prepared and found to exhibit a modest six-fold loss in potency. This result is interesting as 4-aminosalicylic acid is a clinically approved drug for the treatment of tuberculosis, which has been hypothesized to act as an antimetabolite of mycobactin synthesis [48].

As other adenylation enzymes such as DhbE and AsbC activate 2,3-dihydroxybenzoic acid and 3,4-dihydroxybenzoic acid, respectively, inhibitors **32** and **33** were prepared [41, 49]. Compound **32** was a potent nanomolar inhibitor of DhbE with a reported K_I^{app} of 85 nM [39], but displayed a 20-fold loss in potency towards MbtA relative to **1**. In DhbE, the 3-hydroxy hydrogen bonds to Ser240, which is a cysteine in MbtA and therefore a weaker hydrogen-bonding partner. *Bacillus anthracis* incorporates the native substrate 3,4-dihydroxybenzoic acid to create the siderophore petrobactin [50]. Accordingly, compound **33** was prepared and found to inhibit AsbC with an IC_{50} of 250 nM, but was inactive against MbtA, indicating enormous active site differences between the two adenylation enzymes [49].

The incorporation of a basic nitrogen at the 3-position in pyridyl analogue **36** and **37** (Table 14.3) was explored because of the S263C and V352L substitutions. A nitrogen internal to the aryl ring would avoid any steric issues present with a 3-hydroxyl group, and molecular modeling of **36** and **37** showed that Cys263 can donate a hydrogen bond either to the 3-nitrogen or to the backbone of Asn258 (Figure 14.8). The carbon to nitrogen substitutions in 2-fluoropyridyl analogue **36** decreased activity 76-fold relative to the 2-fluorophenyl analogue (**23** versus **36**), suggesting that Cys263 prefers to interact with the Asn258 backbone and causing a desolvation penalty for the inhibitor. In contrast, the activity of the 2-chloropyridyl compound **37** increased 100-fold relative to the 2-chlorophenyl analogue (**27** versus **37**).

Modeling suggested that a 6-substituent might be able to accept a poor-geometry hydrogen bond from the backbone amide of Gly329. Consequently, 6-fluorosalicyl **31** was prepared, which had activity nearly identical to the parent compound. Compounds **38** to **41** were prepared to explore the enzyme active site tolerance for heteroaryl, cycloalkyl, and alkyl groups. Not surprisingly, all four compounds displayed much higher K_I^{app} values, with the cyclohexyl analogue being the best fit and showing a greater than 2600-fold loss of binding affinity to MbtA compared to **1**. Finally, the known phenylalanyl-tRNA synthetase adenylate inhibitor **42** exhibited a 7200-fold loss of activity.

The MIC_{99} values against *M. tuberculosis* for the aryl-modified analogues paralleled the enzyme inhibition data. The *p*-aminosalicyl derivative **34** displayed improved selectivity, showing a 16-fold increase in MIC under iron-deficient conditions, and suggesting that this modification reduced off-target binding. Overall, the collective SAR from this series demonstrated that the aryl domain is poorly tolerant to modification, in accord with the strict specificity of NRPS adenylation enzymes. Although, the 2-hydroxy group is required for optimal activity, the finding that this can be replaced with a fluoro group may be useful for increasing the metabolic stability, as phenols are well-known to undergo glucuronidation.

14.4.3
Role of the Ribose

Modification of the ribose subunit was explored to determine the importance of each atom towards binding affinity [10]. The objective of these modifications was to increase lipophilicity and hence membrane permeability by sequentially removing oxygen atoms without severely compromising binding affinity. Carbocyclic analogue **43** (Table 14.4) displayed a three-fold increase in potency demonstrating that the ribofuranose ring-oxygen is dispensable for activity. The slightly improved activity of **1** may be due to enhanced basicity of the N^6-amine as a result of the missing anomeric effect that enables a stronger H-bond with Val352 (see Figure 14.11) [51]. By contrast, Diederich and coworkers observed a 10^5-fold loss of binding affinity for a bisubstrate nucleoside analogue toward the enzyme catechol methyltransferase when the ribofuranose ring oxygen was replaced by a CH_2 unit (Figure 14.9) [52].

Table 14.4 SAR of the ribose domain.

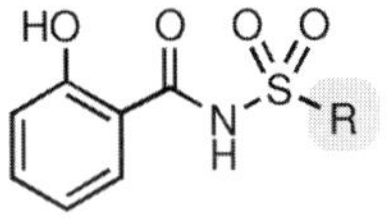

Compound	R =	K_i^{app}(nM)[a)]	MIC_{99} (μM)[b)] (iron-poor)	MIC_{99} (μM)[c)] (iron-rich)
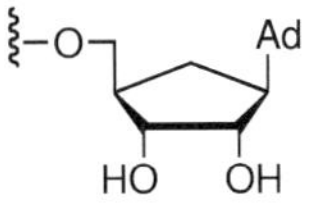	43	2.3 ± 0.4	1.56	6.25
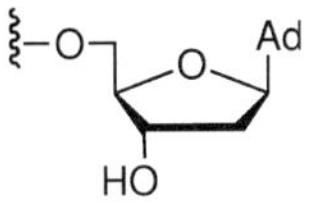	44	3.5 ± 0.4	25.0	100
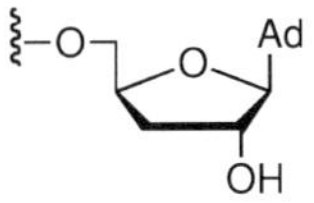	45	3.2 ± 0.5	1.56	12.5
	46	830 ± 78	>200	>200
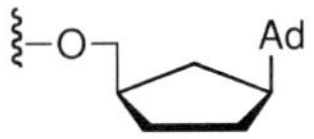	47	61 ± 5	>200	>200
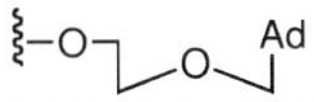	48	16,700 ± 500	>200	>200

a) ATP-PPi exchange assay.
b) Grown in GAST media without Fe^{3+}.
c) Grown in GAST media supplemented with 200 μM Fe^{3+}.

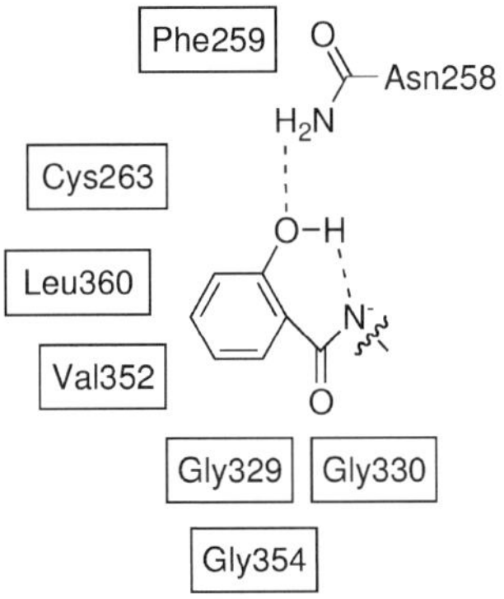

Figure 14.9 Putative binding interactions of the aryl domain of **1** with a MbtA homology model.

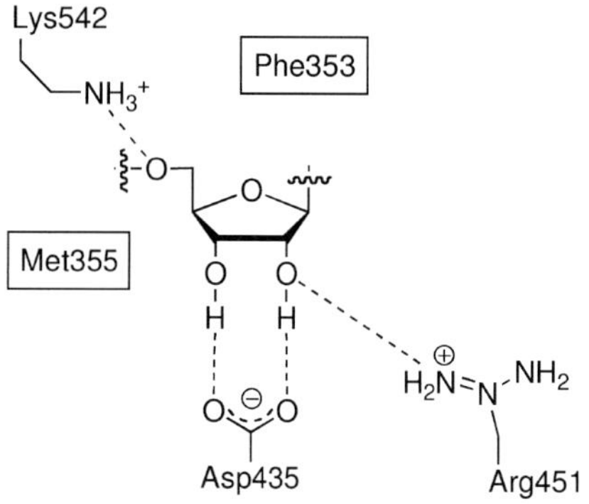

Figure 14.10 Putative binding interactions of the glycosyl domain of **1** with a MbtA homology model.

In order to more precisely map out the structural requirements of the ribofuranose, subunit analogues **44** and **45** (Table 14.4) were evaluated. Deletion of either the 2′-hydroxyl or the 3′-hydroxyl resulted in an approximately two-fold increase in potency. Although Asp435 is positioned to form a bidendate hydrogen with both the 2′- and 3′-hydroxyl groups, and thus fix the sugar in the adenosine binding pocket, the observed SAR demonstrates that only one of these interactions is required to maintain potency (Figure 14.10). Dideoxy dehydro carbocyclic analogue **46** exhibited a pronounced 126-fold loss of potency, while the saturated dideoxy carbocycle **47** resulted in a mere nine-fold loss in binding affinity, despite removal of the 2′ and 3′ alcohols and the ribofuranose ring oxygen. Formation of a salt bridge between Arg451 and Asp435 may compensate for the loss of hydrogen bonds between **47** and MbtA. The attenuated activity of **46** is most likely a result of the rigidity of the cyclopentene moiety that is unable to adopt the required C3′ endo conformation. Acyclo analogue **48** was approximately 2500-fold less active, thereby demonstrating the importance of the conformational rigidity of parent ribose moiety.

Overall, the biological activity of **43–48** paralleled the enzyme inhibition, except for compounds **44** and **47** (Table 14.4). Carbocyclic analogue **43** was noteworthy as this is

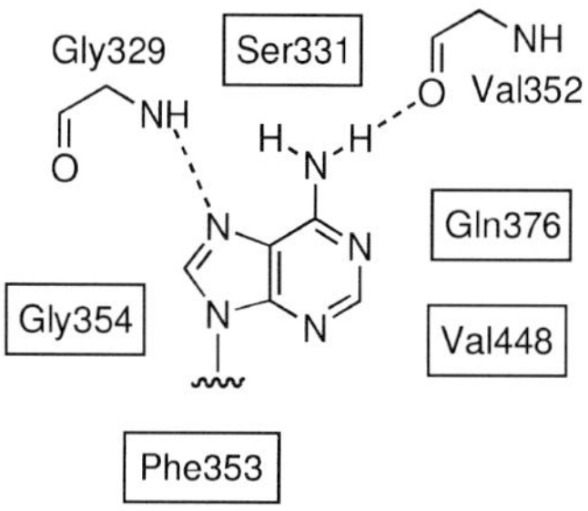

Figure 14.11 Putative binding interactions of the nucleobase domain of **1** with a MbtA homology model.

expected to have significantly improved metabolic stability due to removal of the labile glycosidic linkage, as well as a slight increase in lipophilicity [53]. Removal of the 2-hydroxy group in **44** resulted in a dramatic 64-fold drop in activity compared to **1**, while **47** was inactive. A rationale for the observed discrepancy between enzyme inhibition and the whole-cell biological activity of **44** and **47** is that the sugar is important for recognition by a putative transporter. *M. tuberculosis* encodes 37 ABC (ATP-dependent binding cassette) transporters of which 16 have been unambiguously assigned as importers responsible for the assimilation of amino acids, nucleotides, and other essential cofactors [54]. Additionally, *M. tuberculosis* transports many hydrophilic solutes via a class of proteins known as the porins (the major porin is known as MspA), which have also been shown to play an important role in the transport of antibiotics such as the β-lactams and aminoglycosides [55].

14.4.4 Impact of the Nucleobase

The adenosine scaffold provides several opportunities for improving affinity, increasing lipophilicity, and enhancing metabolic stability; thus, modifications to the nucleobase domain have also been examined. The fairly high $K_M^{(ATP)}$ of 184 μM for MbtA suggests that there are considerable opportunities to improve upon the binding affinity of the inhibitors within the ATP binding pocket [10]. The adenine heterocycles can be conveniently modified at the C-2, C-6, and C-8 positions. Limited sets of modifications at each of the positions have been reported, along with a couple of deaza purine analogues [56].

Initially, modifications to the 8-position of adenosine were explored based on the findings that an 8-aminoadenosine acyladenylate analogue (not shown) developed for methionyl-tRNA synthetase from *E. coli* displayed a greater than 300-fold increase in potency versus the simple unsubsitituted compound [57]. Somu and coworkers reported that the 8-amino derivative **51** (Table 14.5) displayed a 400-fold loss of binding affinity against MbtA, an energetic penalty of almost 6.2 kcal mol^{-1}. Installation of either an azido or bromo function at the 8-position in analogues **50** and **49** respectively also resulted in a substantial loss of binding affinity. Interestingly, molecular modeling has shown this is likely due to unfavorable contacts with the ligands themselves, and not to any protein–ligand interactions. Based on the X-ray co-crystal structure of DhbE with an adenylated 2,3-dihydroxybenzoate, it appears that the prototypical bisubstrate inhibitors given by **1** adopt a unique compact structure wherein the salicyl moiety is folded back towards the adenine base (see Figure 14.7). The well-known preference of simple 8-substituted adenosine derivatives to favor the *syn* conformation around the glycosidic bond could also account in part for the observed decrease in binding affinity, as the ligand must bind in an *anti* conformation. However, the *syn* conformation is strongly disfavored in the folded conformation of the bound ligand.

Replacement of the N^6-amine with an oxygen atom in inosine derivative **52**, or the addition of two methyl substituents in **53**, led to 160-fold and 75-fold loss, respectively, of activity, thereby establishing the requirement for at least one hydrogen bond

Table 14.5 SAR of the base domain.

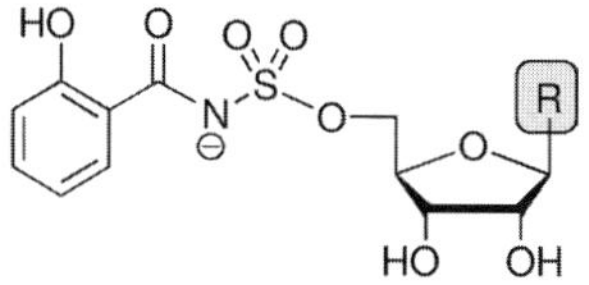

Compound	R =	K_I^{app}(nM)[a]	MIC_{99} (μM)[b] (iron-poor)	MIC_{99} (μM)[c] (iron-rich)
C-8 modifications				
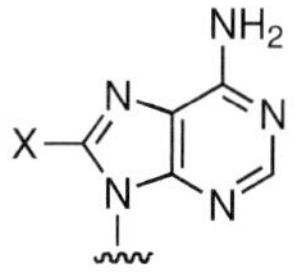	X = **49**, Br	2050 ± 299	>50	>50
	X = **50**, N_3	(42.5 ± 6.2)10^3	>50	>50
	X = **51**, NH_2	(183 ± 19)10^3	>50	>50
C-6 modifications				
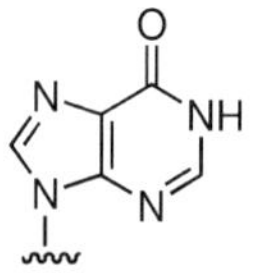	**52**	802 ± 54	>200	>200
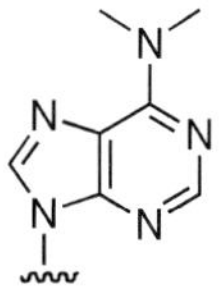	**53**	380 ± 32	50	>50
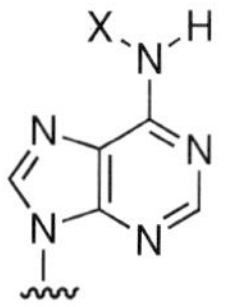	X = **54**, cyclopropyl	1.85 ± 0.13	0.098	6.25
	X = **55**, cyclobutyl	124 ± 12	n.d.	n.d.
	X = **56**, cyclopentyl	9412 ± 520	n.d.	n.d.
	X = **57**, benzyl	8260 ± 820	n.d.	n.d.
	X = **58**, *n*-propyl	3.58 ± 0.21	n.d.	n.d.
C-2 modifications				
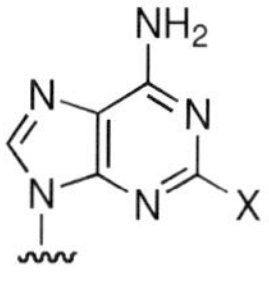	X = **59**, I	3.03 ± 0.34	0.19	3.12
	X = **60**, Ph	0.27 ± 0.07	0.049	0.39
	X = **61**, NHPh	0.94 ± 0.16	0.049	0.39
	X = **62**, CCPh	0.40 ± 0.05	0.049	0.39
	X = **63**, N_3	5.15 ± 0.67	n.d.	n.d.

Table 14.5 (*Continued*)

Compound	R =	K_i^{app}(nM)[a]	MIC_{99} (μM)[b] (iron-poor)	MIC_{99} (μM)[c] (iron-rich)
Deaza analogues				
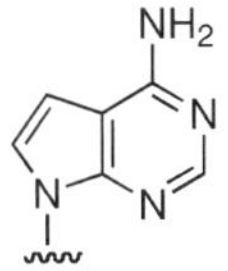	64	24.3 ± 0.7	6.25	50
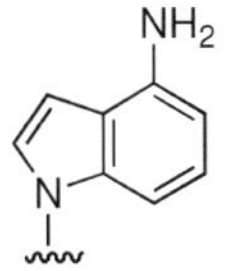	65[d]	20.1 ± 2.3	n.d.	n.d.

[a] ATP-PPi exchange assay.
[b] Grown in GAST media without Fe^{3+}.
[c] Grown in GAST media supplemented with 200 μM Fe^{3+}.
[d] Contains a carbocyclic sugar template wherein the ribofuranose ring oxygen is replaced with a CH_2, see sugar of compound **43** in Table 14.4. n.d., not determined.

donor at this position. A systematic series of mono *N*-substituted alkyl, benzyl, and cycloalkyl analogues was prepared to map out the optimal substituent at N-6 (Table 14.5). Small alkyl and cycloalkyl substituents such as *n*-propyl and cyclopropyl derivatives **58** and **54** respectively led to slight improvements in binding affinity, whereas larger substituents (**55–57**) led to a dramatic reduction in potency.

Modeling had suggested that the C-2 position would be more amenable to modification, because of an accessible binding pocket close to C-2 and N-3 (numbering of adenine base). Thus, the 2-iodo analogue **59** (Table 14.5) was evaluated and found to possess a two-fold increase in potency. The iodo group provided a handle for the introduction of various substituents through Pd-catalyzed cross-coupling reactions, thereby enabling access to several 2-modified analogues, including 2-phenyl **60**, 2-phenylamino **61**, and phenylacetylenyl **62**. These analogues (**60–62**) were five to 19 times more potent than the parent inhibitor.

In order to evaluate the importance of the heteroatoms of the purine base for activity, compounds **64** and **65** were synthesized. 7-Deazaadenine analogue **64** (Table 14.5) displayed a six-fold reduction in potency, indicating that with the putative hydrogen bond with Gly329 is not critical for activity (Figure 14.11). Indole analogue **65** was equipotent to **64**, demonstrating that both N-1 and N-3 of the purine are dispensable for activity.

The relative biological activity of analogues **49** to **65** against whole-cell *M. tuberculosis* H37Rv was consistent with their *in-vitro* enzyme inhibition of MbtA. Notably, C-2 modified derivatives **60–62** exhibited the most potent activity yet observed for this new class of nucleoside antibiotics, with MIC_{99} values of 49 nM under iron-deficient conditions. Even more significantly, cyclopropyl analogue **54** displayed enhanced selectivity, with an MIC_{99} of 98 nM under iron-deficient conditions, but 6.25 μM under iron-rich conditions, representing a selectivity factor of 64.

14.5 Conclusions

Nucleoside inhibitors based on Sal-AMS **1** have great potential as novel antibiotics targeting siderophore biosynthesis. More than 500 structurally different siderophores have been isolated. Thus, the development of a broad-spectrum antibiotic targeting siderophore biosynthesis will not be possible. However, the preparation of an inhibitor for a particular structural class of related siderophores, such as the aryl-capped siderophores, holds promise for providing compounds with a narrow therapeutic window against the targeted pathogen(s). This, in fact, may be advantageous as it is expected that beneficial bacterial flora will not be affected. Installation of the aryl moiety of aryl-capped siderophores is carried out by aryl acid adenylating enzymes. The development of the prototypical intermediate mimetic Sal-AMS of this enzymatic reaction provided exceptionally potent enzyme inhibitors. The excellent whole-cell activity against *M. tuberculosis* is encouraging, but the potential liabilities of this inhibitor should also be recognized, which include: (1) cleavage of acylsulfamate linkage to release 5′-*O*-(sulfamoyl)adenosine; (2) limited oral bioavailability resulting from the highly polar nature of this nucleoside derivative; (3) rapid drug metabolism due to the numerous metabolically liable functions; (4) a limited spectrum of antibiotic activity; and (5) the development of resistance to this chemotherapeutic agent. While it is too early to address all of these concerns, preliminary SAR studies and the measurement of some physico-chemical properties of the inhibitor scaffold have provided useful information to guide future efforts for optimizing the pharmacodynamic and pharmacokinetic properties of these compounds.

References

1 Somu, R. V., Boshoff, H., Qiao, C., Bennett, E. M., Barry, C. E. 3rd, Aldrich, C. C., Rationally designed nucleoside antibiotics that inhibit siderophore biosynthesis of *Mycobacterium tuberculosis*. *J. Med. Chem.* **2006**, *49*, 31–34.

2 Ferreras, J. A., Ryu, J. S., Di Lello, F., Tan, D. S., Quadri, L. E., Small-molecule inhibition of siderophore biosynthesis in *Mycobacterium tuberculosis* and Yersinia pestis. *Nat. Chem. Biol.* **2005**, *1*, 29–32.

3 Crosa, J. H., Walsh, C. T., Genetics and assembly line enzymology of siderophore biosynthesis in bacteria. *Microbiol. Mol. Biol. Rev.* **2002**, *66*, 223–249.

4 Klicic, J. J., Friesner, R. A., Liu, S.- Y., Guida, W. C., Accurate prediction of acidity constants in aqueous solution via density functional theory and self-consistent reaction field methods. *J. Phys. Chem. A* **2002**, *106*, 1327–1335.

5 Blackburn, M. R., Datta, S. K., Wakamiya, M., Vartabedian, B. S., Kellems, R. E.,

Metabolic and immunologic consequences of limited adenosine deaminase expression in mice. *J. Biol. Chem.* **1996**, *271*, 15203–15210.

6 Bloch, A., Coutsogeorgopoulos, C., Inhibition of protein synthesis by 5′-sulfamoyladenosine. *Biochemistry* **1971**, *10*, 4394–4398.

7 Vannada, J., Bennett, E. M., Wilson, D. J., Boshoff, H. I., Barry, C. E. 3rd., Aldrich, C. C., Design, synthesis, and biological evaluation of beta-ketosulfonamide adenylation inhibitors as potential antitubercular agents. *Org. Lett.* **2006**, *8*, 4707–4710.

8 Unpublished results, Daniel Wilson, University of Minnesota, Center for Drug Design.

9 De Voss, J. J., Rutter, K., Schroeder, B. G., Su, H., Zhu, Y., Barry, C. E. 3rd., The salicylate-derived mycobactin siderophores of *Mycobacterium tuberculosis* are essential for growth in macrophages. *Proc. Natl. Acad. Sci. USA.* **2000**, *97*, 1252–1257.

10 Somu, R. V., Wilson, D. J., Bennett, E. M., Boshoff, H. I., Celia, L., Beck, B. J., Barry, C. E. 3rd, Aldrich, C. C., Antitubercular nucleosides that inhibit siderophore biosynthesis: SAR of the glycosyl domain. *J. Med. Chem.* **2006**, *49*, 7623–7635.

11 Trivedi, O. A., Arora, P., Sridharan, V., Tickoo, R., Mohanty, D., Gokhale, R. S., Enzymatic activation and transfer of fatty acids as acyl-adenylates in mycobacteria. *Nature* **2004**, *428*, 441–445.

12 Unpublished results, Brian Beck, Laura Celia, American Type Culture Collection, Manassas, Virginia 20110.

13 Bearden, S. W., Fetherston, J. D., Perry, R. D., Genetic organization of the yersiniabactin biosynthetic region and construction of avirulent mutants in Yersinia pestis. *Infect. Immun.* **1997**, *65*, 1659–1668.

14 Lawlor, M. S., O'Connor, C., Miller, V. L., Yersiniabactin is a virulence factor for *Klebsiella pneumoniae* during pulmonary infection. *Infect. Immun.* **2007**, *75*, 1463–1472.

15 Cendrowski, S., MacArthur, W., Hanna, P., *Bacillus anthracis* requires siderophore biosynthesis for growth in macrophages and mouse virulence. *Mol. Microbiol.* **2004**, *51*, 407–417.

16 Raymond, K. N., Dertz, E. A., Kim, S. S., Enterobactin: An archetype for microbial iron transport. *Proc. Natl. Acad. Sci. USA* **2003**, *100*, 3584–3588.

17 Mihara, K., Tanabe, T., Yamakawa, Y., Funahashi, T., Nakao, H., Narimatsu, S., Yamamoto, S., Identification and transcriptional organization of a gene cluster involved in biosynthesis and transport of acinetobactin, a siderophore produced by *Acinetobacter baumannii* ATCC 19606. *Microbiology* **2004**, *150*, 2587–2597.

18 Quadri, L. E., Sello, J., Keating, T. A., Weinreb, P. H., Walsh, C. T., Identification of a *Mycobacterium tuberculosis* gene cluster encoding the biosynthetic enzymes for assembly of the virulence-conferring siderophore mycobactin. *Chem. Biol.* **1998**, *5*, 631–645.

19 Keating, T. A., Ehmann, D. E., Kohli, R. M., Marshall, C. G., Trauger, J. W., Walsh, C. T., Chain termination steps in nonribosomal peptide synthetase assembly lines: directed acyl-S-enzyme breakdown in antibiotic and siderophore biosynthesis. *Chembiochem* **2001**, *2*, 99–107.

20 Krithika, R., Marathe, U., Saxena, P., Ansari, M. Z., Mohanty, D., Gokhale, R. S., A genetic locus required for iron acquisition in *Mycobacterium tuberculosis*. *Proc. Natl. Acad. Sci. USA* **2006**, *103*, 2069–2074.

21 Stachelhaus, T., Mootz, H. D., Bergendahl, V., Marahiel, M. A., Peptide bond formation in nonribosomal peptide biosynthesis. *J. Biol. Chem.* **1998**, *273*, 22773–22781.

22 Babbitt, P. C., Kenyon, G. L., Martin, B. M., Charest, H., Slyvestre, M., Scholten, J. D., Chang, K.- H., Liang, P.- H., Dunaway-Mariano, D., Ancestry of the 4-chlorobenzoate dehalogenase: analysis of amino acid sequence identities among families of acyl: adenyl ligases, enoyl-CoA hydratases/isomerases, and acyl-CoA

thioesterases. *Biochemistry* **1992**, *31*, 5594–5604.

23 Stachelhaus, T. S., Marahiel, M. A., Rational design of peptide antibiotics by targeted replacement of bacterial and fungal domains. *Science* **1995**, *269*, 69–72.

24 Mootz, H. D., Marahiel, M. A., Biosynthetic systems for nonribosomal peptide antibiotic assembly. *Curr. Opin. Chem. Biol.* **1997**, *1*, 543–551.

25 Schimmel, P., Tao, J., Hill, J., Aminoacyl tRNA synthetases as targets for new anti-infectives. *FASEB J.* **1998**, *12*, 1599–1609.

26 Kim, S., Lee, S. W., Choi, E. C., Choi, S. Y., Aminoacyl-tRNA synthetases and their inhibitors as a novel family of antibiotics. *Appl. Microbiol. Biotechnol.* **2003**, *61*, 278–288.

27 Schimmel, P. R., Soll, D., Aminoacyl-tRNA synthetases: general features and recognition of transfer RNAs. *Annu. Rev. Biochem.* **1979**, *48*, 601–648.

28 Isono, K., Uramoto, M., Kusakabe, H., Miyata, N., Koyama, H., Ubukata, M., Sethi, S. K., McCloskey, J. A., Ascamycin and dealanylascamycin nucleoside antibiotics from *Streptomyces* sp. *J. Antibiot.* **1984**, *37*, 670–672.

29 Ueda, H., Shoku, Y., Hayashi, N., Mitsunaga, J.,in: Y., Doi, M., Inoue, M., Ishida, M. X-ray crystallographic conformational study of 5′-O-[N-L-alanyl)-sulfamoyl]adenosine: A substrate analogue for alanyl-tRNA synthetase. *Biochim. Biophys. Acta.* **1991**, *1080*, 126–134.

30 Belrhali, H., Yaremchuk, A., Tukalo, M., Larsen, K., Berthet-Colominas, C., Leberman, R., Beijer, B., Sproat, B., Als-Nielsen, J., Grübel, G., Legrand, J.-F., Lehmann, M., Cusack, S., Crystal structures at 2.5 angstrom resolution of seryl-tRNA synthetases complexed with two analogs of seryl adenylate. *Science* **1994**, *263*, 1432–1436.

31 Hill, J. M., Yu, G., Shue, Y.- K., Zydowsky, T. M., Rebek, J.,Aminoacyl adenylate mimics as novel antimicrobial and antiparasitic agents, U.S. Patent 5,726,195, May 10, **1998**.

32 Brown, P., Richardson, C. M., Mensah, L. M., O'Hanlon, P. J., Osborne, N. F., Pope, A. J., Walker, G., Molecular recognition of tyrosinyl adenylate analogues by prokaryotic tyrosyl tRNA synthetases. *Bioorg. Med. Chem.* **1999**, 2473–2485.

33 Brown, M. J., Mensah, L. M., Doyle, M. L., Broom, N. J., Osbourne, N., Forrest, A. K., Richardson, C. M., O'Hanlon, P. J., Pope, A. J., Rational design of femtomolar inhibitors of isoleucyl tRNA synthetase from a binding model for pseudomonic acid-A. *Biochemistry* **2000**, *39*, 6003–6011.

34 Koroniak, L., Ciustea, M., Gutierrez, J. A., Richards, N. G. J., Synthesis and characterization of an N-acylsulfonamide inhibitor of human asparagine synthetase. *Org. Lett.* **2003**, *5*, 2033–2036.

35 Lee, J., Kim, S. E., Lee, J. Y., Kim, S. Y., Kang, S. U., Seo, S. H., Chun, M. W., Kang, T., Choi, S. Y., Kim, H. O., N-Alkoxysulfamide, N-hydroxysulfamide, and sulfamate analogues of methionyl and isoleucyl adenylates as inhibitors of methionyl-tRNA and isoleucyl-tRNA synthetases. *Bioorg. Med. Chem. Lett.* **2003**, *13*, 1087–1092.

36 Bernier, S., Dubois, D. Y., Therrien, M., Lapointe, J., Chênevert, R., Synthesis of glutaminyl adenylate analogues that are inhibitors of glutaminyl-tRNA synthetase. *Bioorg. Med. Chem. Lett.* **2000**, *10*, 2441–2444.

37 Heacock, D., Forsyth, C. J., Shiba, K., Musier-Forsyth, K., Synthesis and aminoacyl-tRNA synthetase inhibitory activity of prolyl adenylate analogs. *Bioorg. Chem.* **1996**, *24*, 273–289.

38 Finking, R., Neumuller, A., Solsbacher, J., Konz, D., Kretzschmar, G., Schweitzer, M., Krumm, T., Marahiel, M. A., Aminoacyl adenylate substrate analogues for the inhibition of adenylation domains of nonribosomal peptide synthetases. *Chembiochem.* **2003**, *4*, 903–906.

39 Miethke, M., Bisseret, P., Beckering, C. L., Vignard, D., Eustache, J., Marahiel, M. A., Inhibition of aryl acid adenylation domains involved in bacterial siderophore synthesis. *FEBS J.* **2006**, *273*, 409–419.

40 Copeland, R. A., *Evaluation of Enzyme Inhibitors in Drug Discovery*, Wiley: Hoboken, NJ, **2005**.

41 May, J. J., Kessler, N., Marahiel, M. A., Stubbs, M. T., Crystal structure of DhbE, an archetype for aryl acid activating domains of modular nonribosomal peptide synthetases. *Proc. Natl. Acad. Sci. USA.* **2002**, *99*, 12120–12125.

42 Lee, J., Kang, S. U., Kang, M. K., Chun, M. W., Jo, Y. J., Kwak, J. H., Kim, S., Methionyl adenylate analogues as inhibitors of methionyl-tRNA synthetase. *Bioorg. Med. Chem. Lett.* **1999**, *9*, 1365–1370.

43 Southgate, C. C., Dixon, H. B., Phosphonate analogues of aminoacyl adenylates. *Biochem. J.* **1978**, *175*, 461–465.

44 Callahan, B. P., Lomino, J. V., Wolfenden, R., Nanomolar inhibition of the enterobactin biosynthesis enzyme, EntE: Synthesis, substituent effects, and additivity. *Bioorg. Med. Chem. Lett.* **2006**, *16*, 3802–3805.

45 Qiao, C., Wilson, D. J., Bennett, E. M., Aldrich, C. C., A mechanism-based aryl carrier protein/thiolation domain affinity probe. *J. Am. Chem. Soc.* **2007**, *129*, 6350–6351.

46 Reddick, J. J., Cheng, J., Roush, W. R., Relative rates of Michael reactions of 2′-phenethyl)thiol with vinyl sulfones, vinyl sulfonate esters, and vinyl sulfonamides relevant to vinyl sulfonyl cysteine protease inhibitors. *Org. Lett.* **2003**, *5*, 1967–1970.

47 Qiao, C., Gupte, A., Boshoff, H. I., Wilson, D. J., Bennett, E. M., Barry, C. E., Aldrich, C. C.,Bisubstrate nucleoside inhibitors of siderophore biosynthesis in *Mycobacterium tuberculosis*: Investigation of the aryl domain (unpublished results)

48 Ratledge, C., Brown, K. A., Inhibition of mycobactin formation in *Mycobacterium smegmatis* by *p*-aminosalicylate. A new proposal for the mode of action of *p*-aminosalicylate. *Am. Rev. Respir. Dis.* **1972**, *106*, 774–776.

49 Pfleger, B. F., Lee, J. Y., Somu, R. V., Aldrich, C. C., Hanna, P. C., Sherman, D. H., Characterization and analysis of early enzymes for petrobactin biosynthesis in Bacillus anthracis. *Biochemistry* **2007**, *46*, 4147–4157.

50 Cendrowski, S., MacArthur, W., Hanna, P., *Bacillus anthracis* requires siderophore biosynthesis for growth in macrophages and mouse virulence. *Mol. Microbiol.* **2004**, *51*, 407–417.

51 Froehler, B. C., Terhorst, T., Shaw, J. P., McCurdy, S. N., Triple-helix formation and cooperative binding by oligodeoxynucleotides with a 3(-3(internucleotide junction. *Biochemistry* **1992**, *31*, 1603–1609.

52 Paulini, R., Trindler, C., Lerner, C., Brändli, L., Schweizer, W. B., Jakob-Roetne, R., Zürcher, G., Borroni, E., Diederich, F., Bisubstrate inhibitors of catechol *O*-methyltransferase (COMT): the crucial role of the ribose structural unit for inhibitor binding affinity. *ChemMedChem* **2006**, *1*, 340–357.

53 Vince, R., Hua, M., Synthesis and anti-HIV activity of carbocyclic 2(,3(-didehydro-2(,3(-dideoxy 2,6-disubstituted purine nucleosides. *J. Med. Chem.* **1990**, *33*, 17–21.

54 Braibant, M., Gilot, P., Content, J., The ATP-binding cassette ABC) transport systems of *Mycobacterium tuberculosis*. *FEMS Microbiol. Rev.* **2000**, *24*, 449–467.

55 Niederweis, M., Mycobacterial porins – new channel proteins in unique outer membranes. *Mol. Microbiol.* **2003**, *49*, 1167–1177.

56 Somu, R. V., Wilson, D., Qiao, C., Vannada, J., Bennett, E., Boshoff, H., Barry, C. E. r., Aldrich, C. C. In: Antitubercular nucleosides that inhibit siderophore biosynthesis, ACS National Meeting, Chicago, IL, Chicago, IL, 2007, MEDI-451, **2007**.

57 Blanquet, S., Fayat, G., Poiret, M., Waller, J. P., The mechanism of action of methionyl-tRNA synthetase from *Escherichia* coli. Inhibition by adenosine and 8-aminoadenosine of the amino-acid activation reaction. *Eur. J. Biochem.* **1975**, *51*, 567–571.

15
Synthesis and Biological Activity of Selected Carbocyclic Nucleosides

Adam Mieczkowski and Luigi A. Agrofoglio

15.1
Introduction

Nucleoside analogues are extremely useful for the development of therapeutic agents to control viral diseases and cancer. Replacement the oxygen of furanose ring of a nucleoside by a CH_2 unit results in carbocyclic nucleoside analogues. Carbocyclic nucleosides have emerged as targets of intense investigation due to their potent biological activity and greater metabolic activity and stability to nucleoside phosphorylases than the corresponding carbohydrate counterpart. Among these, aristeromycin (**1**), neplanocin (**2**), abacavir (**3**), lobucavir (**4**), synguanol (**5**) or the newly FDA-approved anti-HBV Baraclude® (**6**, Entecavir) (Figure 15.1) are a few of the carbocyclic nucleosides that are useful antiviral drugs.

The cellular processing and mechanism of action of carbocyclic nucleosides are, in general, similar to those of conventional nucleosides. They may undergo progressive phosphorylation to the corresponding mono-, di-, and triphosphates, and any of these activated forms may interact with natural cellular processes (Figure 15.2). On occasion, the interaction is innocuous and has little or no effect upon cell viability, but very often a significant inhibition of essential biochemical reactions occurs and in these cases measurable effects will be observable.

One class of cellular enzymes to which carbocyclic nucleosides are not susceptible is the nucleoside phosphorylases, as a consequence of which they are extremely stable towards sugar–base cleavage. Most of those carbocyclic nucleosides which have been studied intensively are active as inhibitors of a viral enzyme (mainly a polymerase). Experience has shown that the essential structural features of carbocyclic nucleosides which must be retained are: the ability of the base to engage in base pairing with its DNA partner, and the presence of a 5′-hydroxyl group (or equivalent) which is capable of phosphorylation. The remainder of the sugar component can be regarded as a scaffold which ensures that the base pairing functionality and the 5′-phosphate (or phosphonate) are well distributed in space. The carbocyclic nucleosides do not have the same conformational data as the parent nucleosides; rather, in all cases they

Modified Nucleosides: in Biochemistry, Biotechnology and Medicine. Edited by Piet Herdewijn

ISBN: 978-3-527-31820-9

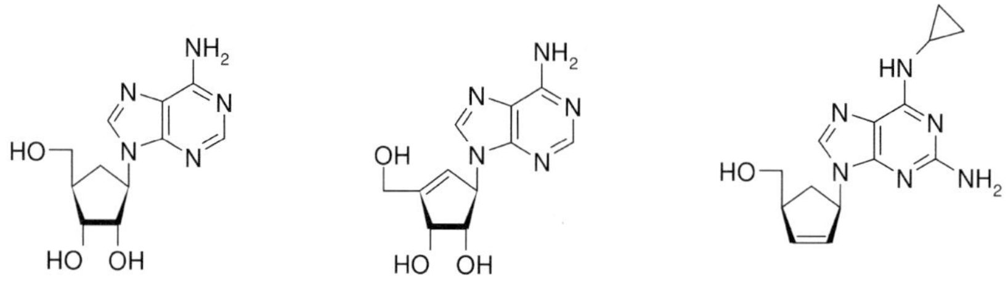

1, Aristeromycin A **2, Neplanocin A** **3, Abacavir**

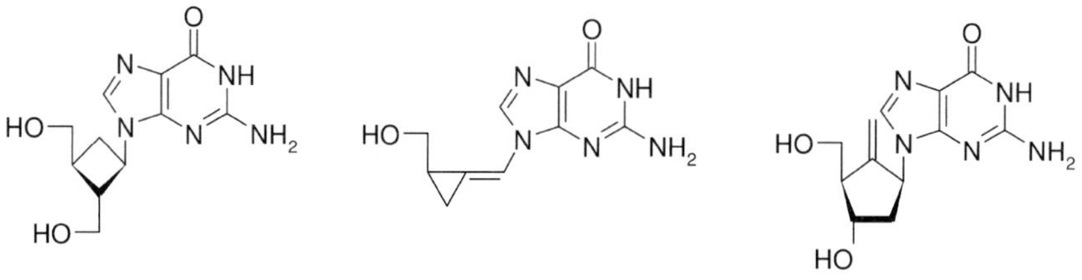

4, Lobucavir **5, Synguanol** **6, Baraclude**

Figure 15.1 Structural formulae of some antiviral carbocyclic nucleosides.

have a preferential hydroxyl group orientation which allows them to interact either with viral or human kinases, or viral DNA polymerases in an efficient manner, leading to potent antiviral agents. The *S*-adenosyl-L-homocysteine hydrolase (AdoHcy *hydrolase*) inhibitors such as aristeromycin, neplanocin and their analogues will not form part of this chapter as they are described elsewhere in the book. The aim of this chapter is not to provide a comprehensive review, but rather to bring together details

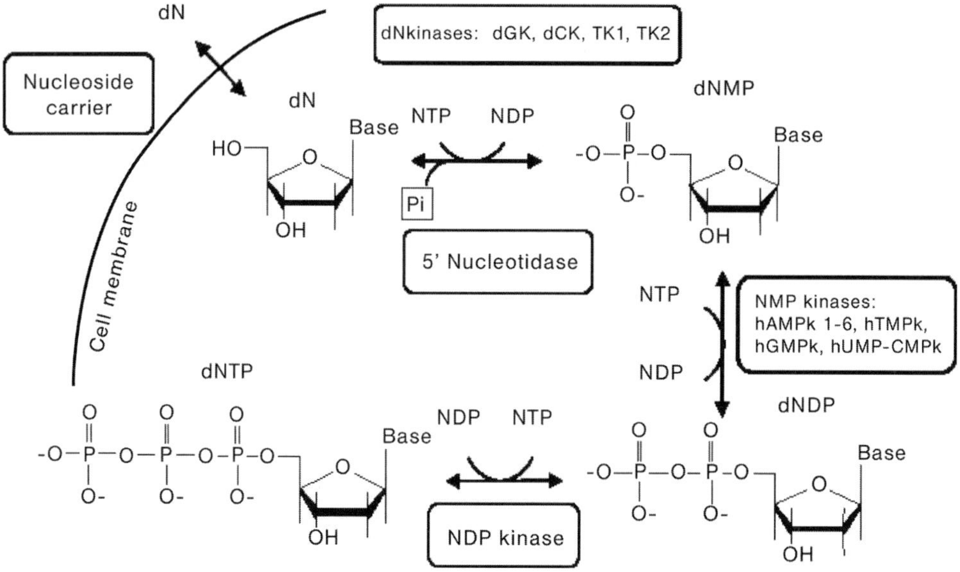

Figure 15.2 Common metabolic pathways of nucleosides.

of recently selected carbanucleosides which have already received FDA approval and will act as leads for antiviral research, or hold great promise as antiviral agents. Thus, attention is focused on cyclopropane, cyclobutane and cyclopentane nucleosides, through selected synthetic pathways. Recently, several excellent reviews have been prepared describing not only the synthetic aspects of the chemistry of carbocyclic nucleosides but also the important pharmacological properties that they exhibit [1].

15.2
A-5021, Synguanol, and Cyclopropane Derivatives

Many nucleosides possessing a cyclopropane ring instead of a sugar ring (Figure 15.3) have demonstrated biological activity. Among these, the cyclopropane guanine – (1′*S*,2′*R*)-9-[[1′,2′-bis(hydroxymethyl)cycloprop-1′-yl]methyl]guanine (A-5021; **7**) exhibits an extremely potent antiherpetic activity against herpes simplex virus and varicella zoster virus (VZV), in addition to a remarkable therapeutic efficiency over the anti-HSV acyclovir in animal models [2]. The anti-HSV and anti-VZV activity of A-5021 (**7**) was found to depend on the phosphorylation by the HSV- and VZV-encoded thymidine kinase (TK). Inhibition of the viral polymerase by the A-5021 triphosphate was stronger than by penciclovir triphosphate [3]. A-5021 has proven also more effective than acyclovir and penciclovir with regard to the reduction of herpetic skin lesions and protection against herpetic encephalitis [4].

With regards to synguanol (**5**), this cyclopropylidene nucleoside is particularly active against human cytomegalovirus (HCMV) (EC_{50} 0.04–2.0 μM) [5]. The antiviral potency of synguanol is enantioselective, with such enantioselectivity reflecting differences in the rates of intracellular phosphorylation [6]. Various analogues of synguanol have been described as antiviral agents (synadenol, syncytol, synthymol) [7], and these are of particular importance because of their ability to inhibit the replication of HCMV and murine cytomegalovirus (MCMV) and Epstein–Barr virus (EBV), as well as human herpes virus 6 (HHV-6). Synadenol (**8**, the adenine analogue) is also an agent against hepatitis B virus (HBV) and VZV. All of these analogues are candidates for clinical investigations. In general, Zemilicka *et al.* [8] have shown that the antiviral potency of the first-generation series resides mostly in purine *Z*-(*cis*)-isomers (Figure 15.4), whereas the *E*-(*trans*)-isomers and pyrimidine analogues are active only on exceptional occasions. The second-generation

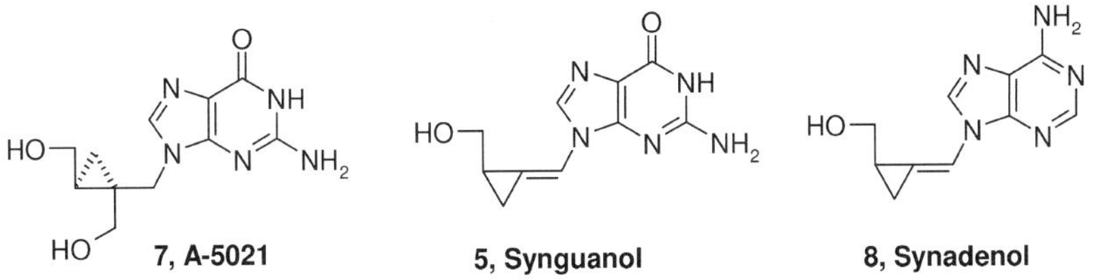

7, A-5021 **5, Synguanol** **8, Synadenol**

Figure 15.3 Some antiviral cyclopropane nucleosides.

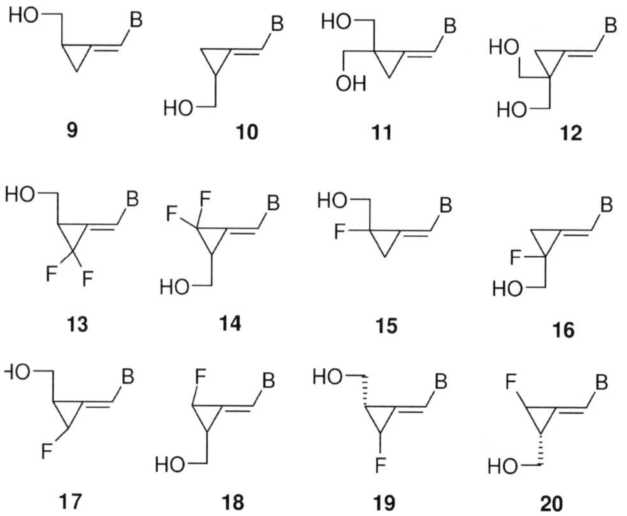

Figure 15.4 Cyclopropylidene analogues of Synguanol.

Z-(*cis*)-isomers have a more narrow antiviral effect, although the guanine analogue cyclopropavir is effective *in vivo* and is currently being developed as a potential agent against HCMV infections. As in the first-generation series, the *E*-(*trans*)-isomers lack anti-HCMV activity, though some EBV potency has been noted.

A practical synthesis of A-5021 (**7**) was proposed by Tsuji *et al.* [9] (Scheme 15.1). The key substrate – bicyclic ester **21**, was synthesized from the epichlorohydrine. In the first step this was selectively reduced with $NaBH_4$ to the alcohol **22**.

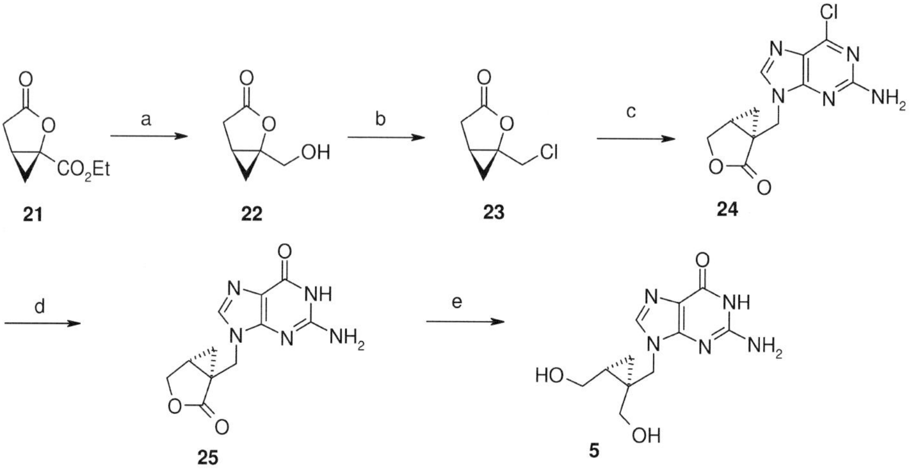

Scheme 15.1 Reagents and conditions: (a) 2.5 equiv. NaOH, r.t., then conc. HCl r.t., then $ClCO_2Et$, Et_3N, THF, −18 °C then $NaBH_4$, THF/H_2O, −18 °C; (b) $SOCl_2$, Et_3N/DCM, r.t.; (c) 2-amino-6-chloropurine, K_2CO_3, DMF, r.t.; (d) 80% HCO_2H, 100 °C, 29% aq. NH_3, r.t.; (e) $NaBH_4$, EtOH, 80 °C.

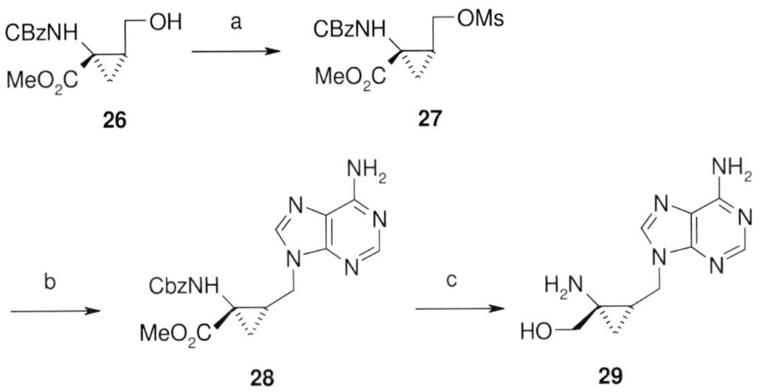

Scheme 15.2 Reagents and conditions: (a) MsCl, Et_3N; (b) Adenine, K_2CO_3, 18-crown-6; (c) $LiBH_4$, then $H_2/Pd(OH)_2/C$.

After treatment with $SOCl_2$, the chloride derivative **23** was obtained and then condensed with 2-amino-6-chloropurine. Treatment with formic acid and aqueous ammonia solution, followed by a second reduction with $NaBH_4$, led to desired compound **5**. The stereoselective synthesis of A-5021 (**5**) was also proposed by Gallos *et al.*, where the D-ribose was used as a starting material [10]. Tricyclic analogues of A-5021 (**7**) were recently investigated by De Clercq *et al.* [11].

Cyclopropyl homo-nucleosides bearing an amino and a methyl group in the cyclopropane ring were synthesized by Ortuno *et al.* [12]. Amino nucleoside **29** was obtained from (−)-(*Z*)-2,3-methanohomoserine ester (**26**) (Scheme 15.2). The substrate was mesylated to intermediate **27**, and then condensed with adenine. The resultant nucleoside **28** was reduced with $LiBH_4$, after deprotection of amine group, whereupon the final nucleoside **29** was obtained.

Cyclopropyl homo-nucleosides bearing a phenyl group in the cyclopropane ring were synthesized by Hong *et al.* [13], starting from 2-hydroxyacetophenone. These compounds showed cytotoxic activity and were tested in several cancer cell lines. Difluorocyclopropyl carbocyclic homo-nucleosides were obtained by Csuk *et al.* (Figure 15.5) [14] from (*Z*)-4-(benzyloxy)-2-butenyl acetate. These compounds also possessed some cytotoxic activity. Cyclopropyl homo-nucleosides bearing hydroxyethyl group were synthesized by Tsuk *et al.* [15] from 3-buten-1-ol (Figure 15.5).

Methylenecyclopropane analogues of nucleosides, synthesized by Zemlicka *et al.* [16], possessed strong antiviral activity against a broad range of viruses. One of the most potent compounds – synadenol (**8**) – could be synthesized from

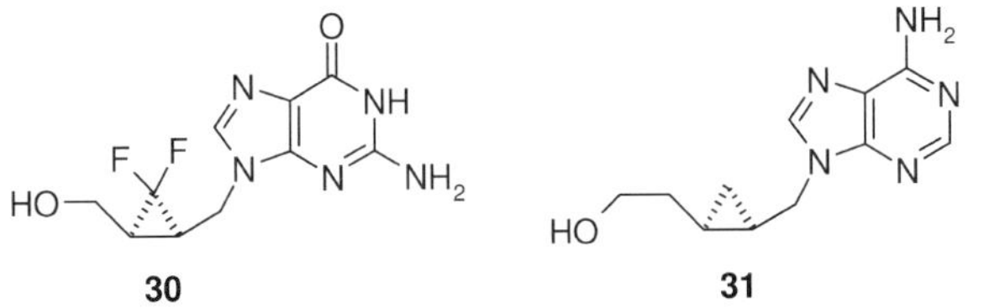

Figure 15.5 Cyclopropane homo-nucleosides.

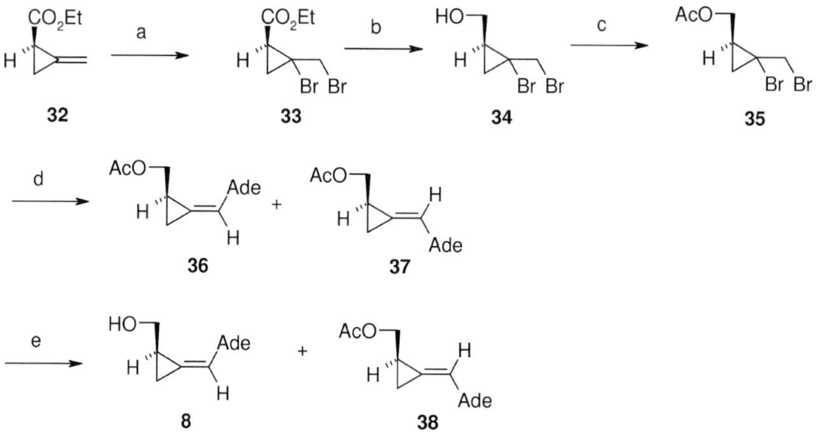

Scheme 15.3 Reagents and conditions: (a) Br_2, CCl_4; (b) DIBAL-H, THF; (c) Ac_2O, pyridine; (d) Adenine, K_2CO_3, DMF; (e) NH_3/MeOH.

cyclopropane ester **32** (Scheme 15.3). Optically pure cyclopropane ester **32**, obtained from a racemic mixture of esters in a four-step procedure, was bromidated to dibromoester **33**. After reduction to dibromoalkohol **34** and acetylation, the cyclopopane ring was coupled with adenine, which led to mixture of isomeric nucleosides **36** and **37**. After deprotection of the hydroxy group, a mixture of two products **8** and **38** was obtained. No racemization for these products was observed.

Methylenecyclopropane analogues of nucleosides possessing a fluorine atom in the cyclopropane ring were also synthesized by Zemlicka *et al.* [17], and their antiviral activity was investigated. Additionally, phosphonate derivatives of methylenecyclopropane nucleosides were recently reported by the same group [18]. All methylene cyclopropane phosphonates were devoid of antiviral activity, with the exception of a guanine derivative that inhibited the replication of VZV and was non-cytotoxic.

15.3 Lobucavir and Cyclobutane Nucleoside Derivatives

Much impetus was provided by the discovery of the naturally occurring oxetanocin-A (**39**), which was isolated from a culture filtrate of *Bacillus megaterium*, and exhibited remarkable antiviral, antitumor, and antibacterial activities [19]. Carbocyclic analogues of oxetanocin-A were subsequently synthesized (Figure 15.6). Two of these, cyclobut-A (**40**) and cyclobut-G (**4**) (lobucavir) – exhibited broad biological activity against many types of virus, including HIV, herpes simplex virus (HSV-1 and HSV-2), cytomegalovirus and hepatitis B [20].

Oxetanocin A isolated from cultures of *Bacillus megaterium* NK84-0218, possesses a unique structure and exhibits potent antiherpes activity, as well as anti-HIV activity. The guanine analogue, oxetanocin-G and its carbocyclic analogue (lobucavir) both have potent antiviral activity against HCMV [21], VZV [22], HBV, and HIV [23].

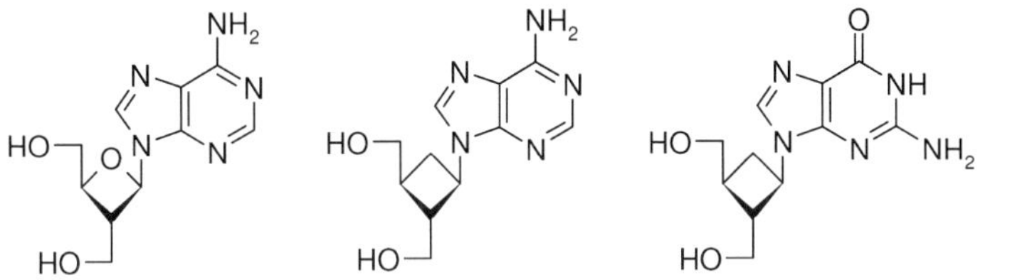

39, Oxetanocin-A **40, Cyclobut-A** **4, Cyclobut-G (Lobucavir)**

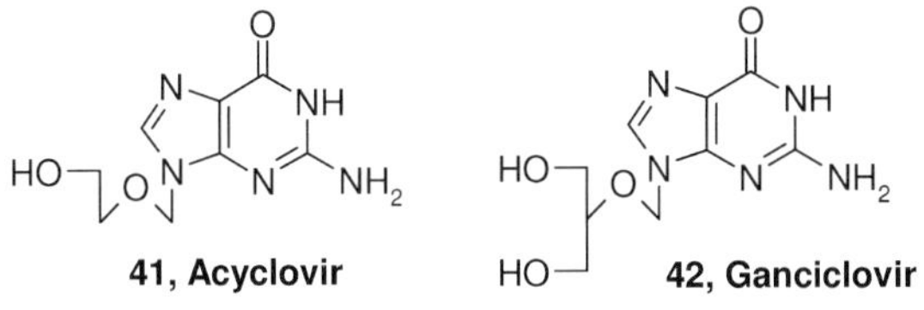

41, Acyclovir **42, Ganciclovir**

Figure 15.6 Cyclobutane analogues of oxetanocin-A.

Among derivatives of carbocyclic-oxetanocin, SQ-32,829 [24], or the fluoro-derivative [25] all exhibited antiviral activity. The carbocyclic guanine analogue exhibits, *in vitro*, anti-HSV activity which is superior to that of acyclovir (**41**) [26] and comparable to that of ganciclovir (**42**) [27], as shown in Table 15.1.

The carbocyclic adenine analogue, cyclobut-A, was more potent than acyclovir against VZV, but less so against HSV-1 and HSV-2. Both, cyclobut-A and cyclobut-G have excellent activities against MCMV and HCMV. Both compounds are also active against HIV, albeit at a lower level than either ddA or ddG. Conformational studies of the compounds showed a close correspondence with the stable conformation of β-nucleoside components of DNA [28], lying within 3 kcal mol^{-1} of each other.

Carbocyclic-oxetanocin-G (lobucavir) therefore appears to have good prospects for treating HCMV and HSV infections. Oxetanocin G itself is synergistic with acyclovir against HSV-1 and HSV-2, but carbocyclic oxetanocin-G is only additive. The latter is a poor substrate for the viral thymidine kinase but, once phosphorylated, the triphosphate form competitively inhibits HSV-1 DNA polymerase with respect to dGTP, acting as a non-obligate chain terminator. Clinical evaluation of carbocyclic oxetanocin G in the treatment of herpes simplex corneal ulcers demonstrated the excellence and safety of this compound, certainly with topical dosing [29]. Interest has also been shown in the chemotherapeutic application of lobucavir for HBV infections [30], with lobucavir reversibly inhibiting HBV production by the HBV-expressing HepG 2.2.15 cell line [31]. In molecular studies, lobucavir nucleotide triphosphate acted as a non-obligate chain terminator of polymerase, inhibiting all three major enzymatic functions in genomic replication: oligodeoxynucleotide primer synthesis, reverse transcriptase, and DNA-dependent DNA (strand) synthesis [32]. This inhibitory mechanism operated against polymerases of other hepadnaviruses, including woodchuck hepatitis virus (WHV) and duck hepatitis virus.

The first racemic synthesis of cyclobut-A was reported by Honjo *et al.* [33] in 1989 (Scheme 15.4). The essential intermediate, the 2,3-bis(benzoyloxymethyl)cyclobuta-

Table 15.1 *In-vitro* antiviral activity of cyclobut-A and cyclobut-G against herpesviruses.

Assay[b]	Cyclobut-A		Cyclobut-G (Lobucavir)		Acyclovir/ Ganciclovir[a]	
	ED_{50}[c]	ID_{50}[d]	ED_{50}	ID_{50}	ED_{50}	ID_{50}
HSV-1[e]						
E-377	1.0	>100	0.05	>100	0.08	>100
HSV-2[e]						
MS	1.6	>100	0.07	>100	0.09	>100
X-79	1.8	>100	0.06	>100	0.10	>100
JEN	2.3	>100	0.05	>100	0.07	>100
HEET	2.2	>100	0.07	>100	0.06	>100
HCMV[e]		>100				
AD169	3.7	>100	6.2	>100	2.7[a]	>100
Davis	0.9	>100	4.9	>100	3.6[a]	>100
EC	2.2	>100	3.9	>100	2.8[a]	>100
LA	1.0	>100	1.2	>100	2.4[a]	>100
CH	0.9	>100			3.1[a]	>100
MCMV[f]						
Smith	0.05	>100	0.10	>100	1.0[a]	>100
VZV[e]						
Ellen	2.1	>100	0.40	>100	2.3	>100
OKA	2.0	>100	0.40	>100	3.7	>100
EBV[g]						
Raji cells			0.01	94	3.8	>100
Cell proliferation[h]		7.7		27.0		165/43[f]

a) Ganciclovir.
b) Mean of two to four assays.
c) Drug concentration ($\mu g\, mL^{-1}$) calculated to reduce plaque formation (or antigen production for EBV) in infected cell monolayers to 50% of unreacted, infected controls.
d) For antiviral assays, the drug concentration ($\mu g\, mL^{-1}$) calculated to reduce uptake of neutral red stain by uninfected cell monolayers to 50% of untreated, uninfected controls; for the cell proliferation assay, the drug concentration ($\mu g\, mL^{-1}$) calculated to reduce proliferation of human foreskin fibroblasts to 50% of untreated controls.
e) Plaque reduction assay in human foreskin fibroblasts.
f) Plaque reduction assay in mouse embryo fibroblasts.
g) Inhibition of diffuse early antigen production assayed by immunofluorescent monoclonal antibody.
h) Human foreskin fibroblats at 25% confluency were incubated in the presence and absence of serial dilutions of drug. After 72 h, the cultures were trypsinized and the number of cells determined with a Coulter counter. The 50% inhibitory dose was calculated from comparison of the number of cells in drug-treated and untreated cultures.

none **45** can be prepared in three steps from diethyl 3,3-diethoxy-1,2-cyclobutane-dicarboxylate **43** by reduction, treatment with benzoyl chloride, and cleavage of the resulting ketal. The conversion of **45** to its oxime **46**, followed by hydrogenation in the presence of platinum oxide, led to a mixture of cyclobutylamines **47** and **48**. Further elaboration of the amino group followed the familiar pattern.

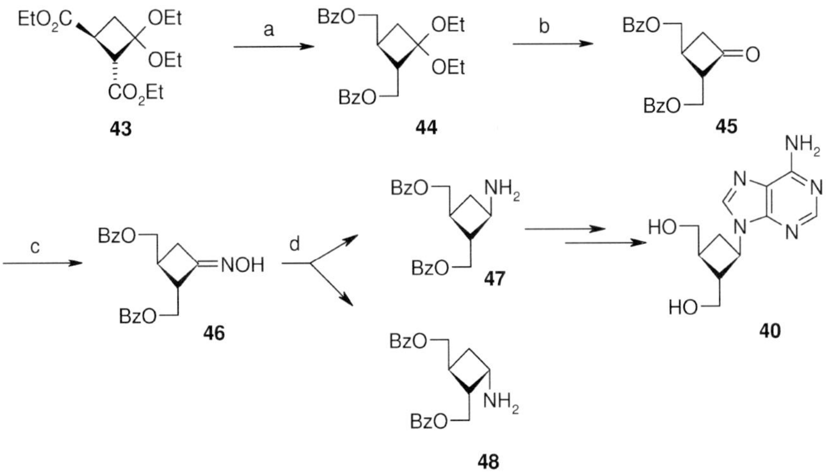

Scheme 15.4 Reagents and conditions: (a) $LiAlH_4$, THF/Et_2O then BzCl, pyridine; (b) pTSA, acetone; (c) NH_2OH; (d) H_2, PtO_2.

Since then, several groups have developed different synthetic pathways, in both racemic and chiral form, to cyclobutyl analogues of oxetanocin-A [34]. Those practicable syntheses have mainly used either cyclobutanones or cyclobutene epoxides as starting materials. In some early studies, chiral resolutions were performed to deliver non-racemic intermediates [35]. Jung *et al.* [36] utilized an enzymatic desymmetrization of a *meso*-cyclobutene as the enantioselective step in their formation of non-racemic cyclobut-A.

Recently, a stereogenic route to optically pure cyclobutyl nucleosides was proposed by Gosh *et al.* [37], starting from the diethyl acetal of acrolein **49** (Scheme 15.5). The key step involved a stereoselective intramolecular [2 + 2] photocycloaddition to provide a trisubstituted cyclobutane derivative with the desired stereochemistry. The nucleoside linkage was established through nucleophilic displacement of an acetate group by adenine.

After an initial *trans*-acetalization with the hexa-2,4-diol, the obtained triene **50** was converted into a mixture of two isomeric photoadducts **51** by irradiation. The photoadduct mixture, on treatment with 80% aqueous acetic acid, led to the thermodynamically more stable lactol mixture, which was oxidized using Jones reagent to **52**. This lactone was then converted to the hydroxy-ketone **53** through a Petasis reagent. Subsequent oxidative cleavage of the olefinic chain with RuO_4, followed by a diazotation, gave keto-diester **54**. The obtained interme-diate **54** was finally transformed into the acetate **55** under Baeyer–Villiger oxidation. The adenine was introduced under standard SN_2 conditions. The obtained diester nucleoside **56** was then treated with $LiBH_4$ to lead the optically pure cyclobut-A **40**.

The (−)-cyclobut-A and (±)-3′-*epi*-cyclobut-A were synthesized by Hegedus *et al.* [38], utilizing photolysis of a (benzyloxymethyl)(methoxy)chromium carbene complex with optically active ene-carbamate to produce the corresponding optically

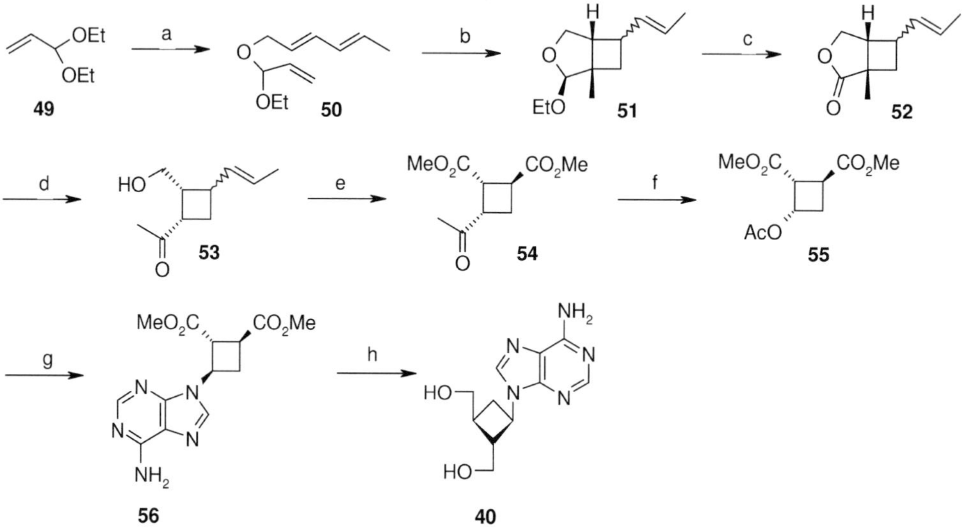

Scheme 15.5 Reagents and conditions: (a) hexa-2,4-dienol, PPTS, C_6H_6; (b) hν, CuOTf, Et_2O; (c) $AcOH/H_2O$, 80 °C, 3.5 h, then Jones reagent; (d) Cp_2TiMe_2, toluene; (e) $RuCl_3$–$NaIO_4$, CCl_4–CH_3CN–H_2O, r.t.; (f) TFAA, urea–H_2O_2, KH_2PO_4, CH_2Cl_2; (g) adenine, K_2CO_3, DMF; (h) $LiBH_4$.

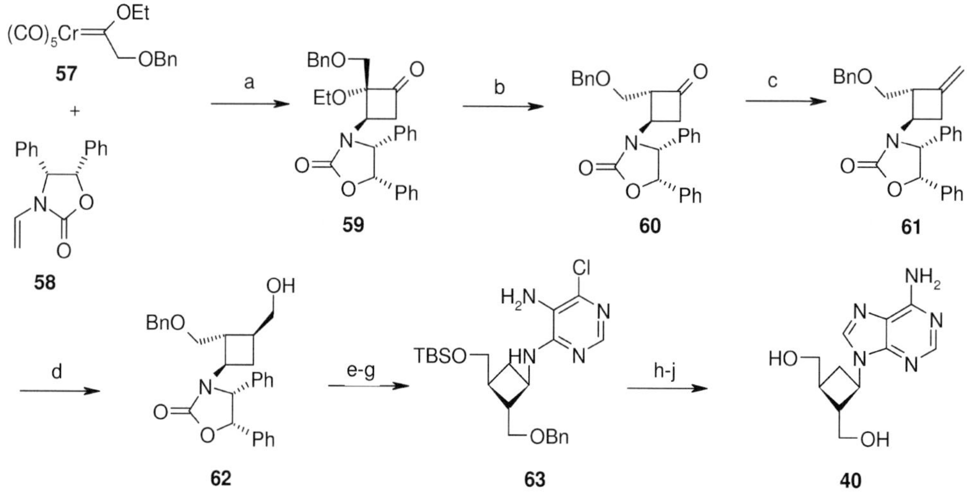

Scheme 15.6 Reagents and conditions: (a) hν, CH_2Cl_2; (b) SmI_2, MeOH, THF; (c) Cp_2TiMe_2 or $Zn/CH_2I_2/TiCl_4$; (d) R_2BH/Ox; (e) TBDMSCl; (f) H_2, $Pd(OH)_2$, Et_3N, EtOH; (g) 5-amino-4,6-chloropyridmine, Et_3N, BuOH; (h) HCl, $HC(OEt)_3$, DMF; (i) $NH_3/MeOH$; (j) BCl_3.

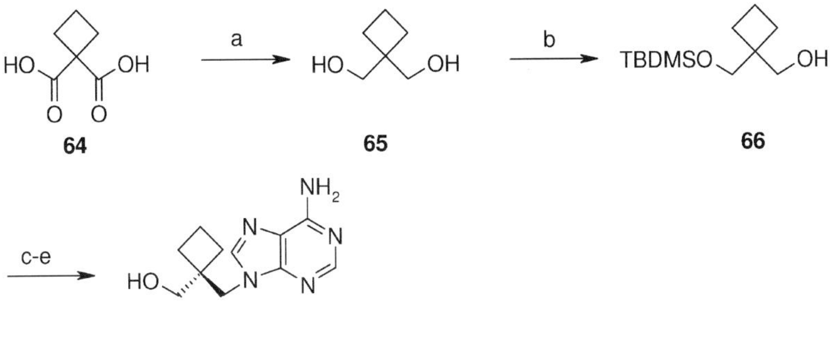

64 65 66

67

Scheme 15.7 Reagents and conditions: (a) $LiAlH_4$, THF; (b) TBDMSCl, imidazole, CH_2Cl_2; (c) MsCl, Et_3N, CH_2Cl_2; (d) Adenine, K_2CO_3, 16-crown-6, DMF; (e) TBAF, THF.

active *R,R*-disubstituted cyclobutanone (Scheme 15.6); this was then converted to the desired compounds. After a reductive cleavage of *R*-heteroatom-substituted carbonyl compounds through SmI_2, the *trans*-disubstituted cyclobutanone was isolated. Methylenation of **60** was successful with both the Petasis reagent, Cp_2TiMe_2 and the Takai reagent, $Zn/CH_2I_2/TiCl_4$. Hydroboration/oxidation of methylenecyclobutane **60** under a variety of conditions always led to the undesired *cis*-dialkyl product as the minor product. Hydrogenolytic cleavage of the oxazolidinone $[Pd(OH)_2/H_2]$ generated the free amine which, by treatment with 5-amino-4,6-dichloropyrimidine, introduced the core of the adenine. Elaboration of cyclobut-A was performed using conventional methodology.

Another example of a synthetic pathway used to synthesize chiral precursors of cyclobutane nucleosides was recently reported by Ortuno *et al.* [39] when, starting from the (−)-(*S*)-vebenone, the hydroxy- and amino-cyclobutane derivatives were obtained.

Hong *et al.* [40] reported on a facile synthesis of cyclobutyl nucleoside **67** (Scheme 15.7), which exhibited significant anti-HCMV activity ($EC_{50} = 11.2\,\mu mol$), without any cytotoxicity up to 100 μmol. The starting material – cyclobutyl dicarboxylic acid **64** – was reduced by $LiAlH_4$ to diol **65**, which was monoprotected with a *tert*-butyldimethylsilyl chloride (TMBDS) group to the intermediate **66**. The remaining hydroxy group was then mesylated, and the obtained diprotected diol coupled with adenine using a standard procedure. The obtained protected nucleoside was treated with tetra-*n*-butylammonium fluoride (TBAF), which led to the free nucleoside **67**.

Using a similar procedure, carbocyclic nucleosides bearing a cyclobutene ring were synthesized from the unsaturated analogue of diol **65** [41], but no significant antiviral activity was identified for these serial compounds. Recently, Zemlicka *et al.* [42] presented a novel synthetic route to methylenecyclobutane nucleosides (Scheme 15.8). Here, ethyl acrylate **68** is reacted with ketene dimethyl thioacetal **69**, giving cyclobutyl disulfide **70**. After reduction of the ester group with $LiAlH_4$, and subsequent benzylation, the thioketal function **71** was hydrolyzed with a NCS/$AgNO_2$ mixture to afford the cyclobutanone **72**. The carbonyl group was then transformed into a double bond by a Wittig reaction and, after treatment with

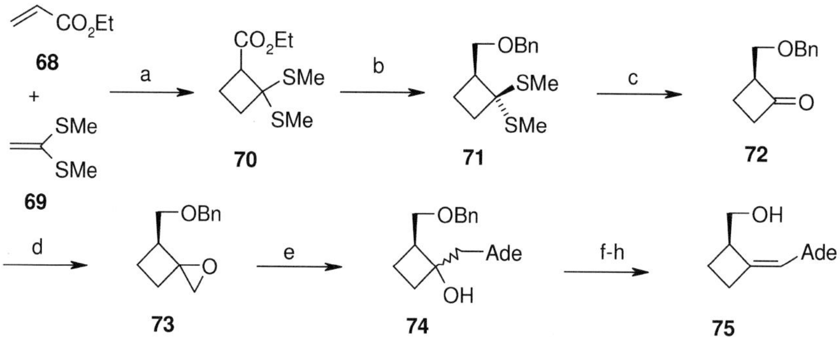

Scheme 15.8 Reagents and conditions: (a) Et_2AlCl, CH_2Cl_2; (b) $LiAlH_4$, THF, then NaH, THF, then BnBr, NBu_4I; (c) NCS, $AgNO_3$, MeCN; (d) [Ph_3PMe]Br, BuLi, THF then m-CPBA, $NaHCO_3$, CH_2Cl_2; (e) Adenine, NaH, DMF, r.t.; (f) MsCl, DMAP, CH_2Cl_2/pyridine; (g) *t*BuOK, THF; (h) BCl_3, CH_2Cl_2, −78 °C.

m-CPBA, gave the epoxide **73**. Opening the oxirane with adenine led to a mixture of two *Z*- and *E*-hydroxyderivatives **75**. After mesylation, treatment with *t*BuOK, and deprotection of the hydroxyl group, the desired unsaturated nucleoside **75** was obtained together with this isomer.

An interesting synthesis which led to nucleosides possessing a dimethylbutyl ring, was presented Fernandez *et al.* [43] (Scheme 15.9). 1*S*-α-pinene (**76**) was used as a starting material, which was oxidized with permanganate to pinonic acid (**77**) and reduced with $NaBH_4$ to pinolic acid (**78**). After acetylation with acetic anhydride in pyridine and treatment with ethyl chlorofomate in the presence of triethylamine and ammonia, acetoxy amide **79** was obtained. Reduction with $LiAlH_4$ led to aminoalcohol **80**, which was coupled with 4,6-dichloropyrimidin-2-amine. The intermediate **81** was transformed into diaza derivative **82** which, after reduction with zinc, gave amine **83**. Treatment with ethyl orthofomate followed by addition of ammonia solution led to the diaminopurine nucleoside **84**.

Some other nucleosides bearing a dimethylbutyl ring were synthesized from 1*S*-α-pinene (**76**) [44] or nopinone [45]. However, except for one compound which exhibited slight anti-herpes properties, these derivatives showed no activity against various types of virus.

In the search for more selective agents, numerous carbocyclic oxetanocins with pyrimidine bases have been synthesized and evaluated. For example, 2-bromovinyl uracil analogues and related compounds were synthesized by Slusarchyk *et al.* [46]. The bromovinyl **85**, the iodovinyl **86**, and the chlorovinyl **87** analogues were found to be potent inhibitors of VZV, but less potent against HCMV and HSV (Table 15.2).

In 2003, Takahashi *et al.* [47] described the inhibition of vertebrate telomerases by the triphosphate derivatives of carbocyclic oxetanocin analogues (Figure 15.7). Telomerase is a cellular endogenous reverse transcriptase that uses its internal RNA as a template for extension of the telomere repeat, thus maintaining telomere length. Both compounds showed potent inhibitory activity. Lineweaver–Burke plot analyses showed that the inhibition mode of these compounds was competitive with dGTP,

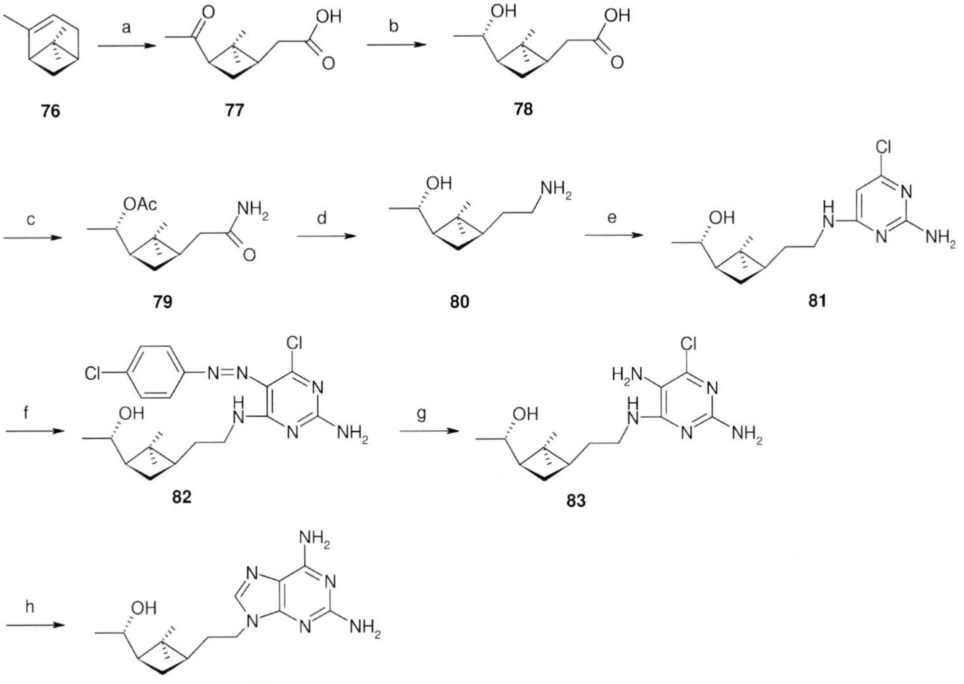

Scheme 15.9 Reagents and conditions: (a) $KMnO_4/H_2O$; (b) $NaBH_4$/EtOH; (c) Ac_2O/pyridine then $ClCO_2Et/Et_3N/NH_3$; (d) $LiAlH_4$/THF; (e) 4,6-dichloropyrimidin-2-amine/Et_3N/1-butanol; (f) $4\text{-}ClC_6H_4N_2/H_2O$; (g) Zn/AcOH; (h) $CH(OEt)_3$/12 M HCl, then 14 M NH_4OH.

the K_i values for C-OXT-GTP and m-C-OXT-GTP being 2.0 µM and 4.9 µM, respectively, and thus smaller than the K_m of dGTP (11 µM).

More recently, several cyclobutyladenine and analogues were synthesized from 1*R*-α-pinene, and their antiviral activity was tested [48]. One of these compounds showed interesting selectivity against both TK^+ and TK^- VZV. However, the remaining cyclobutane nucleosides described to date have not exhibited antiviral activity of note.

15.4
Carbovir and 2′,3′-Unsaturated Nucleoside Derivatives

Carbovir (2′,3′-didehydro-2′,3′-dideoxyguanosine) emerged as a potent and selective anti-HIV agent from a large screening program conducted by the National Cancer Institute [49]. Its hydrolytic stability and ability to inhibit the infectivity and replication of HIV in T cells at concentrations of approximately 200- to 400-fold below toxic concentration made it a prime candidate for development as an anti-HIV agent.

Table 15.2 Antiviral and growth inhibition activities in cell cultures.

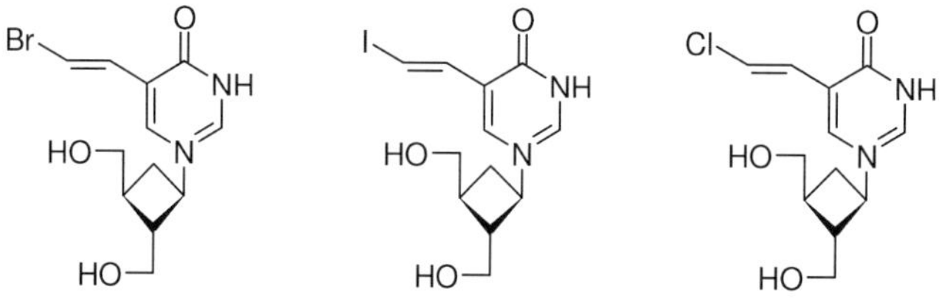

85 86 87

Antiviral[b)]	ID$_{50}$ (μM)[a)]					
	(1′R)-85	(±)-85	(±)-86	(±)-87	BVaraU	Acyclovir
VZV strains						
Ellen	0.06–0.2	0.03–0.15	0.03–0.05	0.2–0.4	0.001–0.003	2–4
Ito	0.015	0.01–0.05	0.05–0.1	0.02–0.05	0.003–0.007	0.4–2
Oka		0.3–0.6			0.001–0.003	1–4
9021	0.06–0.15	0.2–0.6	0.13	0.45–1.1	0.001–0.003	1–4
pplla	0.06–0.2	0.06–0.2	0.13–0.26		0.001–0.003	2–4
40a2 (TK$^-$)		>300				110–220
Kanno-Kohmura (TK$^-$)	75–150	>300	>260	>220	>72	44–110
HSV-1 (Schooler)	1.5–3	<6	5–13	1.7–3.5	0.06–0.14	0.2–0.4
HSV-2 (186)	>300	>300	>260	>280	60–120	0.4–0.8
HCMV (AD169)	>300	>300	130–260	180–350	>290	20–40
WI-38 growth inhibn	≥750	≥750	>800	>400	>75	≥750
Therapeutic index[c)]	$\geq 1.3\times10^4$	$\geq 2.5\times10^4$	$\geq 2.7\times10^4$	$\geq 2.0\times10^3$	$\geq 7.5\times10^4$	$\geq 3.8\times10^2$

a) All ID$_{50}$ values show the range of repeat assays.
b) All plaque reduction assays were performed on WI-38 cell monolayers.
c) ID$_{50}$ for WI-38 cell growth inhibition/ID$_{50}$ anti-VZV (strain Ellen).

Because the brain penetration capability of carbovir was inferior to that of AZT – a factor thought to be important in treating late-stage HIV infections – close analogues of carbovir have been prepared. Of these, the 6-(cyclopropylamino)purine analogue – so-called abacavir – was found to reach levels in the brain of rats and monkeys that were comparable to those of AZT. Finally, abacavir – which is a prodrug of carbovir – was been recently approved by the FDA as an anti-HIV drug with less toxic effects

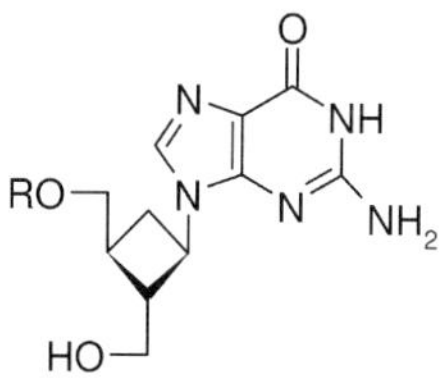

4, R = H
88, R = triphosphate

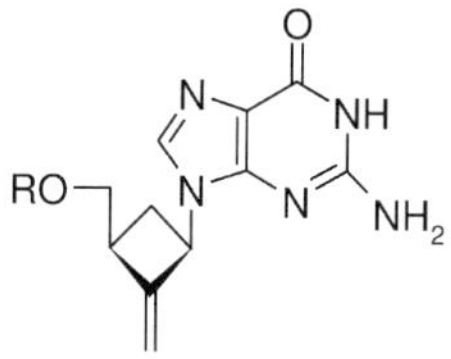

89

Figure 15.7 Carbocyclic oxetanocin analogues.

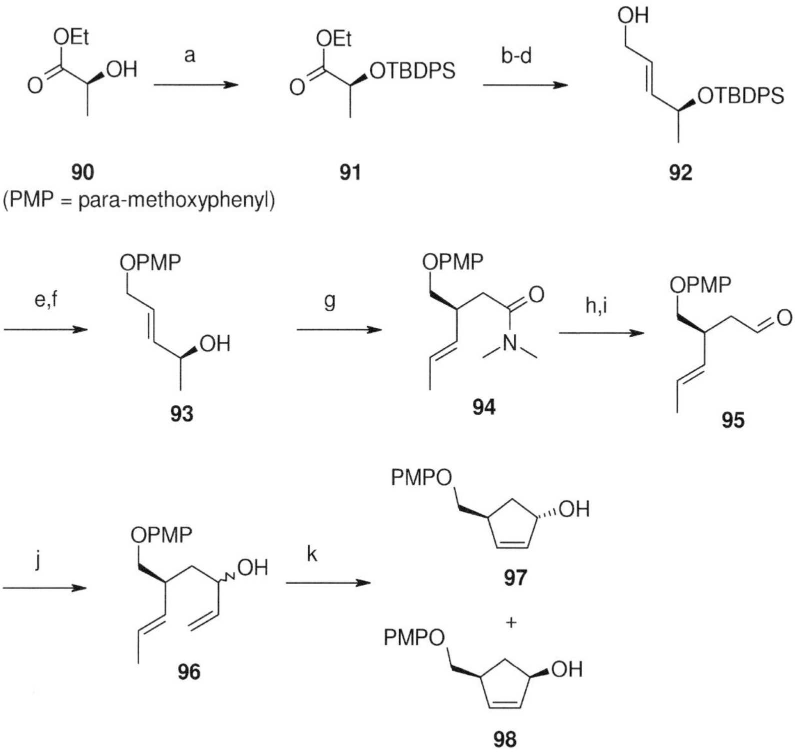

Scheme 15.10 Reagents and conditions: (a) *tert*-$BuPh_2SiCl$, imidazole, CH_2Cl_2, r.t.; (b) *iso*-Bu_2AlH, CH_2Cl_2, −78 °C; (c) $Ph_3P{=}CHCO_2Et$, PhMe, 80 °C; (d) *iso*-Bu_2AlH, THF, −78 °C to 0 °C; (e) DEAD, PPh_3, 4-MeOPhOH, THF, r.t.; (f) TBAF, THF, r.t.; (g) $Me_2N(MeO)_2CCH_3$, decalin, reflux; (h) $LiEt_3BH$, THF, −78 °C to r.t.; (i) $(COCl)_2$, DMSO, CH_2Cl_2, −78 °C, then NEt_3; (j) $BrHC{=}CH_2$, *tert*-BuLi, −78 °C, THF; (k) Grubbs [Ru] = catalyst, CH_2Cl_2, r.t.

than any other nucleoside analogues. As a consequence, related carbocyclic nucleosides have been prepared enantioselectively.

In 2003, Florent *et al.* [50] reported the enantioselective synthesis of (1*S*,4*S*)-4-(4-methoxy-phenoxymethyl)-cyclopent-2-enol **97**, a key intermediate in the synthesis of (−)-carbovir, in 11 steps starting from (*S*)-(−)-ethyl lactate (Scheme 15.10). In the first step, protection of the hydroxy group by a *tert*-butyldiphenylsilyl group led to compound **91**, which was reduced with di-*iso*-butylaluminum hydride. The obtained alcohol was submitted to the Horner–Emmons reaction with (carboethoxymethylene)triphenylphosphorane, and the obtained α,β-unsaturated ethyl ester, was reduced to **92** using di-*iso*-butylaluminum hydride. Protection of the primary hydroxy group as a 4-methoxyphenyl ether, followed by deprotection of the secondary hydroxy group with TBAF, led to the optically pure compound **93**. A Claisen [3 + 3] sigmatropic rearrangement reaction was then applied, which led to intermediate **94**, and this was then transformed in two steps into the appropriate aldehyde **95**. The second ethylenic function was then introduced, and the resulting compound **96** was used in a ring-closing metathesis reaction using Grubb's catalyst. Other examples and aspects

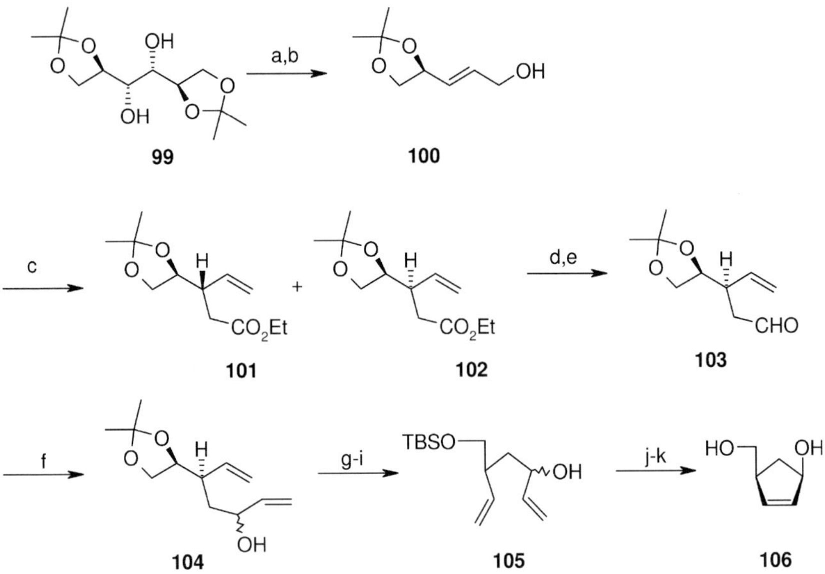

Scheme 15.11 Reagents and conditions: (a) $NaIO_4$, $MeCN/H_2O$, K_2CO_3, $P(O)(OEt)_2CH_2CO_2Et$; (b) $LiAlH_4$, Et_2O; (c) $CH_3CH(OEt)_3$, propionic acid, 140 °C; (d) $LiAlH_4$, Et_2O; (e) oxalyl chloride, DMSO, Et_3N; (f) $CH_2{=}CHMgBr$, THF; (g) 6 M HCl, THF then $NaIO_4$, CH_3CN/H_2O; (h) $LiAlH_4$; (i) TBSCl, DMAP, imidazole, Et_3N, CH_2Cl_2; (j) $(PCy_3)_2Cl_2RuCHPh$, CH_2Cl_2; (k) TBAF, THF, r.t.

of using the ring-closing metathesis (RCM) reaction in the synthesis of carbocyclic nucleosides have been recently reviewed [51].

Recently, Gosh *et al.* [52] reported on a seven-step synthesis of carbovir and its prodrug, including Wittig–Horner reaction, Claisen rearrangement, Swern oxidation and finally ring-closing metathesis reaction. The starting material, 1,2:5,6-di-*O*-cyclohexylidine-D-mannitol was transformed into the silylated (1*S*,4*S*)-4-(4-methoxy-phenoxymethyl)-cyclopent-2-enol and its diastereoisomer (Scheme 15.11).

Olivo *et al.* [53] have used the *endo*-hydroxylactone **107**, obtained by the reaction of glyoxylic acid with cyclopentadiene, separated from its enantiomer using *Pseudomonas fluorescens* lipase, was used in the six-step synthesis of (−)-carbovir. In the first steps, **107** was converted into diol **108**, and then (using triphosgene) transformed into (−)-bicyclic carbonate **109**. Tsuji–Trost palladium-catalyzed cross-coupling with 2-amino-6-chloropurine, followed by hydrolysis of obtained intermediate **109**, led to (−)-carbovir (**111**) (Scheme 15.12).

An enantio- and diastereoselective synthesis of 4-α-alkylcabovir derivatives was proposed by Misayaka *et al.* [54] (Scheme 15.13). In order to synthesize 4′-methyl-carbovir, methyl-2-oxopentanecarboxylate **112** was transformed into the chiral acetal **113**, using (*R*,*R*)-cycloheptanediol. After diastereoselective alkylation with methyl iodide and iodoacetylization of enol ethers **114**, iodoacetal **115** was obtained as a single diastereoisomer. After elimination of iodine with DBU, and acidic hydrolysis,

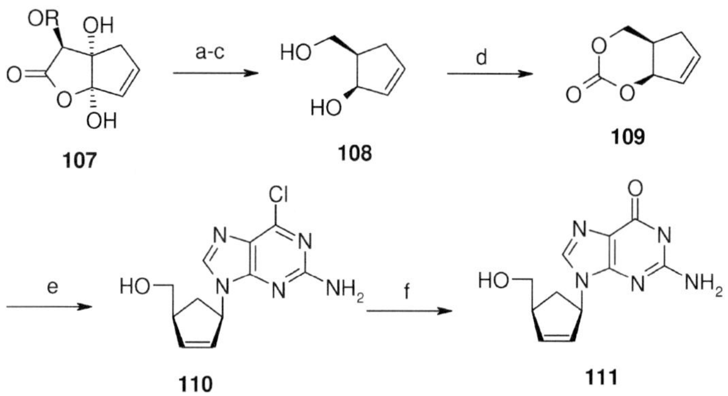

Scheme 15.12 Reagents and conditions: (a) $LiAlH_4$, THF, reflux; (b) $NaIO_4$, $Et_2O–H_2O$; (c) $NaBH_4$, EtOH; (d) triphosgene, Et_3N, CH_2Cl_2; (e) 2-amino-6-chloropurine, $Pd(PPh_3)_4$, DMSO–THF (1 : 1).

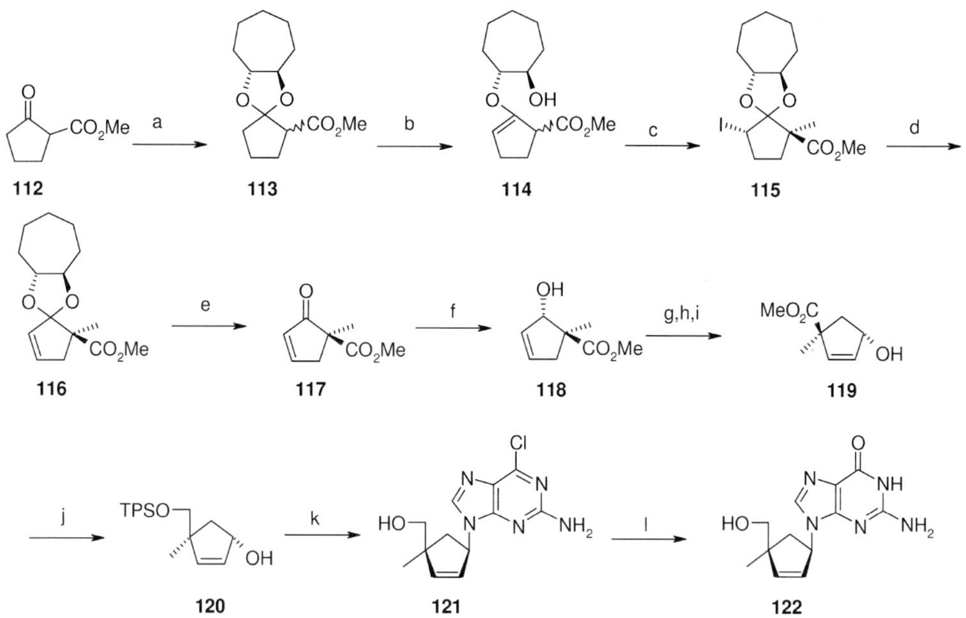

Scheme 15.13 Reagents and conditions: (a) (*R*,*R*)-cycloheptane-1,2-diol, TsOH; (b) LDA, MeI; (c) I_2, Et_3N; (d) DBU; (e) 10% HCl, MeOH; (f) $NaBH_4$, $CeCl_3$; (g) Ac_2O, pyridine; (h) $PdCl_2(MeCN)_2$, benzoquinone; (i) K_2CO_3, MeOH; (j) (i) DPH, (ii) LAH, (iii) TBDPSCl, (iv) PPTS; (k) (i) 2-amino-8-chloropurine, Ph_3P, $EtO_2CN{=}NCO_2Et$, (ii) TBAF, 1 M NaOH.

the obtained intermediate **117**, was applied in a Luche reduction, which gave hydroxy ester **118**. Subsequent acetylation, treatment with a Pd-catalyst, and finally methanolysis gave the rearrangement product **119**. The hydroxy ester **119** was then transformed in a four-step sequence to the key intermediate **120**, which was used directly in a coupling reaction with 2-amino-6-chloropurine. The desired 4′-methylcarbovir **122** was obtained after basic hydrolysis of **121**.

This method was also used for the synthesis of 4′-nonyl and 4′-benzyl derivatives of carbovir. The 4′-methyl analogue of carbovir, together with 4′-cyano analogue, was also synthesized by Hegedus *et al.* [55]. In order to introduced the 4′-substituent, the Michael reaction was used, whereas to introduce the nucleoside base a Pd-catalyzed allylic substitution was applied.

Various carbovir analogues were synthesized by chemists, including 6′-(α)-methyl derivatives [56], 2′,3′-fluoroderivatives [57], *nor*-derivatives [58], and homocarbovir [59]. Among these syntheses, Chu *et al.* [60] presented a convenient preparation of the L-adenine analogue of carbovir, which exhibited moderately potent anti-HIV activity ($EC_{50} = 2.4\,\mu M$), without cytotoxity up to 100 μM. The starting material, (+)-cyclopentanone **123**, which is easily accessible from D-ribose, was transformed into alcohol **124** by regioselective addition and reduction with DIBAl-H. Subsequent benzylation led to intermediate **125**, which was deprotected and treated with trimethyl orthoformate. The resultant cyclic orthoester **126**, after thermal elimination, gave the cyclopentene derivative **127**. After deprotection of the secondary hydroxy group, **127** was condensed with 6-chloropurine and finally transformed into the L-analogue **129**. This method was also used to synthesize the 6-mercaptopurine analogue, which exhibited no anti-HIV activity (Scheme 15.14).

Among many 2′,3′-didehydro-2′,3′-dideoxy carbocyclic nucleoside analogues, a group of branched analogues was designed and synthesized by Hong *et al.* [61]. The 2′,3′,4′-trimethyl unsaturated carbanucleoside **139** (Scheme 15.15) exhibited

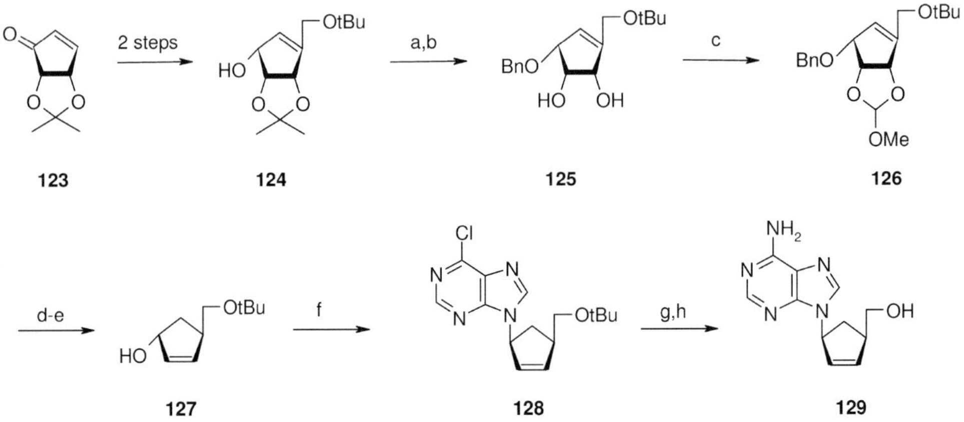

Scheme 15.14 Reagents and conditions: (a) BzCl, pyridine; (b) conc. HCl, MeOH; (c) $CH(OMe)_3$, pyridinium toluene-*p*-sulfonate; (d) Ac_2O; (e) 2 N NaOH/MeOH; (f) 6-chloropurine, Ph_3P, DEAD; (g) NH_3/MeOH; (h) TFA/H_2O, 2 : 1.

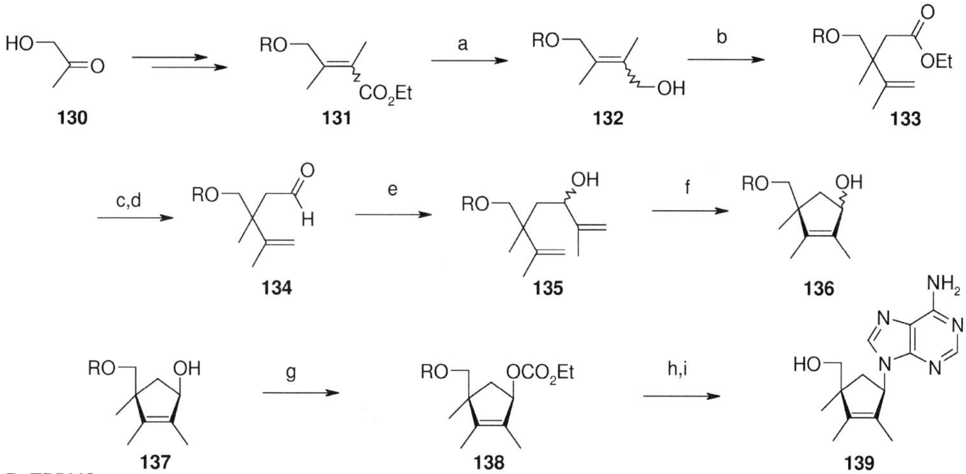

Scheme 15.15 Reagents and conditions: (a) Dibal-H, CH_2Cl_2, −50 °C; (b) $CH_3C(OC_2H_5)_3$, CH_3CH_2COOH, 140 °C; (c) Dibal-H, DCM, 0 °C; (d) PCC, 4 Å MS, CH_2Cl_2, r.t.; (e) $CH_2=C(CH_3)MgBr$, THF, −78 °C; (f) Grubb's catalyst II, reflux, overnight; (g) $ClCO_2Et$, DMAP, pyridine, r.t., overnight; (h) adenine, $Pd_2(dba)_2$, reflux, overnight; (i) TBAF, THF, r.t.

good antiviral activity against HCMV (8.8 μg mL^{-1} in Davis cell), without any cytotoxicity up to 100 μg mL^{-1}. The synthesis was started from acetol **130**, which was transformed into the α,β-unsaturated ester derivative **131** which, after reduction with diisobutylaluminum hydride (DIBAL-H) to allilic alcohol **132** and [3,3]-sigmatropic rearrangement using triethylorthoacetate, gave the γ,δ-unsaturated ester **133**. After further reduction with DIBAL-H, followed by oxidation with PCC, aldehyde **134** was obtained, and used in the next step in Grignard reaction with the appropriate organomagnesium bromide. The obtained olefin **135** was then transformed into an equimolar mixture of two stereoisomeric cyclopentenols **136** and **137**, using Grubb's catalyst II under RCM conditions. Using ethylchloroformate, the β-stereoisomer of **136** was then transformed to the ethoxycabonyl derivative **138**, which was coupled with adenine. Palladium (0)-catalyzed coupling with Pd_2dba_3, followed by deprotection of the primary hydroxy group with TBAF, gave the adenine derivative **139**. The cytosine derivative was also synthesized using this method. When the C′4-methyl group was exchanged into the phenyl group, then obtained analogue exhibited a moderately anti-HCMV activity (30.1 μg mL^{-1}). Additionally, when the C′4 methyl group was exchanged into a second hydroxymethyl group, anti-HCMV activity was not observed, although the compound did show weak anti-HIV-1 activity [62].

Using similar synthetic paths, some other branched derivatives were synthesized, including 2′,4′- [63] and 2′,3′-doubly [64] branched nucleosides. Some of these showed moderate anti-HCMV and anti-CoxB3 activities. Following the discovery of the (−)-bis(hydroxymethyl)-cyclopentenyl adenine (BCA; **149**) [65], which inhibits HIV reverse transcriptase, the interest in 2′,3′-didehydro-2′,3′-dideoxy carbanucleo-

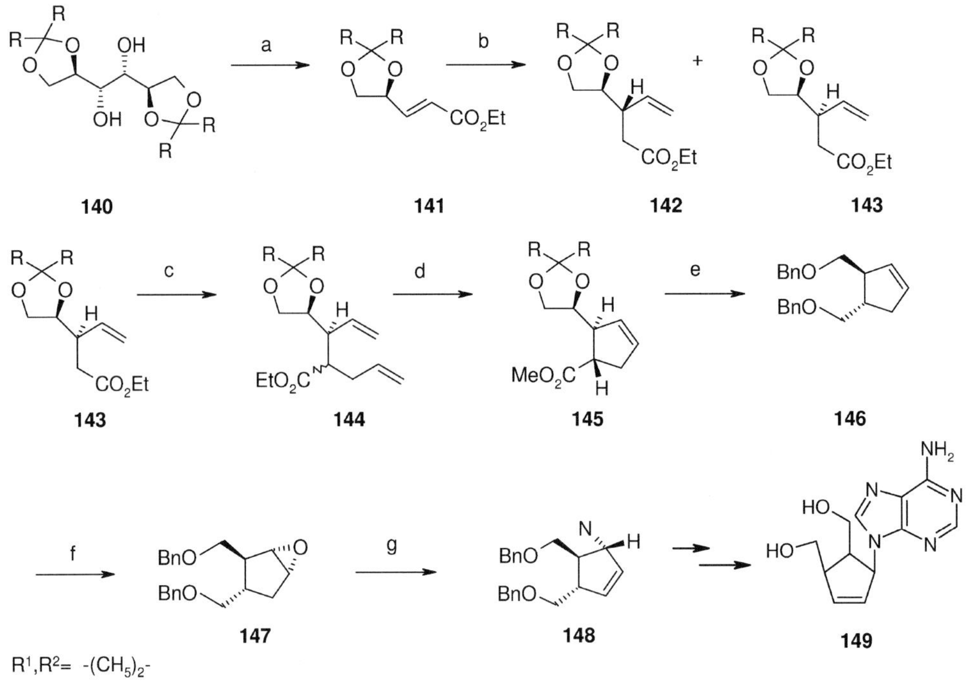

Scheme 15.16 Reagents and conditions: (a) $NaIO_4$, $MeCN/H_2O$ (3:2), K_2CO_3, P(O)$(OEt)_2CH_2CO_2Et$; (b) $LiAlH_4$, Et_2O, −60 °C then $CH_3C(OEt)_3$, $CH_3CH_2CO_2H$, 140 °C, 6 h; (c) LDA, THF, $CH_2{=}CHCH_2Br$, HMPA; (d) Grubb's catalyst, C_6H_6, 60 °C, 20 h, then NaOEt–EtOH, reflux 10 h, then CH_2N_2, Et_2O; (e) 75% AcOH, 18 h, then $NaIO_4$, $MeOH–H_2O$, r.t., 1 h then $NaBH_4$, MeOH then $LiAlH_4$ Et_2O, 0 °C–r.t., 1 h then NaH, $PhCH_2Br$, THF, HMPA, reflux, 6 h; (f) *m*-CPBA, DCE, 0 °C–r.t., 5 h; (g) Ph_2Se_2, $NaBH_4$, then 30% H_2O_2, then *p*-$NO_2C_6H_4COOH$, DEAD, PPh_3, C_6H_6, then KOH, $EtOH–H_2O$, reflux, 30 min, then MsCl, pyridine, 0 °C, then NaN_3, DMF, 100 °C.

sides, bearing hydroxymethyl groups in various positions of the cyclopentene ring, rapidly increased. The synthesis of the carboxylic core of (−)-BCA is outlined in Scheme 15.16 [66]. The synthesis begins with a protected D-mannitol derivative **140** that is transformed into α,β-unsaturated ester **141** by a periodate cleavage of the diol **140**, followed by a Wittig–Horner reaction on the aldehyde with triethyl phosphonoacetate. The unsaturated ester **141** and its (*Z*)-isomer were reduced without separation by using $LiAlH_4$, and the resultant alcohols were applied to an ortho-ester sigmatropic [3 + 3] Claisen rearrangement. The lithium enolate, generated from diastereomer **143**, was alkylated with allyl bromide, which resulted in an inseparable mixture of allylated products **144** that was then submitted to a ring-closing metathesis reaction in the presence of Grubb's catalyst, followed by ester hydrolysis by NaOEt-solution. After diazotation of the resulting carboxylic acid, a single cyclopentene derivative **145** was isolated. Subsequent transformations of **145** resulted in the dibenzyl intermediate **146**, which was treated with *m*-CPBA. The obtained oxirane **147** was converted in a four-step sequence to the key intermediate **148**.

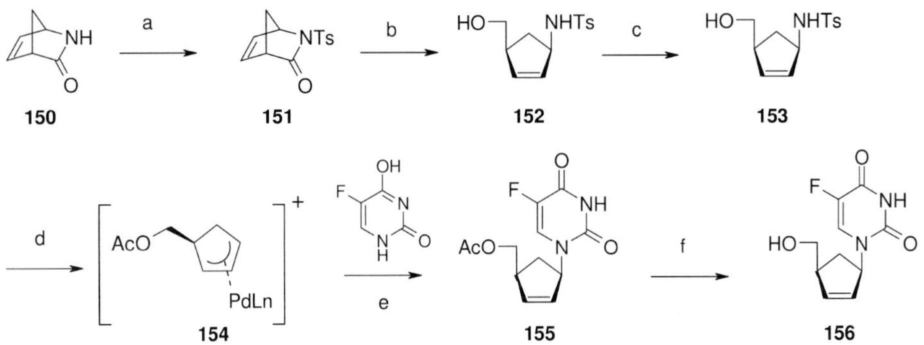

Scheme 15.17 Reagents and conditions: (a) *n*-BuLi, THF, −78 °C, then TsCl; (b) $NaBH_4$, MeOH; (c) Ac_2O, pyridine, then NaH, TsCl; (d) $Pd(OAc)_2$, $P(OiPr)_3$, THF, *n*-BuLi; (e) NaH, DMSO; (f) MeOH, NH_3.

Starting from the 1,2:5,6-di-*O*-isopropylidene-D-glucofuranose, various 4′-hydroxy derivatives were obtained in a stereoselective manner [67]. Starting from solketal, Hong *et al.* [68] presented, in 2003, a stereocontrolled synthesis of 6′-hydroxy analogues of carbovir. Compounds bearing additional hydroxymethyl group in the 3′ and 4′ positions were also obtained by Schmaltz *et al.* [69] and Hong *et al.* [70]. The hydroxyethyl analogues were synthesized by Lundt *et al.* [71], while the synthesis of pyrimidinyl analogues was investigated by Teran *et al.* [72]. No significant activity was found for these compounds.

Chu *et al.* [73] reported an asymmetric synthesis of the carboxylic analogue of the anti-HIV stavudine (d4T), while its 4-fluoro analogue was synthesized by Schinazi *et al.* [74] from the chiral lactam, (1*R*)-(−)-2-azabicyclo[2.2.1]hept-5-en-3-one 150 (*Vince lactam*). The synthetic path (Scheme 15.17) includes tosylation, reduction, acetylation, second tosylation, condensation with heterocyclic base under Trost conditions through a π-allylpalladium complex. A final deprotection led to the desired product **156**. The 4′-ethynyl and 4′-cyano carboxylic analogues of stavudine were synthetized by Kumamoto *et al.* [75], while the phosphono derivatives were investigated by Agrofoglio *et al.* [76]. The carbocyclic analogues of stavudine possessed no significant antiviral activity [77].

15.5 Locked Nucleosides

The analogy between the so-called Northern and Southern pseudorotational conformations of the furanose components of nucleosides and the conformationally fixed bicyclo[3.1.0]hexane systems is illustrated hereafter (Figure 15.8). As highlighted in Section 15.1, nucleoside sugars are required to adopt one or the other of these conformations in order to bind optimally to their target enzymes. However, if the sugar is constrained by structure into the optimal conformation, then it is feasible that better binding to the enzyme might result with the consequence of

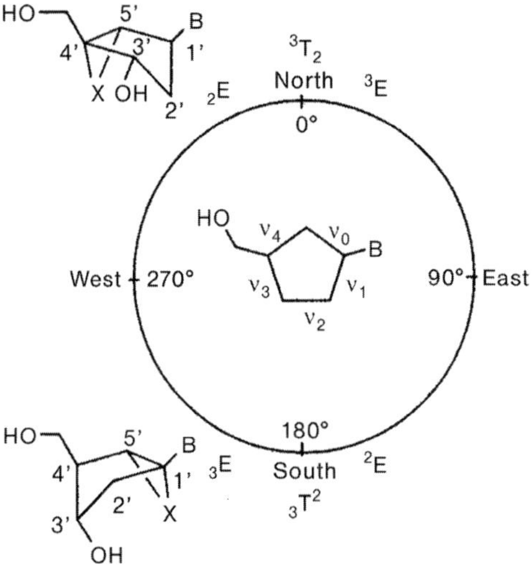

Figure 15.8 Pseudorotational cycle of conformationally rigid nucleosides (from V. E. Marquez *et al.*).

better, or more selective, activity. A bicyclo[3.1.0]hexane system, as a convenient pseudosugar template, exhibits a rigid pseudoboat conformation such that carbanucleosides constructed from it can adopt a fixed conformation that mimics the ring pucker of a true sugar moiety in a specific North or South conformation of the pseudorotational cycle. Because of the exclusive pseudoboat conformation of this template, a rigid North envelope 2*E* conformation can be constructed when the cyclopropane ring is fused between C4′ and C6′. Conversely, fusion of the cyclopropane ring between C1′ and C6′ provides a rigid South 3*E* envelope conformation.

Marquez *et al.* [78] have extended these studies by synthesizing carbocyclic nucleosides with the conformations fixed by the incorporation of fused cyclopropyl rings into the carbocyclic sugars, thus forming bicyclo[3.1.0]hexanes with the five-membered rings corresponding to both Northern and Southern conformations. The compounds (N)-2′-deoxy-methanocarba A **157**, (N)-methanocarba-T **158**, (N)-2′-deoxy-methanocarba-U **159**, (N)-2′-deoxy-methanocarba-C **160**, and (N)-2′-deoxy-methanocarba-G **161** (Figure 15.9), all of which have the "Northern" conformation,

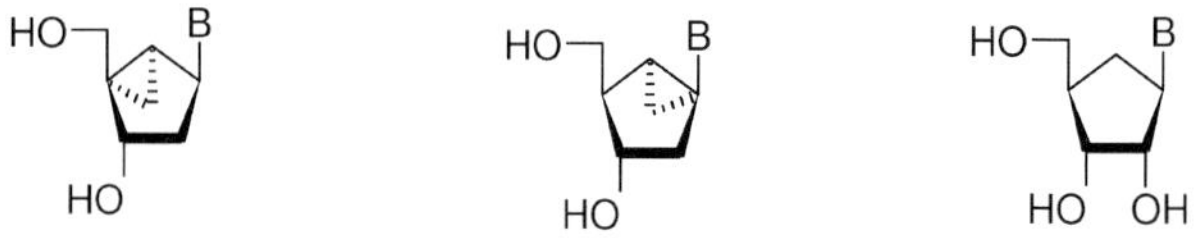

157, B=A
158, B=T
159, B=U
160, B=C
161, B=G

162, B=T

163

Northern conformation

Southern conformation

Figure 15.9 Northern and Southern conformations.

were tested against herpesviruses (HSV-1, HSV-2, HCMV), and the results are summarized in Table 15.3.

As can be seen, the order of potency decreases according to: (N)-2′-deoxy-methanocarba-T > (N)-2′-deoxy-methanocarba-C > (N)-2′-deoxy-methanocarba-G > (N)-2′-deoxy-methanocarba-A. On the other hand, the antiherpetic activity of the compound with the Southern conformation, **162** is poor. The activities of **158** and **162**, along with the unconstrained racemic carba-T, **163**, are summarized in Table 15.3. The racemate, **163** had already been reported to have antiherpetic activity with EC_{50} = 0.8 $\mu g\,mL^{-1}$ (HSV-1), and 7.0 $\mu g\,mL^{-1}$ (HSV-2) in primary rabbit kidney cells, and EC_{50} = 24 $\mu g\,mL^{-1}$ (HSV-1) and 57 $\mu g\,mL^{-1}$ (HSV-2) in Vero cells, and was also non-toxic against host HFF cells. The order of potency against dividing cells was thus (±)-carba-T (CC_{50} > 0.65 $\mu g\,mL^{-1}$) > (N)-methanocarba-T (CC_{50} = 32.9 $\mu g\,mL^{-1}$) > (S)-methanocarba-T (CC_{50} > 100 $\mu g\,mL^{-1}$) (Table 15.4).

These data show that the Northern thymidine analogue is a very potent antiherpetic compound, compared with the reference, acyclovir. It appears likely that, since carbocyclic nucleosides must be activated to the 5′-mono- di-, or triphosphate by viral and/or cellular kinases, this selective antiviral activity of the Northern compounds is directly linked to the Northern isomers being better substrates for the viral

Table 15.3 Antiviral activity of rigid (*N*)-methanocarbocyclic nucleosides according to the cytopathogenic effect (CPE) inhibition assay.

Compound	Virus[a] (HFF cells)	EC_{50}[b] ($\mu g\,mL^{-1}$)	CC_{50}[c] ($\mu g\,mL^{-1}$)	SI[d]	Control[e] (EC_{50}, $\mu g\,mL^{-1}$)
157	HSV-1	72.0	>100	>1.4	ACV (0.80)
	HSV-2	13.9	>100	>7.2	ACV (4.00)
	HCMV	3.1	>100	>32.2	GCV (0.30)
158	HSV-1	0.03	>100	>3333	ACV (0.40)
	HSV-2	0.09	>100	>1111	ACV (0.06)
	HCMV	>20	63.7	<3.2	GCV (0.30)
159	HSV-1	>100	>100	1	ACV (0.60)
	HSV-2	>100	>100	1	ACV (1.50)
	HCMV	>100	>100	1	GCV (0.40)
160	HSV-1	0.14	68	486	ACV (0.70)
	HSV-2	>20	96	<4.8	ACV (6.20)
	HCMV	>4.0	8.8	<2.2	GCV (0.02)
161	HSV-1	4.0	>100	>25	ACV (0.60)
	HSV-2	9.9	>100	<10.1	ACV (1.50)
	HCMV	>20	64.3	<3.2	GCV (0.40)

a) HSV-1 = herpes simplex type 1; HSV-2 = herpes simplex type 2; HCMV = human cytomegalovirus; HFF = human foreskin fibroblasts.
b) EC_{50} = inhibitory concentration required to reduce virus-induced cytopathogenicity by 50%.
c) CC_{50} = cytotoxic concentration that produces 50% cell death.
d) SI = selectivity index (CC_{50}/EC_{50}).
e) ACV = acyclovir; GCV = ganciclovir.

Table 15.4 Antiherpetic activity of rigid (*N*)-methanocarba-T (**158**), rigid (*S*)-methanocarba-T (**162**), and carbocyclic thymidine (**163**), measured with the plaque reduction assay in HFF.

Compound	Virus[a)]	EC_{50}[b)] ($\mu g\,mL^{-1}$)	CC_{50}[c)] ($\mu g\,mL^{-1}$)	SI[d)]	ACV[e)] (EC_{50}, $\mu g\,mL^{-1}$)
163	HSV-1	0.01	>20	>2000	0.30
		0.08[f)]	>20	>250	0.30[f)]
	HSV-2	0.12	>20	>167	0.80
		0.43[f)]	>20	>46.7	1.10[f)]
162	HSV-1	>50	>50	1	0.15
		>20	>20	1	0.30
	HSV-2	>50	>50	1	0.60
		>20	>20	1	1.10
158	HSV-1	>10	>10	1	0.30
	HSV-2	>10	>10	1	0.80

a) HSV-1 = herpes simplex type 1; HSV-2 = herpes simplex type 2.
b) EC_{50} = inhibitory concentration required to reduce the number of virus plaques by 50%.
c) CC_{50} = cytotoxic concentration that produces 50% of cell death.
d) SI = selectivity index (CC_{50}/EC_{50}).
e) ACV = acyclovir control.
f) These values correspond to a drug-pretreated plaque reduction assay.

kinases. It is interesting that the adenosine analogue also displayed anti-HIV activity, though it should be noted that such activity cannot in this case be due to phosphorylation by viral kinases because HIV does not code for them. Almost certainly, the compound requires activation for HIV activity, and since that activity was subsequently found to reside in the enantiomer with the natural configuration, it was hypothesized that the fixed conformation of this compound makes it a good substrate for activating cellular kinases only when the base is adenosine.

Compound **170** is an important intermediate in synthesis of nucleosides with Northern (N)-type conformation, possessing bicyclo[3.1.0]hexane fragment. In 1996, Marquez *et al.* published a seven-step synthesis with an overall yield of 25% [79]. Some three years later, Marquez presented an improved synthesis (Scheme 15.18) of those "locked" nucleosides [80]. An unsaturated dibenzylated diol **164** was treated with silver trifluoroacetate and phenylselenyl chloride to afford **165**. After a basic hydrolysis of trifluoroacetate ester **165**, the obtained cyclopentanol **166** was treated with sodium periodate, which led to the cyclopentenol **166**. After transformation into benzyl ester **168** under Mitsunobu conditions, followed by a basic hydrolysis, the alcohol **169** was obtained. After treatment with Et_2Zn/CH_2I_2 in dichloromethane (DCM), the formed bicyclic intermediate **170** was condensed with purines.

The above method was also applied to the synthesis of carbocyclic nucleosides with an oxirane ring instead of a cyclopropyl ring. Some of these compounds, which possessed an oxabicyclo[3.1.0] fragment, showed potent activity against EBV, with EC_{50} values of 0.34 μg mM (selectivity index >150; guanosine analogue) compared to acyclovir (EC_{50} = 2 μg mM^{-1} [81].

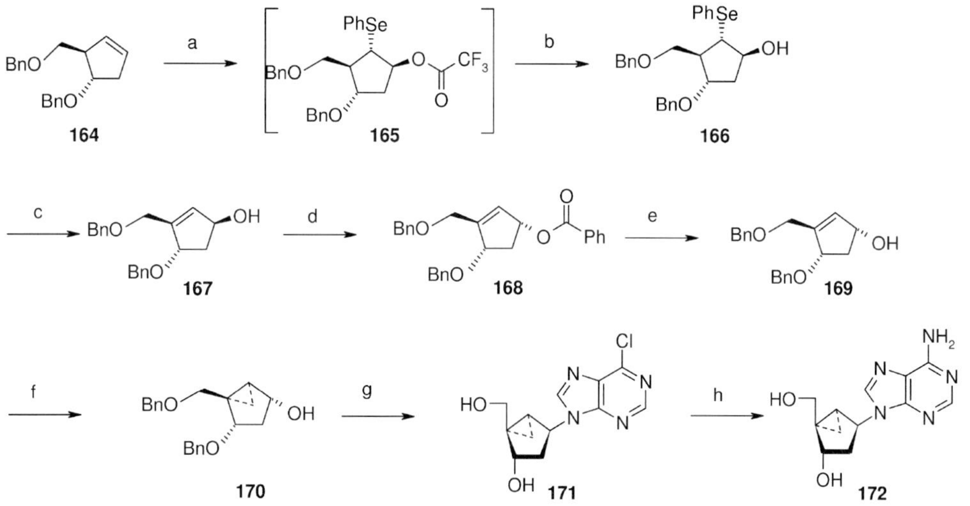

Scheme 15.18 Reagents and conditions: (a) CF_3CO_2Ag, PhSeCl, DMSO; (b) 5% KOH, EtOH; (c) $NaIO_4$ $MeOH/H_2O$; (d) Ph_3P/DEAD, $PhCO_2H$, benzene; (e) K_2CO_3/MeOH; (f) Et_2Zn/CH_2I_2, CH_2Cl_2; (g) 6-chloropurine, Ph_3P, DEAD, THF; (h) NH_3/MeOH then Pd/C, HCO_2H/MeOH.

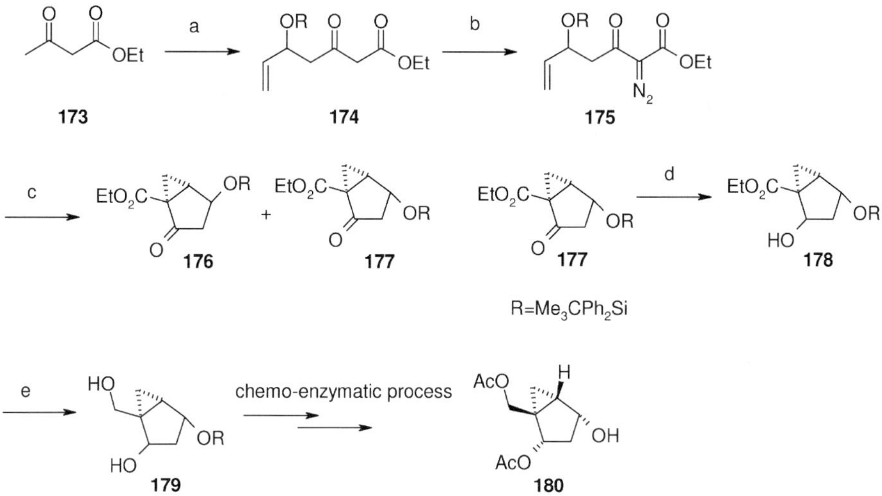

Scheme 15.19 Reagents and conditions: (a) (i) LDA, THF, 0 °C, acrolein, −78 °C, 20 min, (ii) TBDPSCl, imidazole, DCM, r.t., 1 h; (b) TsN_3, Et_3Nm AcCN, r.t., 17 h; (c) $CuSO_4$, cyclohexane, reflux, 35 h; (d) $NaBH_4$, DCM, MeOH, −25 °C, 30 min; (e) $LiAlH_4$, Et_2O, 0 °C, 10 min.

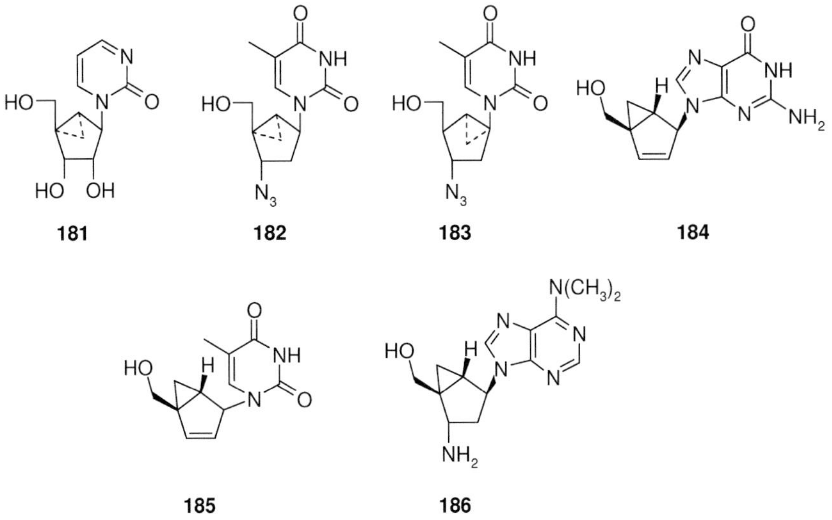

Figure 15.10 Some examples of locked nucleosides.

Bicyclo[3.1.0]hexane analogues of nucleosides could be also obtained from ethyl acetyloacetate **173** (Scheme 15.19) [82]. After treatment with LDA, followed by the addition of acrolein, the obtained allyl alcohol was protected with a *tert*-butyldiphenylsilyl (TBPS) group. Silyl ether **174**, obtained as mixture of keto–enol tautomers, was transformed to the diazo intermediate **175** which, after refluxing in cyclohexane in the presence of copper sulfate, gave a mixture of bicyclic cylopentanons **176** and **177** that could be separated chromatographically. The major product was then transformed to cyclopentanol by reduction with $NaBH_4$. Further reduction with $LiAlH_4$ gave the monoprotected triol **179** as a mixture of epimeric alcohols. Using a chemo-enzymatic process, the optically pure diacetylated triol **180** was obtained used in a direct coupling with an appropriate base.

This method was also applied for the synthesis of bicyclo[3.1.0]hexane analogues of nucleosides functionalized at the tip of the cyclopropane ring [83]. Additionally, some other bicyclic analogues of biologically active nucleosides were synthesized (Figure 15.10).

The locked analogue of Zebularine (1-(β-D-ribofuranosyl)-1,2-dihydropyrimidin-2-one), a potent cytidine deaminase inhibitor **181** did not show any significant activity [84]. Other results showed, that the bicyclic analogue of AZT – (N)-*methano*-carba-AZT **182** – possessed similar potency as AZT, whereas (S)-*methano*-carba-AZT **183** does not inhibit HIV reverse transcriptase [85]. The locked analogue of carbovir **184** showed only weak antiviral activity [86]. The bicyclic analogue of stavudine – D-*N*-MCD4TX **185** – was about four-fold less potent ($EC_{50} = 0.96\ \mu g\ mM^{-1}$ and $EC_{50} = 1.45\ \mu g\ mM^{-1}$) than D4T against both HIV-1 and HIV-2 ($EC_{50} = 0.21\ \mu g\ mM^{-1}$ and $EC_{50} = 0.40\ \mu g\ mM^{-1}$, respectively) [87]. L-*N*-MCD4T was also synthesized, but its activity was weaker than that of the D-analogue ($EC_{50} = 6.76\ \mu g\ mM^{-1}$) [88]. Finally, Marquez presented a carbocyclic version of puromycin **186** which acted as a

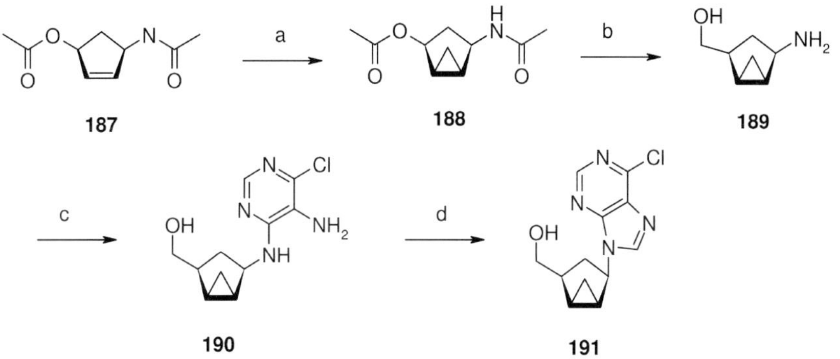

Scheme 15.20 Reagents and conditions: (a) Et_2Zn, CH_2I_2, CH_2Cl_2; (b) 0.5 N $Ba(OH)_2$, reflux; (c) 5-amino-4,6-dichloropyrimidine, Et_3N, *n*-BuOH, reflux; (d) $CH(OCH_2CH_3)_3$.

nucleoside antibiotic [89]. Starting from D-ribose, some other *apio* analogues were synthesized which did not exhibit any significant antiviral activity [90].

2′,3′-methano analogues of nucleosides could be synthesized from 2-azabicyclo [2.2.1] hept-5-en-3-one (ABH) (Scheme 15.20) [91, 92]. Cyclopentene **187** was obtained from ABH in a four-step synthesis, and then transformed into the *endo* bicyclo[3.1.0]hexane derivative **188** using standard Simmons–Smith conditions. The amide and ester groups were removed by basic hydrolysis using barium hydroxide to produce the aminoalcohol **189**; this was then coupled with 5-amino-4,6-dichloropyrimidine to give intermediate **190**. After condensation with triethyl orthoformate, the appropriate purine nucleoside **191** was formed, and this was later transformed to other nucleoside derivatives. Among these, the 6-chloropurine compound **191** was most active against HSV-1 ($EC_{50} = 4.0\,\mu g\,mM^{-1}$), although cytotoxity was observed at a concentration of 40 μM.

Much effort has been directed towards the synthesis of conformationally locked, biologically active nucleosides. Starting from (1*R*,5*S*)-2-oxabicyclo[3.3.0]oct-6-en-3-one, which was prepared in three steps from cyclopentene, Agrofoglio *et al.* reported the stereo synthesis of α-L-bicarbocyclic nucleosides [93]. Together with conformationally locked nucleosides with cyclopropane and oxirane ring, compounds possessing a thiirane ring were also synthesized [94]. Neither group of compounds possessed any antiviral activity. Starting from 3,3-di(isopopoxycarbonyl)-1,1-dimethoxybutane, Wang presented synthesis nucleoside analogues bearing a bicyclo[2.1.1]hexane fragment [95], while Nair *et al.* obtained similar analogues from 5-norbornen-2-ol [96]. Some bicyclic homo-*N*-nucleosides were synthesized from iridoids by Franzyk *et al.* [97].

15.6 Conclusions

During recent years, new nucleosides have been approved to combat many important viral diseases. These compounds are very important, not only because of their biological properties (antiviral, antitumor, or antibiotic), but also because they have induced an explosion among new developments in both heterocycle and nucleoside chemistry. Aristeromycin, Neplanocin, Abacavir, Lobucavir, Synguanol, Entecavir, together with some locked-nucleosides, now show much promise for the therapy of HSV, VZV, CMV, AIDS, and/or hepatitis. Moreover, as well as their anticipated continued contribution to antiviral therapy, carbocyclic nucleosides may also be important for cancer chemotherapy and diseases of the T cells, and in the design and therapeutic applications of modified antisense and antigene agents. The best is yet to come!

References

1 For review on nucleosides, see: (a) Crimmins, M. T., *Tetrahedron* **1998**, *54*, 9229; (b) Huryn, D. M., Okabe, M., *Chem. Rev.* **1992**, *92*, 1745; (c) Borthwick, A. D., Biggadike, K., *Tetrahedron* **1992**, *48*, 571; (d) Agrofoglio, L. A., Suhas, E., Farese, A., Condom, R., Challand, S. R., Earl, R. A., Guedj, R., *Tetrahedron* **1994**, *50*, 10611; (e) Herdewijn, P., *Drug Discov. Today* **1997**, *2*, 235; (f) Crimmins, M. T., *Tetrahedron* **1998**, *54*, 9229; (g) De Clercq, E., *Collect. Czech. Chem. Commun.* **1998**, *63*, 449–479; (h) Agrofoglio, L. A., Challand, S. R., in: *Acyclic, Carbocyclic and L-Nucleosides*, Kluwer Academic, Dordrecht, **1998**; (i) Ichikawa, E., Kato, K., *Synthesis* **2002**, 1; (j) De Clercq, E., Neyts, J., *Rev. Med. Virol.* **2004**, *14*, 289; (k) Schneller, S. W., *Curr. Top. Med. Chem.* **2002**, *2*, 1087; (l) Zemlicka, J., *Pharm. Ther.* **2000**, 251.

2 (a) Sekiyama, T., Hatsuya, S., Tanaka, Y., Uchiyama, M., Ono, N., Iwayama, S., Oikawa, M., Suzuki, K., Okunishi, M., Tsuji, T. J., *J. Med. Chem.* **1998**, *41*, 1284; (b) Iwayama, S., Ono, N., Ohmura, Y., Suzuki, K., Aoki, M., Nakazawa, H., Oikawa, M., Kato, K., Okunishi, M., Nishiyama, Y., Yamanishi, K., *Antimicrob. Agents Chemother.* **1998**, *42*, 1666; (c) Iwayama, S., Ohmura, Y., Suzuki, K., Nakazawa, H., Aoki, M., Tanabe, I., Sekiyama, T., Tsuji, T. J., Okunishi, M., Yamanishi, K., Nishiyama, Y., *Antiviral. Res.* **1999**, *42*, 139.

3 (a) Ini, N., Iwayama, S., Suzuki, K., Sekiyama, T., Nakazawa, H., Tsuji, T., Okunishi, M., Daikoku, T., Nishiyama, Y., *Antimicrob. Agent Chemother.* **1998**, *42*, 2095; (b) De Clercq, E., Andrei, G., Snoeck, R., De Bolle, L., Naesens, L., Degrève, B., Balzarini, J., Zhang, Y., Schols, D., Leyssen, P., Ying, C., Neyts, J., *Nucleosides Nucleotides Nucleic Acids* **2001**, *20*, 271.

4 Iwayama, S., Ohmura, Y., Suzuki, K., Ono, N., Nakazawa, H., Tsuji, T., Okunishi, M., Daikoku, T., Nishiyama, Y., *Antiviral Res.* **1999**, *42*, 139.

5 (a) Qiu, Y.-L., Ksebati, M. B., Ptak, R. G., Fan, B. Y., Bretenbach, J. M., Lin, J. S., Cheng, Y. C., Kern, E. R., Drach, J. C., Zemlicka, J., *J. Med. Chem.* **1998**, *41*, 10; (b) Rybak, R. J., Hartline, C. B., Qiu, Y. L., Zemlicka, J., Harden, E., Marshall, G., Sommadossi, J.-P., Kern, E. R., *Antimicrob. Agents Chemother.* **2000**, *44*, 1506.

6 Zemlicka, J., Qiu, Y. L., Kira, T., Gullen, E., Cheng, Y. C., Ptak, R. G., Breitenbach, J. M., Drach, J. C., Hartline, C. B., Kern, E. R., *Antiviral Res.* **1999**, *41*, A34.

7 Qiu, Y. L., Breitenbach, J. M., Lin, J. S., Cheng, Y. C., Kern, E. R., Drach, J. R., Zemlicka, J., *Antiviral Chem. Chemother.* **1998**, *9*, 341.

8 Preliminary report: (a) Zemlicka, J., Zhou, S., Kern, E. R., Drach, J. C., Mitsuya, H., 19th International Conference on Antiviral Research, San Juan, Puerto Rico, May 7–11. *Antiviral Res.* **2006**, 70, A68; (b) Zemlicka, J., in: *Recent Advances in Nucleosides: Chemistry and Chemotherapy;* Chu, C.K. (Ed.), Elsevier Science, Amsterdam, **2002**; pp. 327–357; (c) Zemlicka, J., Chen, X., in: *Frontiers in Nucleosides and Nucleic Acids;* Schinazi, R. F., Liotta, D. C. (Eds.), IHL Press, Tucker, Georgia, **2004**; pp. 267–307; (d) Zhou, S., Breitenbach, J. M., Borysko, K. Z., Drach, J. C., Kern, E. R., Gullen, E., Cheng, Y.-C., Zemlicka, J., *J. Med. Chem.* **2004**, *47*, 566; (e) Yan, Z., Kern, E. R., Gullen, E., Cheng, Y.-C. Drach, J. C., Zemlicka, J., *J. Med. Chem.* **2005**, *48*, 91; (f) Kern, E. R., Kushner, N. L., Hartline, C. B., Williams-Azziz, S. L., Harden, E. A., Zhou, S., Zemlicka, J., Prichard, M. N., *Antimicrob. Agents Chemother.* **2005**, *49*, 1039; (g) Kern, E. R., Bidanset, D. J., Hartline, C. B., Yan, Z., Zemlicka, J., Quenelle, D. C., *Antimicrob. Agents Chemother.* **2004**, *48*, 4745.

9 Onishi, T., Matsuzawa, T., Nishi, S., Tsuji, T., *Tetrahedron Lett.* **1999**, *40*, 8845.

10 (a) Gallos, J. K., Massen, Z. S., Koftis, T. V., Dellios, C. C., *Tetrahedron Lett.* **2001**, *42*, 7489; (b) Gallos, J., Koftis, T., Massen, Z. S., Dellios, C., Moutzinos, T., Coutouli-Argyropolou, E., Koumbis, A. E., *Tetrahedron* **2002**, *58*, 8043.

11 Ostrowski, T., Golankiewicz, B., De Clercq, E., Balzarini, J., *Bioorg. Med. Chem.* **2006**, *14*, 3535.

12 (a) Muray, E., Rife, J., Branchadel, V., Ortuno, R. M., *J. Org. Chem.* **2002**, *67*, 4520; (b) Rife, J., Ortuno, R., *Org. Lett.* **1999**, *1*, 1221.

13 Wu, Y., Hong, J.-H., *Il Farmaco* **2005**, *60*, 739.

14 Csuk, R., Eversmann, L., *Tetrahedron* **1998**, *54*, 6445.

15 Csuk, R., Kern, A., *Tetrahedron* **1999**, *55*, 8409.

16 Qui, Y.-L., Hempel, A., Camerman, N., Camreman, A., Geiser, F., Ptak, R. G., Breitenbach, J. M., Kira, T., Li, L., Gullen, E., Cheng, Y.-C., Drach, J. C., Zemlicka, J., *J. Med. Chem.* **1998**, *41*, 5257.

17 (a) Zhou, S., Kern, E. R., Gullen, E., Cheng, Y.-C., Drach, C. J., Matsumi, S., Mitsuya, H., Zemlicka, J., *J. Med. Chem.* **2004**, *47*, 6964; (b) Zhou, S., Zemlicka, J., *Tetrahedron* **2005**, *61*, 7112; (c) Zhou, S., Kern, E. R., Gullen, E., Cheng, Y.-C., Drach, J. C., Tamiya, S., Mitsuya, H., Zemlicka, J., *J. Med. Chem.* **2006**, *49*, 6120.

18 (a) Guan, H.-P., Qiu, Y.-L., Ksebati, M. B., Kern, E. R., Zemlicka, J., *Tetrahedron* **2002**, *58*, 6047; (b) Yan, Z., Zhou, S., Kern, E. R., Zemlicka, J., *Tetrahedron* **2006**, *62*, 2608.

19 (a) Hoshino, H., Shimizu, N., Shimada, M., Takita, T., Takeuchi, T., *J. Antibiot.* **1987**, *40*, 1077; (b) Shimada, N., Hasegawa, S., Harada, T., Tomisawa, T., Fujii, A., Takita, T., *J. Antibiot.* **1986**, *39*, 1623; (c) Niitsuma, S., Ichikawa, Y., Takita, T., *Studies in Natural Chemistry* **1992**, *10*, 585.

20 Norbeck, D. W., Kern, E., Hayashi, S., Rosenbrook, W., Sham, H., Herrin, T., Plattner, J. J., Erickson, J., Clement, J., Swanson, R., Shipkowitz, N., Hardy, D., Marsh, K., Arnett, G., Shannon, W., Broder, S., Mitsuya, H., *J. Med. Chem.* **1900**, *33*, 1281.

21 (a) Nishiyama, Y., Yamamoto, N., Takahash, K., Shimada, N., *Antimicrob. Agents Chemother.* **1988**, *32*, 1053; (b) Daikoku, T., Yamamoto, N., Saito, S., Kitagawa, M., Shimada, N., Nishiyama, Y., *Biochem. Biophys. Res. Commun.* **1991**, *176*, 805.

22 Sakuma, T., Saijo, M., Suzutani, T., Yoshida, I., Saito, S., Kitagawa, M., Hasegawa, S., Azuma, M., *Antimicrob. Agents Chemother.* **1991**, *35*, 1512.

23 Nagahata, T., Kitagawa, M., Matsubara, K., *Antimicrob. Agents Chemother.* **1994**, *38*, 707.

24 (a) Gharbaoui, T., Legraverend, M., Ludwig, O., Bisagni, E., Aubertin, A.-M.,

Chertanova, L., *Tetrahedron* **1995**, *51*, 1641; (b) Jacobs, G. A., Tino, J. A., Zahler, R., *Tetrahedron Lett.* **1988**, *30*, 6955; (c) Boumchita, H., Legraverend, M., Guilhem, J., Bisagni, E., *Heterocycles* **1991**, *32*, 867; (d) Pecquet, P., Huet, F., Legraverend, M., Bisagni, E., *Heterocycles* **1992**, *34*, 739.

25 Vite, G. D., Tino, J. A., Zahler, R., Goodfellow, V., Tuomari, A. V., Mc Geever-Rubin, B., Field, A. K., *Bioorg. Med. Chem. Lett.* **1993**, *3*, 1211.

26 (a) Thiers, B. H., *Dermatol. Clin.* **1900**, *8*, 583; (b) O'Brien, J. J., Campoli-Richards, D. M., *Drugs* **1989**, *37*, 233; (c) Campoli-Richards, D. M., *Drugs* **1989**, *37*, 233.

27 Nishiyama, Y., Yamamoto, N., Yamada, Y., Daikoku, T., Ichikawa, Y., Takahashi, K., *J. Antibiot.* **1989**, *42*, 1854.

28 Rao, S. N., *J. Biomol. Struct. Dyn.* **1992**, *9*, 719.

29 Shiota, H., Nitta, K., Naito, T., Mimura, Y., Maruyama, T., *Br. J. Ophthalmol.* **1996**, *80*, 413.

30 Malik, A. H., Lee, W. M., *Ann. Intern. Med.* **2000**, *132*, 723.

31 (a) Innaimo, S. F., Seifer, M., Bisacchi, G. S., Standring, D. N., Zahler, R., Colonno, R. J., *Antimicrob. Agents Chemother.* **1997**, *41*, 1444; (b) Genovesi, E. V., Lamb, L., Medina, I., Taylor, D., Seifer, M., Innaimo, S., Colonno, R. J., Clark, J. M., *Antiviral Res.* **2000**, *48*, 197.

32 Seifer, M., Hamatake, R. K., Colonno, R. J., Standring, R., *Antimicrob. Agents Chemother.* **1998**, *42*, 3200.

33 Honjo, M., Maruyama, T., Sato, Y., Horii, T., *Chem. Pharm. Bull.* **1989**, *37*, 1413.

34 (a) Ichikawa, Y.-I., Narita, A., Shiozawa, A., Hayashi, Y., Narasaka, K., *J. Chem. Soc., Chem. Commun.* **1989**, 1919; (b) Jung, M. E., Sledeski, A. W., *J. Chem. Soc., Chem. Commun.* **1993**, 589; (c) Somekawa, K., Hara, R., Kinnami, K., Muraoka, F., Suishu, T., Shimo, T., *Chem. Lett.* **1995**, 407; (d) Brown, B., Hegedus, L. S., *J. Org. Chem.* **1998**, *63*, 801.

35 (a) Cotterill, I. C., Roberts, S. M., *J. Chem. Soc., Perkin Trans.* **1992**, *1*, 2585; (b) Bissachi, G. S., Braitman, A., Cianci, C. W., Clark, J. M., Field, A. K., Hagen, M. E., Hockstein, D. R., Malley, M. F., Mitt, T., Slusarchyk, W. A., Sundeen, J. E., Terry, B. J., Tuomari, A. V., Weaver, E. R., Young, M. G., Zahler, R., *J. Med. Chem.* **1991**, *34*, 1415.

36 Jung, M. E., Sledeski, A. W., *J. Chem. Soc., Chem. Commun.* **1993**, 589.

37 Panda, J., Ghosh, S., Ghosh, S., *J. Chem. Soc. Perkin Trans.* **2001**, *1*, 3013.

38 Brown, B., Hegedus, L. S., *J. Org. Chem.* **1998**, *63*, 8012.

39 Rouge, P. D., Moglioni, A. G., Moltrasio, G. Y., Ortuño, R. M., *Tetrahedon: Asymm.* **2003**, *14*, 193.

40 Kim, A., Hong, J. H., *Arch. Pham. Chem. Life Sci.* **2005**, *338*, 522.

41 Hubert, C., Alexandre, C., Aubertin, A.-M., Huet, F., *Tetrahedron* **2002**, *58*, 3775.

42 Guan, H.-P., Ksebati, M. B., Kern, E. R., Zemlicka, J., *J. Org. Chem.* **2002**, *65*, 5177.

43 Fernandez, F., Hegueta, A. R., Lopez, C., De Clecq, E., Balzarini, J., *Nucleosides, Nucleotides* **2001**, *20*, 1129.

44 Lopez, C., Balo, C., Blanco, J. M., Fernandez, F., De Clercq, E., Balzarini, J., *Nucleosides, Nucleotides* **2001**, *20*, 1133.

45 Figueira, M. J., Blanco, J. M., Caamano, O., Fernandez, F., Garcia-Mera, X., Lopez, C., Andrei, G., Snoeck, R., Padalko, E., Neyts, J., Balzarini, J., De Clecq, E., *Arch. Pharm. Med. Chem.* **1999**, *332*, 348.

46 Slusarchyk, W. A., Bisacchi, G. S., Field, K., Hockstein, D. R., Jacobs, G. A., McGeever-Rubin, B., Tino, J. A., Tuomari, A. V., Yamanaka, G. A., Young, M. G., Zahler, R., *J. Med. Chem.* **1992**, *35*, 1799.

47 Takahashi, H., Amano, R., Saneyoshi, M., Maruyama, T., Yamaguchi, T., *Nucleic Acids Res.* **2003**, *3*, 285.

48 Lopez, C., Balo, C., Blanco, J. M., Fernandez, F., De Clercq, E., Balzarini, J., *Nucleosides Nucleotides Nucleic Acids* **2001**, *20*, 1133.

49 (a) Vince, R., Hua, M., Brownell, J., Daluge, S., Lee, F. C., Shannon, W. M., Lavelle, G. C., Qualls, J., Weislow, O. S., Kiser, R.,

Canonico, P. G., Schultz, R. H., Narayanan, V. L., Mayo, J. G., Shoemaker, R. H., Boyd, M. R., *Biochem. Biophys. Res. Commun.* **1988**, *156*, 1046; (b) Vince, R., Hua, M., *J. Med. Chem.* **1900**, *33*, 17; (c) Vince, R., *Nucleic Acids Symp. Ser.* **1991**, *25*, 193.

50 Roulland, E., Monneret, C., Florent, J.-C., *Tetrahedron Lett.* **2003**, *44*, 4125.

51 (a) Agrofoglio, L. A., Nolan, S. P., *Curr. Topics Med. Chem.* **2005**, *5*, 1541; (b) Amblard, F., Nolan, S. P., Agrofoglio, L. A., *Tetrahedron* **2005**, *61*, 7067.

52 Nayek, A., Banerjee, S., Sinha, S., Ghosh, S., *Tetrahedron Lett.* **2004**, *45*, 6457.

53 Olivo, H. F., Jiaxin, Y., *J. Chem. Soc. Perkin Trans.* **1998**, *1*, 391.

54 Kato, K., Suzuki, H., Tanaka, H., Misayaka, T., *Tetrahedon: Asymm.* **1998**, *9*, 911.

55 Hegedus, L. S., Cross, J., *J. Org. Chem.* **2004**, *69*, 8492.

56 Jin, W. K., Joon, H. H., *Arch. Pharm. Chem. Life Sci.* **2005**, *338*, 399.

57 Wang, J., Jin, Y., Rapp, K. L., Bennett, M., Schinazi, R. F., Chu, C. K., *J. Med. Chem.* **2005**, *48*, 3736.

58 (a) Koga, M., Uchida, T., Ueda, M., Shigeta, M., Yamamuro, T., Tamai, K., Suzuki, T., Saeki, T., *Nucleic Acid Symposium Series* **2000**, *44*, 117; (b) Koga, M., Kanda, M., Yamamuro, T., Tamai, K., Uchida, T., Suzuki, T., Saeki, T., *Nucleic Acids Res. Suppl.* **2002**, *2*, 15.

59 Balo, C., Blanco, J. M., Fenandez, F., Lens, E., Lopez, C., *Tetrahedron* **1998**, *54*, 2833.

60 Wang, P., Schinazi, R. F., Chu, C. K., *Bioorg. Med. Chem. Lett.* **1998**, *8*, 1585.

61 Aihong, K., Joon, H. H., *Nucleosides, Nucleotides* **2005**, *24*, 63.

62 Ok, H. K., Joon, H. H., *Arch. Pharm. Pharm. Med. Chem.* **2004**, *337*, 579.

63 Aihong, K., Joon, H. H., *Nucleosides, Nucleotides* **2004**, *23*, 813.

64 Ok, H. K., Joon, H. H., *Arch. Pharm. Pharm. Med. Chem.* **2004**, *337*, 457.

65 (a) Katagiri, N., Nomura, M., Sato, H., Kaneko, C., Yusa, K., Tsuruo, T., *J. Med. Chem.* **1992**, *35*, 1882; (b) Katagiri, N., Toyota, A., Shiraishi, T., Sato, H., Kaneko, C., *Tetrahedron Lett.* **1992**, *33*, 3507; (c) Tanaka, M., Norimine, Y., Fujita, T., Suemune, H., Sakai, K., *J. Org. Chem.* **1996**, *61*, 6952.

66 Banerjee, S., Ghosh, S., Sinha, S., Ghosh, S., *J. Org. Chem.* **2005**, *70*, 4199.

67 Gurjar, M. K., Maheshwar, K., *J. Org. Chem.* **2001**, *66*, 7552.

68 Joon, H. H., Chang-Hyun, O., Jung-Hyuck, C., *Tetrahedon* **2003**, *59*, 6103.

69 (a) Velcicki, J., Lex, J., Schmalz, H.-G., *Org. Lett.* **2002**, *4*, 565; (b) Velcicki, J., Lanver, A., Lex, J., Prokop, A., Wieder, T., Schmalz, H.-G., *Chem. Eur. J.* **2004**, *10*, 5087.

70 Ok, H. K., Joon, H. H., *Tetrahedron Lett.* **2002**, *43*, 6399.

71 Johannes, S., Lundt, I., *Eur. J. Org. Chem.* **2001**, 1129.

72 Teran, C. G., Moa, M. J., Mosquera, R., Santana, L., *Nucleosides, Nucleotides* **2001**, *20*, 999.

73 Wang, P., Gulden, B., Newton, M. G., Cheng, Y.-C., Schinazi, R. F., Chu, C. K., *J. Med. Chem.* **1999**, *42*, 3390.

74 Shin, J., Schinazi, R. F., *Nucleosides, Nucleotides* **2001**, *20*, 1367.

75 Kumamoto, H., Haraguchi, K., Tanaka, H., Nitanda, T., Baba, M., Dutschman, G. E., Cheng, Y.-C., Kato, K., *Nucleosides, Nucleotides* **2005**, *24*, 73.

76 Legeret, B., Sarakauskaite, Z., Pradaux, F., Saito, Y., Tumkevicius, S., Agrofoglio, L. A., *Nucleosides, Nucleotides* **2001**, *20*, 661.

77 Shi, J., McAtee, J., Schlueter, J., Wirtz, S., Thanish, P., Joadawlkis, A., Liotta, D. C., Schinazi, R. F., *J. Med. Chem.* **1999**, *42*, 859.

78 (a) Marquez, V. E., Siddiqui, M. A., Ezzitouni, A., Russ, P., Wang, J., Wagner, R. W., Matteuci, M. D., *J. Med. Chem.* **1996**, *39*, 3739; (b) Rodriguez, J. B., Marquez, V. E., Nicklaus, M. C., Mitsuya, H., Barchi, J. J., *J. Med. Chem.* **1994**, *37*, 3389; (c) Altona, C., Sundaralingam, M., *J. Am. Chem. Soc.* **1972**, *94*, 8205.

79 Siddiqui, M. A., Ford, H., George, C., Marquez, V. E., *Nucleosides, Nucleotides* **1996**, *15*, 235.

80 Marquez, V. E., Russ, P., Alonso, R., Siddiqui, M., Hernandez, S., George, C.,

Nicklaus, M. C., Ford, H., *Helv. Chim. Acta* **1999**, *82*, 2119.
81 Comin, M., Rodriguez, J. B., Russ, P., Marquez, V. E., *Tetrahedron* **2003**, *59*, 295.
82 Moon, H. R., Ford, H., Marquez, V. E., *Org. Lett.* **2000**, *28*, 3793; (b) Yoshimura, Y., Moon, H. R., Choi, Y., Marquez, V. E., *J. Org. Chem.* **2002**, *67*, 5938.
83 Comin, M., Parrish, D. A., Deschamps, J. R., Marquez, V. E., *Org. Lett.* **2006**, *8*, 705.
84 Leong, L. S., Buenger, G., McCormack, J. J., Cooney, D. A., Hao, Z. J. R., Marquez, V. E., *J. Med. Chem.* **1998**, *41*, 2572.
85 Marquez, V. E., Ezzitouni, A., Russ, P., Siddiqui, M. A., Ford, H., Feldman, R. J., Mitsuya, H., George, C., Barchi, J. J., *J. Am. Chem. Soc.* **1998**, *120*, 2780.
86 Choi, Y., Sun, G., George, C., Nicklaus, M. C., Kelley, J. A., Marquez, V. E., *Nucleosides, Nucleotides* **2003**, *22*, **2077**.
87 Choi, Y., George, C., Comin, M., Barchi, J. J., Kim, H. S., Jacobson, K. A., Balzarini, J., Mitsuya, H., Boyer, P. L., Hughes, H., Marquez, V. E., *J. Med. Chem.* **2003**, *46*, 3292.
88 Park, A.-Y., Moon, H. R., Kim, K. R., Chun, M. W., Jeong, L. S., *Org. Bioorg. Chem.* **2006**, *4*, 4065.
89 Choi, Y., George, C., Strazewski, P., Marquez, V. E., *Org. Lett.* **2002**, *4*, 589.
90 (a) Moon, H. R., Kim, K. R., Kim, B. T., Hwang, K. J., Chun, M. W., Jeong, L. S. *Nucleosides, Nucleotides* **2005**, *24*, 709; (b) Lee, J. A., Moon, H. R., Kim, H. O., Kim, K. R., Lee, K. M., Kim, B. T., Hwang, K. J., Chun, M. W., Jacobson, A., Jeong, L. S., *J. Org. Chem.* **2005**, *70*, 5006.
91 Katagiri, N., Yamatoya, Y., Ishikura, M., *Tetrahedon Lett.* **1999**, *40*, 9069.
92 Dorr, D. R. Q., Vince, R., *Nucleosides, Nucleotides* **2002**, *21*, 665.
93 Demaison, C., Hourioux, H., Roingeard, P., Agrofoglio, L. A., *Tetrahedon Lett.* **1998**, *39*, 9175.
94 Elhalem, E., Comin, J., Rodriguez, J. B., *Eur. J. Org. Chem.* **2006**, 4473.
95 Wang, G., *Tetrahedon Lett.* **2000**, *41*, 7139.
96 Nair, V., Jeon, G.-S., *Arkivoc* **2004**, *XIV*, 133.
97 Franzyk, H., Rasmussen, J. H., Mazzei, R. A., Jensen, S. R., *Eur. J. Org. Chem.* **1998**, 2931.

16
4′-C-Ethynyl-2′-Deoxynucleosides

Hiroshi Ohrui

16.1
Introduction

4′-*C*-Ethynyl-2′-deoxynucleosides (4′EdNs) are the most promising nucleoside reverse transcriptase inhibitors (NRTIs) among 2′-deoxy-4′-*C*-substituted nucleosides (4′SdNs) [1–3]. Initially, 4′SdN was designed [1] as a highly potent and low-toxicity NRTI that could solve problems such as: (1) the emergence of drug-resistant HIV variants; (2) the need to take large doses of drugs; and (3) problems of adverse side effects – all of which are common during the existing highly active antiretroviral therapy (HAART) using two or more NRTIs and protease inhibitors [4, 5].

The structures and anti-HIV activities against wild-type HIV-1 (EC_{50}), (CC_{50}), and the selectivity index ($SI = CC_{50}/EC_{50}$) of some pyrimidine and purine derivatives of 4′EdNs are listed in Table 16.1. Among the pyrimidine derivatives, 2′-deoxy-4′-*C*-ethynylcytidine (4′EdC) [1, 6] is highly active but, unfortunately, is also highly toxic. Its 5-fluoro derivative – 2′-deoxy-4′-*C*-ethynyl-5-fluorocytidine (4′Ed5FC), which is a nucleoside derivative modified at two positions of physiological 2′-deoxycytidine – is not only one order less active but also far less toxic than 4′EdC; moreover, it has an acceptable SI with MT-4 cells. However, 4′Ed5FC demonstrated toxicity against other cell types [1]. All of the purine derivatives of 4′EdN have both high activity against wild-type HIV, and acceptable SI-values.

The structures and anti-HIV activities of selected 4′SdNs against HIV-1 mutants that are resistant to various clinical NRTIs are listed in Table 16.2. It should be noted that the three cytidine derivatives maintain their activity against the drug-resistant HIV-1 mutants, whilst the activity of 4′-*C*-ethynyl-D-arabinofuranosylcytosine (4′EaraC) and 2′-deoxy-4′-*C*-methylcytidine (4′MedC) each decreased significantly against M184V, M184I, and 41/69/125/SG mutants. The three purine derivatives – 2′-deoxy-4′-*C*-ethynyladenosine (4′EdA), 2′-deoxy-4′-*C*-ethynyl-2-aminoadenosine (4′Ed2AA), and 2′-deoxy-4′-*C*-ethynylguanosine (4′EdG), except for 2′-deoxy-4′-*C*-ethynylinosine (4′EdI) – were all highly potent against all drug-resistant HIV-1 mutants. 4′EdI was much less active than the former three derivatives, especially against M184V.

Modified Nucleosides: in Biochemistry, Biotechnology and Medicine. Edited by Piet Herdewijn

ISBN: 978-3-527-31820-9

Table 16.1 Anti-HIV activity of some 4′EdNs.

Structure	Base	EC_{50}(μM)	CC_{50}(μM)	SI
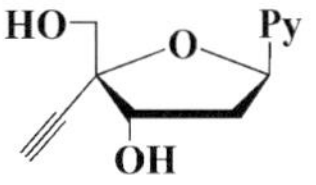	T	0.61	>380	>623
	5-I-U	0.34	>260	>765
	5-Me-C	0.011	0.70	63
	C	0.0048	0.92	192
	5-F-C	0.030	>100	>333
	5-halo-C	>100	>100	ND
	4′-*C*-ethynylarabinofuranosyl-C	0.043	2.0	46.5
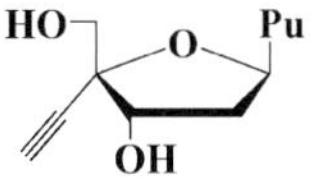	A	0.012	11.7	975
	I	0.15	216	1440
	G	0.0014	1.5	1071
	2,6-di-NH_2-A	0.0003	0.82	2733
	AZT	0.01	>20	>2000

Anti-HIV activity was determined by MTT assay. MT-4 cells and HIV-1 LAI were employed.
5-I-U=5-iodouracil; 5-Me-C=5-methylcytosine; 5-F-C=5-fluorocytocine; 2,6-di-NH_2-A=2,6-diaminoadenine.
ND = not determined.

Additionally, the three were also active against a non-nucleoside reverse transcriptase inhibitor-resistant Y181C. Further, the three purine derivatives were highly potent against the HIV-1s isolated from seven heavily drug-experienced patients with acquired immunodeficiency syndrome (AIDS) as effectively as against wild-type HIV-1 [1, 2]. Thus, 4′EdA, 4′Ed2AA, and 4′EdG were highly potent against all existing HIV-1s. These results suggested that the three purine 4′EdNs might even prevent the emergence of drug-resistant HIV-1s. It should be noted that 4′EdG showed toxicity to HeLa cells at a concentration of 52 μM, and consequently would be highly toxic.

16.2 Murine Toxicity of Purine 4′EdNs [1, 7, 8]

Following the intravenous or oral administration of 4′EdA and 4′EdI (3∼100 mg kg^{-1}), all mice survived, but all died after a single dose (3 mg kg^{-1}) of 4′EdAA and 4′EdG, irrespective of the route of administration (Table 16.3). Thus, it seemed that whereas both 4′EdA and 4′EdI were non-toxic, 4′Ed2AA and 4′EdG were both highly toxic. However, in mice, it was found that 4′EdA and 4′Ed2AA were easily converted to 4′EdI and 4′EdG, respectively, by the action of adenosine deaminase. Hence, these results showed that the actual toxicity of 4′EdA and 4′Ad2AA in animals is difficult to estimate.

Table 16.2 Anti-HIV activities of selected 4′SdNs against wild-type and drug-resistant HIV-1.

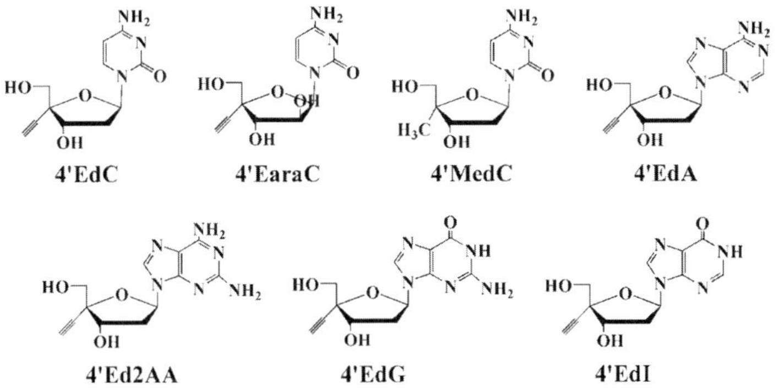

Compound	EC50(μM)									CC50 (μM)
	HXB2[a)]	KH65R	L74V	41/215	M184V	M184I	41/69/ 125/SG	MDR	Y181C	
4′EdC	0.0012	0.0008	0.0013	0.006	0.0024	0.0026	0.015	0.0012	0.0021	>200
4′EaraC	0.0071	0.015	0.026	0.026	0.71	0.48	0.17	0.0079	0.016	>200
4′MedC	0.0058	0.0071	0.0052	ND	0.2	0.74	ND	0.0033	ND	>200
4′EdA	0.008	0.0033	0.004	0.012	0.047	0.022	0.065	0.0062	0.011	>200
4′Ed2AA	0.0014	0.00035	0.0007	0.0017	0.0059	0.0027	0.0041	0.001	0.0008	>200
4′EdG	0.007	0.001	0.0012	0.019	0.008	0.0041	0.0068	0.0048	0.01	52
4′EdI	0.81	0.25	0.61	1.3	1.6	1.5	2.2	0.51	ND	>200
AZT	0.022	0.02	0.02	0.3	0.01	0.017	1.6	15.3	0.014	>100
3TC	0.71	ND	ND	ND	>100	>100	9.9	1.1	ND	>100
ddC	0.2	3.0	1.5	ND	2.2	ND	1.3	5.5	ND	>100
ddI	3.9	12.7	19.5	3.6	10.1	ND	12.2	25	ND	>100

Anti-HIV activity was determined with MAGI assay.
ND = not determined.
a) Wild-type HIV.

16.3
4′EdA Derivatives Stable to Adenosine Deaminase, and their Biological Properties [1]

It has long been known that the introduction of a halogen atom at the 2-position of adenine renders the adenosine derivatives stable towards adenosine deaminase [9, 10]. The structures and anti-HIV activity of 2′-deoxy-4′-*C*-ethynyl-2-fluoroadenosine (4′Ed2FA) and 2′-deoxy-4′-*C*-ethynyl-2-chloroadenosine (4′Ed2ClA) against wild-type HIV-1 are shown in Table 16.4. Both nucleosides are highly active and have acceptable SI-values. Although the M184V mutant is 10- to 15-fold more resistant than wild-type HIV-1 towards these nucleosides, they are both also highly active against drug-resistant HIV-1. Despite 4′Ed2FA being the most active NRTI identified to date, it

Table 16.3 Toxicity of purine derivatives of 4′EdNs towards mice.[a]

	Intravenous administration		Oral administration	
	Dose (mg kg^{-1})	Mortality (%)	Dose (mg kg^{-1})	Mortality (%)
4′Eda and 4′EdI	100	0	100	0
	10	0	10	0
	3	0	3	0
	1	0	1	0
4′Ed2AA	100	100 (1 day)[b]	100	100 (1 day)
	10	100 (2 days)	10	100 (2 days)
	3	0	3	100 (2 days)
	1	0	1	0
4′EdG	100	100 (1 day)	100	100 (1 day)
	10	100 (2 days)	10	100 (4 days)
	3	100 (4 days)	3	100 (4 days)
	1	0	1	0

[a] Six-week-old ICR male mice were used in these studies.
[b] Numbers in parentheses represent survival days of mice after administration.

Table 16.4 Anti-HIV 2′-deoxy-4′-C-ethynyl-2-haloadenosines.

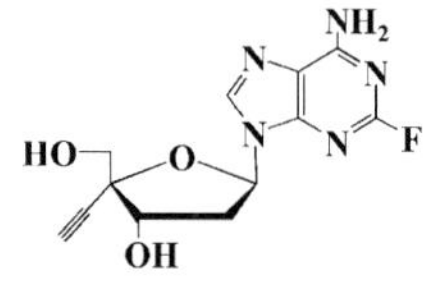

4'Ed2FA

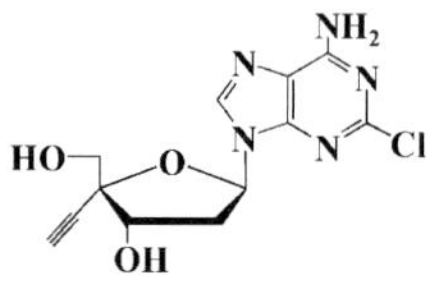

4'Ed2ClA

HIV-1	Compound	EC_{50}(nM)	CC_{50}(nM)	SI
Wild-type	4′Ed2FA	0.068[a]	7500[a]	110 000
	4′Ed2ClA	0.57[a]	18 800[a]	330 000
	AZT	3.2[a]	29 400[a]	9190
M184V	4′Ed2FA	3.1[b]		
	4′Ed2ClA	10.0[b]		
	AZT	10.0[b]		
MDR	4′Ed2FA	0.14[b]		
	4′Ed2ClA	8.4[b]		
	AZT	15 300[b]		

SI = Selectivity index.
[a] MTT-assay used.
[b] MAGI-assay used.

did not demonstrate any acute toxicity in mice when administered at up to 100 mg kg^{-1} by either the oral or intravenous routes [1, 8]. These nucleosides are perfectly stable towards adenosine deaminase under the conditions at which 4′EdA is completely deaminated within 60 minutes. It should also be noted that only 3% of 4′Ed2FA was decomposed after 2 h under the conditions at which ddA was completely decomposed within 10min (pH of gastric juice, 1.06, and a temperature of 36°C).

It is known that the toxicity of NRTIs in animals is due to their inhibitory effects on mitochondrial DNA polymerase γ. The 50% effective concentration (EC_{50}) of 5′-O-triphosphate of 4′Ed2FA (4′Ed2FATP) for inhibiting the incorporation of 2′-deoxyadenosine-5-*O*-triphosphate (dATP) mediated by human mitochondrial DNA polymerase γ was 10 μM, and significantly higher than the value of 0.2 μM for 2,3-dideoxyadenosine-5-*O*-triphosphate (ddATP) [1, 11]. The EC_{50} values of 4′Ed2FATP against DNA polymerases α and β were in excess of 200 μM. These results indicated that whilst the DNA polymerases scarcely recognize 4′Ed2FATP (a derivative modified at two positions of physiological dATP) as a substrate, the reverse transcriptase does afford such recognition.

16.4 4′-*C*-Ethynylnucleosides without 3′-OH

16.4.1 2′,3′-Dideoxy-4′-*C*-ethynylnucleoside

In order to clarify the role of 3′-OH in the biological activity of 4′EdN, 2′,3′-dideoxy-4′-*C*-ethynylcytidine (4′EddC) and its 5′-*O*-triphosphate (4′EddCTP) were prepared and evaluated for their anti-HIV activities [12]. Subsequently, 4′EddC proved to be inactive an *in-vitro* assay against HIV at concentrations between 0.001 and 10 μM in MT-2 and MT-4 cells, whereas 4′EddCTP strongly inhibited HIV reverse transcriptase. These results suggested that the 3′-OH of 4′EdC played an important role in the phosphorylation of 5′-OH of 4′EdC by cellular kinases, and in the cause of toxicity of 4′EdC.

The fact that 2′,3′-dideoxy-4′-*C*-ethynyl-2-fluoroadenosine (4′Edd2FA) is active against HIV-1_{HXB2}, but not against HIV-1_{M184V} and HIV-1_{MDR} (Table 16.5), suggested that 4′Edd2FA (which has no 3′-OH group) may be phosphorylated by cellular kinases. It follows that the 3′-OH group plays a critical role in the activity against drug-resistant HIVs.

16.4.2 2′,3′-Didehydrodideoxy-4′-*C*-ethynylnucleosides

2′,3′-Didehydro-3′-deoxy-4′-*C*-ethynylthymidine (4′Ed4T) was found to be more active, but less toxic, than d4T [13] (Table 16.5). It was possible to rationalize this reduced toxicity to the additional modification of d4T. It was also possible to rationalize the increased activity (to some extent) to the fact that the conversion rate of 4′Ed4T to its 5′-*O*-monophosphate by thymidine kinase is superior to that of d4T

Table 16.5 Anti-HIV activity of 2′,3′-dideoxy-4′-C-ethynyl nucleosides.

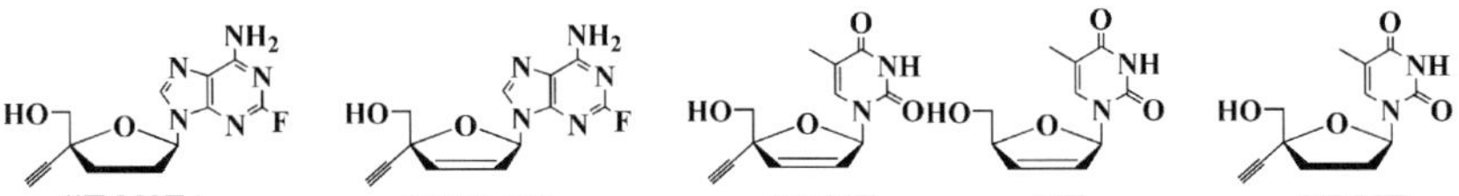

Compound	EC_{50}(μM)		
	HIV-1_{wild}	HIV-1_{MDR}	HIV-1_{M184V}
4′Edd2FA a)	0.94	8.7	97
4′Ed4-2FA a)	0.80	0.15	1.8
4′Ed4T b)	1.5	1.1	17
d4T b)	7.6	64	5.6
4′EddT b)	>100	ND	ND
AZT a)	0.17	74.3	0.13
3TC a)	1.0	2.8	>100

ND = not determined.
a) Refs. [1, 7];
b) Refs. [13, 14].

(with a confidence level of 0.06), and that 4′Ed4T is more resistant to thymidine phosphorylase than d4T [14]. Thus, it appears that the ene–yne system of 4′Ed4T plays only a minor role in the 3′-OH of 4′EdN. Further, the activity profile of 4′Ed4T against drug-resistant HIV-1 strains is different from that of d4T, and 4′Ed4T is equally active against many drug-resistant HIV-1 strains as against wild-type HIV-1. In contrast, the M184V mutation confers some resistance to 4′Ed4T, and additional mutations (PII9WS and T165A) of M184V show high levels of resistance to 4′Ed4T [15].

In contrast, whilst 2′,3′-didehydrodideoxy-4′-C-ethynyl-2-fluoroadenosine (4′Ed42FA) is more active against HIV than 4′Ed4T (Table 16.5), the activity against M184V with additional mutations of PII9S and T165A is not yet known [1, 8].

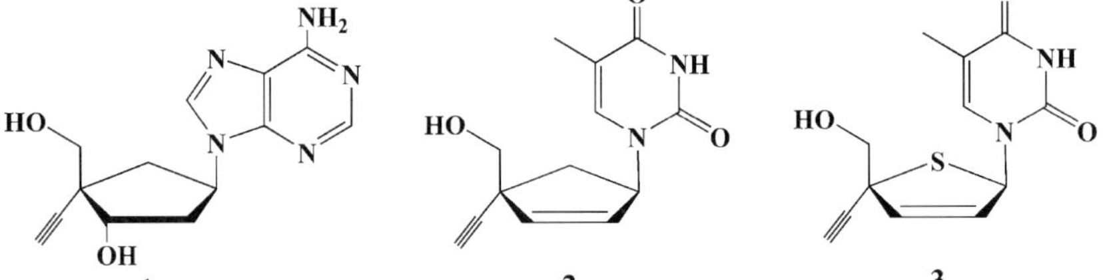

Figure 16.1 Carbocyclic and heterocyclic analogues of 4′-C-ethynylnucleoside.

16.4.3
Carbocyclic and Other Heterocyclic Analogues of 4′-*C*-ethynylnucleoside

The racemic carbocyclic analogues of 4′EdA (**1**) [16] and 4′Ed4T (**2**) [17] (Figure 16.1) have been prepared, but neither demonstrated any remarkable anti-HIV activity.

The racemic 4′-thioanalogue of 4′Ed4T (**3**) (Figure 16.1) showed an anti-HIV activity of $EC_{50} = 0.74\,\mu M$ and low toxicity ($CC_{50} > 100\,\mu M$). Among these 4′-*C*-ethylnucleosides, only **3** has an anti-HIV activity ($EC_{50} = 0.37\,\mu M$) [18]. The absolute configuration of **3** is assumed to be D-, as the L-isomer of d4T does not show any anti-HIV activity [19].

References

1 Ohrui, H., *The Chemical Record* **2006**, *6*, 133.
2 Kodama, E., Kohgo, S., Kitano, K., Machida, H., Gatanaga, H., Shigeta, M., Ohrui, H., Mitsuya, H., *Antimicrob. Agents Chemother.* **2001**, *45*, 1539.
3 Ohrui, H., Mitsuya, H., *Current Drug Targets-Infectious Disorder* **2001**, *1*, 1.
4 Lowe, S., Prins, J. M., Lange, J. M. A., *Neth. J. Med.* **2004**, *62*, 424.
5 Murphy, E. L., Collier, A. C., Kalish, L. A., Mintz, L., Wallach, F. R., Nemo, G. J., *Ann. Intern. Med.* **2001**, *135*, 17.
6 Sugimoto, I., Shuto, S., Mori, S., Shiegeta, S., Matsuda, A., *Bioorg. Med. Chem. Lett.* **1999**, *42*, 2901.
7 Hayakawa, H., Kohgo, S., Kitano, K., Ashida, N., Kodama, E., Mitsuya, H., Ohrui, H., *Antiviral Chem. Chemother.* **2004**, *15*, 169.
8 Patent: International Publication Number, WO2005/090349, PCT/JP2005/005374, Yamasa Corporation.
9 Montgomery, J. A., Hewson, K., *J. Med. Chem.* **1969**, *12*, 498.
10 Brockman, R. W., Schabel, F. M. Jr., Montgomery, J. A., *Biochem. Pharmacol.* **1977**, *26*, 2193.
11 Nakata, H., Amano, M., Koh, Y., Kodama, E., Yang, G., Kohgo, S., Hayakawa, H., Ohrui, H., Matsuoka, M., Mitsuya, H., Abstract of the 13th Conference on Retroviruses and Opportunistic Infections (CROI), F (Antiviral Therapy-Preclinical) (Poster Abstract) December, USA, **2006**, p. 219.
12 Siddiqui, M. A., Hughes, S. H., Boyer, P. L., Mitsuya, H., Van, Q. N., George, C., Saranfinanos, S. G., Marquez, V. E., *J. Med. Chem.* **2004**, *47*, 5041.
13 Haraguchi, K., Takeda, S., Tanaka, H., Nitanda, T., Baba, M., Dutschman, G. E. Cheng, Y. C., *Bioorg. Med. Chem. Lett.* **2003**, *13*, 3775.
14 Tanaka, H., Haraguchi, K., Kumamoto, H., Baba, M., Cheng, Y. C., *Antiviral Chem. Chemother.* **2005**, *16*, 217.
15 Nitanda, T., Wang, X., Kumamoto, H., Haraguchi, K., Tanaka, H., Cheng, Y. C., Baba, M., *Antimicrob. Agents Chemother.* **2005**, *48*, 3355.
16 Miyakoshi, H., Masters Thesis, Tohoku University, **2004**.
17 Kumamoto, H., Haraguchi, K., Tanaka, H., Nitanda, T., Baba, M., Dutchman, G. E., Cheng, Y. C., Kato, K., *Nucleosides, Nucleotides, Nucleic Acids* **2005**, *24*, 73.
18 Kumamoyo, H., Nakai, T., Haraguchi, K., Nakamura, K. T., Tanaka, H., Baba, M., Cheng, Y. C., *J. Med. Chem.* **2006**, *49*, 7861.
19 Mansuri, M. M., Farina, V., Starrett, J. E. Jr., Benigni, D. A., Brankovan, V., Martin, J. C., *Bioorg. Med. Chem. Lett.* **1991**, *1*, 65.

17
Modified Nucleosides as Selective Modulators of Adenosine Receptors for Therapeutic Use

Kenneth A. Jacobson, Bhalchandra V. Joshi, Ben Wang, Athena Klutz, Yoonkyung Kim, Andrei A. Ivanov, Artem Melman, and Zhan-Guo Gao

17.1
Introduction

In addition to its known intracellular actions, adenosine acts as an extracellular signaling molecule by activating adenosine receptors (ARs) present in the plasma membrane. The ARs consist of a family of four G protein-coupled receptor (GPCR) subtypes, designated as A_1, A_{2A}, A_{2B}, and A_3 [1]. Most cells express one or more subtype(s) of these receptors. Adenosine activates the A_{2A} and A_{2B} subtypes to stimulate adenylate cyclase, and activates the A_1 and A_3 subtypes to inhibit adenylate cyclase. Other effector mechanisms are also important in the physiological actions of adenosine, such as ion channels, phosphatidylinositol 3-kinase (PI3K), and mitogen-activated protein kinase (MAPK), which may be activated through both α and $\beta\gamma$ subunits of the G proteins [2]. Structure–activity relationship (SAR) studies have led to the development of chemically modified nucleoside analogues as potent and selective AR agonist ligands [3–6]. While selective agonists for the A_1, A_{2A}, and A_3 receptors have already entered clinical trials, the development of agonists for the A_{2B} receptor has lagged. One reason for the absence of selective agonists for this subtype is that it has the lowest affinity of all the ARs for native adenosine, and therefore a greater design barrier must be overcome in the achievement of selectivity over the other subtypes.

Envisioned therapeutic applications of AR agonists include: for the A_1 AR, treatment of pain [7] and cardiac arrhythmias [8]; for the A_{2A} AR, inflammation, wound healing, diagnostics, psychiatric disorders [9] and sleep disturbances [10]; for the A_{2B} AR, cardioprotection [11]; and for the A_3 AR, treatment of stroke, neurodegenerative diseases [10, 12], cardiac ischemia [11, 13], cancer [14], and rheumatoid arthritis and other autoimmune inflammatory disorders [15]. Some nucleosides act as selective antagonists of the A_3 AR, and are being explored for the treatment of glaucoma [16]. This chapter will focus on the principal molecules that have been recognized as recent clinical candidates by virtue of having selectivity for one of these three subtypes. At this point no attempt will be made systematically to review the history of the design of modified nucleosides as AR ligands.

Modified Nucleosides: in Biochemistry, Biotechnology and Medicine. Edited by Piet Herdewijn

ISBN: 978-3-527-31820-9

17.2
Molecular Targets and Binding Modes

As GPCRs, the overall structural motif of the ARs consists of seven transmembrane helices (TMs) that are connected by three extracellular and three cytoplasmic loops. Since neither nuclear magnetic resonance (NMR) nor crystallographic structures of the ARs have yet been determined, the most widely used method for studying the three-dimensional structures of the receptors is based on homology to the high-resolution crystallographic structure of bovine rhodopsin. Nucleoside ligands have been docked in a putative binding site, which is located within the transmembrane domain in a cleft created by the barrel arrangement of the seven TMs [17]. The overall position of the putative nucleoside ligands in the ARs closely resembles that of the covalently bound ligand retinal in the visual pigment and light transduction protein, rhodopsin.

The molecular recognition of modified nucleosides as AR agonists and the correlation of modeling and mutagenesis data were recently reviewed [17]. The ribose moiety of adenosine and its analogues is proposed to bind in a hydrophilic region defined mainly by conserved polar residues in TMs 3 and 7 of the ARs. A conserved Asn in TM6 acts as a putative H-bond acceptor for the exocyclic amine of the adenine moiety of adenosine analogues, consistent with the need to have at least one H present on that nitrogen in order to bind to the ARs with high affinity. Conserved hydrophobic residues in TMs 5 and 6 delineate the putative binding region for bulky hydrophobic N^6 substituents of adenosine analogues optimized for binding to the A_1 or A_3 AR. Quantitative SAR studies of adenosine agonists have correlated the desired features present on the optimized nucleosides with these regions of the AR proteins [67]. Modeling of the putative binding site of the ARs is intended to guide the development of novel and selective nucleoside ligands.

17.3
AR Agonists as Clinical Candidates

In general, adenosine has a cytoprotective role in the body and, in many physiological systems, has been shown to protect against tissue damage from stress conditions, such as ischemia [18]. When energy is being expended by a particular tissue or organ in an excessive fashion and the ratio of oxygen supply to demand is consequently lowered, adenosine levels in the extracellular medium rise drastically. This may be a function of the utilization of intracellular ATP and its depletion by cell damage. The adenosine level may increase by exiting cell through equilibrative transport proteins, or through the catabolism of extracellular adenine nucleotides. This rise in adenosine activates the ARs to suppress cytotoxic processes, such as apoptosis, calcium overload, mitochondrial damage, and the release of histamine and other inflammatory mediators. Adenosine itself is short-lasting in the circulation and does not possess clear selectivity among AR subtypes (although there is a hierarchy of activation of the ARs as the adenosine concentration increases). For this reason, selective AR agonists that persist for longer periods in the body, are more selective, and are more potent than native adenosine, are being sought as potential therapeutic agents.

17.3.1 Modified Nucleosides as A_1 AR Agonists

The biological action of adenosine was first identified in the cardiovascular system, where it was found to have a significant effect on atrioventricular conduction and heart rate (i.e., the bradycardiac effect of activation of the A_1 AR) and on blood pressure (i.e., to some extent, the vasodilatory effect of activation of the A_{2A} AR). Adenosine (as Adenocard®), when administered by intravenous infusion, is used to restore a normal heart rhythm in patients with paroxysmal supraventricular tachycardia (PSVT) [19]. Unfortunately, however, the endogenous agonist adenosine is not selective for the A_1 AR, and consequently for long-term indications a selective A_1 AR agonist is required in order to avoid adverse side effects such as vasodilation.

The modification of adenosine analogues by attachment of a hydrophobic substituent at the N^6 position has resulted in high selectivity for the A_1 AR (Figure 17.1). Among a series of N^6-cycloalkyl derivatives, CPA (N^6-cyclopentyladenosine) was identified during the 1980s to have the highest affinity and selectivity for the A_1 AR in comparison to the A_{2A} AR. The cyclopentyl group has been optimized in the monosubstituted adenosine derivatives, CVT-510 (Tecadenoson) [20] and GR79236 (*N*-[(1*S*, *trans*)-2-hydroxycyclopentyl] adenosine) [21]. CVT-510 is a full agonist of the A_1 AR, while its 5′-carbamoyl derivative CVT-2759 ((((5-(6-(oxolan-3-yl)amino)purin-9-yl)-3,4-dihydroxyoxolan-2-yl)methoxy)-*N*-methylcarboxamide) is a partial agonist of the A_1 AR [22, 65]. The combined N^6-cyclopentyl and 5′-uronamide substitutions led to RG14202 (Selodenoson, DTI-0009), a clinical candidate by virtue of its potent and selective activation of the A_1 AR [23].

Several of these A_1 AR-selective agonists are currently undergoing clinical trials for cardiac arrhythmias and have been found to be superior to adenosine. RG14202 has completed a Phase II trial to slow the heart rate in atrial fibrillation [24] and was found to have the potential to do this without lowering blood pressure. Previously, RG14202 was investigated for the inhibition of lipolysis, which would indicate its potential for the treatment of diabetes, although at the required doses it was found to have adverse cardiovascular side effects [23]. CVT-510 is currently undergoing Phase III patient trials for its anti-arrhythmic activity, as it was found to terminate PSVT without any significant hypotensive effects. The results of clinical trials suggested that CVT-510 might also have the potential for the treatment of atrial fibrillation. CVT-2759, a partial agonist of the A_1 AR, was found significantly to inhibit isoproterenol-induced arrhythmic activity, but did not cause any potential adverse side effects as are commonly caused by full agonists, such as atrioventricular (AV) block, bradycardia, and atrial arrhythmias, suggesting a potential role for a partial agonist of the A_1 AR in the treatment of cardiac arrhythmias [22].

A_1 AR agonists reduce pain signaling in the spinal cord, where the receptors are highly expressed. In humans, the infusion of adenosine into the spinal cord was effective in decreasing postoperative pain. The results of recent studies have

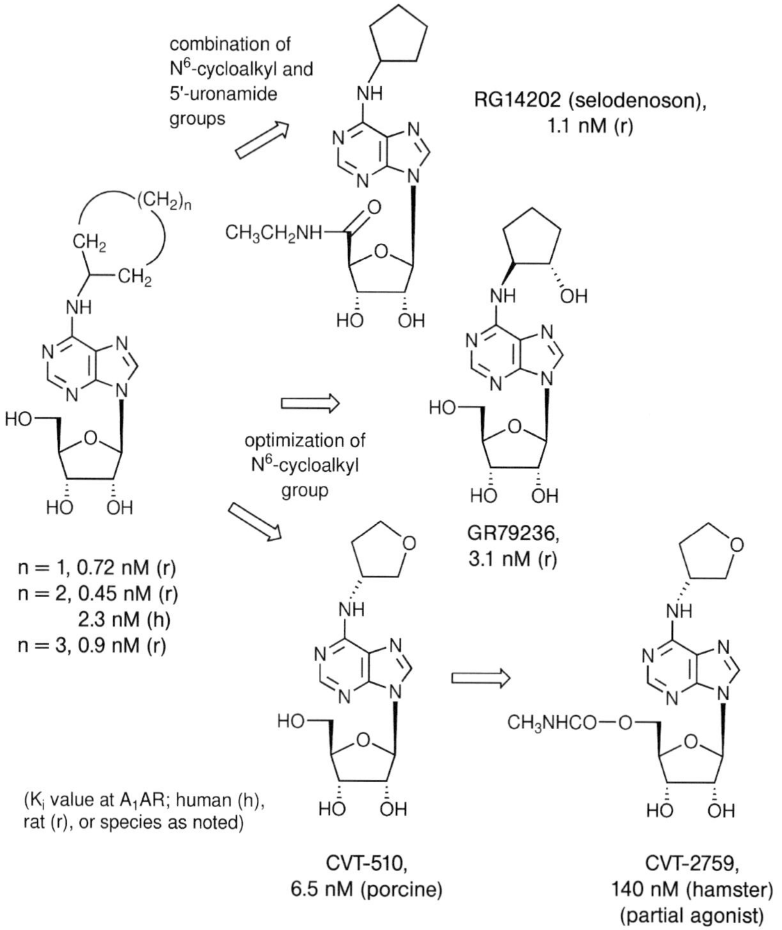

Figure 17.1 Structures of selected A_1 AR agonists and their binding affinity at that subtype. Arrows indicate the conceptual progression of structure–activity relationship studies during the period between 1980 (report of N^6-cyclohexyladenosine) and 2001 (report of CVT-2759) [3, 4, 65].

suggested that A_1 ARs might be more important in chronic pain than in acute pain. Selective activation of the A_1 AR leads to neuronal inhibition without concomitant vasoconstriction, indicating that this might be an effective treatment for migraine and cluster headache. Indeed, the A_1 AR agonists, GR79236 [21] and GW-493838 (structure not disclosed) [25], have been undergoing patient trials for the treatment of pain and migraine. GR79236 is a cardioprotective agent, which also has anti-inflammatory and analgesic actions. In addition, the therapeutic potential of GR79236 in primary headache disorders was successfully demonstrated [21]. GW-493838 is currently undergoing Phase II trials in patients with post-herpetic neuralgia or peripheral nerve injury caused by trauma or surgery.

Adenosine is considered to be an endogenous anti-seizure mediator in the brain, and A_1 AR agonists that enter the brain may eventually be of use in the treatment of epilepsy [26]. The A_1 AR has also been suggested to play a role in neuroprotection, partly due to the direct hyperpolarization of the neurons, and partly because of the inhibition of glutamate release and other mechanisms. Finally, activation of the A_1 AR induces sleep-like electroencephalographic patterns and sleep. However, it has been suggested recently that the A_{2A} AR also has an important role in sleep, such as a tonic role in the regulation of rapid eye movement (REM) sleep [10].

17.3.2
Modified Nucleosides as A_{2A} AR Agonists

The SAR progression leading to current clinical candidates that are selective for the A_{2A} AR are shown in Figure 17.2. Substitutions at the 2 position of adenosine that enhance the A_{2A} AR potency were combined with a favorable 5′-uronamide substitution (either 5′-*N*-ethyl or 5′-*N*-cyclopropyl derivatives) to produce A_{2A} AR-selective agonists. The first somewhat selective and metabolically stable A_{2A} AR agonist was CGS21680 (2-[4-(2-carboxyethyl)phenyl-ethylamino]-5′-*N*-ethylcarboxamidoadenosine). This is selective in binding to the rat A_{2A} AR, but in humans it binds with equal potency to the A_{2A} and A_3 ARs [3]. It also has the advantage (when envisioned for peripheral sites of action) of being largely excluded from crossing the blood–brain barrier. A newer prototypical probe of the A_{2A} AR, ATL-146e (4-(3-[6-amino-9-(5-ethylcarbamoyl-3,4-dihydroxy-tetrahydro-furan-2-yl)-9*H*-purin-2-yl]-prop-2-ynyl)-cyclohexanecarboxylic acid methyl ester) and its closely related derivative ATL-313 (4-(3-[6-amino-9-(5-cyclopropylcarbamoyl-3,4-dihydroxytetrahydrofuran-2-yl)-9*H*-purin-2-yl]prop-2-ynyl)piperidine-1-carboxylic acid methyl ester) [27], are being explored as potential therapeutic agents. Some derivatives of adenosine that are monosubstituted at the 2 position have also been proposed as A_{2A} AR agonists for clinical use; however, these tend to be of only intermediate affinity and selectivity at the A_{2A} AR. For example, MRE0094 (2-([2-(4-chlorophenyl)ethyl)]oxy)-adenosine), although highly A_{2A} AR-selective by functional criteria [28, 64], in binding assays at the human ARs is only six-fold and two-fold selective for the A_{2A} AR in comparison to the A_1 and A_3 ARs, respectively [29].

With respect to their use in the central nervous system, A_{2A} AR agonists also have the potential to be developed as antipsychotic drugs to treat schizophrenia and cocaine addiction. CGS21680 diminishes the affinity of dopamine receptors for dopamine agonists, leading to a decrease in motor activation in rats [30]. Therefore, A_{2A} AR agonists in combination with dopamine antagonists could be used to treat schizophrenia [31]. CGS21680 also decreased the development and expression of cocaine sensitization in rats, as shown by a decline in locomotor activities, thereby opening the door to the development of A_{2A} AR agonists for the treatment of cocaine addiction [32]. However, such agonists will need to be further optimized to guarantee the ability to cross the blood–brain barrier [9].

With respect to the cardiovascular system, one of the most prominent effects of activation of the A_{2A} AR is its vasodilatory effect, both in the aorta and coronary

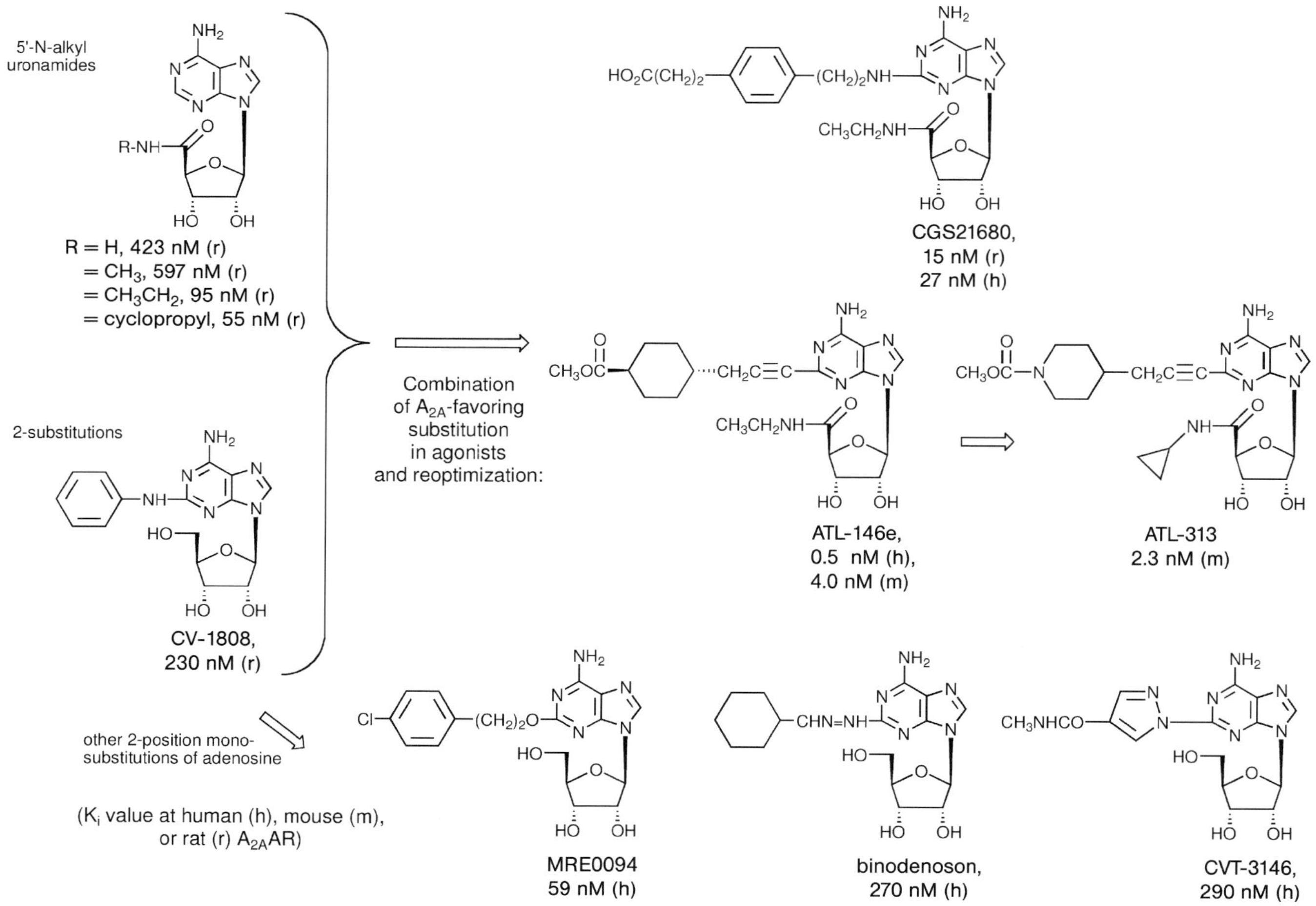

Figure 17.2 Structures of selected A_{2A} AR agonists and their binding affinity at that subtype. Arrows indicate the conceptual progression of structure–activity relationship studies during the period between 1989 (report of CGS21680) and 2005 (report of ATL-313) [3–5, 29, 33, 64].

arteries. The A_{2A} AR-induced vasodilation is the basis for novel therapeutic applications, including imaging. Indeed, adenosine itself (as Adenoscan) is used as an agonist of the A_{2A} AR in cardiac stress imaging to evaluate coronary artery disease. Currently, AMP (adenosine 5′-monophosphate) is being evaluated clinically as a means of delivering adenosine to the tissues. CGS21680 was undergoing clinical trials during the early 1990s for its potential as an antihypertensive agent [33], but was subsequently discontinued, presumably due to the non-selectivity *in vivo*. Thus, more selective A_{2A} AR agonists are required for imaging and other applications. A_{2A} AR agonists have been shown to inhibit platelet aggregation, which suggests that a selective A_{2A} AR agonist might have utility as an antithrombotic agent [34].

Regadenoson (CVT-3146), a potent and selective A_{2A} AR agonist, has already completed Phase III studies for myocardial perfusion imaging [35]. This drug is used as a pharmacological stressor to create vasodilation during myocardial perfusion imaging for patients who are unable to exercise to induce natural vasodilation. Binodenoson (WRC-0470) has entered Phase III clinical trials for cardiac imaging, and seems to be well tolerated as a short-lived coronary vasodilator; it also acts as an adjunct to radiotracers in single-photon emission computed tomography (SPECT) [36]. Apadenoson (ATL-146e) has also entered Phase III clinical trials for coronary imaging. Despite the A_{2A} AR selectivity of these compounds, adverse side effects attributable to other AR subtypes are still observed.

In the immune system, adenosine has been shown to "put the brakes" on excessive inflammation [37]; thus, agonists are being developed as potential anti-inflammatory compounds [38]. Due to the expression of the A_{2A} AR in neutrophils, macrophages, and T lymphocytes, activation of this subtype is a target for the treatment of inflammation [39]. A_{2A} AR agonists suppress the formation of superoxide anion, pro-inflammatory cytokines, and adhesion molecules. ATL-146e reduced ischemic injury in several organs in animal experiments, including the colon and the liver, in which it acts through the inhibition of natural killer (NK) Tcells [27, 40]. Clinical trails of ATL-146e as an anti-inflamatory agent have also been initiated. GW328267 has examined in patient trials for the treatment of asthma and chronic obstructive pulmonary disease (COPD) [41]. MRE0094 has been shown to promote wound healing [28] and is undergoing clinical trials for foot ulcers related to chronic diabetes. In fact, MRE0094 was found to increase angiogenesis and to inhibit the transport of cholesterol from the macrophages, which is a critical step in the formation of pro-atherosclerotic foam cells [42]. ATL-313 has been shown to reduce tissue injury and inflammation in mice with toxin A-induced enteritis [43].

17.3.3 Modified Nucleosides as A_{2B} AR Agonists

Although until recently, the achievement of selectivity of agonists for the A_{2B} AR was unattainable, several groups have now described nucleoside molecules that potently activate the A_{2B} AR, in some cases along with at least one additional subtype (Figure 17.3). Baraldi *et al.* reported the design and synthesis of novel A_{2B} AR agonists consisting of 1-deoxy-1-(6-[((hetero)arylcarbonyl)-hydrazino]-9*H*-purin-9-

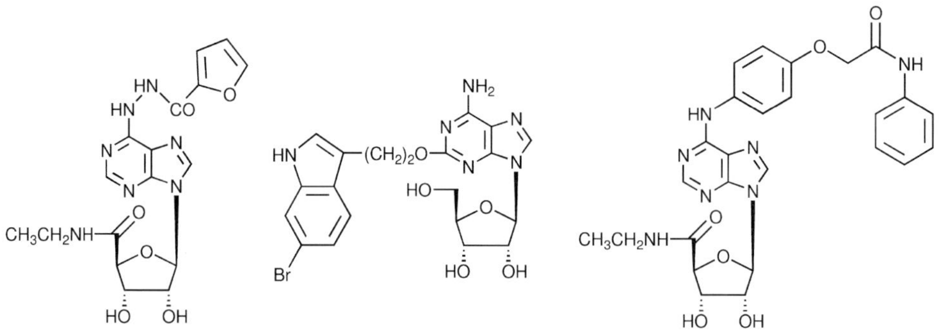

Compound 12b, 82 nM

MRS3997, 128 nM

Compound 24, 7.3 nM

(EC_{50} value at human $A_{2B}AR$)

Figure 17.3 Structures of recently reported nucleosides that potently activate the A_{2B} AR [62, 63, 73].

yl)-*N*-ethyl-β-D-ribofuranuronamide derivatives with EC_{50} values in the nanomolar range. The most effective compound in this series (**12b**; 1-deoxy-1-(6-[N^6-(2-furyl) hydrazino]-9*H*-purin-9-yl)-2,3-*O*-isopropylidene-*N*-ethyl-β-D-ribofuranuronamide, $EC_{50} = 82\,nM$) was two-fold more potent than adenosine-5′-*N*-ethyluronamide (NECA; $EC_{50} = 160\,nM$) [62]. Moreover, this compound was found to be fairly selective versus the A_1 ($K_i = 1050\,nM$), A_{2A} ($K_i = 1550\,nM$), and A_3 ($K_i > 5000\,nM$) ARs. The monosubstituted adenosine derivative MRS3997 [2-(3-(6-bromo-indolyl) ethyloxy)adenosine] fully activates human A_{2A} and A_{2B} ARs equipotently and is a weak partial agonist at both the A_1 and A_3 AR [63]. Compound **24** was recently reported as a highly potent, non-selective A_{2B} agonist [73].

Of all of the four subtypes of AR discovered to date, the functions of the A_{2B} AR have been the most difficult to discern pharmacologically due to the lack of agonists selective for this receptor subtype, as well as its capacity to couple to more than one G protein under different conditions. It has been suggested that A_{2B} AR signaling influences pathways critical for pulmonary inflammation and injury *in vivo*. There is evidence that the A_{2B} AR contributes to the proinflammatory and tissue destructive effects of adenosine in tissues such as lung [44]. The elevation of proinflammatory cytokines and chemokines, as well as mediators of fibrosis and airway destruction, accompanies the activation of the A_{2B} AR. However, a recent study of A_{2B} AR knockout mice suggested that the activation of the A_{2B} AR in macrophages can also have anti-inflammatory effects [45]. Another recent study of the A_{2B} AR knockout mice suggested that mice lacking the A_{2B} AR display increased sensitivity to IgE-mediated anaphylaxis. The A_{2B} AR appears to function as a critical regulator of signaling pathways within the mast cell, which act in concert to limit the magnitude of mast cell responsiveness when an antigen is encountered [46].

Activation of the A_{2B} AR is associated with maintaining vascular integrity and inducing angiogenesis [47]. Depending on species and location, activation of the A_{2B} AR may also have a vasodilatory effect. The relatively high level of expression of the A_{2B} AR in the kidney and in both COS-7 and HEK-293 cell lines suggested its potential role in the regulation of renal cell growth and proliferation [48]. Other potential applications of A_{2B} AR agonists are cardioprotection as a pharmacological post-conditioning agent (in which the agonist would be effective against infarction when administered near the end of a prolonged ischemic period) [69, 70], the prevention of cardiac remodeling and possibly restenosis [49, 50], and the treatment of erectile dysfunction [51]. While A_1 AR appears to trigger the onset of the preconditioned state prior to ischemia, A_{2B} AR has been proposed to participate in mediating the protection early in the reperfusion period. The A_{2B} AR-selective antagonist MRS 1754 blocked protection in preconditioned hearts when administered coincidentally with the restoration of flow after the ischemic insult [71]. Eckle *et al.* [72] found that ischemic preconditioning continued to protect in all AR knockout hearts except in those lacking A_{2B} AR. Ischemic post-conditioning in rabbit cardiac tissue depends on both A_{2B} AR as well as protein kinase C (PKC) activation. Interestingly, PKC could be shown to lie upstream of the A_{2B} AR, suggesting that it increases either endogenous adenosine release or the sensitivity of the heart's A_{2B} AR-dependent pathways to adenosine [69].

17.3.4 Modified nucleosides as A_3 AR Ligands

The progression leading to the design of highly selective nucleoside ligands of the A_3 AR is shown in Figure 17.4. The earliest step was the combination of 5′-*N*-alkyluronamide substitution with a favorable N^6 substitution. N^6-Arylmethyl derivatives of adenosine, such as N^6-benzyl derivatives, were found to have the most favorable selectivity for the A_3 AR. Although N^6-phenylethyladenosine is considerably more potent than N^6-benzyladenosine at the human A_3 AR, it still binds potently at the other AR subtypes. The A_3 agonists IB-MECA [N^6-(3-iodobenzyl)-5′-*N*-methylcarboxamidoadenosine; CF101] and its 2-chloro analogue Cl-IB-MECA (CF102) were the first selective agonists of the A_3 AR. Currently, more selective agonists are available, including the conformationally locked Northern (N) methanocarba derivative, MRS3558 (((1′*R*,2′*R*,3′*S*,4′*R*,5′*S*)-4-(2-chloro-6-[(3-chlorophenylmethyl)amino] purin-9-yl)-1-(methylaminocarbonyl)bicyclo[3.1.0]hexane-2,3-diol)), and the 4′-thio analogue of Cl-IB-MECA, LJ-529 [6, 66]. MRS3558 maintains a receptor-preferred (N) conformation of the ribose-like ring. The tetrahydrothiophene ring in LJ-529 is also expected to be more conformationally constrained than the native ribose moiety in adenosine derivatives. The modified nucleoside MRS1292 ((2*R*,3*R*,4*S*,5*S*)-2-[N^6-3-iodobenzyl)adenos-9′-yl]-7-aza-1-oxa-6-oxospiro[4.4]-nonan-4,5-diol), however, is a moderately potent and selective A_3 AR antagonist in both rat and human [16]. The basis for antagonism of the A_3 AR is reduced flexibility of the 5′-uronamide group in this spirolactam derivative, this being most likely related to its impaired ability to engage in H-bonding with the proper geometry required for A_3 AR activation. Another means of diminishing efficacy at the A_3 AR in nucleoside derivatives is to reduce the H-bonding ability through alkylation of the amide group with a second methyl – an approach that gave rise to the *N*,*N*-5′-dimethyluronamide antagonists, MRS3771 and LJ-1256 [52]. Recently, Jeong *et al.* reported that removing the 5′-uronamide group entirely in the 4′-thio series also produces full antagonists [74].

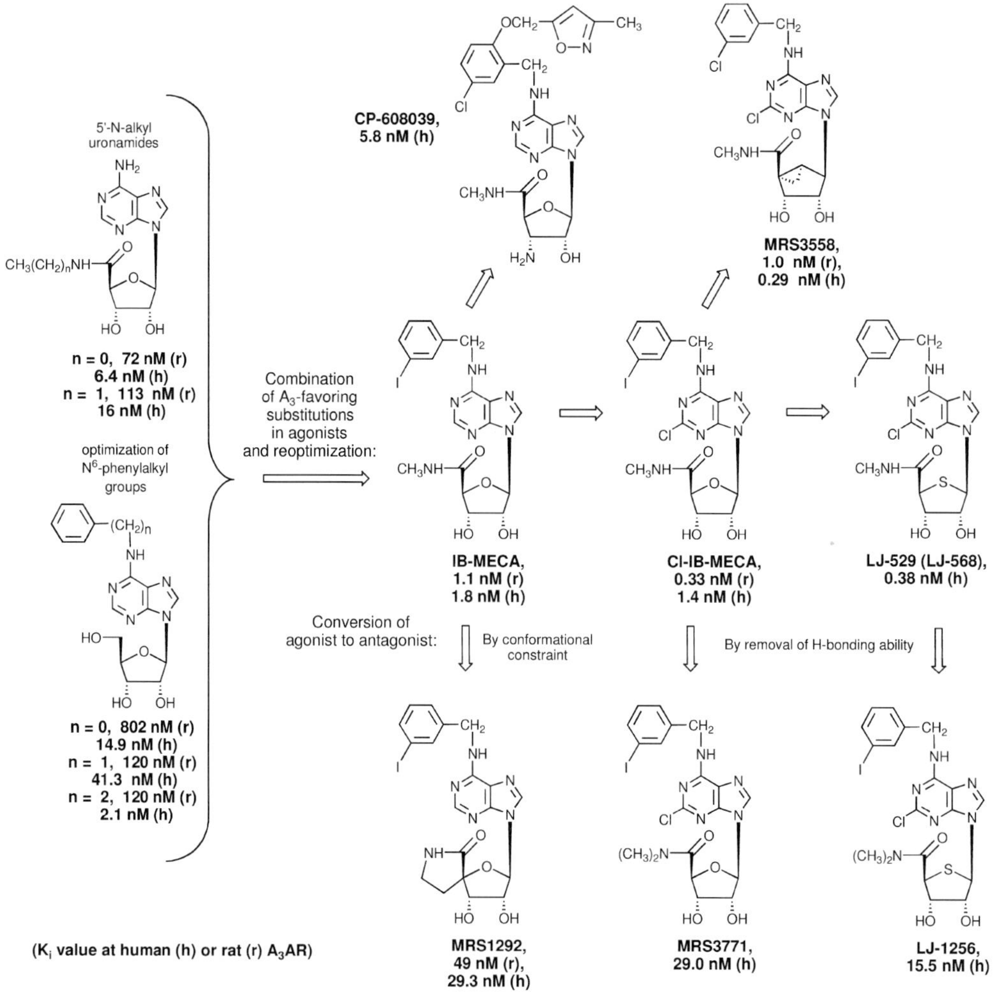

Figure 17.4 Structures of selected A_3 AR agonists and their binding affinity at that subtype. Arrows indicate the conceptual progression of structure–activity relationship studies during the period between 1993 (report of IB-MECA) and 2006 (report of MRS3771 and LJ-1256) [3, 4, 6, 13, 16, 52, 66].

The A_3 AR is highly expressed in some blood cells including eosinophils and activated neutrophils, suggesting a potential function of this receptor. Indeed, it has recently been demonstrated that activation of the A_3 AR is critically involved in inducing neutrophil chemotaxis [53]. The A_3 AR has been suggested to be involved in some inflammation processes, and an A_3 AR agonist IB-MECA has been shown to have anti-inflammatory activity in rheumatoid arthritis patients failing methotrexate therapy [15], as well as in animal models of colitis [54, 55]. IB-MECA is also entering clinical trials for chronic plaque psoriasis, keratoconjunctivitis sicca, and dry eye syndrome.

Although A_3 AR expression levels are low in all regions of the brain, chronic administration of the A_3 AR agonist IB-MECA was highly effective against global ischemia, and deletion of the A_3 AR has a detrimental effect in a model of mild hypoxia, suggesting the possibility of using A_3 AR agonists to treat brain ischemia [56]. The potent and selective A_3 AR agonist MRS3558 reduces damage and apoptosis in a model of traumatic lung injury in the cat [57].

In the heart, the A_3 AR generally has a protective role. A_3 AR agonists administered prior to a period of prolonged cardiac ischemia (preconditioning mode) showed a consistent protective effect, although controversial results have been reported from various studies of AR-knockout animals. It has been shown that activation of either the A_1 AR or the A_3 AR could induce protection of function in preconditioned rat hearts, while maximal preconditioning requires activation of both A_1 AR and A_3 AR. A recent study by Auchampach and colleagues showed that Cl-IB-MECA protects against myocardial ischemia/reperfusion injury in mice via activation of the A_3 AR, and this protection was independent of mast cell degranulation [68]. Overexpression of A_3 ARs decreases heart rate, preserves intracellular energetics, and protects ischemic hearts. Studies of IB-MECA in cardioprotection were recently reviewed [58]. The mechanism of A_3 AR-induced preconditioning involves activation of PKC (downstream from the receptor) and the critical anti-apoptotic enzymatic signal, phosphatidylinositol 3-kinase, leading to an opening of mitochondrial K_{ATP} channels (mitoKATP). An intermediate step in transmitting the protective signal from the cytosol to the mitochondria is activation of guanylyl cyclase and subsequently protein kinase G (PKG). Activated PKG opens mitoKATP, which increases the production of reactive oxygen species. The protective signal at the time of reperfusion is closing of mitochondrial permeability transition pore (MPTP) channels as a result of phosphorylation of glycogen synthase kinase-3β (GSK-3β). GSK-3β is constitutively active and thus keeps the MPTP open; when GSK-3β is phosphorylated it allows the MPTP to close. A_3 AR activation was also demonstrated to protect cardiac myocytes from the damage induced by the cancer chemotherapeutic agent doxorubicin [59]. The highly selective A_3 AR agonist CP-608039 ((2*S*,3*S*,4*R*,5*R*)-3-amino-5-(6-[5-chloro-2-(3-methylisoxazol-5-ylmethoxy)benzylamino]purin-9-yl)-4-hydroxytetrahydrofuran-2-carboxylic acid methylamide) has been under development as a cardioprotective agent [11].

A_3 AR agonists can attenuate apoptosis at low agonist concentrations. However, at higher concentrations (>10 μM), in human eosinophils and human promyelocytic HL-60 cells, the A_3 AR agonist Cl-IB-MECA induces apoptosis, although the mechanism was not purely receptor-mediated. Nevertheless, overexpression of the A_3 AR in transgenic mice resulted in embryonic lethality, suggesting the possible use of selective A_3 AR agonists in some types of cancer. IB-MECA has been undergoing clinical trials for colon carcinoma, and Cl-IB-MECA is being proposed for trials in lung cancer [14]. Combined cancer therapy with an A_3 AR agonist coadministered with other chemotherapeutic agents is envisioned. The administration of IB-MECA has not only a cytostatic effect on tumor growth *in vivo*, but also a myeloprotective effect. The A_3 AR agonist LJ-529 was shown to attenuate human breast tumor growth in mouse xenograft models [60]. It has also been shown that the expression level of the A_3 AR is higher in tumor than in normal cells [61], which may provide additional evidence to justify the A_3 AR as a potential target for cancer therapy. Prolonged A_3 AR receptor activation in tumors is reported to lead to inhibition of PKB/Akt, which consequently deregulates the Wnt signaling pathway, resulting in down-regulation of the transduction factor NF-κB; this in turn inhibits the growth of tumor cells. However, caution must be given with regards to certain species differences of some agonists and some receptor functions. Also, non-A_3 AR mechanisms could account for some of the anticancer effects of the modified nucleosides.

17.4 Summary

After three decades of medicinal chemistry research on adenosine receptors, a considerable number of modified nucleosides have been synthesized, and some have been identified as selective agonists for certain receptor subtypes. At least a dozen of these agonist ligands are now being evaluated clinically. It is envisioned that some of these adenosine derivatives will be used clinically in the near future for the treatment of a variety of diseases.

Acknowledgments

The authors thank the Intramural Research Program of the NIH, National Institute of Diabetes and Digestive and Kidney Diseases (Bethesda, MD) for support. They also thank Prof. James Downey of the University of South Alabama, for helpful discussions.

References

1 Fredholm, B. B., IJzerman, A. P., Jacobson, K. A., Klotz, K. N., Linden, J., International Union of Pharmacology. XXV. Nomenclature and classification of adenosine receptors. *Pharmacol. Rev.* **2001**, *53*, 527–552.

2 Schulte, G., Fredholm, B. B., Signalling from adenosine receptors to mitogen-activated protein kinases. *Cellular Signalling* **2003**, *15*, 813–827.

3 Jacobson, K. A., Gao, Z. G., Adenosine receptors as therapeutic targets. *Nature Reviews Drug Discovery* **2006**, *5*, 247–264.

4 Yan, L., Burbiel, J. C., Maass, A., Müller, C. E., Adenosine receptor agonists: from basic medicinal chemistry to clinical development. *Expert Opinion in Emerging Drugs* **2003**, *8*, 537–576.

5 Cristalli, G.,Lambertucci, C.,Taffi, S., Vittori, S.,Volpini, R.,Medicinal chemistry of A_{2A} adenosine receptor agonists.*Curr. Top. Med. Chem.* **2003**, *3*, 387–401.

6 Joshi, B. V., Jacobson, K. A., Purine derivatives as ligands for A_3 adenosine receptors. *Curr. Top. Med. Chem.* **2005**, *5*, 1275–1295.

7 Sawynok, J., Adenosine and ATP receptors. *Handbook Exp. Pharmacol.* **2007**, *177*, 309–328.

8 Goldstein, R. N., Stambler, B. S., New antiarrhythmic drugs for prevention of atrial fibrillation. *Prog. Cardiovasc. Dis.* **2005**, *48*, 193–208.

9 Fredholm, B. B., Chen, J. F., Masino, S. A., Vaugeois, J. M., Actions of adenosine at its receptors in the CNS: Insights from knockouts and drugs. *Annu. Rev. Pharmacol. Toxicol.* **2005**, *45*, 385–412.

10 Huang, Z. L., Urade, Y., Hayaishi, O., Prostaglandins and adenosine in the regulation of sleep and wakefulness. *Curr. Opin. Pharmacol.* **2007**, *7*, 33–38.

11 Philipp, S., Yang, X. M., Cui, L., Davis, A. M., Downey, J. M., Cohen, M. V., Postconditioning protects rabbit hearts through a protein kinase C-adenosine A_{2B} receptor cascade. *Cardiovasc. Res.* **2006**, *70*, 308–314.

12 Jacobson, K. A., von Lubitz, D. K. J. E., Daly, J. W., Fredholm, B. B., Adenosine receptor ligands: differences with acute and chronic treatment. *Trends Pharmacol. Sci.* **1996**, *17*, 108–113.

13 Tracey, W. R., Magee, W. P., Oleynek, J. J., Hill, R. J., Smith, A. H., Flynn, D. M., Knight, D. R., Novel N^6-substituted adenosine 5′-*N*-methyluronamides with high selectivity for human adenosine A_3 receptors reduce ischemic myocardial injury. *Am. J. Physiol. Heart Circ. Physiol.* **2003**, *285*, H2780–H2787.

14 Fishman, P.,Bar-Yehuda, S.,Madi, L., Adenosine, tumors, and immunity. In *Adenosine Receptors, Therapeutic Aspects for Inflammatory and Immune Diseases* (eds G. Haskó, B. N. Cronstein, C. Szabó), CRC Press, Boca Raton, **2006**, pp. 299–312.

15 Madi, L., Cohn, S., Ochaion, A., Bar-Yehuda, S., Barer, F., Fishman, P., Over-expression of A_3 adenosine receptor in PBMC of rheumatoid arthritis patients: Involvement of NF-κB in mediating receptor level. *J. Rheumatol.* **2007**, *34*, 20–26.

16 Yang, H., Avila, M. Y., Peterson-Yantorno, K., Coca-Prados, M., Stone, R. A., Jacobson, K. A., Civan, M. M., The cross-species A_3 adenosine-receptor antagonist MRS 1292 inhibits adenosine-triggered human nonpigmented ciliary epithelial cell fluid release and reduces mouse intraocular pressure. *Curr. Eye Res.* **2005**, *30*, 747–754.

17 Costanzi, S., Ivanov, A. A., Tikhonova, I. G., Jacobson, K. A., Structure and function of G protein-coupled receptors studied using sequence analysis, molecular modeling, and receptor engineering: Adenosine receptors. *Frontiers Drug Design Discovery* **2007**, *3*, 63–79.

18 Linden, J., Adenosine in tissue protection and tissue regeneration. *Mol. Pharmacol.* **2005**, *67*, 1385–1387.

19 Holdgate, A., Foo, A., Adenosine versus intravenous calcium channel antagonists for the treatment of supraventricular tachycardia in adults. *Cochrane Database Syst. Rev.* **2006**, *4*, CD005154.

20 Peterman, C., Sanoski, C. A., Tecadenoson: a novel, selective A_1 adenosine receptor agonist. *Cardiol. Rev.* **2005**, *13*, 315–321.

21 Giffin, N. J., Kowacs, F., Libri, V., Williams, P., Goadsby, P. J., Kaube, H., Effect of the adenosine A_1 receptor agonist GR79236 on trigeminal nociception with blink reflex recordings in healthy human subjects. *Cephalagia* **2003**, *23*, 287–292.

22 Song, Y.,Wu, L.,Shryock, J. C., Belardinelli, L.,Selective attenuation of isoproterenol-stimulated arrhythmic activity by a partial agonist of adenosine A_1 receptor. *Circulation* **2002**, *105*, 118–123.

23 Cox, B. F., Clark, K. L., Perrone, M. H., Welzel, G. E., Greenland, B. D., Colussi, D. J., Merkel, L. A., Cardiovascular and metabolic effects of adenosine A_1-receptor agonists in streptozotocin-treated rats. *J. Cardiovasc. Pharmacol.* **1997**, *29*, 417–426.

24 Bayes, M., Rabasseda, X., Prous, J. R., Gateways to clinical trials. *Methods*

Find. Exp. Clin. Pharmacol. **2005**, *27*, 331–372.

25 GlaxoSmithKline, The study of GW493838, an adenosine A_1 agonist, in peripheral neuropathic pain. http://clinicaltrials.gov/show/NCT00376454.

26 Zeraati, M., Mirnajafi-Zadeh, J., Fathollahi, Y., Namvar, S., Rezvani, M. E., Adenosine A_1 and A_{2A} receptors of hippocampal CA1 region have opposite effects on piriform cortex kindled seizures in rats. *Seizure* **2006**, *15*, 41–48.

27 Naganuma, M., Wiznerowicz, E. B., Lappas, C. M., Linden, J., Worthington, M. T., Ernst, P. B., Cutting edge: Critical role for A_{2A} adenosine receptors in the T-cell-mediated regulation of colitis. *J. Immunol.* **2006**, *177*, 2765–2769.

28 Victor-Vega, C., Desai, A., Montesinos, M. C., Cronstein, B. N., Adenosine A_{2A} receptor agonists promote more rapid wound healing than recombinant human platelet-derived growth factor (Becaplermin gel). *Inflammation* **2002**, *26*, 19–24.

29 Gao, Z. G., Mamedova, L., Chen, P., Jacobson, K. A., 2-Substituted adenosine derivatives: Affinity and efficacy at four subtypes of human adenosine receptors. *Biochem. Pharmacol.* **2004**, *68*, 1985–1993.

30 Rimondini, R., Ferre, S., Ogren, S. O., Fuxe, K., Adenosine A_{2A} agonists: a potential new type of atypical antipsychotic. *Neuropsychopharmacology* **1997**, *17*, 82–91.

31 Fuxe, K., Canals, M., Torvinen, M., *et al.*, Intramembrane receptor–receptor interactions: a novel principle in molecular medicine. *J. Neural Transm.* **2007**, *114*, 49–75.

32 Filip, M.,Frankowska, M.,Zaniewska, M., Przegalinski, E.,Muller, C. E.,Agnati, L., Franco, R.,Roberts, D. C. S.,Fuxe, K., Involvement of adenosine A_{2A} and dopamine receptors in the locomotor and sensitizing effects of cocaine.*Brain Res.* **2006**, *1077*, 67–80.

33 Hutchison, A. J., Webb, R. L., Oei, H. H., Ghai, G., Zimmerman, M. B., Williams, M., CGS21680C, an A_2 selective adenosine receptor agonist with preferential hypotensive activity. *J. Pharmacol. Exp. Ther.* **1989**, *251*, 47–55.

34 Hourani, S. M., Purinoceptors and platelet aggregation. *J. Auton. Pharmacol.* **1996**, *16*, 349–352.

35 Gordi, T., Frohna, P., Sun, H. L., Wolff, A., Belardinelli, L., Lieu, H., A population pharmacokinetic/pharmacodynamic analysis of regadenoson, an adenosine A_{2A}-receptor agonist, in healthy male volunteers. *Clin. Pharmacokinet.* **2006**, *45*, 1201–1212.

36 Barrett, R. J., Lamson, M. J., Johnson, J., Smith, W. B., Pharmacokinetics and safety of binodenoson after intravenous dose escalation in healthy volunteers. *J. Nucl. Cardiol.* **2005**, *12*, 166–171.

37 Ohta, A., Sitkovsky, M., Role of G-protein-coupled adenosine receptors in downregulation of inflammation and protection from tissue damage. *Nature* **2001**, *414*, 916–920.

38 Cerqueira, M. D., Advances in pharmacologic agents in imaging: new A_{2A} receptor agonists. *Curr. Cardiol. Rep.* **2006**, *8*, 119–122.

39 Lappas, C. M., Sullivan, G. W., Linden, J., Adenosine A_{2A} agonists in development for the treatment of inflammation. *Expert Opin. Investig. Drugs* **2005**, *14*, 797–806.

40 Lappas, C. M., Day, Y. J., Marshall, M. A., Engelhard, V. H., Linden, J., Adenosine A_{2A} receptor activation reduces hepatic ischemia reperfusion injury by inhibiting CD1d-dependent NKT cell activation. *J. Exp. Med.* **2006**, *203*, 2639–2648.

41 van den Berge, M., Hylkema, M. N., Versluis, M., Postma, D. S., Role of adenosine receptors in the treatment of asthma and chronic obstructive pulmonary disease: recent developments. *Drugs in R&D* **2007**, *8*, 13–23.

42 Desai, A., Victor-Vega, C., Gadangi, S., Montesinos, M. C., Chu, C. C., Cronstein, B. N., Adenosine A_{2A} receptor stimulation increases angiogenesis by down-regulating production of the antiangiogenic matrix protein thrombospondin 1. *Mol. Pharmacol.* **2005**, *67*, 1406–1413.

43 Cavalcante, I. C., Castro, M. V., Barreto, A. R., Sullivan, G. W., Vale, M., Almeida, P. R., Linden, J., Rieger, J. M., Cunha, F. Q., Guerrant, R. L., Ribeiro, R. A., Brito, G. A., Effect of novel A_{2A} adenosine receptor agonist ATL 313 on *Clostridium difficile* toxin A-induced murine ileal enteritis. *Infect. Immun.* **2006**, *74*, 2606–2612.

44 Sun, C. X., Zhong, H., Mohsenin, A., Morschl, E., Chunn, J. L., Molina, J. G., Belardinelli, L., Zeng, D., Blackburn, M. R., Role of A_{2B} adenosine receptor signaling in adenosine-dependent pulmonary inflammation and injury. *J. Clin. Invest.* **2006**, *116*, 2173–2182.

45 Yang, D., Zhang, Y., Nguyen, H. G., Koupenova, M., Chauhan, A. K., Makitalo, M., Jones, M. R., Hilaire, C. St., Seldin, D. C., Toselli, P., Lamperti, E., Schreiber, B. M., Gavras, H., Wagner, D. D., Ravid, K., The A_{2B} adenosine receptor protects against inflammation and excessive vascular adhesion. *J. Clin. Invest.* **2006**, *116*, 1913–1923.

46 Hua, X., Kovarova, M., Chason, K. D., Nguyen, M., Koller, B. H., Tilley, S. L., Enhanced mast cell activation in mice deficient in the A_{2B} adenosine receptor. *J. Exp. Med.* **2007**, *204*, 117–128.

47 Feoktistov, I., Ryzhov, S., Zhong, H., Goldstein, A. E., Matafonov, A., Zeng, D., Biaggioni, I., Hypoxia modulates adenosine receptors in human endothelial and smooth muscle cells toward an A_{2B} angiogenic phenotype. *Hypertension* **2004**, *44*, 649–654.

48 Kong, T., Westerman, K. A., Faigle, M., Eltzschig, H. K., Colgan, S. P., HIF-dependent induction of adenosine A_{2B} receptor in hypoxia. *FASEB J.* **2006**, *20*, 2242–2250.

49 Wakeno, M., Minamino, T., Seguchi, O., Okazaki, H., Tsukamoto, O., Okada, K., Hirata, A., Fujita, M., Asanuma, H., Kim, J., Komamura, K., Takashima, S., Mochizuki, N., Kitakaze, M., Long-term stimulation of adenosine A_{2B} receptors begun after myocardial infarction prevents cardiac remodeling in rats. *Circulation* **2006**, *114*, 1923–1932.

50 Peyot, M. L., Gadeau, A. P., Dandre, F., Belloc, I., Dupuch, F., Desgranges, C., Extracellular adenosine induces apoptosis of human arterial smooth muscle cells via A(2b)-purinoceptor. *Circ. Res.* **2000**, *86*, 76–85.

51 Chiang, P. H., Wu, S. N., Tsai, E. M., Wu, C. C., Shen, M. R., Huang, C. H., Chiang, C. P., Adenosine modulation of neurotransmission in penile erection. *Br. J. Clin. Pharmacol.* **1994**, *38*, 357–362.

52 Gao, Z. G., Joshi, B. V., Klutz, A., Kim, S. K., Lee, H. W., Kim, H. O., Jeong, L. S., Jacobson, K. A., Conversion of A_3 adenosine receptor agonists into selective antagonists by modification of the 5′-ribofuran-uronamide moiety. *Bioorg. Med. Chem. Lett.* **2006**, *16*, 596–601.

53 Chen, Y., Corriden, R., Inoue, Y., Yip, L., Hashiguchi, N., Zinkernagel, A., Nizet, V., Insel, P. A., Junger, W. G., ATP release guides neutrophil chemotaxis via $P2Y_2$ and A_3 receptors. *Science* **2006**, *314*, 1792–1795.

54 Mabley, J., Soriano, F., Pacher, P., Haskó, G., Maårton, A., Wallace, R., Salzman, A., Szabó, C., The adenosine A_3 receptor agonist, N^6-(3-iodobenzyl)-adenosine-5′-*N*-methyluronamide, is protective in two murine models of colitis. *Eur. J. Pharmacol.* **2003**, *466*, 323–329.

55 Gurman, J., Yu, J. G., Suntres, Z., Bozarov, A., Cooke, H., Javed, N., Auer, H., Palatini, J., Hassanain, H. H., Cardounel, A. J., Javed, A., Grants, I., Wunderlich, J. E., Christofi, F. L., $ADOA_3R$ as a therapeutic target in experimental colitis: proof by validated high-density oligonucleotide microarray analysis. *Inflamm. Bowel Dis.* **2006**, *12*, 766–789.

56 Fedorova, I. M., Jacobson, M. A., Basile, A., Jacobson, K. A., Behavioral characterization of mice lacking the A_3

adenosine receptor: Sensitivity to hypoxic neurodegeneration. *Cell. Mol. Neurobiol.* **2003**, *23*, 431–447.

57 Matot, I., Weininger, C. F., Zeira, E., Galun, E., Joshi, B. V., Jacobson, K. A., A_3 adenosine receptors and mitogen-activated protein kinases in lung injury following *in vivo* reperfusion. *Critical Care* **2006**, *10*, R65.

58 Xu, Z., Jang, Y., Mueller, R. A., Norfleet, E. A., IB-MECA and cardioprotection. *Cardiovasc. Drug Rev.* **2006**, *24*, 227–238.

59 Shneyvais, V., Mamedova, L., Zinman, T., Jacobson, K. A., Shainberg, A., Activation of A_3 adenosine receptor protects against doxorubicin-induced cardiotoxicity. *J. Mol. Cell. Cardiol.* **2001**, *33*, 1249–1261.

60 Chung, H., Jung, J. Y., Cho, S. D., Hong, K. A., Kim, H. J., Shin, D. H., Kim, H., Kim, H. O., Lee, H. W., Jeong, L. S., Kong, G., The antitumor effect of LJ-529, a novel agonist to A_3 adenosine receptor, in both estrogen receptor-positive and estrogen receptor-negative human breast cancers. *Mol. Cancer Ther.* **2006**, *5*, 685–692.

61 Madi, L., Ochaion, A., Rath-Wolfson, L., Bar-Yehuda, S., Erlanger, A., Ohana, G., Harish, A., Merimski, O., Barer, F., Fishman, P., The A_3 adenosine receptor is highly expressed in tumor vs. normal cells: Potential target for tumor growth inhibition. *Clin. Cancer Res.* **2004**, *10*, 4472–4479.

62 Baraldi, P. G., Preti, D., Tabrizi, M. A., Fruttarolo, F., Romagnoli, R., Carrion, M. D., Cara, L. C. L., Moorman, A. R., Varani, K., Borea, P. A., Synthesis and biological evaluation of novel 1-deoxy-1-6-((hetero)arylcarbonyl)hydrazino-9*H*-purin-9-yl-*N*-ethyl-β-D-ribofuranuronamide derivatives as useful templates for the development of A_{2B} adenosine receptor agonists. *J. Med. Chem.* **2007**, *50*, 374–380.

63 Adachi, H., Palaniappan, K. K., Ivanov, A. A., Bergman, N., Gao, Z. G., Jacobson, K. A., Structure-activity relationships of 2, N^6,5′-substituted adenosine derivatives with potent activity at the A_{2B} adenosine receptor. *J. Med. Chem.* **2007**, *50*, 1810–1827.

64 Ueeda, M., Thompson, R. D., Arroyo, L. H., Olsson, R. A., 2-Aralkyloxyadenosines: Potent and selective agonists at the coronary artery A_2 adenosine receptor. *J. Med. Chem.* **1991**, *34*, 1340–1344.

65 Zablocki, J. A., Wu, L., Shryock, J., Belardinelli, L., Partial A_1 adenosine receptor agonists from a molecular perspective and their potential use as chronic ventricular rate control agents during atrial fibrillation. *Curr. Topics Med. Chem.* **2004**, *4*, 839–854.

66 Jeong, L. S., Lee, H. W., Jacobson, K. A., Kim, H. O., Shin, D. H., Lee, J. A., Gao, Z. G., Lu, C., Duong, H. T., Gunaga, P., Lee, S. K., Jin, D. Z., Chun, M. W., Moon, H. R., Structure–activity relationships of 2-chloro-N^6-substituted-4′-thioadenosine-5′-uronamides as highly potent and selective agonists at the human A_3 adenosine receptor. *J. Med. Chem.* **2006**, *49*, 273–281.

67 Kim, S. K., Jacobson, K. A., Three-dimensional quantitative structure–activity relationship of nucleosides acting at the A_3 adenosine receptor: Analysis of binding and relative efficacy. *J. Chem. Inf. Model.* **2007**, *47*, 1225–1233.

68 Ge, Z. D., Peart, J. N., Kreckler, L. M., Wan, T. C., Jacobson, M. A., Gross, G. J., Auchampach, J. A., Cl-IB-MECA 2-chloro-N^6-(3-iodobenzyl)adenosine-5′-N-methylcarboxamide reduces ischemia/reperfusion injury in mice by activating the A_3 adenosine receptor. *J. Pharmacol. Exp. Ther.* **2006**, *319*, 1200–1210.

69 Philipp, S., Yang, X. M., Cui, L., Davis, A. M., Downey, J. M., Cohen, M. V., Postconditioning protects rabbit hearts through a protein kinase C-adenosine A_{2b} receptor cascade. *Cardiovasc. Res.* **2006**, *70*, 308–314.

70 Albrecht, B., Krahn, T., Philipp, S., Rosentreter, U., Cohen, M., Downey, J., Selective A_{2b} receptor activation mimics postconditioning in a rabbit infarct model. *Circulation* **2006**, *114* (Suppl. II), II-14–II-15.

71 Solenkova, N. V., Solodushko, V., Cohen, M. V., Downey, J. M., Endogenous adenosine protects preconditioned heart during early minutes of reperfusion by activating Akt. *Am. J. Physiol.* **2006**, *290*, H441–H449.

72 Eckle, T., Krahn, T., Grenz, A., Köhler, D., Mittelbronn, M., Ledent, C., Jacobson, M. A., Osswald, H., Thompson, L. F., Unertl, K., Eltzschig, H. K., Cardioprotection by ecto-5′-nucleotidase (CD73) and A_{2B} adenosine receptors. *Circulation*, **2007**, *115*, 1581–1590.

73 Baraldi, P. G., Preti, D., Tabrizi, M. A., Fruttarolo, F., Saponaro, G., Baraldi, S., Romagnoli, R., Moorman, A. R., Gessi, S., Varani, K., Borea, P. A., N^6-(Hetero)aryl/(cyclo)alkyl-carbamoyl-methoxy-phenyl-(2-chloro)-5′-*N*-ethylcarboxamido-adenosines: The first example of adenosine-related structures with potent agonist activity at the human A_{2B} adenosine receptor. *Bioorg. Med. Chem.* **2007**, *15*, 2514–2527.

74 Jeong, L. S., Choe, S. A., Gunaga, P., Kim, H. O., Lee, H. W., Lee, S. K., Tosh, D., Patel, A., Palaniappan, K. K., Gao, Z. G., Jacobson, K. A., Moon, H. R. Discovery of a new nucleoside template for human A_3 adenosine receptor ligands: D-4′-thioadenosine derivatives without 4′-hydroxymethyl group as highly potent and selective antagonists. *J. Med. Chem.* **2007**, *50*, 3159–3162.

18 The Design of Forodesine HCl and Other Purine Nucleoside Phosphorylase Inhibitors

Philip E. Morris and Vivekanand P. Kamath

18.1 Introduction

Purine nucleoside phosphorylase (PNP) is a purine-metabolizing enzyme that catalyzes the reversible phosphorolysis of purine nucleosides such as 2′-deoxyguanosine (dGuo) and 2′-deoxyinosine to their respective bases and deoxyribose-α-1-phosphate [1–5]. In cells, PNP normally acts in the phosphorolytic direction, since the 6-oxopurine metabolic products are further metabolized. The importance of PNP to the integrity of the immune system became apparent with the description of a rare form of immune deficiency found in children who are genetically deficient in the PNP enzyme [6]. Children lacking the PNP enzyme have severe T-cell immunodeficiency while, in most cases, maintaining normal or high B-cell function. The biochemical mechanism underlying the T-cell selective immunosuppression in PNP-deficient patients has been extensively studied. Metabolically, these patients have a low uric acid concentration as they lack the necessary hypoxanthine or guanine substrates. These patients also have elevated levels of the nucleosides 2′-deoxyguanosine, 2′-deoxyinosine, inosine and guanosine as they lack the PNP enzyme necessary to catabolize these materials. While all nucleoside levels are elevated, only dGuo affects the T-cell population. As the concentration of dGuo increases within cells, it accumulates and is phosphorylated to 2′-deoxyguanosine phosphate (dGMP) by the enzyme 2′-deoxycytidine kinase. Subsequently, dGMP is converted to 2′-deoxyguanosine diphosphate (dGDP), which is then further converted to the triphosphate (dGTP) [7, 8]. The elevated dGTP level creates an imbalance in the endogenous nucleotide pool, which in turn induces apoptosis (Figure 18.1).

The unique sensitivity of human T cells to PNP deficiency is attributed to a relatively high level of kinase and a low level of nucleotidase activity compared to other cell types. This unique enzyme ratio is especially characteristic of immature T cells. Based upon these observations, several novel classes of PNP inhibitors have been designed for the treatment of T-cell-mediated diseases. Among the diseases

Modified Nucleosides: in Biochemistry, Biotechnology and Medicine. Edited by Piet Herdewijn

ISBN: 978-3-527-31820-9

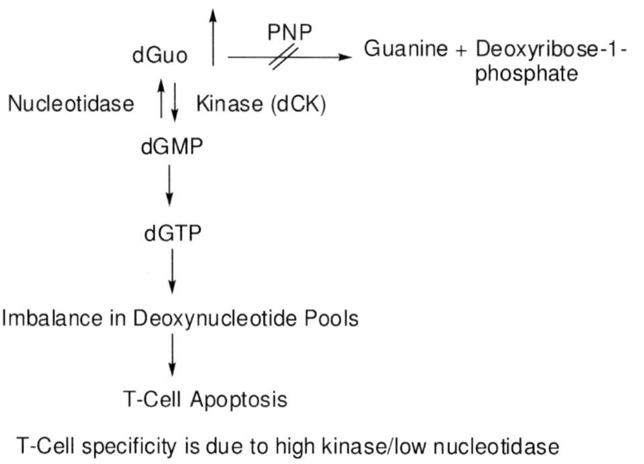

Figure 18.1 Inhibition of purine nucleoside phosphorylase (PNP) prevents the metabolism of 2′-deoxyguanosine (dGuo), which leads to an accumulation of 2′-deoxyguanosine triphosphate (dGTP). The T cells have high kinase activity which converts dGuo to dGTP. These cells are unable to metabolize dGTP because they have low nucleotidase activity. The increased dGTP concentration produces an imbalance in the deoxynucleotide pool, which leads to apoptosis.

which may be amenable to such treatment are T-cell cancers, and lymphomas such as cutaneous T-cell lymphoma (CTCL) and acute lymphoblastic leukemia (ALL). Patients suffering from autoimmune disorders such as psoriasis, Crohn's disease and rheumatoid arthritis resulting from the inappropriate activation of T cells as well as graft-versus-host disease (GVHD) are also possible therapeutic populations.

18.2 Purine Nucleoside Phosphorylase Enzyme Structure

The three-dimensional structure of human erythrocytic PNP was originally determined by X-ray crystallographic analysis at 3.2 Å resolution [9], and more recently the refined structure has been reported at 2.75 Å resolution [10]. In addition to being studied through X-ray crystallographic analyses, human erythrocytic PNP has also been sequenced [11]. The resolved structure shows that PNP is a trimer with a total mass of approximately 97 kDa and consisting of three identical subunits, each of which has an estimated mass of 30 to 33 kDa (Figure 18.2). Each subunit contains an eight-stranded mixed β-sheet and a five-stranded mixed β-sheet, which join to form a distorted β-barrel-type structure. Flanking this core are seven α-helices. Each of the enzyme's subunits contains an active site located near the subunit–subunit boundaries and each site involves five segments from each subunit and a short segment (Phe159) from the adjacent subunit. The active site is composed of three binding regions, namely the purine, sugar, and phosphate binding sites.

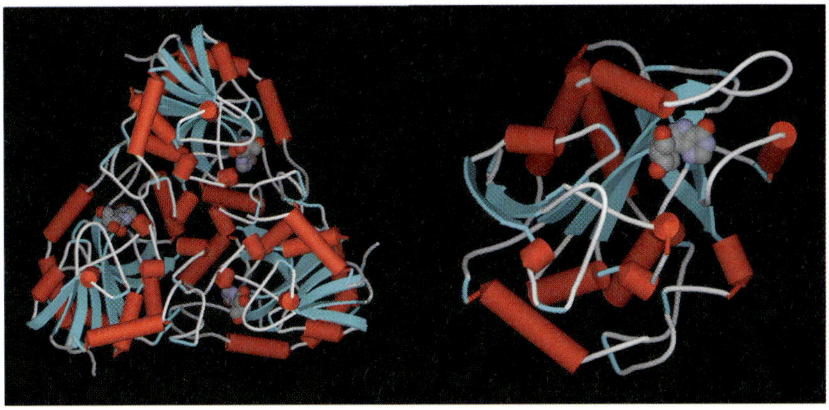

Figure 18.2 Purine nucleoside phosphorylase (PNP) is a trimeric structure with an estimated mass of 97 kDa. Left: A schematic of human PNP at 2.75 Å with inosine in the active site. The PNP trimer has three active sites which are located near the subunit boundaries; note that each active site contains an inosine molecule. Right: A PNP monomer. The α-helices are shown in red, and the β-sheets in blue.

18.2.1 Purine-Binding Site

The purine-binding pocket is the most deeply buried and, in human PNP, is specific for 6-oxopurines. This specificity is driven in part by formation of hydrogen bonds from N-1, 2-NH_2, O-6 and N-7 of the purine with residues Glu201, Thr242, and Asn243. This is clearly demonstrated in Figure 18.3, which compares the binding of the 6-oxopurine, guanine, and the 6-aminopurine, adenine.

The recent crystal structure of PNP at 2.75 Å suggests that the hydroxyl group of Thr242 functions as a hydrogen bond donor to the O-atom of the side-chain carbonyl

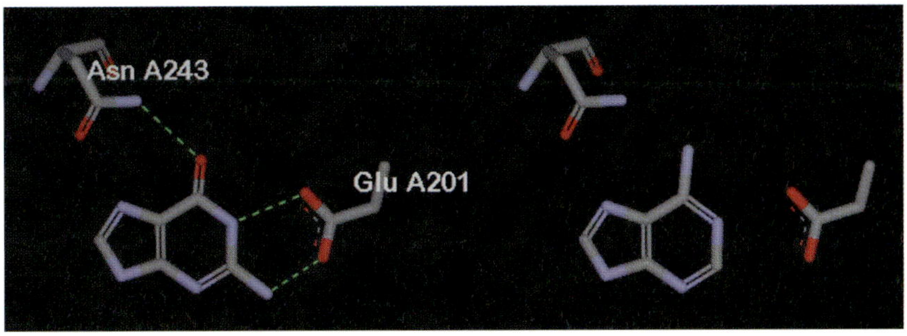

Figure 18.3 6-Oxopurines such as guanine (left) form an extensive hydrogen-bonding network with key residues in the purine binding site. In contrast, 6-aminopurines such as adenine (right) are unable to establish a similar network and, as a consequence, are not bound strongly in the active site.

of Asn243. This allows the amide NH_2 of Asn243 to donate a hydrogen bond to the C-6 carbonyl group, thus imparting specificity. The remaining residues that make up the purine-binding site, Ala116, Phe200, Val217, Met219, Gly218 and Thr242, are largely hydrophobic in nature.

18.2.2 Phosphate-Binding Site

The phosphate-binding site is composed of Ser33, Arg84, His86 and Ser220, and is located near the glycine-rich loop (residues 32–37). In addition, a H_2O molecule (313) is located in this site and, acting in concert with the hydroxyls of Ser33 and Ser220, accepts hydrogen bonds from the phosphate.

18.2.3 Sugar-Binding Site

The sugar-binding site is largely hydrophobic in nature; however, the amino acid residue Tyr88 contains a hydrogen-bonding group which interacts with O-3 of the β-D-ribofuranose ring of nucleosides. Additional hydrogen bonds are formed from the peptide N-atom of Met219 and the NH_2 of His86 with the other ribofuranose hydroxyls. One side of the ribose-binding site is composed of several aromatic amino acids, including Phe159 (from an adjacent subunit), Phe200, His86, and Tyr88. This hydrophobic pocket serves to orient the sugar to facilitate nucleophilic attack by phosphate with subsequent inversion at the anomeric center.

Access to the active site is controlled by a "swinging gate" made up of residues 241 to 260. This gate is closed in the native enzyme, but opens during substrate binding to accommodate the substrate or inhibitor. The maximum movement caused by substrate binding is observed at His257, which can be displaced outward by as much as 5–6 Å. The gate is anchored near the central β-sheet at one end and near the C-terminal helix at the other end. Movement of the gate is complex and, appears to involve a helical transformation near residues 257 to 261 [9, 12].

18.3 First-Generation PNP Inhibitors: Substrate Analogues

An X-ray crystallographic, structure-based drug design approach was used to design several closely related classes of inhibitors [13–16]. Early crystallographic analyses indicated that the N-9 atom of guanine and its derivatives was not involved in binding. As a result, initial synthetic investigations were centered around the closely related 9-deazaguanines (2-amino-3*H*,5*H*-pyrrolo[3,2-*d*]pyrimidin-4-ones), in which the N-9 atom has been replaced by a C-atom. The first-generation PNP inhibitors had the general structure shown in Figure 18.4.

These derivatives containing a 9-deazaguanine ring also bind in the active site of PNP. Although similar in structure, 9-deazaguanines and guanines have different

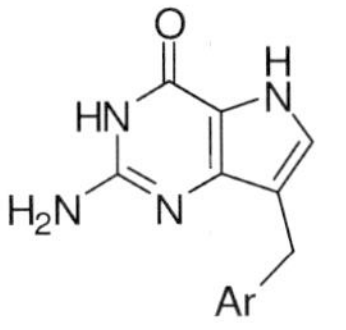

Figure 18.4 The first-generation PNP inhibitors consisted of a 9-deazaguanine ring with aromatic substituents attached via a methylene linker. In this structure, the N-9 atom is replaced by a C-atom to produce a chemically stable inhibitor.

binding motifs in the active site. These differences can be seen in the comparison of soak studies of the thienylmethyl derivatives of guanine and 9-deazaguanine (Figure 18.5) [12]. In the first case, the amino group of Asn243 donates hydrogen bonds with O-6 and N-7 of the purine. Asn243 is anchored in position by a hydrogen bond between its carbonyl moiety and the hydroxyl group of Thr242. In the second case, Asn243 shifts in position to form a more optimal hydrogen bond between its carbonyl oxygen and N-7, which is protonated in the 9-deazapurine analogues, while the hydrogen bond between the amino group of Asn243 and O-6 is maintained. The rearrangement of hydrogen bond donors and acceptors of N-7 results in a shift in the position of the carbonyl oxygen of Asn243 of about 2.5 Å, with a corresponding shift in the Thr242 to maintain a hydrogen bond with Asn243. The introduction of

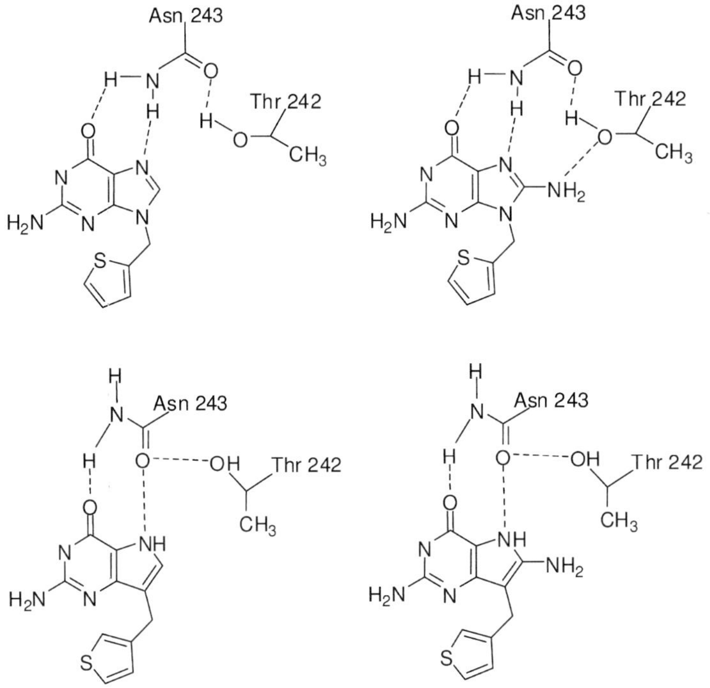

Figure 18.5 Hydrogen-bonding interactions of 9-(2-thienylmethyl)guanine (upper left), 9-(3-thienylmethyl)-9-deazaguanine (lower left), 8-amino-9-(2-thienylmethyl)guanine (upper right), and 8-amino-9-(3-thienylmethyl)-9-deazaguanine (lower right).

an 8-NH_2 group on a guanine ring enhances binding, whilst an 8-NH_2 group on a 9-deazaguanine ring diminishes binding. Soak studies of the thienylmethyl derivatives of 8-aminoguanine and 8-amino-9-deazaguanine reveal the reason for this variation [22, 23]. In the case of 8-aminoguanine, an additional hydrogen bond is formed with the hydroxyl of Thr242. However, in the case of 8-amino-9-deazaguanine, the hydroxyl of Thr242 is no longer available to hydrogen bond to the 8-amino group and, in fact, the methyl group of Thr242 is now in close contact with the 8-amino group – an unfavorable interaction. Consequently, the 8-amino group, which is beneficial for guanine analogues, is detrimental for 9-deazaguanines, with a differential of about three orders of magnitude in IC_{50} values.

Hypoxanthine (2-des-amino) analogues also bind in the PNP active site, but are relatively poor inhibitors when compared to the 2-amino derivatives. The hypoxanthine analogues lack the 2-amino group which plays an important role in binding the ring in the active site with a hydrogen bond to Glu201 (Figure 18.6). As a result, hypoxanthines are generally five- to 10-fold less potent than the corresponding 2-amino compounds when assayed against the human erythrocytic PNP enzyme. These differences in potency are not observed with calf spleen PNP and in this assay, the two ring systems appear to be equipotent [16].

In the first-generation inhibitors, the hydrophobic residues in the sugar pocket of the active site allow replacement of the sugar with a number of aryl and alicyclic groups that are optimally attached to the purine ring through a single methylene linker group. These aromatic groups form particularly strong interactions with the phenyl rings of Phe159 and Phe200 in the active site due to the formation of the classic herringbone (i.e., edge-to-face) arrangement [17] (see Figure 18.7). In this arrangement, the Phe159 residue is from the same subunit containing the inhibitor, whereas Phe200 is from the neighboring subunit. By comparison, acyclic substituents are very poor inhibitors of PNP, thereby underscoring the importance of a rigid group.

Among the first-generation inhibitors studied is BCX-34 (Peldesine), with an $IC_{50} = 35$ nM against human erythrocytic PNP.

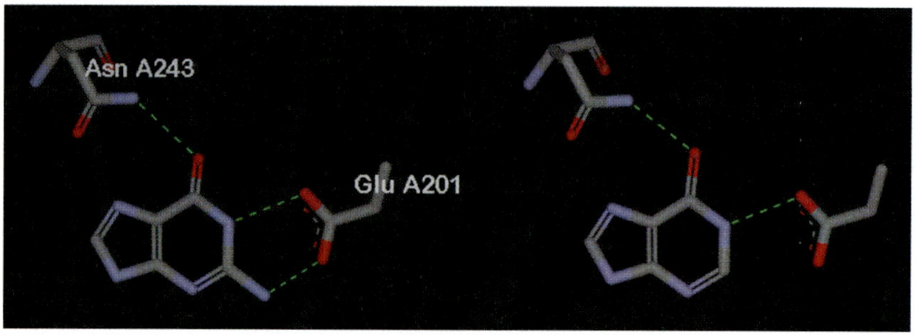

Figure 18.6 The NH_2 group at C-2 of guanines (left) interacts with Glu201. In the corresponding hypoxanthine structure (right), the 2-NH_2 is absent and is replaced by an H atom. As a result, guanines typically bind three- to five-fold stronger than the corresponding hypoxanthine structure.

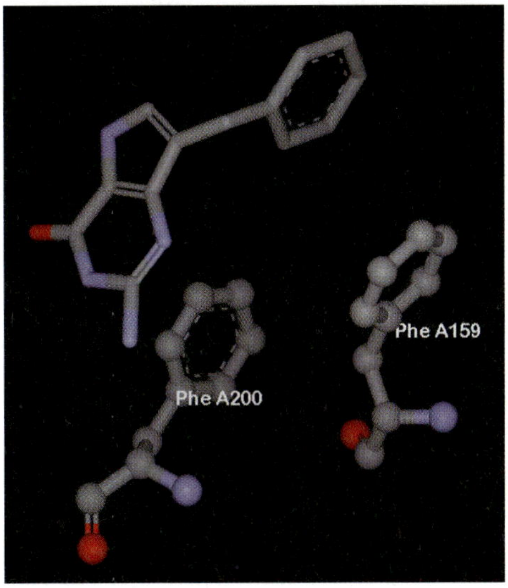

Figure 18.7 The sugar-binding pocket has hydrophobic residues. The first-generation inhibitors were designed with aromatic (or cycloaliphatic) groups attached to the 9-deazapurine rings. A single methylene linker provides the optimal distance between the purine and aromatic group. Note the edge-to-face (herringbone) arrangement of the aromatic group of the inhibitor and residues Phe159 and Phe200.

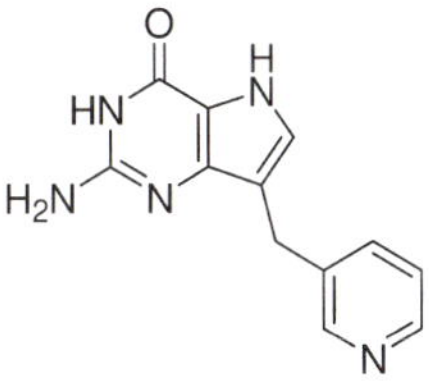

BCX-34 (Peldesine)

As for other inhibitors from this class, soak studies indicate that each monomeric unit of the PNP trimer contains a BCX-34 molecule with the 3-pyridinyl moiety occupying the sugar-binding pocket. Complete inhibition requires three moles of inhibitor per mole of PNP. BCX-34 is the only first-generation PNP inhibitor to have been evaluated for efficacy in human clinical trials. In two separate Phase II clinical trials conducted in patients suffering from cutaneous T-cell lymphoma and in patients with psoriasis, BCX-34 did not demonstrate any clinical benefit [18]. Despite its low (nanomolar) activity, additional studies showed that BCX-34 could produce only a transient increase in the dGuo concentration which was insufficient to produce an accumulation of dGTP. Further measurements of the BCX-34/PNP complex revealed that the inhibitor had a very rapid off-rate [19]. Complete and sustained suppression of the enzyme is required in order for dGuo accumulation to occur. This rapid off-rate with BCX-34, in conjunction with the abundance of

PNP in humans, makes it difficult to mimic what is observed biochemically in the PNP-deficient children who typically have only ≤5% of the normal PNP activity. As a consequence of this, no further development of inhibitors from this class has been conducted. This underscores the need to develop novel inhibitors which are more potent and which have a significantly longer half-life on the enzyme.

18.3.1
Chemistry of First-Generation PNP Inhibitors

An early synthetic route was developed for synthesizing first-generation inhibitors [20]; this route provided a general method for producing analogues with varying aromatic or cycloalkyl substituents in the 9-position (Scheme 18.1). The desired aromatic or alicyclic substituent was introduced by using the corresponding aldehyde as a starting material. Base-catalyzed condensation of the requisite aldehyde **1** with 3,3-dimethoxypropionitrile gave adduct **2**. Typically, these adducts were isolated as oils which were readily converted to the crystalline unsaturated aldehydes **3** in 6 M HCl. Catalytic hydrogenation of **3** gave the corresponding cyanoaldehyde derivatives **4**. This sequence was seen to function well with a variety of substituted aldehydes including heteroaromatic and alicyclic aldehydes, although the yield was poorer in the case of aldehydes containing strong electron-withdrawing groups such as a 4-CF_3 group [21]. This sequence also did not function well in the case of pyridine-3-carboxaldehyde, which was the starting aldehyde for BCX-34 [22]. As a result, the sequence was modified slightly (Scheme 18.2), whereby pyridine-3-carboxaldehyde was condensed with cyanoacetic acid to afford the acrylonitrile intermediate **5**. Hydrogenation of this intermediate gave the nitrile adduct **6** in 54% yield. In this route, the aldehyde group was introduced by formylation of **6** with sodium hydride and ethylformate.

The aldehyde was readily isolated as a stable sodium enolate salt, **7**, and this easily condensed with diethyl aminomalonate to furnish the enamine **8**. Similarly, cyanoaldehydes **4** were converted to the corresponding enamines under the same reaction conditions. Once the enamines were in hand, the sequence was completed in the same manner for all analogues.

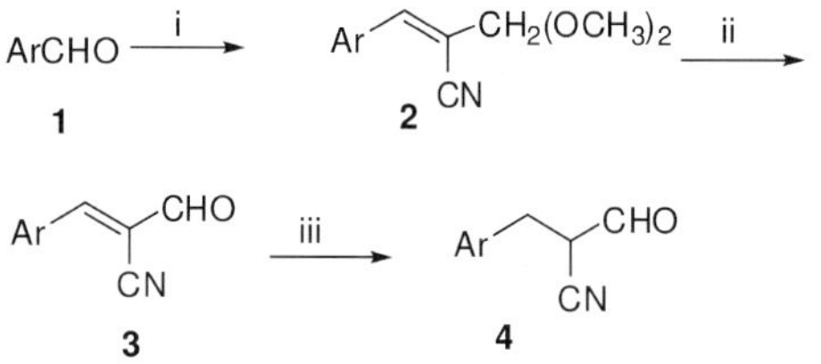

(i) $NCCH_2CH(OCH_3)_2$, NaOMe; (ii) 6N HCl; (iii) H_2, Pd/C.

Ar	Yield (4)*
Phenyl	68%
1-Naphthyl	68%
2-Naphthyl	61%
4-Biphenyl	61%
4-Isopropylbenzyl	60%
4-CF_3-Phenyl	23%
Cyclohexyl	76%
2-Furanyl	67%
3-Thienyl	69%

*Yields not optimized

Scheme 18.1

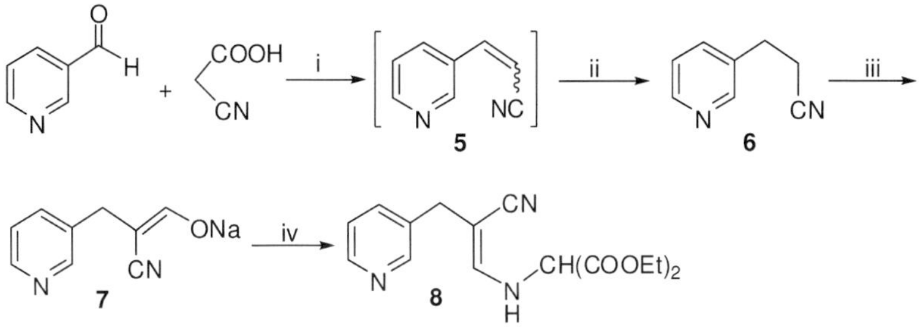

(i) Toluene, pyridine; (ii) H_2, Pd/C; (iii) NaH, HCOOEt; (iv) diethyl aminomalonate.

Scheme 18.2

The enamines were converted to pyrrole **9** by treatment with NaOMe in MeOH at ambient temperature. With the key pyrrole in place, all that remained was annulation to the pyrrolo[3,2-d]pyrimidine ring. The simpler hypoxanthine derivatives (2-desamino) were easily prepared from the corresponding pyrrole **9** by condensation with formamidine acetate in ethanol. Although these reactions worked well, preparation of the guanine analogues (2-NH_2) was less straightforward. The obvious and most direct route for construction of this ring is condensation with an appropriate guanylating reagent such as a guanidine salt, cyanamide, or *S*-methylisothiourea. However, the 3-NH_2 group of pyrrole **9** was found to be surprisingly unreactive towards these reagents. Guanylation was achieved by reaction with the more reactive carbodiimide intermediate, which was readily generated from the treatment of 1,3-dicarbomethoxy-2-methyl-2 methoxypseudourea with mild acid to give **10** (Scheme 18.3). Annulation to **11** was effected by treatment of **10** with NaOMe. Removal of the remaining carbamate group was accomplished with aqueous sodium hydroxide to give **12**.

Other analogues related to the first-generation inhibitors have also been reported by Morris *et al.* [23, 24].

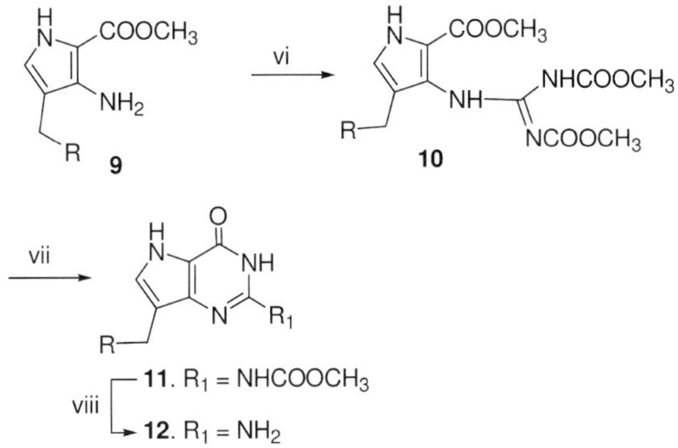

(vi) 1,3-dicarbomethoxy-2-methyl-2-methoxypseudourea;
(Vii) NaOMe; (viii) NaOH.

Scheme 18.3

18.4
Second-Generation PNP Inhibitors: Transition-State Inhibitors

Schramm and Horenstein have used experimentally determined kinetic isotope effects to describe a geometric model of the transition state for the nucleoside hydrolase-catalyzed hydrolysis of inosine [25, 26]. The study revealed that, at the transition state, the ribosyl moiety exists as an oxycarbenium ion with the charge distributed over its surface. Schramm further showed that the oxycarbenium ion character could be mimicked with the stable iminoribitol derivative 1,4-dideoxy-1,4-iminoribitol which, in its protonated form, has a similar charge distribution.

Mechanistically, the nucleoside hydrolase reaction of inosine shares many features with the phosphorolysis of nucleosides by PNP, and as such is an ideal model system for the design of PNP transition-state inhibitors. By applying transition state theory and using kinetic isotope effects, Schramm solved the transition state for PNP and described its two key features [27]. During the transition state, the H-atom attached to N-7 of the purine has an elevated pK_a value. This feature provides an additional interaction within the PNP active site with Asn-243. Similar to nucleoside hydrolase, the ribosyl sugar ring exhibits a strong oxocarbenium ion character at the transition state (Figure 18.8).

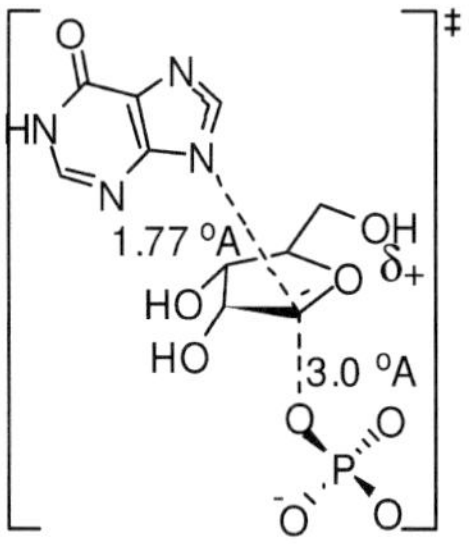

Figure 18.8 Transition-state analysis for the PNP-catalyzed phosphorolysis of inosine. The transition state is characterized by an elevated pK_a at the N-7 position. This arrangement allows the purine base to form a stabilizing interaction with Asn243 in the active site. The ribosyl moiety exhibits a strong oxycarbenium ion character.

Schramm combined these features into stable inhibitors which he termed Immucillin-H (hypoxanthine) and Immucillin-G (guanine).

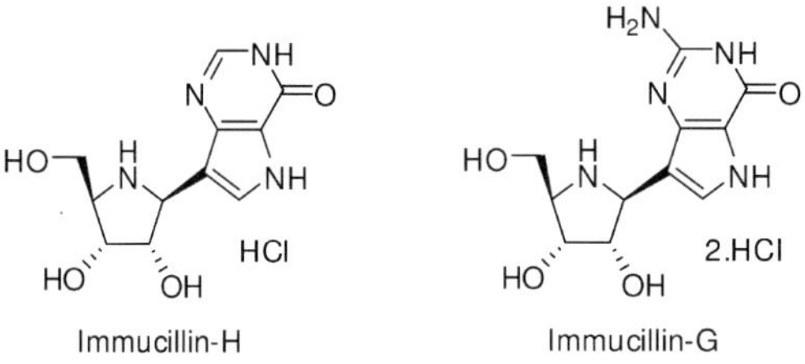

These inhibitors incorporate 9-deazapurines which are also key structural features of the first-generation inhibitors. These rings are protonated at N-7, and this position provides the interaction with Asn243.

In contrast to the first-generation analogues, these inhibitors are slow-onset, tight-binding inhibitors with activities ranging from 42 pM for the 9-deazaguanine derivative to 56 pM for the 9-deazahypoxanthine analogue [28]. Complete inhibition of the PNP enzyme requires only one mole of Immucillin-H or Immucillin-G per mole of PNP, whereas complete PNP inhibition with the first-generation analogues requires three moles of inhibitor per mole of PNP. This one-third-the-sites binding has also been identified in F_1 ATP synthase, which shares some properties with PNP [29]. In contrast to BCX-34, the dissociation of Immucillin-H from the complex is slow [27]. The ability of this class to remain bound to the enzyme for an extended period is one characteristic which makes them useful for clinical development. In a SCID (severe combined immunodeficiency disease) mouse model, Immucillin-H was shown to have immunosuppressive properties similar to cyclosporine [30]. Moreover, the pharmacokinetic properties of Immucillin-H following intravenous and oral dosing to primates have also been reported [31].

18.4.1
Chemistry of Second-Generation PNP Inhibitors

A lengthy linear synthesis (more than 16 steps) was initially developed for the preparation of the Immucillins [32]. In this approach, the iminoribitol moiety was constructed at an early stage, followed by sequential elaboration of the pyrrole and then the pyrimidine ring. As shown in Scheme 18.4 the iminoribitol **13** was derived from D-gulano-1,4-lactone in 10 steps using standard carbohydrate chemistry [33, 34]. Chlorination of **13** with *N*-chlorosuccinimide in pentane gave the *N*-chloro derivative which was dehydrohalogenated at −78 °C with lithium tetramethylpiperidide (LiTMP) to afford the imine **14**. The addition of lithiated acetonitrile to imine **14** occurred selectively from the β-face to give **15**. Unfortunately, the yield of the initial reaction was poor as it generated considerable amounts of the disubstituted cyanomethyl compound; however, the yield was later improved by performing the reaction under dilute conditions with an excess of the lithiated reagent. This was followed by protection of the N-atom as the N-Boc derivative to give compound **16**. Treatment of compound **16** with *tert*-butoxy-bis(*N*,*N*-dimethylamino)methane (Bredereck's reagent) in dimethylformamide (DMF) at 70 °C for 1 h gave the enamine **17**, which was further subjected to mild acid hydrolysis to afford **18**, as a syrup (72%, two-step reaction). Although compounds **17** and **18** can exist as an *E*/*Z* diastereomeric mixture, analysis by ^{1}H NMR indicated that they were single isomers. Compound **18** was treated with ethyl glycinate and sodium acetate at ambient temperature to furnish **19** as a mixture of *E*/*Z* diastereomers in 65% yield. It was first necessary to protect the enamine N prior to cyclization to the pyrrole under basic conditions. The N-atom was protected as the Cbz derivative by treating compound **19** with an excess of benzylchloroformate and 1,8-diazabicyclo(5.4.0)-undec-7-ene (DBU), which introduced the protecting group and promoted cyclization to pyrrole **20**. The Cbz protecting group was readily

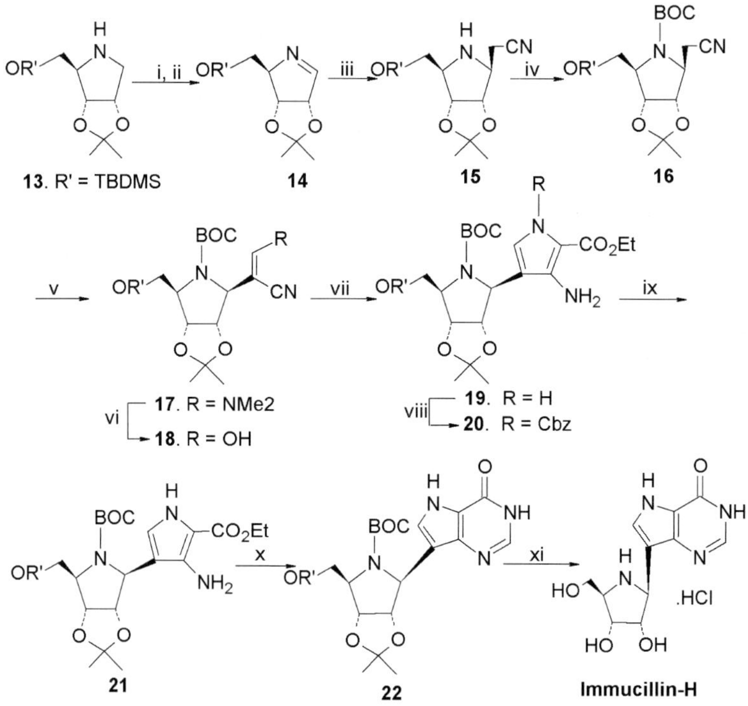

[a]Key: (i) NCS, pentane; (ii) LiTMP, column chromatography, (iii) $LiCH_2CN$, chromatography; (iv) $(BoC)_2O$, DCM; (v) $BuOCH(NMe_2)_2$, DMF; (vi) THF, HOAc, (vii) $H_2NCH_2CO_2EtHCl$, NaOAc, MeOH; (viii) $ClCO_2Bn$, DBU: (ix) H_2, Pd/C, EtOH; (x) formamidine acetate; (xi) conc. HCl.

Scheme 18.4

cleaved by hydrogenolysis to afford **21**, which was a key intermediate for producing Immucillin-H, Immucillin-G, and a limited number of analogues. Treatment of **21** with formamidine acetate in ethanol at reflux gave the deazapurine aza-C-glycoside, **22** in excellent yield (91%), and this intermediate was conveniently deprotected under acidic conditions to give Immucillin-H in 81% yield.

The 9-deazaguanine derivative was also accessible through intermediate **21**. Condensation with benzoyl isothiocyanate in dichloromethane, followed by methylation with methyl iodide and DBU, produced the isothiourea intermediate which was then converted under ammonolysis conditions to the protected 9-dezaguanine derivative. Deprotection under acidic conditions gave Immucillin-G. Additional chemistry with **21** afforded the 5′-deoxy derivatives. Other derivatives such as the 2′-deoxy derivatives were also prepared from this route by making earlier manipulations on the iminoribitol. More recently, the synthesis of the 2′-deoxy 9-deazahypoxanthine analogue and its pharmacokinetic and pharmacodynamic properties has been reported [35].

As a consequence of its low (picomolar) activity and the stability of the Immucillin-H/PNP complex, Immucillin-H was chosen for clinical development. In 2000, BioCryst Pharmaceuticals Inc. began development of Immucillin-H as BCX-1777.

In 2004, the United States Adopted Names Council (USAN) adopted the suffix "*desine*" for PNP inhibitors, and BCX-1777 became known as forodesine hydrochloride. Forodesine is currently undergoing advanced human clinical trials for the treatment of T-cell acute lymphoblastic leukemia (T-ALL) and cutaneous T-cell lymphoma (CTCL). The first "proof of concept" with forodesine was reported by Thomas *et al.* [36], who showed that the intravenous administration of forodesine ($40\,mg\,m^{-2}$ dose on Day 1, followed by $40\,mg\,m^{-2}$ every 12 h for Days 2 to 5) produced an increase in dGuo, with a concomitant increase in intracellular dGTP. As dGTP accumulated, a near-complete depletion of leukemic T cells was observed.

In 2004, forodesine was granted Orphan Drug status in the United States for the treatment of T-cell acute lymphocytic leukemia, T-cell non-Hodgkin lymphoma, chronic lymphocytic leukemia; and related leukemias including prolymphocytic leukemia, adult T-cell leukemia, and hairy cell leukemia. In 2005, forodesine was granted Fast Track designation by the Food & Drug Administration for the treatment of relapsed or refractory T-cell leukemias. In Europe, Orphan Drug Designation was granted by the Committee for Orphan Medicinal Products of the European Medicines Agency (EMEA) in 2006 for the treatment of ALL, following a submission by BioCryst's European partner Mundipharma; and in 2007 the EMEA granted a second Orphan Drug Designation for the treatment of CTCL. The use of PNP inhibitors for treating T-cell malignancies has also been reviewed [37].

18.4.2
Convergent Synthesis of Forodesine HCl

Since the linear approach of more than 16 steps provided only milligram quantities of the desired product, this method was considered to be impractical for scale-up. Therefore, there was a need for the development of an efficient convergent synthesis capable of producing the kilogram quantities of forodesine required for its development. Scheme 18.5 shows the retrosynthetic analysis for the convergent route which was developed [38]. The general method involved lithiation of the 9-bromo-9-deazahypoxanthine **23** and subsequent addition of the anion to imine **14**.

The key 9-bromo-9-deazahypoxanthine intermediate **23** was derived from 9-deazahypoxanthine **28**. The synthesis of **28** from isoxazole has been reported by Tyler *et al.* [39] (Scheme 18.6).

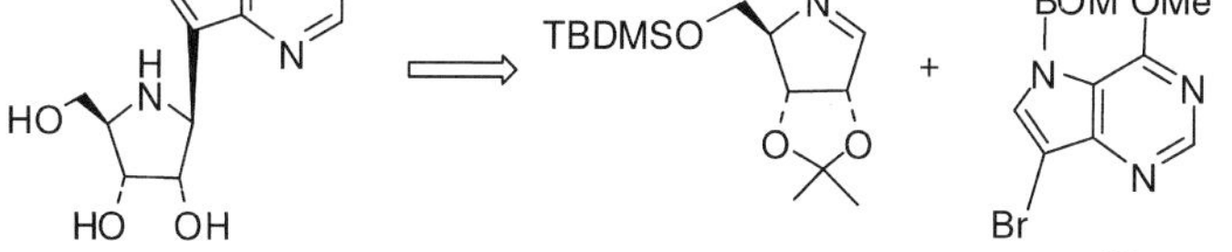

Forodesine

14

23

Scheme 18.5

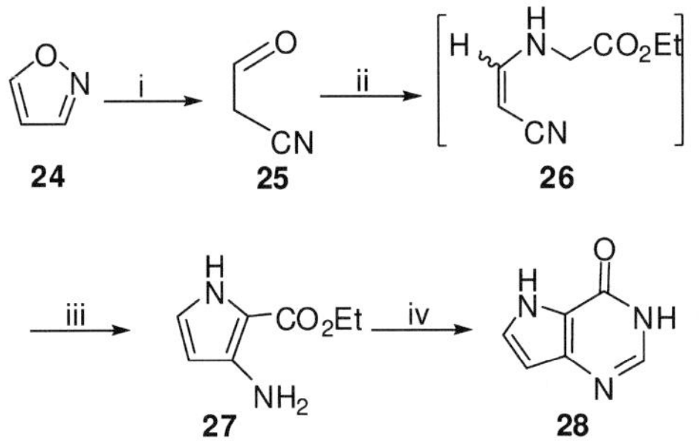

(i) NaOEt, EtOH; (ii) $H_2NCH(COO_2Et)_2$; (iii) NaOEt; (iv) formamidine acetate.

Scheme 18.6

In this synthesis, isoxazole **24** is opened under basic conditions to the isomeric 3-oxopropionitrile **25** which is condensed *in situ* with diethyl aminomalonate to afford an *E*/*Z* mixture of the corresponding enamine **26** as a viscous syrup. Treatment of the enamine **26** mixture with sodium ethoxide afforded the 3-aminopyrrole intermediate **27** as a syrup. Condensation of the 3-aminopyrrole with formamidine acetate at reflux in ethanol gave 9-deazahypoxanthine **28**. Although this route seemed attractive initially, it was could not easily be scaled-up as, surprisingly, isoxazole is not readily available on a commercial basis. In addition, the intermediates were syrups and not readily purified at the kilogram level. Another problem was that the 3-aminopyrrole **27** was air-sensitive, and attempts at using it without stringent purification for the following cyclization step gave poor yields of 9-deazahypoxanthine of varying quality [40].

In an attempt to circumvent these problems a novel scalable procedure (Scheme 18.7) was developed by Morris and Winslow [41]. In this approach, ethyl (ethoxymethylene) cyanoacetate (**29**) was condensed with diethyl aminomalonate and sodium methoxide in methanol at reflux to afford **30** as a tan-colored solid in 79%

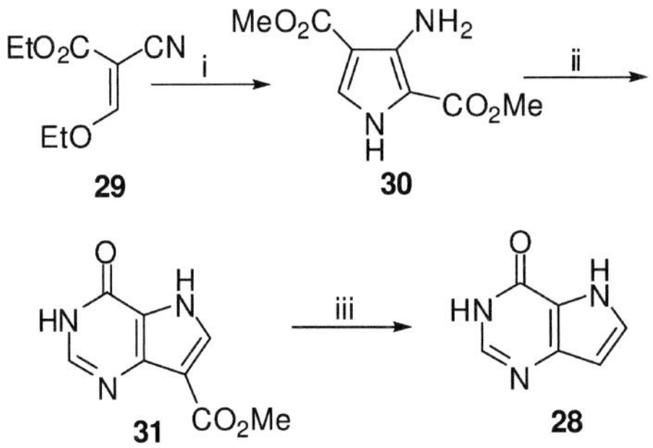

(i) NaOMe, $H_2CH(COO_2Et)_2$; (ii) formamidine acetate; (iii) 10% KOH.

Scheme 18.7

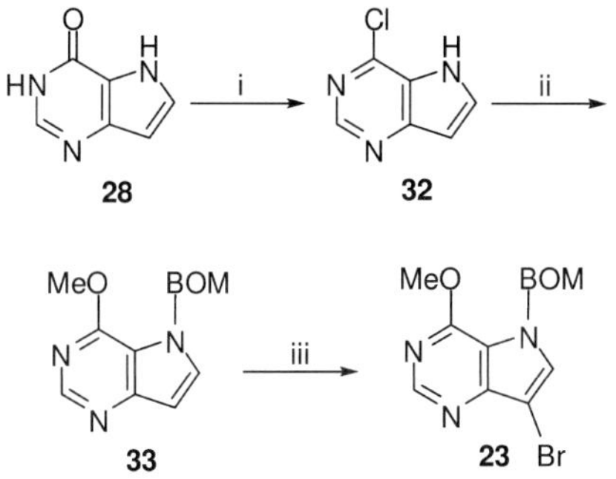

28 32 33 23

(i) $POCl_3$; (ii) a. NaH, BOMCl, THF; b. NaH, MeOH;
(iii) NBS, DCM.

Scheme 18.8

yield. The pyrrole **30** was readily isolated by distilling off most of the methanol and precipitating it by the addition of water to the crude reaction mixture. The isolated pyrrole was sufficiently pure to be used directly in the next step. Condensation of the pyrrole with formamidine acetate in ethanol under reflux conditions gave compound **31** as a solid in 72% yield. Saponification of **31** was efficiently performed with 10% aqueous KOH at reflux, and the resulting carboxylate intermediate spontaneously decarboxylated upon acidification to furnish compound **28** as a solid in 90% yield. By using this approach, more than one metric tonne of **28** has been prepared.

With 9-deazahypoxanthine **28** in hand, 9-bromo-9-deazahypoxanthine **23** was prepared in three steps (Scheme 18.8) [38]. Chlorination with $POCl_3$ gave the 6-chloro-derivative **32** in near-quantitative yield. Subsequent treatment of **32** with sodium hydride and benzyl chloromethyl ether, followed by a methanol quench, gave **33**; this in turn was converted to the requisite 9-bromo-9-dezazhypoxanthine **23** with *N*-bromosuccinimide.

The recently reported synthesis of L-lyxono-1,4-lactone from ribose has rendered the requisite iminoribitol readily available [42]. Although the Fleet method using D-gulono-1,4-lactone works well, the lactone is expensive and not readily available in bulk quantities. The D-ribose route relies on two key conversions which are depicted in Scheme 18.9. The first conversion involves the bromine oxidation of ribose to the corresponding lactone **34**, and its protection as the isopropylidene derivative **35**. This conversion was independently reported by Townsend and Morris [43]. The second key conversion involves a novel epimerization of the protected lactone; this was accomplished by conversion to the mesylate followed by base treatment to afford L-lyxono-1,4-lactone **36**. Standard carbohydrate chemistry based on the Fleet method was employed to convert the L-lyxono-1,4-lactone to the desired iminoribitol in five steps. This method is currently used to generate the iminoribitol, with lots in excess of one metric tonne having been prepared.

Scheme 18.10 describes the condensation of the lithiated purine to imine **14**. Lithiation of **23** with *n*-butyl lithium in ether:anisole at −78 °C gave the lithiated anion. Imine **14** was added to the cold anion and the mixture allowed to warm to 0 °C

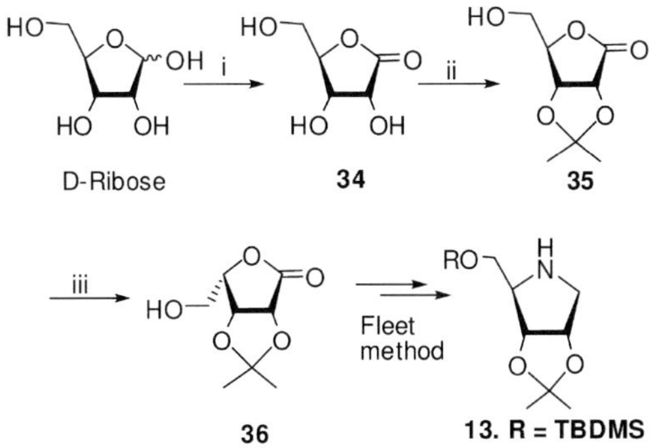

(i) Br_2, H_2O; (ii) acetone, conc. H_2SO_2; (ii) a. MsCl, Et_3N; b. KOH, H_2O

Scheme 18.9

before being quenched with water. Addition of the anion occurs with remarkable stereoselectivity exclusively on the β-face of the imine. This stereoselectivity is largely controlled by the 2,3-*O*-isopropylidene protecting group, which prevents attack from the α-face. The coupled product was Boc-protected (**38**) to facilitate purification. While it is possible to directly convert **37** to forodesine under strongly acidic conditions, it was observed that this procedure produced a tarry byproduct which was difficult to separate from the desired product. Thus, it was found to be more convenient to employ a two-step process in which the benzyloxymethyl (BOM) protecting group is cleaved first under catalytic hydrogenolysis conditions to afford

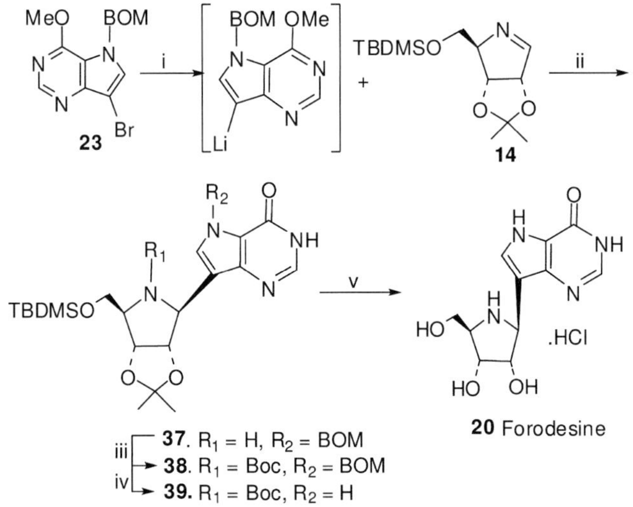

(i) n-BuLi, −78 °C, anisole, ether; (ii) imine addition; −78 °C to 0 °C, (iii) $(Boc)_2O$, DCM; (iv) H_2, $Pd(OH)_2$, $EtOH/NH_3$; (v) conc. HCl

Scheme 18.10

39, followed by treatment of **39** with concentrated HCl to give forodesine as its hydrochloride salt.

Recently, the synthesis and biological activity of the L-enantiomer of forodesine has been reported [44]. This analogue is also a slow-onset, tight-binding inhibitor which is about 15- to 600-fold less active than the D-form.

18.5
Third-Generation PNP Inhibitors: Transition-State Inhibitors

In 2003, Schramm described another novel class of transition-state inhibitors for PNP which he termed the "DADMe-Immucillin" series [28]. In this series, the 9-deazapurine moiety is attached to the N-atom of the iminoribitol via a methylene linker, as shown with DADMe-Immucillin-H and DADMe-Immucillin-G.

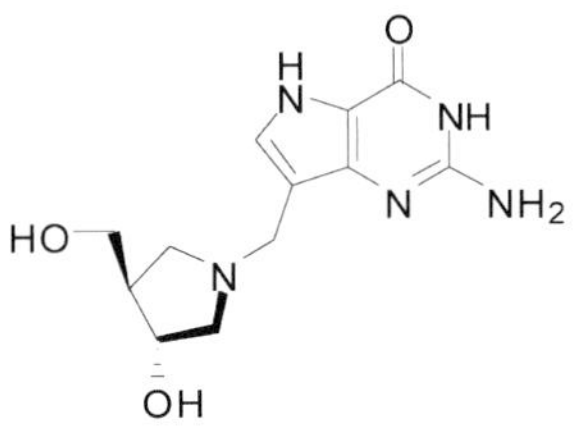

DADMe-Immucillin-G

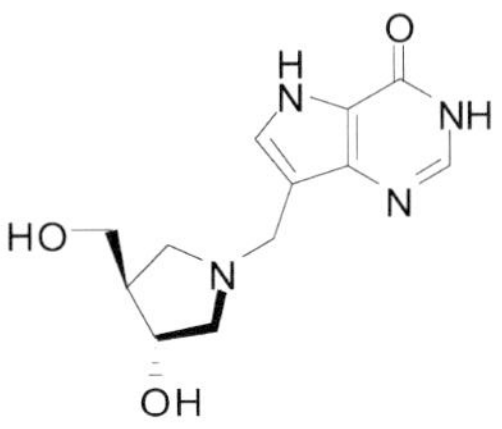

DADMe-Immucillin-H

These inhibitors are the most potent PNP inhibitors identified to date, with $K_d = 7$ pM for the guanine derivative and $K_d = 16$ pM for the hypoxanthine analogue. This arrangement still mimics the oxycarbenium ion character of the transition state and is similar to the second-generation inhibitors; this series also has a long half-life on the PNP enzyme. This has been clearly demonstrated by Schramm in experiments with mice, in which the administration of a single oral dose of DADMe-Imm-H effectively inhibited PNP for the lifetime of the circulating erythrocytes [28]. Currently, DADMe-Imm-H is under clinical development by BioCryst and Roche Pharmaceuticals as BCX-4208.

18.5.1
Chemistry of BCX-4208

The initial synthetic route for the third-generation PNP inhibitors involved a lengthy process, as depicted in Scheme 18.11 [45]. In this route, the 9-bromo-9-dezahypoxanthine derivative **23** was taken in dry anisole and treated with *n*-butyl lithium at −70 °C. The anion was then treated with *N,N*-dimethylformamide to give 9-formyl-9-deazahypoxanthine **40**. With the formyl group in place, a reductive amination with the iminoribitol **41** was performed with sodium cyanoborohydride at ambient temperature. The chiral iminoribitol **41** was derived from D-xylose in 12 steps, using the method of Pedersen *et al.* [46] (Scheme 18.12). Cleavage of the BOM group under

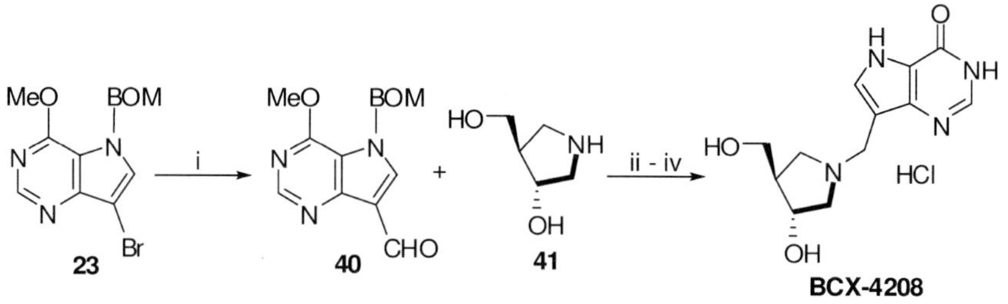

(i) n-BuLi, –70 °C, DMF; (ii) $NaCNBH_4$, MeOH; (iii) H_2, $Pd(OH)_2$; (iv) conc. HCl.

Scheme 18.11

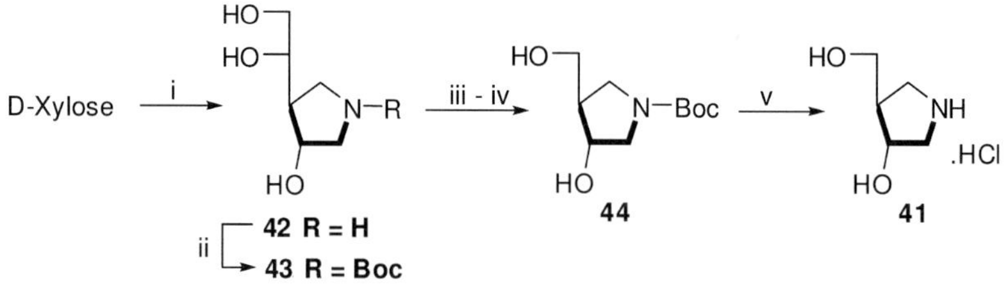

(i) reference 46; (ii) Boc_2O, MeOH: (iii) $NaIO_4$, EtOH; (iv) $NaBH_4$; (v) conc. HCl.

Scheme 18.12

hydrogenolysis conditions, followed by treatment with acid, gave the target compound, as illustrated in Scheme 18.11 for BCX-4208.

A simpler route to the DADMe-Immucillin series has been described by Evans *et al.*, in which the requisite C–C and C–N bonds are formed concomitantly with a Mannich condensation [47] . By using this approach, BCX-4208 was assembled in one step by Mannich condensation of the chiral iminoribitol moiety **41** and 9-deazahypoxanthine **28** with NaOAc and HCHO in water at reflux temperature, as shown in Scheme 18.13. Under similar conditions, Evans *et al.* also reported the synthetic routes for a number of other analogues (see Scheme 18.14 and accompanying table).

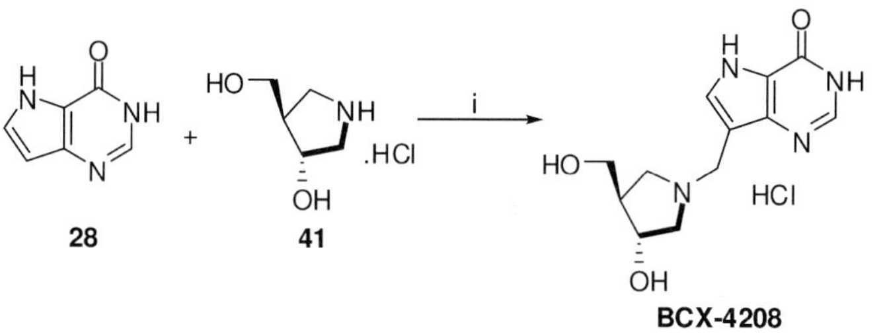

(i) 30% aq HCHO, NaOAc, H_2O, 95 °C.

Scheme 18.13

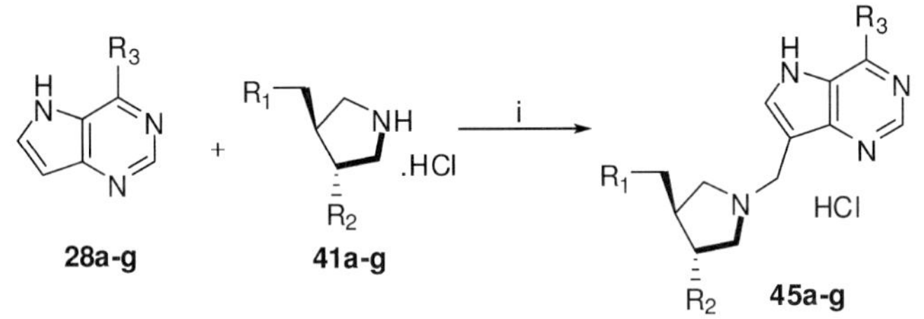

(i) 30% aq HCHO, NaOAc, H_2O, 95 °C.

Entry	Time (h)	R_1	R_2	R_3	Yield (%)
a	16	OH	OH	OH	47
b	1	OH	OH	NH_2	65
c	1	SBn	OH	NH_2	72
d	1	SPhpCl	OH	NH_2	72
e	3	OH	OH	Cl	78
f	3	OH	OH	N_3	65
g	1	OAC	OAc	NH_2	49

Scheme 18.14

Acknowledgments

The authors thank Dr. Krishnan Raman (Crystallography Department, BioCryst Pharmaceuticals) for providing the PNP crystal structures depicted in Figures 18.2, 18.3, 18.6 and 18.7.

References

1 Parks, R. E. Jr., Stoeckler, J. D., Cambor, D., Savarese, T. M., Crabtree, G. W., Chu, S.-H., Purine nucleoside phosphorylase and 5′-methylthioadenosine phosphorylase: targets for chemotherapy. In: *Molecular Actions and Targets for Cancer Chemotherapeutic Agents* (Eds A. Sartoerlli, J. S. Lazo, J. R., Bertino), Academic Press, New York, **1981**, pp. 229–252.

2 Stoeckler, J. D., Purine nucleoside phosphorylase: a target for chemotherapy. In: *Developments in Cancer Chemotherapy* (Ed. R. I. Glazer), CRC Press, Boca Raton, **1984**, pp. 35–60.

3 Erion, M. D., Takabayashi, K., Smith, H. B., Kessi, J., Wagner, S., Honger, S., Shames, S. L., Ealick, S. E., Purine nucleoside phosphorylase. 1. Structure–function studies. *Biochemistry* **1997**, *36*, 11725–11734.

4 Erion, M. D., Stoeckler, J. D., Guida, W. C., Walter, R. L., Ealick, S. E., Purine nucleoside phosphorylase 2. Catalytic mechanism. *Biochemistry* **1997**, *36*, 11735–11748.

5 Stoeckler, J. D., Poirot, A. F., Smith, R. M., Parks, R. E. Jr., Ealick, S. E., Takabayashi, K., Erion, M. D., Purine nucleoside phosphorylase 3. Reversal of purine based specificity site-directed mutagenesis. *Biochemistry* **1997**, *36* 11749–11756.

6 Markert, M. L., Purine nucleoside phosphorylase deficiency. *Immunodeficiency Rev.* **1991**, *3*, 45–81.

7 Sircar, J. C., Gilbertson, R. B., Purine nucleoside phosphorylase (PNP) inhibitors: potentially selective immunosuppressive agents. *Drugs Future* **1988**, *13*, 653–668.

8 Carson, D. A., Kave, J., Seegmiller, J. E., Lymphospecific toxicity in adenosine deaminase deficiency and purine nucleoside phosphorylase deficiency: Possible role of nucleoside kinase(s). *Proc. Natl. Acad. Sci. USA* **1977**, *74*, 5677–5681.

9 Ealick, S. E., Rule, S. A., Carter, D. C., Greenhough, T. J., Babu, Y. S., Cook, W. J., Habash, J., Helliwell, J. R., Stoeckler, J. D., Parks, R. E. Jr., Chen, S. -F., Bugg, C. E., Three-dimensional structure of human erythrocytic purine nucleoside phosphorylase at 3.2 Å resolution. *J. Biol. Chem.* **1990**, *265*, 1812–1820.

10 Narayana, S. V., Bugg, C. E., Ealick, S. E., Refined structure of purine nucleoside phosphorylase at 2.75 Å resolution. *Acta Crystallogr.* **1997**, *D53*, 131–142.

11 Williams, S. R., Goddard, J. M., Martin, D. W. Jr., Human purine nucleoside phosphorylase cDNA sequence and genomic clone characterization. *Nucleic Acids Res.*, **1984**, *12*, 5779–5787.

12 Ealick, S. E., Babu, Y. S., Bugg, C. E., Erion, M. D., Guida, W. C., Montgomery, J. A., Secrist, J. A. III, Application of crystallographic and modeling methods in the design of purine nucleoside phosphorylase inhibitors. *Proc. Natl. Acad. Sci. USA* **1991**, *88*, 11540–11544.

13 Montgomery, J. A., Niwas, S., Rose, J. D., Secrist, J. A. III., Babu, Y. S., Bugg, C. E., Erion, M. D., Guida, W. C., Ealick, S. E., Structure-based design of inhibitors of purine nucleoside phosphorylase. 1. 9-(Arylmethyl) derivatives of 9-deazaguanine. *J. Med. Chem.* **1993**, *36*, 55–69.

14 Secrist, J. A. III., Niwas, S., Rose, J. D., Babu, Y. S., Bugg, C. E., Erion, M. D., Guida, W. C., Ealick, S. E., Montgomery, J. A., Structure-based design of inhibitors of purine nucleoside phosphorylase. 2. 9-Alicyclic and 9-heteroalicyclic derivatives of 9-deazaguanine. *J. Med. Chem.* **1993**, *36*, 1847–1854.

15 Erion, M. D., Niwas, S., Rose, J. D., Ananthan, S., Allen, M., Secrist, J. A. III., Babu, Y. S., Bugg, C. E., Guida, W. C., Ealick, S. E., Montgomery, J. A., Structure-based design of inhibitors of purine nucleoside phosphorylase 3. 9-Arylmethyl derivatives of 9-deazaguanine substituted on the methyl group. *J. Med. Chem.* **1993**, *36*, 3771–3783.

16 Niwas, S., Chand, P. C., Pathak, V. P., Montgomery, J. A., Structure-based design of inhibitors of purine nucleoside phosphorylase 5. 9-Deazahypoxanthines. *J. Med. Chem.* **1994**, *37*, 2477–2480.

17 Burley, S. K., Petsko, G. A., Aromatic-aromatic interaction: a mechanism of protein structure stabilization. *Science* **1985**, *229*, 23–28.

18 Morris, P. E. Jr., Omura, G. A., Inhibitors of the enzyme purine nucleoside phosphorylase as potential therapy for psoriasis. *Current Pharmaceutical Design* **2000**, *6*, 943–959.

19 Zhang, J., Unpublished results from BioCryst Pharmaceuticals.

20 Elliott, A. J., Morris, P. E., Jr., Petty, S. L., Williams, C. H., An improved synthesis of 7-substituted pyrrolo[3,2-*d*]pyrimidines. *J. Org. Chem.* **1997**, *62*, 8071–8075.

21 Morris, P. E. Jr., Elliott, A. J., Unpublished results from BioCryst Pharmaceuticals.

22 Morris, P. E. Jr., Practical synthetic route development for BCX-34, a novel pyrrolo [3,2-*d*]pyrimidine. Presented at the 6th International Conference Organic Process Research and Development, July 10–12, **2002**.

23 Morris, P. E. Jr., Elliott, A. J., Montgomery, J. A., New syntheses of 7-substituted-2-aminothieno- and furo[3,2-*d*]pyrimidines. *J. Heterocyclic Chem.* **1999**, *36*, 423–427.

24 Morris, P. E., Jr., Elliott, A. J., Walton, S. P., Williams, C. H., Montgomery, J. A.,

Synthesis and biological activity of a novel class of purine nucleoside phosphorylase inhibitors. *Nucleosides, Nucleotides Nucleic Acids* **2000**, *19*, 379–404.

25 Horenstein, B. A., Parkin, D. W., Estupinan, B., Schramm, V. L., Transition-state analysis of nucleoside hydrolyase from *Crithidia fasciculata*. *Biochemistry* **1991**, *30*, 10788–10795.

26 Horenstein, B. A. Schramm, V. L., Electronic nature of the transition state for nucleoside hydrolase. A blueprint for inhibitor design. *Biochemistry* **1993**, *32*, 7089–7097.

27 Miles, R. W., Tyler, P. C., Furneaux, R. H., Bagdassarian, C. K., Schramm, V. L., One-thirds-the-sites transition-state inhibitors for purine nucleoside phosphorylase. *Biochemistry* **1998**, *37*, 8615–8621.

28 Lewandowicz, A., Tyler, P. C., Evans, G. B., Furneaux, R. H., Schramm, V. L., Achieving the ultimate physiological goal in transition state analogue inhibitors for purine nucleoside phosphorylase. *J. Biol. Chem.* **2003**, *34*, 31465–31468.

29 Boyer, P. D., The ATP synthase – a splendid molecular machine. *Annu. Rev. Biochem.* **1997**, *66*, 717–749.

30 Bantia, S., Miller, P. J., Parker, C. D., Ananth, S. L., Horn, L. L., Kikpatrick, J. M., Morris, P. E. Jr. Hutchison, T. L., Montgomery, J. A., Sandhu, S., Purine nucleoside phosphorylase inhibitor BCX-1777 (Immucillin-H) – a novel potent and orally active immunosuppressive agent. *Int. Immunopharmacol.* **2001**, *1*, 1199–1210.

31 Kilpatrick, J. M., Morris, P. E. Jr., Serota, D. G. Jr., Phillips, D., Moore, D. R. II, Bennett, J. C., Babu, Y. S., Intravenous and oral pharmacokinetic study of BCX-1777, a novel purine nucleoside phosphorylase transition-state inhibitor. *In vivo* effects on blood 2′-deoxyguanosine in primates. *Int. Immunopharmacol.* **2003**, *3*, 541–548.

32 Evans, G. B., Furneaux, R. H., Gainsford, G. J., Schramm, V. L., Tyler, P. C., Synthesis of transition state analogue inhibitors for purine nucleoside phosphorylase and *N*-riboside hydrolases. *Tetrahedron* **2000**, *56*, 3053–3062.

33 Fleet, G. W. J., Son, J. C., Polyhydroxylated pyrrolidines from sugar lactones: synthesis of 1,4-dideoxy-1,4-imino-D-glucitol from D-galactonolactone and syntheses of 1,4-dideoxy-1,4-imino-D-allitol 1,4-dideoxy-1,4-imino-ribitol, and (2S,3R,4S)-3. 4-dihydroxyproline from D-gulonolactone. *Tetrahedron* **1988**, *44*, 2637–2647.

34 Horenstein, B. A., Zabinshi, R. F., Schramm, V. L., A new class of *C*-nucleoside analogues. 1-(S)-Aryl-1,4-dideoxy-1,4-imino-D-ribitols, transition state analogue inhibitors of nucleoside hydrolase. *Tetrahedron Lett.* **1993**, *34*, 7213–7216.

35 Kezar, H. S. III., Kilpatrick, J. M., Phillips, D., Kellog, D., Zhang, J., Morris, P. E. Jr., Synthesis and pharmacokinetic and pharmacodynamic evaluation of the forodesine HCl analog BCX-3040. *Nucleosides, Nucleotides Nucleic Acids* **2005**, *24*, 1817–1830.

36 Thomas, D., Wierda, W., Faderl, S., O'Brien, S., Kornblau, S., Koller, C., Bantia, S., Kilpatrick, J. M., Bennett, J. C., Kantarjian, H., Gandhi, V., Preliminary activity of intravenous BCX-1777 in aggressive T-cell malignancies. *Blood* **2003**, *102*, Abstract 4772.

37 Bantia, S., Kilpatrick, J. M., Purine nucleoside phosphorylase inhibitors in T-cell malignancies. *Curr. Opin. Drug Discovery Dev.* **2004**, *7*, 243–247.

38 Evans, G. B., Furneaux, R. H., Hutchison, T. L., Kezar, H. S., Morris, P. E. Jr., Schramm, V. L., Tyler, P. C., Addition of lithiated 9-deazapurine derivatives to a carbohydrate cyclic imine: convergent synthesis of the aza-*C*-nucleoside Immucillins. *J. Org. Chem.* **2001**, *66*, 5723–5730.

39 Furneaux, R. H., Tyler, P. C., Improved syntheses of 3*H*,5*H*-pyrrolo[3,2-*d*] pyrimidines. *J. Org. Chem.* **1999**, *64*, 8411–8412.

40 Morris, P. E. Jr., Hutchison, T. L., Unpublished results from BioCryst Pharmaceuticals.

41 Winslow, C. D., Morris, P. E. Jr., Carboxypyrrole, process of preparing and use as precursor US Patent 6,972,331. December 6, **2005**.

42 Batra, H., Moriarty, R. M., Pennmasta, T., Sharma, V., Stanciuc, G., Staszewski, J. P., Tuladhar, S. M., Walsh, D. A., A concise, efficient and production-scale synthesis of a protected L-lyxonolactone derivative: an important aldonolactone core. *Org. Process Res. & Dev.* **2006**, *10*, 484–486.

43 Williams, J. D., Kamath, V. P., Morris, P. E. Jr., Townsend, L. B., D-Ribonolactone and 2,3-isopropylidene(D-ribonolactone). *Organic Synthesis* **2005**, *82*, 75–78.

44 Clinch, K., Evans, G. B., Fleet, G. W. J., Furneaux, R. H., Johnson, S. W., Lenz, D. H., Mee, S. P. H., Rands, P. R., Schramm, V. L., Ringia, E. A. T., Tyler, P. C., Synthesis and bio-activities of the L-enantiomers of two potent transition state analogue inhibitors of purine nucleoside phosphorylases. *Org. Biomol. Chem.* **2006**, *4*, 1131–1139.

45 Evans, G. B., Furneaux, R. H., Lewandowicz, A., Schramm, V. L., Tyler, P. C., Synthesis of second-generation transition state analogues of human purine nucleoside phosphorylase. *J. Med. Chem.* **2003**, *46*, 5721–5276.

46 Filichev, V. V., Brandt, M., Pederson, E. B., Synthesis of an aza analogue of 2-deoxy-D-ribofuranose and its homologues. *Carbohydrate Res.* **2001**, *333*, 115–122.

47 Evans, G. B., Furneaux, R. H., Tyler, P. C., Schramm, V. L., Synthesis of a transition state analogue inhibitor of purine nucleoside phosphorylase via the Mannich reaction. *Org. Lett.* **2003**, *5*, 3639–3640.

19
Formycins and their Analogues: Purine Nucleoside Phosphorylase Inhibitors and their Potential Application in Immunosuppression and Cancer*

Agnieszka Bzowska

19.1
Introduction

The antibiotic formycins are purine nucleosides analogues, in which the pyrazole ring (instead of the imidazole ring) and the carbon–carbon link between the sugar and the heterocycle (instead of the nitrogen–carbon glycosidic bond) are both unusual structural features. Hence, formycins belong to the so-called C-nucleosides, that are analogues of the well-known N-nucleosides. The formycin family includes formycin A (FA), B (FB), and oxoformycin B-structural analogues of adenosine (Ado), inosine (Ino), and xanthosine (Xao), respectively (see Scheme 19.1). The formycins are members of a broader class of so-called "nucleoside antibiotics" – that is, nucleoside analogues isolated from various microorganisms. Such compounds differ from the common nucleosides by modifications to the heterocyclic base moiety or to the sugar portion of the molecule and, as a consequence of these changes they have antibacterial, antitumor, and/or antiviral properties.

Although N-nucleosides are predominant in Nature, many naturally occurring C-nucleosides have also been reported. The first such nucleoside antibiotic – pseudouridine, the C-nucleoside analogue of uridine – was isolated from yeast tRNA in 1957 as an "unknown nucleoside" [1], and later was fully characterized by Cohn [2]. Since then, a number of other C-nucleosides have been isolated. FA and FB were discovered by Hori *et al.* [3] and by Koyama and Umezawa [4], while the oxoformycin B as a metabolite of FA and FB was described by Sawa *et al.* [5]. FA was shown to be a substrate for adenosine deaminase and thus be readily converted to FB [6].

In addition to the interesting and potentially medically useful biological activities, many C-nucleosides represent powerful biochemical and molecular tools. This is especially true in the case of formycins, since these nucleosides – in contrast to their N-nucleoside counterparts – show marked fluorescence under physiological conditions (temperature, pH, and ionic strength) (e.g., Refs. [7, 8]).

*A list of abbreviations may be found at the end of the chapter.

Modified Nucleosides: in Biochemistry, Biotechnology and Medicine. Edited by Piet Herdewijn

ISBN: 978-3-527-31820-9

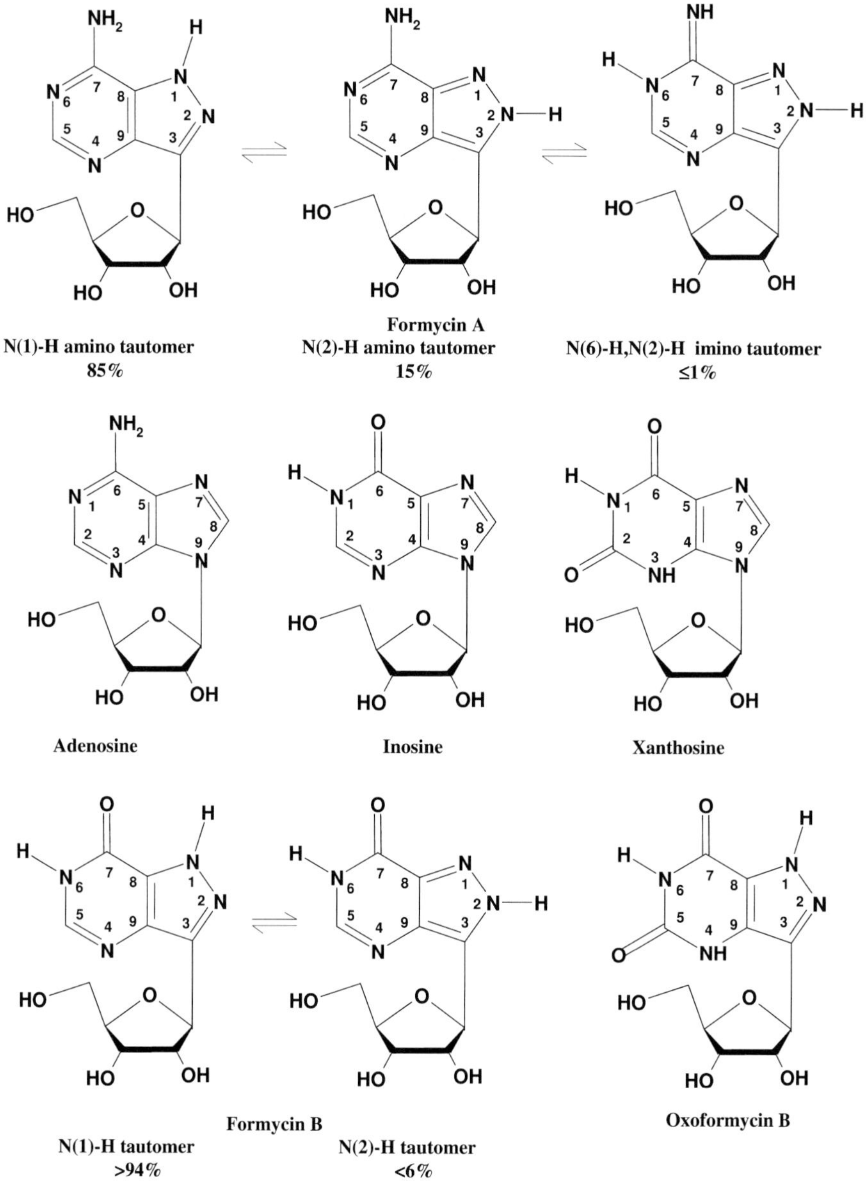

Scheme 19.1 Formycin A, a structural analogue of adenosine, formycin B, a structural analogue of inosine, and oxoformycin B, a structural analogue of xanthosine. Note the difference in the IUPAC numbering for pyrazolo[4,3-d] pyrimidine ring in formycins, and that for the purine ring. In contrast to purine N-nucleosides, FA and FB (C-nucleosides), occur in aqueous media at room temperature as an equilibrium of two tautomeric forms as shown. In the case of FA, also N(6)-H,N(2)-H imino tautomer seems to occur in a very low, but not negligible, amount [8, 25–29].

Their close structural similarity to Ado, Ino and Xao allows formycins to act as alternative substrates of these nucleosides in a variety of enzymatic reactions. However, the C–C glycosidic bond of formycins, which is much stronger than the corresponding glycosidic bond of N-nucleosides, makes them resistant to enzymatic hydrolysis and phosphorolysis. Therefore, these analogues may be expected to act as inhibitors of purine nucleoside phosphorylases (PNP) and purine nucleoside hydrolases (the latter are present only in prokaryotes). In this chapter, a summary is provided of the properties of formycins and their derivatives as non-cleavable ligands of PNP, which has been considered a primary target for chemotherapeutic intervention since it was shown that a child suffering from lymphopenia, and with severely defective T-cell (but normal B-cell) immunity exhibited no PNP activity [9].

The chemistry and biochemistry of nucleoside antibiotics, including various C-nucleosides and formycins, have been the subjects of many reviews [10–17]. At this point, some important facts regarding the chemical and biological properties of formycins we first summarized, after which the application of formycins as molecular tools is briefly discussed.

19.2 Chemical Structure of Formycins and their Analogues

19.2.1 Formycin A and B, Oxoformycin B

The chemical structure of formycin A (FA, also referred to as formycin), formycin B (FB, also referred to as laurusin), and oxoformycin B is shown in Scheme 19.1, together with the structural analogue N-nucleosides. The structures of FA and FB were first determined by a combination of spectroscopic and chemical studies, including ultraviolet (UV), ^{1}H NMR, mass spectra and chemical degradation [18–20], and later confirmed by X-ray crystallography [21–23].

Unlike Ado and Ino [24], the formycins and their analogues exhibit appreciable tautomerism. It has been shown by variety of methods, including ^{13}C NMR and fluorescence, that in aqueous and non-aqueous media FA is in the amino form (like Ado), while FB is in the keto form (like Ino) (see Scheme 19.1). However, both FA and FB exist as an equilibrium mixture of two prototropic tautomers, with the pyrazolo ring proton on either N(1) or N(2) (note the difference in the IUPAC numbering for pyrazolo[4,3-d]pyrimidine ring in FA and FB, and that for the purine ring; see Scheme 19.1). In the case of the neutral form of FA and FB, in aqueous media at room temperature the N(1)-H tautomer predominates; relative population of tautomers N(1)-H/N(2)-H have been reported as $\sim$85 : 15 and >94 : 6 for FA and FB, respectively [8, 25, 26].

While amino forms predominate both in Ado and FA, the latter differs from its N-nucleoside counterpart in that it has small but significant concentration of the imino tautomer. In Ado, the population of the imino tautomer is less than 10^{-6} [24], while in FA and model analogues in which ribose is replaced by methyl or propyl group, the population of the N(6)-H,N(2)-H imino tautomer (see Scheme 19.1) may be up to 1% [27, 28].

The emission properties of FA points to a predominance of protonation (pK_a 4.3–4.4 [7, 29]) on the ring N(4) (formycin ring numbering) [8]. This property distinguishes FA from Ado, which protonates predominantly at N(1) (purine ring numbering) [30]. The monoanion of FA – obtained as the results of deprotonation form the pyrazole ring – is formed with pK_a 9.6 [30]. FB and oxoformycin B also first deprotonate from the pyrazole ring with pK_a 8.6, identical for both nucleosides. This mechanism distinguishes them from their N-nucleoside counterparts, Ino and Xao, which deprotonate with pK_a 9.0 and 5.7, respectively, with both forming position N(1) of the six-membered ring (purine ring numbering). Dianions of FB and oxoformycin B, in which the proton from position N(6) is also removed (formycin ring numbering), are formed with pK_a > 11 [31].

Crystallographic studies revealed two other important features of the formycins structure. First, the glycosidic C–C bond (1.55 Å) is significantly longer than the 1.47 Å bond of C–N glycosidic linkage. Second, a higher flexibility in torsion angles of formycins is observed, which causes the conformations of FA and FB to differ significantly from their N-nucleoside counterparts. Formycins adopt preferentially either the *syn* (formycin hydrobromide; protonated) or the intermediate between *syn* and *anti* conformation (formycin monohydrate; neutral form) in the crystal state [19, 21, 32]. Later, it was also shown that in solution FA and FB prefer the *syn* conformation [33], this conformation most likely being stabilized by an intramolecular hydrogen bond between O(5′) and N(4) (formycin ring numbering).

The increased conformational flexibility of formycins has important biological implications. It is probably responsible for the fact that FA can substitute for Ado, even in enzymatic systems such as adenosine deaminase (ADA), which are known to require nucleosides substrate in the *anti* conformation. On the other hand, polymers containing formycins may demonstrate many unusual features (see Section 19.2.3) [7, 34–36].

19.2.2 Structural Modifications of Formycins

As many of naturally occurring C-nucleosides show antitumor, antiviral and antibacterial activities, a variety of their unnatural analogues were synthesized and many of these were seen to display interesting biological properties.

19.2.2.1 N-Methyl and N-Substituted Analogues

The synthesis of methyl derivatives of FA was prompted by suggestions that a restriction of the rotameric freedom around the glycosidic bond to the *syn* conformation would confer resistance to deamination. N(1)- and N(2)-methyl formycin A derivatives (m^1FA, m^2FA) were prepared by variety of treatments [37, 38]. Although m^2FA exists predominantly in the *syn* conformation, it exhibits no enhanced activity when compared with FA. Nevertheless other possible N-methylformycins were prepared, namely m^4FA, m^6FA and m^7FA, as well as m^1FB, m^4FB, and m^6FB [39].

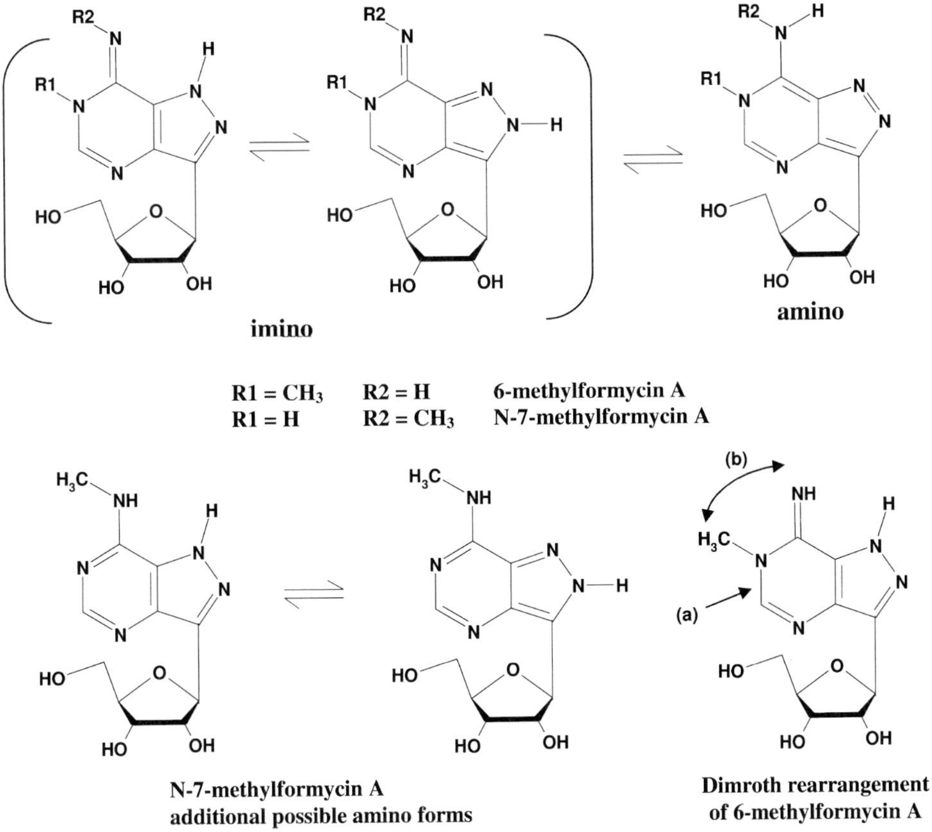

Scheme 19.2 N-methyl derivatives of FA and their tautomeric forms. For *N*(6)-methylformycin A (m^6FA), there are one amino and two imino tautomers; in aqueous medium, the amino and imino forms occur in comparable populations, whereas in low-polar solvents the imino form predominates [39, 114, 168]. *N*(7)-methylformycin A (m^7FA) may in theory occur in three amino and two imino forms, but this has not yet been studied experimentally. As the pK_a of this analogue is 4.6 (see Table 19.1) (i.e., similar to FA and lower than expected for protonation of the imino derivative), the amino forms presumably predominate. In aqueous solution, m^6FA undergoes rearrangement to four different formycin derivatives; the main product, m^7FA, is formed by Dimroth rearrangement, as shown.

Although not more chemotherapeutically active, some of these proved to be very useful in studies of the mechanism of purine nucleoside phosphorylase (see Section 19.6.2.2). This is especially true for m^6FA, which in aqueous solution undergoes rearrangement to different formycin derivatives. The main product, m^7FA, is formed by Dimroth rearrangement (Scheme 19.2) [39]. m^1FA, m^2FA and m^7FA show significant cytotoxicity, and are incorporated into erythrocytic nucleotide pools, while m^6FA displays only very weak cytotoxic activity [40].

19.2.2.2 Other Base-Modified Analogues

Among these compounds, the 7-substituted, 5-substituted, and 5,7-disubstituted derivatives dominate [41–43], some with the 2′-deoxyribose [44], as these are analogues of the 6-thiopurine nucleosides and 2-chloro-2′-deoxyadenosine which exhibit antitumor activity [45, 46]. The chemical and biological properties of 7-substituted analogues were reviewed by Daves and Cheng [13], although the data did not include any impressive antitumor and antiviral activities. The 5-substituted derivatives (e.g. 5-fluoro- and 5-aminoFA) were found to be no more active than FA [43]. An analogue with no substituent at position 7 (so-called deaminoformycin) was synthesized and shown to be a good herbicide [47]. Recently, the preparation of ring-modified FA analogues (4-deazaformycins) was also reported [48].

19.2.2.3 Sugar-Modified Analogues

Although 2′-deoxy- and 3′-deoxyFA were synthesized [49–52], their antitumor activities (with one exception, against S49 lymphoma cells) were no better than that of FA. The synthesis of 2′3′-dideoxyFA was also achieved [50], whereby the structure showed – as in the case of FA and FB (see above) – a *syn* conformation about the glycosidic bond, stabilized by an intramolecular hydrogen bond between the O(5′) and N(4) atoms [53]. More recently, the synthesis of H-phosphonates and phosphoramidites of 2′-deoxy-N(1)-methylFA (formycin ring system) was also reported [54].

5′-Deoxy-5′-iodoFB, 5′-deoxy-5′-phenylthioFB and 5′-[(*p*-fluorosulfonyl)benzoyl] FB [55, 56] cannot be 5′-phosphorylated, and their potential biological activity is connected with inhibitory properties versus purine nucleoside phosphorylase (see Sections 19.6.3 and 19.6.4). More recently, carbocyclic formycin analogues (i.e., with the ribofuranose oxygen replaced by methylene) were prepared [57], in addition to C-4′ truncated carbocyclic racemic forms of FB and FA (i.e., additionally lacking the C-4’ hydroxylmethylene moiety), although an antiviral analysis of the latter compound did not disclose any such activity [58].

19.2.3 Formycin Phosphates and Polyformycin Phosphates

The crystal structure of 5′-phosphate of FA (FMP) revealed the key structural features: the N(4) of the base forms an intramolecular hydrogen bond to the phosphate oxygen O(1); the glycosyl torsion angle is *syn* with O(4′)-C(1′) relative to C(9)-C(4) being −6.43 degrees; and the furanose ring pucker is C(3′)-endo, with a pseudorotation angle of 20.3 degrees. The major difference between the AMP and FMP structures is that the glycosyl torsion angles differ by 190 degrees. The computed conformational energy necessary to distort AMP so that it has the same glycosyl torsion angle as FMP is 4.6 kcal mol^{-1} [59]. These differences are most likely responsible for atypical behavior of RNA containing FA (see below).

Copolymers of formycin are readily synthesized by the polymerizing enzymes, and they also efficiently direct protein synthesis *in vitro*, while the polymer of formycin (poly FA) is difficult to synthesize for RNA polymerase and fails to code for any polypeptide. This is in contrast to poly A, which is known to direct the synthesis of polylysine. A third anomaly of poly FA concerns its interactions with nucleases. It is degraded much more slowly than poly A by nucleases known not to possess base specificity. On the other hand, unlike poly A, poly FA is a true substrate for pancreatic RNase, a pyrimidine-specific enzyme, and is degraded at a rate similar to that found for poly C [34–36]. All of the above findings reflect the unique conformational properties of the poly FA polymer. Individual formycin residues in ribopolymers may exist either in the *syn* conformation, as in poly FA, or in the *anti* conformation, for example in ordered structures with the Watson–Crick base-pairing [34, 36].

19.3 Spectral Properties of Formycins

In contrast to their N-nucleoside counterparts, which are practically non-fluorescent, the neutral form and the cation of FA, FMP, and some of the N-methylated analogues of FA, are fluorescent at room temperature, with good quantum yield (Table 19.1) [7, 8, 29]. The absorption spectra of formycins are red-shifted with comparison to their N-nucleoside analogues, and also with comparison to fluorescent protein residues (see Table 19.1 and Figure 19.1), which makes possible their selective excitation without excitation of the intrinsic protein fluorescence, even in the high excess of typical nucleosides. The absorption and fluorescence spectra of FA and FMP differ considerably, which allows a spectrometric distinction to be made between these compounds (Figure 19.1).

FA exists as a N(1)-H and N(2)-H 85 : 15 tautomeric mixture in the ground and excited electronic states, and the relative contribution of the emission from both tautomers depends on the excitation and emission wavelengths. Absorption and emission spectra of the N(2)-H form are red-shifted relative to the spectra of the N(1)-H form, and excitation at 315 nm goes almost exclusively to the N(2)-H tautomer, while selective observation of its fluorescence is possible at wavelengths above 360 nm [60].

The fluorescence of FA is quenched several-fold on its incorporation into polymers, and further quenching occurs when the residues are base-paired in helical, double-stranded structures [7]. It is also possible to quench the protein-bound formycin fluorescence by the resonance energy transfer mechanism if some of the protein residues are labeled with properly chosen resonance energy acceptors. Resonance energy transfer occurs also from FA to terbium ions (e.g., Ref. [61]), the latter being able to substitute for calcium and magnesium ions in various proteins.

The fluorescence spectra of the cation of FA consists of two bands [7], that may be separated. The presence of a 445 nm band is the effect of proton migration in the excited state and formation of the rare tautomeric form N(4)-H (formycin ring numbering) [8]. This is one of the few examples of phototautomerism in heterocyclic systems.

By contrast, FB is practically non-fluorescent at neutral pH, and only slightly so in alkaline media [62]. Both, the neutral and monoanionic forms of oxoformycin B are strongly fluorescent, and their fluorescence may be excited at wavelengths where normal nucleic acid bases do not absorb [31].

Table 19.1 Spectral properties of formycins and some analogues in aqueous media at room temperature.

Compound	pK_{a1} protonation	pK_{a2} deprotonation	Absorption		Emission			Ref.
			λ_{max} [nm]	ε_{max} [M^{-1} cm^{-1}]	λ_{exc} [nm]	λ_{max} [nm]	Φ	
Formycin A neutral form	4.4	9.6	294	10 300				[7]
					280 305 315[a)]	338 338 350	0.058 0.052	[8] [8] [60]
Formycin A cation			295 232	10 400 8400				[29]
					280 320	380[b)] 445[b)] 380[b)] 445[b)]	0.0045 0.017 0.010 0.017	[8]
Formycin B neutral form	0.9[c)]	8.6[d)]	278	7300				[29]
						340	<0.005	[62]
Formycin B anion			282	8800				[62]
					290	335	0.03	[62]
Oxoformycin B neutral form	<1	8.6[d)]	287	5650	300	370		[31]
Oxoformycin B monoanion			297	4400	322	392		[31]
m^1FA	4.0		302 332	10 000 7700				[29]
					[e)]	351	0.087	[8]

m^2FA	4.9		306 234	11 000 5200				[29]
					e)	354	0.084	[8]
m^4FA	6.4		314 270 246	6500 9700 10 500				[39]
					e)	440	0.083	[8]
m^6FA	6.9	10.3	289 278 241	6300 6300 18 600				[28] [114]
					300 320 f)	340,440 440	0.002 0.007	[8] [168]
m^7FA	4.6		298	17 100				[128]
FMP	4.9	10.2	295	10 000				[62]
					295 318	340 355	0.053 –	[62]

a) Selective excitation of the *N*(2)-H tautomeric form.
b) The fluorescence spectra of the cation of formycin consists of two bands (see text, Section 19.3 for discussion).
c) Ref. [207] reports p*K* 1.3.
d) Second deprotonation occurs at pH > 11 [31].
e) Independent of wavelength of excitation.
f) Selective excitation of the amino tautomeric form.

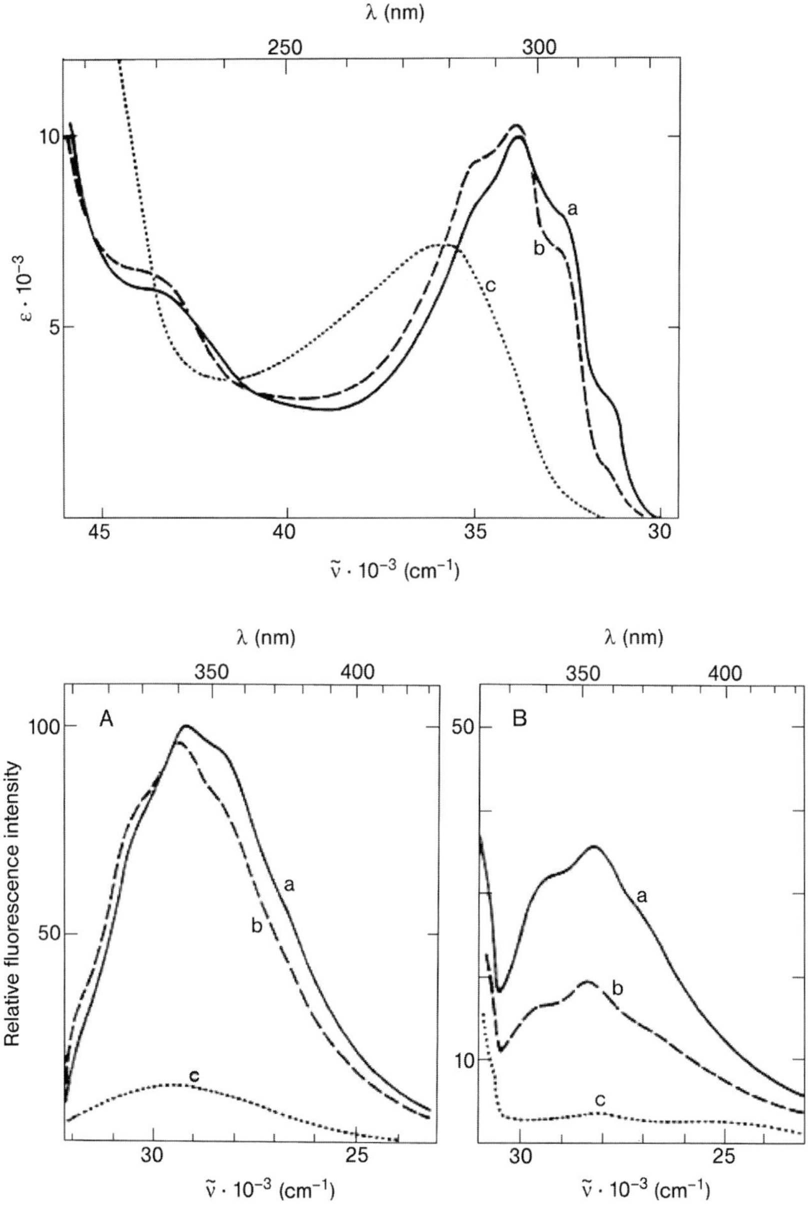

Figure 19.1 Absorption (upper panel) and emission (lower panel) spectra of 5′-FMP (a, solid line), FA (b, dashed line) and FB (c, dotted line). Excitation was at 308 nm (A) and 318 nm (B), pH was 7.6 and 8.6 for absorption and emission spectra, respectively. ε is in ($M^{-1}cm^{-1}$). (Reprinted from [62], © Elsevier).

Due to their favorable spectral and fluorescence properties, formycins and their phosphates are very suitable for studies of nucleic acid structure and functions, and for biophysical studies of enzyme/ligand interactions. Such studies are summarized briefly in Sections 19.8.1 and 19.8.3.

19.4 Sources of Formycins

19.4.1 Natural Sources and Biosynthesis

FA, FB, and oxoformycin B are all natural products of the Actinomycetes. Originally, FA was isolated from culture filtrates of the rice mold, *Nocardia interforma*, at the Institute of Microbial Chemistry in Tokyo [3] during a screening program for antibiotics with antitumor activity. One year later, a report was made of the isolation of FB from *Nocardia interforma* and the relationship of this new antibiotic to FA [4]. Both FA and FB were also obtained from *Streptomyces lavendulae* [63], while oxoformycin B was shown to be a metabolite of FA and FB in *Nocardia interforma* [5]. To date, the C-nucleoside analogue of Guo obtained from natural sources has not been reported in the literature.

In view of the structural similarity of formycins and their N-nucleoside counterparts, it was of great interest to determine whether the biosynthesis occurs via similar pathways. On the basis of published data regarding biosynthesis in *Nocardia interforma* and *Streptomyces* sp. MA406-A-1, it appeared that the intact ribose (as in the case of naturally occurring purine nucleosides) is incorporated into the ribosyl moiety of formycins. By contrast, biosynthesis of the heterocyclic portion is different, with the carbon and nitrogen atoms being derived from sources other than that for purines [64–67]. Both, L-lysine and L-glutamate appear to be precursors of the aglycone, with the former donating nitrogen atoms N(3), N(7), and N(8), and the latter being incorporated into four contiguous positions C(9), C(4), C(5), and C(6) (purine ring numbering system) [68, 69]. The final steps of formycin biosynthesis were proposed to be as follows: formycin B 5′-monophosphate → FMP → FA → FB → oxoformycin B [5]. The first step in this sequence is very similar to that observed in the purine nucleoside synthesis (IMP → AMP).

19.4.2 Synthesis

The interesting biological properties of naturally occurring C-nucleosides have prompted intensive trials to identify synthetic routes leading to these analogues. Oxoformycin B was the first of the formycins to be synthesized (Scheme 19.3). This was achieved, and preliminary reports made, by Bobek, Farkas and Sorm [70], whereas the full synthetic procedure was published some two years later [71]. The starting compound [72] was the nitrile of the O-benzyl-protected D-ribose

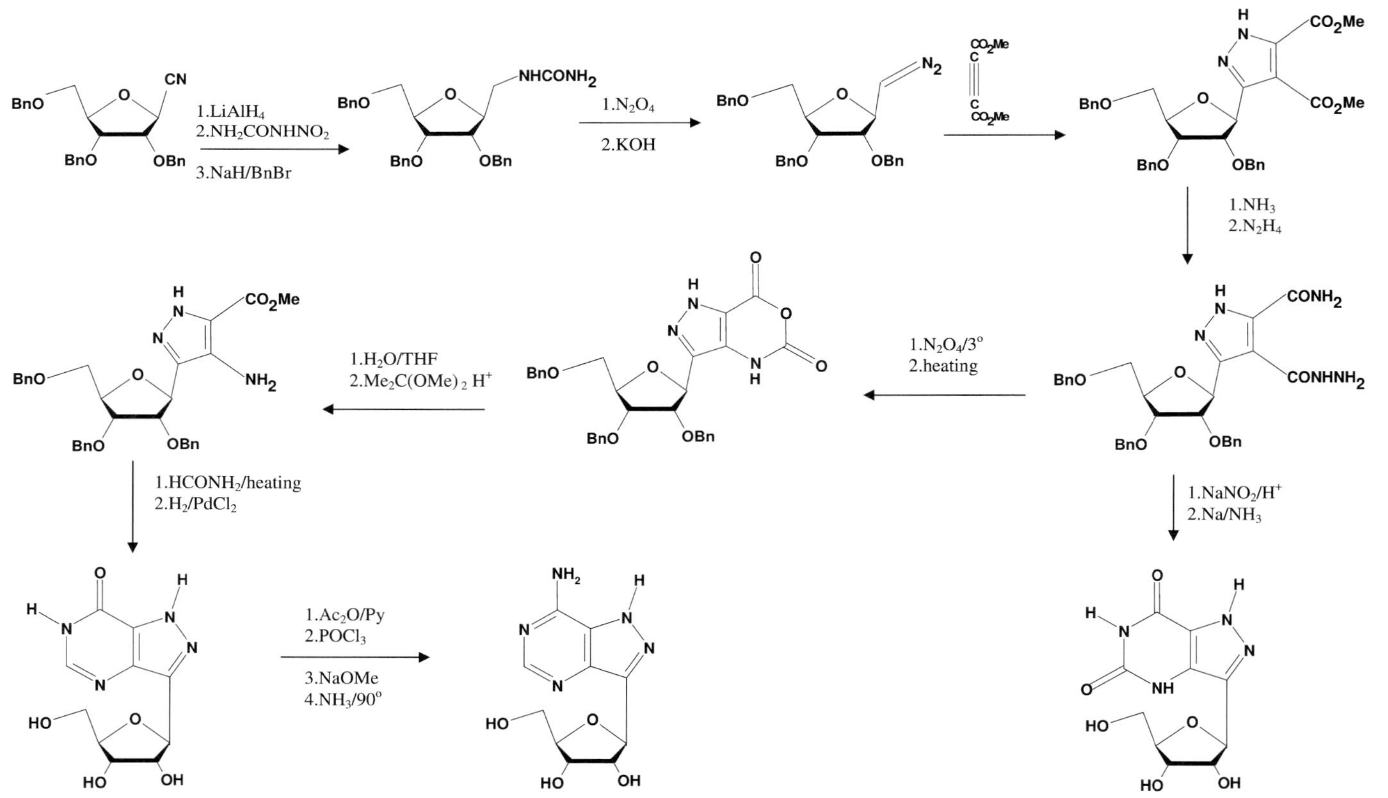

Formycin B
Acton et al., 1971 [73]

Formycin A
Lang et al., 1971 [42]

Oxoformycin B
Bobek et al., 1970 [70]
Farkaš and Šorm, 1972 [71]

Scheme 19.3 The first synthetic procedures leading to oxoformycin B, FB, and FA from the latter compound. Abbreviations: Bn, benzyl group; Me, methyl group; Ac, acetyl group.

(Scheme 19.3). The nitrogen heterocycle was constructed in several steps on this precursor, via the dicarboxylate pyrazole derivative, as the key intermediate. The next intermediate of the reaction leading to oxoformycin B – amide hydrazine (see Scheme 19.3) – was later used to obtain FB [73]. FA was first prepared by Lang *et al.* [42] from FB (Scheme 19.3) via the 7-chloro derivative (note the difference in the IUPAC numbering for pyrazolo(4,3-d)pyrimidine ring in FA and FB, and that for the purine ring; see Scheme 19.1).

Later, several optimized synthetic routes leading directly to formycin were reported, for example, by Kalvoda [74]. This approach, in which the starting material is also the nitrile, required fewer steps than the previous method, and made possible a direct synthesis of both FA and FB as shown (Scheme 19.4). The elegant method of Buchanan *et al.* [75] rendered the synthetic route leading to FA even more efficient. In this case, the *O*-benzyl-protected ribofuranosylalkyne was used as the precursor (Scheme 19.5), which was converted to the nitropyrazole derivative with the ribose protected by acetic groups, as described previously by the same authors [76]. The resultant tri-*O*-acetyl derivative was converted to FA in four steps (Scheme 19.5), followed by deprotection, with the yields of the steps ranging from 77% to 90%.

19.5
The Biological Activity of Formycins: A Brief Summary

As an antitumor agent, FA displays activity against several experimental tumors, including Ehrlich and Yoshida carcinomas, mouse leukemia L-1210 and HeLa cells [77–80]. By contrast, FB has little antitumor activity, except towards L-5178Y in mice [81]. Both analogues are active against the organism responsible for a rice plant disease, *Xanthomonas oryzae*, although FB shows a higher activity than FA [3, 4, 63]. FB owes its activity to the prevention of the uptake of nucleosides from the medium [82].

Both, FA and FB possess some antiviral activity [83], for example against influenza A1 virus [84], myxovirus [85], and tobacco mosaic virus [86, 87]. FA also inhibits Rous sarcoma virus production [88], and has some antibacterial properties, for example against *Sphaerotilus natans* and *Beggiatoa* sp. [89]. An insulinotropic action of FA has also been reported [90] in which FA was shown to enhance insulin output in rat pancreatic islets; later, FA was also found to display a cytotoxic potential in tumor islet cells [91]. Within the islets, FA is converted to formycin-5′-triphosphate (FTP), thereby increasing insulin secretion in the absence and presence of glucose [92].

FA, with the exception of some molecular processes (see below) must be phosphorylated before it can become an effective antitumor and antiviral agent since, as noted [93], cell lines that lack adenosine kinase activity are resistant to formycin. FA undergoes phosphorylation to the mono-, di-, and triphosphate by a soluble multi-enzyme system (e.g., in mouse liver [94]), but it is also deaminated to FB by ADA [95]. However, the succino-AMP synthetase/lyase system provides a pathway for the metabolic regeneration of FMP from the formycin B monophosphate [96].

To date, the mode of action of FA against tumors has not been generally elucidated. It is probable that FA inhibits *de-novo* purine synthesis in tumor cells, since it was

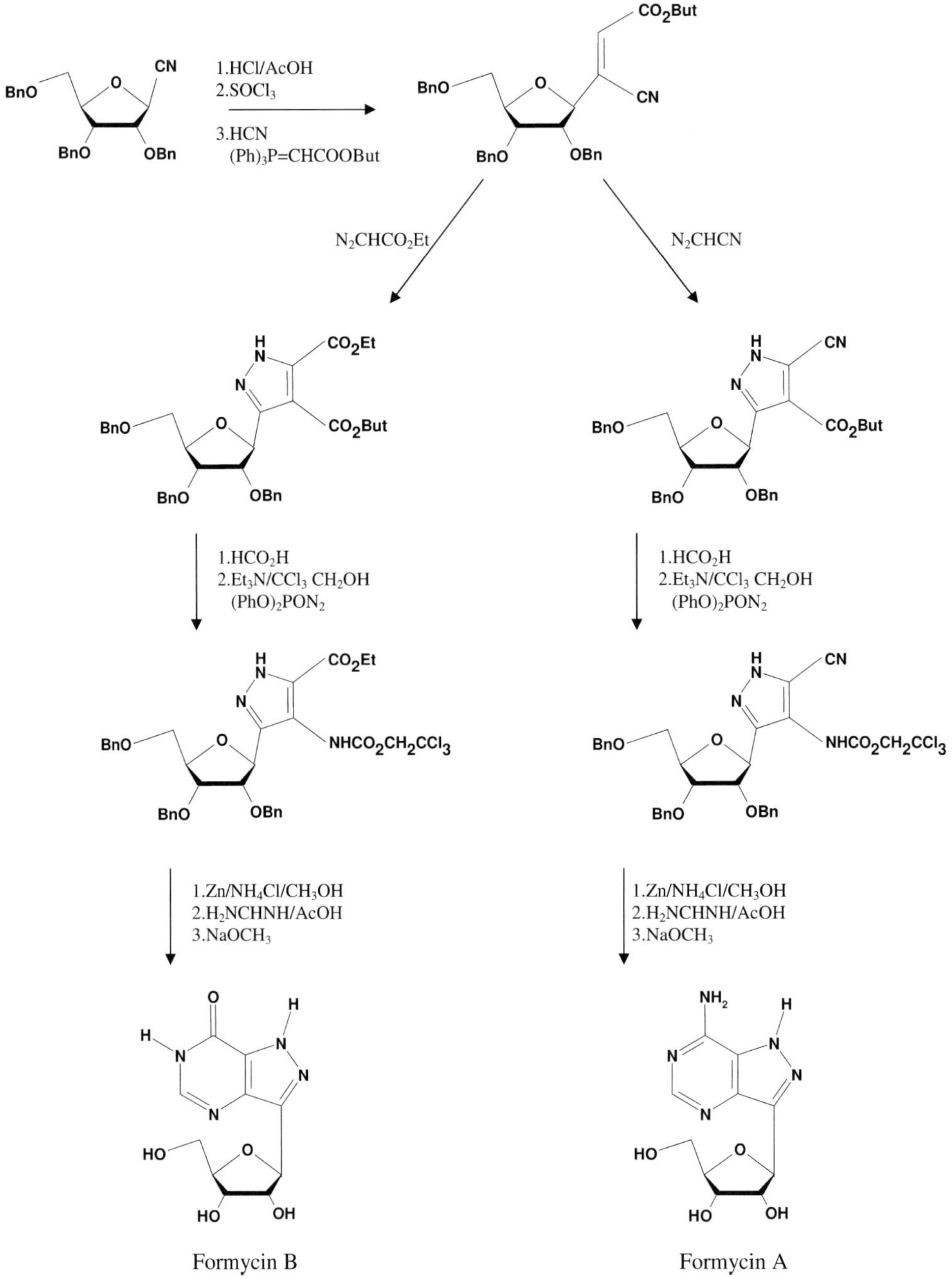

Scheme 19.4 Direct synthetic procedure by Kalvoda [74] leading to FA and FB. Abbreviations: Bn, benzyl group; But, *tert*-butyl group; Et, ethyl group; Ph, phenyl group.

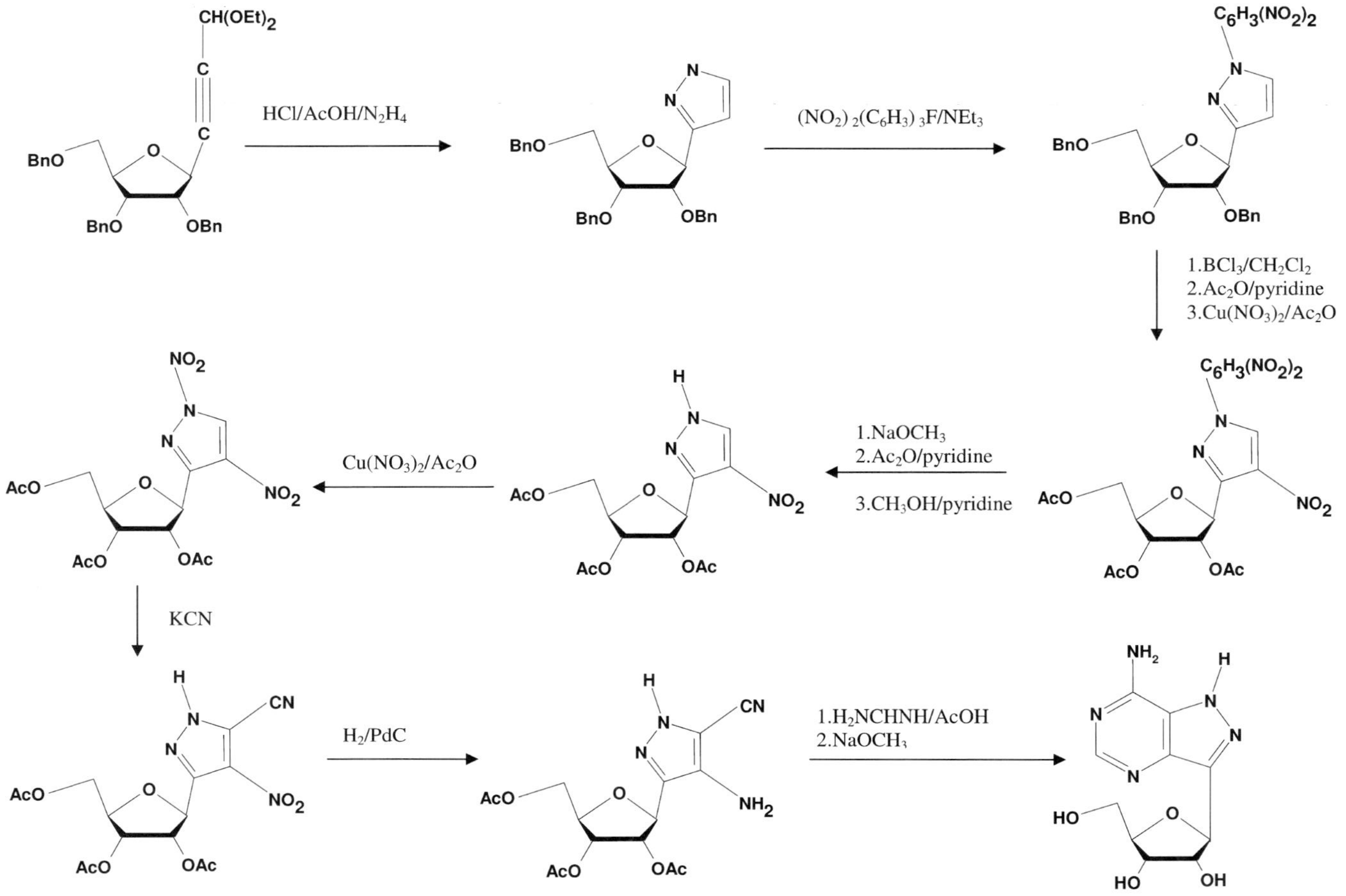

Scheme 19.5 Synthesis of FA according to Buchanan *et al.* [75].
Abbreviations: Ac, acetyl group; Bn, benzyl group; But, *tert*-butyl group; Et, ethyl group; DNPhe, 2,4-dinitrophenyl group.

shown that FMP inhibits the synthesis of 5′-phosphoribosylpyrophosphate, the important pathway of *de-novo* biosynthesis of purine and pyrimidine nucleotides [97]. Other reports have suggested that the incorporation of FA into nucleic acids, particularly into DNA, correlates closely with its lethal effect on cell viability in human colon carcinoma cells in culture [98].

By contrast, FB does not undergo efficient phosphorylation to the 5′-nucleotide in most of the systems analyzed [16]. The initial studies reported that this includes also mammalian cells [79], but it was shown later that FB-resistant mutants of the Chinese hamster ovary cell extracts contained no detectable adenosine kinase activity [99]. Subsequently, it was documented that FB (and also 9-deazaIno) were phosphorylated by at least two different routes in the mouse L cells [100], whereas in Chinese hamster ovary cells FB was phosphorylated by adenosine kinase [101].

Due to biochemical differences between some parasites and their mammalian hosts, FB is efficiently converted to the 5′-monophosphate only in the former, because it is activated by the nucleoside phosphotransferase that is found in these parasites but is absent from mammals. It is subsequently converted to cytotoxic adenosine nucleotide analogues of FA (by the succino-AMP synthetase/lyase system [96]) that become incorporated into RNA [102]. This is probably the reason why FB is cytotoxic towards some parasitic protozoa. FB is also a potent inhibitor of growth of the promastigote forms of *Leishmania tropica, L. mexicana, L. braziliensis,* and *L. donovani* [102] and of several *Trypanosoma* species [103–105]. Recently, FB was found to be toxic to the parasite *Cryptosporidium parvum* and to induce a marked decrease in the gamont stages of this organism [106]. However, the *in-vivo* use of FB is limited by its toxicity to humans due to the fact that, as discussed above, it is also converted (though not very efficiently) to 5′-phosphate in mammalian cells.

Due to the fact that tautomerism of formycin is mainly restricted to the pyrazole ring, and hydrogen bond donor/acceptors pattern on the six-membered ring of this nucleoside is the same as in the case of Ado (see Scheme 19.1), FA functions as the analogue of Ado in base pairing, including nucleic acid replication, while both FA and FB and their nucleotides effectively replace Ado and Ino, and corresponding nucleotides in a wide variety of enzymatic reactions. In the following sections, the properties of formycins will be summarized, and the use of their derivatives as potential immunosuppressive and anticancer agents *via* the inhibition of human purine nucleoside phosphorylases (PNP) will be discussed.

19.6
Formycins and Analogues as Purine Nucleoside Phosphorylase Inhibitors

19.6.1
Molecular Target of Formycins: Purine Nucleoside Phosphorylase (PNP)

PNP (purine nucleoside : orthophosphate ribosyl transferase; EC 2.4.2.1) is a ubiquitous enzyme of the nucleoside salvage pathway in which purine bases from degraded nucleic acids are rescued by the cell. PNP catalyzes the reversible phosphorolytic cleavage of the glycosidic bond of purine nucleosides [107], as follows:

$$\text{purine nucleoside} + \text{orthophosphate} \leftrightarrow \text{purine base} + \text{pentose-1-phosphate}$$

Depending on the enzyme source, natural PNP substrates include 6-oxopurine nucleosides (i.e., Ino and Guo) or both 6-oxo and 6-aminopurine nucleosides (hence also Ado). More specific, low-molecular-mass PNPs are found mainly in mammals, and human PNP belongs to this class. Less-specific, high-molecular-mass phosphorylases occur mainly in bacteria [107]. Both classes show no sequence homology, although X-ray studies have revealed similar topologies of the monomers, with the active site located in the same region of a subunit relative to secondary structure elements, and a similar geometric location of bound ligands in the active site (see Section 19.6.2). By contrast, different amino acid residues constitute the binding sites for the purine base, the ribose moiety, and inorganic phosphate in the low- and high-molecular-mass PNPs.

PNPs from various sources are considered as target enzymes for a variety of diseases and metabolic disorders. Potent inhibitors of the human enzyme are expected to be potential immunosuppressive and anticancer drugs (see Section 19.6.3). Less-specific, high-molecular-mass phosphorylases were proposed to be useful in anticancer gene therapy. In such an approach, tumor cells are transduced with a gene encoding *Escherichia coli* PNP, which differs in specificity from the corresponding human enzyme. The application of a prodrug (a nucleoside analogue, such as 6-methylpurine 2′-deoxyriboside, which is non-toxic to the organism), the cleavage of which at the glycosidic bond releases the heterocyclic base (which is an active cytotoxic agent) leads to the selective destruction of tumor cells [108]. In order to reduce toxicity resulting from the activation of 6-methylpurine 2′-deoxyriboside and other prodrugs by the intestinal tract flora, a redesigned *E. coli* PNP was constructed that is able to cleave those prodrugs which are not cleaved by wild-type *E. coli* PNP [109].

FA and FB are close structural analogues of the natural substrates of PNP, corresponding purine N-nucleosides, Ado, and Ino. However, due to the C–C glycosidic bond, formycins cannot be cleaved by the enzyme. Therefore, formycins and their base-modified analogues were used in X-ray studies of the three-dimensional (3-D) structures of PNPs from various sources, and also in the elucidation of the reaction mechanism.

19.6.2
Formycins and Analogues in Studies of the Molecular Mechanism of Catalysis: The 3-D Structure of PNPs

FB is a moderate inhibitor of mammalian PNP, with an inhibition constant (K_i) of about 100 μM versus the human enzyme [110] and 280 μM versus murine PNP [111]. FA is totally inactive versus the mammalian PNP at a 100 μM concentration [112]. By contrast, PNPs from other sources, such as the malarial parasite, *Plasmodium lophurae* and from *E. coli*, interact much more strongly with formycins. The K_is for

FB are 0.4 μM and 4.6 μM, respectively [113, 114], whereas that for FA versus *E. coli* PNP is 5.3 μM [114]. An examination of a series of N-methyl analogues of FA and FB led to the finding that *N*(6)-methylformycin A (m^6FA), which is virtually inactive versus to the human enzyme, is the most potent inhibitor of *E. coli* PNP, with a K_i of approximately 0.3 μM at neutral pH [114]. Phosphorylases bind their natural nucleoside substrates (albeit very weakly) with dissociation constants of several hundreds of μM [115–117]. Hence, whilst changing from the imidazole ring to the pyrazole ring enhances binding by orders of magnitude with *Plasmodium lophurae* and *E. coli* PNPs, it has only minimal effect in the case of the human enzyme. The reasons for such differences between mammalian phosphorylase and those from other sources became clear when the 3-D structures of PNPs became available.

19.6.2.1 **Low-Molecular-Mass PNPs**

Low-molecular-mass PNPs proved to be homotrimes (e.g., [118–122]) with one complete active site per PNP monomer. The base-binding site is located deeply in the active site pocket, and has three side-chains that are capable of interacting via hydrogen bonds with the base (Figures 19.2 and 19.3). In all of the complexes examined, the side chain of Glu201 forms a hydrogen bond with the base position N(1)-H, and thus restricts the specificity of trimeric PNPs to those purines and their nucleosides that have a hydrogen atom at position N(1). The flexible side chain of Asn243 forms a strong hydrogen bond only with those substrates (e.g., hypoxanthine)

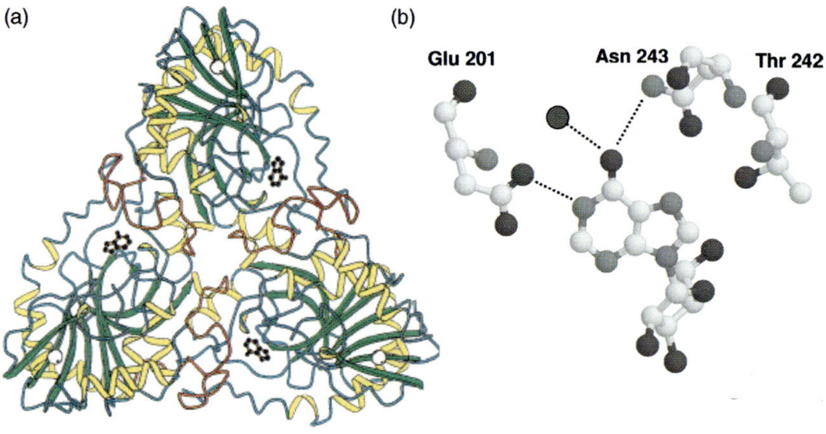

Figure 19.2 (a) A trimer of calf spleen PNP complexed with hypoxanthine. (Reprinted from [119]. © Elsevier). (b) Mode of binding Ino and sulfate in the active site of trimeric (bovine spleen) PNP, derived from the X-ray structure of the PNP-Ino-sulfate ternary complex (PDB entry 1A9S) [120]. Oxygen atoms are black, nitrogen atoms are dark gray. The water molecule present in the active site is shown as a black sphere. The possible hydrogen bond (broken line) of Asn243N^{δ} to O^6 of the base (3.44 Å) is derived from data in the PDB file. Due to the fact that classical X-ray analysis cannot distinguish nitrogen and oxygen atoms, the side chain of Asn243 may be also rotated in such a way that hydrogen bonds from Asn243N^{δ} to N(7) of the base and to Thr242O^{γ}, 3.2 Å and 3.1 Å, respectively, are formed, as shown in [120].

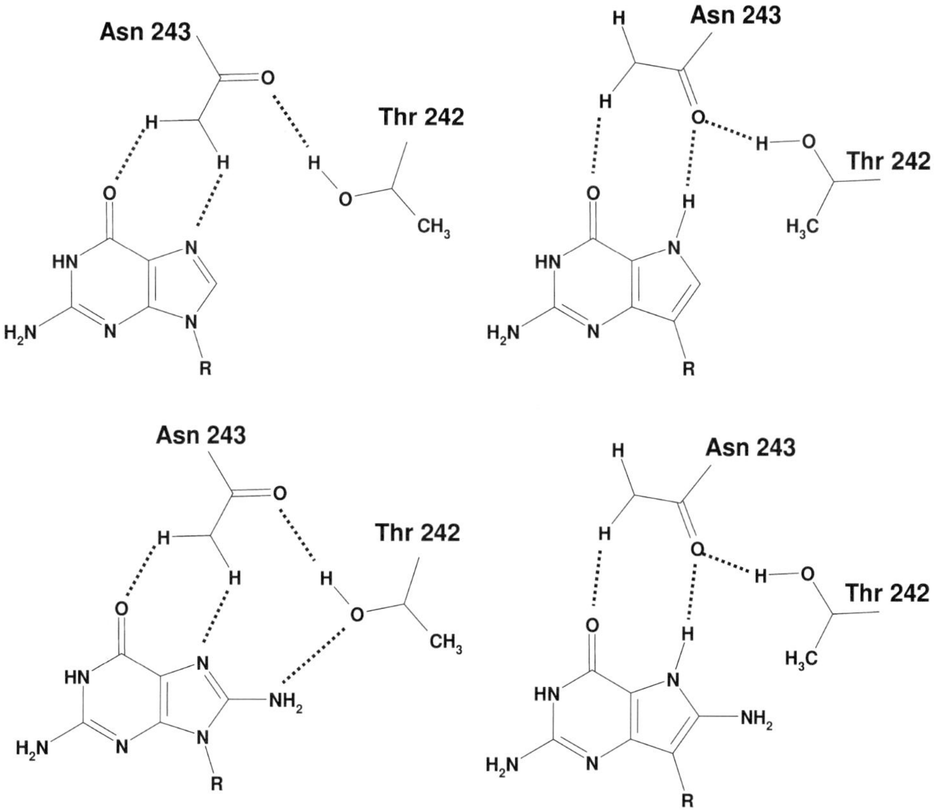

Figure 19.3 Comparison of the effects of the 9-deaza and 8-amino substituents on the mode of binding of the base in the active site of trimeric PNP; R group is 2-thienylmethyl or 3-thienylmethyl. (See Scheme 19.8; according to [118]).

and substrate analogues (e.g., 9-deazaIno) that have a hydrogen bond donor at position N(7), whereas with naturally occurring nucleoside substrates (Ino and Guo) it interacts weakly or not at all (several hydrogen bond patterns were observed; see Figures 19.2 and 19.3). Finally, Thr242 may form an additional bond with those analogues, such as 8-aminoguanine, that have a hydrogen bond donor at position C(8). However, in this case some shift of Thr242 is necessary which results in an inability of Asn243 to interact strongly with N(7)-H. Therefore, analogues with hydrogen bond donors at both positions, N(7) and C(8), bind less tightly than those with a donor at only one position. This is also the most likely reason why FB, which is a mixture of two tautomers with hydrogen either at N(7) or N(8) (purine ring numbering, see Scheme 19.1), is not able to stabilize one of the strong binding conformations and thus somehow binds better than natural nucleoside substrates, though still not very strongly.

The observation that the presence of a hydrogen bond donor at base position N(7) is beneficial for binding with trimeric PNPs was later exploited in attempts to construct inhibitors with nanomolar dissociation constants as potential drugs (see Sections 19.6.3 and 19.6.4).

19.6.2.2 High-Molecular-Mass PNPs

High-molecular-mass PNPs, for example from *E. coli* [123, 124], from the parasites *Plasmodium falciparum* [125] and *Trichomonas vaginalis* [126], and from *Bacillus anthracis* [127], were characterized as homohexamers with one complete active site per monomer (Figure 19.4). However, the minimal functional unit seems to be a dimer because each active site contains two residues contributed by the neighboring monomer (Figure 19.5). Hence, enzymes belonging to this class may be considered as a trimer of dimers. In contrast to trimeric PNPs, the base-binding site in hexameric enzymes is exposed to solvent, and only one residue – Asp204 – forms a hydrogen bond with the base. Two X-ray structures of *E. coli* PNP, one with FB and one with FA derivatives, m^6FA and m^7FA (formycin ring numbering is used in this section) obtained from the former via hydrolysis during the course of crystallization, were extremely helpful in the clarification of the reaction mechanism [124, 128, 129]. Two conformations of the active site were identified – open binding loosely, and closed binding tightly – as verified by parallel solution studies [128]. In the proposed

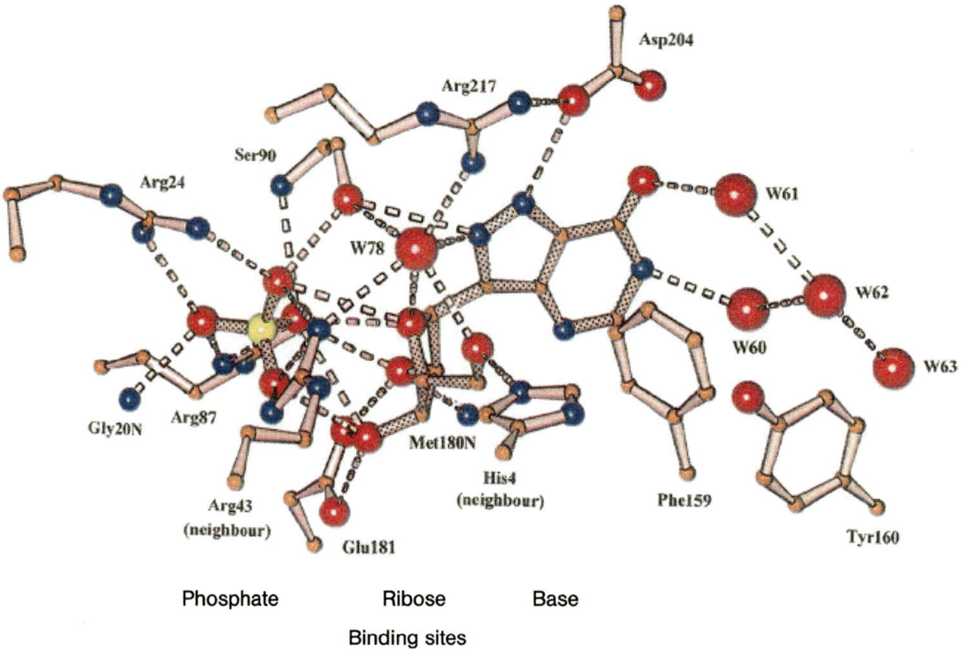

Figure 19.4 Formycin B and phosphate (or sulfate) ion bound in the active site of *E. coli* PNP. (Reprinted from [124], © Elsevier).

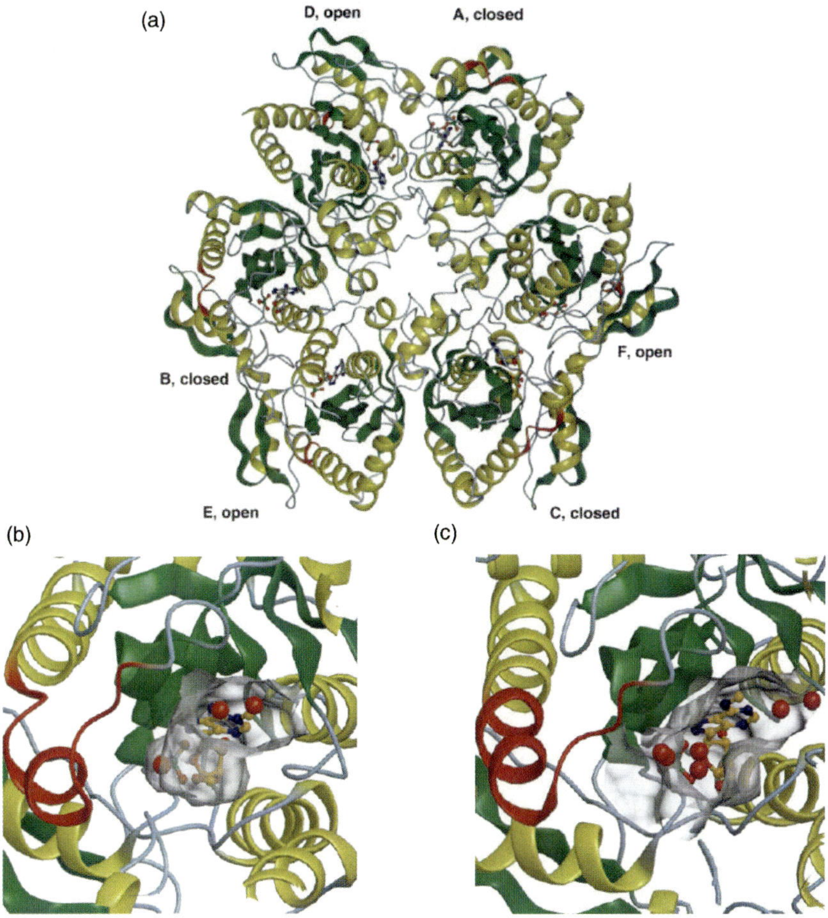

Figure 19.5 (a) A hexamer of *E. coli* PNP complexed with *N*(6)-methylformycin A (m^6FA) and *N*(7)-methylformycin A (m^7FA), and phosphate (or sulfate). The six monomers are symmetry-independent, and form three unsymmetrical dimers (A–D, B–E, and C–F). One monomer in each dimer (A, B, C) binds m^6FA, whereas the other monomers (D, E, F) bind a hydrolysis product, m^7FA (see Scheme 19.2 for chemical structures of the bound ligands). (b, c) The binding pocket of *E. coli* PNP as described by the solvent-accessible surface calculated without the ligands (MSMS; [202]). Middle panel: "Open" or "loose-binding" conformation in monomer D, binding m^7FA. Right panel: "Closed" or "tight-binding" conformation in monomer A, binding m^6FA. Residues in helix H8 undergoing conformational changes are highlighted. Water molecules are shown as balls. (Reprinted from [128], © Elsevier).

mechanism, substrate binding occurs in the open, with the catalytic action taking place in the closed conformation (Figure 19.6). The catalytic process of phosphorolysis resembles the well-known acid-catalyzed hydrolysis – that is, the process which is dependent on prior nucleoside protonation at position N(7) (purine ring numbering). In the case of hexameric PNPs, protonation is achieved with the help of the side chain of Asp204, which is initially in the protonated form. The side chain of Arg217, which

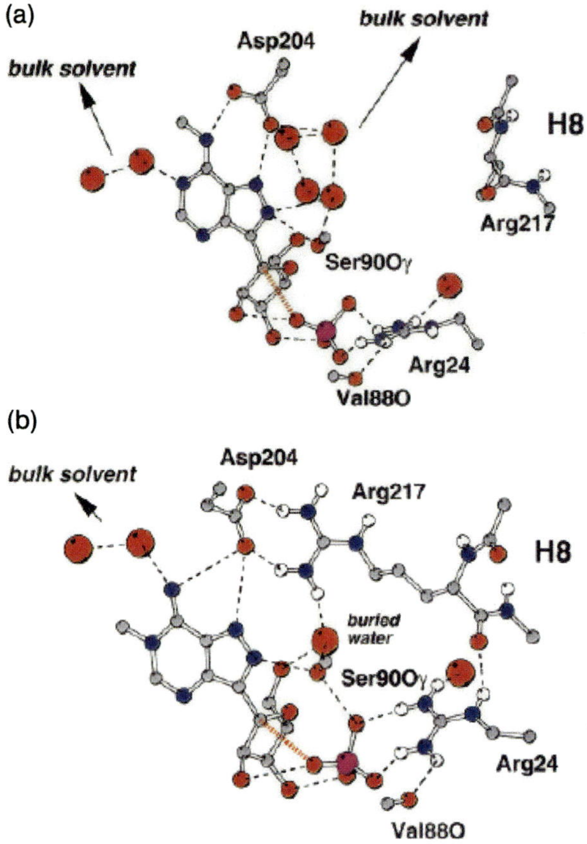

Figure 19.6 Sections of the base binding modes observed in the *E. coli* PNP complexed with m^6FA and phosphate (sulfate). (a) "Open" or "loose-binding" conformation of the active site (observed in monomers D–F) binding a hydrolysis product of m^6FA, m^7FA. (b) The "closed" or "tight-binding" conformation of the active site (observed in monomers A–C) binding m^6FA. (Drawn with MOLSCRIPT [203], reprinted from [128], © Elsevier).

moves close to Asp204 as a consequence of the conformational change from the open to close active site forms, triggers the proton transfer (Figure 19.6). The oxocarbenium ion structure of the ribose ring and the positive charge of the purine base are the key features of the transition state of the reaction.

Formycins bind excellently with hexameric PNPs because they form additional hydrogen bonds between the N(2)-H of the pyrazole ring and the Ser90 O [124]. The second hydrogen bond is formed between the formycin N(1) position (formycin ring numbering) and Asp204, which is protonated [128]. Thus, the tautomeric equilibrium in the PNP/formycin complex is shifted from N(1)-H (which is predominant in solution) to the N(2)-H form, as confirmed also by spectroscopic studies, including enzyme–ligand fluorescence and phosphorescence spectra, and fluorescence intensity decays [60, 130, 131], and fully in line with the proposed catalytic mechanism.

19.6.3
PNP Deficiency and the Potential Role of PNP Inhibitors

PNP deficiency – that is, a lack of PNP activity – is a rare genetic disorder that was first discovered by Gibblet *et al.* [9]. The condition results in abnormalities in purine metabolism and a selective cellular, but not humoral, immunodeficiency. The mechanism of action for cell death in the case of PNP deficiency has been attributed to increased phosphorylation of dGuo. This results in an intracellular accumulation of dGTP – a potent inhibitor of ribonucleotide reductase – and the subsequent inhibition of DNA synthesis in T cells, but not in B cells [107]. However, it has also been suggested [132, 133] that dGTP accumulation is not the only pathway of PNP deficiency leading to T-cell dysfunction (see [107]). Indeed, the mode of MOLT-4 T-cell death following PNP inhibition was recently shown to involve apoptosis with a contribution of a caspase-3-like proteases [134].

The symptoms of PNP deficiency suggested possible chemotherapeutic applications of the potent inhibitors of human PNP. Such compounds are expected to act as selective immunosuppressive drugs, causing selective cellular immunodeficiency with normal B-cell function. They could be used, for example, against the host versus graft reaction in organ transplantation, for the treatment of T-cell-mediated autoimmune diseases such as lupus erythematosus and rheumatoid arthritis, and also against T-cell tumors [107, 134]. They also may be employed for potentiation of the antitumor and antiviral activities of purine nucleoside analogues which, following administration, must reach the target cell and undergo phosphorylation in order to manifest their activities. Many of these analogues are substrates of human PNP and are inactivated by this enzyme.

Due to the high activity of PNP in human tissues, a PNP inhibitor must have an inhibition constant in the nanomolar range in order to be considered for clinical applications. For example, to attain 99.9% inhibition *in vivo* requires an inhibitor concentration of 10 μM and a K_i of 10 nM [55].

In contrast to the high-molecular-mass PNPs, the mechanism of catalysis in the case of low-molecular-mass phosphorylases is still not fully elucidated. An oxocarbenium ion structure of the ribose ring [136], which has been proposed on the basis of structural [120, 137] and kinetic isotope effect studies [138], is generally accepted as the transition state of the reaction. However, uncertainty persists with regards to the extent to which the catalytic process of phosphorolysis resembles the acid-catalyzed hydrolysis; that is, the possible protonation [135, 139]) of the imidazole ring N(7) by Asn243 is not fully documented, and alternative mechanisms have been proposed [121, 137, 140, 141]. Elucidation of the catalytic mechanism of low-molecular-mass PNPs is difficult due to the unusual kinetic behavior of these enzymes caused by molecular phenomena such as the random binding of substrates and the slow, rate-limiting release of some products [117].

19.6.4
Formycins and Analogues as Inhibitors of Mammalian PNPs

Sheen *et al.* [110] first noted the inhibitory activity of FB with human erythrocyte PNP, with a K_i value shown to be 100 µM. The K_i values for FA and FB with another trimeric PNP, murine enzyme, were determined as 670 µM and 280 µM, respectively [111]; hence, formycins are rather weak inhibitors of trimeric PNPs (see Section 19.6.2). Although it was demonstrated that FB inhibits normal circulating lymphocytes and lymphoblastoid cell lines (LCL) with T- and B-cell characteristics, the primary effect was found to be due to a mechanism other than PNP inhibition [142]. In fact, it was suggested that this effect was due to the formation of FB nucleotides and action at other loci (see previous sections).

In 1985, Stoeckler *et al.* [55, 143, 144] found that another class of C-nucleosides – the 9-deazapurine ribonucleosides, 9-deazaGuo and 9-deazaIno – are good inhibitors of the human erythrocyte PNP, with K_i values of 2.9 µM and 2.0 µM, respectively. Hence, 9-deaza purine nucleosides are PNP inhibitors with an at least 30-fold greater affinity for the human enzyme than the corresponding C-nucleosides of the FB series. Moreover, the combination of a 5′-halogeno substitution with a 9-deaza modification leads to even more potent inhibitors. These include 5′-deoxy-5′-iodo-FB, with a K_i of 7 µM as against 100 µM for FB; and 5′-deoxy-5′-iodo-9-deazaIno, with a K_i of 0.18 µM as against 2.0 µM for the parent 9-deazaIno (Scheme 19.6).

As outlined above, the structural changes of formycins as compared with corresponding N-nucleosides, are restricted to atoms located at the 8 and 9 positions of the base (according to the purine ring numbering; see Scheme 19.1). These changes result in differences in hydrogen bond donors and acceptors positions in the five-membered ring (compare Scheme 19.1). As shown later by X-ray crystallography and by modeling studies using the original atomic coordinates of PNPs [e.g., 118–120,122,145,146], enhanced binding by the trimeric PNP of 9-deazaIno, 9-deazaGuo, FB and other 9-deaza analogues, each with a hydrogen at base position N(7) [N(1) in the formycin ring numbering; see Scheme 19.1], is a consequence of better hydrogen bonding of purine substituent O^{δ} to the N^{δ}-H_2, and position N(7)-H to the O^{δ}, of flexible side chain of Asn243, when compared with N-nucleosides lacking a hydrogen bond donor at position N(7) (see Section 19.6.2.1).

The enhanced binding of C-nucleosides when compared with N-nucleosides, and potentiation of this effect in the 9-deaza, five-membered ring (instead of 8-aza-9-deaza) and by additional modification of the pentose portion of the molecule, prompted intensive trials to synthesize other 9-deazapurine nucleosides and acyclonucleosides as potential PNP inhibitors.

X-ray structural analysis and intensive solution studies of complexes of various PNP ligands bound to human and calf phosphorylases showed as mentioned above that 9-deaza analogues acquire inhibitory potency through the donation of a hydrogen bond to the flexible side chain of Asn243 (see Section 19.6.2). Moreover, substitution of the hydrophobic group or a hydrophobic chain at the base position N(9) may enhance binding due to hydrophobic interactions with the pentose binding site [118]. Finally, if such a group or a chain were to be flexible and sufficiently long, its terminal part may

Nucleoside	Chemical structure		K_i [µM] *vs* trimeric PNPs
formycin B	see Scheme 1		100 [a]; 280[b]
formycin A	see Scheme 1		670[b]
5'-deoxy-5'-iodoformycin B	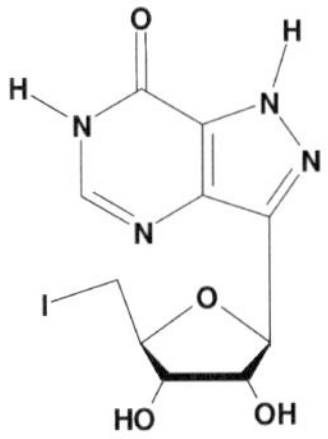		7.0
9-deazainosine	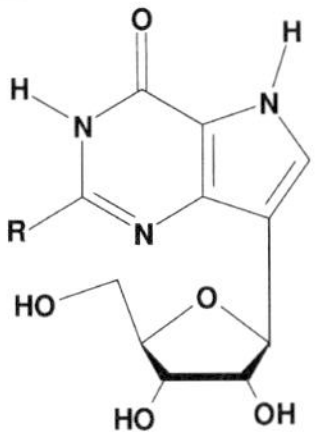	R = H	2.0
9-deazaguanosine		R = NH_2	2.9
5'-deoxy-5'-iodo-9-deazainosine	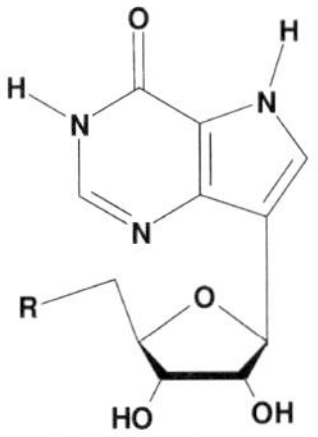	R = I	0.18
5'-deoxy-5'-chloro-9-deazainosine		R = Cl	0.20

Scheme 19.6 Properties of formycins, 9-deazapurine nucleosides and their 5′-halogenated derivatives as inhibitors of trimeric PNPs. Unless otherwise indicated, the inhibition constants (K_i) are from [55] and were determined versus human erythrocyte PNP in the presence of 50 mM phosphate.
a) From [110]
b) Murine PNP; data from [111]

additionally interact with the phosphate binding site, thereby competing for this site with the orthophosphate which is one of the PNP substrates (see the reaction scheme in Section 19.6.1). In line with the latter finding, the IC_{50} and K_i values for many inhibitors depend on the concentration of phosphate in the assay mixture. The 1 mM concentration, which is close to the intracellular concentration [147], is now routinely employed,

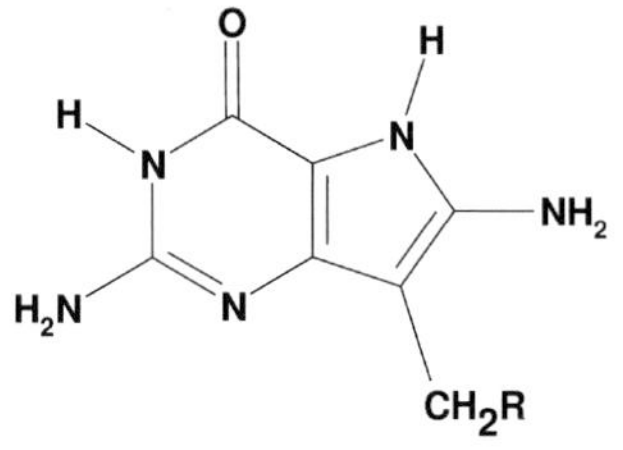

R	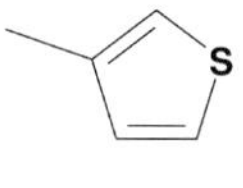	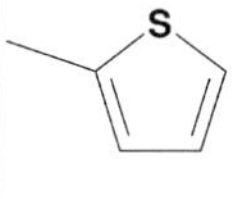	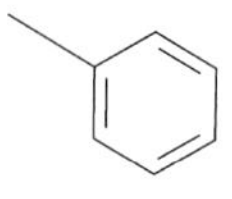	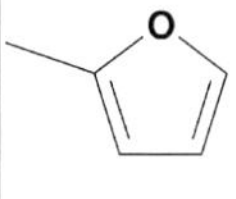	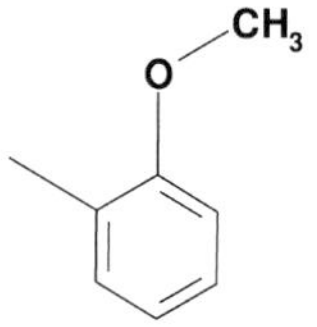
IC_{50} [µM] 50 mM P_i	0.9	1.0	1.6 (0.85)	5.4 (0.75)	31.4

Effect of the base (IC_{50} [µM] vs human PNP in the presence of 50 mM phosphate)

	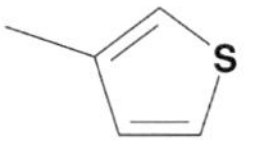	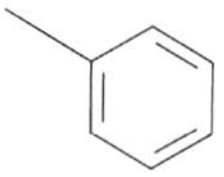	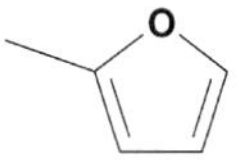
guanine	11.0	29.6	17.9
8-aminoguanine	0.16	0.47	0.25
9-deazaguanine	0.03	0.085	0.057
8-amino-9-deazaguanine	1.0	0.85 (1.6)	0.75 (5.4)

Scheme 19.7 Upper panel: Inhibitors of trimeric PNP with 8-amino-9-deazaguanine aglycone and the pentose ring replaced by the arylmethyl substituent. IC_{50} were determined versus human erythrocyte enzyme in the presence of 50 mM P_i [148, 204]; data in brackets are from [145]. The analogue 8-amino-9-deaza-9-(3-thienylmethyl) Gua (first from left) was designated CI-972 and under this name was tested as a potential T-cell-selective immunosuppressive agent [e.g., [148, 151–153]]. Lower panel: Effect of base on the inhibitory properties. Data are from [145] for 9-deazaguanine analogues, and from [206] for guanine analogues and data shown in brackets.

so that measured IC_{50} and K_i values under these conditions may more adequately reflect the intracellular effectiveness of an inhibitor.

In 1991, two groups reported first PNP inhibitors with 9-deazapurine aglycone and various groups replacing ribose. Sircar *et al.* [148] synthesized several 9-deaza-8-amino-9-arylmethylguanine derivatives (Scheme 19.7), while Ealick *et al.* [118], Montgomery *et al.* [145], Secrist *et al.* [149] and Erion *et al.* [150] all reported a much broader series of 9-substituted 9-deazapurine analogues with various aromatic, heteroaromatic, and cyclic aliphatic 9-substituents (Scheme 19.8). 8-Amino-9-deazaguanine analogues are relatively poor inhibitors, with IC_{50} (at 50 mM phosphate) of approximately 1 µM; this is due to the fact that, as shown by X-ray studies [118], an additional 8-amino substituent prevents formation of the strong hydrogen bond via

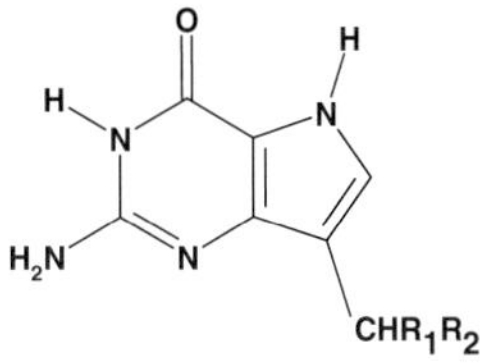

R_1	(*S*)-3-chlorophenyl 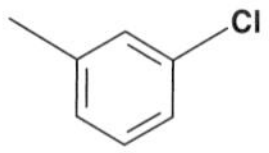	(*R*,*S*) 3-chlorophenyl 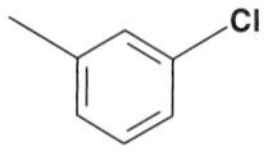	2-tetrahydrothienyl 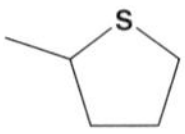	3,4-dichlorophenyl 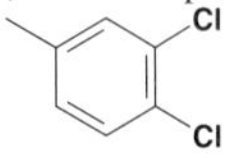
R_2	CH_2COOH	CH_2CN	H	H
IC_{50} (μM) 1 mM P_i	0.0059	0.010	0.011	0.012
K_i (μM) 1 mM P_i	0.004	0.011	0.009	–
IC_{50} (μM) 50 mM P_i	0.031	1.8	0.22	0.25

R_1	3-thienyl 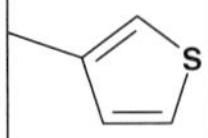	cyclopentyl 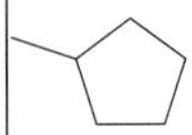	3-pyridinyl 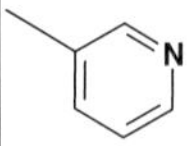	cycloheptyl 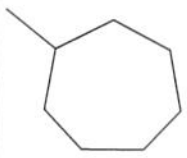	cyclohexyl 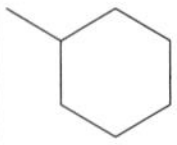	(*R*)-3-chlorophenyl 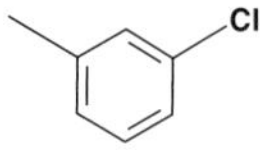
R_2	H	H	H	H	H	CH_2COOH
IC_{50} (μM) 1 mM P_i	0.020	0.029	0.030	0.030	0.043	0.160
K_i (μM) 1 mM P_i	0.016	0.009	–	–	0.013	0.149
IC_{50} (μM) 50 mM P_i	0.08	1.8	0.20	0.86	2.0	0.90

Scheme 19.8 Inhibitors of trimeric PNPs with 9-deazaguanine aglycone and the pentose ring replaced by other cyclic moieties. IC_{50} were determined versus calf spleen enzyme in the presence of 50 mM and 1 mM P_i [118]; K_i values versus Ino, at 1 mM P_i are from Farutin *et al.* [205]. Two of the analogues, 9-deaza-9-(3-thienylmethyl)Gua (lower table, first from left) and 9-deaza-9-(3-pyridinylmethyl)Gua (lower table, third from left) were tested under the names PD 141955 (also CI-1000 and BCX-5) and BCX-34 (peldesine), respectively as potential T-cell-selective immunosuppressive agents (e.g., [151–155]).

position N(7) of the 9-deazaguanine ring (see Figures 19.2 and 19.3). Several compounds with the 9-deazapurine aglycone exhibit good affinity for trimeric PNPs, with IC_{50} values (at 50 mM phosphate) about 10-fold lower. The best in this series, versus calf spleen PNP, was (with 50 mM phosphate) 9-(3-thienylmethyl)-9-deazaguanine, with $IC_{50} = 80$ nM (Scheme 19.8, lower table, first entry). However, some other compounds showed a higher IC_{50} (at 1 mM phosphate)/IC_{50} (at 50 mM phosphate) ratio, so at 1 mM phosphate the most potent inhibitor in this series was the (*S*)-9-[1-(3-chlorophenyl)-2-carboxyethyl]-9-deazaguanine, with $IC_{50} = 5.9$ nM (Scheme 19.8, upper table, first entry).

Two of such analogues, 9-deaza-9-(3-thienylmethyl)Gua and 9-deaza-9-(3-pyridinylmethyl)Gua (Scheme 19.8, lower table, first and third from the left, respectively) were tested under the names PD 141955 (also CI-1000 and BCX-5) and BCX-34 (peldesine) as potential T-cell-selective immunosuppressive agents, for the treatment of cutaneous T-cell lymphoma, acute lymphoblastic leukemia, HIV infections, and psoriasis [151–155]. BCX-34 was shown to inhibit the proliferation of normal human peripheral blood mononuclear cells (PBMC), and CD4 and CD8 cells, variously stimulated to induce proliferation, with IC_{50} values in the range 0.2 to 15 μM [154]. The antiproliferative effect on T- (but not B-) cell lines was not due to suppression of interleukin-2 production [156, 157]. This suggested that BCX-34 might serve as a selective T-cell immunosuppressive drug for use in combination, not competition, with cyclosporine A. The suppression of proliferation of the human T-lymphoblastic leukemia (CCRF-CEM) cell line by BCX-34 was accompanied by an accumulation of intracellular dGTP and a reduction of GTP. The mechanism appears to be dGuo-dependent, in contrast to normal human T cells, where it is both dGuo-dependent and dGuo-independent [154, 158]. BCX-34 is orally bioavailable (76%) in rats, and pharmacologically active in rodents after oral dosing, with no apparent toxicity or adverse side effects (e.g., [155, 158]). However, following these initial somewhat optimistic reports, BCX-34 proved not to be sufficiently active for consideration as a drug [159]. The oxocarbenium ion structure of the ribose ring proposed for the PNP transition state, led to the development of stable transition-state analogues, the immucillins. These iminoribitol C-nucleosides were seen to interact with trimeric PNPs, with picomolar inhibition constants [136], and to date are the most effective PNP inhibitors known (see Chapter 8). This finding subsequently led to the synthesis of novel ground-state PNP inhibitors being much less intensively investigated.

19.7
Formycins as Inhibitors of Parasitic PNPs and Hydrolases

Nucleoside hydrolases, which non-phosphorolytically and irreversibly cleave the glycosidic bond of not only purine but also, in some instances, pyrimidine nucleosides are widely distributed in protozoan parasites, yet are conspicuously absent from host mammalian cells (e.g., [125, 160]). Protozoan parasites lack *de-novo* purine synthesis, and so are dependent on purine salvage to replenish their nucleotide pools. Hence, the purine salvage system, including purine nucleoside phosphorylase and/or purine nucleoside hydrolase, presents several potential targets for the design of selective antiparasitic drugs.

FB is a competitive inhibitor of PNP from the protozoan *Dictyostelium discoideum* (a soil-living amoeba), with a K_i of 14 μM [161], and also of PNP from human and bird malarial parasites, *Plasmodium lophurae* and *Plasmodium falciparum*, with K_i values of 0.4 μM and 1.1 μM, respectively [113, 162]. FA, an analogue of Ado which is known to be an inhibitor of *E. coli* PNP without any effect on mammalian PNPs, was shown to inhibit PNP from a parasitic protozoan *Trichomonas vaginalis* with a K_i of 2.3 μM by competing with Ado [163]. FB inhibits purine nucleoside hydrolase from *Trypanosoma gambiense* [164] and *Trypanosoma brucei brucei*, in the latter case with a K_i of 13 μM [165].

Formycins should be modified to become useful antiparasitic drugs, as both are toxic (FA much more than FB) not only to parasite but also to mammalian cells due to the 5′-phosphorylation discussed in Section 19.5. The metabolism of pyrazolopyrimidines in parasitic protozoa, which was reviewed some time ago by Ullman [166], included the possible use of FA and FB as chemotherapeutic agents. The potential role of purine and pyrimidine nucleoside analogues (including FB) in the chemotherapy of leishmaniasis was summarized by Ram and Nath [167].

19.8 Actual and Potential Applications of Formycins

19.8.1 Formycin and Analogues in Assays of Enzyme Activity

As discussed previously, FA is a substrate for ADA from various sources. In fact, it is an excellent substrate, with a rate of deamination which is several-fold higher than that for adenosine. Such deamination may easily be followed spectrophotometrically due to the marked differences between the UV spectra of FA and FB. It is also possible to use fluorescence detection since the product, FB, is non-florescent at neutral pH and only slightly so at alkaline pH (see Figure 19.1). FA is therefore a superior substrate both for routine assays in crude extracts and for kinetic studies of the purified ADA [168].

The continuous fluorimetric assay for AMP deaminase makes use of FMP which undergoes deamination to formycin B 5′-phosphate, which does not fluoresce at neutral pH. The deamination of FMP may also be followed spectrophotometrically at 306 nm, thus permitting better assays of crude extracts [169].

The modification in UV absorption and fluorescence emission spectra accompanying dephosphorylation of FMP may be employed in the assay of 5′-nucleotidase activity. Moreover, the sensitivity of this method may be enhanced, especially when fluorescence monitoring is used, by coupling the reaction with ADA, which deaminates FA more quickly than Ado. As the final product, FB, is non-florescent at neutral pH (see Figure 19.1), this method permits the use of substrate concentration as low as 1 μM [62].

FTP is also a useful substrate of the membrane-bound adenylate cyclase, the reaction product of which – 3′,5′-cyclic formycin monophosphate – is separated from the substrate by reverse-phase high-performance liquid chromatography (HPLC), after which its presence is detected by fluorimetry. Such an approach provides an alternative, non-radioactive direct assay of adenylate cyclase catalytic activity [170]. A similar chromatographic assay procedure for several other enzymatic activities, using FA analogues of adenosine 5′-mono, 5′-tri, and cyclic 3′,5′-monophosphate as substrates, has also been proposed [171].

19.8.2
Formycin and Analogues as Protein Ligands for X-Ray Structural Studies

Formycins and its nucleotides with C–C glycosidic bonds resistant to enzymatic cleavage were used as ligands in the X-ray structure determination of various proteins, and not only from the PNP family. For example, FA was complexed with 5′-methylthioadenosine phosphorylase (MTAP) from the thermostable archea *Sulfolobus solfataricus* and from human sources [172, 173], and with methylthioadenosine/*S*-adenosylhomocysteine nucleosidase (MTAN) from *E. coli* [174] to obtain the 3-D structure of these enzymes. The hexameric structure of MTAP is similar to that of PNP from *E. coli*, although only SsMTAP accepts 5′-deoxy-5′-methylthioadenosine as a substrate. Hypoxanthine phosphoribosyltransferase of *Trypanosoma cruzii* was crystallized in a complex with FB [175]. Another example is that of AMP nucleosidase (this enzyme is present only in prokaryotes) which, by solving the structure of its complex with FMP, was shown to be a new member of the high-molecular-mass PNP family (see Section 19.6.2.2) [176]. The structure of several ribosome-inactivating enzymes which, in the N-ribohydrolase reaction remove a specific adenine residue from the ribosome, was obtained in the complex with FMP. These included alpha-momorcharin from *Momordica charantia* [177], ricin A-chain [178], and pokeweed (*Phytolacca americana*) antiviral protein from leaves [179] and from seeds [PDB entry 1J1S, to be published].

19.8.3
Formycins and Analogues as Molecular Probes

FA and its nucleosides have been widely used in both biochemical and biophysical studies of numerous macromolecules. In many cases, such approaches profited from changes in the fluorescence of FA and its nucleotides upon binding to proteins due to resonance energy transfer to appropriately labeled protein residues (e.g., tyrosines and lysines), or upon incorporation into polymers and base-pairing in helical, double-stranded structures (see Section 19.3). FA fluorescence in many cases is also sensitive to conformational changes of the binding macromolecule, a point utilized in the elucidation of the structure and function of myosin. This included probing myosin's ATPase activity with FTP (which replaces ATP as a substrate), and probing the binding properties with the non-cleavable analogue formycin A 5′-[βγ-imido]triphosphate (e.g., [180, 181]). In another series of experiments, the spatial relationship between the nucleotide-binding site and Lys-61 in actin and a conformational change induced by myosin subfragment-1 binding, and the distance separating Tyr69 from the high-affinity nucleotide in actin, were studied by the resonance energy transfer using FTP as the fluorescent ATP analogue [182, 183].

The fluorescence of FA was also used to monitor the conformational changes at the 3'-terminus of tRNA caused by aminoacylation and hydrolysis of aminoacyl residue from aminoacyl-tRNAs. Fluorescent tRNAs species with FA in the 3′-terminal position (tRNA-CCF) were derived from *E. coli* tRNA(Val), and *Thermus thermophilus* tRNA(Asp) and tRNA(Phe) [184]. The kinetics of the product ternary complex formation by adenine phosphoribosyltransferase from *Leishmania donovani* was investigated using FMP as a fluorescent probe [185].

As mentioned earlier, resonance energy transfer occurs from FA to terbium ions, and it was shown that the terbium–FTP complex binds with – but is not a substrate for – numerous ATPases. Thus, the complex used, for example, to investigate the nucleotide-binding site of the yeast *Schizosaccharomyces pombe* plasma membrane H^+-ATPase, and to characterize the change of conformation triggered by H^+ in the yeast plasma membrane H^+-ATPase [61, 186]. Evidence of a calcium-induced structural change in the ATP-binding site of the sarcoplasmic reticulum Ca^{2+}-ATPase was also presented using terbium–FTP as an analogue of Mg-ATP [187]. The use of formycin nucleotides to measure the conformational states of Na^+,K^+-ATPase was reviewed by Karlish [188]. The kinetics of nucleotide binding to pyruvate carboxylase have also been studied by measuring the fluorescence changes that occur on the binding and release of FTP and FDP [189]. FA was also incorporated into the dinucleoside cap structure, m^7GpppAdo (where m^7G is 7-methylguanosine) [190].

In other experimental approaches, the enzyme structure and mechanisms were investigated by utilizing the fact that FA and it nucleotides are analogues of Ado and its nucleotides, with an elevated pK_a for protonation at the five-membered ring (see Section 19.2.1), and with the C–C glycosidic bond being resistant to hydrolysis and phosphorolysis (see Sections 19.6 and 19.7). For example, the mechanism of the ricin-A chain, which catalyzes the depurination of a single adenine at position 4324 of 28S rRNA in a N-ribohydrolase reaction, was studied by the incorporation of FA at the depurination site [191]. Interactions of the MutY enzyme (the DNA repair enzyme that removes misincorporated adenine residues) with DNA duplexes containing 2′-deoxyformycin A were studied. Subsequently, 2′-deoxyformycin A was found effectively to mimic the recognition properties of 2′-deoxyadenosine, but was resistant to the glycosylase activity of MutY, owing to structural properties [192].

19.8.4
Formycins as Ligands in Affinity Chromatography

As non-cleavable substrate analogues of PNP, formycins proved to be very useful in affinity chromatography. The use of FB, with a K_i of approximately 100 μM versus human PNP [110], permitted the purification of this enzyme to homogeneity in two steps, and with an overall yield of about 60% [193]. FB in this case was linked to Sepharose 4B. Later, the same nucleoside was linked with Sepharose 6B to provide an excellent affinity resin for PNP for the enzyme from *E. coli* [194]. These authors reported that after five years' of use, and exposure to a variety of biological preparations, the resin showed no detectable decrease in its ability to bind PNP. In the present authors' group, a column with formycin B linked to epoxy-activated Sepharose 6B was used for almost 10 years, with no change in PNP binding properties. The capacity of the 50-mL column was estimated to be to 80 mg of *E. coli* PNP (i.e., 160 units mL^{-1}) [195, 196].

19.8.5
Formycin B as a Tool to Study Nucleoside Transport

As an analogue of inosine (Ino), FB is not catabolized by mammalian cells and is only very inefficiently phosphorylated in them (see Section 19.5). Nonetheless, this relative inertness allows the transport of FB into and out of the cells to be measured, as noted by Plagemann and Wolfendin [197]. FB was first used as an unmetabolized analogue of Ino to study sodium-dependent, concentrative nucleoside transport in cultured intestinal epithelial cells [198]. Since then, it has become a well-established substrate for nucleoside transporters including equlibrative *es* and *ei* (equilibrative inhibitor-sensitive and equilibrative inhibitor-insensitive) as well as concentrative Na^+-dependent transporter types. It has also been widely used to study the properties of nucleoside transporters in various normal and tumor mammalian cells (e.g., [199–201]).

Abbreviations

FA	formycin A, formycin, 9-deaza-8-azaAdo, 7-Amino-3-β-D-ribofuranosyl-1*H*-pyrazolo[4,3-*d*]pyrimidine
FB	formycin B, laurusin, 9-deaza-8-azaIno, 1,4-Dihydro-3-β-D-ribofuranosyl-7*H*-pyrazolo[4,3-*d*]pyrimidin-1-one
m^1FA	*N*(1)-methyl formycin A, with similar connotations for other methylated formycin analogues
FTP	formycin-5′-triphosphate, with analogous abbreviations for monophosphate FMP, and diphosphate, FDP
Ado	adenosine
Ino	inosine
Guo	guanosine
Xao	xanthosine
PNP	purine nucleoside phosphorylase
ADA	adenosine deaminase

References

1 Davies, F. F., Allen, F. W., *J. Biol. Chem.* **1957**, *227*, 907–915.
2 Cohn, W. E., *J. Biol. Chem.* **1960**, *235*, 1488–1498.
3 Hori, M., Ito, E., Takita, T., Koyama, G., Takeuchi, T., Umezawa, H., *J. Antibiot. (Tokyo)* **1964**, *17*, 96–99.
4 Koyama, G., Umezawa, H., *J. Antibiot. (Tokyo)* **1965**, *18*, 175–177.
5 Sawa, T., Fukagawa, Y., Homma, I., Wakashiro, T., Takeuchi, T., *J. Antibiot. (Tokyo)* **1968**, *21*, 334–339.
6 Sawa, T., Fukagawa, Y., Homma, I., Takeuchi, T., Umezawa, H., *J. Antibiot. (Tokyo)* **1967**, *20*, 1–4.
7 Ward, D. C., Reich, E., Stryer, L., *J. Biol. Chem.* **1969**, *244*, 1228–1237.

8 Wierzchowski, J., Shugar, D., *Photochem. Photobiol.* **1982**, *35*, 445–458.

9 Giblett, E. R., Ammann, A. J., Wara, D. W., Sandman, R., Diamond, L. K., *Lancet* **1975**, *1*, 1010–1013.

10 Suhadolnik, R. J., *Nucleoside Antibiotics*, J. Wiley, New York, 1970.

11 Suhadolnik, R. J., *Prog. Nucleic Acid Res. Mol. Biol.* **1979**, *22*, 193–291.

12 Suhadolnik, R. J., *Nucleosides as Biological Probes*, J. Wiley, New York, **1979**.

13 Daves, G. D., Cheng, C. C., *Prog. Med. Chem.* **1976**, *13*, 304–349.

14 Hanessian, S., Pernet, A. G., *Adv. Carbohydr. Chem. Biochem.* **1976**, *33*, 111–188.

15 Goodchild, J., The biochemistry of nucleoside antibiotics, in *Topics in Antibiotic Chemistry, Vol.6* (ed. P. G. Sammes), Ellis Horwood Ltd. Chichester, UK, **1982**, pp. 99–228.

16 Buchanan, J. G., *Fortschr. Chem. Org. Naturst.* **1983**, *44*, 243–299.

17 Hacksell, U., Daves, G. D. Jr. *Prog. Med. Chem.* **1985**, *22*, 1–65.

18 Robins, R. K., Townsend, L. B., Cassidy, F., Gerster, J. F., Lewis, A. F., Miller, R. L., *J. Heterocyclic Chem.* **1966**, *3*, 110–114.

19 Koyama, G., Maeda, K., Umezawa, H., Iitaka, Y., *Tetahedron Lett.* **1966**, 597–602.

20 Kawamura, K., Fukatsu, S., Murase, G., Koyama, G., Maeda, K., Umezawa, H., *J. Antibiot. (Tokyo)* **1966**, *19*, 91–92.

21 Prusiner, P., Brennar, T., Sundaraligam, M., *Biochemistry* **1973**, *12*, 1196–1202.

22 Koyama, G., Umezawa, H., Iitaka, Y., *Acta Crystallogr. Sect. B* **1974**, *30*, 1511–1516.

23 Koyama, G., Nakamura, H., Umezawa, H., Iitaka, Y., *Acta Crystallogr. Sect. B* **1976**, *32*, 813–820.

24 Shugar, D., Psoda, A., Tautomerism of purines and pyrimidines, their nucleosides and various analogues, in *Landoldt-Börnstein, Vol. VII/1d*, Springer-Verlag, Heidelberg, **1990**, pp. 308–348.

25 Chenon, M. T., Panzica, R. P., Smith, J. C., Pugimire, R. J., Grant, D. M., Townsend, L. B., *J. Am. Chem. Soc.* **1976**, *98*, 4736–4745.

26 Cho, B. P., McGregor, M. A., *Nucleosides Nucleotides* **1994**, *13*, 481–490.

27 Caesar, G. P., Greene, J. J., *J. Med. Chem.* **1974**, *17*, 1122–1134.

28 Wierzchowski, J., *PhD Thesis*, University of Warsaw, Warsaw, Poland, **1980**.

29 Giziewicz, J., Shugar, D., *Acta Biochim. Pol.* **1977**, *24*, 231–246.

30 Markowski, V., Sullivan, G. B., Roberts, D. J., *J. Am. Chem. Soc.* **1977**, *99*, 714–718.

31 Davies, R. J., *Z. Naturforsch. [C]* **1975**, *30*, 835–837.

32 Abola, E. M., Sims, M. J., Abraham, D. J., Lewis, A. F., Townsend, L. B., *J. Med. Chem.* **1974**, *17*, 62–65.

33 Ludemann, H. D., Westhof, E., *Z. Naturforsch. [C]* **1977**, *32*, 528–538.

34 Ward, D. C., Reich, E., *Proc. Natl. Acad. Sci. USA* **1968**, *61*, 1494–1501.

35 Ward, D. C., Fuller, W., Reich, E., *Proc. Natl. Acad. Sci. USA* **1969**, *62*, 581–585.

36 Ward, D. C., Cerami, A., Reich, E., Acs, G., Altwreger, L., *J. Biol. Chem.* **1969**, *244*, 3243–3250.

37 Townsend, L. B., Long, R. A., McGraw, J. P., Miles, D. W., Robins, R. K., Eyring, H., *J. Org. Chem.* **1974**, *39*, 2023–2027.

38 Makabe, O., Nakamura, N., Umezawa, S., *J. Antibiot. (Tokyo)* **1975**, *28*, 492–495.

39 Lewis, A. F., Townsend, L. B., *J. Am. Chem. Soc.* **1980**, *102*, 2817–2823.

40 Crabtree, G. W., Agarwal, R. P., Parks, R. E. Jr., Lewis, A. F., Wotring, L. L., Townsend, L. B., *Biochem. Pharmacol.* **1979**, *28*, 1491–1500.

41 Watanabe, S., Matsuhashi, G., Fukatsu, S., Koyama, G., Maeda, K., Umezawa, H., *J. Antibiot. (Tokyo)* **1966**, *19*, 93–96.

42 Long, R. A., Lewis, A. F., Robins, R. K., Townsend, L. B., *J. Chem. Soc. Perkin Trans. I* **1971**, *13*, 2443–2446.

43 Secrist, J. A. III, Shortnacy, A. T., Montgomery, J. A., *J. Med. Chem.* **1985**, *28*, 1840–1842.

44 Upadhya, K. G., Sanghvi, Y. S., Robins, R. L., Revankar, G. R., Ugarkar, B. G., *Nucleic Acids Res.* **1986**, *14*, 1747–1764.

45 Montgomery, J. A., *Prog. Med. Chem.* **1970**, *7*, 69–123.

46 Montgomery, J. A., *Cancer Res.* **1982**, *42*, 3911–3917.

47 Lindell, S. D., Moloney, B. A., Hewitt, B. D., Ernshaw, C. G., Dudfield, P. J., Dancer, J. E., *Bioorg. Med. Chem. Lett.* **1999**, *9*, 1985–1990.

48 Kourafalos, V. N., Marakos, P., Pouli, N., Townsend, L. B., *J. Org. Chem.* **2003**, *68*, 6466–6469.

49 Robins, M. J., McCarthy, J. R. Jr., Jones, R. A., Mengel, R., *Can. J. Chem.* **1973**, *51*, 1313.

50 Jain, T. C., Russell, A. F., Moffat, J. G., *J. Org. Chem.* **1973**, *38*, 3179–3189.

51 Roskovsky, A., Solna, V. C., Gudas, L. A., *J. Med. Chem.* **1985**, *28*, 1096–1099.

52 De Clercq Balzarini, J., Madej, D., Hansske, F., Robins M. J., *J. Med. Chem.* **1987**, *30* 481–486.

53 Neidle, S., Urpi, L., Serafinowski, P., Whitby, D., *Biochem. Biophys. Res. Commun.* **1989**, *161*, 910–916.

54 Seela, F., Chen, Y. M., Melenewski, A., Rosemeyer, H., Wei, C. F., *Acta Biochim. Pol.* **1996**, *43*, 45–52.

55 Stoeckler, J. D., Ealick, S. E., Bugg, C. E., Parks, R. E. Jr. *Fed. Proc.* **1986**, *45*, 2773–2778.

56 Chern, J. W., Lee, Y. H., Chen, C. S., Schewach, D. S., Daddona, P. E., Townsend, L. B., *J. Med. Chem.* **1993**, *36*, 1024–1031.

57 Zhou, J., Yang, M. M., Schneller, S. W., *Tetrahedron Lett.* **2004**, *452*, 8233–8234.

58 Zhou, J., Yang, M. M., Akdag, A., Schneller, S. W., *Tetrahedron* **2006**, *62*, 7009–7013.

59 Giranda, V. L., Berman, H. M., Schramm, V. L., *Biochemistry* **1988**, *27*, 5813–5818.

60 Kierdaszuk, B., Modrak-Wojcik, A., Wierzchowski, J., Shugar, D., *Biochim. Biophys. Acta* **2000**, *1476*, 109–128.

61 Ronjat, M., Lacapere, J. J., Dufour, J. P., Dupont, Y., *J. Biol Chem.* **1987**, *262*, 3146–3153.

62 Wierzchowski, J., Lassota, P., Shugar, D., *Biochim. Biophys. Acta* **1984**, *786*, 170–178.

63 Aizawa, S., Hidaka, T., Otake, N., Yonehara, H., Isono, K., Igarashi, N., Suzuki, S., *Agric. Biol. Chem.* **1965**, *29*, 375–376.

64 Kunimoto, T., Sawa, T., Wakashiro, T., Hori, M., Umezawa, H., *J. Antibiot. (Tokyo)* **1971**, *204*, 253–258.

65 Ochi, K., Iwamoto, S., Hayase, E., Yashima, S., Okami, Y., *J. Antibiot. (Tokyo)* **1974**, *27*, 909–916.

66 Ochi, K., Yashima, S., Eguhi, Y., *J. Antibiot. (Tokyo)* **1975**, *28*, 965–973.

67 Ochi, K., Kikuchi, S., Yashima, S., Eguchi, Y., *J. Antibiot. (Tokyo)* **1976**, *29*, 638–645.

68 Ochi, K., Yashima, S., Eguchi, Y., Matsushita, K., *J. Biol. Chem.* **1979**, *254*, 8819–8824.

69 Buchanan, J. G., Hamblin, M. R., Sood, G. R., Wightman, R. H., *Chem. Commun.* **1980**, 917.

70 Bobek, M., Farkas, J., Sorm, F., *Tetrahedron Lett.* **1970**, *52*, 4611–4614.

71 Farkas, J., Sorm, F., *Collect. Czech. Chem. Commun.* **1972**, *37*, 2798–2803.

72 Bobek, M., Farkas, J., *Coll. Czech. Chem. Commun.* **1968**, *34*, 247.

73 Acton, E. M., Ryan, K. J., Henry, W., Goodman, L., *Chem. Commun.* **1971**, 986–988.

74 Kalvoda, L., *Collect. Czech. Chem. Commun.* **1978**, *43*, 1431–1437.

75 Buchanan, J. G., Edgar, A. R., Stobie, A., Wightman, R. H., *J. Chem. Soc. Perkin Trans. I* **1980**, 2567–2571.

76 Buchanan, J. G., Stobie, A., Wightman, R. H., *Can. J. Chem.* **1980**, *58*, 2624–2627.

77 Ishizuka, M., Takeuchi, T., Nitta, K., Koyama, G., Hori, M., Umezawa, H., *J. Antibiot. (Tokyo)* **1964**, *17*, 124–126.

78 Ishizuka, M., Sawa, T., Hori, S., Takazama, H., Takeuchi, T., *J. Antibiot. (Tokyo)* **1968**, *21*, 5–12.

79 Umezawa, H., Sawa, T., Fukagawa, Y., Homa, I., Ishizuka, M., *J. Antibiot. (Tokyo)* **1967**, *20*, 308–316.

80 Kunimoto, T., Hori, M., Umezawa, H., *J. Antibiot. (Tokyo)* **1967**, *20*, 277–281.

81 Muller, W. E. G., Rohde, H. J., Steffen, R., Maidhof, A., Lachmann, M., Zahn, R. K.,

Umezawa, H., *Cancer Res.* **1975**, *35*, 3673–3681.

82 Hori, M., Wakshiro, T., Ito, E., Sawa, T., Takeuchi, T., *J. Antibiot. (Tokyo)* **1968**, *21*, 264–271.

83 Ishida, N., Homma, M., Kumagai, K., Shimizu, Y., Matsumoto, S., *J. Antibiot. (Tokyo)* **1967**, *20*, 49–52.

84 Takeuchi, T., Iwanaga, J., Aoyagi, T., Umezawa, H., *J. Antibiot. (Tokyo)* **1966**, *19*, 286–287.

85 Ishida, N., Izawa, A., Homma, M., Kumagai, K., Shimizu, Y., *J. Antibiot. (Tokyo)* **1967**, *20*, 129–131.

86 Huang, K. T., Katagiri, M., Mistao, T., *J. Antibiot. (Tokyo)* **1966**, *19*, 75–77.

87 Tezeuka, N., Hirai, T., *Jpn. J. Microbiol.* **1969**, *13*, 367–374.

88 Sarich, L., Agoutin, B., Lecoq, O., Weill, D., Jullien, P., Heyman, T., *Virology* **1985**, *145*, 171–175.

89 Takiguchi, Y., Yoshikawa, H., Terao, M., *Appl. Environ. Microbiol.* **1978**, *36*, 658–661.

90 Malaisse, W. J., Sener, A., Gruber, H. E., Erion, M. D., *Biochem. Med. Metab. Biol.* **1994**, *53*, 22–27.

91 Raschaert, J., Tanigawa, K., Mallaisse, W. J., *Biochem. Mol. Med.* **1995**, *54*, 138–141.

92 Sato, Y., Henquin, J. C., *Diabetes* **1998**, *47*, 1713–1721.

93 Caldwell, I. C., Henderson, J. F., Paterson, A. R., *Can. J. Biochem.* **1969**, *47*, 901–908.

94 Dye, F. J., Rossomando, E. F., *Biosci. Rep.* **1982**, *2*, 229–2234.

95 Ishizuka, M., Sawa, T., Koyama, G., Takeuchi, T., Umezawa, H., *J. Antibiot. (Tokyo)* **1968**, *21*, 1–4.

96 Spector, T., Jones, T. E., LaFon, S. W., Nelson, D. J., Berens, R. L., Marr, J. J., *Biochem. Pharmacol.* **1984**, *33*, 1611–1617.

97 Henderson, J. F., Paterson, A. R., Caldwell, I. C., Hori, C., *Cancer Res.* **1967**, *27*, 715–719.

98 Glazer, R. I., Lloyd, L. S., *Biochem. Pharmacol.* **1982**, *31*, 3207–3214.

99 Metha, K. D., Gupta, R. S., *Biochem. Biophys. Res. Commun.* **1985**, *130*, 910–917.

100 LaFon, S. W., Nelson, D. J., Berens, R. L., Marr, J. J., *J. Biol. Chem.* **1985**, *260*, 9660–9665.

101 Metha, K. D., Gupta, R. S., *Mol. Cell. Biol.* **1983**, *3*, 1468–1477.

102 Rainer, P., Santi, D. V., *Proc. Natl. Acad. Sci. USA* **1983**, *80*, 288–292.

103 Carson, D. A., Chang, K. P., *Biochem. Biophys. Res. Commun.* **1981**, *100*, 1377–1383.

104 McCabee, R. E., Remington, J. S., Araujo, F. G., *Antimicrob. Agents Chemother.* **1985**, *27*, 491–494.

105 Berman, J. D., Hanson, W. L., Lovelace, J. K., Waits, V. B., Jackson, J. E., Chapman, W. L. Jr., Klein, R. S., *Antimicrob. Agents Chemother.* **1987**, *31*, 111–113.

106 Lawton, P., Hejl, C., Mancassola, R., Niciri, M., Petavy, A. F., *FEMS Microbiol. Lett.* **2003**, *226*, 39–43.

107 Bzowska, A., Kulikowska, E., Shugar, D., *Pharmacol. Ther.* **2000**, *88*, 349–425.

108 Sorscher, E. J., Peng, S., Bebok, Z., Allan, P. A., Bennett, L. L. Jr., Parker, W. B., *Gene Ther.* **1994**, *1*, 233–238.

109 Zhang, Y., Parker, W. B., Sorscher, E. J., Ealick, S. E., *Curr. Top. Med. Chem.* **2005**, *5*, 1259–1274.

110 Sheen, M. R., Kim, B. K., Parks, R. E., *Mol. Pharmacol.* **1968**, *4*, 293–299.

111 Maynes, J. T., Yam, W., Jenuth, J. P., Gang Yuan, R., Litster, S. A., Phipps, B. M., Snyder, F. F., *Biochem. J.* **1999**, *344*, 585–592.

112 Bzowska, A., Kulikowska, E., Shugar, D., *Z. Naturforsch. [C]* **1990**, *45*, 59–70.

113 Schimandle, C. M., Tanigoshi, L., Mole, L. A., Schermann, I. W., *J. Biol. Chem.* **1985**, *260*, 4455–4460.

114 Bzowska, A., Kulikowska, E., Shugar, D., *Biochem. Biophys. Acta* **1992**, *1120*, 239–247.

115 Doskočil, J., Holý, A., *Coll. Czech. Chem. Commun.* **1977**, *42*, 370–383.

116 Porter, D. J., *J. Biol. Chem.* **1991**, *267*, 7342–7351.

117 Bzowska, A., *Biochim. Biophys. Acta* **2002**, *1596*, 293–317.

118 Ealick, S. E., Babu, S., Bugg, C. E., Erion, M. D., Guida, W. C., Montgomery, J. A.,

Secrist, J. A. III, *Proc. Natl. Acad. Sci. USA* **1991**, *88*, 11540–11544.

119 Koellner, G., Luič, M., Shugar, D., Saenger, W., Bzowska, A., *J. Mol. Biol.* **1997**, *265*, 202–216.

120 Mao, C., Cook, W. J., Zhou, M., Federov, A. A., Almo, S. C., Ealick, S. E., *Biochemistry* **1998**, *37*, 7135–7146.

121 Tebbe, J., Bzowska, A., Wielgus-Kutrowska, B., Kazimierczuk, Z., Schröder, W., Shugar, D., Saenger, W., Koellner, G., *J. Mol. Biol.* **1999**, *294*, 1239–1255.

122 de Azevedo, W. F. Jr., Canduri, F., dos Santos, D. M., Pereira, J. H., Bertacine Dias, M. V., Silva, R. G., Mendes, M. A., Bassso, L. A., Palma, M. S., Santos, D. S., *Biochem. Biophys. Res. Commun.* **2003**, *312*, 767–772.

123 Mao, C., Cook, W. J., Zhou, M., Koszalka, G. W., Krenitsky, T. A., Ealick, S. E., *Structure* **1997**, *5*, 1373–1383.

124 Koellner, G., Luič M., Shugar, D., Saenger, W., Bzowska, A., *J. Mol. Biol.* **1998**, *280*, 153–166.

125 Shi, W., Schramm, V. L., Almo, S. C., *J. Biol. Chem.* **1999**, *274*, 21114–21120.

126 Zang, Y., Wang, W. H., Wu, S. W., Ealick, S. E., Wang, C. C., *J. Biol. Chem.* **2005**, *280*, 22318–22325.

127 Grenha, R., Levdikov, V. M., Fogg, M. J., Blagova, E. V., Brannigan, J. A., Wilkinson, A. J., Wilson, K. S., *Acta Crystallogr. Sect. F Struct. Biol. Cryst. Commun.* **2005**, *61*, 459–462.

128 Koellner, G., Bzowska, A., Wielgus-Kutrowska, B., Luić, M., Steiner, T., Saenger, W., Stępiński, J., *J. Mol. Biol.* **2002**, *315*, 351–371.

129 Bennett, E. M., Li, C., Allan, P. W., Parker, W. B., Ealick, S. E., *J. Biol. Chem.* **2003**, *278*, 47110–47118.

130 Wlodarczyk, J., Stoychev Galitonov, G., Kierdaszuk, B., *Eur. Biophys. J.* **2004**, *33*, 377–385.

131 Wlodarczyk, J., Kierdaszuk, B., *Biophys. Chem.* **2006**, *123*, 146–153.

132 Duan, D. S., Nagashima, T., Hoshino, T., Waldman, F., Pawlak, K., Sadee, W., *Biochem. J.* **1990**, *268*, 725–731.

133 Boehncke, W. H., Gilbertsen, R. B., Hemmer, J., Sterry, W., *Scand. J. Immunol.* **1994**, *39*, 327–332.

134 Posmantur, R., Wang, K. K., Nath, R., Glibertsen, R. B., *Immunopharmacology* **1997**, *37*, 231–244.

135 Stoeckler, J. D., Purine nucleoside phosphorylase: a target for chemotherapy, in *Developments in Cancer Chemotherapy* (ed. R. J. Glazer), CRC Press Inc. Boca Raton, FL, **1984**, pp. 35–60.

136 Miles, R. W., Tyler, P. C., Furneaux, R. H., Bagdassarian, C. K., Schramm, V. L., *Biochemistry* **1998**, *37*, 8615–8621.

137 Erion, M. D., Stoeckler, J. D., Guida, W. C., Walter, R. L., Ealick, S. E., *Biochemistry* **1997**, *36*, 11735–11748.

138 Kline, P. C., Schramm, V. L., *Biochemistry* **1995**, *34*, 1153–1162.

139 Fedorov, A., Shi, W., Kicska, G., Fedorov, E., Tyler, P. C., Furneaux, R. H., Hanson, J. C., Gainsford, G. J., Larese, J. Z., Schramm, V. L., Almo, S. C., *Biochemistry* **2001**, *40*, 853–860.

140 Wierzchowski, J., Bzowska, A., Stepniak, K., Shugar, D., *Z. Naturforsch.* **2004**, *59c*, 713–725.

141 Canduri, F., Fadel, V., Basso, L. A., Palma, M. S., Santos, D. S., de Azevero, W. F. Jr., *Biochem. Biophys. Res. Commun.* **2005**, *327*, 646–649.

142 Osborne, W. R., Sullivan, J. L., Scott, C. R., *Immunol. Commun.* **1980**, *9*, 257–267.

143 Stoeckler, J. D., Ryden, J. B., Chu, M.-Y., Lim, M. -I., Ren, W. -Y., Klein, R. S., Parks, R. E. Jr. *Proc. Am. Assoc. Cancer Res.* **1985**, *26*, 953.

144 Stoeckler, J. D., Ryden, J. B., Parks, R. E. Jr., Chu, M.-Y., Lim, M.-I., Ren, W.-Y., Klein, R. S., *Cancer Res.* **1986**, *46*, 1774–1778.

145 Montgomery, J. A., Niwas, S., Rose, J. D., Secrist, J. A. III, Babu, Y. S., Bugg, C. E., Erion, M. D., Guida, W. C., Ealick, S. E., *J. Med. Chem.* **1993**, *36*, 55–69.

146 Erion, M. D., Takabayashi, K., Smith, H. B., Kessi, J., Wagner, S., Honger, S., Shames, S. L., Ealick, S. E., *Biochemistry* **1997**, *36*, 11725–11734.

147 Traut, T. W., *Mol. Cell. Biol.* **1994**, *140*, 1–22.

148 Sircar, J. C., Kostlan, C. R., Gilbertsen, R. B., Dong, M. K., Cetenko, W. J., *Adv. Exp. Med. Biol.* **1991**, *309A*, 45–48.

149 Secrist, J. A. III, Niwas, S., Rose, J. D., Babu, Y. S., Bugg, C. E., Erion, M. D., Guida, W. C., Ealick, S. E., Montgomery, J. A., *J. Med. Chem.* **1993**, *36*, 1847–1854.

150 Erion, M. D., Niwas, S., Rose, J. D., Ananthan, S., Allen, M., Secrist, J. A. III, Babu, Y. S., Bugg, C. E., Guida, W. C., Ealick, S. E., Montgomery, J. A., [published erratum in J. *Med. Chem.* **1994**, **37** 1034] *J. Med. Chem.* **1993**, *36*, 3771–3783.

151 Gilbertsen, R. B., Dong, M. K., Kossarek, L. M., Sircar, J. C., Kostlan, C. R., Conroy, M. C., *Biochem. Biophys. Res. Commun.* **1991**, *178*, 1351–1358.

152 Gilbertsen, R. B., Dong, M. K., Wilburn, D. J., Kossarek, L. M., Sircar, J. C., Kostlan, C. R., Conroy, M. C., *Adv. Exp. Med. Biol.* **1991**, *309A*, 41–44.

153 Gilbertsen, R. B., Josyula, U., Sircar, J. C., Dong, M. K., Wum, W. S., Wilburn, D. J., Conroy, M. C., *Biochem. Pharmacol.* **1992**, *44*, 996–999.

154 Conroy, R. M., Bantia, S., Turner, H. S., Barlow, D. L., Allen, K. O., LoBuglio, A. F., Montgomery, J. A., Walsh, G. M., *Immunopharmacology* **1998**, *40*, 1–9.

155 Morris, P. E. Jr., Elliott, A. S., Walton, S. P., Williams, C. H., Montgomery, J. A., *Nucleosides, Nucleotides Nucleic Acids* **2000**, *19*, 379–404.

156 Wada, Y., Yagihashi, A., Terasawa, K., Miyao, N., Hirata, K., Cicciarelli, J., Iwaki, Y., *Artificial Organs* **1996**, *20*, 849–852.

157 Iwata, H., Wada, Y., Walsh, M., Montgomery, J. A., Hirose, H., Mendez, R., Cicciarelli, J., Iwaki, Y., *Transplant. Proc.* **1998**, *30*, 983–986.

158 Bantia, S., Montgomery, J. A., Johnson, H. G., Walsh, G. M., *Immunopharmacology* **1996**, *35*, 53–63.

159 Duvic, M., Olsen, E. A., Omura, G. A., Maize, J. C., Vonderheid, E. C., Elmets, C. A., Shupack, J. L., Demeirre, M. F., Kuzel, T. M., Sanders, D. Y., *J. Am. Acad. Dermatol.* **2001**, *44*, 940–947.

160 Miller, R. L., Sabourin, C. L., Krenitsky, T. A., Berens, R. L., Marr, J. J., *J. Biol. Chem.* **1984**, *259*, 5073–5077.

161 Cohen, A., Sussman, M., *Proc. Natl. Acad. Sci. USA* **1975**, *72*, 4479–4482.

162 Daddona, P. E., Wiesmann, W. P., Milhouse, W., Chern, J. -W., Townsend, L. B., Hershfield, M. S., Webster, H. K., *J. Biol. Chem.* **1986**, *261*, 11667–11673.

163 Mungala, N., Wang, C. C., *Biochemistry* **2002**, *41*, 10382–10389.

164 Schmidt, G., Walter, R. D., Konigk, E., *Tropenmed. Parasitol.* **1975**, *26*, 19–26.

165 Parkin, D. W., *J. Biol. Chem.* **1996**, *271*, 21713–21719.

166 Ullman, B., *Pharmaceut. Res.* **1984**, *1*, 194–203.

167 Ram, V. J., Nath, M., *Curr. Med. Chem.* **1996**, *3*, 303–316.

168 Wierzchowski, J., Shugar, D., *Collect. Czech. Chem. Commun.* **1993**, *58*, 14–17.

169 Bzowska, A., Shugar, D., *Z. Naturforsch.* **1989**, *44c*, 581–589.

170 Rossamondo, E. F., Jahngen, J. H., Eccleston, J. F., *Proc. Natl. Acad. Sci. USA* **1981**, *78*, 2278–2282.

171 Rossamondo, E. F., Jahngen, J. H., Eccleston, J. F., *Anal. Biochem.* **1981**, *116*, 80–88.

172 Appleby, T. C., Mathews, I. I., Porcelli, M., Cacciapuoti, G., Ealick, S. E., *J. Biol. Chem.* **2001**, *276*, 39232–39242.

173 Lee, J. E., Settembre, E. C., Cornell, K. A., Riscoe, M. K., Sufrin, J. R., Ealick, S. E., Howell, P. L., *Biochemistry* **2004**, *43*, 5159–5169.

174 Lee, J. E., Cornell, K. A., Riscoe, M. K., Howell, P. L., *J. Biol. Chem.* **2003**, *278*, 8761–8770.

175 Focia, P. J., Craig, S. P. III, Nieves-Alicea, R., Fletterick, R. J., Eakin, A. E., *Biochemistry* **1998**, *37*, 15066–15075.

176 Zhang, Y., Cottet, S. E., Ealick, S. E., *Structure* **2004**, *12*, 1383–1394.

177 Ren, J., Wang, Y., Dong, Y., Stuart, D. I., *Structure* **1994**, *2*, 7–16.

178 Weston, S. A., Tucker, A. D., Thatcher, D. R., Derbyshire, D. J., Pauptit, R. A., *J. Mol. Biol.* **1994**, *244*, 410–422.

179 Monzingo, A. F., Collins, E. J., Ernst, S. R., Irvin, J. D., Robertus, J. D., *J. Mol. Biol.* **1993**, *223*, 705–715.

180 Jackson, A. P., Bagshaw, C. R., *Biochem. J.* **1988**, *257*, 527–540.

181 Olney, J. J., Sellers, J. R., Cremo, C. R., *J. Biol. Chem.* **1996**, *271*, 20375–20384.

182 Barden, J. A., Miki, M., *Biochem. Int.* **1986**, *12*, 321–329.

183 Miki, M., dos Remedios, C. G., Barden, J. A., *Eur. J. Biochem.* **1987**, *168*, 339–345.

184 Schlosser, A., Nawrot, B., Grillenbeck, N., Sprinzl, M., *J. Biomol. Struct. Dyn.* **2001**, *19*, 285–291.

185 Bashor, C., Denu, J. M., Brennan, R. G., Ullman, B., *Biochemistry* **2002**, *41*, 4020–4031.

186 Blanplain, J. P., Ronjat, M., Supply, P., Duffour, J. P., Goffeau, A., Dupont, Y., *J. Biol. Chem.* **1992**, *267*, 3735–3740.

187 Girardet, J. L., Dupont, Y., Lacapere, J. J., *Eur. J. Biochem.* **1989**, *184*, 131–140.

188 Karlish, S. J., *Methods Enzymol.* **1988**, *156*, 271–277.

189 Geeves, M. A., Branson, J. P., Attwood, P. V., *Biochemistry* **1995**, *34*, 11846–11854.

190 Stepinski, J., Zuberek, J., Jemielity, J., Kalek, M., Stolarski, R., Darzynkiewiecz, E., *Nucleosides Nucleotides Nucleic Acids* **2005**, *24*, 629–33.

191 Chen, X. Y., Link, T. M., Schramm, V. L., *Biochemistry* **1998**, *37*, 11605–11613.

192 Porello, S. L., Williams, S. D., Kuhn, H., Michaels, M. L., David, S. S., *J. Am. Chem. Soc.* **1996**, *118*, 10684–10692.

193 Zanis, V., Doyle, D., Martin, D. W., *J. Biol. Chem.* **1978**, *253*, 504–510.

194 Hall, W. W., Krenitsky, T. A., *Prep. Biochem.* **1990**, *20*, 75–85.

195 Bzowska, A., Kazimierczuk, Z., Seela, F., *Acta Biochim. Polon.* **1998**, *45*, 755–768.

196 Modrak-Wójcik, A., Stepniak, K., Akoev, V., Żółkiewski, M., Bzowska, A., *Prot. Sci.* **2006**, *15*, 1794–1800.

197 Plagemann, P. G., Wolffendin, C., *Biochim. Biophys. Acta* **1989**, *1010*, 7–15.

198 Jakobs, E. S., Paterson, A. R., *Biochem. Biophys. Res. Commun.* **1986**, *140*, 1028–1035.

199 Plagemann, P. G., Aran, J. M., Wolffendin, C., *Biochim. Biophys. Acta* **1990**, *1022*, 93–102.

200 Dagnino, L., Bennett, L. L. Jr., Paterson, A. R., *J. Biol. Chem.* **1991**, *266*, 6308–6311.

201 Stolk, M., Cooper, E., Vilk, G., Litchfield, D. W., Hammond, J. R., *Biochem. J.* **2005**, *386*, 281–289.

202 Sanner, M. F., Olson, A. J., Spehner, J. C., *Biopolymers* **1996**, *38*, 305–320.

203 Kraulis, P. J., *J. Appl. Crystallogr.* **1991**, *24*, 946–950.

204 Sircar, J. C., Kostlan, C. R., Gilbertsen, R. B., Bennett, M. K., Dong, M. K., Cetenko, W. J., *J. Med. Chem.* **1992**, *35*, 1605–1609.

205 Farutin, V., Masterson, L., Andricopulo, A. D., Cheng, J., Riley, B., Hakimi, R., Frazer, J. W., Cordes, E. H., *J. Med. Chem.* **1999**, *42*, 2422–2431.

206 Sircar, J. C., Glibertsen, R. B., *Drugs Future* **1988**, *13*, 653–668.

207 Luyten, I., Thibaudeau, C., Chattopahyaya, J., *Tetrahedron* **1997**, *53*, 6903–6906.

20
1-(3-*C*-Ethynyl-β-D-*ribo*-pentofuranosyl)cytosine (ECyd)

Akira Matsuda

20.1
Introduction

Tumor cells are well known to be heterogeneous, and in this respect the cell cycle of certain tumor tissues is not synchronized. If the mechanism of action of an antitumor agent is solely to inhibit DNA synthesis, then such an agent functions only during the S phase of the cell cycle. Although solid tumor cells grow more rapidly than normal cells, their growth rate is usually much slower than that of blood cells, and consequently the antitumor agent is able to kill only a small portion of the tumor cells. However, it has been reported that some nucleoside antimetabolites also inhibit RNA synthesis to some extent. Although this inhibition of RNA synthesis would not serve as a major mechanism of tumor cell death, it is conceivable that such inhibition might contribute to the antitumor efficacy against slow-growing solid tumors. Therefore, in the quest for more potent inhibitors of tumor cell growth, the target has been a nucleoside antimetabolite which inhibits both DNA and RNA syntheses. Hence, the compound 1-(3-*C*-ethynyl-β-D-*ribo*-pentofuranosyl)cytosine (ECyd; Figure 20.1) was designed, whereby it was expected that ECyd 5′-diphosphate (ECDP) would inhibit ribonucleotide reductase, while ECyd 5′-triphosphate (ECTP) would inhibit RNA polymerases in order to cause an inhibition of both DNA and RNA synthesis [1].

20.2
Synthesis of ECyd and its Analogues

In the original preparation of ECyd, the synthesis was effected by the condensation of 1-*O*-acetyl-2,3,5-tri-*O*-benzoyl-3-*C*-ethynyl-α,β-D-*ribo*-pentofuranose and pertrimethylsilylated cytosine in the presence of $SnCl_4$ as a Lewis acid in CH_3CN, followed by debenzoylation [2]. However, for clinical trials of ECyd (TAS-106), the synthetic method had to be improved in a more practical manner. Therefore, a practical method for the large-scale preparation of ECyd from 1,2-*O*-isopropylidene-D-xylofuranose (**1**)

Modified Nucleosides: in Biochemistry, Biotechnology and Medicine. Edited by Piet Herdewijn

ISBN: 978-3-527-31820-9

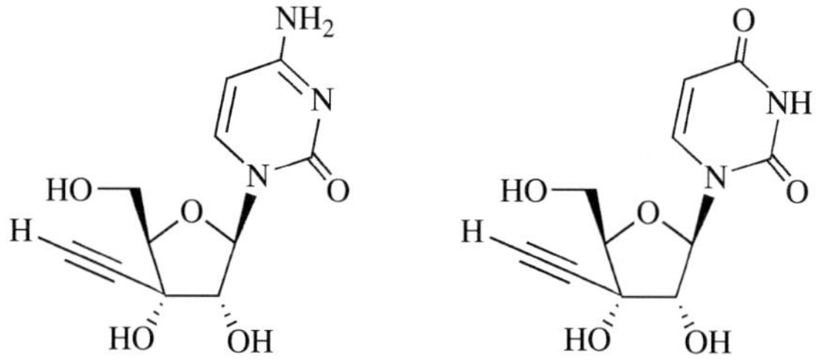

ECyd (TAS-160)

EUrd

Figure 20.1 The chemical structures of 1-(3-*C*-ethynyl-β-D-*ribo*-pentofuranosyl)cytosine (ECyd; TAS-106) and 1-(3-*C*-ethynyl-β-D-*ribo*-pentofuranosyl)uracil (EUrd).

has been developed, as shown in Scheme 20.1 [3]. As most of the intermediates were obtained as crystals, the target ECyd was obtained without any chromatographic purification in 31% overall yield from **1** (seven steps). An isobutyryloxy group was found to be a most effective leaving group at the anomeric position of the 3-β-*C*-ethynyl glycosyl donor **6** in the key Vorbrüggen glycosylation reaction [3]. Using a similar procedure, but without chromatographic purification, the uracil congener EUrd [1-(3-*C*-ethynyl-β-D-*ribo*-pentofuranosyl)uracil], which also has a potent antitumor effect [2, 4], was synthesized from **1** in 39% overall yield. This method was also applicable to the glycosylation of less-reactive sugars [5]. EUrd has also been prepared from 3′-keto-uridine derivatives [6, 7].

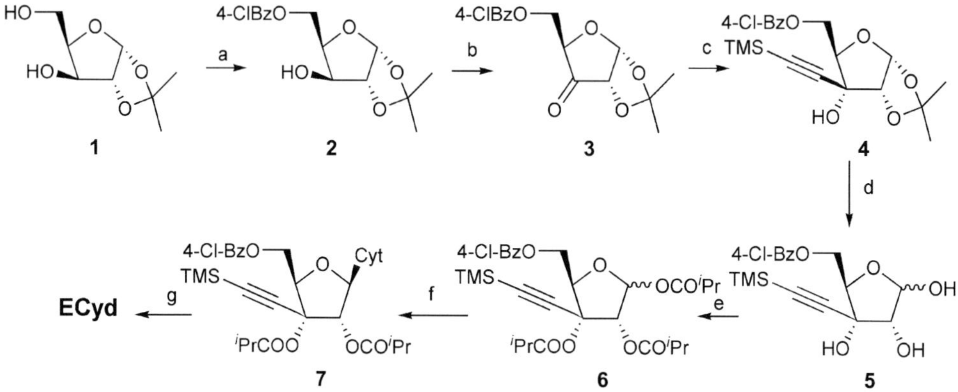

(a) 4-ClBzCl, Et_3N, CH_2Cl_2, 0 °C, 2 h, 74%; (b) TEMPO, NaOCl, CH_2Cl_2-aq $NaHCO_3$, 0 °C, 15 min, 77%; (c) TMSC≡CMgBr; THF, 4 °C, 91%, (d) 50% aq HCO_2H, reflux, 45 min, 97%; (e) isobutyryl chloride, Et_3N, DMAP, CH_2Cl_2, rt, 5 h; (f) bis-TMScytosine, $SnCl_4$, MeCN, 30 °C, 3 h, 81%; (g) DBU, MeOH, 30 °C, 3h, 68%

Scheme 20.1 The practical synthesis of ECyd.

20.3
Cytotoxic Activity and Structure–Activity Relationships of ECyd Analogues *In Vitro*, and *In Vitro* Antitumor Activity

Among the various 3′-*C*-ethynyl *ribo*-nucleosides with different nucleobases (uracil, 5-fluorouracil, thymine, cytosine, 5-fluorocytosine, adenine, guanine) [2], and cytosine and uracil nucleosides bearing substituents other than the ethynyl group at the 3′-β-position [8], and an ethynyl group other than the 3′-β position [8–10], both ECyd and EUrd showed the most potent cytotoxicity *in vitro* against various human tumor cell lines, with IC_{50} values of nanomolar magnitude (Table 20.1). These data would reflect the substrate specificity of the first activation enzymes, such as uridine-cytidine kinase (UCK), which would recognize the bulky substituents near the 5′-position in space. However, the cytotoxic spectrum of ECyd and EUrd differed somewhat from those of DNA synthesis inhibitors, such as araC [1]. ECyd exhibited strong antitumor activity against a variety of human xenografts implanted into nude mice and nude rats, after intravenous administration (Table 20.2) [2, 11–13]. Indeed, the antitumor potency of ECyd was rather schedule-independent, and did not produce severe toxic effects such as diarrhea, myelosuppression, or loss in body weight.

The preparation was also conducted of a 4′-thio congener of ECyd, 1-(3-*C*-ethynyl-4-thio-β-D-*ribo*-pentofuranosyl)cytosine (4′-thio-ECyd) [14]. However, whilst this nucleoside did not show any significant cytotoxicity against mouse L1210 and human KB cells *in vitro* at a cell medium concentration of 100 μg mL^{-1}, ECyd proved to be a strong inhibitor of tumor cell proliferation, with IC_{50} values of 16 and 28 nM, respectively, against the same cell lines. One explanation for this finding might be the difference in substrate specificity towards a nucleoside kinase. As described above, ECyd is phosphorylated by UCK, and successively converted to the active metabolite, ECTP. Since for these metabolic activations, the first phosphorylation is thought to be the most important step, the relative susceptibility to the kinase was compared between 4′-thio-ECyd, ECyd, and cytidine. Consequently, whilst ECyd was seen to be phosphorylated 26% relative to cytidine, no phosphorylation of 4′-thio-ECyd was observed under the same conditions.

20.4
Structural Features of ECyd and 4′-Thio-ECyd

The structures of ECyd and 4′-thio-ECyd were initially confirmed by X-ray analyses (Figure 20.2) [14], and some important conformational characteristics of both structures are summarized in Table 20.3. Among these data, striking differences in the bond lengths and angles were observed in C1′–O4′ (S4′) and C4′–O4′ (S4′), and C4′–O4′ (S4′)–C1′. Thus, the bond lengths C1′–O4′ and C4′–O4′ in ECyd were 1.408 (3) and 1.449 (3) Å, respectively, while the corresponding bond lengths of 4′-thio-ECyd were much longer (i.e., 1.818 (2) and 1.840 (3) Å, respectively). The other bond lengths, including that of the glycosidic bond (C1′–N1) were quite

Table 20.1 Inhibitory effects of ECyd and EUrd on the growth of human tumor cell lines *in vitro*.[a)]

Cell line	Origin	IC_{50} (μM)	
		ECyd	EUrd
MKN-28	Stomach adenocarcinoma	0.021	0.065
MKN-45		0.0088	0.052
KATO-III		0.031	0.031
NUGC-4		0.024	0.22
ST-KM		0.047	0.086
KKLS		0.021	0.23
STSA-1		0.15	1.9
NAKAJIMA		0.17	0.78
Colo320DM	Colon adenocarcinoma	0.02	0.06
HCT-15		0.0082	0.075
SW-48		>1.9	>1.9
SW-480		0.26	0.34
PC-8	Lung adenocarcinoma	0.090	0.33
PC-9		0.11	0.19
PC-10		0.16	>1.9
QG-56	Lung squamous-cell carcinoma	0.045	0.15
QG-95		0.21	0.28
Lu-65	Lung large-cell carcinoma	0.032	0.089
QG-90	Lung small-cell carcinoma	0.13	1.0
QG-96		0.039	0.38
NCI-H-82		0.12	0.23
NCI-H-417		0.043	0.11
MDA-MB-321	Breast adenocarcinoma	0.024	>1.9
YMB-1-E		0.16	>1.9
MCF-7		0.069	0.2
PANC-1	Pancreas adenocarcinoma	>1.9	>1.9
Mia-PaCa-2		0.015	0.054
T24	Bladder carcinoma[b)]	0.028	0.21
KK-47		0.12	0.089
HOS	Osteosarcoma	1.0	1.8
MG-63		0.15	>1.9
A431	Leiomyosarcoma	0.033	0.29
HT-1080	Fibrosarcoma	0.073	0.16
A375	Melanoma	0.026	0.021
SK-MEL-28		0.047	0.086
KB	Pharyngeal carcinoma	0.028	0.029

a) Tumor cells (2×10^3 cells per well) were incubated in the presence or absence of the compounds for 72 h. MTT reagent was added to each well, and the plate incubated for an additional 4 h. The resulting MTT formazan was dissolved in DMSO, and the OD (540 nm) was measured. Percentage inhibition was calculated as follows: % inhibition = [1 − OD(540 nm) of sample well/OD(540 nm) of control well] × 100. IC_{50} (μM) represents the concentration at which cell growth was inhibited by 50%.

b) Transitional cell.

Table 20.2 Anti-tumor effects of ECyd, EUrd, and 5-fluorouracil (5-FU) on human tumor xenografts in nude mice.[a)]

Tumor	Origin	Tumor inhibition ratio (%)		
		ECyd[b)]	EUrd[c)]	5-FU[d)]
		$0.25\ mg\ kg^{-1} \times 10$	$2.0\ mg\ kg^{-1} \times 10$	$15\ mg\ kg^{-1} \times 14$
AZ-521	Stomach	92	81	25
H-81	Stomach	84	92	83
NUGC-3	Stomach	80	54	ND
CO-3	Colon	83	90	49
KM12C	Colon	88	81	59
DLD-1	Colon	78	73	50
JRC-11	Renal	73	75	69
H-31	Breast	76	82	68
Hucc-T1	Bile duct	55	64	1
PAN-12	Pancreas	75	81	40
BxPC-3	Pancreas	59	35	50

a) Tumor mass ($2\ mm^3$) was transplanted subcutaneously into nude mice. The tumor inhibition ratio was evaluated at day 15 and calculated as: inhibition ratio (%) = $(A - B)/A \times 100$, where A is the average tumor weight in the control group and B is that in the treated group. Each group consisted of eight nude mice.
b) ECyd was administered intravenously for 10 consecutive days when the tumor volume reached $60–100\ mm^3$.
c) EUrd was administered intravenously for 10 consecutive days when the tumor volume reached $60–100\ mm^3$.
d) 5-FU was administered intravenously for 14 consecutive days when the tumor volume reached $60–100\ mm^3$.

similar to each other. In contrast to the longer bond length of 4′-thio-ECyd, the bond angle C4′–S4′–C1′ in the thio-sugar is 94.9 degrees, which is some 15.8 degrees less than that of ECyd. The other bond angles in the two sugar moieties do not differ markedly. In spite of the partial structural differences between ECyd and 4′-thio-

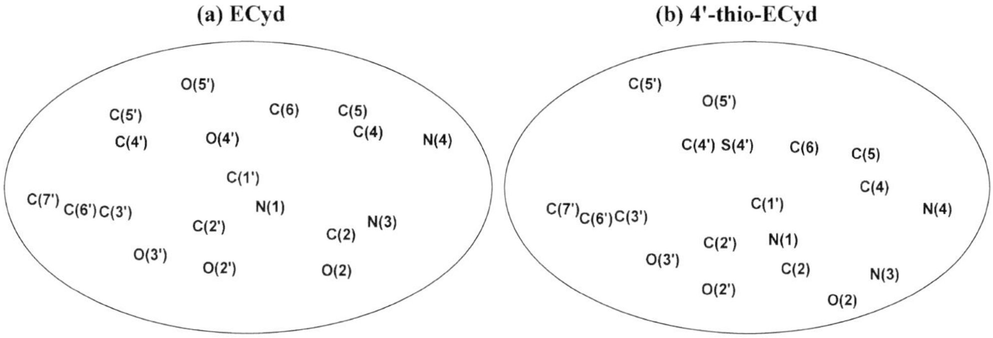

Figure 20.2 The crystal structures of (a) ECyd and (b) 4′-thio-ECyd.

Table 20.3 Geometric parameters: bond lengths, angles, and torsion angles that represent important structural features of ECyd and 4′-thio-ECyd.

	ECyd	4′-thio-ECyd
Bond lengths (Å)		
C1′–C2′	1.523 (3)	1.527 (3)
C2′–C3′	1.551 (3)	1.532 (3)
C3′–C4′	1.547 (3)	1.546 (3)
C1′–O4′ (S4′)	1.408 (3)	1.818 (2)
C4′–O4′ (S4′)	1.449 (3)	1.840 (3)
C1′–N1′	1.458 (3)	1.473 (3)
Bond angles (degrees)		
C1′–C2′–C3′	101.7 (2)	108.6 (2)
C2′–C3′–C4′	101.4 (2)	107.2 (2)
C3′–C4′–O4′ (S4′)	106.1 (2)	106.3 (2)
C4′–O4′ (S4′)–C1′	110.7 (2)	94.9 (1)
O4′ (S4′)–C1′–C2′	106.3 (2)	107.4 (2)
O4′ (S4′)–C1′–N1′	109.1 (2)	112.2 (2)
Torsion angles (degrees)		
C4′–O4 (S4′)–C1′–C2′ (ν_0)	−19.5 (3)	−10.2 (2)
O4′ (S4′)–C1′–C2′–C3′ (ν_1)	34.5 (2)	31.7 (2)
C1′–C2′–C3′–C4′ (ν_2)	−35.2 (2)	−42.8 (3)
C2′–C3′–C4′–O4′ (S4′) (ν_3)	25.0 (2)	33.7 (2)
C3′–C4′–O4′ (S4′)–C1′ (ν_4)	−4.0 (3)	−13.7 (2)
O4′ (S4′)–C1′–N1′–C2′ (χ)	−139.2 (2)	−134.6 (2)
O5′–C5′–C4′–C3′ (γ)	58.6 (3)	58.1 (3)
Pseudoroation parameters (degrees)		
Phase angle (P)	167.5	182.2
Puckering amplitude (ν_m)	36.1	42.8

ECyd, their overall structures are quite similar. Thus, the cytosine bases are both in the *anti* conformation with the glycosidic torsion angle χ (O4′–C1′–N1–C2) −139.2° and χ (S4′–C1′–N1–C2) = −134.6°. The furanose ring of ECyd exhibits a South-type conformation with a pseudorotation phase angle $P = 167.5°$ and maximum puckering amplitude $\nu_m = 36.1°$, respectively. The 4′-thiosugar of 4′-thio-ECyd is also found in a South-type puckered conformation with the values of $P = 182.2°$ and $\nu_m = 42.8°$. This type of conformation was also maintained in both structures in solution. The conformation of each of the sugar rings was analyzed on the basis of the coupling constant $J_{1',2'}$ in DMSO-d_6. The J-value of ECyd was found to be 6.6 Hz, while that of 4′-thio-ECyd was 8.6 Hz. These data show that both compounds prefer a South-type puckered conformation both in solid state and in solution.

From these structural features, UCK2 was able to discriminate these slight changes in the sugar conformation to be a substrate of the phosphorylation.

20.5
Metabolism and Mechanism of Action

ECyd was first activated by UCK [15, 16], being converted to its 5′-monophosphate (ECMP). As UCK activity in human tumor tissues is well known to be relatively high compared to its activity in normal tissues, a tumor-selective growth inhibitory action of ECyd could, therefore, be expected to manifest clinically. Recently, a UCK family consisting of two members – UCK1 and UCK2 – has been reported in human cells [17, 18], and consequently for further development it is important to determine which of these isoenzymes is responsible for the phosphorylation of ECyd. Initially, investigations were conducted, using a panel of 10 human tumor cell lines, into the relationship between the expression of UCK1 and UCK2 at both mRNA and protein levels, and also of ECyd phosphorylation activity [15]. In fact, the UCK activity was found to correlate well with the cells' sensitivity to ECyd. Moreover, the mRNA or protein expression level of UCK2 was closely correlated with UCK activity in these cell lines, although neither the level of expression of UCK1 mRNA nor that of protein was correlated with the enzyme activity. A comparison of the protein expression level of UCK2 in several human tumor tissues, and the corresponding normal tissues, showed that the expression of UCK2 protein was barely detectable in four out of the five human tissues, but tended to be high in pancreatic tumor tissue. In any of the normal tissues, the expression could not be detected at all, and consequently the expression of UCK2 appeared to be correlated with cellular sensitivity to ECyd, and may contribute to the tumor-selective growth inhibitory activity of ECyd.

The crystal structures of human UCK2, both alone and complexed with a substrate (cytidine), a feedback inhibitor (CTP or UTP), and with phosphorylation products (CMP and ADP), respectively, were also solved [19].

Further phosphorylations to the 5′-diphosphate (ECDP) and 5′-triphosphate (ECTP) were found to occur concomitantly [20]. ECTP, the dead-end metabolite of ECyd, was shown to accumulate in tumor cells, and was rather stable in mouse mammary FM3A cells, with a half-life of about 81 h, whereas that of araCTP was less than 10 min. Therefore, intracellular ECTP was some 500-fold more stable than araCTP [12, 21]. RNA polymerase was inhibited competitively by ECTP in the isolated nuclei of FM3A cells, with a K_i value for ECTP of 21 nM, and an apparent K_m value of RNA polymerase for CTP of 8 μM. Since the IC_{50} value of ECyd in the growth of FM3A cells *in vitro* was 30 nM, the target enzyme responsible for the compound's cytotoxicity would be RNA polymerases, of which three (I, II and III) occur in eukaryotes. ECTP was found not to selectively inhibit any of these polymerases. Ultimately, this RNA inhibition may lead the tumor cells to undergo apoptosis (Figure 20.3) [22, 23].

Although ECyd was found to be a very poor substrate of cytidine deaminase (CDA) from mouse kidney *in vitro*, its deaminated analogue, EUrd, exhibited very similar cytotoxicity to ECyd, which was also metabolized to its 5′-triphosphate (EUTP) and inhibited RNA polymerase with a K_i value of 84 nM (the apparent K_m of UTP was 13 μM) [24].

Initially, although ECDP was anticipated to inhibit ribonucleotide reductase, it was found not to inhibit the enzyme from *Escherichia coli*, even at millimolar concentrations.

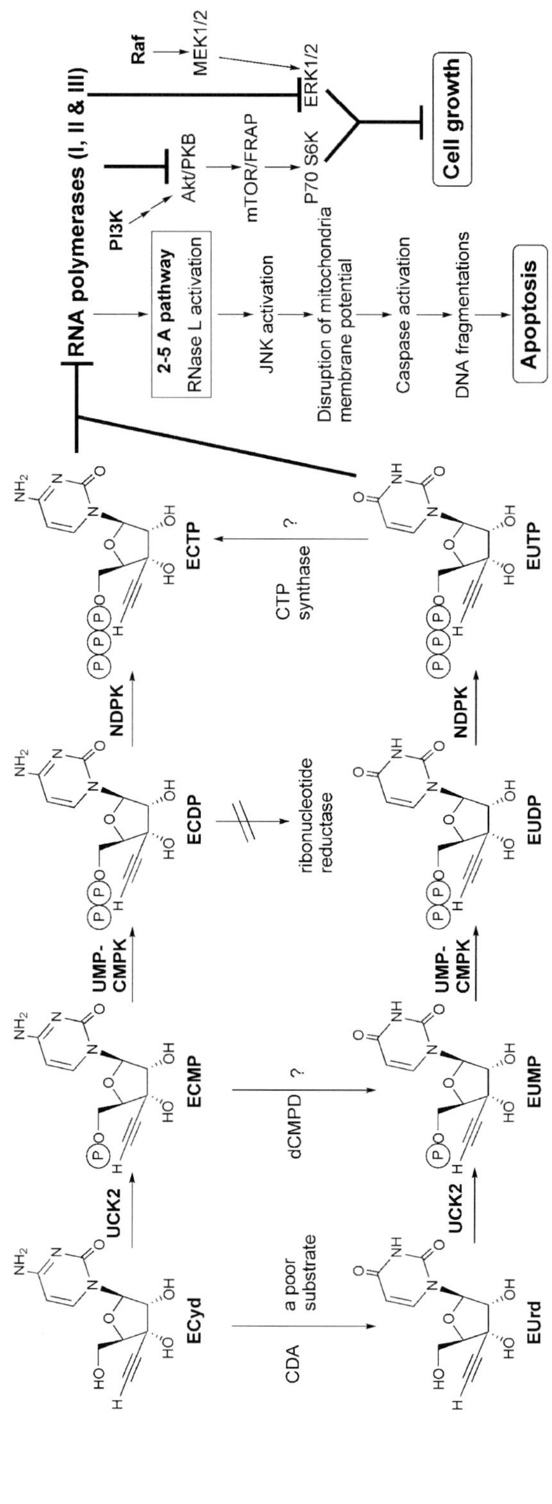

Figure 20.3 The metabolism and mechanisms of action of ECyd.

20.6
An Apoptotic Pathway Involving the Action of ECyd

ECyd-induced cell death in FM3A cells [23] and in MKN-45 [13] cells with the wild-type *p53* gene was accompanied by the release of approximately 100- to 200-kilo-base pair-sized DNA and internucleosomal DNA fragments, as confirmed by gel electrophoresis. However, in tumor cells which had a mutated *p53* gene or which were deficient in the *p53* gene – for example, MKN-28 (point mutation) and KATO III (deficient) cells – treatment with ECyd caused cell death by necrosis, but not by apoptosis.

As a result of this RNA synthesis inhibition, ribosomal RNA fragmentations occurred in the D8 domain of 28S rRNA in FM3A cells [15]. The fragmentation pattern was quite similar to, and the cleavage sites were identical, to that produced by RNase L (an endoribonuclease). When the expression of RNase L by its siRNA was suppressed, however, the apoptosis induced by ECyd was also inhibited in a concentration-dependent manner. Moreover, at the same time, the level of the mitochondrial membrane potential was greatly reduced. Therefore, it appears that RNase L may be involved in the mitochondria-caspase-dependent apoptotic cell death induced by ECyd (Figure 20.3). RNase L is known to be activated by 2′-5′-oligoadenylate (2-5A) synthetases, which are activated by not only double-stranded (ds) but also single-stranded (ss) RNA [26]. Hence, the inhibition of RNA synthesis in the nucleus by ECTP would produce short-length ssRNAs, as well as specific rRNA cleavage products, which might activate the 2-5A synthetases. It was also reported that the c-Jun NH_2-terminal kinases (JNK) family of MAP kinases is essential for mitochondrion-dependent apoptosis in response to RNase L [27]. The apoptotic cell death induced by ECyd correlates in part through the 2-5A/RNase L pathway (Figure 20.3).

20.7
Combination of ECyd with Low-Dose X-Irradiation

Ionizing radiation is known to cause cell cycle arrest in the G_2/M phase, as well as cell death. DNA double-strand breaks induced by radiation are recognized by ataxia telangiectasia mutated (ATM), which inhibits the passage of DNA-damaged cells from G_2 into M phase. ATM is required for the activation of Chk1/2 in response to DNA damage, followed by the phosphorylation of Cdc25C phosphatase on Ser216. The phosphorylation of Cdc25C phosphatase creates a binding site for 14-3-3 proteins, after which the 14-3-3-bound phosphatase is sequestered outside of the nucleus and cannot dephosphorylate and activate the meiosis-promoting Cdc-cyclin B1 complex [28].

The abrogation of the G_2/M checkpoint often leads to a marked increase in the sensitivity of cells towards ionizing radiation and to some types of chemotherapy. A low dose of ECyd induced the radiosensitization of caspase-dependent apoptosis and reproductive cell death in cells of the human gastric tumor cell lines MKN45 and MKN28 and murine rectum adenocarcinoma Colon26 [29, 30]. Flow cytometry studies subsequently showed that ECyd induced the abrogation of the X-ray-induced G_2/M checkpoint. Western blot analysis also showed that X rays increased the expression of cyclin B1, phospho-Cdc2 and Wee1, whereas co-treatment with ECyd and X-rays decreased the expression of these cell-cycle proteins associated with the G_2/M checkpoint. Furthermore, ECyd was shown to decrease the radiation-induced expression of survivin but not Bcl2 and $BclX_L$, regardless of TP53 status and cell type. The overexpression of wild-type surviving in MKN45 cells inhibited the induction of apoptosis induced by co-treatment with ECyd and X-rays. These results suggest that ECyd enhances X-ray-induced cell death through the down-regulation of survivin and abrogation of the cell-cycle machinery.

Investigations of the *in-vivo* antitumor efficacy of a low dose of X-irradiation (2 Gy) combined with low-level ECyd administration (0.1 $mg\,kg^{-1}$ for Colon26 and 0.5 $mg\,kg^{-1}$ for MNK45, injected intraperitoneally) was also carried out using Colon26 murine rectal adenocarcinoma cells and MNK45 human gastric adenocarcinoma cells inoculated into the footpad of BALB/c mice and severe combined immunodeficient (SCID) mice, respectively [31]. A significant reduction in tumor growth was observed in both types of tumor compared to that in mice treated with X-irradiation or ECyd alone. A near-complete remission of tumors was observed in half of the mice that received the combined treatment on three occasions at two-day intervals. An immunohistochemical analysis of the apoptotic and proliferative cells showed that large numbers of apoptotic and Ki-67-negative cells were induced by the combined treatment. In addition, the mRNA expression of anti-apoptotic proteins and phosphorylation of ERK1/2 and Akt were found to be inhibited by ECyd.

20.8
ECyd is Effective against Gemcitabine-Resistant Human Pancreatic Cancer Cells

Currently, pancreatic cancer is recognized as one of the most intractable cancers, and has been associated with an increasing mortality rate in recent years. Although gemcitabine is currently the standard chemotherapeutic drug for metastatic and advanced pancreatic cancer, it provides only a modest improvement in the patient's quality of life and survival. A variant of MIAPaCa-2 human pancreatic cancer cells has been shown to be about 2500-fold more resistant to gemcitabine than the wild-type cell line, the mechanism of resistance being a decrease in the intracellular pool of the drug's active metabolites and a subsequently reduced incorporation of gemcitabine 5′-triphosphate into the DNA. This would in turn be related to a decreased dCK activity, increased cytidine deaminase (CDA) and ribonucleotide reductase activities, and increased 5′-nucleotidase mRNA expression. The cytotoxicity of ECyd was very similar in both parental and gemcitabine-resistant cells, with IC_{50} values of 6.25 and 6.27 nM, respectively. The *in-vivo* antitumor activity of ECyd against MIAPaCa-2 and gemcitabine-resistant cells implanted into nude mice was also monitored, with tumor growth inhibited by weekly ECyd treatment (7 $mg\,kg^{-1}$, i.v.) in parental cells and gemcitabine-resistant tumors by 73% and 76%, respectively. By comparison, twice-weekly gemcitabine (240 mg kg^{-1}, i.v.) caused 84% and 34% growth inhibition, respectively [32]. Such results suggest that ECyd might indeed be valuable for treating patients with advanced pancreatic carcinoma in whom gemcitabine-based chemotherapy had failed.

20.9 Conclusions

Although Phase I clinical studies of intravenously administered ECyd (TAS-106) against various solid tumors have recently been completed in the USA, Phase II clinical trials in which ECyd is administered either as a single drug or in combination with other chemotherapeutic agents, are planned to determine its clinical potential.

References

1 Matsuda, A., Sasaki, T., Antitumor activity of sugar-modified cytosine nucleosides. *Cancer Sci.* **2004**, *95*, 105–111.

2 Hattori, H., Tanaka, M., Fukushima, M., Sasaki, T., Matsuda, A., 1-(3-*C*-Ethynyl-β-D-*ribo*-pentofuranosyl)cytosine (ECyd), 1-(3-*C*-ethynyl-β-D-*ribo*-pentofuranosyl) uracil (EUrd), and their nucleobase analogues as new potential multifunctional antitumor nucleosides with a broad spectrum of activity. *J. Med. Chem.* **1996**, *39*, 5005–5011.

3 Nomura, M., Sato, T., Washinosu, M., Tanaka, M., Asao, T., Shuto, S., Matsuda, A., Practical large-scale synthesis of 1-(3-*C*-ethynyl-β-D-*ribo*-pentofuranosyl)cytosine (ECyd), a potent antitumor nucleoside. Isobutyryloxy group as an efficient anomeric leaving group in the Vorbruggen glycosylation reaction. *Tetrahedron* **2002**, *58*, 1279–1288.

4 Matsuda, A., Hattori, H., Tanaka, M., Sasaki, T., 1-(3-*C*-Ethynyl-β-D-*ribo*-pentofuranosyl)uracil as a potential broad spectrum multifunctional antitumor nucleoside. *Bioorg. Med. Chem. Lett.* **1996**, *6*, 1887–1892.

5 Mochizuki, M., Kondo, Y., Abe, H., Tovey, S. C., Dedos, S. G., Taylor, C. W., Paul, M., Potter, B. V. L., Matsuda, A., Shuto, S., Synthesis of adenophostin A analogues conjugating an aromatic group at the 5′-position as potent IP_3 receptor ligands. *J. Med. Chem.* **2006**, *49*, 5750–5758.

6 Jung, P. M. J., Burger, A., Biellmann, J.-F., Rapid and efficient stereocontrolled synthesis of C-3′-ethynyl ribo and xylonucleosides by organocerium addition to 3′-ketonucleosides. *Tetrahedron Lett.* **1995**, *36*, 1031–1034.

7 Jung, P. M. J., Burger, A., Biellmann, J.-F., Diastereofacial selective addition of ethynylcerium reagent and Barton-McCombie reaction as the key steps for the synthesis of *C*-3′-ethynylribonucleosides and of *C*-3′-ethynyl-2′-deoxyribonucleosides. *J. Org. Chem.* **1997**, *62*, 8309–8314.

8 Hattori, H., Nozawa, E., Iino, T., Yoshimura, Y., Shuto, S., Shimamoto, Y., Nomura, M., Fukushima, M., Tanaka, M., Sasaki, T., Matsuda, A., The structural requirements of the sugar moiety for the antitumor activities of new nucleoside antimetabolites, 1-(3-*C*-ethynyl-β-D-*ribo*-pentofuranosyl)cytosine and -uracil. *J. Med. Chem.* **1998**, *41*, 2892–2902.

9 Yoshimura, Y., Iino, T., Matsuda, A., Stereoselective radical deoxygenation of

tert-propargyl alcohols in sugar moiety of pyrimidine nucleosides: Synthesis of 2′-*C*-alkynyl-2′-deoxy-1-β-D-arabinofuranosylpyrimidines. *Tetrahedron Lett.* **1991**, *32*, 6003–6006.

10 Iino, T., Yoshimura, Y., Matsuda, A., Synthesis of 2′-*C*-alkynyl-2′-deoxy-1-β-D-arabinofuranosylpyrimidines via radical deoxygenation of tert-propargyl alcohols in the sugar moiety. *Tetrahedron* **1994**, *50*, 10397–10406.

11 Shimamoto, Y., Fujioka, A., Kazuno, H., Murakami, Y., Ohshimo, H., Kato, T., Matsuda, A., Sasaki, T., Fukushima, M., Antitumor activity and pharmacokinetics of TAS-106, 1-(3-C-ethynyl-β-D-ribo-pentofuranosyl)cytosine. *Jpn. J. Cancer Res.* **2001**, *92*, 343–351.

12 Takatori, S., Kanda, H., Takenaka, K., Wataya, Y., Matsuda, A., Fukushima, M., Shimamoto, Y., Tanaka, M., Sasaki, T., Antitumor mechanisms and metabolism of novel antitumor nucleoside analogues, 1-(3-*C*-ethynyl-β-D-*ribo*-pentofuranosyl) cytosine and 1-(3-*C*-ethynyl-β-D-*ribo*-pentofuranosyl)-uracil. *Cancer Chemother. Pharmacol.* **1999**, *44*, 97–104.

13 Tabata, S., Tanaka, M., Matsuda, A., Fukushima, M., Sasaki, T., Antitumor effect of a novel multifunctional antitumor nucleoside, 3′-ethynylcytidine, on human cancers. *Oncology Rep.* **1996**, *3*, 1029–1034.

14 Minakawa, N., Kaga, D., Kato, Y., Kobayashi, K., Tanaka, M., Sasaki, T., Matsuda, A., Synthesis and structural elucidation of 1-(3-*C*-ethynyl-4-thio-β-D-ribofuranosyl)cytosine (4′-thioECyd). *J. Chem. Soc. Perkin Trans. 1*, **2002**, 2182–2189.

15 Shimamoto, Y., Koizumi, K., Okabe, H., Kazuno, H., Murakami, Y., Nakagawa, F., Matsuda, A., Sasaki, T., Fukushima, M., Sensitivity of human cancer cells to the new anticancer *ribo*-nucleoside TAS-106 is correlated with expression of uridine-cytidine kinase 2. *Jpn. J. Cancer Res.* **2002**, *93*, 825–833.

16 Murata, D., Endo, Y., Obata, T., Sakamoto, K., Syouji, Y., Kadohira, M., Matsuda, A., Sasaki, T., A crucial role of uridine/cytidine kinase 2 in antitumor activity of 3′-ethynyl nucleosides. *Drug Metab. Dispos.* **2004**, *32*, 1178–1182.

17 Van Rompay, A. R., Norda, A., Linden, K., Johansson, M., Karlsson, A., Phosphorylation of uridine and cytidine nucleoside analogs by two human uridine-cytidine kinases. *Mol. Pharmacol.* **2001**, *59*, 1181–1186.

18 Koizumi, K., Shimamoto, Y., Azuma, A., Wataya, Y., Matsuda, A., Sasaki, T., Fukushima, M., Cloning and expression of uridine/cytidine kinase cDNA from human fibrosarcoma cells. *Int. J. Mol. Med.* **2001**, *8*, 273–278.

19 Suzuki, N. N., Koizumi, K., Fukushima, M., Matsuda, A., Inagaki, F., Structural basis for the specificity, catalysis and regulation of human uridine-cytidine kinase. *Structure* **2004**, *12*, 751–764.

20 Shimamoto, Y., Kazuno, H., Murakami, Y., Azuma, A., Koizumi, K., Matsuda, A., Sasaki, T., Fukushima, M., Cellular and biochemical mechanisms of the resistance of human cancer cells to a new anticancer *ribo*-nucleoside, TAS-106. *Jpn. J. Cancer Res.* **2002**, *93*, 445–452.

21 Azuma, A., Matsuda, A., Sasaki, T., Fukushima, T., 1-(3-*C*-Ethynyl-β-D-*ribo*-pentofuranosyl)cytosine (ECyd, TAS-106): Antitumor effect and mechanism of action. *Nucleosides, Nucleotides, Nucleic Acids* **2001**, *20*, 609–619.

22 Tabata, S., Tanaka, M., Endo, Y., Obata, T., Matsuda, A., Sasaki, T., Anti-tumor mechanisms of 3′-ethynyluridine and 3′-ethynylcytidine as RNA synthesis inhibitors: development and characterization of 3′-ethynyluridine-resistant cells. *Cancer Lett.* **1997**, *116*, 225–231.

23 Takatori, S., Tsutsumi, S., Hidaka, M., Kanda, H., Matsuda, A., Fukushima, M., Wataya, Y., The characterization of cell death induced by 1-(3-*C*-ethynyl-β-D-*ribo*-pentofuranosyl)cytosine (ECyd) in FM3A cells. *Nucleosides Nucleotides* **1998**, *17*, 1309–1317.

24 Yokogawa, T., Kanda, H., Takatori, S., Takenaka, K., Naito, T., Sasaki, T., Matsuda, A., Fukushima, M., Kim, H.-S., Wataya, Y., Inhibitory mechanism of 1-(3-*C*-ethynyl-β-D-*ribo*-pentofuranosyl) uracil (EUrd) on RNA synthesis. *Nucleosides, Nucleotides Nucleic Acids* **2005**, *24*, 227–232.

25 Naito, T., Yokogawa, T., Kim, Hye-Sook, Matsuda, A., Sasaki, T., Fukushima, M., Kitade, Y., Wataya, Y., An apoptotic pathway of 3′-ethynylcytidine (ECyd) involving the inhibition of RNA synthesis mediated RNase L. *Nucleic Acids Symposium Ser.* **2006**, *50*, 103–104.

26 Hartman, R., Norby, P. L., Martensen, P. M., Jorgensen, P., James, M. C., Jacobsen, C., Moestrup, S. K., Clemens, M. J., Justesen, J., Activation of 2′-5′ oligoadenylate synthetase by single-stranded and double-stranded RNA aptamers. *J. Biol. Chem.* **1998**, *273*, 3236–3246.

27 Li, G., Xiang, Y., Sabapathy, K., Silverman, R. H., An apoptotic signaling pathway in the interferon antiviral response mediated by RNase L and c-Jun NH_2-terminal kinase. *J. Biol. Chem.* **2004**, *279*, 1123–1131.

28 Khanna, K. K., Lavin, M. F., Jackson, S. P., Mulhern, T. D., ATM, a central controller of cellular responses to DNA damage. *Cell Death Differ.* **2001**, *8*, 1052–1065.

29 Inanami, O., Iizuka, D., Iwahara, A., Yamamori, T., Kon, Y., Asanuma, T., Matsuda, A., Kashiwakura, I., Kitazato, K., Kuwabara, M., A novel anticancer ribonucleoside, 1-(3-*C*-ethynyl-β-D-*ribo*-pentofuranosyl)cytosine, enhances radiation-induced cell death in tumor cells. *Radiat. Res.* **2004**, *162*, 635–645.

30 Iizuka, D., Inanami, O., Matsuda, A., Kashiwakura, I., Asanuma, T., Kuwabara, M., X irradiation induces the proapoptotic state independent of the loss of clonogenic ability in Chinese hamster V79 cells. *Radiation Res.* **2005**, *164*, 36–44.

31 Yasui, H., Inanami, O., Asanuma, T., Iizuka, D., Nakajima, T., Kon, Y., Matsuda, A., Kuwabara, M., Treatment combining X irradiation and a ribonucleoside anticancer drug, 1-(3-*C*-ethynyl-β-D-*ribo*-pentofuranosyl) cytosine, synergistically suppresses the growth of tumor cells transplanted in mice. *Int. J. Rad. Oncol. Biol. Phys.* **2007**, *68*, 218–228.

32 Kazuno, H., Sakamoto, K., Fujioka, A., Fukushima, M., Matsuda, A., Sasaki, T., Possible antitumor activity of 1-(3-*C*-ethynyl-β-D-*ribo*-pentofuranosyl)cytosine (ECyd, TAS-106) against an established gemcitabine-resistant human pancreatic cancer cell line. *Cancer Sci.* **2005**, *96*, 295–302.

21
Syntheses and Biological Activity of Neplanocin and Analogues

Dilip K. Tosh, Hea Ok Kim, Shantanu Pal, Jeong A. Lee, and Lak Shin Jeong

21.1
Introduction

Neplanocin A (**1**) [1] is a carbocyclic nucleoside which is isolated from *Ampullariella regularis* and possesses excellent chemical and metabolic stability of the glycosidic bond (Figure 21.1). Neplanocin A also exhibits potent biological activity such as antiviral and antitumor activities, which result from the inhibition of *S*-adenosyl-L-homocysteine (SAH) hydrolase [2, 3].

SAH hydrolase catalyzes the interconversion of SAH into adenosine and L-homocysteine [4], and inhibition of this enzyme leads to an accumulation of SAH and a negative inhibition of cellular *S*-adenosyl-L-methionine (SAM)-dependent methyltransferase (Figure 21.2).

As SAM-dependent methyltransferase is responsible for formation of the 5′-terminal methylated N^7-methylguanosine cap of the mRNA in most animal-infecting viruses, SAH hydrolase is essential for viral replication [4, 5]. Thus, the inhibition of SAH hydrolase can exhibit a broad spectrum of antiviral activity [6].

Neplanocin A is one of the most potent inhibitors of SAH hydrolase, and shows a strong antiviral activity against various RNA and DNA viruses. Neplanocin A inhibits SAH hydrolase by virtue of a cofactor-depleting mechanism [5, 7, 8], and is oxidized to its 3′-keto form by the enzyme-bound cofactor NAD^+, which is tightly bound to the reduced cofactor NADH, resulting in the depletion of NAD^+. However, despite its potent enzyme inhibitory activity, neplanocin A could not be developed as a clinically useful antiviral agent because it proved to be too toxic to the host cells. This toxicity was derived from the triphosphorylation of the 5′-hydroxyl group of neplanocin A by cellular kinases (including adenosine kinase [9, 10]), with the resultant triphosphate being incorporated into RNA and causing the inhibition of RNA synthesis [10].

Aristeromycin (**2**) [11] is another natural product isolated from *Streptomyces citricolor*, and has a similar carbocyclic skeleton as neplanocin A, but is devoid of the C4′=C6′ double bond. Like neplanocin A, aristeromycin also shows potent inhibitory activity against SAH hydrolase, but again also proved to be cytotoxic and could not be developed as an antiviral agent. The cellular toxicity of aristeromycin was

Modified Nucleosides: in Biochemistry, Biotechnology and Medicine. Edited by Piet Herdewijn

ISBN: 978-3-527-31820-9

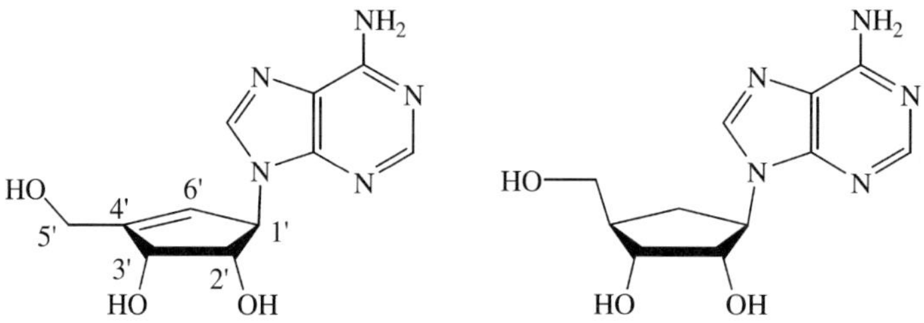

Neplanocin A (**1**)　　Aristeromycin (**2**)

Figure 21.1 Structures of neplanocin A (**1**) and aristeromycin (**2**).

similar to that of neplanocin A. Compound **2** was converted to the corresponding triphosphate [12], which inhibits RNA synthesis. In addition, aristeromycin was also metabolized intracellularly to the carbocyclic guanosine monophosphate (GMP), which inhibited hypoxanthine-guanine phosphoribosyltransferase (HGPRT), a key enzyme in purine biosynthesis [13, 14].

Following the discovery of neplanocin A and aristeromycin, many carbocyclic nucleosides have been synthesized as potential inhibitors of SAH hydrolase [6], although very few were identified as potent inhibitors of SAH hydrolase, partly because of the synthetic problems of the carbocyles. However, the recent development of a synthetic methodology [15] for the carbocycles using ring-closing metathesis (RCM) made it possible to monitor thorough structure–activity relationships (SARs) for the carbocyclic nucleosides, and hence to discover promising carbocyclic nucleosides with potent antiviral and antitumor activities. Although many excellent reviews on the carbocyclic nucleosides have been prepared [16], this chapter will focus on the recent developments of neplanocin A and its analogues, by describing and emphasizing details of biological activity and new synthetic metho-

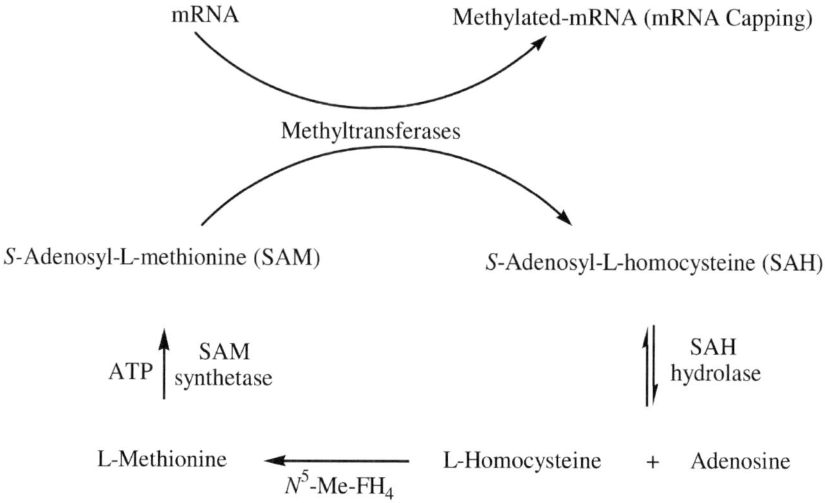

Figure 21.2 The *S*-adenosylhomocysteine (SAH) hydrolase cycle.

dologies. The chapter comprises two sections: the first section describes recent development in the syntheses of neplanocin A, while the second section provides details of the neplanocin A analogues and their biological activities.

21.2 New Methodologies in the Synthesis of Neplanocin A

At this point, the most recent methods used to synthesize the key intermediates via which neplanocin A is produced are described. For example, Jeong and coworkers [17] have reported a highly elegant synthesis of the key intermediate, D-cyclopentenone derivative **8** with a 3-hydroxymethyl side chain (Scheme 21.1). Here, D-ribose was converted to the bulky acetonide **3** which was treated with methyl ylide to give **4**. A Swern oxidation of **4** afforded the ketone **5**, which was subjected to a Grignard addition to give the desired diene **6** as a single stereoisomer. In the Grignard reaction, bulky protecting groups such as *tert*-butyldiphenylsilyl (TBDPS) and trityl (Tr) controlled the facial selectivity to give the desired diene **6**, but use of a less bulky benzyl protecting group yielded the undesired diastereomeric diene as a major product. A RCM reaction of diene **6** in the presence of a Grubbs second-generation catalyst gave the β-allylic alcohol **7**, which underwent a smooth oxidative rearrangement in the presence of pyridinium dichromate (PDC) to give the key intermediate **8**

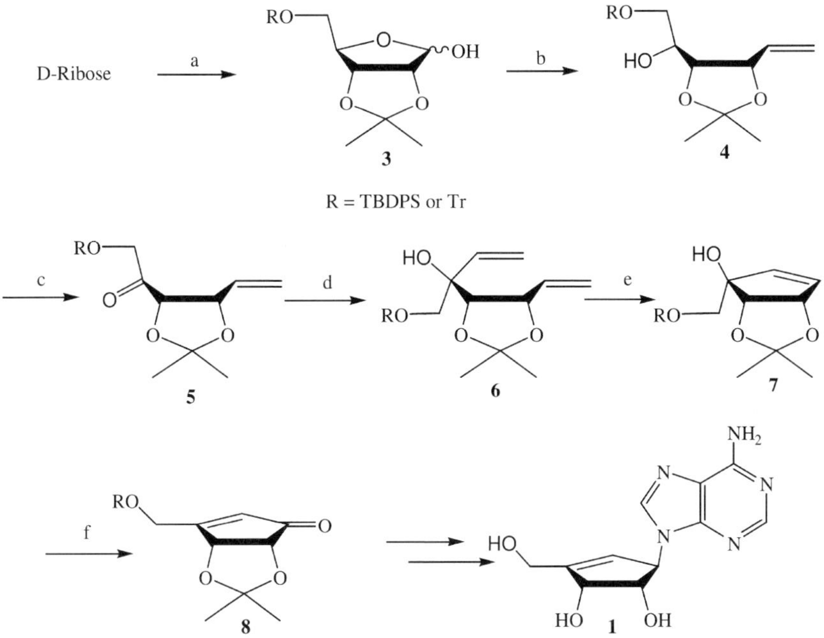

Reagents: (a) (i) acetone, H_2SO_4; (ii) R-Cl, pyridine; (b) ph_3PCH_3Br, KO^tBu; (c) $(COCl)_2$, DMSO; (d) $CH_2=CHMgBr$, THF; (f) PDC, DMF.

Scheme 21.1 Synthesis of the key intetmediate **8** for the synthesis of neplanocin A [17].

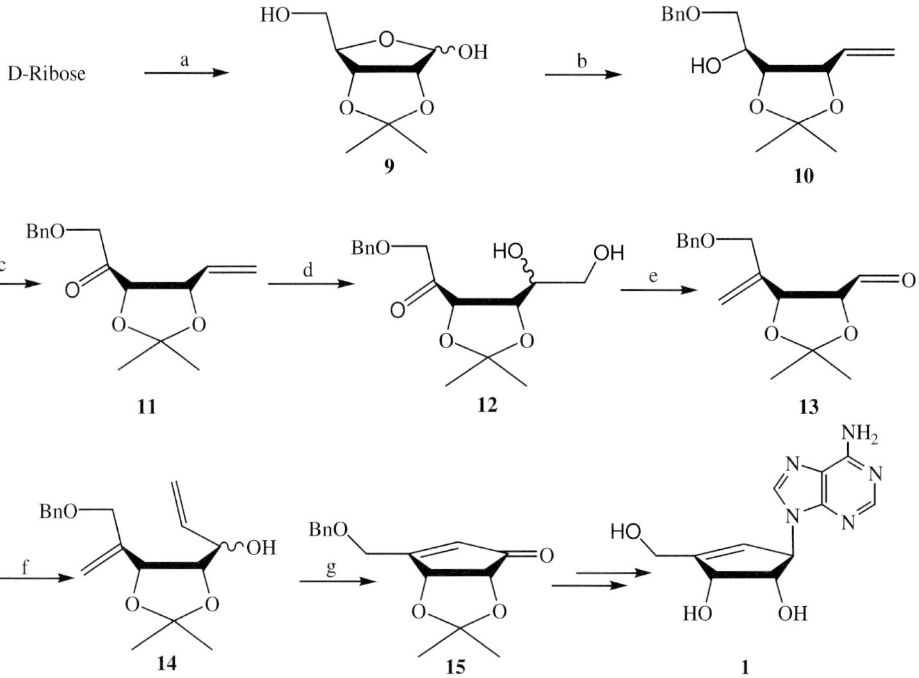

Reagents: (a) (i) acetone, H_2SO_4; (b) (i) Ph_3PCH_3Br, KO^tBu; (ii) Bu_2SnO, TBAI, BnBr; (c) $(COCl)_2$, DMSO; (d) OsO_4, NMO; (e) (i) Ph_3PCH_3Br, KO^tBu; (ii) $NaIO_4$; (f) CH_2=CHMgBr, THF; (g) (i) Grubbs catalyst; (ii) TPAP, NMO.

Scheme 21.2 An alternative synthesis of the key intermediate **15** [19].

for the synthesis of neplanocin A. However, the diastereomeric tertiary α-allylic alcohol formed in case of the benzyl protecting group failed to give the same intermediate **8** due to steric hindrance by the isopropylidene group. Thus, the key intermediate **8** was synthesized from D-ribose in seven steps, with an overall yield of 45–50%. Subsequently, the intermediate **8** could be elaborated to neplanocin A by conventional methods [18].

Jeong's group [19] has also reported an alternative synthesis of the key intermediate cyclopentenone **15** with a benzyl protecting group from D-ribose (Scheme 21.2). D-Ribose was converted to the isopropylidene derivative **9** under acidic conditions, after which a Wittig reaction of **9** followed by selective benzyl protection of the primary alcohol using organotin chemistry gave **10**. Swern oxidation of **10** followed by dihydroxylation with OsO_4 in the presence of NMO provided the diol **12** as a diastereomeric mixture. A Wittig reaction of **12** followed by oxidative cleavage gave **13**, the treatment of which with vinyl magnesium bromide yielded the diene **14** as a diastereomeric mixture. A RCM reaction of **14** and subsequent oxidation of the resulting cyclopentenol with tetrapropylammonium perruthenate (TPAP) in the presence of NMO afforded the key intermediate **15**, which was elaborated to neplanocin A according to the known procedure [20].

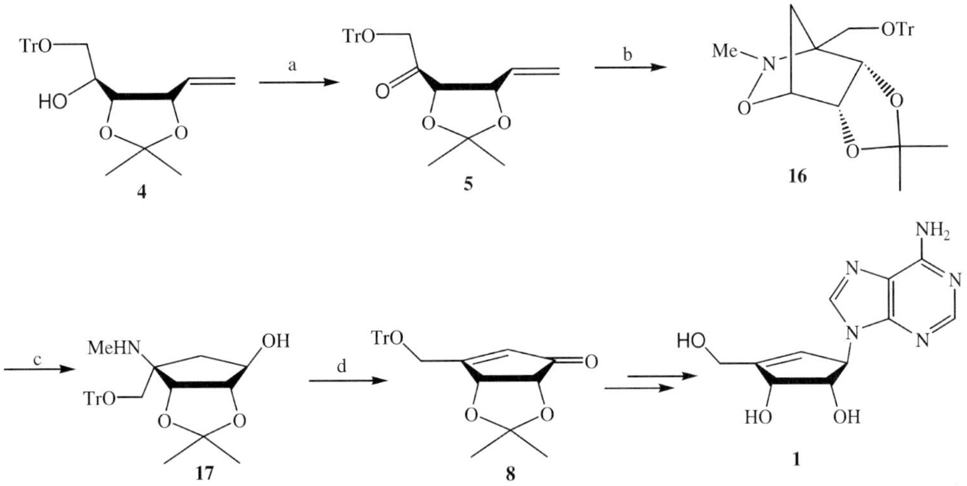

Reagents: (a) DMSO, DCC, THA; (b) (i) MeNHOH · HCl, pyridine; (ii) PhCl, Δ; (c) Zn, AcOH; (d) (i) MeI, K_2CO_3; (ii) Ag_2O, H_2O; (iii) PDC, CH_2Cl_2.

Scheme 21.3 An alternative synthesis of the key intermediate **8** using intramolecular cyclocaddition [21].

Gallos and coworkers [21] have disclosed a total synthesis of neplanocin A, using intramolecular nitrone cycloaddition and reductive cleavage of the N–O bond as key steps (Scheme 21.3). Moffatt oxidation of compound **4**, prepared from D-ribose by a reported method [17] gave the ketone **5**, the treatment of which with MeNHOH in pyridine and subsequent refluxing with chlorobenzene gave the intramolecular nitron cycloaddition adduct **16**. Reductive N–O cleavage of the cycloadduct **16** in the presence of zinc in acetic acid provided the methylamine **17**. Quaternization of **17** with excess MeI, and E2 elimination of the quaternary ammonium ion with Ag_2O, yielded cyclopentenol, which was oxidized with PDC to furnish the key intermediate **8**. The cyclopentenone **8** was converted to neplanocin A (**1**) according to the known method [18].

Jeong and coworkers [22] have reported a highly efficient synthesis of D-3-unsubstituted cyclopentenone derivative **22** and its enantiomer, using RCM as a key step, which can also serve as a versatile intermediate for the synthesis of neplanocin A (Scheme 21.4). The synthesis began with 2,3-*O*-isopropylidene-D-erythrono-γ-lactone (**18**). DIBAL reduction of **18**, followed by vinyl Grignard addition to the resulting lactol, afforded vinyl diol **19** as a single stereoisomer. Selective oxidation of the primary alcohol of **19** with TPAP in the presence of *N*-methylmorpholine *N*-oxide (NMO) gave the lactone **20**. Reduction of **20** with DIBAL, followed by Wittig reaction of the resulting lactol with methyl ylide, yielded the RCM precursor **21**. Exposure of **21** to Grubbs catalyst gave the cyclopentenol which, upon oxidation with MnO_2, furnished the key intermediate **22** from which neplanocin A could be synthesized [23].

Jeong's group [24] has also reported an improved and alternative synthesis of the same intermediate **22**, starting from a cheap and readily available D-ribose (Scheme 21.5). Treatment of the lactol **9** with vinyl magnesium bromide gave the triol **23**, the oxidative cleavage of which with sodium metaperiodate provided

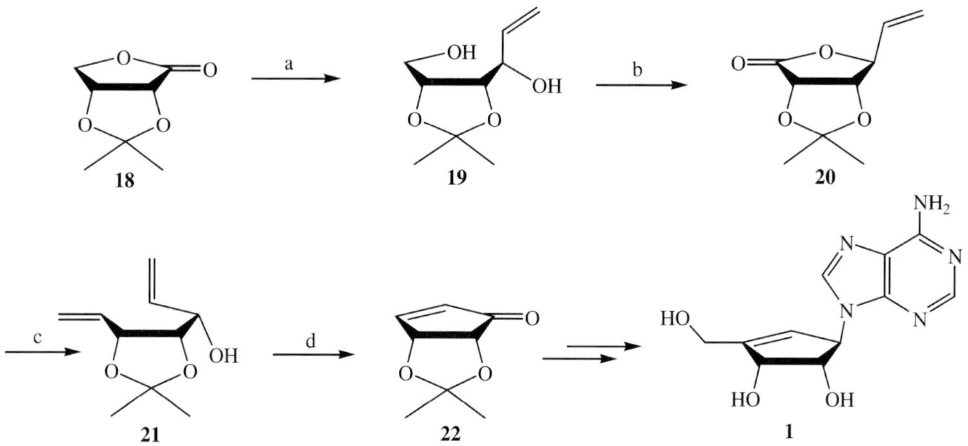

Reagents: (a) (i) DIBAL; (ii) CH_2=CHMgBr; (b) TPAP, NMO; (c) (i) DIBAL; (ii) Ph_3PCH_3Br; NaH, DMSO; (d) (i) Grubbs catalyst; (ii) MnO_2.

Scheme 21.4 Synthesis of D-3-unsubstitued cyclopentenone **22** for the synthesis of neplanocin A [22].

the lactol **24**. Wittig olefination of **24** with methyl ylide produced the diene **21**. A RCM reaction of **21**, followed by allylic oxidation with manganese dioxide, furnished the key intermediate **22**. This intermediate was synthesized in six steps and 45% overall yield from D-ribose. Ultimately, this synthetic procedure was found to be much more improved in terms of overall yield and number of steps when compared to that described in Scheme 21.4.

Chu and coworkers [25] have also reported an efficient and practical synthesis of the same intermediate **22**, using the similar synthetic approach as described in

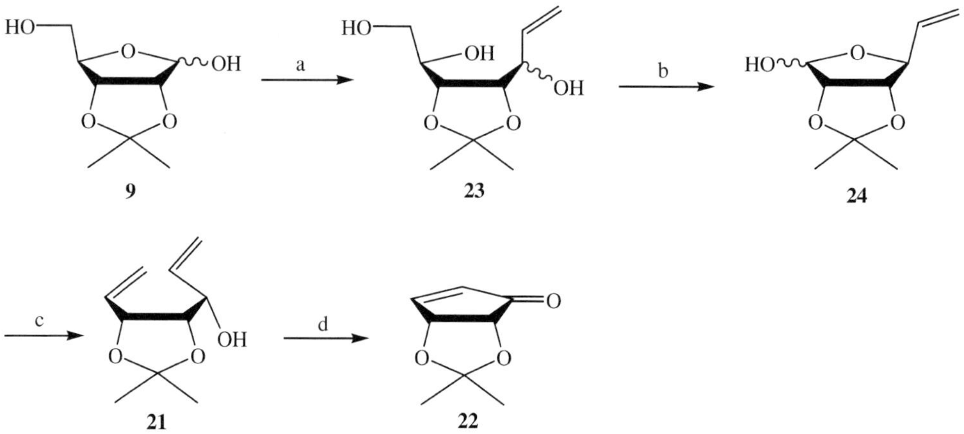

Reagents: (a) CH_2=CHMgBr; (b) $NaIO_4$, H_2O; (c) Ph_3PCH_3Br; NaH; (d) (i) Grubbs catalyst; (ii) MnO_2.

Scheme 21.5 An alternative and practical synthesis of D-cyclopentenone intermediate **22** [24].

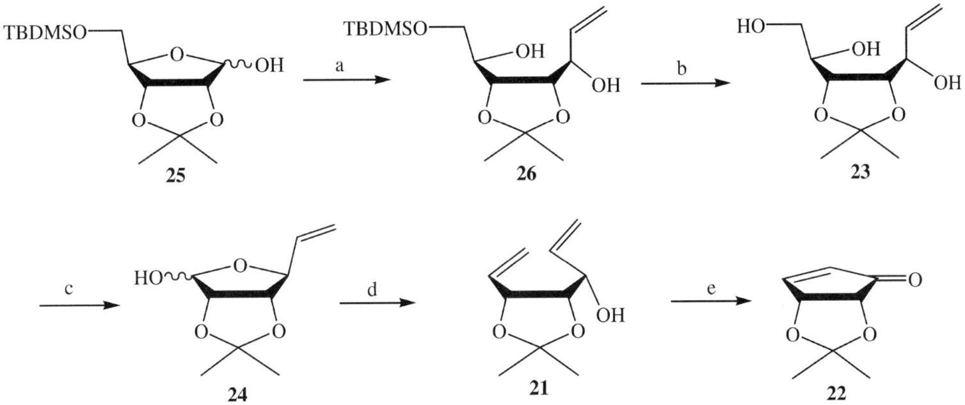

Reagents: (a) CH_2=CHMgBr; (b) TBAF; (c) $NaIO_4$, H_2O; (d) Ph_3PCH_3Br; NaH; (e) (i) Grubbs catalyst; (ii) PCC, CH_3COOH.

Scheme 21.6 An alternative synthesis of D-cyclopentenone intermediate **22** [25].

Scheme 21.5. The compound **25**, prepared from D-ribose, was converted to the diol **26** (Scheme 21.6). Removal of the TBDMS protecting group of **26** with tetra-*n*-butyl ammonium fluoride (TBAF) gave the triol **23**, which underwent an oxidative cleavage to give the lactol **24**. The latter compound was then converted to the same intermediate **22**, using the same procedure as described in Scheme 21.5.

Trost and coworkers [26] have disclosed an enantioselective total synthesis of neplanocin A, using palladium-catalyzed enantioselective base condensation as a key step (Scheme 21.7). The condensation of carbasugar **27** with 6-chloropurine in the presence of palladium catalyst and chiral ligand **28** provided exclusively the β-isomer **29**. In order to introduce the 4′-hydroxymethyl group, the compound **29** underwent a second palladium-catalyzed alkylation with phenylsulfonyl nitromethane to give the nitrosulfone derivative **30**, which was then elaborated to compound **31**. *Cis*-dihydroxylation of **31** with osmium tetroxide gave the tetraol, which was converted to the 2,3;4,5-di-*O*-acetonide **32** by treatment with 2,2-dimethoxy propane in the presence of *p*-toluenesulfonic acid, indicating that *cis*-dihydroxylation took place exclusively from the β face because acetonide will not be formed a *trans* fused bicyclo [3.3.0]ring system.

Selective deprotection of the 4′,5′-acetonide group in **32**, followed by selective protection of the 5′-hydroxyl group of the resulting **33** with a bulky pivaloyl group, gave **34**. Dehydration of the remaining 4′-hydroxyl group of **34** with thionyl chloride and pyridine in the presence of dimethylformamide (DMF) afforded the major *endo*-dehydrated product **35**, along with minor amount of *exo*-elimination product. The compound **35** was routinely transformed to neplanocin A.

Recently, Paquette and coworkers [27] have reported a total synthesis of neplanocin A, which involved a zirconocene-promoted ring contraction of vinyl-substituted pyranoside as a key step (Scheme 21.8). The vinyl-substituted pyranoside **36** was prepared from D-glucose in 13 steps, after which ring contraction of the pyranoside **36**

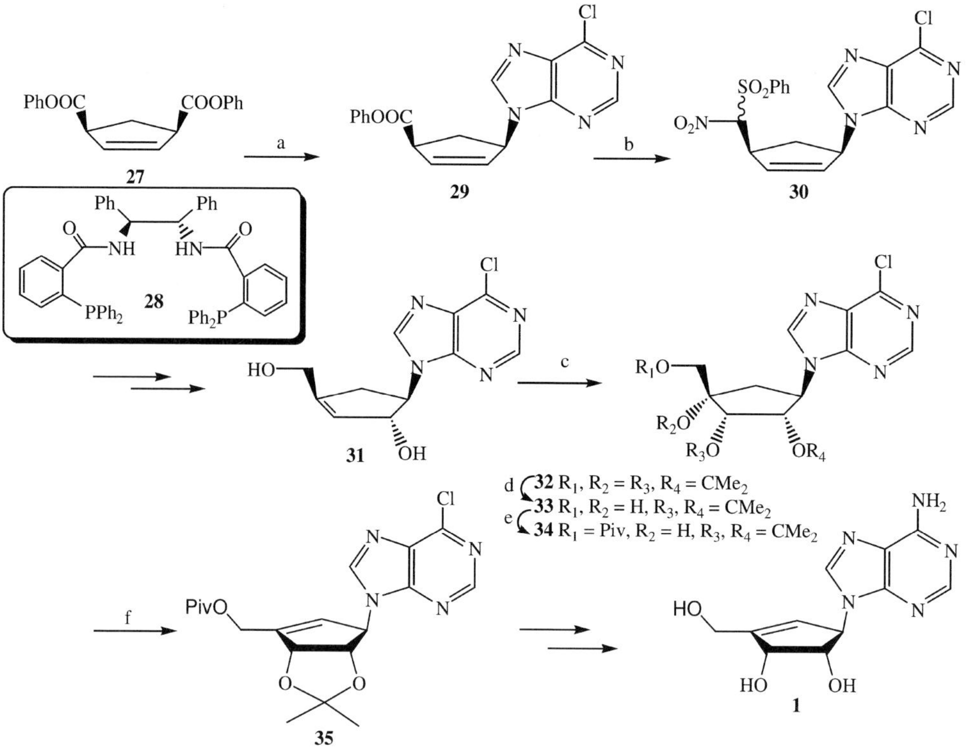

Reagents: (a) Pd_2dba_3. $CHCl_3$, **28**, Et_3N; (b) Pd_2dba_3. $CHCl_3$, PPh_3, phenylsulfonyl-nitromethane, Et_3N; (c) (i) OsO_4, NMO; (ii) $(MeO)_2CMe_2$, *p*-TSA; (d) $FeCl_3.H_2O$; (e) PivCl, pyridine; (f) SO_2Cl, pyridine, DMF.

Scheme 21.7 Total synthesis of neplanocin A, using palladium-catalyzed enantioselective base condensation [26].

with $ZrCl_2(Cp)_2$ and BuLi afforded the cyclopentane derivative **37**. The latter compound was converted to **39** by mesylation, debenzylation, and acylation. Ozonolysis of **39** followed by an elimination reaction in the presence of Hunig's base gave the aldehyde **40**. A chemoselective Luche reduction of aldehyde **40**, followed by TBS protection, yielded compound **41**. As the palladium-catalyzed base condensation did not succeed with **41**, it was transformed to the key intermediate **42** by DIBAL reduction and PDC oxidation. The intermediate **42** was elaborated to neplanocin A by the conventional method. This entire synthesis was completed in 25 steps from D-glucose.

21.3
Modifications on Neplanocin A and Aristeromycin

In this section are described the various modified analogues of neplanocin A and aristeromycin, together with details of their biological activities. The section is further subdivided according to the position of the modification made on neplanocin A and

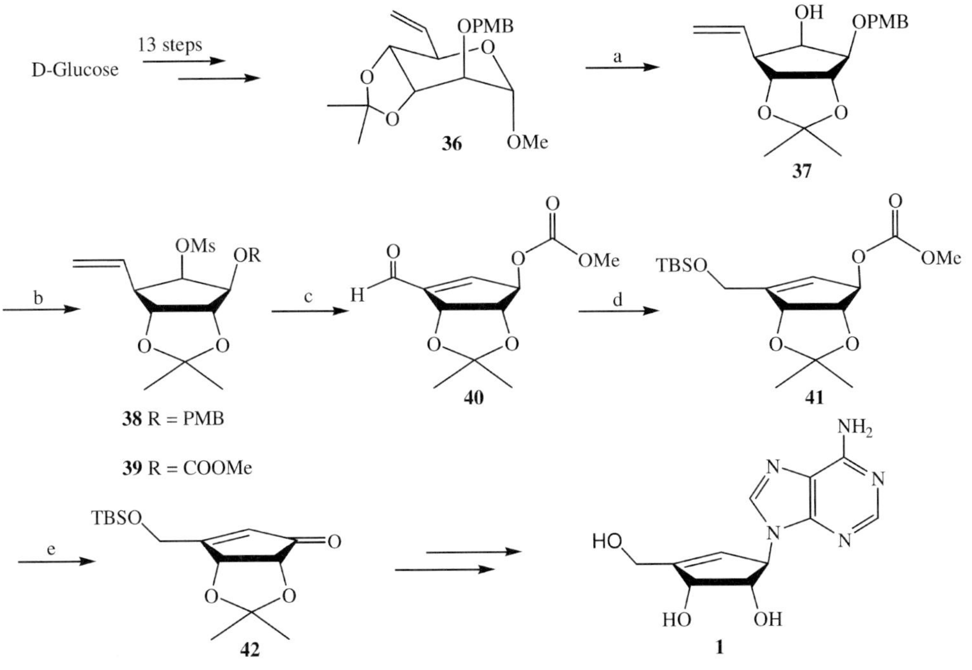

Reagents: (a) $ZrCl_2(Cp)_2$, BuLi; (b) (i) MsCl, Et_3N; (ii) DDQ; (iii) methyl carbonocyanidate, Et_3N, DMAP; (c) (i) O_3, Me_2S; (ii) $^iPr_2NEt$; (d) (i) $NaBH_4$, $CeCl_3.7H_2O$; (ii) TBSOTf, 2,6-lutidine; (e) (i) DIBAL; (ii) PDC.

Scheme 21.8 Total synthesis of neplanocin A, using a zirconocene-promoted ring contraction [27].

aristeromycin. These details commence with the C2′ position, as no reports exist on modification of the C1′ position of either neplanocin A or aristeromycin.

21.3.1 C2′ Modification

Schneller and coworkers [28] have reported the synthesis of a 2′-deoxy-neplanocin A analogue having a C1′=C6′ double bond (Scheme 21.9). The key feature of their method was to place a leaving group at the C6′ position of a requisite adenine derivative for 1,2-elimination in order to generate the double bond between C1′ and C6′. The synthesis began with tritylated cyclopentenol **43**, the epoxidation of which, followed by *p*-methoxybenzyl protection, gave the epoxide **44**.

Nucleophilic opening of the epoxide **44** with an adenine anion resulted in the adenine derivative **45** as a single regioisomer. In order to produce the 1′,2′-double bond in **45**, the N^6-amino group of adenine was first protected with *N,N*-dimethylformamide dimethyl acetal, with subsequent mesylation, elimination, and deprotection of acetal group to give the compound **46**; this was then converted to the final 2′-deoxy-neplanocin A analogue **47** under acidic conditions. The biological activity of this compound was not reported.

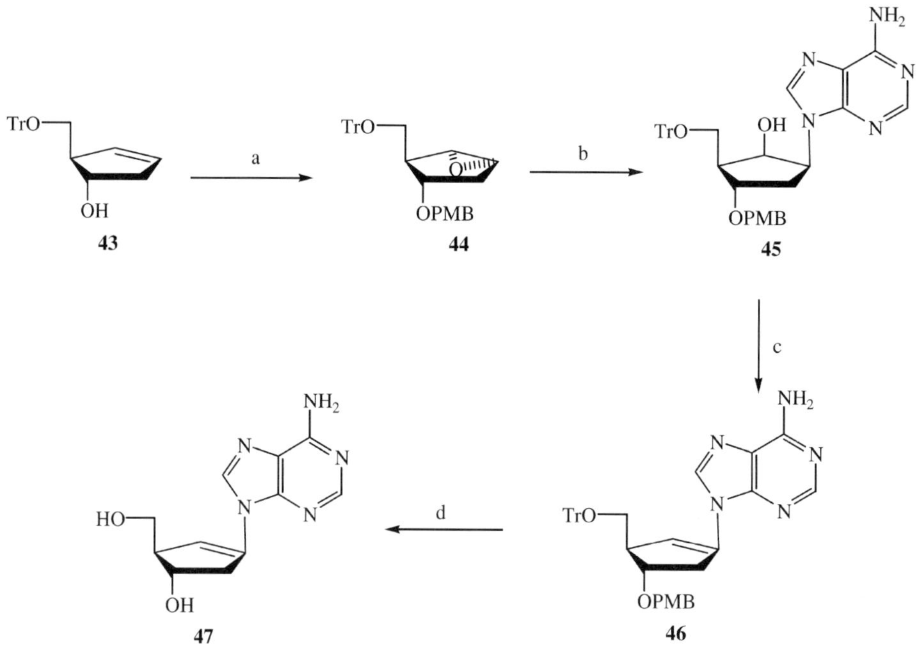

Reagents: (a) (i) *m*CPBA, CH_2Cl_2; (ii) PMBCl, NaH, DMF; (b) adenine, NaH, 15-crown-6; (c) (i) $HC(OMe)_2NMe_2$, DMF; (ii) MsCl, DMAP, CH_2Cl_2; (iii) NaOMe, THF/MeOH; (iv) MeOH, reflux; (d) (i) 10% TFA, CH_2Cl_2; (ii) 1 *N*, HCl.

Scheme 21.9 Synthesis of a 2′-deoxy-neplanocin A analogue having a C-1=C-6 double bond [28].

Meillon *et al.* [29] have reported the synthesis of 2′-*C*-methyl-aristeromycin (**53**), which involved a stereoselective base condensation with the epoxy alcohol **51** as a key step (Scheme 21.10). The methyl cyclopentenone **48** was alkylated with BOM bromide to give **49** which, upon reduction with $LiBH_4$, gave the diastereomeric mixture of **50a** and **50b** in a 2 : 1 ratio. Epoxidation of the desired alcohol **50a** with *m*CPBA provided **51** as a sole epoxide, which was condensed with adenine anion to give the β-nucleoside derivative **52** exclusively. The 2′-methyl group created the perfect regio- and stereoselectivity during attack of the adenine anion to the epoxide to give the exclusive product **52**. Debenzylation of **52** afforded the final analogue **53**. Similarly, the stereoisomer **50b** was also used to synthesize the 4′-epimer of **53**. The antiviral activities of all synthesized nucleosides were tested against various viruses such as bovine viral diarrhea, hepatitis B virus (HBV), human immunodeficiency virus (HIV), Dengue, and West Nile virus. However, none of these 2′-branched nucleosides showed any antiviral activity.

Jacobson and coworkers [30] have synthesized various cyclopropyl-fused 2′-deoxy-neplanocin A analogues (**56–58**), and investigated their binding affinity for adenosine receptors (Scheme 21.11). The key carbasugar **54** was synthesized from the known cyclopentenone derivative **15** [31]. The Mitsunobu condensation of **54** with 2,6-dichloropurine, followed by debenzylation with BCl_3, provided the compound **55**. The N^6-chloro group of **55** was treated with cyclopentylamine and 3-iodo-benzylamine

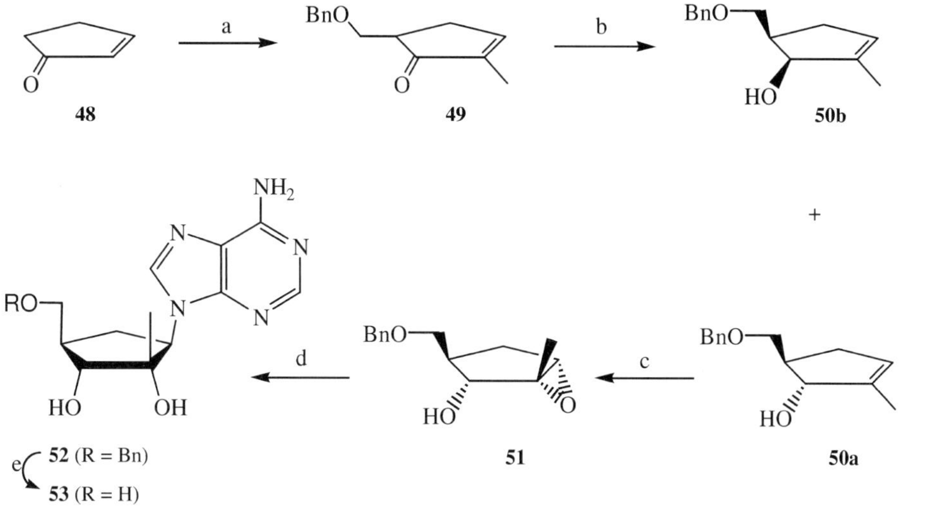

Reagents: (a) LDA, BOM-Br, THF; (b) $LiBH_4$, ether (**50a** major) or $NaBH_4$, $CeCl_3$-$7H_2O$, MeOH (**50b** major); (c) *m*CPBA, CH_2Cl_2; (d) NaH, adenine, DMF; (e) $Pd(OH)_2$/C, cyclohexene, MeOH

Scheme 21.10 Synthesis of 2′-*C*-methyl-aristeromycin (**53**) [29].

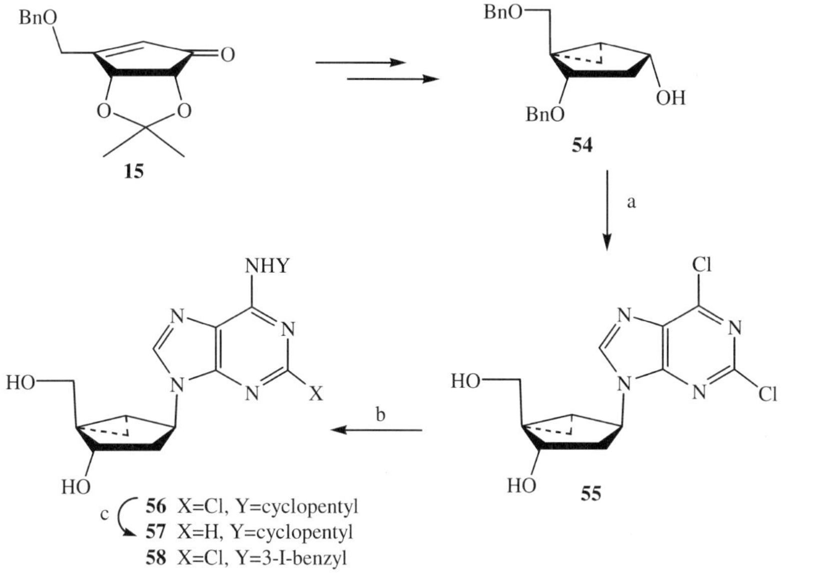

Reagents: (a) (i) 2,6-dichloropurine, DEAD, PPh_3, THF; (ii) BCl_3, CH_2Cl_2; (b) cyclopentylamine (for **56**) or 3-I-benzylamine-HCl, TEA (for **58**) MeOH; (c) H_2/Pd-C, MeOH.

Scheme 21.11 Synthesis of cyclopropyl-fused 2′-deoxy-neplanocin A analogues [30].

to give **56** and **58**, respectively. Catalytic hydrogenation of the 2-chloro group in **56** yielded **57**. The radioligand binding assay of **56–58** indicated that compounds **56** and **57** were weakly and non-selectively bound to the adenosine receptors, whereas compound **58** displayed a greatly reduced affinity and selectivity for A_3 adenosine receptors. The K_i values for **56** and **57** at the A_1 AR were $2.28 \pm 0.99\,\mu M$ ($40 \pm 8\%$ efficacy) and $2.89 \pm 0.14\,\mu M$ ($30 \pm 1\%$ efficacy), respectively, and found to be partial agonists at the A_1 AR.

21.3.2
C3′ Modification

Although very few reports exist on 3′-modified neplanosin A derivatives, Jeong and colleagues [32] have investigated a novel stereoselective method for synthesis of the apio analogue of neplanocin A, using RCM and stereoselective hydroxymethylation as key steps (Scheme 21.12). The intermediate **24** [24] underwent an aldol condensation with formaldehyde in the presence of K_2CO_3 to give the aldol **59** as a sole product. Subsequent Wittig olefination of **59** with methyl ylide gave the diene **60**, which underwent a smooth RCM reaction upon treatment with Grubbs second-generation catalyst to give the key cyclopentenol **61**. The latter compound was elaborated to the final apio-neplanocin A (**62**) by a routine procedure. Compound **62** was shown not to possess any inhibitory activity against human recombinant SAH hydrolase, possibly due to the presence of a tertiary hydroxyl group at the C3′ position, which could not be oxidized by cofactor-bound NAD^+. When tested for its binding affinity to the adenosine receptors (by using competitive radioligand binding assays),

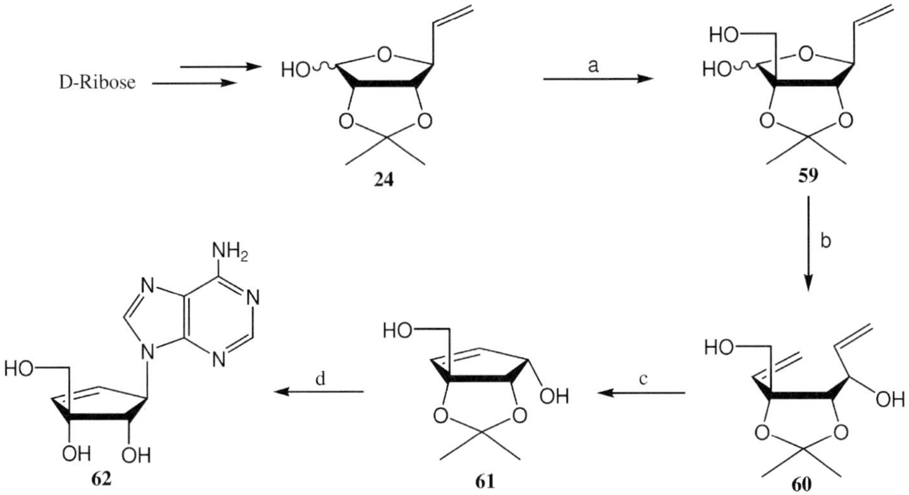

Reagents: (a) K_2CO_3, 37% HCHO, MeOH; (b) Ph_3PCH_3Br, $KOBu^t$, THF; (c) Grubbs catalyst (2nd generation), CH_2Cl_2; (d) (i) TrCl, DMAP, pyridine; (ii) MsCl, Et_3N, CH_2Cl_2 (iii) adenine, K_2CO_3, 18-crown-6, DMF; (iv) aqueous TFA, THF.

Scheme 21.12 Stereoselective synthesis of apio-neplanocin A [32].

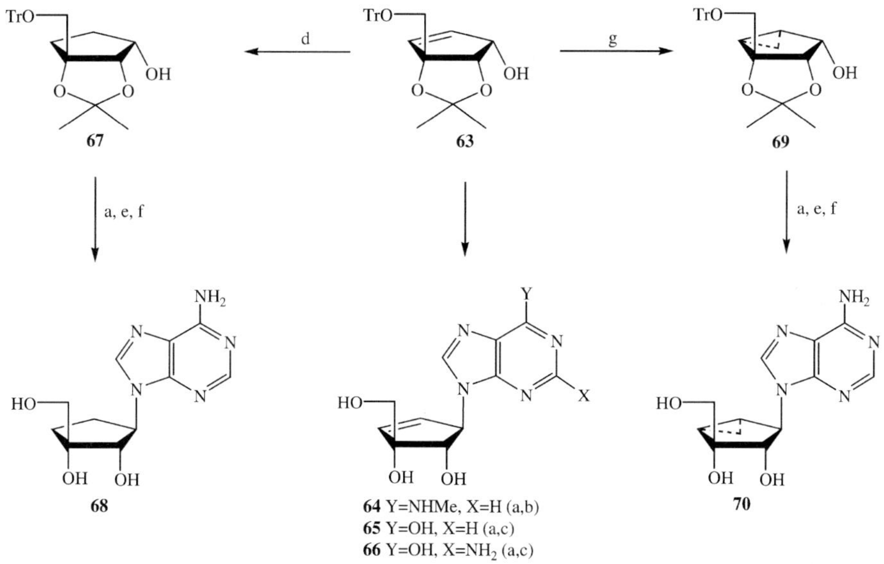

Reagents: (a) 6-chloropurine or 2-amino-6-chloropurine, Ph_3P, DEAD, THF; (b) 40% $MeNH_2$, MeOH; (c) 3 *N* HCl, THF; (d) 10% Pd/C, H_2; (e) NH_3, MeOH; (f) 30% aq. CF_3COOH, THF; (g) CH_2I_2, Et_2Zn, CH_2Cl_2.

Scheme 21.13 Synthesis of various apio-neplanocin A analogues [33].

compound **62** was found to be a potent and highly selective agonist at A_3 AR ($K_i = 628 \pm 69$ nM), but did not show any binding affinity at human A_1 and A_2 ARs (8% and 0% inhibition at 10 μM, respectively) [33].

Jeong's group [33] has further reported the synthesis of various apio-neplanocin A analogues, and studied their activity against SAH hydrolase (Scheme 21.13). Condensation of the glycosyl donor **63** [16] with 6-chloropurine under the Mitsunobu condition, followed by treatment with methyl amine, gave **64** after acid hydrolysis with 3 M HCl. The 6-chloropurine product was converted to the inosine derivative **65** by treating with 3 M HCl. Similarly, the guanine derivative **66** was prepared from **63**. The apio aristeromysin analogue **68** and apio-(*N*)-methanocarba-adenosine analogue **70** were each prepared, using a similar method, from the glycosyl donors **67** and **69**, respectively. Although all of the final compounds were tested against SAH hydrolase, none showed any inhibitory activity.

Recently, Jeong and coworkers [34] have reported the synthesis of a 3′-modified-neplanocin A derivative (homoapio-neplanocin A), using chemoselective hydroboration as a key step (Scheme 21.14). Selective oxidation of the diene precursor **60** [16] with TEMPO, followed by RCM and Wittig olefination, provided the intermediate **71**. Chemoselective hydroboration of compound **71** was achieved with Sia_2BH to give the diol **72** after subsequent oxidation with sodium perborate. Selective protection of the primary hydroxyl group with a trityl group, followed by tosylation, furnished the glycosyl donor **73**. The latter compound was elaborated to the homoapio-neplanocin A **74** using the conventional method. Compound **74** did not exhibit any inhibitory activity against SAH hydrolase.

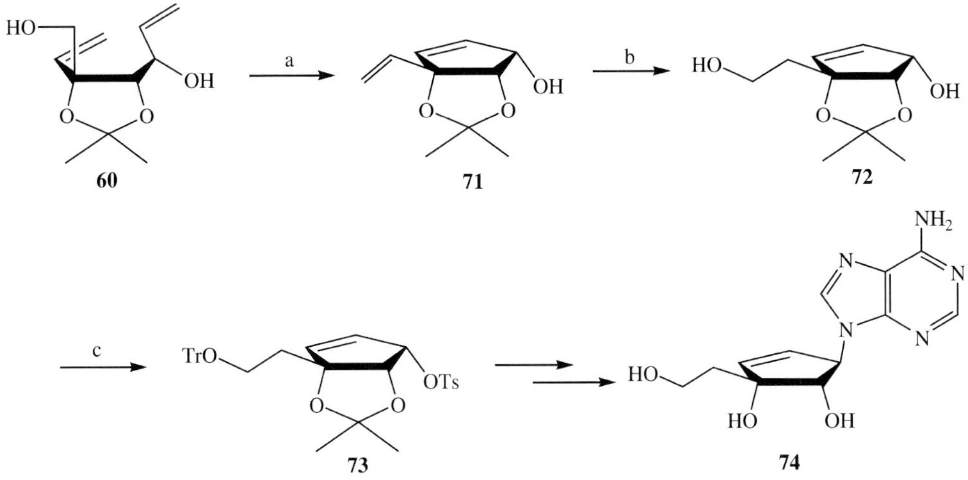

Reagents: (a) (i) TEMPO, TBACl, NCS; (ii) Grubbs catalyst; (iii) $MePh_3PBr$, KO^tBu; (b) (i) Sia_2BH; (ii) $NaBO_3.H_2O$; (c) (i) TrCl, pyridine; (ii) TsCl, DMAP.

Scheme 21.14 Synthesis of homoapio-neplanocin A using chemoselective hydroboration [34].

21.3.3
C4′ Modification

As the C4′ position of neplanocin A is a quaternary center, there is very little opportunity to modify this position, and only a few examples of such modification for neplanocin A have been reported. Hence, the 4′-modification of aristeromycin rather than neplanocin A will be described in this section.

Jeong and coworkers [35] have reported an efficient method for the synthesis of the fluoro analogue of 9-(*trans*-2′-, *trans*-3′-dihydroxycyclopent-4′-enyl)adenine (DHCeA), which is a neplanocin A derivative devoid of the 4′-hydroxymethyl group (Scheme 21.15). The cyclopentenone **22** [24] was iodinated using iodine and pyridine to produce the iodo cyclopentenone **75**, the reduction of which with sodium borohydride under Luche's condition, followed by silylation with TBDPSCl, yielded the silyl ether **76**. The electrophilic fluorination of **76** was achieved by treatment with *n*-BuLi and *N*-fluorobenzenesulfonimide (NFSI) to give the fluoro derivative **77** in excellent yield. Removal of the TBDPS group in **77** with TBAF, followed by mesylation of the resulting alcohol, provided the glycosyl donor **78**. Condensation of **78** with the adenine anion gave the condensed product which was treated with 10% trifluoroacetic acid to yield the fluoro-DHCeA **79**. The inhibitory activity ($IC_{50} = 8.9\,\mu M$) of **79** against SAH hydrolase was almost equipotent as that of DHCeA ($IC_{50} = 8.7\,\mu M$).

Schneller *et al.* [36] have reported a simple method for the synthesis of 4′-*C*-methylaristeromycin (**85**) (Scheme 21.16), whereby the monoacetate **80** was converted to the cyclopentanone derivative **81** in five steps [37]. The compound **81** was treated with MeMgBr to give the 4′-methyl derivative **82** which, upon Mitsunobu-type elimination, followed by desilylation and subsequent oxidation with PCC, yielded the

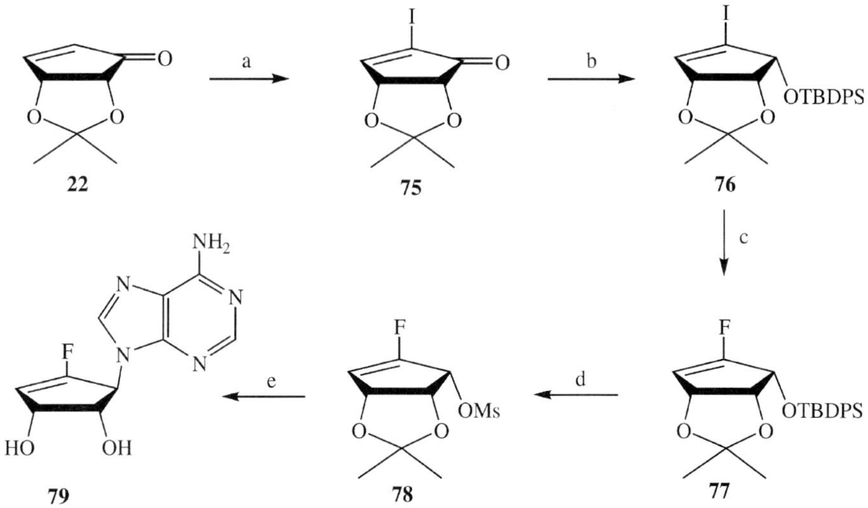

Reagents: (a) I_2, pyridine; (b) (i) $NaBH_4$, $CeCl_3$-$7H_2O$; (ii) TBDPSCl, Imidazol; (c) NFSI, *n*-BuLi; (d) (i) TBAF; (ii) MsCl, Et_3N; (e) (i) adenine, K_2CO_3, 18-crown-6-ether; (ii) 10% TFA.

Scheme 21.15 Synthesis of fluorinated DHCeA using electrophilic vinyl fluorination as a key step [35].

enone **83**. Treatment of **83** with vinyl magnesium bromide in the presence of CuBr·Me_2S afforded the 1,4-adduct **84**, which in turn was elaborated to the 4′-*C*-methylaristeromycin (**85**).

Schneller *et al.* [38] have also reported the synthesis of 5′-noraristeromycin possessing a C-1′/C-6′ double bond (Scheme 21.17). The epoxidation of cyclopentenone

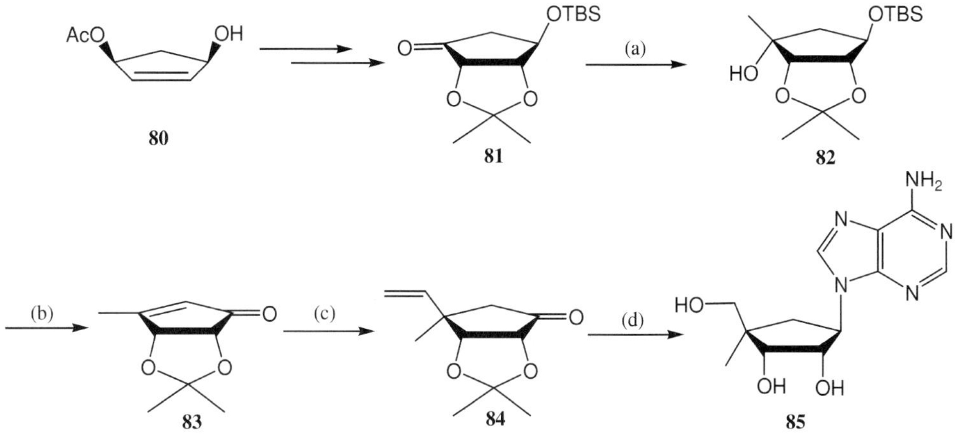

Reagents: (a) MeMgBr, THF; (b) (i) TPP, DIAD, toluene; (ii) TBAF, THF; (iii) PCC, Celite, CH_2Cl_2; (c) CH_2=CHMgBr, HMPA, TMSCl, CuBr-Me_2S, THF; (d) (i) DIBAL, THF; (ii) TPP, DIAD, 6-chloropurine, THF; (iii) OsO_4, $NaIO_4$, MeOH; (iv) $NaBH_4$, MeOH; (v) NH_3, MeOH; (vi) HCl, MeOH.

Scheme 21.16 Synthesis of 4′-*C*-methylaristeromycin [36].

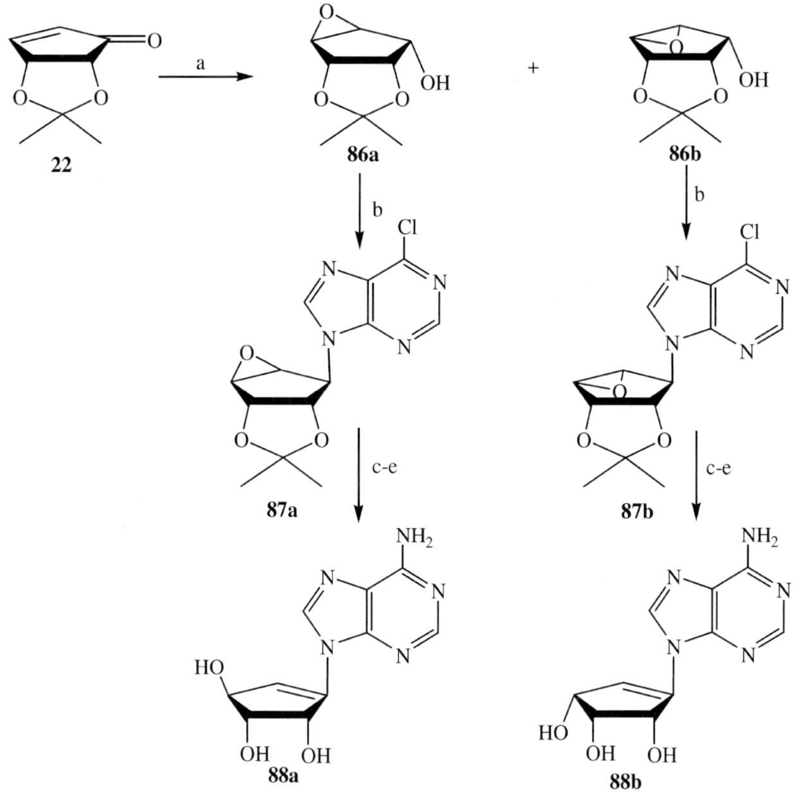

Reagents: (a) (i) ^tBuOOH, Triton B, THF, MeOH; (ii) $NaBH_4$, $CeCl_3$-$7H_2O$, MeOH; (b) 6-chloropurine, TPP, DIAD, THF; (c) NaOMe, MeOH; (d) NH_3, MeOH; (e) 0.5 *N* HCl, MeOH.

Scheme 21.17 Synthesis of 5′-noraristeromycin possessing a C-1′/C-6′ double bond [38].

22 with *tert*-butyl hydrogen peroxide in the presence of Triton B provided both α and β epoxides which, upon stereoselective reduction with sodium borohydride, afforded the glycosyl donors **86a** and **86b**. Mitsunobu condensation of **86a** and **86b** with 6-chloropurine yielded the condensed products **87a** and **87b**, which underwent an epoxy opening with sodium methoxide, with the resultant alcohols being elaborated to **88a** and **88b**, respectively, by conventional methods. Both, **88a** and **88b** were tested for their antiviral activity against Daudi host cells, but proved to be inactive.

21.3.4 C5′ Modification

The C5′ hydroxyl group is one of the most important moieties for the modification of neplanocin A, mainly because the compound's toxicity results from phosphorylation of the primary hydroxyl group at the C5′ position by adenosine kinase and subsequent

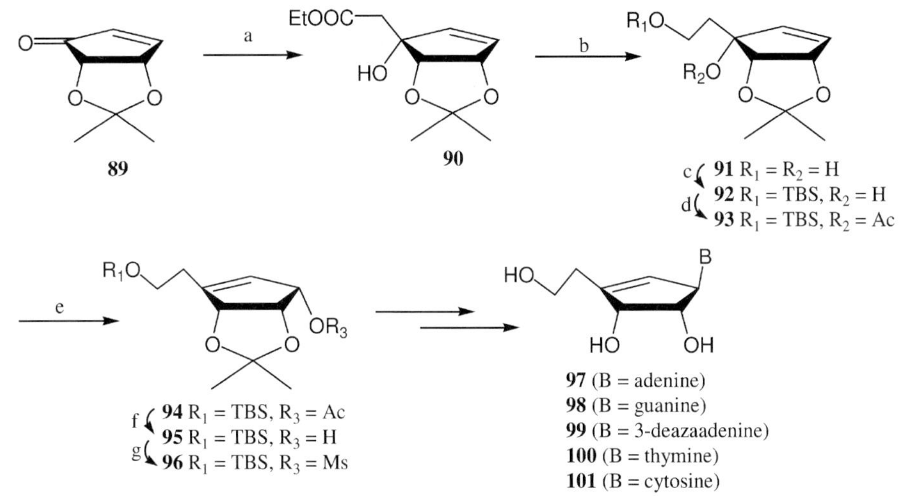

Reagents: (a) $(TMS)_2NH$, BuLi, EtOAc, THF; (b) $LiBH_4$, THF; (c) TBSCl, Imidazole, THF; (d) Ac_2O, DMAP, Et_3N, THF; (e) $PdCl_2(MeCN)_2$, *p*-benzoquinone, THF; (f) K_2CO_3, MeOH; (g) MsCl, DMAP, CH_2Cl_2

Scheme 21.18 Synthesis of various 5′-homologated neplanocin A derivatives [40].

metabolism by cellular enzymes [9, 10, 39]. Although to date, many reports have been made on modifications of the C5′ position, only those compounds of biological interest will be discussed at this point.

Matsuda and coworkers [40] have reported the synthesis of various 5′-homologated neplanocin A derivatives, together with details of their biological activities (Scheme 21.18). In order to synthesize the 5′-homologated glycosyl donor **96**, the cyclopentenone **89** was treated with EtOAc in the presence of $LiN(TMS)_2$ to give the 1,2-adduct **90**. Reduction of the ester **90**, followed by selective silyl protection of the primary hydroxyl group and subsequent acetylation of secondary hydroxyl group, provided the compound **93**. Palladium-catalyzed allylic rearrangement of the acetoxy group of **93** was achieved with a $PdCl_2(MeCN)_2$ catalyst to yield the allylic acetate **94**, which was then converted to the glycosyl donor **96** after deacetylation followed by mesylation. Compound **96** was then elaborated to various homo-neplanocin A derivatives (**97–101**) by conventional methods. All of the final nucleosides were tested for antiviral activities against vesicular stomatitis virus (VSV) and parainfluenza virus (PV-3). Homo-neplanocin A (**97**) showed significant antiviral efficacy against VSV ($IC_{50} = 1.0\,\mu g\,mL^{-1}$) and PV-3 ($IC_{50} = 0.35\,\mu g\,mL^{-1}$) without cytotoxicity up to $500\,\mu g\,mL^{-1}$, when compared with neplanocin A ($IC_{50} = 1.0\,\mu g\,mL^{-1}$ for VSV; $0.74\,\mu g\,mL^{-1}$ for PV-3) with cytotoxicity ($CC_{50} = 152\,\mu g\,mL^{-1}$). The 3-deazaadenine derivative **99** also showed significant antiviral activity against VSV ($IC_{50} = 3.9\,\mu g\,mL^{-1}$), but was inactive against PV-3. The other compounds, namely **98**, **100**, and **101**, were each inactive against both viruses.

Recently, Jeong and coworkers [41] have reported the synthesis of various 5′-substituted fluoro-neplanocin A analogues, in which the substitution of a hydrogen-bonding donor at the C5′ position was essential for the inhibitory activity of SAH

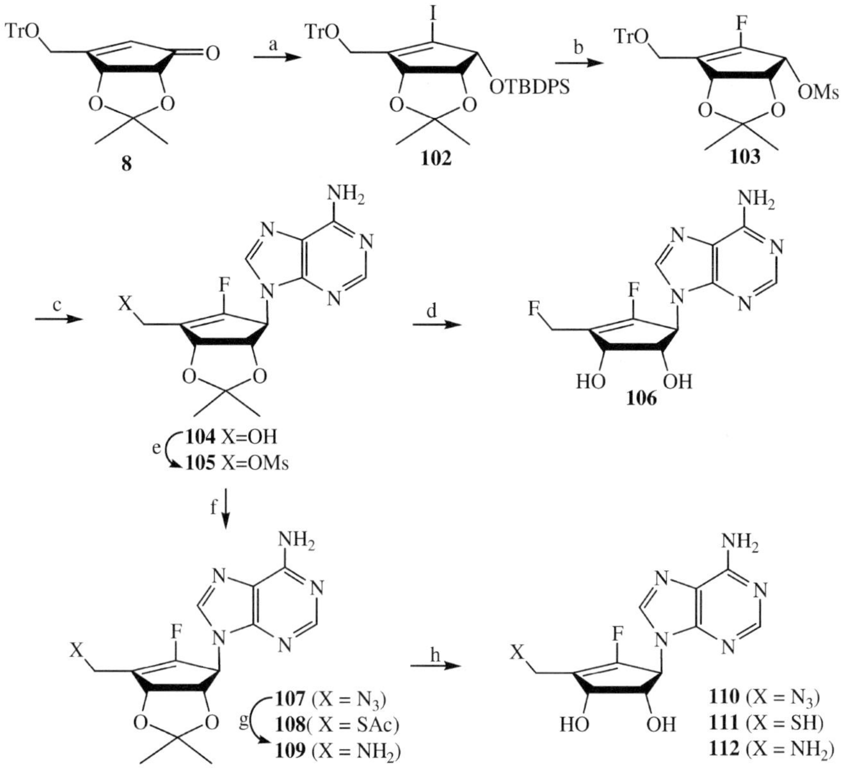

Reagents: (a) (i) I_2, Pyridine, CCl_4; (ii) $NaBH_4$, $CeCl_3$, MeOH; (iii) TBDPSCl, Imidazole, DMF; (b) (i) NFSI, BuLi; (ii) TBAF, THF; (iii) MsCl, Et_3N; (c) (i) adenine, K_2CO_3, 18-crown-6; (ii) *p*-TsOH, MeOH; (d) (i) DAST, CH_2Cl_2; (ii) aq. TFA; (e) MsCl; (f) NaN_3, DMF for **107** or KSAc, DMF for **108**; (g) $Pd(OH)_2$, H_2; (h) aq. TFA then NH_4OH, MeOH.

Scheme 21.19 Synthesis of various 5′-substituted fluro-neplanocine A analogues [41].

hydrolase (Scheme 21.19). The cyclopentenone **8** [17] was converted to the glycosyl donor **103** using the similar procedure as described in Scheme 21.15. Condensation of **103** with adenine anion, followed by deprotection of the 5′-trityl group, provided the 5′-OH derivative **104**. Treatment of the latter compound with DAST furnished the 5′-fluoro-neplanocin A derivative **106**, after removal of the acetonide group. For the synthesis of 5′-azido- and 5′-mercapto derivatives of fluoro-neplanocin A, compound **104** was mesylated and then treated with NaN_3 and KSAc to give **107** and **108**, respectively. Removal of the protecting groups of **107** and **108** yielded the 5′-azido derivative **110** and 5′-mercapto derivative **111**, respectively. The 5′-azido derivative **107** was converted to the 5′-amino derivative **112** by reduction with Pd $(OH)_2$, followed by deprotection with aqueous trifluoroacetic acid. Neither compound **106** nor **110** showed any inhibitory activity against SAH hydrolase; in contrast, compound **112** exhibited a potent inhibitory activity ($IC_{50} = 12.68\,\mu M$) but was less potent than the parent fluoro-neplanocin A ($IC_{50} = 0.48\,\mu M$). Compound **111** showed very weak inhibitory activity ($IC_{50} = 97.27\,\mu M$). Based on the results of these

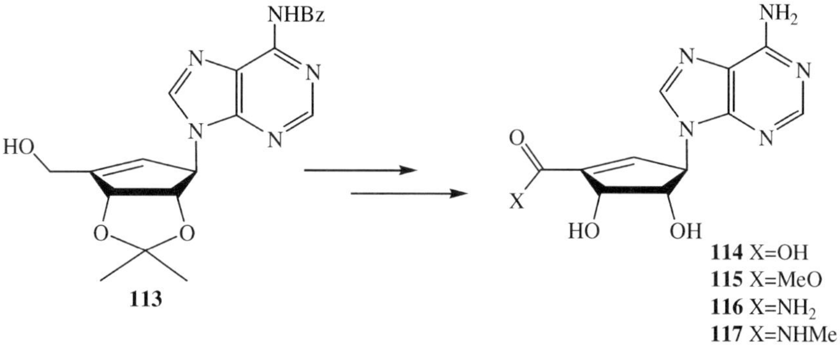

Scheme 21.20 Synthesis of various 5′-acid derivatives of neplanocin A [43].

studies, it was concluded that compounds with hydrogen-bonding acceptors such as N_3 and F at the 5′-position of fluoro-neplanocin A had no enzyme inhibitory activity, whereas those with hydrogen donors such as NH_2 and OH showed a restoration of enzyme inhibitory activity. All of these 5′-substituted compounds were much less cytotoxic than fluoro-neplanocin A in human colon cancer and lung cancer cell lines.

It is known that the 3′-hydroxyl group of neplanocin A is oxidized by NAD^+ bound to SAH hydrolase to form the 3′-keto intermediate, which has an enone structure and binds tightly to the enzyme. However, this intermediate could not be isolated due to its instability [42]. On the basis of this idea, Matsuda *et al.* [43] have designed and synthesized various acid derivatives (**114**–**117**) of neplanocin A which modified the C5′ position of neplanocin A (Scheme 21.20).

The inhibitory effects on SAH hydrolase from rabbit erythrocytes and antiviral activities of all the synthesized compounds **114**–**117** were monitored. Among these, the amide derivative **116** had anti-RNA-virus activities against parainfluenza-3, measles or mumps within the IC_{50} range of 0.14 to 4.88 $\mu g\,mL^{-1}$. Carboxylic acid derivative **114** and amide derivative **116** were both potent inhibitors of SAH hydrolase ($IC_{50} = 0.33\,\mu g\,mL^{-1}$ and $2.56\,\mu g\,mL^{-1}$, respectively). Although compound **114** showed significant inhibitory effects on SAH hydrolase, it did not show any anti-RNA-virus activities, though this may be due to its inability to pass through the cell membrane because of its high polarity.

Matsuda and coworkers [44] have also studied the biological effects of different substituents at the 5′-position of neplanocin A (Scheme 21.21).

The intermediate **118** was treated with ethynylmagnesium bromide to provide two diastereomers, (*R*)-**119a** and (*S*)-**119b**, after debenzoylation and deprotection of the acetonide. The catalytic hydrogenation of (*R*)-**119a** and (*S*)-**119b** gave ethynyl derivatives **120a** and **120b** or ethyl derivatives **121a** and **121b**, respectively. The inhibitory activities of all synthesized compounds were examined on murine L929 cells, in which the (*R*)-stereoisomers were found to be more potent than the (*S*)-stereosiomers, with their inhibitory effects decreasing in the following order: **119a** > **120a** > **119b** > **121a** > **120b** > **121b**. These compounds were sensitive to vaccinia virus (VV) and to VSV. The order of antiviral activity was approximately similar to the inhibitory effect for SAH hydrolase. For example,

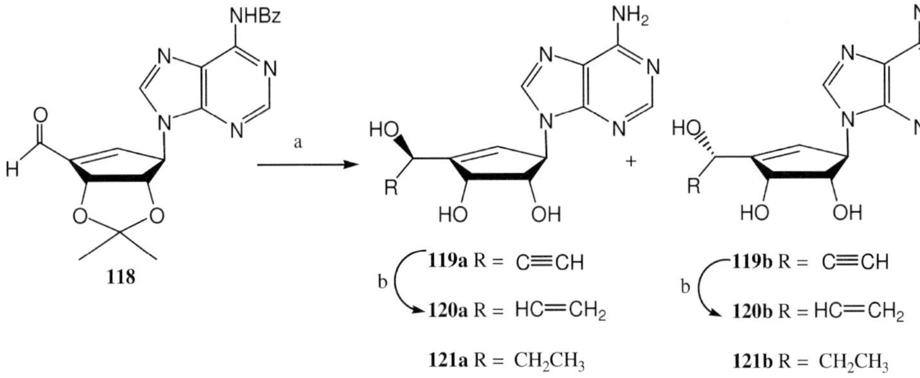

Reagents: (a) (i) ethynylmagnesium bromide; (ii) NaOMe, MeOH; (iii) 50% HCOOH; (b) Pd/$CaCO_3$, H_2, MeOH.

Scheme 21.21 Synthesis of 5′-substituted-neplanocin A analogues [44].

(5′*R*)-5′-*C*-ethynyl-neplanocin A (REyNPA) had significant antiviral effects (IC_{50} = 0.07 μg mL^{-1} for VV and 0.2 μg mL^{-1} for VSV) respectively, and was ten-fold more potent than neplanocin A (IC_{50} = 0.7 μg mL^{-1} for VV; 0.2 μg mL^{-1} for VSV). It was suggested that neither REyNPA nor (5′*R*)-5′-*C*-methyl-neplanocin A (RMNPA) would be phosphorylated at the 5′-position in the cell because the hydroxyl group would be sterically hindered against phosphorylation by adenosine kinase. From a comparative study on the growth of wild-type and tubercidin-resistant FM3A (FM3A Tub^{r4}) cells, neplanocin A was seven-fold less inhibitory to the proliferation of FM3A Tub^{r4} cells than for wild-type cells, while tubercidin was more than 100-fold less cytotoxic towards the Tub^{r4} FM3A cells. However, REyNPA and RMNPA were equipotent against the wild-type and Tub^{r4} FM3A cells.

Daelemans and coworkers [45] also conducted a comparative investigation on the inhibitory effects of both *R*- and *S*-isomers of various 5′-*C*-substituted analogues of neplanocin A (Figure 21.3).

The inhibitory effects on SAH hydrolase were compared with those on HIV-1 replication and Tat transactivation. When a series of compounds (**119–122**) were tested against HIV-1 transactivation, RMNA **122a** (IC_{50} = 0.32 μM) and REyNPA **119a** (IC_{50} = 0.9 μM) were highly potent inhibitors. The *R*-isomer generally showed much higher inhibitory activities than the *S*-isomer against SAH hydrolase

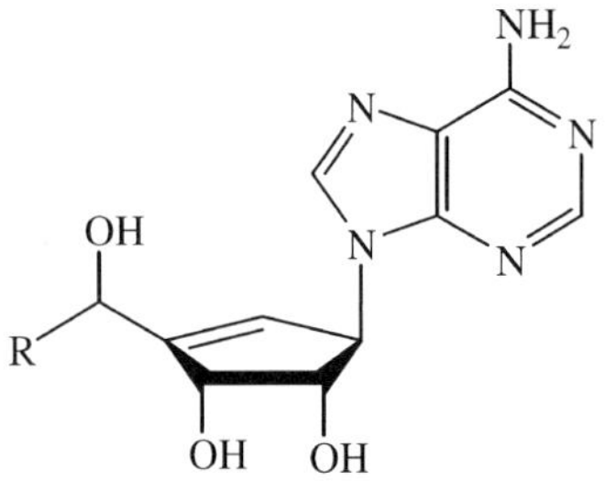

119a: (5'R)-5'-C-ethynyneplanocin A (REyNPA)
120a: (5'R)-5'-C-ethenylneplanocin A (REnNPA)
121a: (5'R)-5'-C-ethylneplanocin A (REtNPA)
122a: (5'R)-5'-C-methylneplanocin A(RMNPA)
119b: (5'S)-5'-C-ethynyneplanocin A (SEyNPA)
120b: (5'S)-5'-C-ethenylneplanocin A (SEnNPA)
121b: (5'S)-5'-C-ethylneplanocin A (SEtNPA)
122b: (5'S)-5'-C-methylneplanocin A(SMNPA)

Figure 21.3 *R* and *S* isomers of various 5′-C-substituted analogues of neplanocin A [45].

and HIV-1 LTR transactivation. For example, RMNPA **122a** ($IC_{50} = 0.32\,\mu M$) was more than 110-fold more potent against HIV-1 Tat transactivation than SMNPA **122b** ($IC_{50} > 36\,\mu M$). A similar differential inhibitory activity was observed for the *R*- and *S*-isomers of 5′-C-ethynyl-neplanocin A **119**. However, the REtNPA **121a** did not exhibit any inhibitory effects on HIV-1 Tat transactivation.

The inhibition of HIV-1 replication by these analogues in HeLa-CD4 cells with stably integrated LTR-*LacZ* was also investigated. In this study, RMNPA **122a** ($IC_{50} = 3.5\,\mu M$), REtNPA **121a** ($IC_{50} = 4.6\,\mu M$), and REyNPA **119a** ($IC_{50} = 6.1\,\mu M$) were each found to be potent anti-HIV inhibitors, with the same differential activity of the *R*- and *S*-isomers being observed as for the HIV-1 Tat transactivation assay. The compounds were also not toxic towards the HeLa-CD4 cells. The differential inhibitory effects of the compounds on HIV-1 replication also followed a similar pattern as their inhibitory effects on SAH hydrolase activity and Tat-dependent transactivation, except that REyNPA **119a** was less active and REtNPA **121a** more active against HIV-1 replication.

(5′*R*)-5′-*C*-Methyl-neplanocin A (RMNPA, **122a**) was a potent inhibitor of SAH hydrolase, and also exhibited better antiviral activity than neplanocin A [46]. Based on these results, Matsuda's group [47] have reported an improved method for the synthesis of **122a**, and studied its antimalarial activity both *in vitro* and *in vivo* (Scheme 21.22).

The treatment of aldehyde **123** with $MeTiCl_3$ at −50 °C provided the (5′*S*)-5′-*C*-methyl-neplanocin A derivative **124** as the major product. A chelation-controlled mechanism resulted in the stereoselective addition of $MeTiCl_3$ to the aldehyde **123** from the less-hindered *Si*-face to give the 5′*S*-product **124** stereoselectively. Debenzoylation of **124** followed by a Mitsunobu inversion of the C5′ hydroxyl group and subsequent deprotection of the benzoyl and silyl groups furnished RMNPA, **122a**. The antimalarial activity of **122a** and neplanocin A was first evaluated *in vitro*, whereby a significant antimalarial activity ($IC_{50} = 0.1\,\mu M$) was identified which was greater than that of the parent compound neplanocin A ($IC_{50} = 0.2\,\mu M$), and 5.6-fold higher than that of chloroquine. Compound **122a** exhibited a lower cytotoxicity against the proliferation of KATO III cells ($IC_{50} = 1.1\,\mu M$) than neplanocin A (IC_{50} $0.22\,\mu M$), and was completely inactive in Vero cells in the stationary phase, being

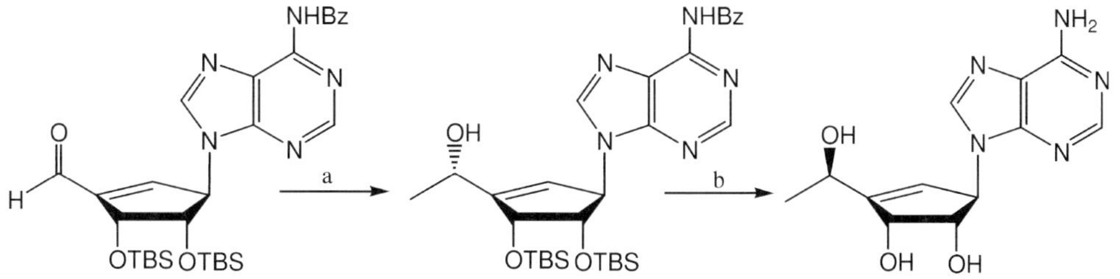

123 **124** **122a**

Reagents: (a) $MeTiCl_3$, CH_2Cl_2; (b) (i) NH_3, MeOH; (ii) TPP, DIPAD, $ClCH_2COOH$, THF; (iii) NH_3, MeOH; NH_4F, MeOH.

Scheme 21.22 Synthesis of RMNPA **122a** [47].

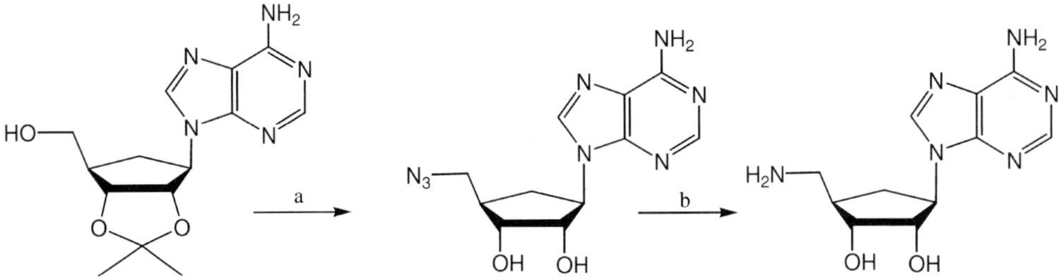

125 126 127

Reagents: (a) (i) $(PhO)_2P(O)N_3$, TPP, DIAD, THF; (ii) TFA, rt; (b) H_2, 10% Pd/C

Scheme 21.23 Synthesis of 5′-amino-5′-deoxy-aristeromycin [48].

non-cytotoxic up to 1800 μM. The *in-vivo* antimalarial activity of **122a** indicated that it effectively inhibited the growth of *Plasmodium berghei* in mice ($ED_{50} = 1.0\,mg\,kg^{-1}$ per day), being more potent than both chloroquine ($ED_{50} = 1.8\,mg\,kg^{-1}$ per day) and neplanocin A ($ED_{50} > 5.0\,mg\,kg^{-1}$ per day).

Schneller and coworkers [48] have synthesized 5′-amino-5′-deoxy-aristeromycin (**127**) and studied its antiviral effects (Scheme 21.23).

The 2′,3′-isopropylidene-aristeromycin (**125**) was treated with diphenylphosphoryl azide (DPPA) under Mitsunobu conditions to give the azido derivative **126** after acid-catalyzed hydrolysis. Compound **126** was then reduced to give the desired compound **127**, which exhibited antiviral activity against vaccinia virus ($EC_{50} = 76.5\,\mu g\,mL^{-1}$, CPE inhibition in HFF cells; Cidofovir, $EC_{50} = 2\,\mu g\,mL^{-1}$), against HSV-2 ($EC_{50} = 27.3\,\mu g\,mL^{-1}$, CPE inhibition in HFF cells; Acyclovir, $EC_{50} = 1.6\,\mu g\,mL^{-1}$), and against human cytomegalovirus ($EC_{50} = 17.5\,\mu g\,mL^{-1}$, CPE inhibition in HFF cells; Ganciclovir, $EC_{50} = 0.3\,\mu g\,mL^{-1}$). Unfortunately, however, compound **127** showed significant cytotoxicity towards the host cells.

Robins and coworkers [49] have reported the synthesis and the inhibitory effects on SAH hydrolase of the halovinyl aristeromycin analogues **130** and **131** (Scheme 21.24).

Moffatt oxidation of 5′-hydroxyl group of **125**, followed by Wittig olefination of the resulting aldehyde with [(*p*-toluoylsulfonyl)methylene]triphenylphosphorane gave the (*E*)-homovinyl sulfone **128**. Stannodesulfonylation of **128** with tributyltin hydride in the presence of AIBN produced a partially separable mixture of vinylstannanes **129**. Treatment of **129** with *N*-iodosuccimide (NIS) and subsequent deprotection of acetonide furnished the (*E*)-5′-iodovinyl analogue **130a** and (*Z*)-5′-iodovinyl analogue **130b**. Subsequent fluorodestannylation of **129** was achieved by treating with XeF_2 and AgOTf to give the (*E*)-5′-fluorovinyl derivative **131a** and its (*Z*)-isomer **130b** after acid hydrolysis. The inhibitory activity of the synthesized compounds on SAH hydrolase obtained from human placenta was measured. The enzyme was completely inactivated in 10 min by compound **130a** at a concentration of 10 μM, whereas both **130b** and **131a** proved to be less potent than **130a** (70% and 90%, respectively). The inhibitory activity of these compounds towards SAH hydrolase was in the following order: **130a** > **131a** > **130b**.

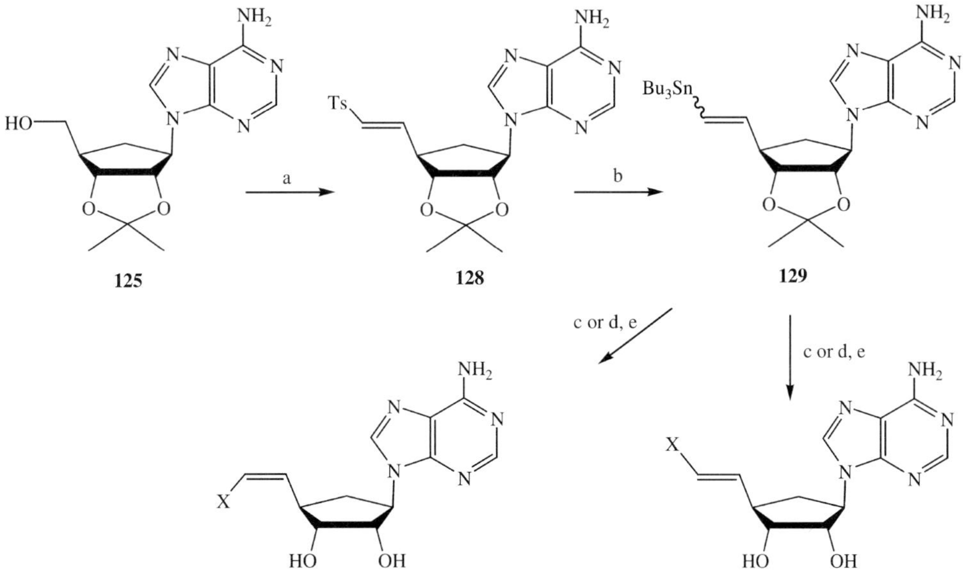

125 128 129

130b X = I **130a** X = I

131b X = F **131a** X = F

Reagents: (a) (i) DMSO, DCC, $Cl_2CHCCOH$; (ii) $Ph_3P{=}CHTs$; (b) Bu_3SnH, AIBN, toluene; (c) NIS, CH_2Cl_2, CCl_4; (d) XeF_2, AgOTf, THF; (e) aq. TFA.

Scheme 21.24 Syntheses of halovinyl aristeromycin analogues [49].

Since the hydroxylamine (HO−NH−) moiety at the 5′-position could serve as a bioisostere of hydroxymethyl ($HO{-}CH_2{-}$) group of aristeromycin, which is needed for recognition by viral kinase, Miller's group [50] has synthesized the 5′-hydroxylamino-aristeromycin derivative **136** (Scheme 21.25). A Mitsunobu condensation of cyclopentenol acetate **132** with 6-chloropurine, followed by hydrolysis of the acetate with catalytic KCN in methanol, provided the compound **133**. Treatment of the latter compound with BocNHOBn under the Mitsunobu conditions afforded the hydroxylamino derivative **134**, the reaction of which with $RuCl_3{\cdot}3H_2O$ and $NaIO_4$ in ethyl acetate, acetonitrile, and water afforded a diastereomeric mixture of *cis*-diols **135a** and **135b** in 15 : 1 ratio, whereas no other method gave such good diastereoselectivity. The desired β-isomer **135a** was converted to the final derivative **136** by amination and subsequent deprotection of the protecting groups. The biological activity of **136** was not reported.

Schneller's group [51] has reported the synthesis of 5′-monomethyl- and dimethyl-aristeromycin analogues, employing Corey–Bakshi–Shibata (CBS) reduction as a key step (Scheme 21.26). Oxidative cleavage of the intermediate **138** gave the ketone **139** which, upon treatment with (*R*)-MeCBS and diethylaniline borane (DEANB) under CBS conditions, afforded compound **140** as a single stereoisomer.

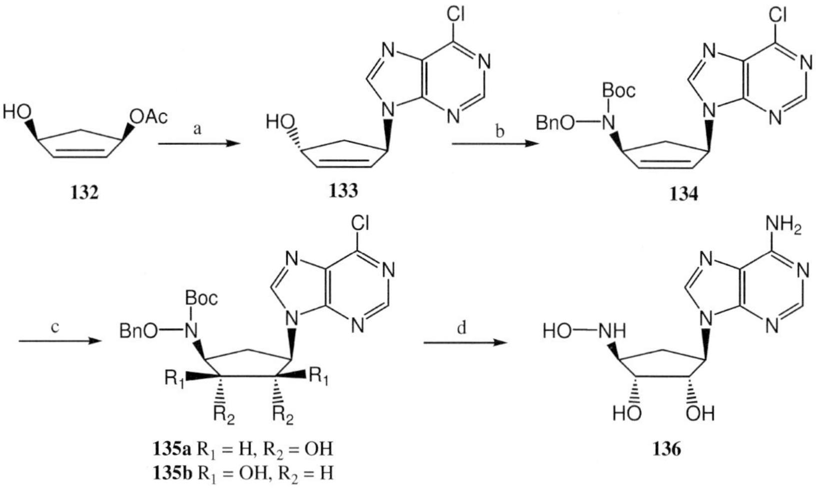

Reagents: (a) (i) TPP, DIAD, 6-chloropurine, THF; (ii) KCN, MeOH; (b) DBAD, TPP, Boc-NH-OBn, THF; (c) $RuCl_3$-$3H_2O$, $NaIO_4$; (d) (i) NH_3, THF; (ii) H_2, $Pd(OH)_2/C$, MeOH; (iii) AcCl, MeOH, THF.

Scheme 21.25 Synthesis of 5′-hydroxylamino-aristeromycin derivative **136** [50].

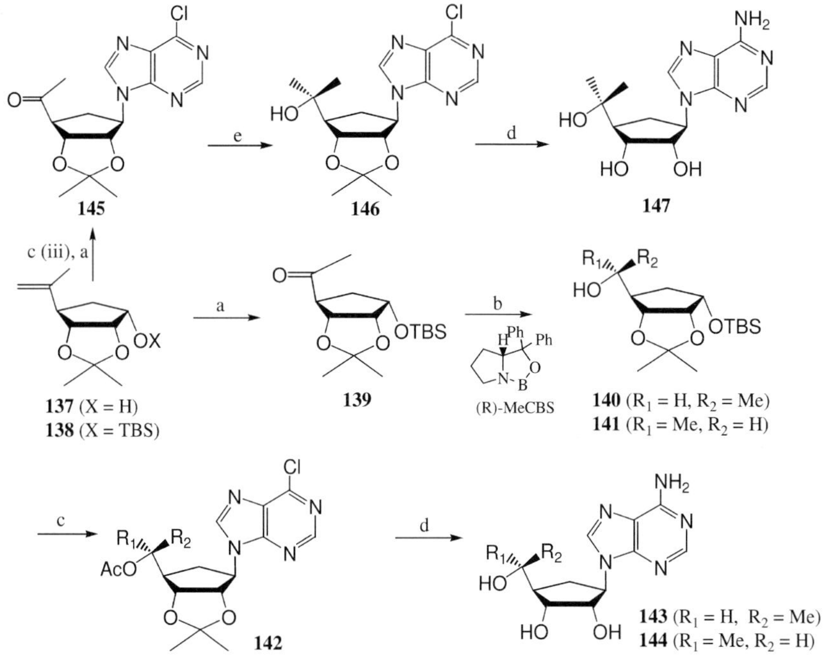

Reagents: (a) OsO_4, $NaIO_4$; (b) for **140**, (R)-MeCBS, diethylaniline; for **141**, (S)-MeCBS; (c) (i) Ac_2O, pyridine; (ii) TBAF; (iii) TPP, 6-chloropurine, DIAD; (d) (i) NH_3/MeOH; (ii) 1 *N* HCl, MeOH; (e) MeMgBr, THF.

Scheme 21.26 Synthesis of 5′-monomethyl- and dimethyl-aristeromycin analogues [51].

Similarly, treatment of ketone **139** with (*S*)-MeCBS gave the single diastereomer **141**, of which the structure was confirmed by X-ray single-crystal analysis. Desilylation of **140**, followed by acetylation and Mitsunobu condensation with 6-chloropurine, yielded the compound **142**. Treatment of **142** with methanolic ammonia followed by acid hydrolysis furnished the (5′*S*)-5′-methyl-aristeromycin **143**. Similarly, the diastereomer (5′*R*)-5′-methyl-aristeromycin **144** was prepared from compound **141**. In order to synthesize the 5′-dimethyl-aristeromycin, compound **137** was converted to the protected nucleoside **145** by a Mitsunobu condensation with 6-chloropurine and subsequent oxidative cleavage. Treatment of ketone **145** with methylmagnesium bromide gave the tertiary alcohol **146**, which was elaborated to 5′,5′-dimethyl-aristeromycin **147**. Compound **144** exhibited antiviral activity ($EC_{50} = 0.32\,\mu g\,mL^{-1}$; CPE inhibition in Vero cell; positive drug control $EC_{50} = 55\,\mu g\,mL^{-1}$) towards yellow fever, and without cytotoxicity.

Miller and coworkers [52] have disclosed a novel asymmetric synthesis of 5′-aza noraristeromycin siderophore conjugate **155** which involved a cycloaddition reaction with a nitroso compound to produce the carbasugar as a key step (Scheme 21.27).

A Diels–Alder reaction of cyclopentadiene with an acylnitroso compound generated *in situ* by the oxidation of *N*-protected L-amino acid hydroxamate **148** gave the cyclo adduct **149** as the major product. A molybdenum-mediated N−O reduction of adduct **149** followed by acetylation of the resulting alcohol gave the cyclopentene

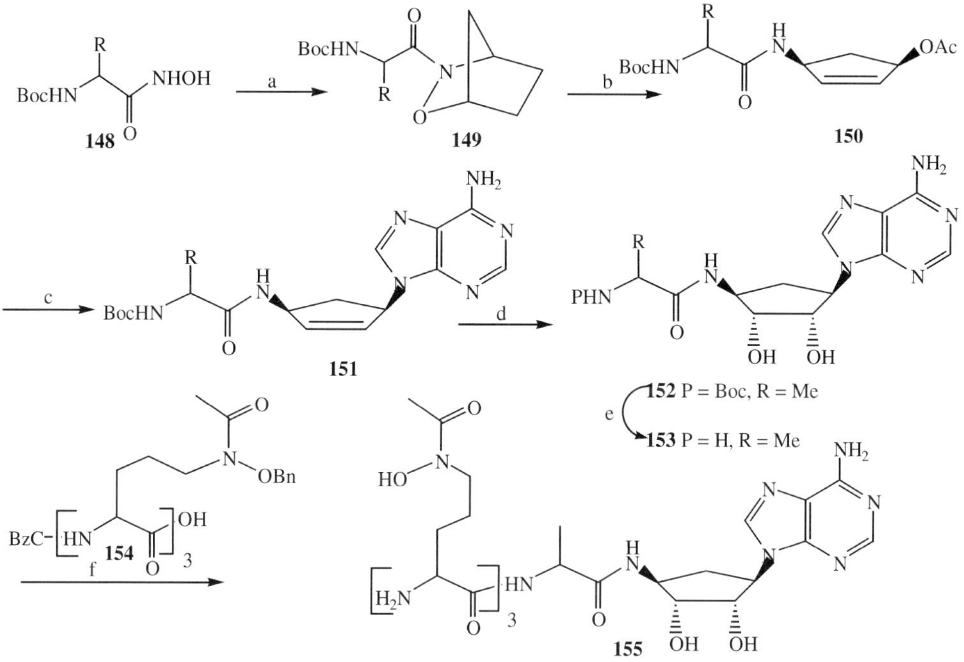

Reagents: (a) $NaIO_4$, cyclopentadiene; (b) $Mo(CO)_6$, Ac_2O; (c) adenine, Pd(0); (d) OsO_4, NMO; (e) TFA; (f) (i) **147**, DCC, NHOSu; (ii) H_2, Pd/C.

Scheme 21.27 Asymmetric synthesis of 5′-aza-noraristeromycin siderophore congugate **155** [52].

derivative **150**. Palladium-mediated condensation of the allylic acetate **150** with adenine anion provided the condensed product **151** as a single stereoisomer. *Cis*-dihydroxylation of **151** with OsO_4 gave the desired diol **152** as a major product which, upon deprotection of the Boc group, yielded the amine derivative **153**. In order to produce the conjugate, the free amine **153** was coupled with protected trihydroxamate tripeptide **154** in the presence of DCC, and then deprotected under catalytic hydrogenation to provide the conjugate **155**. When the compounds **152**, **153**, and **155** were tested for biological activity, all were inactive and non-selective against herpes simplex virus (HSV), HIV-1, and other viruses. However, notable inhibitory effects were found for these compounds against several viruses such as reovirus, parainfluenza virus, vaccinia virus, and cytomegalovirus. Among these compounds, **155** was active at a concentration ranging from 0.2 to 20 $\mu g\,mL^{-1}$ against the above viruses that are known to be inhibited by SAH hydrolase inhibitors.

21.3.5
C6′ Modification

Very few reports have been made on the modification of 6′-position of neplanocin A. Jeong and coworkers [53] have recently proposed a new mechanism for the irreversible inhibition of 6′-halo-analogues of neplanocin A (Figure 21.4).

As shown in Figure 21.4, the SAH hydrolase-catalyzed oxidation of neplanocin A by the cofactor NAD^+ led to the formation of its keto form II which, by virtue of

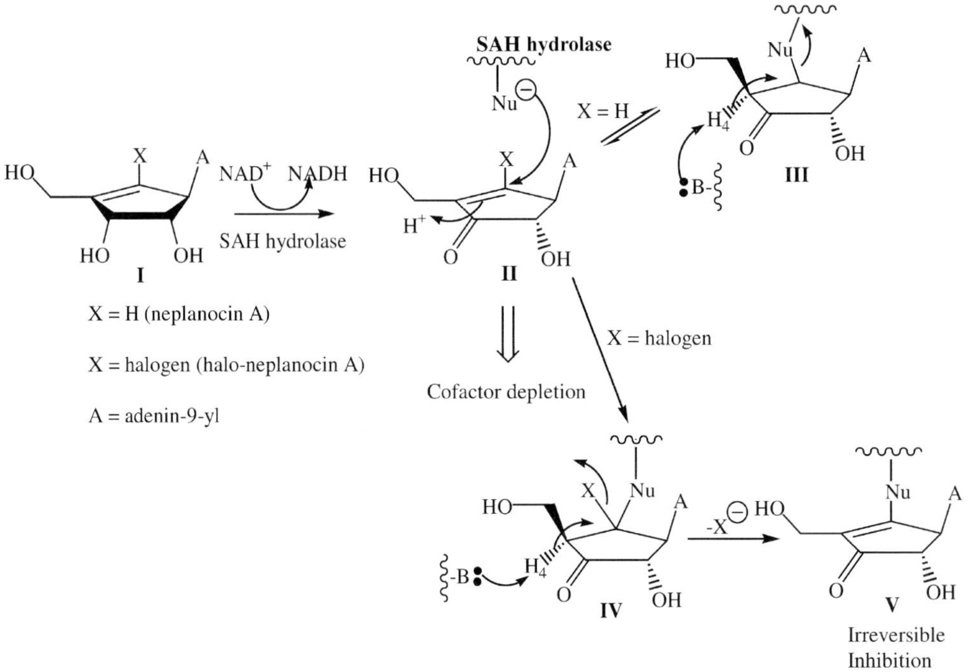

Figure 21.4 Proposed inhibitory mechanism of neplanocin A and halo-naplanocin A [53].

being a dead-end product, maintained the cofactor permanently in its reduced form (NADH), resulting in cofactor depletion. In addition to this well-known cofactor depletion mechanism, it has been hypothesized that the keto derivative **II** is well poised to form a covalent complex **III** with a nucleophile in the active site of the enzyme via Michael addition. However, formation of this intermediate has never been demonstrated because of the expected reversibility of the Michael adduct **III** due to the presence of the acidic α-keto hydrogen (H_4). On the basis of this hypothesis, an attempt was made to demonstrate one likely mechanism by designing an appropriate substrate that would provide the enzyme with an alternative leaving group for the addition–elimination reaction. The substrate designed for this purpose was the 6′-halo analogue of neplanocin A which, after Michael addition, would have the option of eliminating the halide ion to trap the intermediate **IV** by forming an irreversible covalent complex. The synthesis of various 6′-halo neplanocin A analogues to prove the proposed mechanism is illustrated in Scheme 21.28 [53].

The intermediate **15** [19] was treated with halogen in presence of pyridine to give the halogenated ketones **156**, the reduction of which with sodium borohydride, followed by mesylation, gave the mesylate **157**. Mitsunobu condensation of **157** with adenine anion followed by deprotection afforded the 6′-chloro-, bromo-, and iodo-neplanocin A derivatives **158a–c**. For the synthesis of 6′-fluoro-neplanocin A, the iodo derivative **156** was reduced and protected with a TBDPS group to give **159**. Electrophilic vinyl fluorination of **159** was achieved by treating with *N*-fluorobenzenesulfonimide

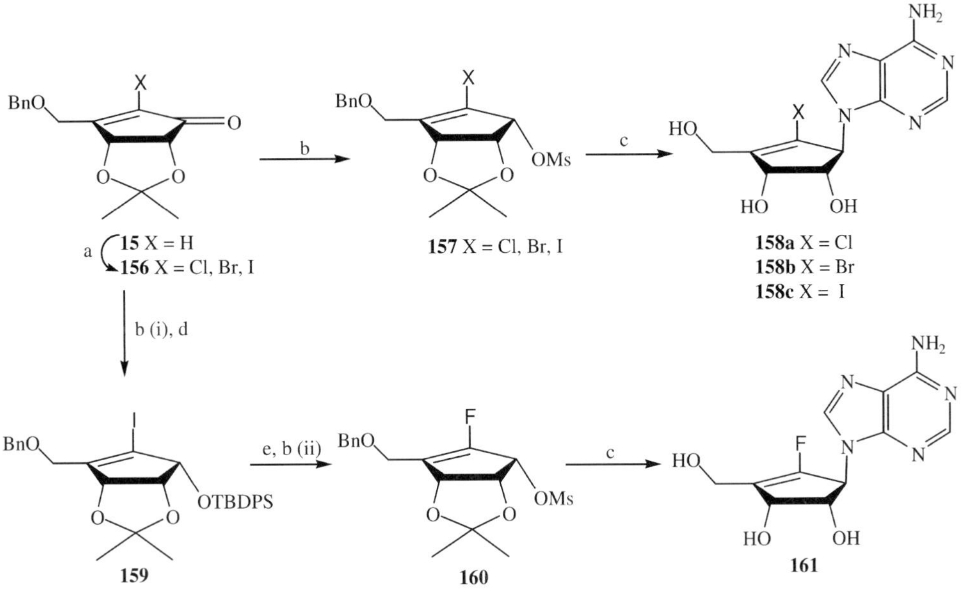

Reagents: (a) X_2, pyridine; (b) (i) $NaBH_4$, $CeCl_3$-$7H_2O$; (ii) MsCl, Et_3N; (c) (i) adenine, 18-crown-6; (ii) BCl_3, CH_2Cl_2; (d) TBDPSCl, DMF; (e) (i) n-BuLi, NFSI; (ii) TBAF, THF.

Scheme 21.28 Synthesis of various 6′-halo-neplanocin A analogues [53].

(NFSI) in the presence of *n*-BuLi, while subsequent mesylation provided the glycosyl donor **160**. According to the similar procedure used for **158**, the fluoro intermediate **160** was transformed to 6′-fluoro-neplanocin A **161**. The SAH hydrolase inhibitory activity of the 6′-halo-neplanocin A analogues, **158a–c** was measured, whereby the 6′-fluoro analogue **161** ($IC_{50} = 0.48\,\mu M$) was found to be two-fold more potent than the parent neplanocin A ($IC_{50} = 0.87\,\mu M$). However, the 6′-chloro derivative **158a** ($IC_{50} = 36.46\,\mu M$) and 6′-bromo derivative **158b** ($IC_{50} = 60.17\,\mu M$) were found to be less potent than neplanocin A, while the 6′-iodo derivative **158c** was inactive ($IC_{50} = 1.0\,mM$). The irreversible nature of inhibition with 6′-fluoro-neplanocin A **161** was demonstrated using dialysis, incubation with excess NAD^+ or adenosine, and ^{19}F NMR experiments, with results indicating that **161** is the mechanism-based inhibitor of SAH hydrolase operated by the proposed mechanism. 6′-Fluoro-neplanocin A **161** exhibited a more potent antiviral activity ($EC_{50} = 0.43\,\mu M$) against VSV than the control, ribavirin ($IC_{50} = 59.0\,\mu M$), without cytotoxicity up to 40 μM. However, the compound showed toxicity-dependent antiviral activities against other viruses such as HIV-1, HSV-1, and HSV-2.

21.3.6
Base Modification

Chu and co-workers [54] have reported the synthesis and antiviral activity of neplanocin A analogues having different nucleobases (Scheme 21.29).

D-Ribose was converted to the diastereomeric mixture **164** according to the same procedure [17] described in Scheme 21.1. Desilylation of **164** followed by oxidative cleavage of the resulting diol with sodium periodate provided the key intermediate L-cyclopent-2-enone **89**. The addition of a carbanion of Me_3COCH_3 to **89**, followed by acetylation, gave **165**. A palladium-catalyzed allylic rearrangement of **165** afforded the cyclopentenol derivative **166** after deacetylation, which was elaborated to neplanocin

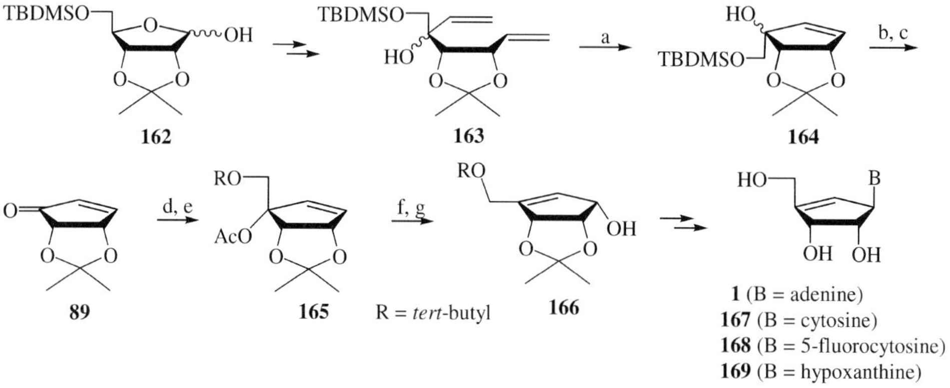

Reagents: (a) Grubbs catalyst, anhydrous CH_2Cl_2; (b) TBAF, THF; (c) $NaIO_4$, H_2O; (d) Me_3COMe, *t*-BuOK, *sec*-BuLi; (e) Ac_2O; (f) $PdCl_2(CH_3CN)_2$, (g) K_2CO_3, MeOH.

Scheme 21.29 Synthesis of neplanocin A analogues having different nucleobases [54].

A (**1**) and its purine and pyrimidine analogues **167–169**. All of the synthesized nucleosides were evaluated for their antiviral activity against smallpox (7124, BSH), monkeypox (MPX), cowpox (CPX), and vaccinia (VAC) viruses, and also for their cytotoxicity in Vero and MK2 cells. Among these compounds, the adenine **1**, cytosine **167**, and 5-fluorocytosine **168** analogues showed potent antiviral activity against smallpox-7124 (EC_{50} = 0.10, 0.08, and 1.73 μg mL^{-1}, respectively), smallpox-BSH (EC_{50} = 0.10, 0.03, and 0.51 μg mL^{-1}, respectively), monkeypox (EC_{50} = 0.26, 0.1, and 0.61 μg mL^{-1}, respectively), cowpox (EC_{50} = 100, 0.06, and 0.43 μg mL^{-1}, respectively), and vaccinia (EC_{50} = 2.62, 0.12, and 0.53 μg mL^{-1}, respectively) viruses in MK2 cells. Each of the analogues also showed antiviral activity against smallpox-7124 (EC_{50} = 0.03, <0.05, and 2.63 μg mL^{-1}, respectively), smallpox-BSH (EC_{50} = 0.14, >100, and 1.17 μg mL^{-1}, respectively), monkeypox (EC_{50} = 0.21, <0.05, and 2.05 μg mL^{-1}, respectively), cowpox (EC_{50} = >100, 0.05, and 1.35 μg mL^{-1}, respectively), and vaccinia (EC_{50} = >100, 0.04, and 1.46 μg mL^{-1}, respectively) viruses in Vero cells.

Chu and coworkers [55] have extended the structure–activity relationship shown in Scheme 21.29 by synthesizing various D- and L-cyclopentenyl purine and pyrimidine nucleosides exhibiting anti-HIV-1 and anti-West Nile virus activities (Scheme 21.30).

The L-glycosyl donor **170** was also prepared from the D-cyclopentenone **22**. The 1,4-conjugated addition of lithium bis(*tert*-butyloxymethyl)cuprate to **22**, followed by

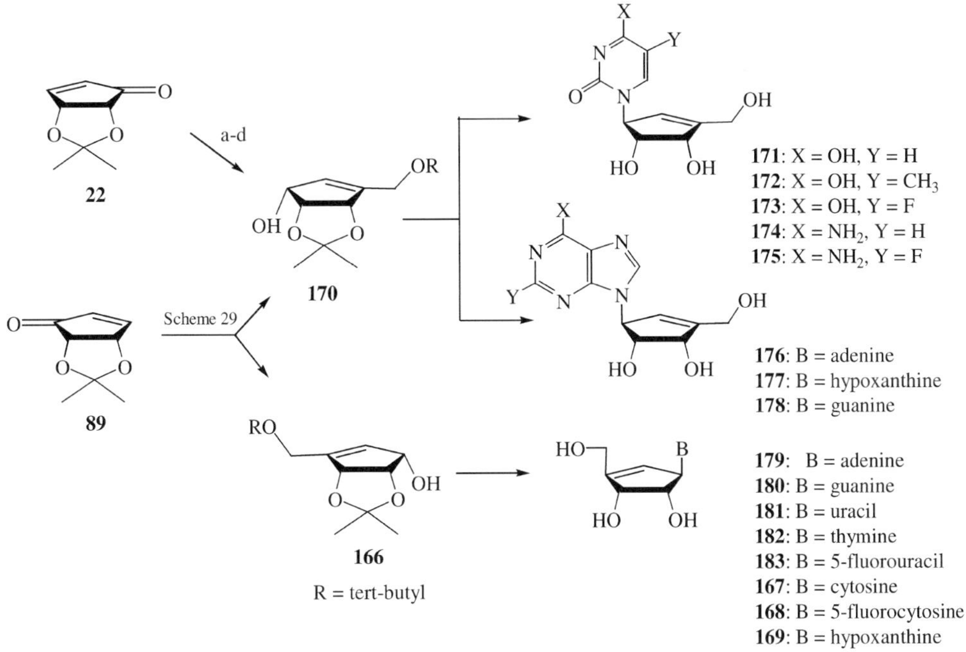

Reagents: (a) $(t\text{-BuOCH}_2)_2\text{CuLi}$, ^{t}BuOMe; (b) PhSeBr, LDA; (c) H_2O_2/H_2O; (d) $NaBH_4$, $CeCl_3$.

Scheme 21.30 Synthesis of D- and L-cyclopentenyl nucleosides with various purine and pyrimidine bases [55].

the generation of a double bond, gave the key intermediate **170** after reduction with $NaBH_4$. The glycosyl donors **166** and **170** were elaborated to D-nucleoside analogues, **167** to **169** and **179** to **183** and L-nucleoside analogues **171** to **178**, respectively. The anti-HIV-1 activity of all synthesized nucleosides was evaluated on human peripheral blood mononuclear (PBM) cells infected with HIV-1. Among the D-nucleosides, adenine analogue **179** and cytosine analogue **167** exhibited potent anti-HIV-1 activity (EC_{50} = 0.1 and 0.06 μM, respectively) and 5-fluorocytosine analogue **168** showed moderate anti-HIV-1 activity (EC_{50} = 5.34 μM), but showed significant cytotoxicity in PBM, CEM, and Vero cells. The 5-fluorouracil **183** displayed weak anti-HIV-1 activity (EC_{50} = 93.1 μM).

Among the L-nucleosides, only the cytosine analogue **174** showed weak anti-HIV-1 activity (EC_{50} = 58.9 μM). Anti-West Nile virus activity of all synthesized nucleosides was also evaluated with, among the D-nucleosides, cytosine analogue **167** and 5-fluorocytosine analogue **168** exhibiting the most potent antiviral activities (EC_{50} 0.2–3.0 and 15–20 μM, respectively), whereas the L-nucleosides did not exhibit any significant antiviral activity.

Matsuda and coworkers [56, 57] have reported the antiviral activity of various 2-halo derivatives of neplanocin A (Figure 21.5).

All 2-halo derivatives of neplanocin A were completely resistant to adenosine deaminase. The antiviral activities of these compounds were measured against VSV, mumps virus, and measles virus. Among the compounds tested, 2-fluoro-neplanocin A (**184a**) showed significant antiviral activities against VSV (IC_{50} = 0.25 μg mL^{-1}), measles (IC_{50} = 0.03 μg mL^{-1}), and mumps (IC_{50} = 1.2 μg mL^{-1}) without showing cytotoxicity towards the host cells. Another 2-fluoro-derivative **186** also exhibited weak antiviral activities against VSV (IC_{50} = 16 μg mL^{-1}) and mumps (IC_{50} = 7.0 μg mL^{-1}). However, neither 2-fluoro-DHCA **188** nor all of the 2-chloro derivatives exhibited any significant antiviral activity against these viruses.

In a comparative study of 2-fluoro-neplanocin A (**184a**) with neplanocin A, the former compound was found to be more active than the latter against vaccinia virus (VV) and VSV, to be equally active against parainfluenza virus, reovirus, and arenaviruses (Junin, Tacaribe), and less active against human cytomegalovirus (HCMV). The antiviral activity was derived from an inhibition of the SAH hydrolase. Compound **184a** was also 10-fold more inhibitory than neplanocin A to the growth of HEL cells, but 10-fold less inhibitory to the growth of L1210 and FM3A

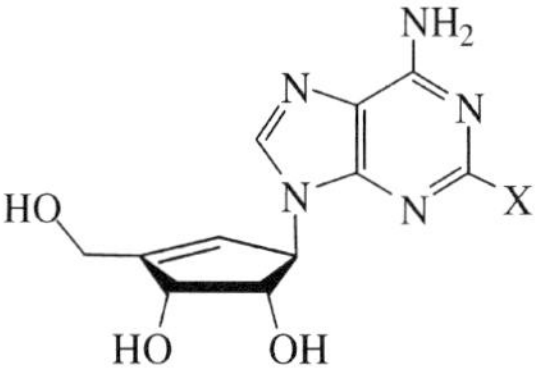

1 (X = H; NPA)
184a (X = F; 2-F-NPA)
184b (X = Cl; 2-Cl-NPA)

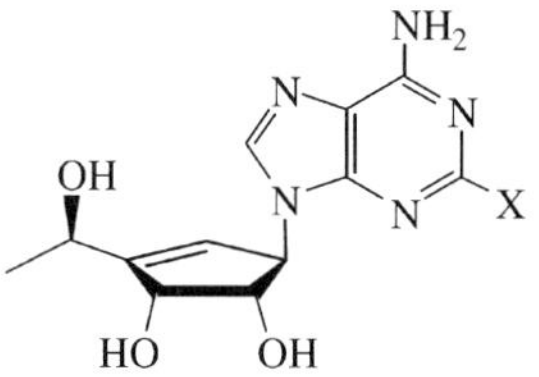

185 (X = H; RMNPA)
186 (X = F)

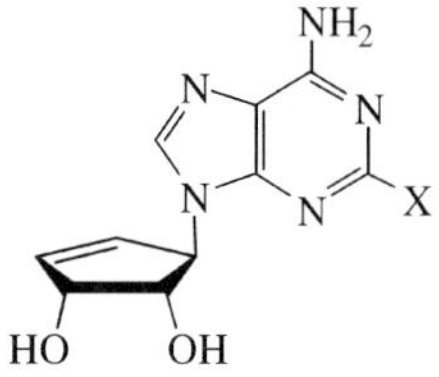

187 (X = H; DHCA)
188 (X = F)
189 (X = Cl)

Figure 21.5 Various 2-halo derivatives of neplanocin A [56, 57].

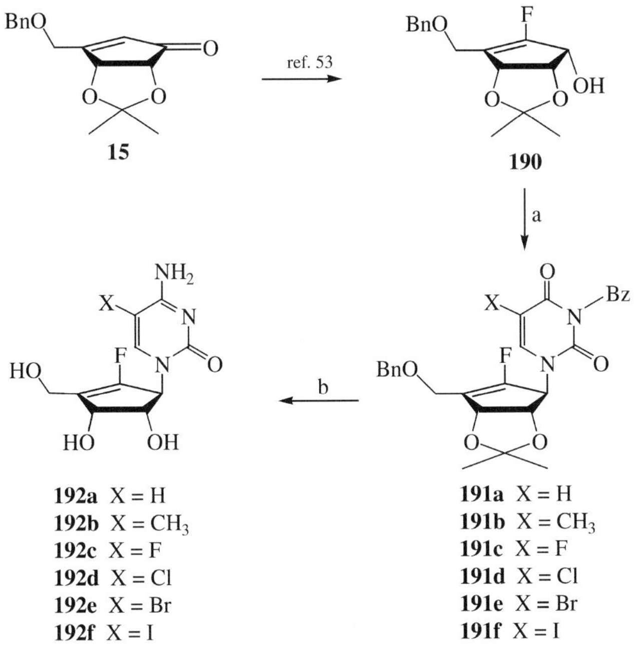

Reagents: (a) N^3-benzoylpyrimidines, DEAD, PPh_3P; (b) (i) NH_3/MeOH; (ii) BCl_3, CH_2Cl_2; (iii) Ac_2O, pyridine; (iv) 1,2,4-triazole, $POCl_3$; (v) NH_4OH, 1,4-dioxane; (v) NH_3/MeOH.

Scheme 21.31 Synthesis of 6′-fluorocyclopentenyl-5′-substituted pyrimidine nucleosides [58].

cells. However, **184a** did not exhibit any antiviral activity against HIV-1 and HIV-2 in CEM cells.

Jeong and coworkers [58] have disclosed the synthesis of 6′-fluorocyclopentenyl-5′-substituted pyrimidine nucleosides and their potent antitumor activity (Scheme 21.31).

The ketone **15** was converted to fluorocyclopentenol **190** according to the same procedure [53] described earlier. The glycosyl donor **190** was condensed with N^3-benzoylpyrimidines under Mitsunobu conditions to give **191a–f** which, upon deprotection and conversion of uracil to cytosine moiety by the standard conditions, furnished the final nucleosides **192a–f**. Among the synthesized compounds **192a–f**, cytosine derivative **192a** exhibited the most potent inhibitory activity on the cell growth of broad ranges of human cancer cell lines: 0.80 μM in human OVCAR-3 (ovary), 0.34 μM in MCF-7 (breast, hormone-dependent), 0.18 μM in MDA-MB-231 (breast), 1.35 μM in HeLa (cervix), 0.63 μM in PC3 (prostate), 2.67 μM in LNCap (prostate), 0.79 μM in HepG2 (liver), 0.50 μM in A549 (lung), 0.25 μM in NCI-/H226 (lung), 0.28 μM in HT-29 (colon), 0.19 μM in HCT116 (colon), 1.38 μM in SK-MEL-28 (melanoma), 0.62 μM in PANC-1 (pancreas), 0.83 μM in U251 (brain), 0.34 μM in MKN-45 (stomach), and 0.83 μM in UMRC2 (kidney).

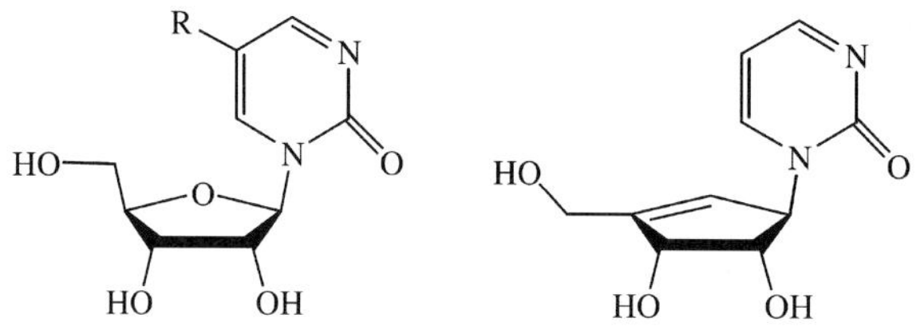

193a R = H (Zebularine)
193b R = F

194

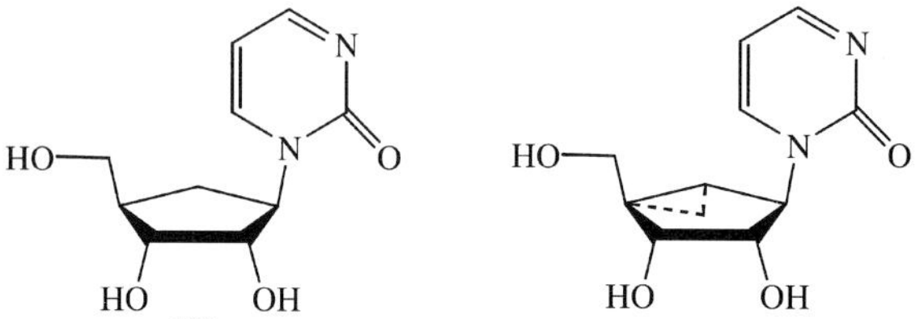

195

196

Figure 21.6 Zebularine derivatives as cytidine deaminase inhibitors [59].

Zebularine **193a** and 5-fluorozebularine **193b** are potent cytidine deaminase inhibitors, which have been proposed as antitumor agents in combination with cytosine arabinoside (ara-C) or 5-aza-2′-deoxycytidine (Figure 21.6).

However, zebularine and its fluoro derivative were extremely sensitive under basic conditions, resulting in a complex decomposition that destroyed their biological activity. In order to solve this problem, Marquez and coworkers [59] have designed and synthesized carbocyclic analogues **194**, **195**, and **196** of zebularine, which were stable under basic conditions. Biological results indicated that whilst neither **194** nor **196** showed any significant inhibitory activity on mouse or human cytidine deaminase, compound **195** exhibited significant inhibitory activity, although it was 16-fold less potent than zebularine ($K_i = 38\,\mu M$ versus $K_{i\ (apparent)} = 2.3\,\mu M$). This result indicated that replacement of the electronegative oxygen with the less-electronegative carbon presumably reduced the capacity of the pyrimidin-2(1*H*)-one ring to form a covalent hydrate, a step considered crucial for the compound to function as a transition-state inhibitor of the enzyme [59].

Schneller and coworkers have reported the synthesis of 7-deaza-5′-noraristeromycin (**197**) [60] and 8-aza-7-deaza-5′-noraristeromycin derivatives **198** to **203** [61], as well as their antitrypanosomal activity against *Trypanosoma brucei* (Figure 21.7). The *in-vitro* activity of the (+)- and (−)-enantiomers of **197** against *T. brucei* indicated that the (+)-enantiomer only showed strong inhibitory activity ($IC_{50} = 0.165–5.3\,\mu M$). (+)-**197** was also found to be a weak inhibitor ($IC_{50} > 500\,\mu M$) of SAH hydrolase, albeit without cytotoxicity.

Compounds **198** to **203** were also evaluated against bloodstream forms of *T. brucei* and *T. brucei rhodesiense* grown *in vitro*, with compound **200** being shown as the most potent ($IC_{50} = 12.2–16.8\,\mu M$) in three test strains. None of the compounds was shown to display cytotoxicity against HEL, E_6SM, HeLa, and Vero cells.

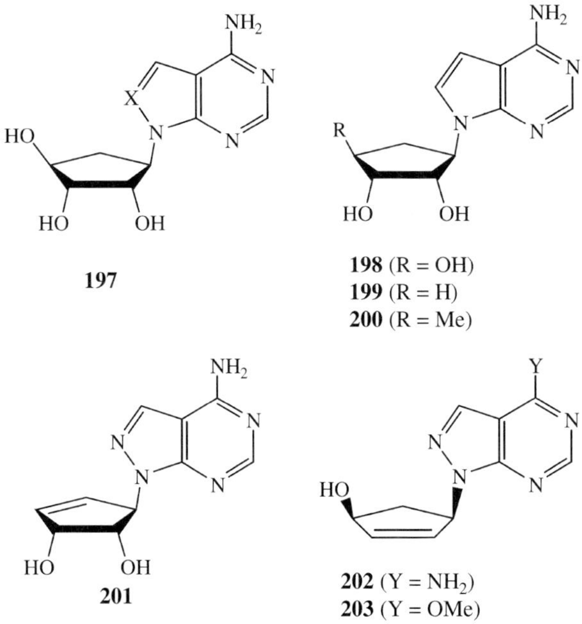

Figure 21.7 7-Deaza-5′-noraristeromycin **197** and 8-aza-7-deaza-5′-noraristeromycin derivatives **198–203** as anti-trypanosomal agents [60, 61].

Previously, Chu and coworkers [62] has reported the synthesis of cyclopentenyl triazole nucleosides and their antiviral activity (Scheme 21.32). The cyclopentenol **204** [17, 54, 55] was condensed with methyl-1*H*-1,2,4-triazole-3-carboxylate and methyl imidazole-4-carboxylate under the Mitsunobu conditions to give the products **205** and **206**, respectively.

Treatment of **205** and **206** with methanolic ammonia, followed by removal of the protecting groups, furnished the 1,2,4-triazole and imidazole nucleosides **207** and **208**, respectively. For the synthesis of 1,2,3-triazole nucleoside **211**, the intermediate **204** was converted to the azide **209**, which was subjected to 1,3-dipolar cycloaddition with methyl propiolate to afford **210** that, upon removal of the protecting groups, yielded the 1,2,3-triazole nucleoside **211**. Compounds **207**, **208**, and **211** were tested for antiviral efficacy, whereby the 1,2,4-triazole derivative **207** exhibited moderate activity ($EC_{50} = 21\ \mu M$, $SI > 4.8$) against SARS virus. The 1,2,3-triazole derivative **211** exhibited the most potent and selective antiviral activity ($EC_{50} = 0.4\ \mu M$, $SI > 750$) against vaccinia virus, but only moderate activity against cowpox virus ($EC_{50} = 39\ \mu M$, $SI > 7.7$) as well as marginal activity against SARS virus ($EC_{50} = 47\ \mu M$, $SI > 2.1$). The imidazole derivative **208** did not exhibit any significant antiviral activity.

Recently, Chu and coworkers [63] have also reported the antiviral activity of various purine and pyrimidine *C*-nucleosides **212–215** of neplanocin A (Figure 21.8).

Among compounds tested, only 9-deazaneplanocin A (**212**) exhibited moderate anti-HIV activity ($EC_{50} = 2.0\ \mu M$), while the other compounds did not show any

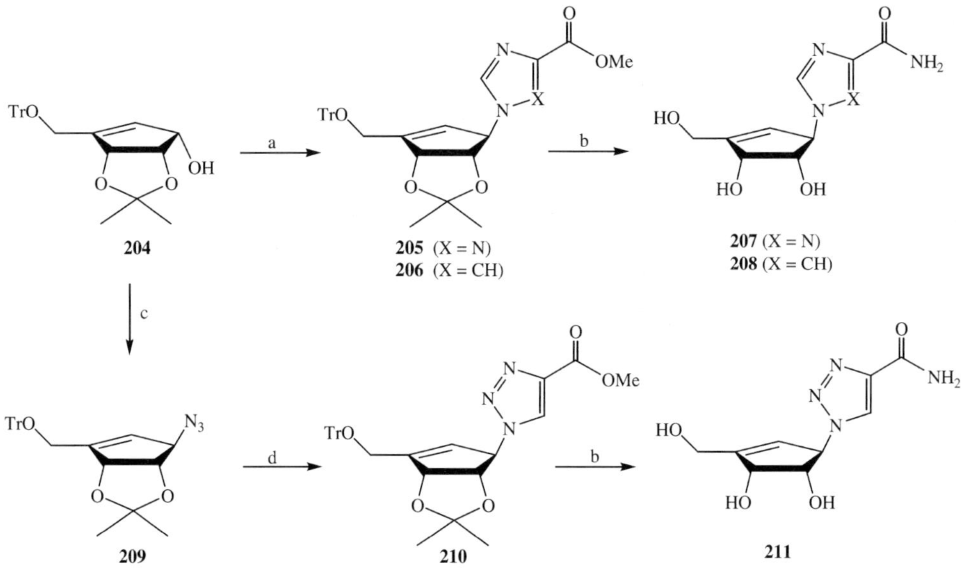

Reagents: (a) DIAD, Ph_3P, methyl-1*H*-1,2,4-triazole-3-carboxylate or methyl imidazole-4-carboxylate; (b) (i) NH_3, MeOH; (ii) 0.1 N HCl, MeOH; (c) (i) MsCl, TEA (ii) NaN_3, DMF; (d) methyl propiolate, CuI, TEA.

Scheme 21.32 Synthesis of cyclopentenyl- imidazole and triazole nucleosides [62].

anti-HIV activity. Compounds **213** and **215** displayed moderate antiviral activity against Punta Toro virus (EC_{50} = 2.5 μM and 65 μM, respectively), without any cytotoxicity. Compound **214** was slightly active against West Nile virus (EC_{50} = 11 μM), while **215** showed marginal activity against SARS CoV (EC_{50} = 49 μM).

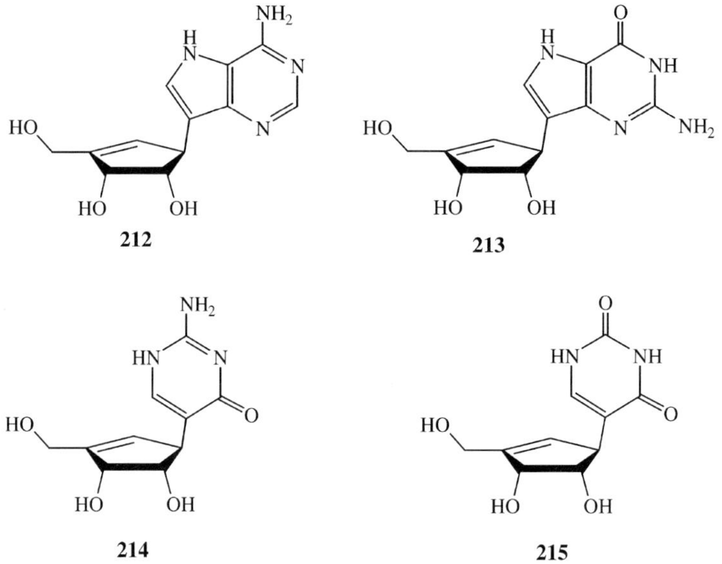

Figure 21.8 Structures of cyclopentenyl C-nucleosides [53].

21.3.7 Miscellaneous

In this section, the syntheses and biological activity of several neplanocin A and aristeromycin derivatives modified at more than one position of the carbasugar are described.

Marquez and coworkers [64] have designed and synthesized *C*4′–*C*6′ epoxy-constrained 2′-deoxy-neplanocin A derivatives in which the epoxy group could fix the sugar puckering to the desired Northern conformation (Scheme 21.33).

The sugar intermediate **217**, prepared from compound **216** using a known method [64b] was treated with *m*CPBA to afford the desired α-epoxy derivative **218** as a single stereoisomer. Mitsunobu condensations of glycosyl donor **217** with 6-chloropurine and 2-amino-6-benzyloxypurine gave the protected nucleosides **219** and **220**, which were further elaborated to the final nucleosides **221** and **222**, respectively. Similarly,

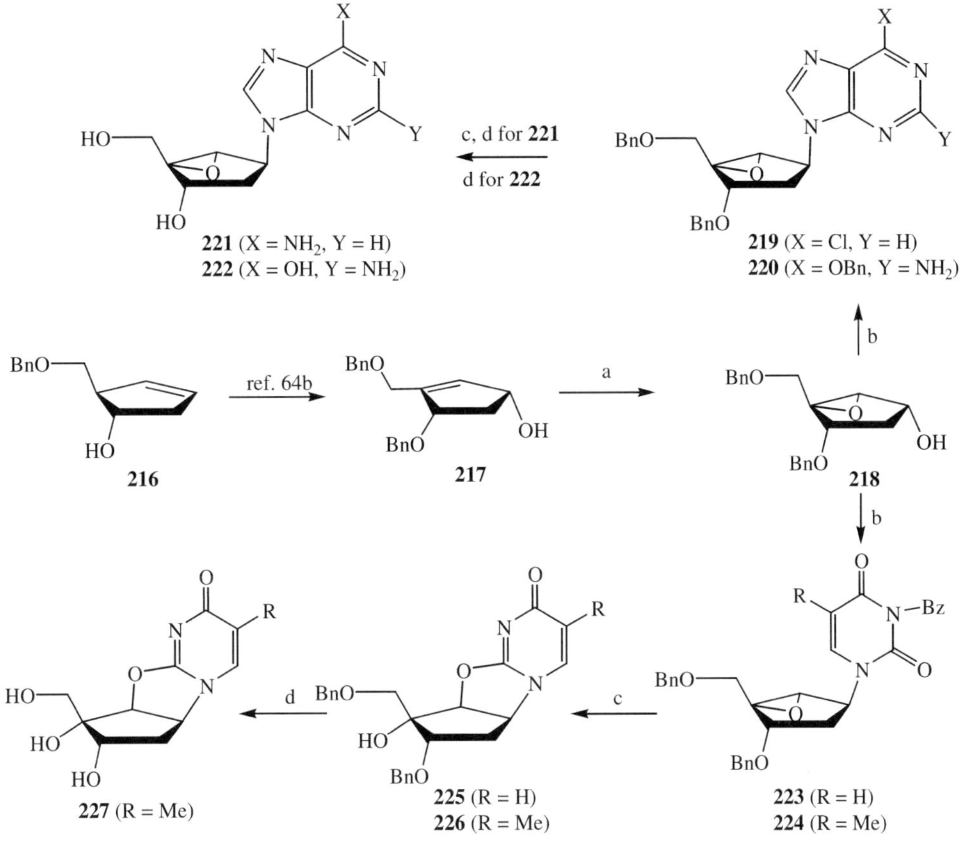

Reagents: (a) *m*-CPBA; (b) 6-chloropurine, 2-amino-6-benzyloxypurine, N^3-benzoylthymine, or N^3- benzoyluracil; (c) NH_3/MeOH; (d) Pd/C, H_2.

Scheme 21.33 Synthesis of C4′-C6′ epoxy-constrained 2′-deoxy-neplanocin A derivatives [64].

compounds **223** and **224** were prepared from the glycosyl donor **218**. However, when **223** and **224** were treated with methanolic ammonia, they underwent an intramolecular nucleophilic opening of the epoxide, resulting in an exclusive formation of the corresponding anhydrides **225** and **226**, respectively. The compound **225** could not be converted to the final nucleoside due to instability, but **226** was smoothly converted to the anhydride compound **227**. All final compounds were tested for antiviral activity, whereby the guanosine derivative **222** ($EC_{50} = 0.34\,\mu g\,mL^{-1}$, SI > 150) was more potent than compound **227** against Epstein–Barr virus (EBV), with less cytotoxicity. Compound **227** also showed good antiviral activity against EBV ($EC_{50} = 1.2\,\mu g\,mL^{-1}$, SI > 40). However, compounds **221** and **222** were both inactive against HSV-1 and HSV-2.

Jeong's group [65] has recently disclosed a novel stereoselective synthesis of 2′-*C*-methyl-cyclopropyl fused neplanocin A analogues as potential anti-HCV agents (Scheme 21.34).

The known cyclopentenone **8** [17] was reduced and subjected to the modified Simmons–Smith cyclopropanation to afford the cyclopropyl-fused bicyclic compound **228** as a single stereoisomer. Regioselective cleavage of the isopropylidene group of **228**, followed by selective silyl protection and subsequent Swern oxidation, afforded the ketone **229**. Grignard reaction of **229** with methylmagnesium iodide gave a tertiary alcohol **230** as a single stereoisomer which, upon deprotection of silyl groups and selective re-protection of the 5′-primary hydroxyl group as TBS ether, provided the compound **231**. Treatment of the diol **231** with $SOCl_2$ and subsequent oxidation with $RuCl_3$ in the presence of $NaIO_4$ yielded the cyclic sulfate **232**. Condensation of **232** with adenine and 2-amino-6-chloropurine in the presence of

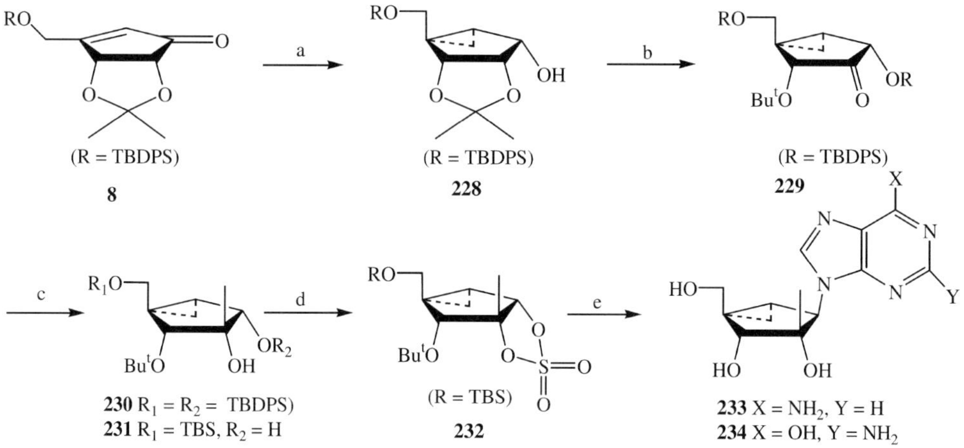

Reagents: (a) (i) $NaBH_4$, $CeCl_3$; (ii) Et_2Zn, CH_2I_2; (b) (i) Me_3Al; (ii) TBDPSCl, imidazole; (iii) Swern oxidation; (c) (i) MeMgI; (ii) TBAF; (iii) TBSCl, imidazole; (d) (i) SO_2Cl, Et_3N; (ii) $RuCl_3.3H_2O$, $NaIO_4$; (e) (i) adenine or 2-amino-6-chloropurine, DMF; (ii) acid hydrolysis.

Scheme 21.34 Stereoselective synthesis of 2′-C-methyl-cyclopropyl-fused neplanocin A analogues [65].

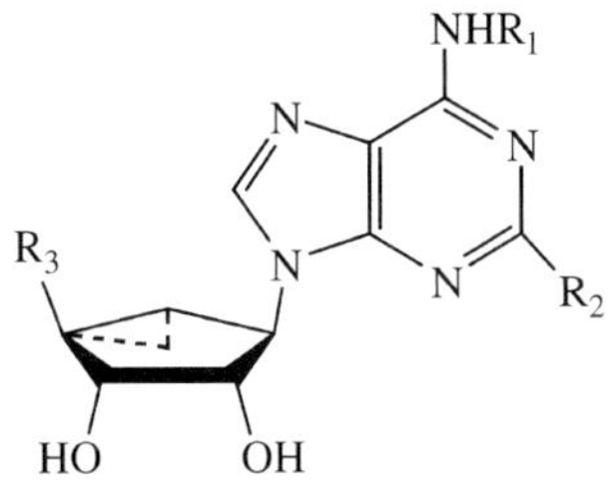

235 $R_1 = H, R_2 = H, R_3 = CH_2OH$
236 $R_1 = H, R_2 = Cl, R_3 = CH_2OH$
237 $R_1 = H, R_2 = H, R_3 = CONHEt$
238 $R_1 = Me, R_2 = H, R_3 = CH_2OH$
239 $R_1 = Me, R_2 = Cl, R_3 = CH_2OH$
240 $R_1 = Me, R_2 = Cl, R_3 = CONHMe$
241 $R_1 =$ Cyclopentyl, $R_2 = H, R_3 = CH_2OH$
242 $R_1 =$ Cyclopentyl, $R_2 = Cl\ R_3 = CH_2OH$
243 $R_1 =$ 3-iodobenzyl, $R_2 = H, R_3 = CH_2OH$
244 $R_1 =$ 3-iodobenzyl, $R_2 = H, R_3 = CONHMe$
245 $R_1 =$ 3-iodobenzyl, $R_2 = Cl, R_3 = CH_2OH$
246 $R_1 =$ 3-iodobenzyl, $R_2 = Cl, R_3 = CONHMe$

Figure 21.9 Ring-constrained (*N*)-methanocarba nucleosides as A_3 adenosine receptor agonists [30].

sodium hydride gave adenine derivative **233** and guanine derivative **234**, after acidic hydrolysis. The antiviral activities of **233** and **234** were measured against hepatitis C virus (HCV), but unfortunately none of these compounds showed significant antiviral activity in a cell-based HCV replication assay.

Jacobson and coworkers [30] have reported a series of ring constrained (*N*)-methanocarba nucleosides (**235–246**), which are considered as a bioisostere of neplanocin A, and monitored their binding affinity to adenosine receptors (ARs) (Figure 21.9).

The binding assays of compounds **235** to **246** at A_1, A_2, and A_3 ARs indicated that adenine derivative **237** with 5′-ethyluronamide enhanced the binding affinity at the A_1 AR by 53-fold and at the A_3 AR by 14-fold, when compared with adenine derivative **235** having the 4′-hydroxymethyl group. In the N^6-methyl series, the binding at A_{2A} AR was absent, and compounds **238** and **239** with a 4′-hydroxymethyl group were selective in binding at A_3 AR versus A_1 AR by 12- and 39-fold, respectively.

The group of Rodriguez [66] has disclosed an enantioselective synthesis of (+)-neplanocin F, which is a member of neplanocin family (Scheme 21.35).

The cyclopentene derivative **247** was prepared from 1,4-D-ribonolactone by a reported method [67], and hydrolyzed with 60% acetic acid to give the diol **248**. Regioselective protection of the unhindered hydroxyl group of diol **248** by treatment with trimethyl orthoformate, followed by reductive cleavage with DIBAL, gave the mono-protected MOM derivative **249**. The benzylation of **249**, followed by deprotection of the MOM group with trifluoroacetic acid, provided the glycosyl donor **251**, which was converted to the neplanocin F (**252**) according to the conventional method.

Mandal and coworkers [68] have reported the synthesis of an aristeromycin analogue **257**, employing intramolecular nitrone cycloaddition as the key step (Scheme 21.36).

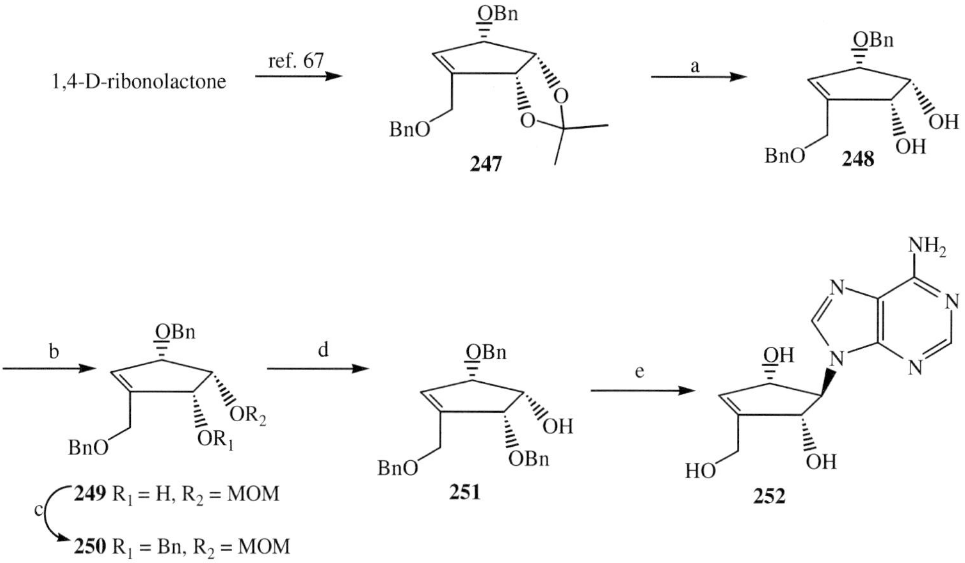

Reagents (a) 60% AcOH; (b) (i) trimethyl orthoformate, CAN; (ii) DIBAL; (c) BnBr, NaH; (d) CF_3COOH; (e) (i) 6-chloropurine, TPP, DEAD; (ii) NH_3/MeOH; (iii) BCl_3, CH_2Cl_2.

Scheme 21.35 Enantioselective synthesis of (+)-neplanocin F [66].

Compound **253**, prepared from D-glucose was hydrolyzed under acidic conditions, and the resulting diol treated with *N*-benzyl hydroxylamine to give the isoxazolidine derivative **254**; this was formed by intramolecular cycloaddition between the *N*-benzyl nitrone of the masked aldehyde at C1 and the vinyl group at the C5 position.

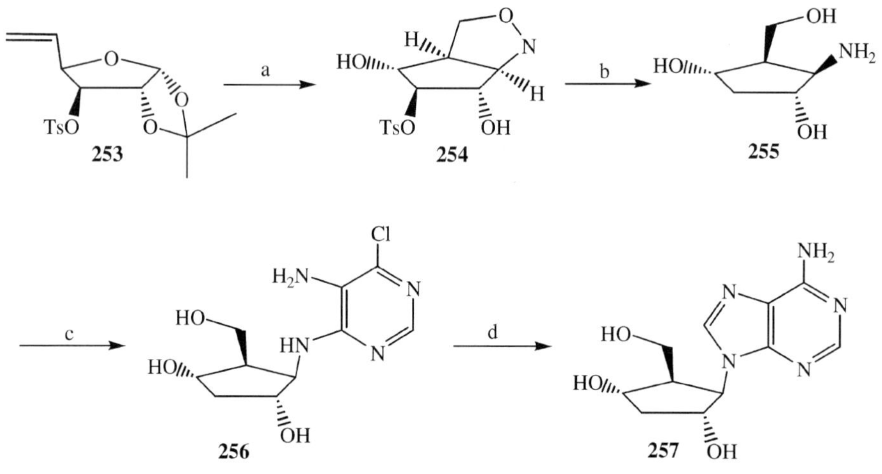

Reagents: (a) (i) aq. H_2SO_4; (ii) $PhCH_2NHOH$; (b) (i) $LiAlH_4$; (ii) Pd/C, cyclohexene; (c) 5-amino-4,6-dichloropyrimidine, Et_3N; (d) (i) $HC(OEt)_3$, *p*-TSA; (ii) NH_3/MeOH.

Scheme 21.36 Synthesis of an aristeromycin analogue **257** employing intramolecular nitrone cycloaddition [68].

Reductive elimination of the tosyl group in **254** with $LiAlH_4$, followed by reductive cleavage of the isoxazolidine ring, furnished the key precursor aminocyclopentane **255**. Treatment of the latter compound with 5-amino-4,6-dichloropyrimidine afforded **256**, which underwent cyclization with triethyl orthoformate; subsequent amination yielded the isoaristeromycin **257**.

21.4
Conclusions

It has been observed from recent discussions that neplanocin A and its analogues are highly popular targets for medicinal and organic chemists alike, due to their novel architecture and promising biological activities. In this chapter, several recently developed methods have been described for the synthesis not only of neplanocin but also of a variety of new analogues of neplanocin A. The recent development of aristeromycin analogues was also included. It is of interest to note that various modified neplanocin A and aristeromycin analogues exhibit promising biological activities, and consequently several biologically active modified analogues of neplanocin A and aristeromycin are undergoing further study in the development of new drugs. It is important, however, to continue exploring chemical and biological approaches for the development of neplanocin A analogues, as the existing agents are limited in terms of their toxicity and antiviral efficacy.

References

1 (a) Yaginuma, S., Muto, N., Tsujino, M., Sudate, Y., Hayashi, M., Otani, M., *J. Antibiot. (Tokyo)* **1980**, *34*, 359; (b) Hayashi, M., Yaginuma, S., Yoshioka, H., Nakatsu, K., *J. Antibiot. (Tokyo)* **1981**, *34*, 675.

2 De Clercq, E., *Biochemical Pharmacol.* **1987**, *36*, 2567.

3 (a) Hasobe, M., McKee, J. G., Borchardt, R. T., *Antimicrob. Agents Chemother.* **1989**, *33*, 828; (b) Cools, M., De Clercq, E., *Biochemical Pharmacol.* **1990**, *40*, 2259.

4 (a) Cantoni, G. L., The centrality of *S*-adenosylhomocysteinase in the regulation of the biological utilization of *S*-adenosylmethionine, In *Biological Methylation and Drug Design* (eds R. T. Borchardt, C. R. Creveling, P. M. Ueland), Humana Press, Clifton, NJ, **1986**, pp. 227–238; (b) Turner, M. A., Yang, X. D., Kuczera, K., Borchardt, R. T., Howell, P. L., *Cell Biochem. Biophys.* **2000**, *33*, 101.

5 (a) Borchardt, R. T., Wolfe, M. S., *J. Med. Chem.* **1991**, *34* 1521, and references cited therein; (b) Liu, S., Wolfe, M. S., Borchardt, R. T., *Antiviral Res.* **1992**, *19*, 247.

6 De Clercq, E., *Nucleosides, Nucleotides & Nucleic Acids* **2005**, *24*, 1395.

7 Matuszewska, B., Borchardt, R. T., *Arch. Biochem. Biophys.* **1987**, *256*, 50.

8 Paisley, S. D., Wolfe, M. S., Borchardt, R. T., *J. Med. Chem.* **1989**, *32*, 1415.

9 Saunders, P. P., Tan, M. T., Robins, R. K., *Biochem. Pharmacol.* **1985**, *34*. 2749.

10 Glazer, R. I., Knode, M. C., *J. Biol. Chem.* **1984**, *259*, 12964.

11 (a) Kusaka, T., Yamomoto, H., Shibata, M., Muroi, M., Kishi, T., *J. Antibiot.* **1968**, *21*, 255; (b) Kishi, T., Muroi, M., Kusaka, T., Nishikawa, M., Kamiya, K., Mizuno, K., *Chem. Pharm. Bull.* **1972**, *20*, 940.

12 Bennett, L. L. Jr., Allan, P. W., Hill, D. L., *Mol. Pharmacol.* **1968**, *4*, 208.

13 Bennett, L. L. Jr., Brockman, R. W., Rose, L. M., Allan, P. W., Shaddix, S. C., Shealy, Y. F., Claton, J. D., *Mol. Pharmacol.* **1985**, *27*, 666.

14 Bennett, L. L. Jr., Allan, P. W., Rose, L. M., Comber, R. N., Secrest, J. A. III, *Mol. Pharmacol.* **1986**, *29*, 383.

15 Jeong, L. S., Lee, J. A., *Antiviral Chem. Chemother.* **2004**, *15*, 235.

16 (a) Piperno, A., Chiacchio, M. A., Iannazzo, D., Romeo, R., *Curr. Med. Chem.* **2006**, *13*, 3675; (b) Melroy, J., Nair, V., *Curr Pharm Des.* **2005**, *11*, 3847; (c) Rodriguez, J. B., Comin, M. J., *Mini Rev. Med. Chem.* **2003**, *3*, 95; (d) Schneller, S. W., *Curr. Top. Med. Chem.* **2002**, *2*, 1087; (e) Ferrero, M., Gotor, V., *Chem. Rev.* **2000**, *100*, 4319; (f) De Clercq, E., Andrei, G., Snoeck, R., De Bolle, L., Naesens, L., Degreve, B., Balzarini, J., Zhang, Y., Schols, D., Leyssen, P., Ying, C., Neyts, J., *Nucleosides Nucleotides Nucleic Acids* **2001**, *20*, 271.

17 Choi, W. J., Moon, H. R., Kim, H. O., Yoo, B. N., Lee, J. A., Shin, D. H., Jeong, L. S., *J. Org. Chem.* **2004**, *69*, 2634.

18 Ohira, S., Sawamoto, T., Yamato, M., *Tetrahedron Lett.* **1995**, *36*, 1537.

19 Moon, H. R., Choi, W. J., Kim, H. O., Jeong, L. S., *Chem. Lett.* **2004**, *33*, 506.

20 (a) Marquez, V. E., Lim, M.-I., Tseng, C. K. H., Markovac, A., Priest, M. A., Khan, M. S., Kaskar, B., *J. Org. Chem.* **1998**, *53*, 5709; (b) Wolfe, M. S., Anderson, B. L., Borcherding, D. R., Borchardt, R. T., *J. Org. Chem.* **1990**, *55*, 4712.

21 Gallos, J. K., Stathakis, C. I., Kotoulas, S. S., Koumbis, A. E., *J. Org. Chem.* **2005**, *70*, 6884.

22 Choi, W. J., Park, J. G., Yoo, S. J., Kim, H. O., Moon, H. R., Chun, M. W., Jung, Y. H., Jeong, L. S., *J. Org. Chem.* **2001**, *66*, 6490.

23 (a) Bestmann, H. J., Roth, D., *Synlett*, **1999**, 751; (b) Borcherding, D. R., Scholtz, S. A., Borchardt, R. T., *J. Org. Chem.* **1987**, *52*, 5457.

24 Moon, H. R., Choi, W. J., Kim, H. O., Jeong, L. S., *Tetrahedron Asymm.* **2002**, *13*, 1189.

25 Jin, Y. H., Chu, C. K., *Tetrahedron Lett.* **2002**, *43*, 4141.

26 Trost, B. M., Madsen, R., Guile, S. D., Brown, B., *J. Am. Chem. Soc.* **2000**, *122*, 5947.

27 Paquette, L. A., Tian, Z., Seekamp, C. K., Wang, T., *Helv. Chim. Acta* **2005**, *88*, 1185.

28 Xue-qing, Y., Schneller, S. W., *Tetrahedron Lett.* **2005**, *46*, 1927.

29 Gosselin, G., Griffe, L., Meillon, J. -C., Storer, R., *Tetrahedron* **2006**, *62*, 906.

30 (a) Lee, K., Ravi, G., Ji, X.-D., Marquez, V. E., Jacobson, K. A., *Bioorg. Med. Chem. Lett.* **2001**, *11*, 1333; (b) Nandanan, E., Jang, S. Y., Moro, S., Kim, H. O., Siddiqui, M. A., Russ, P., Marquez, V. E., Busson, R., Herderwijn, P., Harden, T. K., Boyer, J. L., Jacobson, K. A., *J. Med. Chem.* **2000**, *43*, 829.

31 Siddiuui, M. A., Ford, H. Jr., George, C., Marquez, V. E., *Nucleosides Nucleotides* **1996**, *15*, 235.

32 (a) Moon, H. R., Kim, H. O., Lee, K. M., Chun, M. W., Kim, J. H., Jeong, L. S., *Org. Lett.* **2002**, *4*, 3501; (b) Moon, H. R., Kwon, S. H., Lee, J. A., Yoo, B. N., Kim, H. O., Chun, M. W., Kim, H.-D Kim, J. H., Jeong, L. S., *Nucleoside Nucleotides Nucleic Acids* **2003**, *22*, 1475.

33 Lee, J. A., Moon, H. R., Kim, H. O., Kim, K. R., Lee, K. M., Kim, B. T., Hwang, K. J., Chun, M. W., Jacobson, K. A., Jeong, L. S., *J. Org. Chem.* **2005**, *70*, 5006.

34 Kim, J. H., Kim, H. O., Lee, K. M., Chun, M. W., Moon, H. R., Jeong, L. S., *Tetrahedron* **2006**, *62*, 6339.

35 (a) Kim, H. O., Yoo, S. J., Ahn, H. S., Choi, W. J., Moon, H. R., Lee, K. M., Chun, M. W., Jeong, L. S., *Bioorg. Med. Chem. Lett.* **2004**, *14*, 2091; (b) Jeong, L. S., Park, J. G., Choi, W. J., Moon, H. R., Lee, K. M., Kim, H. O., Kim, H.-D., Chun, M. W., Park, H.-Y., Kim, K., Sheen, Y. Y., *Nucleosides Nucleotides Nucleic acids* **2003**, *22*, 919.

36 Yin, X., Schneller, S. W., *Tetrahedron Lett.* **2006**, *47*, 4057.

37 (a) Deardorff, D. R., Myles, D. C., *Org. Synth.* **1989**, *67*, 114; (b) Siddiqi, S. M., Chen, X., Schneller, S. W., *Nucleoside Nucleotides* **1993**, *12*, 267.

38 Yin, X., Schneller, S. W., *Tetrahedron* **2004**, *60*, 3451.

39 Hoshi, A., Yoshida, M., Iigo, M., Tokugen, R., Fukukawa, K., Ueda, T., *J. Pharmacobio-Dyn.* **1986**, *9*, 202.

40 Shuto, S., Obara, T., Saito, Y., Andrei, G., Snoeck, R., De Clercq, E., Matsuda, A., *J. Med. Chem.* **1996**, *39*, 2392.

41 (a) Moon, H. R., Lee, H. J., Kim, K. R., Lee, K. M., Lee, S. K., Kim, H. O., Chun, M. W., Jeong, L. S., *Bioorg. Med. Chem. Lett.* **2004**, *14*, 5641; (b) Moon, H. R., Lee, K. M., Lee, H. J., Lee, S. K., Park, S. B., Chun, M. W., Jeong, L. S., *Nucleoside Nucleotides Nucleic Acids* **2005**, *24*, 707.

42 Paisesley, S. D., Wolfe, M. S., Borchardt, R. T., *J. Med. Chem.* **1989**, *32*, 1415.

43 Obara, T., Shuto, S., Saito, Y., Toriya, M., Ogawa, K., Yaginuma, S., Shigeta, S., Matusda, A., *Nucleosides Nucleotides* **1996**, *15*, 1157.

44 Shuto, S., Obara, T., Saito, Y., Yamashita, K., Tanaka, M., Sasaki, T., Andrei, G., Snoeck, R., Neyts, J., Padalko, E., Balzarini, J., Clercq, E. D., Matsuda, A., *Chem. Pharm. Bull.* **1997**, *45*, 1163.

45 Daelemans, D., Vandamme, A.-M., Shuto, S., Matsuda, A., Clercq, E. D., *Nucleosides Nucleotides* **1998**, *17*, 479.

46 (a) Shuto, S., Obara, T., Toriya, M., Hosoya, M., Snoeck, R., Andrei, G., Balzarini, J., De Clercq, E., *J. Med. Chem.* **1992**, *35*, 324; (b) Shigeta, S., Mori, S., Baba, M., Ito, M., Honzumi, K., Nakamura, K., Oshitani, H., Numazaki, Y., Matsuda, A., Obara, T., Shuto, S., De Clercq, E., *Antimicrob. Agents Chemother.* **1992**, *36*, 435.

47 Shuto, S., Minakawa, N., Niizuma, S., Kim, H.-S., Wataya, Y., Matsuda, A., *J. Med. Chem.* **2002**, *45*, 748.

48 Rajappan, V. P., Schneller, S. W., *Bioorg. Med. Chem.* **2003**, *11*, 5199.

49 Wnuk, S. F., Yuan, C.-S., Borchardt, R. T., Robins, M. J., *Nucleosides Nucleotides* **1998**, *17*, 99.

50 Mulvihill, M. J., Miller, M. J., *Tetrahedron* **1998**, *54*, 6605.

51 Ye, W., Schneller, S. W., *J. Org. Chem.* **2006**, *71*, 8641.

52 Ghosh, A., Miller, M. J., De Clercq, E., Balzarini, J., *Nucleosides Nucleotides* **1999**, *18*, 217.

53 (a) Jeong, L. S., Yoo, S. J., Lee, K. M., Koo, M. J., Choi, W. J., Kim, H. O., Moon, H. R., Lee, M. Y., Park, J. G., Lee, S. K., Chun, M. W., *J. Med. Chem.* **2003**, *46*, 201; (b) Jeong, L. S., Moon, H. R., Park, J. G., Shin, D. H., Choi, W. J., Lee, K. M., Kim, H. O., Chun, M. W., Kim, H.-D., Kim, J. H., *Nucleosides Nucleotides Nucleic Acids* **2003**, *22*, 589.

54 Chu, C. K., Jin, Y. H., Baker, R. O., Huggins, J., *Bioorg. Med. Chem. Lett.* **2003**, *13*, 9.

55 Song, G. Y., Paul, V., Choo, H., Morrey, J., Sidwell, R. W., Schinazi, R. F., Chu, C. K., *J. Med. Chem.* **2001**, *44*, 3985.

56 Shuto, S., Obara, T., Itoh, H., Kosugi, Y., Saito, Y., Toriya, M., Yaginuma, S., Shigeta, S., Matsuda, A., *Chem. Pharm. Bull.* **1994**, *42*, 1688.

57 Obara, T., Shuto, S., Saito, Y., Snoeck, R., Andrei, G., Balzarini, J., Clercq, E. D., Matsuda, A., *J. Med. Chem.* **1996**, *39*, 3847.

58 Zhao, L. X., Yun, M., Kim, H. O., Lee, J. A., Choi, W. J., Lee, K. M., Lee, S. K., Lee, Y. B., Ahn, C. H., Jeong, L. S., *Nucleic Acid Symposium Series* **2005**, *49*, 107.

59 Jeong, L. S., Buenger, G., McCormack, J. J., Cooney, D. A., Hao, Z., Marquez, V. E., *J. Med. Chem.* **1998**, *41*, 2572.

60 Seley, K. L., Schneller, S. W., *J. Med. Chem.* **1997**, *40*, 622.

61 Seley, K. L., Schneller, S. W., *J. Med. Chem.* **1997**, *40*, 625.

62 Cho, J. H., Bernard, D. L., Sidwell, R. W., Kern, E. R., Chu, C. K., *J. Med. Chem.* **2006**, *49*, 1140.

63 Rao, J. R., Schinazi, R. F., Chu, C. K., *Bioorg. Med. Chem.* **2007**, *15*, 839.

64 (a) Comin, M. J., Rudriguez, J. B., Russ, P., Marquez, V. E., *Tetrahedron* **2003**, *59*, 295;

(b) Marquez, V. E., Russ, P., Alonso, R., Siddiqui, M. A., Hernandez, S., George, C., Nicklaus, M. C., Dai, F., Ford, H. Jr. *Helv. Chim. Acta* **1999**, *82*, 2119.

65 Lee, J. A., Kim, H. O., Tosh, D. K., Moon, H. R., Kim, S., Jeong, L. S., *Org. Lett.* **2006**, *8*, 5081.

66 Coumin, M. J., Leitofuter, J., Rodriguez, J. B., *Tetrahedron* **2002**, *58*, 3129.

67 Coumin, M. J., Rodriguez, J. B., *Tetrahedron* **2000**, *56*, 4639.

68 Roy, A., Patra, R., Achari, B., Mandal, S. B., *Synlett* **1997**, *11*, 1237.

22
Clitocine and Its Analogues

Hyunik Shin and Changhee Min

22.1
Clitocine: Isolation, Synthesis, and Biological Activity

22.1.1
Isolation

Guided by the *insecticidal* activity against the agricultural pest insect *Pectinophora gossypiella*, Kubo and Kim isolated clitocine (**1a**) [1], along with adenosine, from *Clitocybe inversa*, a medium-sized, buff-colored mushroom found in western North America. The NMR spectral data for clitocine was found to be very similar to that of adenosine, with the exception of the absence of the C-8 carbon peak of adenosine, while the low chemical shift of the 4-NH proton suggested strong hydrogen bonding between the 5-NO_2 and 4-NH groups. Later, Moss and coworkers [2] confirmed its structure through X-ray crystallographic analysis: as postulated earlier by Kubo *et al.*, the entire aglycone moiety resembles a tricyclic ring system as a result of the hydrogen bonding between the 4-NH and 6-NH protons and the NO_2 group.

22.1.2
Synthesis

The interesting biological activity of clitocine, together with its biogenetically close relationship to adenosine, triggered an immediate attempt to synthesize the molecule *de novo*. Two years after its initial isolation, two independent syntheses have now been completed. The first synthesis [3] is based upon the condensation of 2,3,5-tribenzoyl-ribofuranosylamine (**2**) with 4,6-dichloro-5-nitropyrimidine (**3**) in the presence of a base to give a mixture of the α- and β-anomers in a 7 : 1 ratio (Scheme 22.1), although the yield (19%) is relatively poor. Deprotection of the benzoate group of **4** using a methanolic ammonia solution produced clitocine with a 71% yield. In order to improve the efficiency and reproducibility of the coupling reaction, an acetonide-protected amino sugar (**5**) was employed for the coupling reaction with 4-chloro-5-

Modified Nucleosides: in Biochemistry, Biotechnology and Medicine. Edited by Piet Herdewijn

ISBN: 978-3-527-31820-9

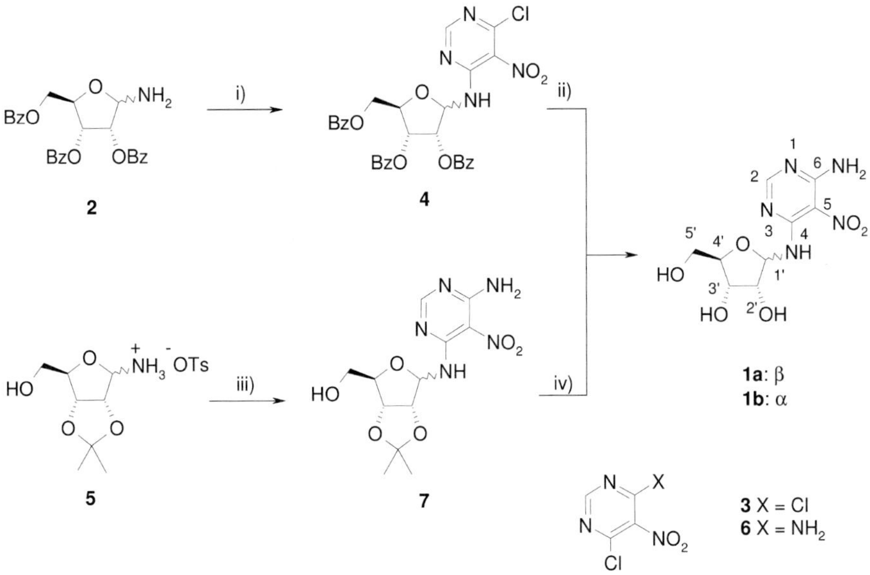

Scheme 22.1 Reagents and conditions: (i) **3**, Et_3N, DMF, room temp.; (ii) NH_3, MeOH, 0 °C; (iii) **6**, Et_3N, DMF, room temp.; (iv) CF_3CO_2H-H_2O (9:1), room temp., 6 min.

nitro-6-aminopyrimidine (**6**). With this modification, the α/β isomer ratio of **7** was improved to 1 : 2.8. However, the selective cleavage of the acetonide group in the presence of the *N*-glycosidic linkage was not trivial. After careful screening of the reaction parameters, it was found that deprotection with 90% aqueous trifluoroacetic acid at room temperature represent the optimal conditions for the production of clitocine, with a yield of 43%.

The second synthesis [4] is more efficient than the first with respect to both the stereocontrol at the anomeric center and the yield (Scheme 22.2). The reaction of 4,6-diamino-5-nitropyrimidine (**8**) with hexamethyldisilazane (HMDS) in the presence of a catalytic amount of sulfuric acid produced the silylated 4,6-diamino-5-nitropyrimidine (**9**), which was coupled with 1-*O*-acetyl-2,3,5-tri-*O*-benzoyl-D-ribofuranose (**10**) in the presence of trimethylsilyl trifluoromethanesulfonate (TMSOTf) to produce the β-product (**14**) with virtually no sign of its α-isomer. However, deprotection using a saturated solution of ammonia in methanol or ethanol caused significant epimerization of the anomeric center resulting in an α/β isomer ratio of 1 : 2.8. After considerable optimization, the use of 0.2 equivalents of sodium methoxide in methanol was determined to be optimal for the production of **1a**, with less than 10% epimerization at the anomeric center.

Recently, Choi *et al.* identified a new intermediate in Kini's synthesis [5]. In a repetition of the latter procedure, the NMR data of the crude coupling reaction mixture of **9** and **10** revealed no evidence of any formation of the desired product (**14**). However, column chromatography of the crude mixture using silica gel yielded the desired product (**14**) and its *endo*-isomer (**12**) in a 3 : 1 ratio. From this observation,

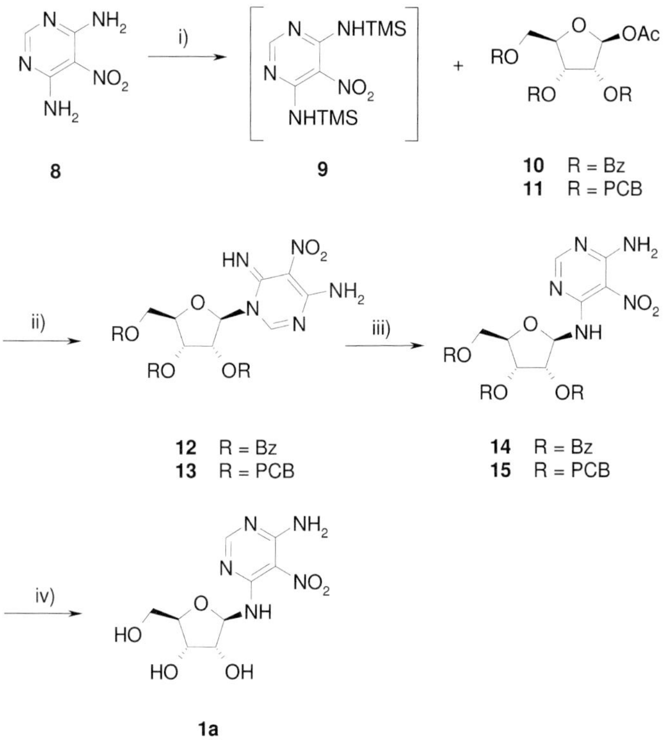

Scheme 22.2 Reagents and conditions: (i) HMDS, pyridine, H_2SO_4, $(NH_4)_2SO_4$; (ii) TMSOTf; (iii) silica gel or acetic acid; (iv) NaOMe, MeOH/dioxane.

it was postulated that the *endo*-isomer (**12**) was formed kinetically, isomerized to its *exo*-isomer (**14**) through a 1,3-N (*endo*) to N (*exo*)-migration. In order to minimize the epimerization of the anomeric center at the final deprotection stage, Choi *et al.* introduced a *p*-chlorobenzoate group instead of a benzoate group as in **11**; this modification was based on the assumption that the more electron-deficient *p*-chlorobenzoate group would facilitate the final deprotection reaction. The coupling reaction of **11** with **9** gave **13**, which was readily solidified with high purity by trituration with diethyl ether, thus making the purification very facile. The subsequent isomerization of **13** by silica gel or acetic acid produced **15** in quantitative yield; this molecule, as expected, was deprotected at a considerably faster rate to give **1a**, without any sign of epimerization.

22.1.3
Biological Activity

Motivated by the observed adenosine kinase (AKase) inhibitory effect of nucleosides structurally related to **1a** [6], Moss and collaborators tested its activity towards AKase and discovered that it served as a substrate and an inhibitor of AKase, with a K_i value of 3 μM [4]. In a growth inhibition study using human B lymphoblast-derived WI-L2 cells, these authors detected mono-, di-, and triphosphate derivatives of clitocine, indicating an efficient phosphorylation of the molecule after its transport into the cell. Clitocine inhibited the growth of L1210 (murine lymphocytic leukemia cells), WI-L2 (human B-lymphoblastic leukemia cells), and CCRF-CEM (human T-lymphoblastic leukemia cells) cell lines, with an ID_{50} of 30 nM. However, no significant antiviral activity was detected in cell culture against parainfluenza virus type 3, measles and vaccinia viruses, and herpes simplex virus (HSV) type 2.

Fortin and coworkers isolated clitocine fractions in various α/β ratios from *Lepista inverse* – the pure β-anomer clitocine, mixture A (α/β = 60 : 40), and mixture B (α/β = 20 : 80) [7]. Interestingly, all of these fractions exhibited similar cytotoxic activities, with IC_{50} values ranging from 20.5 to 42 nM in murine cancer cell lines (3LL and L1210), and from 185 to 578 nM in human cancer cell lines (DU145, K-562, MCF7, and U251). Based on this information, a detailed *in-vivo* study of the readily available mixture B was carried out using 3LL- and L1210-tumor-bearing mice. Treatment with clitocine at 5 mg kg^{-1} did not significantly increase the survival rate and lifespan of the 3LL-tumor-bearing mice (Figure 22.1A). In contrast, clitocine exhibited a meaningful antitumor activity in the L1210-tumor-bearing mice, with a 50% increase in life span and a decrease in the development of ascites at 3 mg kg^{-1} administration. These findings are in agreement with those obtained from cell-cycle analysis by flow cytometry, whereby clitocine induced the appearance of a sub-G_1 peak only in the L1210 cells; this represented the presence of aneuploid cells and possibly cells with fragmented DNA. However, in light of the increased mortality at a dose of 5 mg kg^{-1} and a loss of activity at 0.5 mg kg^{-1}, a narrow therapeutic window was evident (Figure 22.1B). A further search for clitocine analogues as potential anticancer reagents [8] should be directed towards improving this aspect.

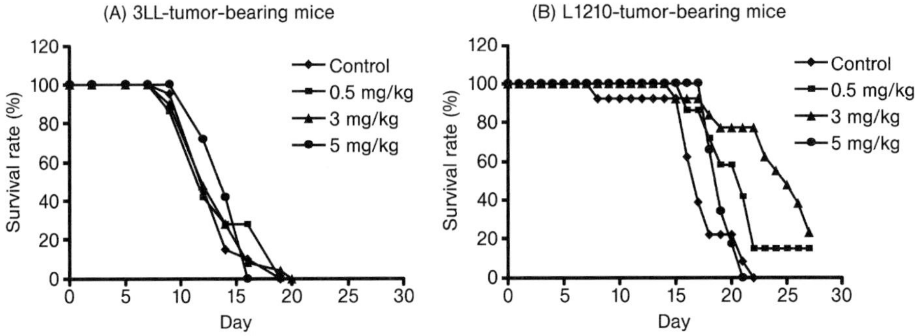

Figure 22.1 (A) Effects of mixture B on the survival rate of 3LL-tumor-bearing mice. The animals received daily intraperitoneal injections of mixture B over 14 days at the indicated dose. Data shown are as follows: control, n = 20 (♦); 0.5 mg kg^{-1}, n = 7 (■); 3 mg kg^{-1}, n = 21 (▲); and 5 mg kg^{-1}, n = 7 (●). (B) Effects of mixture B on the survival rate of L1210-tumor-bearing mice. Mice were administered daily intraperitoneal injections of mixture B at the indicated dose. Data shown are as follows: control, n = 13 (♦); 0.5 mg kg^{-1}, n = 7 (■); 3 mg kg^{-1}, n = 13 (▲); and 5 mg kg^{-1}, n = 6 (●).

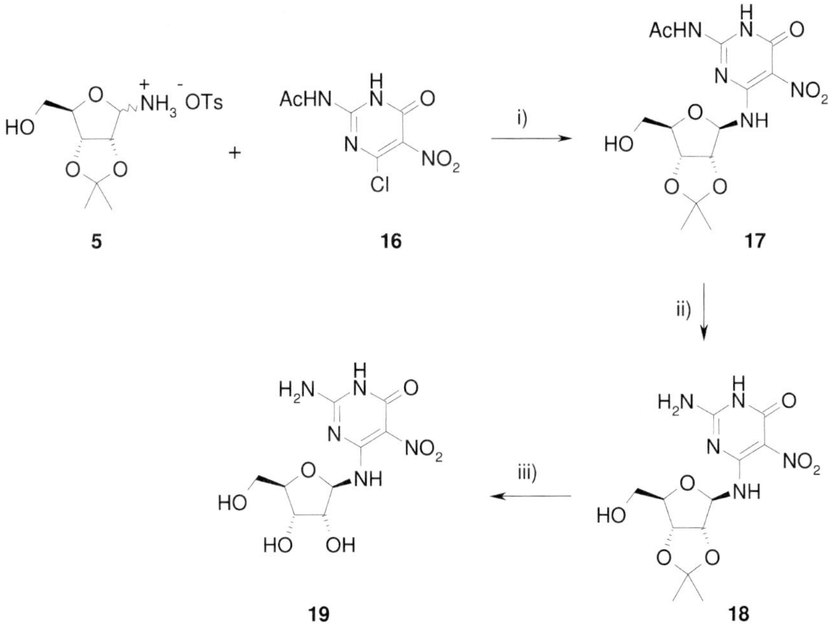

Scheme 22.3 Reagents and conditions: (i) (a) NaOEt, EtOH, (b) recrystallization; (ii) NH_2NH_2, EtOH, 15 min; (iii) CF_3CO_2H-H_2O (95:5), −10 °C, 10 min.

22.2
Clitocine Analogues

22.2.1
Aglycone Modifications

As an extension of the structural resemblance between clitocine and adenosine, Lhomme and collaborators published details of the synthesis of a clitocine derivative that corresponded to guanosine (Scheme 22.3) [9]. Following Kamikawa's synthetic protocol [3] for clitocine, the coupling reaction of **5** and **16** was initially accomplished in the presence of triethylamine in dimethylformamide (DMF) to produce **17**, albeit with a low yield and poor reproducibility. The yield and reproducibility were, however, significantly improved by using sodium ethoxide in absolute ethanol as a base. Recrystallization in water provided the pure β-anomer (**17**), the acetyl group of which was deprotected by treatment with hydrazine to give **18**. Final deprotection using 95% trifluoroacetic acid in water yielded the guanosine analogue of clitocine (**19**) without any noticeable anomerization.

In 1991, the synthesis of the 3-deaza analogue of clitocine was reported by Franchetti *et al.* (Scheme 22.4) [10]. Compound **20** was silylated and reacted with 1-*O*-acetyl-2,3,5-tri-*O*-benzoyl-D-ribofuranose (**10**) in the presence of TMSOTf to yield two major products (**21** and **22**) via either the *exo*-N attack or *endo*-N attack

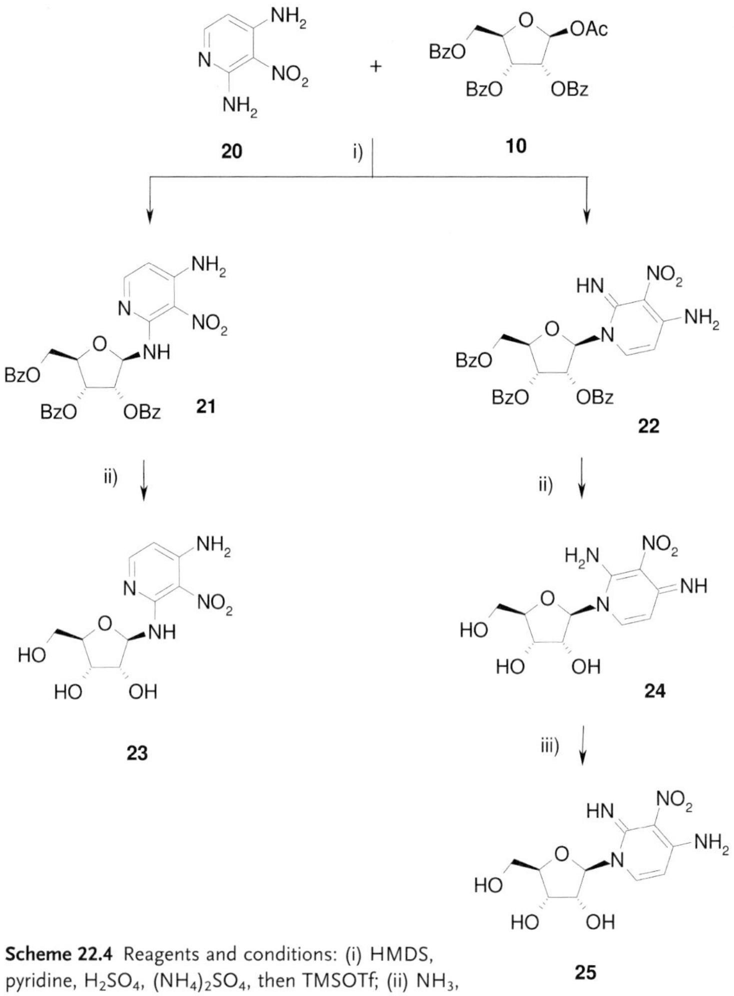

Scheme 22.4 Reagents and conditions: (i) HMDS, pyridine, H_2SO_4, $(NH_4)_2SO_4$, then TMSOTf; (ii) NH_3, MeOH; (iii) recrystallization in MeOH.

of **20**. Deprotection of the benzoate group produced **23** and **24**, and the latter was isomerized to **25** upon recrystallization in methanol. Compounds **23** and **25** exhibited a growth-inhibitory effect on murine leukemia P388 and human promyelocytic leukemia HL60 cell lines. The former compound was less active than clitocine (ID_{50} 8.8×10^{-5} M and 7.5×10^{-5} M, respectively), whereas the latter exhibited more potent activity (ID_{50} 2.8×10^{-5} M and 8.5×10^{-6} M).

Recently, a solid-phase parallel synthesis was developed for the construction of a nucleoside library (Scheme 22.5) [11]. The anchoring of **26** to polystyrene monomethoxytrityl chloride (MTT-Cl) resin provided the key intermediate (**27**), the azide group of which was reduced and subsequently condensed with 4,6-dichloro-5-nitropyrimidine (**3**) to yield **29**. Substitution of its 6-chloro group with various amine

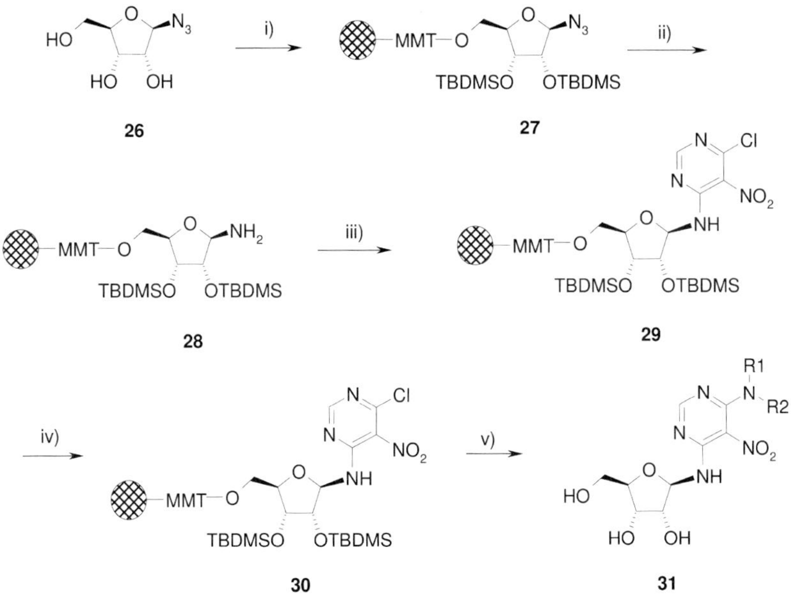

Scheme 22.5 Reagents and conditions: (i) (a) polystyrene MMT-Cl resin, DMAP, pyridine, (b) *t*-butyldimethylsilyl chloride, imidazole; (ii) PMe_3; (iii) 4,6-dichloro-5-nitropyrimidine, diisopropylethylamine; (iv) amine building blocks, diisopropylethylamine; (v) (a) 1 M tetrabutylammonium fluoride, 16 h, (b) 1.5% CF_3CO_2H, room temp., 1.5 min.

building blocks in a parallel manner, followed by deprotection of the silyl group and detachment from the resin, produced various clitocine derivatives with high purity (65–100%).

22.2.2 Carbocyclic Analogues

Palmer *et al.* reported the synthesis of a carbocyclic analogue of clitocine (±)-**38** (Scheme 22.6) [12]. Epoxidation of the allylic alcohol (**32**) by *m*-chloroperbenzoic acid (mCPBA) produced the epoxide (**33**) with a 50% yield and with high diastereoselectivity. Regioselective epoxide ring opening (93 : 7 selectivity), followed by protection of the formed vicinal diol group as an acetonide, produced **34** contaminated with its regioisomer (**35**) (Scheme 22.6). After purification by column chromatography, the azide group of **34** was reduced to the amine group under hydrogenation conditions. The coupling reaction of **36** with 4,6-dichloro-5-nitropyrimidine (**3**) yielded **37**, the chloro group of which was transformed to the amino group by treatment with methanolic ammonia solution. Subsequent sequential deprotections of the silyl and the acetonide groups produced (±)-**38** with a 50% yield.

In a similar manner, Baxter and collaborators prepared carbocyclic clitocine (±)-**38** in a single step (Scheme 22.7) using the coupling reaction of the readily available

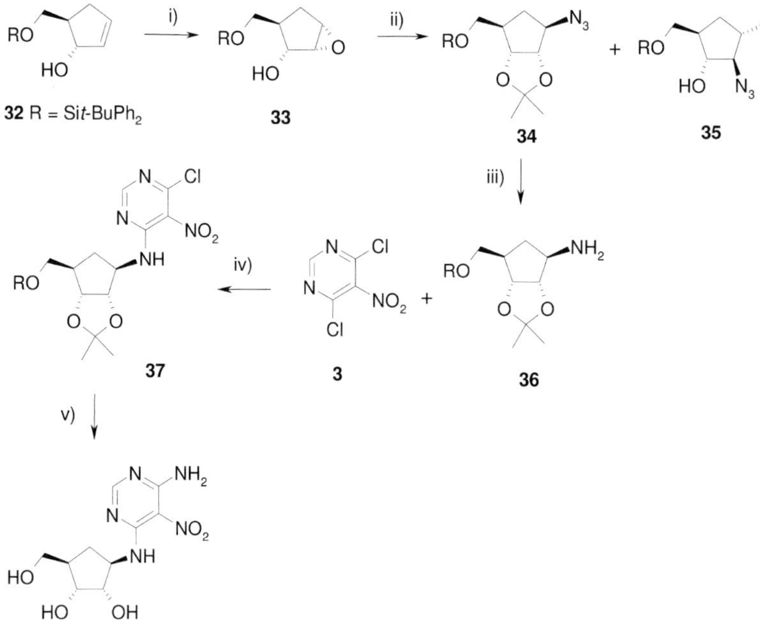

38

Scheme 22.6 Reagents and conditions: (i) mCPBA; (ii) (a) NaN_3, NH_4Cl, DMF, 100 °C, 4 days, (b) 2,2-dimethoxypropane, *p*-TsOH, room temp. 18 h; (iii) H_2, Lindlar catalyst, MeOH; (iv) 4,6-dichloro-5-nitropyrimidine, Et_3N, CH_2Cl_2; (v) (a) NH_3, MeOH, (b) tetrabutylammonium fluoride, THF, (c) H^+.

aminotriol (**39**) [13] with 4-chloro-5-nitro-6-aminopyrimidine (**6**) to give a 76% yield [14]. Iteration of the same transformation with 6-chloro-4,5-dihydro-5-nitro-4-oxo-2-aminopyrimidine (**40**) produced the guanosine-type carbocyclic analogue (**41**). The introduction of a hydroxyl group, by replacing the chloro group of the known intermediate (**42**) by sodium hydroxide in ethylene glycol, produced the 5-denitro derivative (**43**).

The antiviral activity of compounds **38**, **41**, and **43** was tested against influenza virus, herpesvirus, retroviruses and myxoviruses, among which only the influenza virus was susceptible (Table 22.1). A noteworthy observation was the significant phosphorylation of **38** by AKase and its selective and potent *in-vitro* activity against influenza A (Singapore/1/57).

The incorporation of neplanocin [15] glycone into clitocine has been reported by Marquez *et al.* (Scheme 22.8) [16]. The selective reduction of the azide group of **44** using the Staudinger reaction with trimethylphosphine produced **45**, which was *in-situ*-coupled with 4,6-dichloro-5-nitropyrimidine (**3**) or with 6-chloro-3,4-dihydro-5-nitro-4-oxo-2-aminopyrimidine (**40**) to produce the intermediates **46** and **47**, respectively. The chloride group of **46** was replaced by ammonia to produce **50**, which was treated with boron trichloride to induce simultaneous deprotections of the

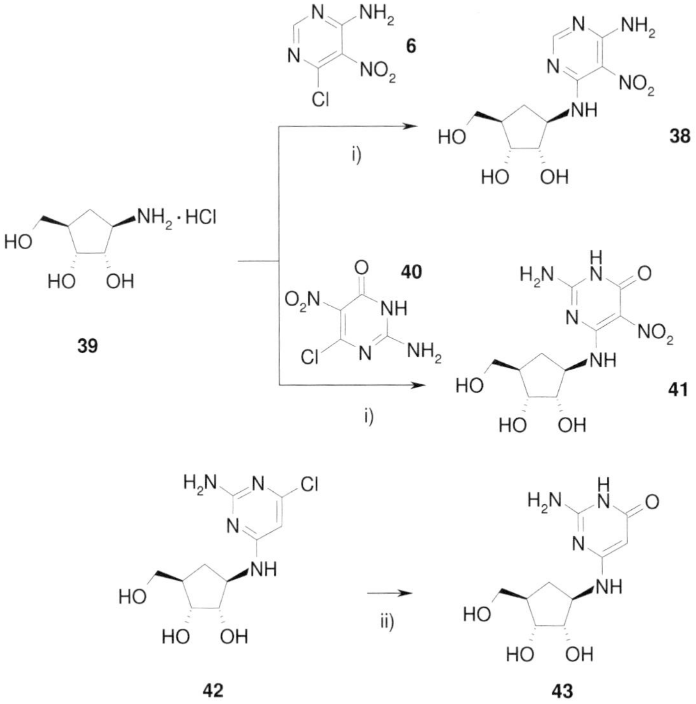

Scheme 22.7 Reagents and conditions: (i) Et_3N, EtOH, reflux; (ii) NaOH, $(CH_2OH)_2$, 150 °C, 51%.

benzyl and acetonide groups to yield the cyclopentenyl analogue of clitocine (**51**). In contrast to clitocine **51** was unstable, and an attempted recrystallization in water failed, resulting instead in the formation of a dark, tarry material. In contrast, when the neutralization procedure was omitted after deprotection by boron trichloride, the HCl salt of **51** was isolated in a pure and stable form. Deprotections of the benzyl and

Table 22.1 *In-vitro* biological activity of clitocine and its analogues.

Compound	Extent of phosphorylation by adenosine kinase (A = 1)	Inhibition of virus replication (IC_{50}): ELISA assay (μM) Influenza A (Singapore/1/57)	Plaque reduction assay [toxicity] (μM) Influenza A (Singapore/1/57)	Influenza B (Victoria/102/85)	Toxicity to L1210 cells (μM) (mouse DBA/2)
38	1.48	70	5.2 [>50.0]	105.2 [>500]	1.75
41	0.09	99	165 [>500]	>500 [>500]	>50
43	0.11	>1 mM	–	–	>50

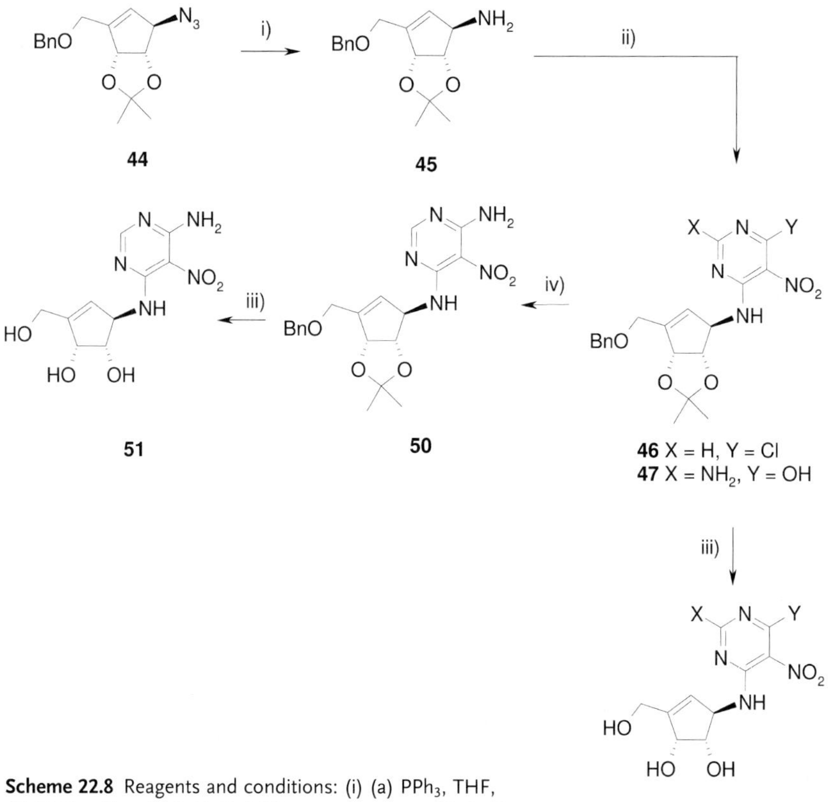

Scheme 22.8 Reagents and conditions: (i) (a) PPh_3, THF, (b) H_2O, reflux; (ii) **3**, Et_3N, THF, room temp., or **40**, Et_3N, THF, reflux; (iii) BCl_3, CH_2Cl_2; (iv) NH_3, MeOH.

the acetonide groups of **46** using boron trichloride provided the inosine derivative (**48**) via the concomitant introduction of a hydroxyl group; treatment of **47** with boron trichloride gave the guanosine derivative (**49**).

22.2.3
Acyclic Analogues

Franchetti and his collaborators prepared new nucleotide analogues [17] of the anti-HIV agents PMEA (Adefovir; 9-[2-(phosphonomethoxy)ethyl]adenine), (*R*)-PMPA (9-[2-(phosphonomethoxy)propyl]adenine), PMEDAP (9-[2-(phosphonomethoxy)ethyl]-2,6-diaminopurine), and (*R*)-PMPDAP (9-[2-(phosphonomethoxy)propyl]-2,6-diaminopurine), by replacing their adenine fragment with 4,6-diamino-5-nitropyrimidine (**67** and **68**), 2,4,6-triamino-5-nitropyrimidine (**69** and **70**), and 1-deaza-, and 3-deaza-4,6-diamino-5-nitropyrimidine (**71** and **72**) (Scheme 22.9). The same group

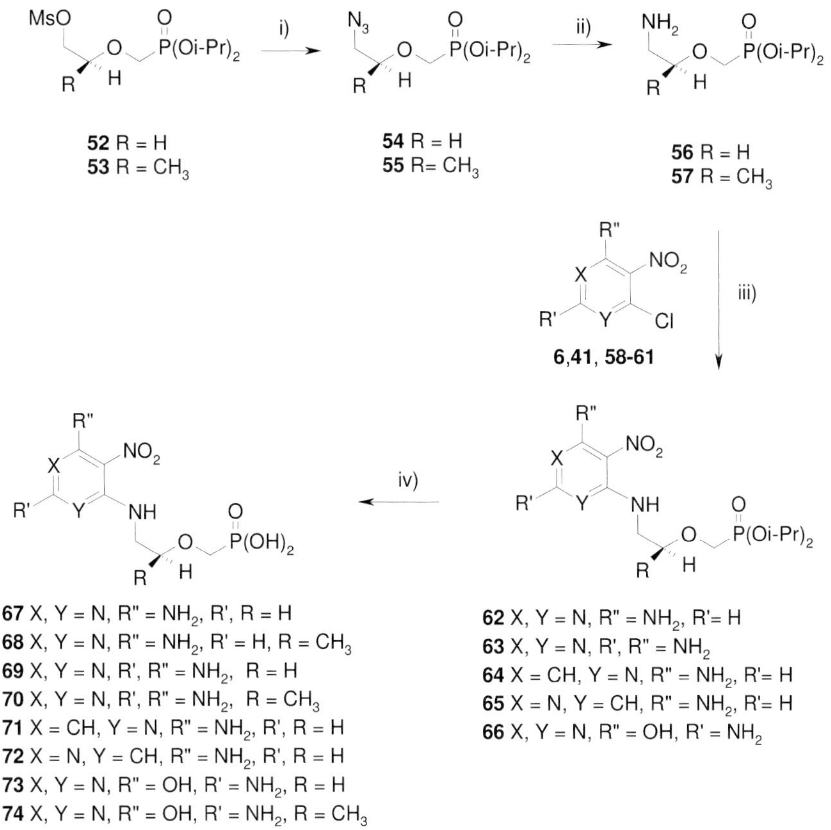

Scheme 22.9 Reagents and conditions: (i) NaN_3; (ii) H_2, Pd/C; (iii) **6**, **41**, **58–61**, Et_3N; (iv) TMSBr.

also prepared analogues of PMEG (9-[2-(phosphonomethoxy)ethyl]guanine) and (*R*)-PMPG (9-[2-(phosphonomethoxy)propyl]guanine) in which the guanine moiety was replaced by 2,6-diamino-5-nitro-4(3H)-pyrimidinone (**73** and **74**).

The protection of cell cultures against HIV-1- and HIV-2-induced cytopathogenicity was observed with clitocine analogues **67** to **70**, albeit with slightly lower selectivity and activity than PMEA and PMEDAP (Table 22.2). Analogous with the increased activity of PMEDAP compared to PMEA [18], **69** and **70** – the analogues of PMEDAP and (*R*)-PMPDAP, respectively – exhibited a four-fold higher activity against HIV-1 than **67** and **68** (analogues of PMEA and (*R*)-PMPA, respectively). The introduction of a CH group instead of an N at position 1 or 3 (as in **71** and **72)** led to the loss of activity. The observed biological data strongly supported the supposition that monocyclic 4,6-diaminopyrimidine and 2,4,6-triamino-5-nitropyrimidine function as bioisosteres of bicyclic adenine and 2-amino-adenine, respectively. In contrast, **73** and **74** (analogues of PMEG and (*R*)-PMPG, respectively) exhibited no

Table 22.2 Comparative cytotoxicity and anti-HIV activity of phosphonates **67–70**.

Compound	CC_{50} [a)]	HIV-1		HIV-2	
		EC_{50} [b)]	SI [c)]	EC_{50} [b)]	SI [c)]
67	>342	118	>2.9	120	>2.8
68	>326	111	>2.9	115	>2.8
69	>325	28	>11.6	63	>5.2
70	>310	26	>12.5	60	>5.2
PMEA	229	5.3	43	26	8.8
PMEDAP	48	2.6	18.4	ND	–
PMEG	2.4	0.19	12.6	1.1	2.2
(*R*)-PMPG	>333	4.5	>74.0	5.0	>67

a) Compound dose (μM) required to reduce the viability of mock-infected MT-4 cells by 50%.
b) Compound dose (μM) required to achieve 50% protection of MT-4 and C8166 cells against the cytopathic effect of HIV-1 and HIV-2, respectively.
c) Selective index: CC_{50}/EC_{50} ratio.

activity, presumably due to either poor cell penetration or a low affinity towards phosphorylating enzymes.

A combination of the aglycone of clitocine and a 4-hydroxybutylamino or 3-hydroxypropoxyamino group led to carboacyclic and 1′-oxaacyclic clitocine derivatives, respectively (Scheme 22.10). The former derivatives were prepared by the condensation of 4-aminobutanol (**75**) with 4,6-dichloro-5-nitropyrimidine (**3**) in the presence of triethylamine, with subsequent substitution of the 6-chloro group by cyclopropylamino, 2′-ethylimidazolyl, or amino groups to produce **77**, **78**, and **79**, respectively [19]. The reaction of the hydroxyl group of the intermediate (**76**) with pivaloyl chloride in the presence of triethylamine and *N,N*-dimethylaminopyridine (DMAP) yielded the *O*-pivaloyl derivative, which was further transformed with amine building blocks to produce pivaloate derivatives **80**, **81**, and **82**. In a similar manner, 1′- derivatives were obtained via the condensation of 3-[(*tert*-butyldimethylsilyl)oxy]propoxyamine (**83**) with **3** in the presence of *N*-methylmorpholine, followed by transformation of the 6-chloro group to the amino or cyclopropylamino moiety and subsequent deprotection of the silyl group to give **85** and **86**, respectively.

Compounds **77**, **79**, and **80** exhibited anti-Epstein–Barr virus (EBV) activity (EC_{50} 0.86 $\mu g\,mL^{-1}$, 6.6 $\mu g\,mL^{-1}$, and 6.4 $\mu g\,mL^{-1}$, respectively), whereas the other 1′-derivatives did not exhibit any activity towards HCMV, varicella zoster virus (VZV), HSV-1, HSV-2, EBV, and HIV-1.

22.2.4
5′-Amino Analogues

Lee and collaborators reported the synthesis and screening of several 5′-deoxyclitocine analogues (Scheme 22.11) [20]. The 5′-deoxy-5′-amino analogue of clitocine

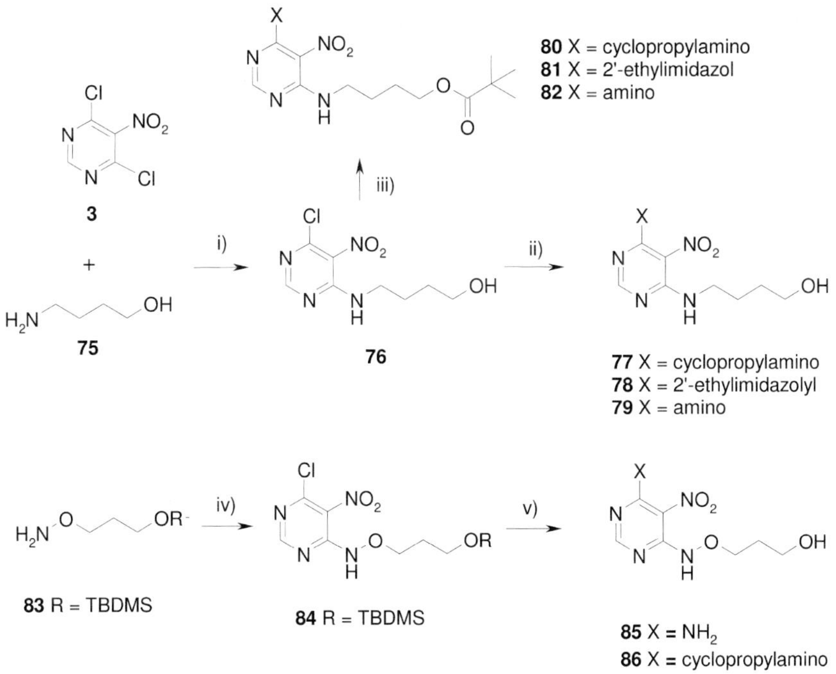

Scheme 22.10 Reagents and conditions: (i) Et_3N, CH_2Cl_2, room temp. 4 h; (ii) XH, CH_2Cl_2; (iii) PivCl, DMAP, pyridine, then XH, CH_2Cl_2; (iv) NMM, **3**, CH_2Cl_2; (v) XH, in CH_2Cl_2 or MeOH followed by EtOH/1% HCl.

(**94**) was prepared by functional group modification of the 5′-hydroxy moiety. Bromination of the 5′-hydroxy group of **87** by treatment with triphenylphosphine bromine and subsequent displacement of the 5′-bromo group with potassium phthalimide gave the key intermediate (**92**). Sequential deprotections of the phthalimide by hydrazine and the acetonide group by trifluoroacetic acid yielded **94**. On the assumption that the H-bond between the 5-nitro group and the 6-NH proton of clitocine is crucial, a series of analogues possessing a 5-electron-withdrawing group that could form an H-bond was introduced to provide clitocine analogues such as **96** to **100**, starting from intermediates **87** to **91** and its 5′-dexoy-5′-amino-5-methoxycarbonyl derivative **95** from **93**. Analogues unsubstituted at the 4-position, such as **102** and **103**, were prepared from a common intermediate (**88**).

Starting from D-ribono-1,4-lactone (**104**), 5′-deoxy-clitocine as well as 5′-deoxy-5-carboxylate analogues were prepared as shown in Scheme 22.12. The reaction of **104** with triphenylphosphine and *N*-iodosuccinimide (NIS), followed by reduction by diisobutylaluminum hydride (DIBAL), produced 5-deoxylactol, which was converted to **105** via the formation of a 2-azido intermediate by treatment with aluminum chloride and sodium azide, and its subsequent reduction under hydrogenation conditions. The coupling of **105** with **6** or 4,6-dichloro-5-methoxycarbonylpyrimidine

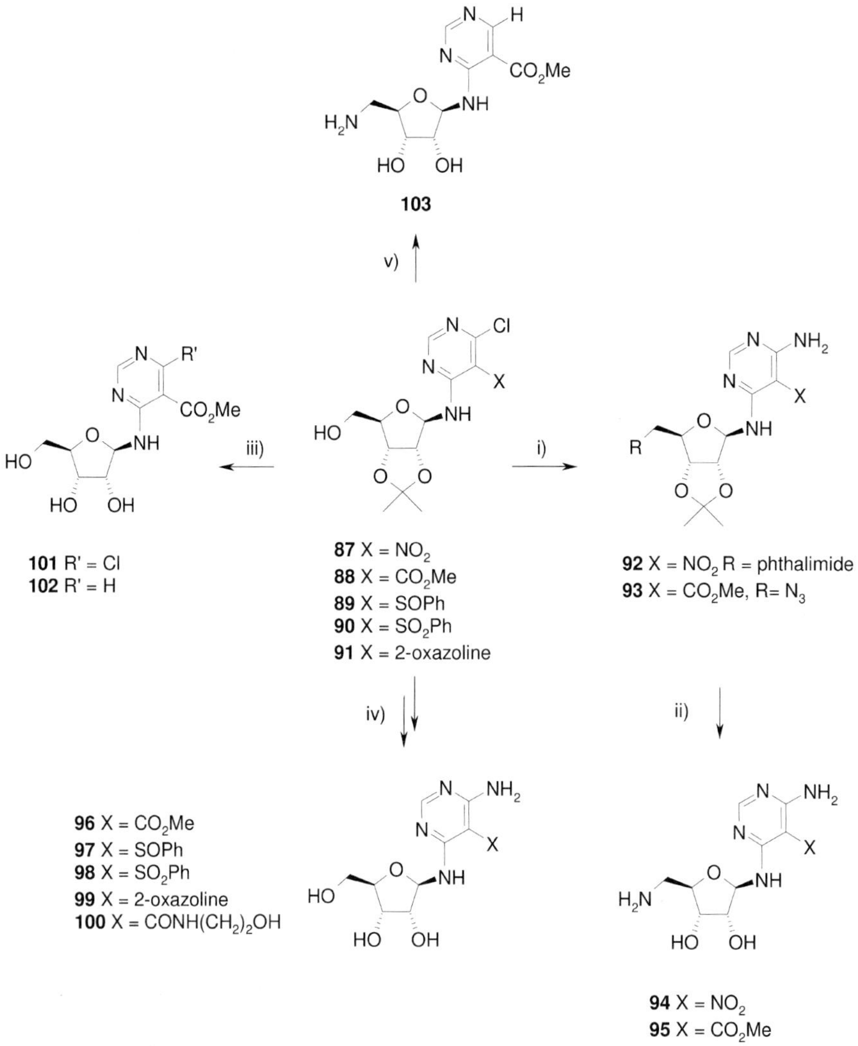

Scheme 22.11 Reagents and conditions: (i) For **92** from **87**: (a) NH_3, MeOH, (b) PPh_3P/Br_2, (c) potassium phthalimide. For **93** from **88**: (a) NH_3, MeOH, (b) $(PhO)_2PON_3$. (ii) For **94** from **92**: N_2H_4 followed by CF_3CO_2H-H_2O; for **95** from **93**: PPh_3/pyridine followed by CF_3CO_2H-H_2O. (iii) For **101** from **88**: CF_3CO_2H-H_2O, for **102** from **88**: (a) H_2, Pd/C, (b) CF_3CO_2H-H_2O. (iv) (a) NH_3, MeOH, (b) CF_3CO_2H-H_2O. (v) For **103** from **88**: (a) H_2, Pd/C, (b) $(PhO)_2PON_3$, (c) PPh_3/pyridine, (d) CF_3CO_2H-H_2O.

yielded **106** or **107**, respectively. Deprotection of the acetonide group of **106** gave **108**; the introduction of an amino group to **107**, followed by deprotection, gave **109**.

The *in-vitro* biological activity of the 5′-deoxy derivatives and the 5-electron-withdrawing group-modified analogues are summarized in Table 22.3. By replacing the

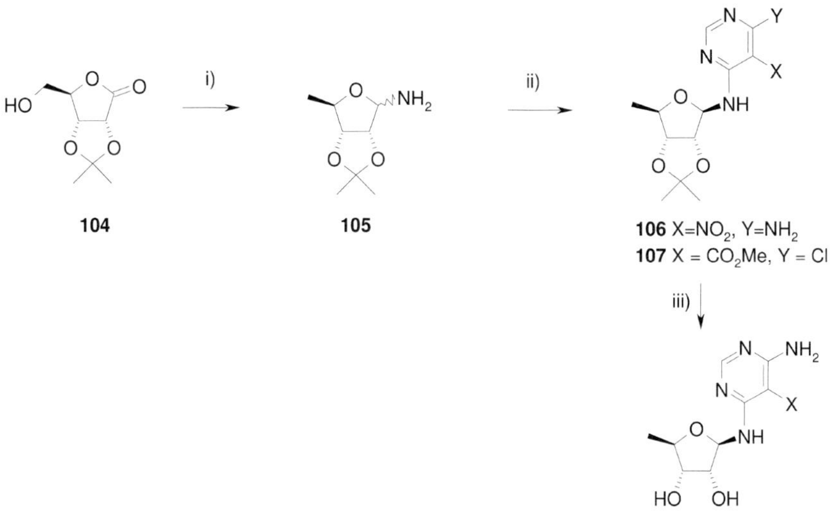

104 **105** **106** X=NO_2, Y=NH_2 **107** X = CO_2Me, Y = Cl **108** X=NO_2 **109** X= CO_2Me

Scheme 22.12 Reagents and conditions: (i) (a) PPh_3/NIS, (b) DIBAL, (c) $AlCl_3$, (d) NaN_3, (e) PtO_2/H_2, 44% overall yield; (ii) Et_3N, DMF, **6** for **106**, 4,6-dichloro-5-methoxycarbonylpyrimidine for **107**; (iii) CF_3CO_2H-H_2O for **108**, NH_3-MeOH followed by CF_3CO_2H-H_2O for **109**.

Table 22.3 *In-vitro* biological activity of adenosine kinase inhibitors.

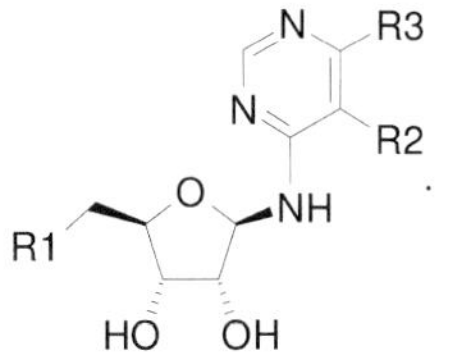

Compound	R^1	R^2	R^3	Adenosine kinase inhibition IC_{50} (nM)[a)]
Clitocine	OH	NO_2	NH_2	2000 (±1000)
94	NH_2	NO_2	NH_2	35 (±12)
95	NH_2	CO_2CH_3	NH_2	2 (±1)
96	OH	CO_2CH_3	NH_2	1000 (±857)
97	OH	SOPh	NH_2	>10 000
98	OH	SO_2Ph	NH_2	>10 000
99	OH	2-Oxazoline	NH_2	4000 (±3420)
100	OH	$CONHC_2H_4OH$	NH_2	>10 000
101	OH	CO_2CH_3	Cl	>10 000
102	OH	CO_2CH_3	H	600 (±323)
103	NH_2	CO_2CH_3	H	4 (±1)

a) All values are the mean ± SEM of at least three separate observations run in triplicate.

5′-hydroxy substituent with an amino group, Lee and coworkers gained a 50-fold improvement in the enzyme inhibitory activity ($IC_{50} = 35$ nM for **94** versus 2 μM for clitocine). The same trend was observed with the 5-methyl carboxylate analog **96** ($IC_{50} = 1$ μM): changing its 5′-hydroxyl group to an amino group (as in **95**) again led to a significant improvement in enzyme inhibition ($IC_{50} = 2$ nM). Interestingly, the analogue with no substituent at the 4-position (**103**) exhibited equally high activity ($IC_{50} = 4$ nM). A similar correlation was observed between **102** ($IC_{50} = 0.6$ μM) and **96** ($IC_{50} = 1$ μM). The introduction of phenylsulfoxide or phenylsulfone (as in **97** and **98**) instead of the 5-nitro group of clitocine led to the loss of activity. The oxazoline analogue (**99**) exhibited a potency which was comparable to that of clitocine, whereas the hydroxyethyl amide analog (**100**), obtained as a minor product, was found to be inactive.

By using solid-phase reaction technologies, various 5′-deoxy-5′-amino derivatives of nucleosides were prepared in a parallel manner (Scheme 22.13) [21]. After attaching the amino sugar (**110**) to polystyrene monomethoxytrityl chloride, reduction of the azido group provided **112**, which was condensed with 4,6-dichloro-5-nitropyrimidine to yield the key intermediate (**113**). Parallel derivatization at the

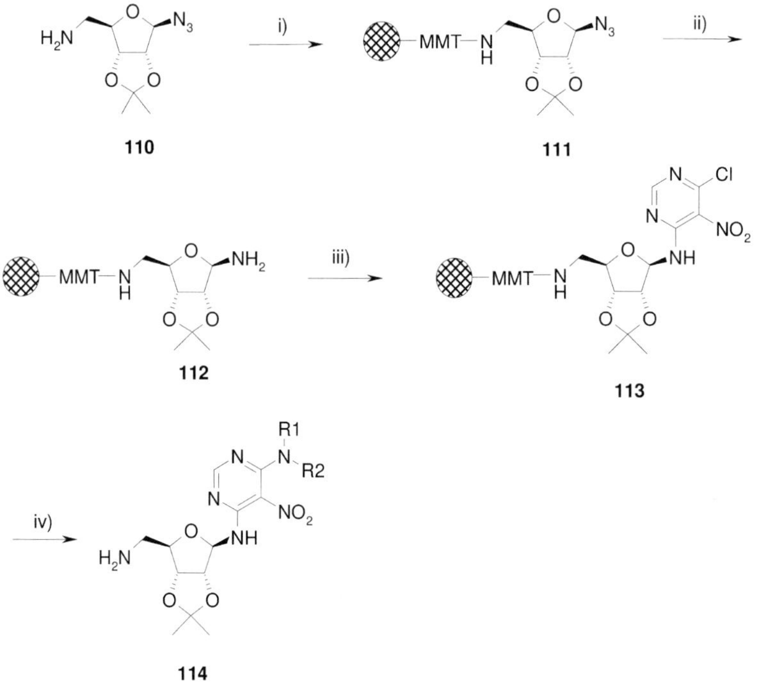

Scheme 22.13 Reagents and conditions: (i) polystyrene MMT-Cl resin, DMAP, pyridine; (ii) Me_3P, THF-H_2O; (iii) 4,6-dichloro-5-nitropyrimidine, diisopropylethylamine; (iv) (a) amine, DMAP, (b) 90% CF_3CO_2H-H_2O.

6-position by various primary and secondary amines and detachment from the resin provided 5′-deoxy-5′-amino-clitocine derivatives (**114**) with a purity of 60 to 93%.

Acknowledgments

The authors thank Drs. Joo-Yong Yoon, Hee Bong Lee, and Mr. Jayhyok Chang for their critical reading of this chapter.

References

1 Kubo, I., Kim, M., Wood, W. F., Naoki, H., *Tetrahedron Lett.* **1986**, *27*, 4277.

2 Larson, S. B., Moss, R. J., Kini, G. D., Robins, R. K., *Acta Crystallogr.* **1988**, *C44*. 1076.

3 Kamikawa, T., Fujie, S., Yamagiwa, Y., Kim, M., Kawaguchi, H., *J. Chem. Soc., Chem. Commun.* **1988**, 195.

4 Moss, R. J., Petrie, C. R., Meyer, R. B., Jr., Nord, L. D., Willis, R. C., Smith, R. A., Larson, S. B., Kini, G. D., Robins, R. K., *J. Med. Chem.* **1988**, *31*, 786.

5 Choi, H.-W., Choi, B. S., Chang, J. H., Lee, K. W., Nam, D. H., Kim, Y. K., Lee, J. H., Heo, T., Shin, H., Kim, N.-S., *Synlett* **2005**, 1942.

6 Bennett, L. L., Jr., Hill, D. L., Allan, P. W., *Biochem. Pharmacol.* **1978**, *27*, 83.

7 Fortin, H., Tomasi, S., Delcros, J.-G., Bansard, J.-Y., Boustie, J., *ChemMedChem* **2006**, *1*, 189.

8 (a) Wilde, R. G., Almstead, N. G., Welch, E. M., Beckmann, H., (PTC Therapeutics Inc.), WO A/2004009610, *Use of nucleoside compounds for nonsense suppression and T diseases;* **2004**. (b) Wilde, R. G., Kennedy, P. D., Almstead, N. G., Welch, E. M., Takasugi, J. J., Friesen, W. J., (PTC Therapeutics Inc.), WO A/ 2004009609, *Nucleosides compounds and their use for treating cancer and diseases associated with somatic mutations,* **2004**.

9 Espie, J. C., Lhomme, M. F., Morat, C., Lhomme, J., Tetrahedron Lett., **1990**, *31*, 1423.

10 Franchetti, P., Cappellacci, L., Cristalli, G., Grifantini, M., Vittori, S., *Nucleosides Nucleotides,* **1990**, *10*, 543.

11 Varaprasad, V. C., Habib, Q., Li, D. Y., Huang, J., Abt, J. W., Rong, F., Hong, Z., An, Z. H., *Tetrahedron* **2003**, *59*, 2297.

12 Palmer, C. F., Parry, K. P., Roberts, S. M., *Tetrahedron Lett.* **1990**, *31*, 279.

13 Cermak, R. C., Vince, R., *Tetrahedron Lett.* **1981**, *22*, 2331.

14 Baxter, A. D., Penn, C. R., Storer, R., Weir, N. G., Woods, J. M., *Nucleosides Nucleotides,* **1991**, *10*, 393.

15 (a) Marquez, V. E., Lim, M. -I., Tseng, C. K.- H., Markovac, A., Priest, M. A., Khan, M. S., Kaskar, B., *J. Org. Chem.* **1988**, *53*, 5709; (b) Yaginuma, S., Muto, N., Tsujino, M., Sudate, Y., Hayashi, M., Otani, M., *J. Antibiot.* **1981**, *34*, 359.

16 Marquez, V. E., Lim, B. B., Driscoll, J. S., Snoek, R., Balzarini, J., Ikeda, S., Andrei, G., De Clercq, E., *J. Heterocyclic Chem.* **1993**, *30*, 1393.

17 Franchetti, P., Cappellacci, L., Sheikha, G. A., Grifantini, M., Loi, A. G., De Montis, A., Spiga, M. G., La Colla, P., *Nucleosides Nucleotides,* **1995**, *14*, 607.

18 Holy, A., De Clercq, E., Votruba, I., ACS Symposium Series No. 40, *Nucleotide Analogues as Antiviral Agents,* American Chemical Society, USA, **1989**, pp. 51–71.

19 Bacchelli, C., Condom, R., Patino, N., Aubertin, A. -M., *Nucleosides Nucleotides* **2000**, *19*, 567.

20 Lee, C.-H., Daanen, J. F., Jiang, M., Yu, H., Kohlhaas, K. L., Alexander, K., Jarvis, M. F., Kowaluk, E. L., Bhagwat, S. S., *Bioorg. Med. Chem. Lett.* **2001**, *11*, 2419.

21 Varaprasad, V. C., Habib, Q., An, H., Hong, Z., *Nucleosides Nucleotides Nucleic Acids* **2006**, *25*, 61.

Part IV
Antitumorals and Antivirals

23
Capecitabine Preclinical Studies: From Discovery to Translational Research

Hideo Ishitsuka and Nobuo Shimma

23.1
Introduction

The most desirable formulation for anti-cancer drugs requiring frequent administration would be a tablet, similar to the treatment of many other diseases. Oral treatment would not only reduce medical treatment costs but also allow cancer patients to spend less time in hospital and more time with their family and friends. In fact, cancer patients prefer oral treatment if it is as effective as parenteral treatment [1]. In addition, a tablet is most appropriate for the long-term frequent treatment of cancer cells. Although many cytotoxic drugs effectively inhibit tumor cells only in particular processes of the cell cycle, such as the S phase and M phase, not all tumor cells are present in the tissues during such processes. In order to be effective, cytotoxic drugs should be administered continuously over a period of time which is longer than one cell cycle. However, the daily administration of cytotoxic drugs is limited because of the accumulation of non-tumor-selective toxicity. When given orally, cytotoxic drugs generally lead to high concentrations of the drug, predominantly at the local sites, such as the intestine and liver, and this results in treatment-associated adverse effects on these organs. On the other hand, oral doses of cytotoxic drugs are generally larger than parenteral doses, because drugs given orally are often subjected to first-pass metabolic effects in the intestinal tract and liver. Unfortunately, a reduction in dose to spare such local organ toxicity results in an insufficient efficacy compared with that of parenteral administration.

In 1986, the present authors commenced studies into the design and synthesis of orally available cytotoxic compounds with high tumor selective activity and low myelotoxicity or intestinal toxicity, and which therefore would be safe for daily oral treatment over prolonged periods of time in the comfort of the home. The strategy employed, specifically identifying new cytotoxic compounds with high tumor-selective action, included the design and synthesis of a prodrug that, from oral administration, generates an active cytotoxic component from enzymes

Modified Nucleosides: in Biochemistry, Biotechnology and Medicine. Edited by Piet Herdewijn

ISBN: 978-3-527-31820-9

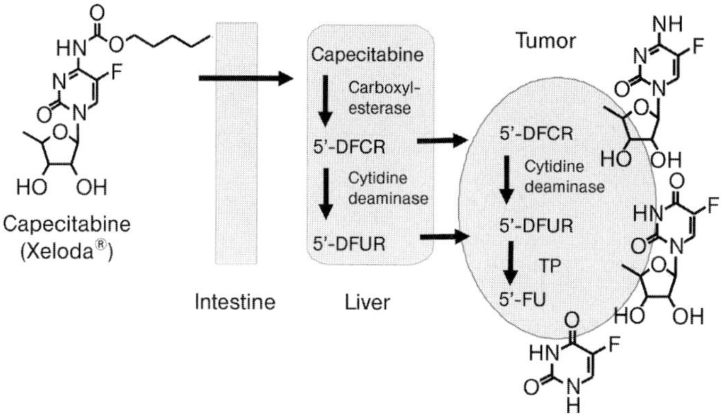

Figure 23.1 The metabolic pathway of capecitabine. Capecitabine, when administered orally, yields 5-FU preferentially within tumors via a cascade of three enzymes.

located preferentially in tumors. Previously, a prodrug – 5′-deoxy-5-fluorouridine (5′-DFUR; doxifluridine; Furtulon®) – had been developed which is metabolized to the active drug 5-fluorouracil (5-FU) by an enzyme which is located preferentially in tumors [2]. The prodrug had demonstrated better efficacy than 5-FU in many experimental tumor models [2, 3] and was approved in Japan, China, and Korea for the treatment of breast, colorectal, and gastric cancers. However, the efficacy of 5′-DFUR is not strictly tumor-selective, and it may cause dose-limiting intestinal toxicity [4]. Therefore, an attempt was made rationally to synthesize a novel fluoropyrimidine which would be characterized by improved efficacy and safety profiles compared to those of 5′-DFUR and 5-FU. From these studies, the compound capecitabine (N^4-pentyloxycarbonyl-5′-deoxy-5-fluoro-cytidine; Xeloda®; Figure 23.1) [5, 6] was identified, which showed improved pharmacodynamics, efficacy, and safety profiles compared with 5-FU and 5′-DFUR [5, 6, 8, 9]. Preclinical studies were also conducted to optimize capecitabine efficacy in both combination and personalized therapies [10–18]. In this chapter the discovery of capecitabine is described, and its superiority to 5-FU and 5′-DFUR is demonstrated, with details of subsequent clinical studies. Today, capecitabine, which has proved to be provide clinical benefits [19–23], is well accepted for the treatment of breast and colorectal cancers.

23.2
Drug Design and Discovery of Capecitabine

23.2.1
5′-DFUR as a Lead Compound of Capecitabine

5′-DFUR was selected as the lead compound of the precursor that generates 5-FU selectively in tumors. The compound was originally synthesized by Cook *et al.* [24], and subsequently developed by the present authors in Japan. It was found that 5′-DFUR generates 5-FU by the action of thymidine phosphorylase (TP) or uridine phosphorylase, which are located preferentially in the tumor tissues of cancer patients and in many mouse tumor models, respectively [2, 25, 26]. 5′-DFUR was also found to be more effective and safer than 5-FU in some tumor models [2, 3]. However, it was also found to be converted to 5-FU by enzymes within normal tissues, thus causing dose-limiting toxicity such as diarrhea when passing through the intestine at high concentrations [4, 27]. Orally administered 5′-DFUR also causes myelotoxicity when given at doses higher than the approved dose ranges. Hence, the initial investigations focused on the design and synthesis of novel 5′-DFUR prodrugs that would pass intact through the intestine, be first metabolized to 5′-DFUR by enzymes located in the liver, and ultimately converted to 5-FU by TP in the tumors. The expectation was that such prodrugs could be given at higher doses with improved efficacy over 5-FU and 5′-DFUR

23.2.2
5′-DFCR Derivatives

In order to minimize the myelotoxicity and increase the tumor selective activity of 5′-DFUR, 5′-deoxy-5-fluorocytidine (5′-DFCR) [24] was selected as the lead from among the many 5′-DFUR derivatives synthesized. 5′-DFCR is metabolized to 5′-DFUR by cytidine (Cyd) deaminase, the enzyme responsible for the metabolism of cytosine arabinoside (Ara C) – a cytotoxic drug used to treat leukemia – to the inactive molecule, uracil arabinoside. The enzyme is highly expressed in the liver, kidney [28], and solid tumors [29] of humans, whereas it is minimally expressed in immature, growing marrow cells and leukemic myeloblasts, compared to that in mature, normal granulocytes [30]. This unique localization of the enzyme explains the clinical efficacy of Ara C for leukemia and its dose-limiting adverse effect towards bone marrow. The unique tissue enzyme localization [5] and low expression in the granulocyte progenitor cells of both human bone marrow and umbilical cord blood was confirmed by the present authors (unpublished results). 5′-DFCR was thus selected as a potential lead compound based on the rationale that the specific tissue distribution of Cyd deaminase would result in the generation of 5′-DFUR at higher concentrations in liver and tumors, but not in growing bone marrow cells. Capecitabine, a derivative of 5′-DFUR, demonstrated minimal myelotoxicity in clinical studies [31].

23.2.3
N^4-Acyl-5′-DFCR Derivatives

Because certain levels of Cyd deaminase and TP activities have been found in the human intestinal tract [5], it was hypothesized that 5-FU could be generated from 5′-DFCR to some extent within the intestinal mucosa and thus cause gastrointestinal toxicity (particularly diarrhea) when given orally. In addition, the oral absorbability of 5′-DFCR was poor in mice, possibly as a result of its hydrophilic nature. Thus, a series of N^4-acyl-5′-DFCR derivatives was synthesized that would pass through the intestine as intact molecules and then be converted to 5′-DFCR by hepatic acylamidase [4]. From the derivatives synthesized, N^4-trimethoxybenzoyl-5′-DFCR (galocitabine) was selected because it generated levels of 5-FU which were equal to those by 5′-DFUR in tumors, but much lower in the intestine [4]. Although the therapeutic indices of galocitabine were several-fold greater than those of 5′-DFUR in mouse tumor models [4], in the human liver extract galocitabine was not efficiently metabolized to 5′-DFCR because of a much lower susceptibility to acylamidase compared with mouse liver (3.7 versus 280 nmol mg^{-1} protein h^{-1}) (unpublished results). In Japan, therefore, the development of galocitabine was terminated at clinical Phase 2.

23.2.4 N^4-Alkoxycarbonyl-5′-DFCR Derivatives

Subsequently, a large series of N^4-substituted 5′-DFCR derivatives was synthesized which are converted to 5′-DFCR by the action of human and monkey hepatic enzymes, but not by intestinal enzymes [6]. In parallel, hepatic enzymes that catalyze the metabolic conversion to 5′-DFCR were investigated for rational drug design [5]. These studies identified one particular series of 5′-DFCR derivatives, N^4-alkoxycarbonyl-5′-DFCR, which convert to 5′-DFCR by the action of an isoenzyme of the 60-kDa carboxylesterase family, which exists preferentially in the liver but not in the intestine in humans [5, 6]. Among the hundred or so N^4-alkoxycarbonyl-5′-DFCR derivatives synthesized, those that were chemically stable at acidic pH, susceptible to human hepatic carboxylesterase, effective in human cancer xenografts in mice, and orally available in monkeys, were screened [6]. Capecitabine was moderately susceptible to carboxylesterase (20-fold more so than galocitabine), yielded the largest area-under-concentration curve (AUC) for 5′-DFUR in the plasma (4.2-fold higher than that of galocitabine), and was efficiently metabolized to 5-FU and its further metabolites by 66% when given orally to monkeys [6]. Capecitabine was also highly effective in human cancer xenograft models, and demonstrated much less intestinal toxicity and myelotoxicity than 5′-DFUR in monkeys [5, 6]. Finally, the improved pharmacokinetic profile of capecitabine was confirmed in an exploratory human pharmacokinetic study before starting Phase 1 studies in 1994 simultaneously in the United States, European Union, and Japan.

23.3 Preclinical Studies

23.3.1 Tumor-Selective Delivery of the Active 5-FU

Capecitabine selectively generates 5-FU in tumors, as rationally intended. Therefore, it can be safely administered at higher doses, which lead to higher 5-FU concentrations in tumors than is possible with either 5-FU or 5′-DFUR [7]. When 5-FU was

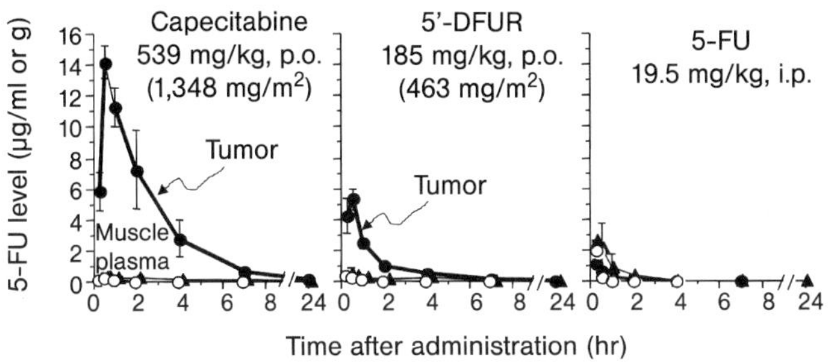

Figure 23.2 Tumor-selective conversion of capecitabine to 5-fluorouracil (5-FU) in a human colon cancer xenograft model. Capecitabine (1.5 mmol kg^{-1}, p.o.) and 5-FU (0.15 mmol kg^{-1}, i.p.) were given at their maximal tolerated dose (MTD) in multiple treatments to mice bearing the HCT116 human cancer xenograft. Data points represent the mean ± SD of 5-FU concentrations in plasma (□), tumor (●), and muscle (▲) (n = 3).

given (intraperitoneally; i.p.) at the maximum tolerated dose (in long-term treatment) to mice bearing the HCT116 human colon cancer xenograft, it yielded generally uniform concentrations of 5-FU in plasma, muscle, and tumors (Figure 23.2). In contrast, after oral administration of capecitabine at equitoxic doses, the concentration of 5-FU in tumor tissue was considerably higher relative to that in plasma or muscle: the intratumor AUCs for 5-FU were 114- and 209-fold greater than the plasma and muscle AUCs, respectively [7]. The administration of capecitabine (per os; p.o.) also resulted in a 5.5- to 36-fold and 2.8- to 4.3-fold higher AUC for 5-FU within tumors compared to 5-FU (i.p.) and 5′-DFUR (p.o.), respectively, in four human cancer colon xenograft models studied (HCT116, COLO205, CXF280, and WiDr). This 5-FU generation in tumors at higher concentrations clearly explains why capecitabine was more effective than 5-FU in these and other tumor models, despite the fact that the active principle, 5-FU, is the same. The high 5-FU level in human tumor tissues after capecitabine administration was predicted in a physiologically based pharmacokinetics analysis, in which 5-FU levels were simulated based on plasma pharmacokinetic profiles and the activities of four enzymes involved in the activation of capecitabine to 5-FU and its elimination in the gastrointestinal tract, liver, and tumor tissues [15]. The tumor-selective 5-FU delivery was later demonstrated in a pharmacodynamic study in patients with colorectal cancer [32].

The pharmacokinetic studies of capecitabine in patients demonstrated rapid gastrointestinal absorption, followed by extensive conversion into 5′-DFCR and 5′-DFUR, with low systemic 5-FU levels, as expected [33]. Among capecitabine and its metabolites, 5-FU showed the greatest level of cytotoxicity in cultures of human cancer cell lines, followed by 5′-DFUR, while capecitabine and 5′-DFCR showed only slight cytotoxic activity with median IC_{50}-values of 3.7, 67, >1000, and 1000 μM, respectively (cell lines, $n = 10$) [5]. The cytotoxic activity of 5′-DFCR and 5′-DFUR was much reduced by the inhibition of Cyd deaminase and TP [5]. These results supported the concept that capecitabine becomes effective only after its conversion to 5-FU by the enzymes involved in its activation. By selectively generating high concentrations of 5-FU within tumor tissue through non-cytotoxic intermediates, capecitabine was expected to be more effective than 5-FU within the clinical setting.

23.3.2 Anti-Tumor Activities

Capecitabine was shown to be more effective and to have a wider spectrum of anti-tumor activity than 5-FU, 5′-DFUR, or UFT (a 1:4 combination of tegafur and uracil) against 24 randomly selected human cancer xenograft models, which included colon, breast, gastric, cervical, bladder, ovarian, and prostate cancers [5, 8]. In these experiments, capecitabine administered orally at the maximal tolerated dose (MTD) was effective (defined as >50% growth inhibition) in 18 of 24 models (75%), and inhibited tumor growth by more than 90% in seven models [8]. In contrast, 5′-DFUR was effective in 15 models (63%) and inhibited tumor growth by >90% in only one model. 5-FU and UFTwere each effective in one model (4.1%) and five (21%) models, respectively. Neither of these compounds inhibited the growth of any of the tumor models tested by more than 90%. Thus, capecitabine showed activity against tumors that are resistant to 5-FU and UFT *in vivo*.

Capecitabine also showed anti-tumor activity in dose ranges much broader than those of 5-FU, 5′-DFUR, and UFT in 12 human cancer xenograft models, including HCT116 and CXF280 [9], which express respectively the intermediate and highest levels of susceptibility to the fluoropyrimidines among the tumor models studied. The therapeutic index (defined as the ratio of the lowest toxic dose to ED_{50}, the minimum dose inhibiting tumor growth by 50%) in the CXF280 model was 94 for capecitabine, versus 2.7 for 5-FU measured in the only model susceptible to 5-FU [9]. The safety margin of capecitabine was also confirmed from the ratio of the dose causing intestinal toxicity to the dose showing the minimum efficacy (ED_{50}) [5, 9]. In the HCT116 model, this ratio for capecitabine was 6.4, compared to 1.5 for 5′-DFUR. Capecitabine was the only fluoropyrimidine demonstrating efficacy at doses associated with no intestinal toxicity [5].

23.3.3 Dose Fractionation and Schedule

The optimal capecitabine dosing regimen was then investigated from a pharmacodynamic viewpoint. 5-FU, the principal active product of capecitabine, acts upon tumor cells after metabolic conversion to FUTP and FdUMP, the latter of which inactivates thymidylate synthase (TS) resulting in DNA synthesis inhibition [34]. Capecitabine generated 5-FU and FdUMP in tumors shortly after its administration to mice bearing HCT116 human colon cancer xenograft. 5-FU and FdUMP levels in the tumor peaked at 1 h after administration, and then decreased quickly within a few hours (unpublished data). TS inactivation also occurred at 2 h after capecitabine administration; however, the concentration of free TS (i.e., TS not bound to FdUMP) subsequently recovered very slowly and remained at less than 50% of the control at 24 h after capecitabine administration. Therefore, consecutive daily administration of capecitabine, employing commonly used dosing schedule, would be expected to result in a similar level of cumulative TS inactivation and consequently a similar level of efficacy. In fact, the degree of cumulative TS inhibition and tumor growth inhibition by capecitabine given once- and twice-daily for 10 days was similar when the total dose amount was the same [14]. Capecitabine was also shown to be equally effective independent of its dosing schedule in prolonged administration in a human colon cancer xenograft model. It was equally effective when given uninterrupted daily for 5 days per week, or when given for 14 consecutive days every three weeks (2 weeks on/1 week off) during the 12-week treatment period (unpublished data). These results provided the rationale for determining the dose fractionation and schedule for capecitabine in clinical studies of twice-daily for 14 consecutive days every three weeks.

23.3.4 Safety (Dose Range and Mild Myelotoxicity)

The efficacy and safety margin of capecitabine was greater compared with that of 5-FU and 5′-DFUR in human cancer xenograft models, as described above. Capecitabine was the only fluoropyrimidine to demonstrate efficacy at doses associated with little or marginal myelosuppression or intestinal toxicity [5, 9]. Shindoh *et al.*, at the present authors' institute, also demonstrated a high safety margin for capecitabine in cynomolgus monkeys [35]. Capecitabine given orally daily for four weeks showed no toxicity at 0.1 mmol kg^{-1} day^{-1}, and slight toxicity in hematopoietic and intestinal organs at 0.5 mmol kg^{-1} day^{-1}, with 5′-DFUR AUCs of 5.8 and 26.1 μg· h mL^{-1}, respectively. On the other hand, 5′-DFUR given at 0.25 mmol kg^{-1} day^{-1} showed slight to moderate toxicity, with a 5′-DFUR AUC of only 3.2 μg· h mL^{-1}. Thus, a high safety margin was suggested from the results of their studies of capecitabine in monkeys.

23.4 Translational Research for Optimizing Capecitabine Efficacy

23.4.1 Factors that Influence Capecitabine Efficacy

Two main factors affect the efficacy of capecitabine. The first factor is TP, which generates 5-FU from 5′-DFUR in tumors; and the second factor is dihydropyrimidine dehydrogenase (DPD), which catabolizes 5-FU to inactive molecules. Hence, a series of pharmacodynamic studies were performed in mouse xenografts bearing human colon cancer lines that were either susceptible (HCT116) or refractory (WiDr) to capecitabine, in order to demonstrate the importance of TP in determining the susceptibility of tumor to this drug [7]. When capecitabine was given orally, intratumoral concentrations of 5-FU in the refractory WiDr line were lower than those in the susceptible HCT116 line (4.8-fold lower AUC). In contrast, 5′-DFUR concentrations were higher (4.9-fold higher AUC). The molar ratio of the 5′-DFUR to 5-FU AUC in the WiDr tumors was 11, as opposed to 0.47 in the susceptible xenograft (HCT116). These data suggested that 5′-DFUR was not efficiently converted to 5-FU by TP in WiDr tumors, which could be the basis of drug resistance to capecitabine. It was also demonstrated that the intratumoral TP level or TP/DPD ratio was a factor which determines susceptibility to capecitabine treatment [9]. In the studies with 24 human cancer xenograft models, tumor susceptibility to capecitabine therapy correlated directly with TP activity ($p = 0.0164$), and inversely with DPD activity to a minor extent ($p = 0.125$). The TP/DPD ratio correlated best with the susceptibility of tumor cell lines to capecitabine ($p = 0.0015$) (Figure 23.3A). No similar correlation could be established between the efficacy of UFT (which is converted to 5-FU mainly in the liver) and the tumor levels of either enzyme. On the other hand, no significant correlation was found between susceptibility and TS levels in this study with the 24 tumor models (unpublished data). The correlation established for capecitabine to levels of the two enzymes in tumors supports the concept that its antitumor effects are mediated through 5-FU generated within tumors, as opposed to 5-FU generated in the peripheral tissues. These results provided the insight for optimizing the efficacy of capecitabine.

23.4.2 Combination Therapy with Rational Partners

23.4.2.1 Combination with TP Up-Regulators

One approach to optimize capecitabine therapy is either to up-regulate TP or to down-regulate DPD activity in tumor tissues. Previously, it was found that cell lines transfected with the human TP gene became more susceptible to 5′-DFUR, the intermediate metabolite of capecitabine, and that inflammatory cytokines (such as TNFα, IFNγ, IL-1α) up-regulate TP gene expression in human cancer cells [36]. Since paclitaxel was known to up-regulate TNFα, a potential TP up-regulator, in cultures of macrophages [37], various cytotoxic drugs were examined (including paclitaxel) to determine if they affected TP levels, thus enhancing the efficacy of capecitabine in mice bearing human cancer xenografts [10]. Paclitaxel and docetaxel indeed up-regulated TP, and consequently improved the efficacy of capecitabine [10], as did cyclophosphamide [11]. It was also observed that X-ray irradiation up-regulated TP in tumor tissues but not in the liver, and thereby X-ray in combination with capecitabine showed additive to the synergistic activity *in vivo* [12]. These cytotoxic drugs and X-ray irradiation simultaneously increased the level of human TNFα in tumor, which could in turn up-regulate TP in the tumor cells. The up-regulation of TP by taxanes and radiation has since been confirmed by other investigators [38, 39]. In clinical trials, capecitabine combined with either taxanes [22] or with radiation treatment [40] has shown promising outcomes.

23.4.2.2 Combination with DPD Down-Regulators

DPD levels in tumors influence the efficacy of capecitabine, which generates 5-FU in tumor tissues. The role of DPD was confirmed in reversed susceptibility to

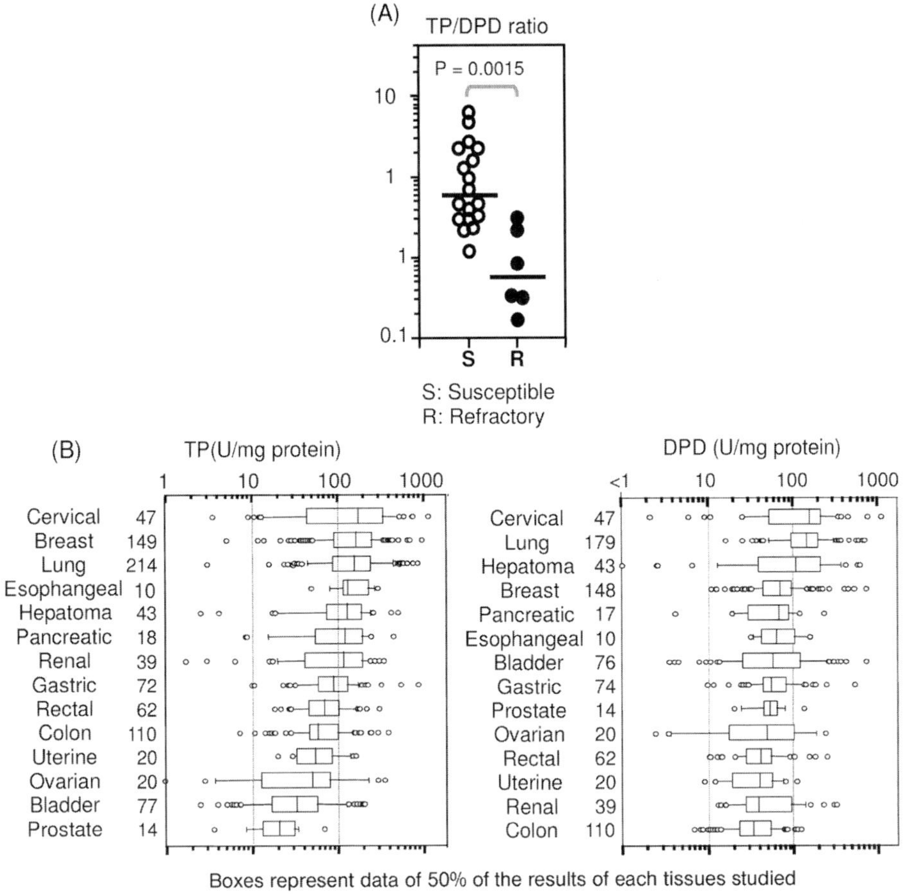

Figure 23.3 Thymidine phosphorylase (TP)/dihydropyrimidine dehydrogenase (DPD) ratios in human cancer xenografts, and TP and DPD activities in human cancer tissues. (A) *In-vivo* susceptibility to capecitabine correlates with the intratumoral ratio of TP to DPD activity (human cancer xenograft models, n = 24). The activity of the enzymes affecting 5-FU concentrations following capecitabine administration, TP and DPD, was measured. The correlation between enzyme levels and *in-vivo* susceptibility to capecitabine was analyzed, in which the susceptibility was defined as defined as >50% growth inhibition by capecitabine *in vivo*. (B) TP and DPD levels in various human tumor tissues. TP and DPD levels in tumor tissues were determined using ELISA. The boxes in the figures indicate the range where one half of the data is included (between upper and lower quartile). The line in a box indicates the median value; bars at both sides of a box indicate the range of lower 10% (right bar) and upper 90% (left bar); open circles indicate values beyond these ranges.

capecitabine in a human colon cancer xenograft transfected with human DPD genes. The efficacy of capecitabine was reduced in such cancer xenografts compared with non-transfected counterpart xenografts [17]. In humans, DPD activity exists mainly in the liver, even though high levels are also found in many types of human cancer [13]. Therefore, when combined with capecitabine, only tumor-selective

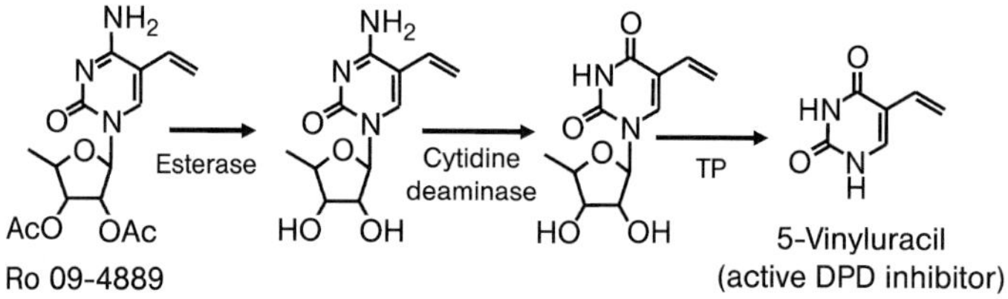

Figure 23.4 The metabolic pathway of Ro 09-4889. When administered orally, Ro 09-4889 yields 5-vinyluracil, a DPD inhibitor, preferentially within tumors via a cascade of three enzymes [18].

and/or preferential DPD inhibitors should enhance capecitabine efficacy, whereas non-tumor-selective DPD inhibitors should enhance levels of 5-FU in the liver and consequently distribute it into the whole body, resulting in increased toxicity. Based on these assumptions, the orally available prodrug RO-09-4889 (2′,3′-O-diacetyl-5′-deoxy-5-vinylcytidine) was designed and synthesized. This prodrug generates a known DPD inhibitor, 5-vinyluracil (5-VU), preferentially in tumor tissues, similar to capecitabine generating 5-FU from enzymes including cytidine deaminase and TP (Figure 23.4) [17, 18]. Capecitabine in combination with RO-09-4889 gave higher levels of 5-FU in tumors, and was thereby more effective in those human cancer xenografts with at least some level of DPD than capecitabine alone. Rational target diseases for the combination would therefore be cancer types with high DPD activity, such as non-small cell lung cancer (NSCLC), hepatoma, pancreatic cancer, and individual cancer patients who have tumors with high DPD activity (see Figure 23.3A) [13].

23.4.3
Personalized Therapy of Rational Patient Populations

Another approach towards optimizing capecitabine efficacy is rationally to select the most appropriate patient populations. As described above, it has been shown that tumors with higher levels of TP or with a higher TP/DPD ratio were more susceptible to capecitabine treatment in human cancer xenografts models (see Figure 23.3A) [8]. Enzyme levels in tumors vary among human cancer types and individual cancer patients (Figure 23.3B) [13]. Cancer types with high TP levels or a high TP/DPD ratio, and individual patients who have tumors with such enzyme expression, would therefore be the rational target patient population for capecitabine therapy. Hence, monoclonal antibodies to human TP [41] and DPD [13] were prepared, and methods for measuring the levels of enzymes established using ELISA and immunohistochemical methods. With these antibodies, Nishina *et al.* showed that gastric cancer with a higher TP/DPD ratio was treated more effectively by 5′-DFUR combination therapy than was such cancer with a lower TP/DPD ratio (overall response of 64% versus 9.1%) [42]. Meropol *et al.* also showed that colorectal cancer with higher levels of TP had a survival benefit from treatment with capecitabine and irinotecan [43].

23.5 Conclusions

In these studies, capecitabine was rationally designed as an orally available cytotoxic drug such that it would generate 5-FU preferentially within human tumors through three sequential enzymatic reactions. The most critical issue experienced in the discovery of capecitabine was interspecies difference in the types, activities, and tissue distribution patterns of metabolic enzymes necessary for prodrug design. The first step of the study involved the development of galocitabine (N^4-trimethoxybenzoyl-5′-DFCR), which had shown improved efficacy and safety profiles in mice compared with its parent drugs, 5-FU and 5′-DFUR. However, due to interspecies differences, galocitabine was not well biotransformed to 5′-DFUR or 5-FU in patients. Subsequently, capecitabine was designed and identified based mainly on its susceptibility to human metabolic enzymes and to its improved pharmacokinetic and safety profiles in monkeys, in which the tissue distribution patterns of metabolic enzymes useful for prodrug design were most similar to those in humans. The metabolic enzymes involved in prodrug activation include: (1) a carboxylesterase isoenzyme, Cyd deaminase, which is located preferentially in the liver and in various types of cancer; and (2) thymidine phosphorylase (TP), which shows higher concentrations in tumor tissues than in healthy tissues. Capecitabine, when given orally, is sequentially converted to 5-FU preferentially in tumors through the non-cytotoxic intermediate metabolites 5′-DFCR and 5′-DFUR. Consequently, the drug can be safely given at higher doses, thus generating much higher levels of 5-FU in tumors and improving the efficacy and safety of the parent drug 5-FU or 5′-DFUR. Capecitabine has indeed been proven to show improved safety profiles compared with intravenous 5-FU. In particular, capecitabine proved to be less myelosuppressive than 5-FU in humans, most likely due to the unique tissue distribution of the three enzymes, including Cyd deaminase, which is poorly expressed in growing marrow cells. Thus, the strategy for the prodrug approach was successful, and could be applied to the improvement of other cytotoxic drugs. Today, with data available from genome, proteome, and metaborome databases of both human and experimental animals, it is possible to identify the metabolic enzymes for prodrug design [44].

A rational approach to optimizing capecitabine therapy was also pursued. The results of preclinical studies have provided a scientific rationale for conducting subsequent capecitabine clinical trials, such as combination therapies with radiation or with anti-cancer drugs, including taxanes. Findings that the DPD inhibitor RO 09-4889 improved the efficacy and safety of capecitabine, and that tumors with a higher TP/DPD ratio were more susceptible to capecitabine, also provide a rationale for additional clinical trials and an opportunity to optimize the efficacy of the drug. Indeed, combination therapy with a tumor-selective DPD inhibitor may broaden the range of indications for capecitabine, and personalized therapy might further increase its efficacy, thus avoiding unnecessary treatment.

Acknowledgments

The authors thank their colleagues at Nippon Roche Research Center, Nippon Roche K.K., who were involved in the discovery, development, and optimization of capecitabine.

References

1 Liu, G., Pranssen, E., Fitch, E. F., Warnew, E., *J. Clin. Oncol.* **1997**, *15*, 110–115.

2 Ishitsuka, H., Miwa, M., Takemoto, K., Fukuoka, K., Itoga, A., Maruyama, H. B., *Gann (Jpn. J. Cancer Res.)* **1980**, *71*, 112–123.

3 Bollag, W., Hartmann, H. R., *Eur. J. Cancer* **1990**, *16*, 427–432.

4 Ninomiya, Y., Miwa, M., Eda, H., Sahara, H., Fujimoto, K., Ishida, M., Umeda, I., Yokose, K., Ishitsuka, H., *Jpn. J. Cancer Res.* **1990**, *81*, 188–195.

5 Miwa, M., Ura, M., Nishida, M., Sawada, N., Ishikawa, T., Mori, K., Shimma, N., Umeda, I., Ishitsuka, H., *Eur. J. Cancer* **1998**, *34*, 1274–1281.

6 Shimma, N., Umeda, I., Arasaki, M., Murasaki, C., Masubuchi, K., Kohchi, Y., Miwa, M., Ura, M., Sawada, N., Tahara, H., Kuruma, I., Horii, I., Ishitsuka, H., *Bioorg. Med. Chem.* **2000**, *8*, 1697–1706.

7 Ishikawa, T., Utoh, M., Sawada, N., Nishida, M., Fukase, Y., Sekiguchi, F., Ishitsuka, H., *Biochem. Pharmacol.* **1998**, *55*, 1091–1097.

8 Ishikawa, T., Sekiguchi, F., Fukase, Y., Sawada, N., Ishitsuka, H., *Cancer Res.* **1998**, *58*, 685–690.

9 Ishikawa, T., Fukase, F., Yamamoto, T., Sekiguchi, F., Ishitsuka, H., *Biol. Pharm. Bull.* **1998**, *21*, 713–717.

10 Sawada, N., Ishikawa, T., Fukase, Y., Nishida, M., Yoshikubo, T., Ishitsuka, H., *Clin. Cancer Res.* **1998**, *4*, 1013–1019.

11 Endoh, M., Shiratori, N., Fukase, Y., Sawada, N., Ishikawa, T., Ishitsuka, H., Tanaka, Y., *Int. J. Cancer* **1999**, *83*, 127–134.

12 Sawada, N., Ishikawa, T., Sekiguchi, F., Tanaka, Y., Ishitsuka, H., *Clin. Cancer Res.* **1999**, *5*, 2948–2953.

13 Mori, M., Hasegawa, M., Nishida, M., Toma, H., Fukuda, M., Kubota, T., Nagasue, N., Yamane, H., *et al.*, *Int. J. Oncol.* **2000**, *17*, 33–38.

14 Ishitsuka, H., *Investigational New Drugs* **2000**, *18*, 343–354.

15 Tsukamoto, Y., Kato, Y., Ura, M., Horii, I., Ishikawa, T., Ishitsuka, H., Sugiyama, Y., *Biopharm. Drug. Dispos.* **2001**, *22*, 1–14.

16 Fujimoto-Ouchi, K., Sekiguchi, F., Tanaka, Y., *Cancer Chemother. Pharmacol.* **2002**, *49*, 211–216.

17 Endoh, M., Miwa, M., Eda, H., Ura, M., Tanimura, H., Ishikawa, T., Miyazaki-Nose, T., Hattori, K., Shimma, N., Yamada-Okabe, H., Ishitsuka, H., *Int. J. Cancer* **2003**, *106*, 799–805.

18 Hattori, K., Kohchi, Y., Oikawa, N., Suda, H., Ura, M., Ishikawa, T., Miwa, M., Endoh, M., Eda, H., Tanimura, H., Kawashima, A., Horii, I., Ishitsuka, H., Shimma, N., *Bioorganic Medicinal Chemistry Lett.* **2003**, *13*, 867–872.

19 Blum, J. L., Jones, S. E., Buzdar, A. U., LoRusso, P. M., Kuter, I., Vogel, C., Osterwalder, B., Burger, H.-U., Brown, C. S., Griffin, T., *J. Clin. Oncol.* **1999**, *17*, 485–493.

20 Hoff, P. M., Ansari, R., Batist, G., Cox, J., Kocha, W., Kuperminc, M., Maroun, J., Walde, D., Weaver, C., Harrison, E., Burger, H. U., Osterwalder, B., Wong, A.G., Wong, R., *J. Clin. Oncol.* **2001**, *19*, 2282–2292.

21 Van Cutsem, E., Twelves, C., Cassidy, J., Allman, D., Bajetta, E., Boyer, M., Bugat, R., Findlay, M., Frings, S., Jahn, M., McKendrick, J., Osterwalder, B., Perez-Manga, G., Rosso, R., Rougier, P., Schmiegel, W. H., Seitz, J.-F., Thompson, P., Vieitez, J. M., Weitzel, C., Harper, P., *J. Clin. Oncol.* **2001**, *19*, 4097–4106.

22 O'Shaughnessy, J., Miles, D., Vukelja, S., Moiseyenko, V., Ayoub, J. P., Cervantes, G., Fumoleau, P., Jones, S., Lui, W.-Y., Mauriac, L., Twelves, C., Hazel, G. V., Verma, S., Leonard, R., *J. Clin. Oncol.* **2002**, *20*, 2812–2823.

23 Twelves, C., Wong, A., Nowacki, M. P., Abt, M., Burris, H. III, Carrato, A., Cassidy, J., Cervantes, A., Fagerberg, J., Georgoulias, V., Husseini, F., Jodrell, D., Koralewski, P., Kröning, H., Maroun, J., Marschner, N., McKendrick, J., Pawlicki, M., Rosso, R., Schüller, J., Seitz, J.-F., Stabuc, B., Tujakowski, J., Van Hazel, G., Zaluski, J., Scheithauer, W., *N. Engl. J. Med.* **2005**, *352*, 2696–2704.

24 Cook, A. F., Holman, M. J., Kramer, M. J., Trown, P. W., *J. Med. Chem.* **1979**, *22*, 1330–1335.

25 Kono, A., Hara, Y., Sugata, S., Karube, Y., Matsushima, Y., Ishitsuka, H., *Chem. Pharm. Bull.* **1983**, *31*, 175–178.

26 Miwa, M., Cook, A., Ishitsuka, H., *Chem. Pharm. Bull. (Tokyo)* **1986**, *34*, 4225–4232.

27 Taguchi, T., Sakai, K., Terasawa, T., Irie, K., Okajima, K., Yamamoto, M., Kawahara, T., Satomi, T., Tomita, K., Yamaguchi, A., Shiratori, T., Enomoto, I., Kimura, M., Ito, A., Satani, M., Shimoyama, T., Okamoto, E., Okuno, G., Kurabori, T., Okumura, T., *Gan To Kagaku Ryoho* **1985**, *12*, 2179–2184. (Japanese).

28 Camiener, G. W., Smith, C. G., *Biochem. Pharmacol.* **1965**, *14*, 1405–1416.

29 Giusti, G., Mangoni, C., De Petrocellis, B., Scarano, E., *Enzym. Biol. Clin.* **1970**, *11*, 375–383.

30 Chabner, B. A., Johns, D. G., Coleman, C. N., Drake, J. C., Evans, W. H., *J. Clin. Invest.* **1974**, *53*, 922–931.

31 Scheithauer, W., McKendrick, J., Begbie, S., Borner, M., Burns, W. I., Burris, H. A., Cassidy, J., Jodrell, D., Koralewski, P., Levine, E. L., Marschner, N., Maroun, J., Garcia-Alfonso, P., Tujakowski, J., Van Hazel, G., Wong, A., Zaluski, J., Twelves, C., *Ann. Oncol.* **2003**, *14*, 1735–1743.

32 Schuller, J., Cassidy, J., Dumont, E., Roos, B., Durston, S., Banken, L., Utoh, M., Mori, K., Weidekamm, E., Reigner, B., *Cancer Chemother. Pharmacol.* **2000**, *45*, 291–297.

33 Mackean, M., Planting, A., Twelves, C., Schellens, J., Allman, D., Osterwalder, B., Reigner, B., Griffin, T., Kaye, S., *J. Clin. Oncol.* **1998**, *16*, 2977–2985.

34 Pinedo, H. M., *J. Clin. Oncol.* **1988**, *6*, 1653–1664.

35 Shindoh, H., Kawashima, A., Shishido, N., Nakano, K., Kobayashi, K., Horii, I., *J. Toxicol. Sci.* **2006**, *31*, 265–285.

36 Eda, H., Fujimoto, K., Watanabe, S., Ura, M., Hino, A., Tanaka, Y., Wada, K., Ishitsuka, H., *Cancer Chemother. Pharmacol.* **1993**, *32*, 333–338.

37 Bogdan, C., Ding, A., *J. Leukocyte Biol.* **1992**, *52*, 119–121.

38 Hata, K., Osaki, M., Kanasaki, H., Nakayama, K., Fujiwaki, R., Ito, H., Miyazaki, K., *Anticancer Res.* **2003**, *23*, 2665–2669.

39 Blanquicett, C., Gillespie, G. Y., Nabors, L. B., Miller, C. R., Bharara, S., Buchsbaum, D. J., Diasio, R. B., Johnson, M. R., *Mol. Cancer Therapeutics* **2002**, *1*, 1139–1145.

40 Glynne-Jones, R., Dunst, J., Sebag-Montefiore, D., *Ann Oncol.* **2006**, *17*, 361–371.

41 Nishida, M., Hino, A., Mori, K., Matsumoto, T., Yoshikubo, T., Ishitsuka, H., *Biol. Pharm. Bull.* **1996**, *19*, 1407–1411.

42 Nishina, T., Hyodo, I., Miyaike, J., Inaba, I., Suzuki, S., Shiratori, Y., *Eur. J. Cancer* **2004**, *40*, 1566–1571.

43 Meropol, N., Gold, P. J., Diasio, R. B., Andria, M., Dhami, M., Godfrey, T.,

Kovatich, A. J., Lund, K. A., Mitchell, E., Schwarting, R., *J. Clin. Oncol.* **2006**, *24*, 4069–4077.

44 Kohchi, Y., Hattori, K., Oikawa, N., Mizuguchi, E., Isshiki, Y., Aso, K., Yoshinari, K., Shirai, H., Miwa, M., Inagaki, Y., Ura, M., Ogawa, K., Okabe, H., Ishitsuka, H., Shimma, N., *Bioorganic Medicinal Chemistry Lett.* **2007**, *17*, 2241–2245.

24
Tenofovir and Adefovir as Antiviral Agents

Tomas Cihlar, William E. Delaney IV, and Richard Mackman

24.1
Introduction

Over the past decade, nucleotide analogues have emerged as a novel class of clinically effective antiviral agents. Cidofovir was the first antiviral nucleotide to reach commercial approval for the treatment of cytomegalovirus retinitis in 1996. Subsequently, prodrugs of two closely related adenine nucleotide analogues – tenofovir and adefovir – have been licensed for the treatment of human immunodeficiency virus (HIV) and chronic hepatitis B virus (HBV) infections, respectively. Both tenofovir [(*R*)-9-(2-phosphonomethoxypropyl)adenine; PMPA] and adefovir [9-(2-phosphonomethoxyethyl)adenine; PMEA] (Figure 24.1) are acyclic nucleoside phosphonates (ANPs), a structurally unique class of nucleotide analogues containing aliphatic sugar-like moieties, covalently attached to phosphonate groups.

The concept of ANPs emerged during the mid-1980s [1] as a result of combining the early pursuits of bioisosteric nucleoside phosphonates [2, 3] with clinically validated antiviral acyclic nucleoside analogues such as acyclovir or ganciclovir. In contrast to antiviral nucleosides that require phosphorylation to their corresponding triphosphates in order to become active, two phosphorylation steps are sufficient for the metabolic activation of ANPs. Unlike the phosphate in natural nucleotides and phosphorylated nucleoside analogues ($-CH_2-O-P$ linkage), the bioisosteric phosphonate moiety present in ANPs ($-O-CH_2-P$ linkage) is resistant to enzymatic hydrolysis by phosphatases. Consequently, the active metabolites of ANPs (i.e., ANP diphosphates), which act through potent inhibition of viral polymerases, exhibit prolonged intracellular half-lives [4] and provide persistent antiviral effects [5]. The aliphatic linker mimicking the sugar part of nucleotides possesses multiple rotatable bonds, and offers a greater degree of flexibility to allow ANPs to maintain their potency against genetically diverse viral strains, including a range of drug-resistant viruses.

Contrary to the vast majority of therapeutics, ANPs are quite hydrophilic due to two negative charges on their phosphonate moiety present at physiological pH. Although

Modified Nucleosides: in Biochemistry, Biotechnology and Medicine. Edited by Piet Herdewijn

ISBN: 978-3-527-31820-9

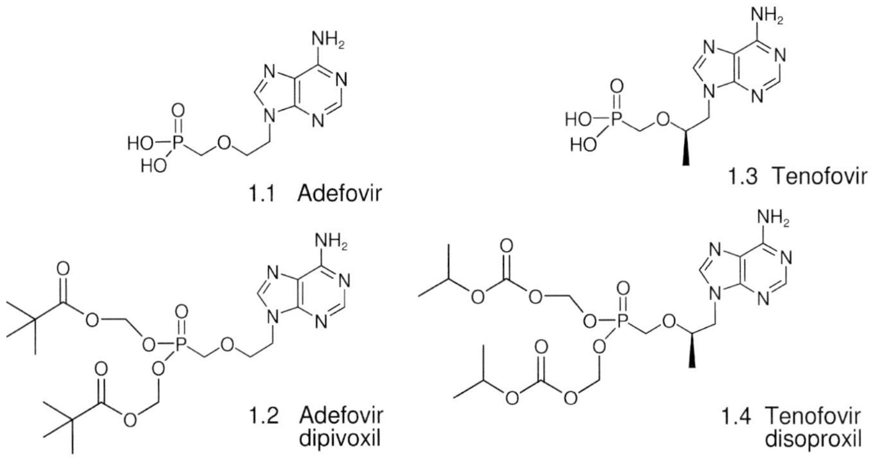

Figure 24.1 Structures of adefovir and tenofovir together with their prodrugs adefovir dipivoxil and tenofovir disoproxil (fumarate salt omitted).

this property does not diminish their antiviral activity, it becomes a limiting factor for intestinal absorption due to low cellular permeability. Hence, the oral administration of ANPs requires the use of prodrugs such as lipophilic diesters that have been extensively explored on adefovir and tenofovir, leading to the design, clinical development, and subsequent commercial approval of tenofovir disoproxil fumarate (TDF; Viread®) in 2001 and adefovir dipivoxil (ADV; Hepsera®) in 2002 for the treatment of HIV and HBV infection, respectively (Figure 24.1, compounds **1.4** and **1.2**, respectively). Because of their efficacy, convenient once-daily dosing regimen, activity against drug-resistant viral strains, and favorable long-term safety profiles, these two drugs have markedly improved the clinical management of HIV and HBV infections. The most illustrative example of progress in the therapeutic application of ANPs is the recent commercial approval of Atripla™, the first fixed-dose combination of three antiretrovirals, which contains TDF as one of the active components. Although several comprehensive reviews on ANPs have been recently published [6–9], this chapter summarizes the breadth of tenofovir- and adefovir-related research in the areas of chemical synthesis, virology, and molecular pharmacology, and highlights key results from the preclinical and clinical development of TDF and ADV.

24.2 Synthesis

Although the clinical approval of ADV followed that of TDF, adefovir and its prodrug were synthesized and characterized prior to tenofovir. Both, the initial synthesis and subsequent larger-scale production of tenofovir and TDF were derived from the synthetic pathways originally developed for adefovir and ADV.

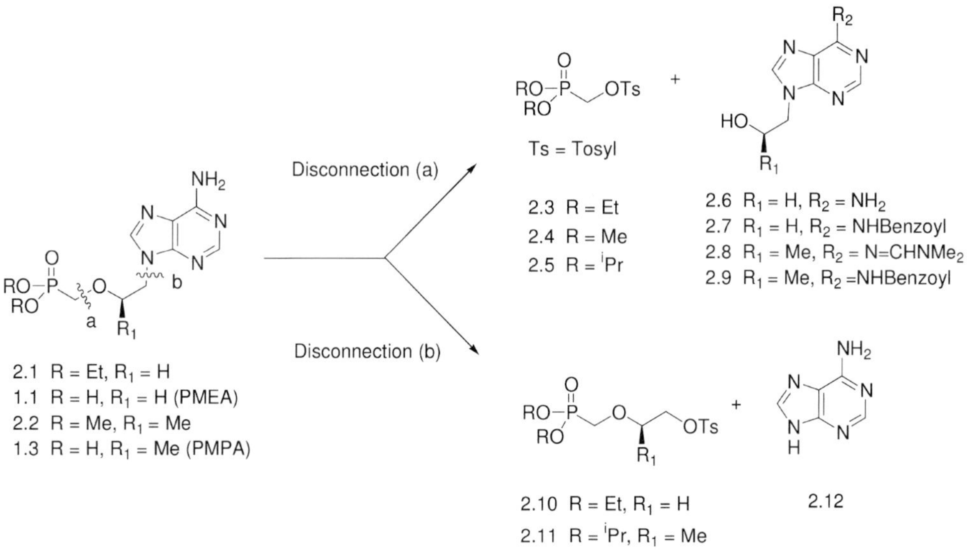

Figure 24.2 Synthesis strategies for adefovir (PMEA) and tenofovir (PMPA).

24.2.1
Adefovir and Adefovir Dipivoxil

The first synthesis of PMEA (adefovir; **1.1**), was described by Holy and Rosenberg in 1986 [10]. The synthesis was based on the disconnection strategy (route a in Figure 24.2) in which phosphonate synthon **2.3** was coupled with alkoxy adenine synthon (**2.6** or **2.7**) to generate the phosphonate diester precursor **2.1**. The specific method described was the reaction of benzoyl-protected hydroxyl alkyladenine **2.7** [11, 12] under basic (sodium hydride) conditions with diethyl tosyl phosphonate **2.3** [13]. Later, an alternative approach based on disconnection (route b in Figure 24.2) was published [12]. This approach utilized an Arbuzov reaction to generate a phosphonate synthon **2.10** that contained the complete linear chain of PMEA. Thus, similar to disconnection strategy (a), treatment of the tosyl phosphonate synthon **2.10** with adenine **2.12** under basic conditions (sodium hydride) yielded the diethyl phosphonate **2.1**. Strategy (b) avoided the need for a benzoyl protecting group on the adenine synthon, and was ideally suited to modifications of the nucleobase to other purines and pyrimidines for SAR studies [14].

For the preparation of greater than 100 kg quantities of PMEA and the corresponding bis(pivaolyloxymethyl) [bis(POM)] prodrug (ADV) required for oral delivery, several differences to the early routes can be noted. The synthesis is depicted in Figure 24.3A, and follows the original disconnection strategy (a) [15]. Generation of alkoxy adenine **3A.3** was achieved in high yields by condensing adenine with ethylene carbonate **3A.1** in hot dimethylformamide (DMF) under basic conditions [16]. In the next step, the reaction between the unprotected alkoxy adenine **3A.3** and phosphonate **2.3** in the presence of sodium hydride as the deprotonating base of choice, was

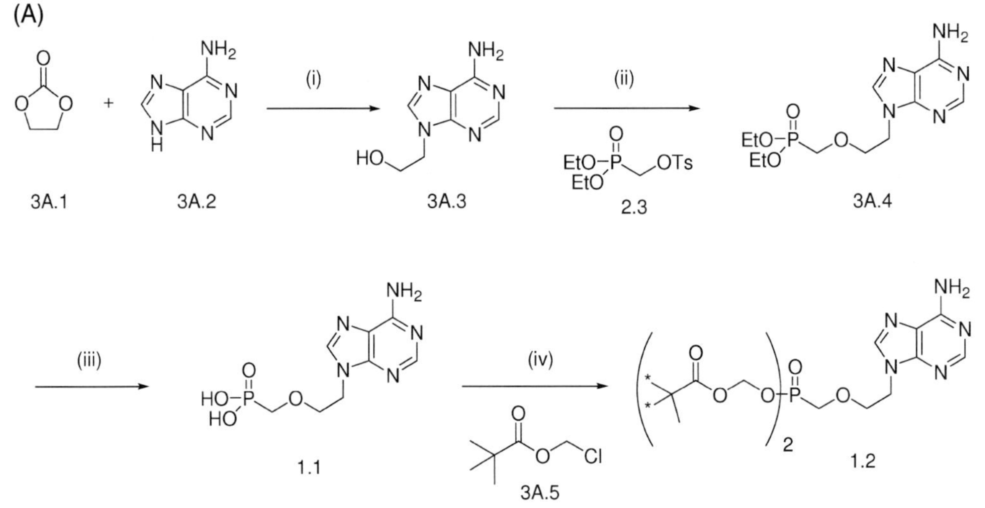

(i) NaOH, DMF, 120 °C; (ii) DMF, NaO^{t}Bu (1.75 eq.), Tosylate **2.3** (1.25 eq.), -10–0 °C; (iii) TMSBr, CH_3CN, rt; (iv) chloromethyl pivaloate **3A.5**, Et_3N, NMP, 60 °C.

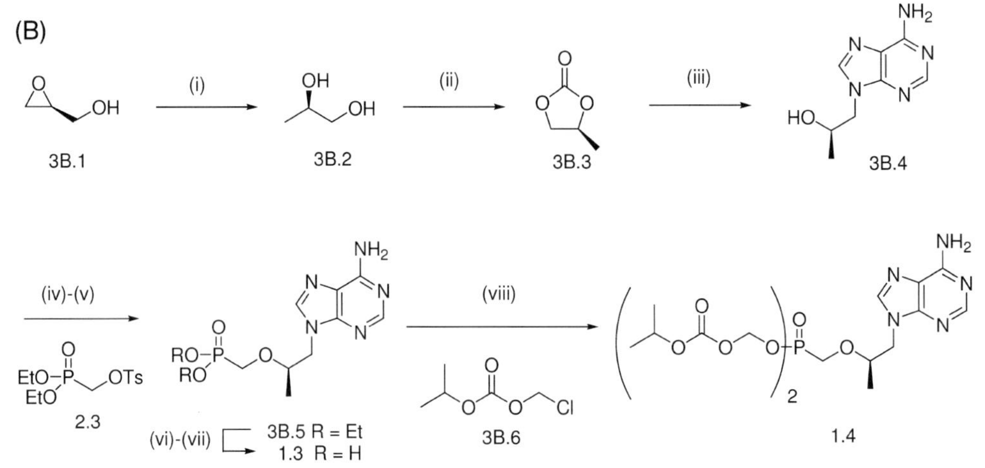

(i) H_2, 5%Pd/C, cat NaOH, EtOH; (ii) $(EtO)_2CO$, cat NaOEt; (iii) NaOH, DMF, adenine; (iv) LiO^{t}Bu, tosylate **2.3**; (v) AcOH, then CH_2Cl_2, H_2O; (vi) TMSBr, CH_3CN; (vii) NaOH, H_2O; (viii) Et_3N, DMF, chloromethyl carbonate, **3B.6**, 50 °C.

Figure 24.3 Synthesis of adefovir dipivoxil (A) and tenofovir disoproxil (B).

plagued with several issues. These included variable yields, a lack of homogeneity of the reaction mixture, and safety concerns regarding the evolution of hydrogen gas. Sodium *tert*-butoxide was found to be an effective replacement for sodium hydride that solved these issues and also allowed for reduced equivalents of reagent, thus leading to fewer side products [15]. Kilo-scale optimization resulted in the isolation of

3A.4 in consistent yields of 42%. More recently, magnesium *tert*-butoxide has been reported to be another potential base for this reaction [17]. The dealkylation of precursor **3A.4** to form adefovir (**1.1**) was achieved by treatment with dealkylating agent, trimethylsilyl bromide. The bis(POM) prodrug **1.2** was initially prepared by reaction of adefovir with chloromethyl pivaloate **3A.5** in the presence of a hindered base, for example *N,N* dicyclohexylmorpholine carboxamidine [18]. However, for the kilo-scale production of **1.2**, triethylamine base was used.

24.2.2
Tenofovir and Tenofovir Disoproxil Fumarate

Not surprisingly, the synthetic methods for the generation of (*R*)-PMPA (tenofovir) **1.3** are closely related to those of adefovir. Although both disconnection strategies (a) and (b) are once again applicable (Figure 24.2), it should be noted that the additional methyl group in the alkyl side chain introduced a stereocenter into the molecule and therefore required the use of chiral pool reagents or a resolution step to isolate the individual stereoforms of PMPA. The first synthesis of PMPA prepared the racemate through disconnection strategy (a) [19–21]. Coupling of the racemic, adenine-protected 9-(*R,S*)-2-hydroxypropyl adenine **2.8**, with dimethyl tosyl phosphonate **2.4** was achieved in the presence of sodium hydride to afford diester **2.2**. Dealkylation of the methyl ester groups readily afforded racemic PMPA. The first stereocontrolled synthesis was based on chiral pool reagents, esters of D- and L- lactate, which resulted in the preparation of (*R*)-PMPA and (*S*)-PMPA respectively [22]. The chiral synthon **2.9** that was prepared from D-isobutyl-lactate was coupled with bis-isopropyl phosphonate ester **2.5**, followed by removal of the isopropyl groups to afford (*R*)-PMPA **1.3**. An alternative approach based on disconnection (b) was also pursued, starting from a suitable ester of D-lactate [23]. In this approach, chiral synthon **2.11** was prepared in seven steps and then alkylated with adenine **2.12**.

The kilo-scale preparation of (*R*)-PMPA followed disconnection (a), and was closely based on the kilo-scale preparation of PMEA described above (Figure 24.3B) [24]. By analogy, the chiral propylene carbonate **3B.3** served as the chiral synthon, and was readily derived in two steps from the chiral pool reagent (*S*)-glycidol **3B.1** (86% *e.e.*) in an overall yield of 75–80%. The coupling of adenine to chiral carbonate **3B.3** could be effected without protection of the adenine, and resulted in the preparation of the desired chiral alkoxy adenine precursor **3B.4**. The reaction of phosphonate **2.3** with **3B.4** was performed in the presence of lithium *tert*-butoxide. Diethyl phosphonate ester **3B.5** was dealkylated using trimethylsilyl bromide to afford (*R*)-PMPA **1.3**.

The orally bioavailable prodrug form of (*R*)-PMPA is a carbonate ester, bis (isopropyloxymethylcarbonyl) [bis(POC)] PMPA **1.4**. Although this is different from the prodrug designed for PMEA, the synthesis of the carbonate ester and other PMPA prodrugs followed similar procedures. On a small scale, the bis(POC) prodrug of (*R*)-PMPA **1.3** was prepared in a modest 10% yield by treatment of (*R*)-PMPA with triethylamine and chloromethyl carbonate **3B.6** [25]. An example of a large-scale synthesis of **1.4** was published in the patent literature using a similar procedure, followed by isolation of **1.4** as the fumarate salt [26].

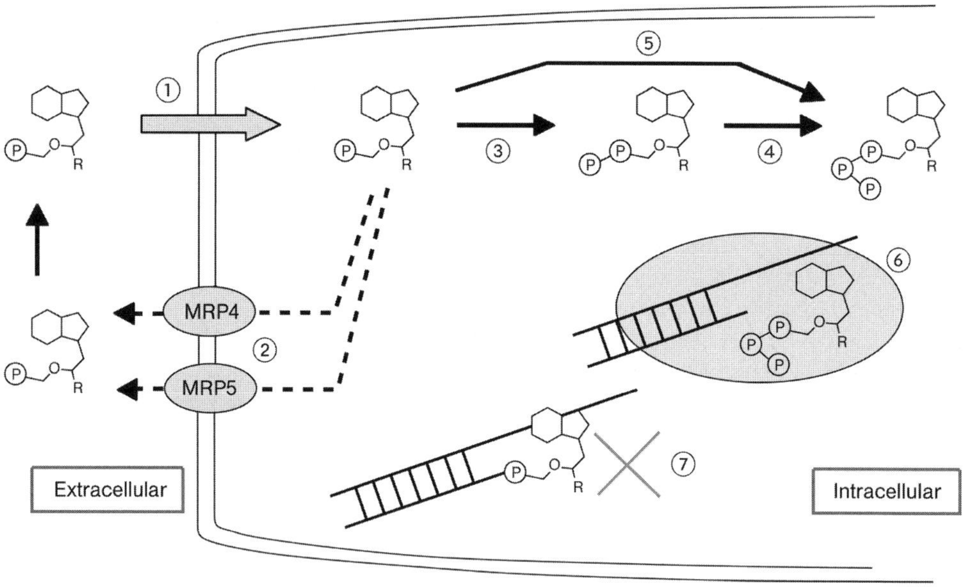

① Uptake into cell
- Specific transporters (epithelial cells)
- Endocytosis (lymphocytes)

② Potential active efflux
- MRP4, 5 efflux pumps

③ First phosphorylation
- AMP kinase

④ Second phosphorylation
- AMP kinase
- NDP kinase
- Creatine kinase

⑤ Potential diphosphorylaton
- PRPP synthetase

⑥ Binding to viral polymerase

⑦ Termination of viral DNA synthesis following ANP incorporation

Figure 24.4 Membrane transport, intracellular activation, and mechanism of action of adefovir and tenofovir.

24.3 Mechanism of Action

The manifestation of antiviral activity of ANPs consists of multiple distinct steps, including cellular transport and retention, intracellular conversion to active diphosphate metabolites, and the interaction of diphosphate with a target viral polymerase (Figure 24.4). The combined efficiency of these steps determines the overall potency and spectrum of antiviral activity for each specific ANP.

24.3.1 Membrane Transport and Intracellular Metabolism

Nucleoside analogues are transported across the plasma membrane either via facilitated diffusion or by membrane nucleoside transporters [27, 28]. In contrast, the cellular uptake of ANPs is likely slower and less efficient, due to the negative charge of their phosphonate moieties. A range of *in-vitro* studies with various ANPs have demonstrated the cell type-dependent mechanism of their membrane transport. In H-9 [29] and CEM [30] T cells, non-specific fluid-phase endocytosis has been suggested as the primary mechanism of adefovir transport, whereas epithelial HeLa cells take up adefovir via receptor-mediated endocytosis that exhibits a narrow substrate specificity for only certain types of ANPs [31]. One important and clinically relevant aspect of the active cellular transport of ANPs is their selective accumulation in renal proximal tubules that may cause renal dysfunction in a proportion of patients; this was predominantly observed in patients treated with cidofovir or high-dose adefovir originally explored for the treatment of HIV [32]. Recent studies have established the role of human organic anion transporters types 1 and 3 in the uptake of ANPs into renal cells [33–35]. It should be noted that not only cellular uptake, but also active elimination via cellular efflux, affects the intracellular concentration of ANPs and their metabolites. Both, adefovir and tenofovir are substrates for multidrug resistance proteins (MRP) types 4 and 5 [36–38] – two membrane efflux pumps that actively eliminate a wide range of small organic anions from various cell types. In contrast, neither of the two ANPs is transported by MRP2 or P-glycoprotein [36, 37]. The up-regulation of MRP4 expression has been identified as a mechanism of *in-vitro* cellular resistance to the cytotoxic effects of adefovir [39]. Although the active interaction of ANPs with MRPs may affect their distribution to specific tissues and body compartments, it is unlikely to be a relevant mechanism for the development of antiviral resistance, since supratherapeutic concentrations of ANPs are required to up-regulate MRP4 [39].

After entering cells, tenofovir and adefovir are activated by phosphorylation to their respective diphosphate metabolites (Figure 24.4). *In-vitro* enzymatic studies have suggested at least two different pathways for adefovir phosphorylation. For example, Merta *et al.* showed that AMP kinase is capable of phosphorylating adefovir and several other adenine ANPs to their diphosphates in two distinct steps [40]. As shown by cellular fractionation, the mitochondrial form of AMP kinase (AK2) phosphorylates adefovir more efficiently than the cytosolic enzyme (AK1) [41]. In addition, creatine kinase is capable of converting adefovir monophosphate to its diphosphate [40]. It is likely that this phosphorylation step can also be catalyzed by nucleoside diphosphate kinase, which is known to have broad substrate specifi-city [42]. The results of independent studies have suggested that 5-phosphoribosyl-1-pyrophosphate (PRPP) synthetase may activate various adenine ANPs directly to diphosphates [43]. Although the characterization of these enzymatic phosphorylation pathways has been conducted primarily with adefovir, it is likely that the results are also applicable to tenofovir, because of its close structural similarity.

In metabolic studies conducted *in vitro* with radiolabeled adefovir or tenofovir, only parental nucleotides together with their mono- and diphosphate forms have been detected in various cell types [4, 41, 44, 45], indicating that neither adefovir nor tenofovir undergo deamination or any other type of intracellular metabolic conversion, aside from phosphorylation. One important general aspect of the metabolism of ANPs is the intracellular persistence of their active diphosphate metabolites. The intracellular half-lives of adefovir and tenofovir diphosphates in T cells are approximately 16 h [44] and 21 h [4], respectively, which compares favorably with the shorter half-lives of active triphosphates of nucleoside analogues [46]. The efficient conversion of adefovir and tenofovir into their respective diphosphates in hepatic cells (HepG2 or primary hepatocytes) has also been confirmed [47, 48]. The half-lives of adefovir diphosphate (75 h) and tenofovir diphosphate (95 h) are significantly longer in hepatic cells than in T cells [48]. Overall, the prolonged intracellular retention of active adefovir and tenofovir diphosphates in target cells most likely contributes to their sustained activities against HIV and HBV, and has enabled these drugs to be developed as once-daily therapeutics.

24.3.2
Inhibition of Viral Polymerases

The diphosphates of ANPs are analogues of natural deoxynucleoside triphosphates, and can thus act as competitive inhibitors or substrates for viral polymerases. An essential aspect of the mechanism of action of adefovir and tenofovir, which both lack the 3′-like hydroxyl in their acyclic structures, is their ability to function as obligate chain terminators of DNA synthesis upon their incorporation into viral DNA by target polymerases [49]. The inhibitory activity of adefovir and tenofovir diphosphates against viral polymerases, including HIV-1 reverse transcriptase (RT) and HBV DNA polymerase – the two therapeutically most relevant target enzymes – are summarized in Table 24.1.

Similar to nucleoside HIV RT inhibitors, active metabolites of ANPs inhibit the HIV reverse transcription step (RNA-dependent synthesis of negative strand DNA) more efficiently than the DNA polymerization step (DNA-dependent synthesis of positive strand DNA) [50]. The overall efficacy of any DNA chain-terminating inhibitor against HIV is affected by the ability of RT to catalyze the reverse (excision) reaction in the presence of pyrophosphate or ATP, which leads to a removal of the incorporated chain terminator and allows the viral DNA synthesis to resume [51]. It has been shown that HIV RT is able to excise ANPs (including tenofovir) from the terminated DNA, albeit with less efficiency than some nucleosides such as zidovu-

Table 24.1 Inhibition of viral target enzymes and host DNA polymerases by diphosphates of tenofovir and adefovir.

Enzyme	K_i [μM]		References
	Tenofovir diphosphate	Adefovir diphosphate	
HIV reverse transcriptase			
• RNA template	0.022	0.012	[50]
• DNA template	1.55	0.98	[50]
HBV DNA polymerase	0.18	0.10	[48, 53]
HSV-1 DNA polymerase		0.11	[56]
HCMV DNA polymerase		0.45	[57]
Human DNA polymerase α	5.2	1.18	[50]
Human DNA polymerase β	81.7	70.4	[50]
Human DNA polymerase γ	59.5	0.97	[50]

dine [52]. This difference is significant as it likely compensates for the lower binding affinity of tenofovir diphosphate to HIV RT compared to zidovudine triphosphate.

Adefovir and tenofovir also inhibit HBV DNA replication, which is not surprising given that HIV and HBV replicate by related mechanisms that include a reverse transcription step and have structurally related polymerases. The mechanism of HBV polymerase inhibition is similar to that of HIV, as both adefovir and tenofovir diphosphate can be incorporated into replicating HBV DNA and serve as chain terminators. This has been confirmed by enzymatic studies with isolated HBV polymerase [53], in intact HBV core particles [54], and by experiments with the related duck hepatitis B virus (DHBV) polymerase [55]. Adefovir and tenofovir diphosphates have inhibition constants (K_i) of 0.1 and 0.18 μM against recombinant HBV polymerase, respectively. These K_i values are two- to four-fold lower than the K_m of the natural substrate (dATP, $K_m = 0.38$ μM) [48, 53].

Consistent with the antiherpesviral activity of adefovir, its diphosphate is a potent competitive inhibitor of DNA polymerases from herpes simplex virus type 1 (HSV-1) [56] and human cytomegalovirus (HCMV) (Table 24.1) [57]. The inhibition of HSV-1 ribonucleotide reductase by adefovir metabolites, particularly adefovir diphosphate, has also been described and may contribute to the overall antiherpesviral activity of adefovir [58].

As substrate analogues, diphosphates of ANPs also have the potential to interact with host DNA polymerases (Table 24.1). Among the polymerases tested *in vitro*, mitochondrial DNA polymerase γ was the most sensitive to adefovir diphosphate [50]. Tenofovir diphosphate is both a less effective substrate [59] and a weaker inhibitor [50] of human DNA polymerases α, β, and γ compared to adefovir diphosphate. Despite the substantial difference in their interaction with mitochondrial DNA polymerase γ (see Table 24.1), neither tenofovir nor adefovir affect the replication of mitochondrial DNA *in vitro* [60, 61] or *in vivo* [62] at therapeutically relevant levels. Tenofovir diphosphate is also inefficiently incorporated into DNA by replicative nuclear DNA polymerases δ and ε [63], and its weak interaction with host DNA polymerases is the likely explanation for the low *in-vitro* cytotoxicity of tenofovir compared to various nucleoside analogues [64].

24.3.3 Spectrum of Antiviral Activity

Adefovir and tenofovir are both potent inhibitors of the *in-vitro* replication of HIV-1 and HBV, which represent their therapeutically most relevant antiviral activities. Both ANPs inhibit HIV-1 in a variety of host cell types including T-cell lines [65], primary peripheral blood mononuclear cells (PBMCs) [66] and macrophages [67]. While the anti-HIV activity of nucleosides such as zidovudine or stavudine is compromised in resting PBMCs, the activity of neither adefovir nor tenofovir is affected by cell activation status [68]. This difference is due to the fact that the activation of ANPs requires only two phosphorylation steps (see Section 24.3.1), and thus is not dependent on the activity of nucleoside kinases that are often tightly regulated in distinct phases of the cell cycle. Although tenofovir differs from adefovir

Table 24.2 *In-vitro* antiviral activity of adefovir and tenofovir.

Virus	EC_{50} [μM]				References
	Tenofovir	TDF	Adefovir	ADV	
Retroviruses					
HIV-1					
T-cell lines	4.2		11		[67]
			3.7	0.18	[83]
	5.0	0.05			[82]
PBMCs	0.3–4.5		0.8–10.5		[66]
	0.37		2.5		[67]
	0.18	0.005			[4]
Macrophages	0.04		0.02		[67]
HIV-2	2.0		5.6		[71]
			6.0	0.25	[83]
Hepadnaviruses					
HBV	1.2		0.2		[74]
	1.1	0.02	0.8	0.1	[48]
DHBV			0.2		[75]
	0.11		0.14		[198]
WHBV			<1		[76]
Herpesviruses					
HSV-1	>300		26	1.7	[84, 85, 199]
HSV-2			26	0.6	[199, 200]
HCMV	>300		28–70		[57, 84, 86]
EBV			1.1		[87]

TDF: tenofovir disoproxil fumarate (Viread®); ADV: adefovir dipivoxil (Hepsera®).

only by the addition of a single methyl group in the sugar-mimicking linker, it exhibits improved *in-vitro* antiretroviral activity and lower cytotoxicity. In PBMCs, tenofovir and adefovir are active against various subtypes of HIV-1 clinical isolates, with EC_{50} values of 0.3–4.5 μM and 0.8–10.5 μM, respectively (Table 24.2) [66]. Importantly, tenofovir has been found less cytotoxic than adefovir in various cell types, including renal proximal tubule cells [34], a profile that was key in supporting its development after adefovir failed to meet the clinical safety and efficacy criteria for the treatment of HIV infection. Both, tenofovir and adefovir retain their activity against multiple HIV-1 strains resistant to zidovudine, abacavir, lamivudine and other nucleosides [66, 69]. Subsequent to its approval, the favorable profile of tenofovir has been confirmed against a large pool of both nucleoside-sensitive and -resistant HIV-1 strains [70]. Based on results of these analyses, TDF has become extensively used in patients infected with nucleoside-resistant HIV-1.

The sensitivity of HIV-2 to the inhibition by tenofovir and adefovir is similar to that of HIV-1 [71]. However, the antiretroviral spectrum of adefovir and tenofovir is not limited only to human pathogens as they both are active *in vitro* against related animal viruses, including simian immunodeficiency virus (SIV), Moloney murine sarcoma virus, sheep visna virus, and other retroviruses [72, 73]. These activities have proven useful for the initial proof-of-concept efficacy studies in animal models (see Section 24.4).

The cell-based activity of adefovir and tenofovir against HBV is comparable (Table 24.2), which is in agreement with the similar phosphorylation rates observed in hepatic cells and the similar K_i values of active metabolites against HBV polymerase [48]. Yokota and colleagues first demonstrated the selective antiviral activity of adefovir and tenofovir against HBV in a human liver cell line stably replicating the virus [74]. These findings were subsequently confirmed by several other investigators, although adefovir has been more extensively studied in cell culture. In addition to HBV, adefovir also has antiviral activity against related hepadnaviruses including DHBV, in primary duck hepatocytes [75] and woodchuck hepatitis virus (WHV) in primary woodchuck hepatocytes [76].

Adefovir maintains its potency against a variety of clinical isolates, multiple HBV genotypes [77, 78], and clinically relevant nucleoside-resistant HBV mutants [79, 80]. One of the most promising *in-vitro* properties of adefovir was its ability to inhibit drug-resistant viruses selected by other anti-HBV agents (e.g., famciclovir or lamivudine). Lamivudine-selected HBV polymerase mutations of rtM204V/M204I, with or without secondary mutations of rtL180M and/or rtV173L, cause high level cross-resistance to multiple other nucleosides, but the corresponding mutant viruses are still effectively inhibited by adefovir [79–81]. Recent studies have confirmed that tenofovir shares adefovir's ability to inhibit lamivudine-resistant strains of HBV [79].

Dialkoxyester prodrugs of adefovir and tenofovir exhibit enhanced cell membrane permeability which, coupled with the efficient intracellular hydrolysis of the prodrug moieties, markedly increases the *in-vitro* antiviral potency of parent ANPs. In cell-based assays, TDF and ADV are 20- to 100-fold more potent against HIV-1 compared to their respective parent nucleotides [4, 82, 83]. Similarly, the potency of adefovir and tenofovir against HBV in cell culture models increases by 10- and 100-fold, respectively, after their conversion to the dialkoxyester prodrugs (Table 24.2) [48].

Among nucleoside and nucleotide analogues, adefovir is a unique example of an agent with broad-spectrum antiviral activity. In addition to its antiretroviral and antihepadnaviral activity, adefovir is active *in vitro* against multiple types of human herpesviruses [57, 84–86] with the highest potency shown against Epstein–Barr virus (EBV) [87] (Table 24.2). Unlike the antiherpesviral nucleosides acyclovir or ganciclovir, adefovir does not require virus-encoded enzymes for its intracellular activation, and therefore it retains its antiviral activity against thymidine kinase and UL97 kinase mutants of HSV and HCMV, respectively [57, 85]. Despite its structural similarity with adefovir, tenofovir does not possess any activity against human or animal herpesviruses, and neither of the two ANPs inhibits any non-retroviral RNA viruses.

Unlike other inhibitors of viral polymerases, ANPs have also been shown to possess *in-vitro* immunomodulatory effects in addition to their direct antiviral activity [88, 89]. The mechanism of these effects involves enhancing the production of various cytokines, probably through the activation of multiple purine P(1) receptors [90]. Although this type of ANP-associated activity has been demonstrated and characterized in multiple cell culture models, its relevance in the clinic remains to be established.

24.4
Activity in Animal Models

The antiviral spectrum of adefovir and tenofovir, as defined *in vitro* in various cell culture systems, has been confirmed *in vivo* in many animal models, in which both nucleotides were found to be effective and selective agents for the suppression, treatment, or prophylaxis of different viral infections. Many of these models played critical roles in the development of adefovir and tenofovir by providing early *in-vivo* proof-of-concept data. A comprehensive review covering in detail various *in-vivo* models used to study ANPs has been published [91].

24.4.1
Models for Retroviral Infections

One of the first models used to evaluate the *in-vivo* antiretroviral activity of adefovir was the Moloney murine sarcoma virus (MSV) infection in newborn mice. In this model, adefovir suppressed MSV-induced tumor development with 25-fold higher efficiency and five-fold better selectivity than zidovudine, and was effective even when given prophylactically [92]. In the same model, oral treatment with ADV exhibited efficacy comparable to that of subcutaneous adefovir [83]. Subsequently, the antiretroviral effect of adefovir was confirmed in mice infected with Friend leukemia virus [93] or LP-BM5 retrovirus complex [94]. In the latter model, adefovir showed dual antiviral activity as the treatment not only inhibited retroviral replication, but also suppressed various herpesviral infections in co-infection experiments [94]. In addition, the *in-vivo* efficacy of adefovir was demonstrated against feline immunodeficiency virus [95]. Visna virus infection of sheep has long been used as an *in-vivo* model for the replication of neurotropic retroviruses. Infected lambs treated with adefovir showed reductions in brain lesions and an effective suppression of viral replication in the central nervous system [96], suggesting that at higher doses, adefovir is capable of crossing the blood–brain barrier.

Both, adefovir and tenofovir have been extensively tested against various types of SIV infection in macaques, which is considered to be one of the most relevant animal models for HIV infection in humans. Both ANPs were found to be highly effective in adult macaques with established chronic SIV infection [97, 98] and the potent antiviral activity of tenofovir has also been demonstrated in newborn macaques infected with SIV [99]. Multiple recent studies have identified a severe depletion of $CD4^+$ lymphocytes in gut-associated lymphatic tissue as one of the early effects of retrovirus-induced immunodeficiency, both in animals and humans [100]. When initiated early after primary infection, tenofovir treatment of SIV-infected monkeys prevented the destruction of the gut-associated pool of $CD4^+$ lymphocytes [101].

Simultaneously with the (thus far) unsuccessful development of an effective HIV-1 vaccine, chemoprophylaxis is being explored as an alternative means of preventing HIV-1 infection. Initial studies indicated a significant pre-exposure prophylactic effect of subcutaneous adefovir treatment in SIV-infected macaques, with an efficiency greater than 80% [102], whereas parallel experiments with zidovudine showed the prevention of SIV infection only in 6% of experimental animals. The higher potency of adefovir in this model may be due to the long intracellular half-life of its active metabolite and/or the retention of its antiretroviral activity in resting lymphocytes, in which zidovudine undergoes only limited metabolic activation [68]. Subsequently, a promising prophylactic efficacy of tenofovir was demonstrated when it was administered subcutaneously both pre- and post-SIV exposure [103]. Even with treatment starting 24 h after the inoculation of animals, subcutaneously administered tenofovir (30 mg kg^{-1}) showed complete protection against SIV acute infection, without any toxic effects. Tenofovir also effectively prevents SIV infection in a macaque model for vaginal transmission [104] and reduces the oral transmission of SIV in newborn macaques [105]. More recently, tenofovir in combination with emtricitabine effectively blocked rectal infection in monkeys following repeated exposure to a low-titer SIV inoculum [106]. Hence, results from multiple animal studies investigating different mechanisms of virus transmission suggest that tenofovir and/or TDF may be useful agents for the chemoprophylaxis of HIV-1 infection.

24.4.2 Models for Hepadnavirus Infections

In-vivo proof of efficacy for adefovir was first established in the DHBV model by Heijtink and colleagues [75]. DHBV-positive ducklings raised from a stock of congenitally infected ducks were treated with 15 or 30 mg kg^{-1} adefovir by intraperitoneal injection every other day for a period of three weeks. A significant reduction of serum DHBV DNA was observed within a few days in the animals treated with 30 mg kg^{-1} adefovir, but not in those treated with 15 mg kg^{-1}, or in control animals. Importantly, no toxicity was observed following adefovir treatment, thereby establishing that adefovir could elicit a selective short-term antiviral effect *in vivo*. Following the discontinuation of adefovir treatment, serum DHBV DNA levels rebounded to pre-treatment values, indicating that short-term treatment was suppressive, but not curative. Additional studies subsequently confirmed that adefovir markedly suppresses DHBV replication in congenitally infected [107] or experimentally infected [108] ducks, without toxicity.

Two studies have established that adefovir inhibits the replication of WHV in woodchucks when administered orally as ADV. Cullen *et al.* first demonstrated this effect in an 18-week study wherein woodchucks chronically infected with wild-type WHV were treated with ADV daily for 12 weeks. At the end of treatment there were reductions of >2.5 and 1.6 $\log_{10}$ copies of serum WHV DNA mL^{-1} in the 15 and 5 mg kg^{-1} groups, respectively [109]. After a six-week follow-up period without drug, serum WHV DNA returned to baseline levels in both dose groups. A subsequent study was performed in chronically infected woodchucks that had developed lamivudine-resistant WHV as a result of prolonged lamivudine therapy [110]. In this study, lamivudine-resistant woodchucks were treated with a combination of 15 mg kg^{-1} lamivudine plus 15 mg kg^{-1} ADV, and compared to woodchucks maintained on 15 mg kg^{-1} lamivudine monotherapy as a control. Woodchucks receiving

ADV had a mean 4.5 $\log_{10}$ reduction in virus titer after 12 weeks of therapy, compared to a 0.8 $\log_{10}$ reduction in control animals. The observed efficacy of ADV in lamivudine-resistant woodchucks was consistent with the activity of adefovir against lamivudine-resistant HBV in both cell culture studies and in patients (see Section 24.6.2). No adefovir-associated toxicities were observed during the study.

The antiviral efficacy of ADV against HBV was also confirmed *in vivo* in two strains of transgenic mice replicating the virus [111, 112]. Kajino *et al.* showed that oral administration of ADV (100 mg kg^{-1}) for 14 days reduced serum HBV DNA to undetectable levels [113]. Similarly, Julander *et al.* treated HBV transgenic mice with 100 mg kg^{-1} of ADV orally for 10 days, and observed an approximate 3 $\log_{10}$ reduction in serum HBV DNA copies mL^{-1} [114]. In both transgenic studies, ADV also reduced levels of replicating HBV DNA in the livers of the mice. No adefovir-related toxicities were observed during either transgenic mouse study.

The antiviral activity of tenofovir against hepadnaviruses in animals has only been tested in the WHV model. Menne *et al.* studied the antiviral efficacy of 0.5, 1.5, 5, and 15 mg kg^{-1} daily oral doses of TDF in woodchucks over a four-week treatment period [115]. The higher doses produced reductions in serum WHV DNA of 1.1–1.5 $\log_{10}$ copies mL^{-1}. However, during a 12-week drug-free follow-up period, the viral load returned to baseline in all dose groups, and there were no signs of toxicity in any of the treated woodchucks.

24.4.3
Herpes Models

An *in-vivo* antiherpesviral activity of adefovir was initially demonstrated in nude mice infected intracutaneously, intraperitoneally, or intracerebrally with HSV-1 and, to a lesser extent, in animals infected with HSV-2. In these studies, adefovir exhibited activity when administered both topically and systemically [116]. Unlike acyclovir, adefovir was efficacious against both TK^+ and TK^- HSV-1 in a rabbit keratitis model, in which it afforded complete healing of eye lesions when applied topically as a 0.2% solution [117]. An *in-vivo* antiviral activity of systemically administered adefovir was also observed in irradiated immunodeficient mice infected with mouse cytomegalovirus [118]. However, in all explored herpesvirus animal models, adefovir was less potent than some other ANPs, including hydroxyphosphonomethoxypropyl derivatives of adenine and cytosine (cidofovir).

24.5
Clinical Experience

24.5.1
Tenofovir Disoproxil Fumarate (Viread®)

Clinical efficacy of TDF (300 mg once daily) against HIV-1 was initially established in treatment-experienced patients. When added to a stable suboptimal antiretroviral regimen, TDF showed a safety profile similar to that of placebo, and produced a significant plasma viral load reduction in two cohorts, in which more than 90% of patients had multiple nucleoside resistance mutations in RT at baseline [119, 120]. Subsequently, the durable efficacy and safety of TDF in combination with lamivudine (nucleoside RT inhibitor) and efavirenz (non-nucleoside RT inhibitor) was demonstrated in treatment-naïve patients through 144 weeks [121]. This study was performed as a head-to-head comparison with stavudine combined with lamivudine and efavirenz, and its analysis established the TDF/laminudine/efavirenz combination as one of the most efficacious antiretroviral regimens in treatment-naïve patients [6]. In 2004, a co-formulation of emtricitabine (a close structural analogue of lamivudine) and TDF (Truvada®) was approved, representing the first once-daily, fixed dose combination of two antiretrovirals [122]. A recent study in treatment-naïve patients demonstrated the superior efficacy and safety of TDF/emtricitabine once-daily compared to zidovudine/lamivudine twice-daily, both in combination with efavirenz [123]. A fixed-dose combination of TDF/emtricitabine/efavirenz is currently available (approved in 2006) as the only once-a-day, one-pill therapy for HIV infection (Atripla™). TDF also showed durable efficacy and safety in combination with HIV protease inhibitors (PIs), both in treatment-naïve [124] and experienced patients [125]. Taken together, the results from multiple studies indicate that TDF can be successfully used in combination with most available antiretrovirals. One exception, however, is the nucleoside analogue didanosine, which requires dose-reduction due to its pharmacokinetic drug interaction with TDF [126]. The increase in didanosine exposure observed in combination with TDF is thought to be due to the inhibitory effect of tenofovir metabolites on purine nucleoside phosphorylase, an enzyme involved in the systemic clearance of didanosine [127]. Regimens containing TDF in combination with dose-adjusted didanosine were found to be suboptimal in both naïve [128] and experienced [129] patients. Some additional triple nucleoside/nucleotide regimens such as TDF/abacavir/lamivudine also showed inferior efficacy in HIV patients [130]. As no antagonistic drug–drug interactions were found with this combination [131, 132], convergent resistance pathways appear to be the most likely explanation for the diminished clinical efficacy of this treatment strategy [133].

In comparison with some nucleosides (particularly stavudine), chronic treatment with TDF causes significantly fewer lipid abnormalities [121]. Several studies have demonstrated improvements in dyslipidemia and overall lipid metabolism profiles in patients who switched from stavudine to TDF [134, 135]. The likely mechanistic explanation of this favorable effect is the lower potential of tenofovir to induce mitochondrial toxicity [61]. Renal tubular dysfunction is the primary adverse effect found in a small subset of TDF treated patients [32]. This is likely due to an effective transport of tenofovir into the proximal tubules via renal organic anion transporters (see Section 24.3.1) [33, 34]. Although no difference in renal safety between placebo and TDF was reported in control trials and some patient cohorts [136, 137], analyses of other observational cohorts have indicated a relatively low incidence of renal dysfunction among patients treated with TDF [138]. Primary confounding factors include low $CD4^+$ cell counts, decreased baseline renal function, and diabetes. However, the overall favorable long-term safety and tolerability profile in comparison with some anti-HIV nucleosides is one of the major reasons for extensive clinical use of TDF in combination with many other antiretrovirals.

Based on a number of successful studies in animal models (see Section 24.4.1), tenofovir and TDF are being considered as potential agents for chemoprophylaxis of HIV infection [139]. Recent exploratory clinical studies established the pharmacokinetics of tenofovir in the genital tract [140] and addressed the safety/tolerability of a tenofovir vaginal gel in both HIV-positive and negative women [141]. Several pre-exposure prophylaxis clinical trials with TDF are currently ongoing.

In addition to demonstrating potent and durable efficacy against HIV, TDF was also shown effectively to suppress the replication of HBV. The clinical activity of TDF against HBV was first reported by Van Bommel *et al.*, who observed serum HBV DNA suppressions after treating lamivudine-resistant HIV/HBV co-infected patients [142]. Shortly thereafter, the anti-HBV activity of TDF was confirmed in multiple studies conducted in lamivudine-resistant HBV patients (with or without HIV co-infection) [143–149], as well as in co-infected patients who participated in TDF registrational trials for HIV infection [150]. The results of several of these studies suggested that the viral load reduction with 300 mg TDF was greater than that of the 10 mg approved dose of ADV. This difference was further demonstrated in two small head-to-head efficacy studies, wherein 300 mg TDF produced a $\geq$10-fold greater antiviral suppression compared to 10 mg ADV [151, 152]. These findings are not surprising given the 30-fold difference in dose and the similar *in-vitro* potency of tenofovir and adefovir against HBV (see Section 24.3.3).

TDF entered development for the treatment of chronic hepatitis B, and completed enrollment of two pivotal Phase III studies in 2006. These studies have been conducted in patients with $HBeAg^+$ and $HBeAg^-$ chronic hepatitis B, respectively, and compared the 300 mg TDF to the 10 mg ADV for one year in order to provide a head-to-head comparison in a greater number of patients. Results of these studies indicated that 71% and 67% of $HBeAg^+$ and $HBeAg^-$ patients, respectively, had complete response to TDF. Complete response were achieved in a significantly greater proportion of patients in the TDF arms compared to the ADV arms, in which complete responses to ADV were observed in 49% of $HBeAg^+$ patients and 12% of $HBeAg^-$ patients. Based on these data, TDF received marketing approval for the treatment of chronic HBV in multiple countries in 2008.

24.5.2
Adefovir Dipivoxil (Hepsera®)

ADV was initially explored in clinical studies as a novel antiretroviral agent for the treatment of HIV-1 infection. However, oral treatment of nucleoside-experienced patients with 60–120 mg doses of ADV once daily showed only moderate antiretroviral efficacy, and its clinical utility was limited by a significant incidence of drug-related nephrotoxicity [153]. Taking advantage of the improved potency of adefovir against HBV, ADV was subsequently explored in a wide range of clinical studies for the treatment of chronic HBV infection. Gilson and colleagues reported the first clinical study of ADV against HBV, wherein a 125 mg dose elicited a substantial reduction in serum HBV DNA [154]. Safety and efficacy observations from this trial, coupled with clinical data from HIV trials, prompted the investigation of lower doses in HBV-infected patients. In subsequent Phase II studies, doses of 5–60 mg once daily resulted in significant antiviral activity, with up to 4 $\log_{10}$ reductions in plasma viremia [155]. Based on efficacy and safety data, 30 mg and 10 mg doses were selected for Phase III studies.

Two pivotal Phase III studies were conducted in $HBeAg^{+}$ and $HBeAg^{-}$ chronic hepatitis B patients, respectively [156, 157]. These studies later served as the basis for drug approval in the United States in 2002, and in the European Union in 2003. During these studies, ADV 10 mg was significantly better than placebo for the primary endpoint of histological improvement (i.e., significantly reduced hepatic necroinflammation and fibrosis) in both $HBeAg^{+}$ and $HBeAg^{-}$ patients. Secondary clinical endpoints of viral load reduction and normalization of serum transaminase levels were also significantly improved in ADV-treated patients. Although the viral load suppression was greater in patients receiving 30 mg of ADV than in those who received 10 mg, the 30 mg dose was associated with increases in serum creatinine (a biomarker of nephrotoxicity) in 8% of patients and was consequently not pursued further. It is important to note that nephrotoxicity induced by ADV is reversible and resolves once therapy is discontinued. In contrast, the 10 mg ADV dose was not associated with clinical evidence of nephrotoxicity during the placebo-controlled part of these studies. Patients who completed the two-year Phase III studies were followed for up to three additional years on open-label ADV 10 mg to study the long-term safety and efficacy. The results of these studies indicated that continuing ADV therapy resulted in the maintenance or further improvement in necroinflammation, fibrosis, serum HBV suppression, and biochemical response in the majority of patients [158]. The clinical benefit of ADV is substantially more durable than that of lamivudine, this being a direct result of the greatly reduced frequency of viral resistance to ADV compared to lamivudine (see Section 24.6.2).

Based on the promising *in-vitro* activity of adefovir against lamivudine-resistant HBV mutants described above, the efficacy of ADV was tested in patients failing lamivudine therapy. Studies were conducted in several lamivudine-resistant patient populations, including those with compensated chronic hepatitis B, those wait-listed for or having received orthotopic liver transplantation, and those with HIV co-infection. In lamivudine-resistant patients with compensated liver disease, the clinical effects of switching from lamivudine to ADV monotherapy, adding ADV to lamivudine, or maintaining lamivudine alone were compared in a one-year study [159]. The ADV monotherapy and ADV plus lamivudine arms indicated that ADV was as effective in lamivudine-resistant patients as in the wild-type $HBeAg^{+}$ and $HBeAg^{-}$ patients studied during Phase III. There was no significant difference in efficacy between the patients switched to ADV monotherapy and those who received ADV plus lamivudine. Patients who remained on lamivudine only did not have any median change in serum HBV DNA.

In liver transplant studies, lamivudine-resistant patients received a 10 mg dose of ADV individually adjusted on the basis of the patients' renal function, while lamivudine therapy was either maintained or discontinued based on the treating physician's discretion. In both pre- and post-transplant patients, the antiviral efficacy of ADV was similar to that in patients with compensated liver disease with wild-type or lamivudine-resistant HBV infection [160, 161]. This result was particularly important as it confirmed that significant (decompensated) liver disease did not compromise the metabolic activation and efficacy of ADV. Currently, ADV is the only therapy approved for patients with decompensated liver disease and lamivudine-resistant HBV. The efficacy of ADV against lamivudine-resistant HBV DNA has also been demonstrated in HIV co-infected patients. For example, Benhamou and colleagues have treated a cohort HBV/HIV co-infected patients for over five years, and observed antiviral suppression and improvement in liver histology similar to that in other ADV-treated patient populations [162–164].

24.6 Drug Resistance

24.6.1 HIV Resistance

Similar to nucleoside antiretroviral inhibitors, the susceptibility of HIV to ANPs can be reduced by specific mutations in the viral reverse transcriptase (RT). Early studies established K65R as the primary RT mutation selected by both adefovir and tenofovir *in vitro* [165, 166]. Low-frequency emergence of K65R was found in multiple clinical studies both in treatment-naïve and NRTI-experienced patients exposed to TDF [121, 123, 167]. Notably, less than 3% of patients treated for three years with a TDF-containing first-line regimen developed K65R [121]. Although K65R can also be selected by some nucleosides such as abacavir or didanosine, its overall prevalence in treatment-experienced patients remains less than 5% [168], most likely because the mutation substantially reduces virus replicative fitness [169] and confers only partial resistance to various nucleosides and tenofovir [52]. Consequently, a subset of patients with the K65R mutation still respond to TDF treatment [170].

More recent reports have described the emergence of a K70E mutation in RT, primarily in patients treated with TDF/abacavir/lamivudine [171]. This mutation was originally found in association with exposure to adefovir *in vitro* [172] and in patients [173]. Similar to K65R, K70E also confers reduced susceptibility to tenofovir and several nucleosides, but the levels of resistance are generally lower [174] and the overall prevalence is substantially less frequent compared to K65R (<1% versus 3.5–4%) [175]. The effect of K70E on the clinical response to TDF is currently unknown.

In addition to K65R, multiple (three or more) thymidine analogue mutations (TAMs) containing primarily M41L and/or L210W are associated with reduced *in-vitro* susceptibility to tenofovir [176] and diminished clinical response to TDF [167]. In contrast, the presence of an M184V mutation in RT increases the susceptibility to tenofovir and slightly enhances the clinical response to TDF [167]. Results from extensive phenotypic analyses have shown that many HIV strains with mutations in

RT exhibit less resistance to tenofovir compared to most antiretroviral nucleosides [66, 69, 70]. Because of its favorable resistance profile, tenofovir is frequently used for the treatment of nucleoside-experienced patients with various mutations in RT. A comprehensive review on tenofovir resistance profile and efficacy in various populations of experienced patients has been published [176].

24.6.2 HBV Resistance

Since HBV does not infect cells *in vitro*, it was not possible to select adefovir-resistant HBV and to identify resistance mutations prior to clinical use. Adefovir resistance was thus identified during Phase III studies and mapped to two mutations in the polymerase gene of HBV: rtN236T and rtA181V. These mutations are distinct from those that emerge during therapy with other approved anti-HBV drugs (e.g., lamivudine, entecavir, telbivudine). The Phase III/open label extension study in HBeAg$^-$ patients provides the most comprehensive data on the frequency of viral resistance during long-term ADV therapy. During this study, no resistance mutations were observed following the first year of therapy. The cumulative probability of resistance emergence (rtN236T or rtA181V mutations) was 3%, 11%, 18%, and 29% after two, three, four, and five years of therapy, respectively [158]. In comparison, lamivudine resistance is considerably more frequent, with approximately 23% of patients developing resistance after one year, and 71% of patients after four years [177]. Due to orthogonal resistance patterns, the emergence of adefovir resistance is rare when ADV is used in combination with lamivudine. In the transplant and HIV co-infected populations described above, no patients receiving the combination of ADV and lamivudine developed resistance during up to five years of therapy [161, 163]. Additional studies by Lampertico *et al.* in larger patient populations have also demonstrated very low resistance rates in combination therapy (0.8% in ADV plus lamivudine patients after two years of therapy) [178].

The rtN236T and rtA181V resistance mutations result in relatively small changes in phenotypic susceptibility to adefovir (seven- to 14- fold and three- to four-fold, respectively) [179, 180]. These mutations show varying levels of *in-vitro* cross-resistance to several other nucleosides [180]. The susceptibilities to tenofovir of viruses containing rtN236T and rtA181V are reduced four-fold and three-fold, respectively; these values are slightly less than those of adefovir in side-by-side assays. Several preliminary clinical reports have indicated that patients with either rtN236T or rtA181V respond to TDF [181, 182], entecavir [183], or lamivudine [184].

24.7 Novel Antiviral Nucleotides and Nucleotide Prodrugs

Following the successful introduction of TDF and ADV into clinical practice, the focused search for novel prodrug approaches, as well as novel types of nucleotide

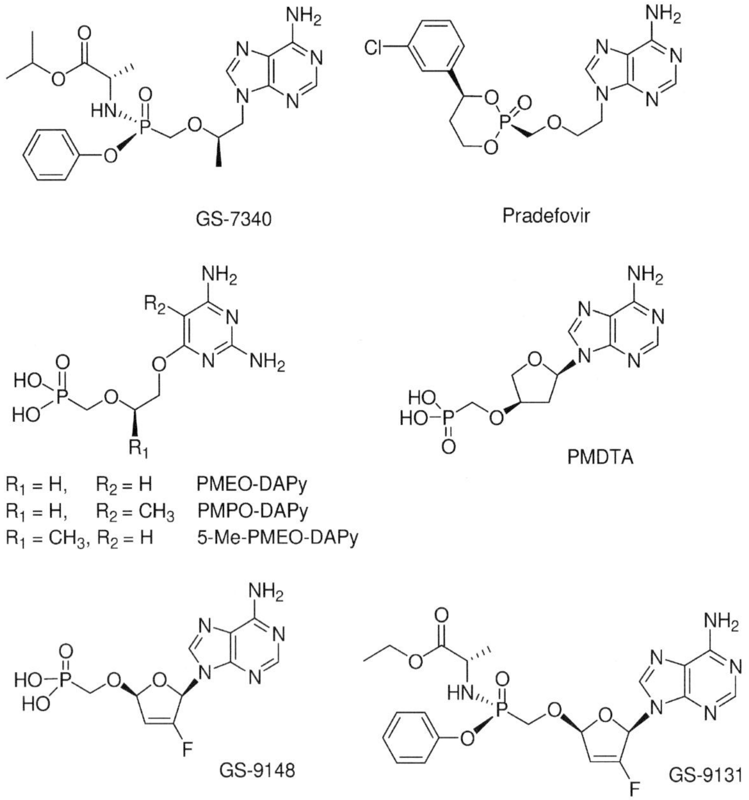

Figure 24.5 Novel antiviral nucleotides and their prodrugs.

analogues, has continued, yielding a number of new drug candidates with attractive pharmacological profiles.

In order to enhance the *in-vivo* delivery of parent tenofovir into target lymphocytes, various novel mono- and bis-amidate prodrugs have recently been explored. Among these, the mono-alaninyl mono-phenyl ester of tenofovir (GS-7340; Figure 24.5) showed the most favorable pharmacological properties. Compared to tenofovir itself, GS-7340 is up to 1000-fold more potent against HIV-1 *in vitro* [82]. Importantly, GS-7340 is substantially more stable in blood and plasma than TDF, but undergoes rapid hydrolysis in lymphocytes, resulting in an enhanced intracellular accumulation of tenofovir and tenofovir diphosphate [82, 185]. When administered orally to dogs, GS-7340 was seen to distribute more favorably into peripheral lymphocytes and lymphatic tissues than TDF, increasing the cell- and tissue-associated tenofovir levels by 15- to 30-fold but without changing the liver, kidney, and systemic exposure to tenofovir [82]. Results of multiple studies with GS-7340 have established amidate prodrugs as a promising strategy for effective *in-vivo* intracellular delivery of ANPs.

For the treatment of hepatitis B, a novel adefovir prodrug with a cyclic phosphonate promoiety (pradefovir; Figure 24.5), which is selectively activated by CYP3A4 in the liver, is being developed with the goal of reducing the systemic exposure of adefovir [186]. This liver-targeting strategy would potentially allow higher and more efficacious doses of adefovir to be delivered, without the risk of nephrotoxicity. Pharmacokinetic studies have confirmed the conversion of pradefovir to adefovir *in vivo* [187], while preliminary data from a Phase II study have indicated that doses of 10–30 mg pradefovir are more efficacious compared to 10 mg of ADV, without a significant changes in kidney function markers [188]. However, carcinogenicity was observed in rats and mice after prolonged dosing with pradefovir and it is unclear whether the clinical development will continue.

The extensive exploration of structurally simple ANPs yielded a new series of phosphonates containing a 6-substituted 2,4-diaminopyrimidine (DAPy) base that mimics the natural purine base [189]. Similar to adefovir and tenofovir, their respective DAPy analogues PMEO-DAPy and PMPO-DAPy (Figure 24.5) are both active against HIV and HBV [189, 190]. More recently, PMEO derivatives with various 5-substituted DAPy bases were prepared, several of which (e.g., 5-CN, 5-CH_3, 5-Br, and 5-CHO derivatives) were identified as potent inhibitors of HIV-1 and/or HBV replication [191, 192]. 5-CH_3-PMEO-DAPy in particular has shown promising activity against some of the nucleoside-resistant strains of HIV-1, and was more effective than adefovir in inhibiting murine retroviruses *in vivo* [71].

In contrast to ANPs, nucleotides with a cyclic sugar moiety remain less explored, primarily because of synthetic challenges, and consequently comparatively fewer active cyclic nucleoside phosphonates have been identified to date. Among recent examples, nucleotides containing a 2-deoxythreose sugar are of note [193] with the adenine derivative (PMDTA; Figure 24.5) being particularly active against HIV with *in-vitro* potency and selectivity comparable to that of tenofovir. Among the series of cyclic phosphonates containing dideoxy-didehydro-ribose (d4) characterized in the past, the adenine derivative (d4AP) is a potent inhibitor of HIV-1 [194]. However, its diphosphate metabolite is an effective substrate for DNA polymerase γ, suggesting a potential for mitochondrial toxicity [59]. A variety of d4AP analogues have been explored, yielding the 2′-fluoro substituted nucleotide GS-9148 (Figure 24.5), which has a significantly diminished interaction with DNA polymerase γ [195] and thus a low potential for mitochondrial toxicity [196]. GS-9148 exhibits a favorable *in-vitro* resistance profile, retaining its activity against viruses with single or multiple nucleoside resistance mutations [196]. Its mono-amidate prodrug GS-9131 (ethylalaninyl phenyl ester; Figure 24.5) markedly enhances the cell-based antiviral potency of GS-9148 and allows for a substantial accumulation and a prolonged retention of the active metabolite GS-9148 diphosphate in peripheral lymphocytes *in vivo*, a profile which is suggestive (potentially) of infrequent oral dosing [197].

24.8 Conclusions

The successful clinical development of TDF and ADV, followed by their extensive use across multiple populations of HIV- and HBV-infected patients, has firmly validated ANPs as a novel class of clinically efficacious antiviral agents. Although the application of prodrug strategies is necessary for the effective oral delivery of ANPs, they possess unique and attractive features such as improved catabolic stability that enhances their intracellular retention relative to nucleosides, providing the advantage of less-frequent dosing. In addition, the aliphatic linker mimicking the nucleotide sugar moiety offers additional structural flexibility, allowing ANPs to maintain their activity against a broad range of genetically diverse viruses, including strains that are resistant to multiple nucleoside analogues. The concept of ANPs that emerged during the mid-1980s as a unique approach to the design of viral polymerase inhibitors, and stemming from the earlier successful development of acyclic nucleoside analogues, has culminated some 20 years later (in 2006) with the commercial approval of Atripla™, the first available three-drug fixed-dose combination of antiretroviral agents containing TDF co-formulated with emtricitabine and efavirenz. This one-pill, once-a-day product with potent and durable clinical efficacy has set a new standard for the management of HIV-infected patients. Currently, both ADV and TDF are among the most frequently prescribed drugs in their respective therapeutic areas, and additional applications – particularly with TDF and tenofovir in the treatment of chronic HBV and prophylaxis of HIV, respectively– are being extensively explored. Importantly, the clinical benefit provided by ANPs continues to stimulate drug discovery efforts not only in academia but also within the pharmaceutical industry. This, in turn, should lead to the identification of novel structural types of both acyclic and cyclic nucleotides, which may enter clinical development in near future and expand the class of clinically efficacious antiviral nucleotide therapeutics.

References

1 De Clercq, E., Holy, A., Rosenberg, I., Sakuma, T., Balzarini, J., Maudgal, P. C., *Nature* **1986**, *323*, 464–467.

2 Holy, A., *Tetrahedron Lett.* **1967**, *10*, 881–884.

3 Jones, G. H., Moffatt, J. G., *J. Am. Chem. Soc.* **1968**, *90*, 5336–5338.

4 Robbins, B., Srinivas, R., Kim, C., Bischofberger, N., Fridland, A., *Antimicrob. Agents Chemother.* **1998**, *42*, 612–617.

5 Ledford, R., Vela, J., Ray, A., Callebaut, C., Miller, M., McColl, D., *Abstract PE 4.1/7 10th European AIDS Conference*, Dublin, Ireland, **2005**.

6 De Clercq, E., *Expert Rev. Anti-Infect. Ther.* **2003**, *1*, 21–43.

7 De Clercq, E., Holy, A., *Nat. Rev. Drug Discov.* **2005**, *4*, 928–940.

8 Holy, A., *Curr. Pharm. Des.* **2003**, *9*, 2567–2592.

9 Lee, W. A., Martin, J. C., *Antiviral Res.* **2006**, *71*, 254–259.

10 Holy, A., Rosenberg, I., European Patent Application 0206 459 A2, **1986**.

11 Holy, A., Kohoutova, J., Merta, A., Votruba, I., *Collect. Czech. Chem. Commun.* **1986**, *51*, 459–477.

12 Holy, A., Rosenberg, I., *Collect. Czech. Chem. Commun.* **1987**, *52*, 2801–2809.

13 Holy, A., Rosenberg, I., *Collect. Czech. Chem. Commun.* **1982**, *47*, 3447–3463.

14 Holy, A., Rosenberg, I., Dvorakova, H., *Collect. Czech. Chem. Commun.* **1989**, *54*, 2190–2210.

15 Yu, R., Schultze, L. M., Rohloff, J. C., Dudzinski, P. W., Kelly, D. E., *Org. Proc. Res. Dev.* **1999**, *3*, 53–55.

16 Kondo, K., Iwasaki, H., Ueda, N., Takemoto, M., Imoto, M., *Makromol. Chem.* **1968**, *120*, 21.

17 Becker, M., Chapman, H., Cihlar, T., Eisenberg, E., He, G., Kernan, M., Lee, W., Prisbe, E., Rohloff, J., Sparacino, M., WO0208241A2, **2002**.

18 Starrett, J. E. Jr., Tortolani, D. R., Russell, J., Hitchcock, M. J. M., Whiterock, V., Martin, J. C., Mansuri, M. M., *J. Med. Chem.* **1994**, *37*, 1857–1864.

19 Holy, A., *Collect. Czech. Chem. Commun.* **1975**, *40*, 187–214.

20 Holy, A., Rosenberg, I., Dvorakova, H., DeClercq, E., *Nucleosides Nucleotides* **1988**, *7*, 667–670.

21 Rosenberg, I., Holy, A., Masojidkova, M., *Collect. Czech. Chem. Commun.* **1988**, *53*, 2753–2777.

22 Holy, A., Masojidkova, M., *Collect. Czech. Chem. Commun.* **1995**, *60*, 1196–1212.

23 Holy, A., Dvorakova, H., Masojidkova, M., *Collect. Czech. Chem. Commun.* **1995**, *60*, 1390–1409.

24 Schultze, L., Chapman, H., Dubree, N., Jones, R., Kent, K., Lee, T., Louie, M., Postich, M., Prisbe, E., Rohloff, J., Yu, R., *Tetrahedron Lett.* **1998**, *39*, 1853–1856.

25 Arimilli, M. N., Kim, C. U., Dougherty, J., Mulato, A., Oliyai, R., Shaw, J. P., Cundy, K. C., Bischofberger, N., *Antiviral Chem. Chemother.* **1997**, *8*, 557–564.

26 Arimilli, M., Cundy, K., Dougherty, J., Kim, C., Oliyai, R., Stella, V., WO9804569, **1998**.

27 Yao, S. Y., Ng, A. M., Sundaram, M., Cass, C. E., Baldwin, S. A., Young, J. D., *Mol. Membr. Biol.* **2001**, *18*, 161–167.

28 Zimmerman, T. P., Mahony, W. B., Prus, K. L., *J. Biol. Chem.* **1987**, *262*, 5748–5754.

29 Palu, G., Stefanelli, S., Rassu, M., Parolin, C., Balzarini, J., De Clercq, E., *Antiviral Res.* **1991**, *16*, 115–119.

30 Olsanska, L., Cihlar, T., Votruba, I., Holy, A., *Collect. Czech. Chem. Commun.* **1997**, *62*, 821–828.

31 Cihlar, T., Rosenberg, I., Votruba, I., Holy, A., *Antimicrob. Agents Chemother.* **1995**, *39*, 117–124.

32 Izzedine, H., Launay-Vacher, V., Deray, G., *Am. J. Kidney Dis.* **2005**, *45*, 804–817.

33 Cihlar, T., Bleasby, K., Roy, A., Pritchard, J., *44th Interscience Conference on Antimicrobial Agents and Chemotherapy*, Washington DC, USA, Abstract A-443, **2004**.

34 Cihlar, T., Ho, E. S., Lin, D. C., Mulato, A. S., *Nucleosides Nucleotides Nucleic Acids* **2001**, *20*, 641–648.

35 Ho, E. S., Lin, D. C., Mendel, D. B., Cihlar, T., *J. Am. Soc. Nephrol.* **2000**, *11*, 383–393.

36 Imaoka, T., Kusuhara, H., Adachi, M., Schuetz, J. D., Takeuchi, K., Sugiyama, Y., *Mol. Pharmacol.* **2007**, *71*, 619–627.

37 Ray, A. S., Cihlar, T., Robinson, K. L., Tong, L., Vela, J. E., Fuller, M. D., Wieman, L. M., Eisenberg, E. J., Rhodes, G. R., *Antimicrob. Agents Chemother.* **2006**, *50*, 3297–3304.

38 Wijnholds, J., Mol, C. A., van Deemter, L., de Haas, M., Scheffer, G. L., Baas, F., Beijnen, J. H., Scheper, R. J., Hatse, S., De Clercq, E., Balzarini, J., Borst, P., *Proc. Natl. Acad. Sci. USA* **2000**, *97*, 7476–7481.

39 Schuetz, J. D., Connelly, M. C., Sun, D., Paibir, S. G., Flynn, P. M., Srinivas, R. V., Kumar, A., Fridland, A., *Nature Med.* **1999**, *5*, 1048–1051.

40 Merta, A., Votruba, I., Jindrich, J., Holy, A., Cihlar, T., Rosenberg, I., Otmar, M., Herve, T. Y., *Biochem. Pharmacol.* **1992**, *44*, 2067–2077.

41 Robbins, B. L., Greenhaw, J., Connelly, M. C., Fridland, A., *Antimicrob. Agents Chemother.* **1995**, *39*, 2304–2308.

42 Miller, W. H., Miller, R. L., *Biochem. Pharmacol.* **1982**, *31*, 3879–3884.

43 Balzarini, J., Nave, J. F., Becker, M. A., Tatibana, M., De Clercq, E., *Nucleosides Nucleotides* **1995**, *14*, 1861–1871.

44 Aduma, P., Connelly, M. C., Srinivas, R. V., Fridland, A., *Mol. Pharmacol.* **1995**, *47*, 816–822.

45 Balzarini, J., Hao, Z., Herdewijn, P., Johns, D. G., De Clercq, E., *Proc. Natl. Acad. Sci. USA* **1991**, *88*, 1499–1503.

46 Balzarini, J., *Adv. Exp. Med. Biol.* **1994**, *370*, 459–464.

47 Ray, A. S., Vela, J. E., Olson, L., Fridland, A., *Biochem. Pharmacol.* **2004**, *68*, 1825–1831.

48 Delaney, W. E. t., Ray, A. S., Yang, H., Qi, X., Xiong, S., Zhu, Y., Miller, M. D., *Antimicrob. Agents Chemother.* **2006**, *50*, 2471–2477.

49 Suo, Z., Johnson, K. A., *J. Biol. Chem.* **1998**, *273*, 27250–27258.

50 Cherrington, J. M., Allen, S. J. W., Bischofberger, N., Chen, M. S., *Antiviral Chem. Chemother.* **1995**, *6*, 217–221.

51 Meyer, P. R., Matsuura, S. E., So, A. G., Scott, W. A., *Proc. Natl. Acad. Sci. USA* **1998**, *95*, 13471–13476.

52 White, K. L., Margot, N. A., Ly, J. K., Chen, J. M., Ray, A. S., Pavelko, M., Wang, R., McDermott, M., Swaminathan, S., Miller, M. D., *AIDS* **2005**, *19*, 1751–1760.

53 Xiong, X., Flores, C., Yang, H., Toole, J. J., Gibbs, C. S., *Hepatology* **1998**, *28*, 1669–1673.

54 Gaillard, R. K., Barnard, J., Lopez, V., Hodges, P., Bourne, E., Johnson, L., Allen, M. I., Condreay, P., Miller, W. H., Condreay, L. D., *Antimicrob. Agents Chemother.* **2002**, *46*, 1005–1013.

55 Seigneres, B., Aguesse-Germon, S., Pichoud, C., Vuillermoz, I., Jamard, C., Trepo, C., Zoulim, F., *J. Hepatol.* **2001**, *34*, 114–122.

56 Merta, A., Votruba, I., Rosenberg, I., Otmar, M., Hrebabecky, H., Bernaerts, R., Holy, A., *Antiviral Res.* **1990**, *13*, 209–218.

57 Xiong, X., Flores, C., Fuller, M. D., Mendel, D. B., Mulato, A. S., Moon, K., Chen, M. S., Cherrington, J. M., *Antiviral Res.* **1997**, *36*, 131–137.

58 Cerny, J., Votruba, I., Vonka, V., Rosenberg, I., Otmar, M., Holy, A., *Antiviral Res.* **1990**, *13*, 253–264.

59 Cihlar, T., Chen, M. S., *Antiviral Chem. Chemother.* **1997**, *8*, 187–195.

60 Birkus, G., Gibbs, C. S., Cihlar, T., *J. Viral Hepatol.* **2003**, *10*, 50–54.

61 Birkus, G., Hitchcock, M. J., Cihlar, T., *Antimicrob. Agents Chemother.* **2002**, *46*, 716–723.

62 Biesecker, G., Karimi, S., Desjardins, J., Meyer, D., Abbott, B., Bendele, R., Richardson, F., *Antiviral Res.* **2003**, *58*, 217–225.

63 Birkus, G., Hajek, M., Kramata, P., Votruba, I., Holy, A., Otova, B., *Antimicrob. Agents Chemother.* **2002**, *46*, 1610–1613.

64 Cihlar, T., Birkus, G., Greenwalt, D. E., Hitchcock, M. J., *Antiviral Res.* **2002**, *54*, 37–45.

65 Mulato, A. S., Cherrington, J. M., *Antiviral Res.* **1997**, *36*, 91–97.

66 Palmer, S., Margot, N., Gilbert, H., Shaw, N., Buckheit, R. Jr., Miller, M., *AIDS Res. Hum. Retroviruses* **2001**, *17*, 1167–1173.

67 Balzarini, J., Aquaro, S., Perno, C. F., Witvrouw, M., Holý, A., De Clercq, E., *Biochem. Biophys. Res. Commun.* **1996**, *219*, 337–341.

68 Shirasaka, T., Chokekijchai, S., Yamada, A., Gosselin, G., Imbach, J. L., Mitsuya, H., *Antimicrob. Agents Chemother.* **1995**, *39*, 2555–2559.

69 Miller, M. D., Margot, N. A., Hertogs, K., Larder, B., Miller, V., *Nucleosides, Nucleotides, Nucleic Acids* **2001**, *20*, 1025–1028.

70 Harrigan, P. R., Miller, M. D., McKenna, P., Brumme, Z. L., Larder, B. A., *Antimicrob. Agents Chemother.* **2002**, *46*, 1067–1072.

71 Balzarini, J., Schols, D., Van Laethem, K., De Clercq, E., Hockova, D., Masojidkova, M., Holy, A., *J. Antimicrob. Chemother.* **2007**, *59*, 80–86.

72 Balzarini, J., Naesens, L., Slachmuylders, J., Niphuis, H., Rosenberg, I., Holy, A., Schellekens, H., De Clercq, E., *AIDS* **1991**, *5*, 21–28.

73 Thormar, H., Balzarini, J., Debyser, Z., Witvrouw, M., Desmyter, J., De Clercq, E., *Antiviral Res.* **1995**, *27*, 49–57.

74 Yokota, T., Mochizuki, S., Konno, K., Mori, S., Shigeta, S., De Clercq, E., *Antimicrob. Agents Chemother.* **1991**, *35*, 394–397.

75 Heijtink, R. A., De Wilde, G. A., Kruining, J., Berk, L., Balzarini, J., De Clercq, E.,

Holy, A., Schalm, S. W., *Antiviral Res.* **1993**, *21*, 141–153.

76 Dandri, M., Burda, M. R., Will, H., Petersen, J., *Hepatology* **2000**, *32*, 139–146.

77 Yang, H., Xiong, S., Delaney, W. E. IV *J. Hepatol.* **2003**, *38*, 8.

78 Chen, R. Y., Edwards, R., Shaw, T., Colledge, D., Delaney, W. E. IV, Isom, H., Bowden, S., Desmond, P., Locarnini, S. A., *Hepatology* **2003**, *37*, 27–35.

79 Yang, H., Qi, X., Sabogal, A., Miller, M., Xiong, S., Delaney, W. E., *Antiviral Ther.* **2005**, *10*, 625–633.

80 Ono, S. K., Kato, N., Shiratori, Y., Kato, J., Goto, T., Schinazi, R. F., Carrilho, F. J., Omata, M., *J. Clin. Invest.* **2001**, *107*, 449–455.

81 Delaney, W. E. t., Edwards, R., Colledge, D., Shaw, T., Torresi, J., Miller, T. G., Isom, H. C., Bock, C. T., Manns, M. P., Trautwein, C., Locarnini, S., *Antimicrob. Agents Chemother.* **2001**, *45*, 1705–1713.

82 Lee, W. A., He, G. X., Eisenberg, E., Cihlar, T., Swaminathan, S., Mulato, A., Cundy, K. C., *Antimicrob. Agents Chemother.* **2005**, *49*, 1898–1906.

83 Naesens, L., Balzarini, J., Bischofberger, N., De Clercq, E., *Antimicrob. Agents Chemother.* **1996**, *40*, 22–28.

84 Balzarini, J., Holy, A., Jindrich, J., Naesens, L., Snoeck, R., Schols, D., De Clercq, E., *Antimicrob. Agents Chemother.* **1993**, *37*, 332–338.

85 De Clercq, E., Sakuma, T., Baba, M., Pauwels, R., Balzarini, J., Rosenberg, I., Holy, A., *Antiviral Res.* **1987**, *8*, 261–272.

86 Snoeck, R., Andrei, G., De Clercq, E., *Eur. J. Clin. Microbiol. Infect. Dis.* **1996**, *15*, 574–579.

87 Lin, J. C., DeClercq, E., Pagano, J. S., *Antimicrob. Agents Chemother.* **1987**, *31*, 1431–1433.

88 Zidek, Z., Frankova, D., Holy, A., *Int. J. Immunopharmacol.* **2000**, *22*, 1121–1129.

89 Zidek, Z., Frankova, D., Holy, A., *Antimicrob. Agents Chemother.* **2001**, *45*, 3381–3386.

90 Kmonickova, E., Potmesil, P., Holy, A., Zidek, Z., *Eur. J. Pharmacol.* **2006**, *530*, 179–187.

91 Naesens, L., Snoeck, R., Andrei, G., Balzarini, J., Neyts, J., De Clercq, E., *Antiviral Chem. Chemother.* **1997**, *8*, 1–23.

92 Balzarini, J., Naesens, L., Herdewijn, P., Rosenberg, I., Holy, A., Pauwels, R., Baba, M., Johns, D. G., De Clercq, E., *Proc. Natl. Acad. Sci. USA* **1989**, *86*, 332–336.

93 Naesens, L., Neyts, J., Balzarini, J., Holy, A., Rosenberg, I., De Clercq, E., *J. Med. Virol.* **1993**, *39*, 167–172.

94 Gangemi, J. D., Cozens, R. M., De Clercq, E., Balzarini, J., Hochkeppel, H. K., *Antimicrob. Agents Chemother.* **1989**, *33*, 1864–1868.

95 Hartmann, K., Kuffer, M., Balzarini, J., Naesens, L., Goldberg, M., Erfle, V., Goebel, F. D., De Clercq, E., Jindrich, J., Holy, A., Bischofberger, N., Kraft, W., *J. Acquir. Immune Defic. Syndr. Hum. Retrovirol.* **1998**, *17*, 120–128.

96 Thormar, H., Georgsson, G., Palsson, P. A., Balzarini, J., Naesens, L., Torsteinsdottir, S., De Clercq, E., *Proc. Natl. Acad. Sci. USA* **1995**, *92*, 3283–3287.

97 Tsai, C. C., Follis, K. E., Beck, T. W., Sabo, A., Bischofberger, N., Dailey, P. J., *AIDS Res. Hum. Retroviruses* **1997**, *13*, 707–712.

98 Tsai, C. C., Follis, K. E., Sabo, A., Grant, R., Bischofberger, N., *J. Infect. Dis.* **1995**, *171*, 1338–1343.

99 Van Rompay, K. K., Cherrington, J. M., Marthas, M. L., Berardi, C. J., Mulato, A. S., Spinner, A., Tarara, R. P., Canfield, D. R., Telm, S., Bischofberger, N., Pedersen, N. C., *Antimicrob. Agents Chemother.* **1996**, *40*, 2586–2591.

100 Li, Q., Duan, L., Estes, J. D., Ma, Z. M., Rourke, T., Wang, Y., Reilly, C., Carlis, J., Miller, C. J., Haase, A. T., *Nature* **2005**, *434*, 1148–1152.

101 George, M. D., Reay, E., Sankaran, S., Dandekar, S., *J. Virol.* **2005**, *79*, 2709–2719.

102 Tsai, C. C., Follis, K. E., Grant, R., Sabo, A., Nolte, R., Bartz, C., Bischofberger, N.,

Benveniste, R., *J. Med. Primatol.* **1994**, *23*, 175–183.

103 Tsai, C. C., Follis, K. E., Sabo, A., Beck, T. W., Grant, R. F., Bischofberger, N., Benveniste, R. E., Black, R., *Science* **1995**, *270*, 1197–1199.

104 Otten, R. A., Smith, D. K., Adams, D. R., Pullium, J. K., Jackson, E., Kim, C. N., Jaffe, H., Janssen, R., Butera, S., Folks, T. M., *J. Virol.* **2000**, *74*, 9771–9775.

105 Van Rompay, K. K., McChesney, M. B., Aguirre, N. L., Schmidt, K. A., Bischofberger, N., Marthas, M. L., *J. Infect. Dis.* **2001**, *184*, 429–438.

106 Garcia-Lerma, J., Otten, R., Qari, S., Jackson, E., Luo, W., Monsour, M., Schinazi, R., Janssen, R., Folks, T., Heneine, W., *13th Conference on Retroviruses and Opportunistic Infections, Abstract 32LB,* Denver, Colorado, **2006**.

107 Nicoll, A. J., Colledge, D. L., Toole, J. J., Angus, P. W., Smallwood, R. A., Locarnini, S. A., *Antimicrob. Agents Chemother.* **1998**, *42*, 3130–3135.

108 Delmas, J., Schorr, O., Jamard, C., Gibbs, C., Trepo, C., Hantz, O., Zoulim, F., *Antimicrob. Agents Chemother.* **2002**, *46*, 425–433.

109 Cullen, J. M., Brown, C., Li, D., Eisenberg, G., Cundy, K. C., Gibbs, C. S., *Antiviral Ther.* **2000**, *5*, B.67.

110 Jacob, J. R., Korba, B. E., Cote, P. J., Toshkov, I., Delaney, W. E. t., Gerin, J. L., Tennant, B. C., *Antiviral Res.* **2004**, *63*, 115–121.

111 Araki, K., Miyazaki, J., Hino, O., Tomita, N., Chisaka, O., Matsubara, K., Yamamura, K., *Proc. Natl. Acad. Sci. USA* **1989**, *86*, 207–211.

112 Guidotti, L. G., Matzke, B., Schaller, H., Chisari, F. V., *J. Virol.* **1995**, *69*, 6158–6169.

113 Kajino, K., Kamiya, N., Yuasa, S., Takahara, T., Sakurai, J., Yamamura, K., Hino, O., *Biochem. Biophys. Res. Commun.* **1997**, *241*, 43–48.

114 Julander, J. G., Sidwell, R. W., Morrey, J. D., *Antiviral Res.* **2002**, *55*, 27–40.

115 Menne, S., Cote, P. J., Korba, B. E., Butler, S. D., George, A. L., Tochkov, I. A., Delaney, W. E. t., Xiong, S., Gerin, J. L., Tennant, B. C., *Antimicrob. Agents Chemother.* **2005**, *49*, 2720–2728.

116 De Clercq, E., Holy, A., Rosenberg, I., *Antimicrob. Agents Chemother.* **1989**, *33*, 185–191.

117 Maudgal, P. C., De Clercq, E., *Curr. Eye Res.*, **1991**, 10 Suppl, 139–142.

118 Neyts, J., Stals, F., Bruggeman, C., De Clercq, E., *Eur. J. Clin. Microbiol. Infect. Dis.* **1993**, *12*, 437–446.

119 Schooley, R. T., Ruane, P., Myers, R. A., Beall, G., Lampiris, H., Berger, D., Chen, S.-S., Miller, M. D., Isaacson, E., Cheng, A. K., *AIDS* **2002**, *16*, 1257–1263.

120 Squires, K., Pozniak, A. L., Pierone, G., Steinhart, C. R., Berger, D., Bellos, N. C., Becker, S. L., Wulfsohn, M., Miller, M. D., Toole, J. J., Coakley, D. F., Cheng, A., Team, S., *Ann. Intern. Med.* **2003**, *139*, 313–320.

121 Gallant, J. E., Staszewski, S., Pozniak, A. L., DeJesus, E., Suleiman, J. M., Miller, M. D., Coakley, D. F., Lu, B., Toole, J. J., Cheng, A. K., Group, S., *JAMA* **2004**, *292*, 191–201.

122 Dando, T. M., Wagstaff, A. J., *Drugs* **2004**, *64*, 2075–2082.

123 Gallant, J., DeJesus, E., Arribas, J., Pozniak, A., Gazzard, B., Campo, R., Lu, B., McColl, D., Chuck, S., Enejosa, J., Toole, J., Cheng, A., Group, S., *N. Engl. J. Med.* **2006**, *354*, 251–260.

124 Johnson, M. A., Gathe, J. C. Jr., Podzamczer, D., Molina, J. M., Naylor, C. T., Chiu, Y. L., King, M. S., Podsadecki, T. J., Hanna, G. J., Brun, S. C., *J. Acquir. Immune Defic. Syndr.* **2006**, *43*, 153–160.

125 Johnson, M., Grinsztejn, B., Rodriguez, C., Coco, J., Dejesus, E., Lazzarin, A., Lichtenstein, K., Wirtz, V., Rightmire, A., Odeshoo, L., McLaren, C., *AIDS* **2006**, *20*, 711–718.

126 Kearney, B. P., Sayre, J. R., Flaherty, J. F., Chen, S. S., Kaul, S., Cheng, A. K., *J. Clin. Pharmacol.* **2005**, *45*, 1360–1367.

127 Ray, A., Olson, L., Fridland, A., *Antimicrob. Agents Chemother.* **2004**, *48*, 1089–1095.

128 Maitland, D., Moyle, G., Hand, J., Mandalia, S., Boffito, M., Nelson, M., Gazzard, B., *AIDS* **2005**, *19*, 1183–1188.

129 Leon, A., Mallolas, J., Martinez, E., De Lazzari, E., Pumarola, T., Larrousse, M., Milincovic, A., Lonca, M., Blanco, J. L., Laguno, M., Biglia, A., Gatell, J. M., *AIDS* **2005**, *19*, 1695–1697.

130 Gallant, J. E., Rodriguez, A. E., Weinberg, W. G., Young, B., Berger, D. S., Lim, M. L., Liao, Q., Ross, L., Johnson, J., Shaefer, M. S., Study, E., *J. Infect. Dis.* **2005**, *192*, 1921–1930.

131 Hawkins, T., Veikley, W., St Claire, R. L., Guyer, B., Clark, N., Kearney, B. P., *J. Acquir. Immune Defic. Syndr.* **2005**, *39*, 406–411.

132 Ray, A. S., Myrick, F., Vela, J. E., Olson, L. Y., Eisenberg, E. J., Borroto-Esoda, K., Miller, M. D., Fridland, A., *Antiviral Ther.* **2005**, *10*, 451–457.

133 Delaunay, C., Brun-Vezinet, F., Landman, R., Collin, G., Peytavin, G., Trylesinski, A., Flandre, P., Miller, M., Descamps, D., *J. Virol.* **2005**, *79*, 9572–9578.

134 Domingo, P., Labarga, P., Palacios, R., Guerro, M. F., Terron, J. A., Elias, M. J., Santos, J., Ruiz, M. I., Llibre, J. M., Group, R. S., *AIDS* **2004**, *18*, 1475–1478.

135 Schewe, C. K., Maserati, R., Wassmer, G., Adam, A., Weitner, L., *Clin. Infect. Dis.* **2006**, *42*, 145–147.

136 Izzedine, H., Hulot, J. S., Vittecoq, D., Gallant, J. E., Staszewski, S., Launay-Vacher, V., Cheng, A., Deray, G., Team, S., *Nephrol. Dial. Transplant.* **2005**, *20*, 743–746.

137 Izzedine, H., Isnard-Bagnis, C., Hulot, J. S., Vittecoq, D., Cheng, A., Jais, C. K., Launay-Vacher, V., Deray, G., *AIDS* **2004**, *18*, 1074–1076.

138 Gallant, J. E., Parish, M. A., Keruly, J. C., Moore, R. D., *Clin. Infect. Dis.* **2005**, *40*, 1194–1198.

139 Roland, M. E., *Curr. Opin. Infect. Dis.* **2007**, *20*, 39–46.

140 Vourvahis, M., Tappouni, H., Patterson, K., Chen, Y. C., Rezk, N., Fiscus, S., Kearney, B., Rooney, J., Cohen, M., Kashuba, A., *J. Acquir. Immune Defic. Syndr.* **2008**, *47* 329–333.

141 Mayer, K. H., Maslankowski, L. A., Gai, F., El-Sadr, W. M., Justman, J., Kwiecien, A., Masse, B., Eshleman, S. H., Hendrix, C., Morrow, K., Rooney, J. F., Soto-Torres, L., *AIDS* **2006**, *20*, 543–551.

142 van Bommel, F., Wunsche, T., Schurmann, D., Berg, T., *Hepatology* **2002**, *36*, 507–508.

143 Cecil, B. D., *Hepatology* **2002**, 630A.

144 Nunez, M., Perez-Olmeda, M., Diaz, B., Rios, P., Gonzalez-Lahoz, J., Soriano, V., *AIDS* **2002**, *16*, 2352–2354.

145 Ristig, M. B., Crippin, J., Aberg, J. A., Powderly, W. G., Lisker-Melman, M., Kessels, L., Tebas, P., *J. Infect. Dis.* **2002**, *186*, 1844–1847.

146 Nelson, M., Portsmouth, S., Stebbing, J., Atkins, M., Barr, A., Matthews, G., Pillay, D., Fisher, M., Bower, M., Gazzard, B., *AIDS* **2003**, *17*, F7–F10.

147 Benhamou, Y., Tubiana, R., Thibault, V., *N. Engl. J. Med.* **2003**, *348*, 177–178.

148 Bruno, R., Sacchi, P., Zocchetti, C., Ciappina, V., Puoti, M., Filice, G., *AIDS* **2003**, *17*, 783–784.

149 Kuo, A., Dienstag, J. L., Chung, R. T., *Clin. Gastroenterol. Hepatol.* **2004**, *2*, 266–272.

150 Dore, G. J., Cooper, D. A., Pozniak, A. L., DeJesus, E., Zhong, L., Miller, M. D., Lu, B., Cheng, A. K., *J. Infect. Dis.* **2004**, *189*, 1185–1192.

151 van Bommel, F., Wunsche, T., Mauss, S., Reinke, P., Bergk, A., Schurmann, D., Wiedenmann, B., Berg, T., *Hepatology* **2004**, *40*, 1421–1425.

152 Peters, M. G., Andersen, J., Lynch, P., Liu, T., Alston–Smith, B., Brosgart, C. L., Jacobson, J. M., Johnson, V. A., Pollard, R. B., Rooney, J. F., Sherman, K. E., Swindells, S., Polsky, B., *Hepatology* **2006**, *44*, 1110–1116.

153 Kahn, J., Lagakos, S., Wulfsohn, M., Cherng, D., Miller, M., Cherrington, J.,

Hardy, D., Beall, G., Cooper, R., Murphy, R., Basgoz, N., Ng, E., Deeks, S., Winslow, D., Toole, J. J., Coakley, D., *JAMA* **1999**, *282*, 2305–2312.

154 Gilson, R. J., Chopra, K. B., Newell, A. M., Murray-Lyon, I. M., Nelson, M. R., Rice, S. J., Tedder, R. S., Toole, J., Jaffe, H. S., Weller, I. V., *J. Viral Hepatol.* **1999**, *6*, 387–395.

155 Heathcote, E. J., Jeffers, L., Wright, T., Sherman, M., Perrillo, R., Sacks, S., Carithers, R., Rustgi, V., Di Bisceglie, A., Balan, V., Murray, A., Rooney, J., Jaffe, H. S., *Hepatology* **1998**, *28*, 317A.

156 Hadziyannis, S. J., Tassopoulos, N. C., Heathcote, E. J., Chang, T. T., Kitis, G., Rizzetto, M., Marcellin, P., Lim, S. G., Goodman, Z., Wulfsohn, M. S., Xiong, S., Fry, J., Brosgart, C. L., *N. Engl. J. Med.* **2003**, *348*, 800–807.

157 Marcellin, P., Chang, T.-T., Lim, S. G., Tong, M. J., Sievert, W., Shiffman, M. L., Jeffers, L., Goodman, Z., Wulfsohn, M. S., Xiong, S., Fry, J., Brosgart, C. L., and the Adefovir Dipivoxil 437 Study Group, *N. Engl. J. Med.* **2003**, *348*, 808–816.

158 Hadziyannis, S. J., Tassopoulos, N. C., Heathcote, E. J., Chang, T. T., Kitis, G., Rizzetto, M., Marcellin, P., Lim, S. G., Goodman, Z., Ma, J., Brosgart, C. L., Borroto-Esoda, K., Arterburn, S., Chuck, S. L., *Gastroenterology* **2006**, *131*, 1743–1751.

159 Peters, M. G., Hann Hw, H., Martin, P., Heathcote, E. J., Buggisch, P., Rubin, R., Bourliere, M., Kowdley, K., Trepo, C., Gray, D., Sullivan, M., Kleber, K., Ebrahimi, R., Xiong, S., Brosgart, C. L., *Gastroenterology* **2004**, *126*, 91–101.

160 Schiff, E. R., Lai, C. L., Hadziyannis, S., Neuhaus, P., Terrault, N., Colombo, M., Tillmann, H. L., Samuel, D., Zeuzem, S., Lilly, L., Rendina, M., Villeneuve, J. P., Lama, N., James, C., Wulfsohn, M. S., Namini, H., Westland, C., Xiong, S., Choy, G. S., Van Doren, S., Fry, J., Brosgart, C. L., *Hepatology* **2003**, *38*, 1419–1427.

161 Schiff, E., Lai, C. L., Hadziyannis, S., Neuhaus, P., Terrault, N., Colombo, M., Tillmann, H. L., Samuel, D., Zeuzem, S., Villeneuve, J. P., Arterburn, S., Borroto-Esoda, K., Brosgart, C., Chuck, S., *Liver Transplant.* **2007**, *13*, 349–360.

162 Benhamou, Y., Bochet, M., Thibault, V., Calvez, V., Fievet, M. H., Vig, P., Gibbs, C. S., Brosgart, C., Fry, J., Namini, H., Katlama, C., Poynard, T., *Lancet* **2001**, *358*, 718–723.

163 Thibault, V., Benhamou, Y., Valantin, M. A., Brosgart, C. L., Xiong, S., *Hepatology* **2005**, *42*, 581.

164 Benhamou, Y., Thibault, V., Vig, P., Calvez, V., Marcelin, A. G., Fievet, M. H., Currie, G., Chang, C. G., Biao, L., Xiong, S., Brosgart, C., Poynard, T., *J. Hepatol.* **2006**, *44*, 62–67.

165 Foli, A., Sogocio, K. M., Anderson, B., Kavlick, M., Saville, M. W., Wainberg, M. A., Gu, Z., Cherrington, J. M., Mitsuya, H., Yarchoan, R., *Antiviral Res.* **1996**, *32*, 91–98.

166 Wainberg, M. A., Miller, M. D., Quan, Y., Salomon, H., Mulato, A. S., Lamy, P. D., Margot, N. A., Anton, K. E., Cherrington, J. M., *Antiviral Ther.* **1999**, *4*, 87–94.

167 Miller, M. D., Margot, N., Lu, B., Zhong, L., Chen, S. S., Cheng, A., Wulfsohn, M., *J. Infect. Dis.* **2004**, *189*, 837–846.

168 Segondy, M., Montes, B., *J. Acquir. Immune Defic. Syndr.* **2005**, *38*, 110–111.

169 Weber, J., Chakraborty, B., Weberova, J., Miller, M. D., Quinones-Mateu, M. E., *J. Clin. Microbiol.* **2005**, *43*, 1395–1400.

170 McColl, D. J., Miller, M. D., *J. Antimicrob. Chemother.* **2003**, *51*, 219–223.

171 Ross, L., Gerondelis, P., Liao, Q., Wine, B., Lim, M., Shaefer, M., Rodriguez, A., Limoli, K., Huang, W., Parkin, N., Gallant, J., Lanier, R., *Antiviral Ther.* **2005**, *10*, S102.

172 Cherrington, J. M., Mulato, A. S., Fuller, M. D., Chen, M. S., *Antimicrob. Agents Chemother.* **1996**, *40*, 2212–2216.

173 Mulato, A. S., Lamy, P. D., Miller, M. D., Li, W.-X., Anton, K. E., Hellmann, N. S.,

Cherrington, J. M., *Antimicrob. Agents Chemother.* **1998**, *42*, 1620–1628.

174 Sluis-Cremer, N., Sheen, C. W., Zelina, S., Torres, P. S., Parikh, U. M., Mellors, J. W., *Antimicrob. Agents Chemother.* **2007**, *51*, 48–53.

175 Stanford HIV Drug Resistance Database. http://hivdb.stanford.edu/ **2006**.

176 Miller, M. D., *AIDS Rev.* **2004**, *6*, 22–33.

177 Lok, A. S., Lai, C. L., Leung, N., Yao, G. B., Cui, Z. Y., Schiff, E. R., Dienstag, J. L., Heathcote, E. J., Little, N. R., Griffiths, D. A., Gardner, S. D., Castiglia, M., *Gastroenterology* **2003**, *125*, 1714–1722.

178 Lampertico, P., Marzano, A., Levrero, M., Santantonio, T., Di Marco, V., Brunetto, M., Andreone, P., Sagnelli, E., Fagiuoli, S., Mazzella, G., Raimondo, G., Gaeta, G., Ascione, A., *Hepatology* **2006**, *44*, 639A.

179 Angus, P., Vaughan, R., Xiong, S., Yang, H., Delaney, W., Gibbs, C., Brosgart, C., Colledge, D., Edwards, R., Ayres, A., Bartholomeusz, A., Locarnini, S., *Gastroenterology* **2003**, *125*, 292–297.

180 Qi, X., Xiong, S., Yang, H., Miller, M., Delaney, W. E. IV *Antiviral Ther.* **2007**, *12*, 255–362.

181 Ratziu, V., Thibault, V., Benhamou, Y., Poynard, T., *Comp. Hepatol.* **2006**, *5*, 1.

182 Villeneuve, J.-P., Willems, B., Zoulim, F., *Hepatology* **2005**, *42*, 588A.

183 Fung, S. K., Andreone, P., Han, S. H., Rajender Reddy, K., Regev, A., Keeffe, E. B., Hussain, M., Cursaro, C., Richtmyer, P., Marrero, J. A., Lok, A. S., *J. Hepatol.* **2005**, *43*, 937–943.

184 Lim, S. G., Hadziyannis, S., Tassopoulos, N., Chang, T. T., Heathcote, J., Kitis, G., Rizzetto, M., Marcellin, P., Arterburn, S., Ma, J., Xiong, S., Qi, X., Brosgart, C. L., Currie, G.,and Adefovir Dipivoxil 438 Study Group, *40th Annual Meeting of the European Association for the Study of the Liver,* Paris, France, **2005**.

185 Eisenberg, E. J., He, G. X., Lee, W. A., *Nucleosides Nucleotides Nucleic Acids* **2001**, *20*, 1091–1098.

186 Lin, C. C., Fang, C., Benetton, S., Xu, G. F., Yeh, L. T., *Antimicrob. Agents Chemother.* **2006**, *50*, 2926–2931.

187 Lin, C. C., Xu, C., Teng, A., Yeh, L. T., Peterson, J., *J. Clin. Pharmacol.* **2005**, *45*, 1250–1258.

188 Lee, K. S., Lim, S. G., Chuang, W. L., Hwang, S. G., Cho, M., Lai, M., Chao, Y. C., Chang, T. T., Han, K., Lee, C., Um, S., Yeon, J. E., Yang, S. S., Teo, E. K., Peng, C. Y., Lin, H. H., Yang, S. S., Huo, T. I., Nguyen, T., Chen, T. Y., Hu, K. Q., Xu, Y., Sullivan-Bolyai, J. Z., *J. Hepatol.* **2006**, *44* S16.

189 Balzarini, J., Pannecouque, C., De Clercq, E., Aquaro, S., Perno, C. F., Egberink, H., Holy, A., *Antimicrob. Agents Chemother.* **2002**, *46*, 2185–2193.

190 De Clercq, E., Andrei, G., Balzarini, J., Leyssen, P., Naesens, L., Neyts, J., Pannecouque, C., Snoeck, R., Ying, C., Hockova, D., Holy, A., *Nucleosides Nucleotides Nucleic Acids* **2005**, *24*, 331–341.

191 Hockova, D., Holy, A., Masojidkova, M., Andrei, G., Snoeck, R., De Clercq, E., Balzarini, J., *J. Med. Chem.* **2003**, *46*, 5064–5073.

192 Ying, C., Holy, A., Hockova, D., Havlas, Z., De Clercq, E., Neyts, J., *Antimicrob. Agents Chemother.* **2005**, *49*, 1177–1180.

193 Wu, T., Froeyen, M., Kempeneers, V., Pannecouque, C., Wang, J., Busson, R., De Clercq, E., Herdewijn, P., *J. Am. Chem. Soc.* **2005**, *127*, 5056–5065.

194 Kim, C., Luh, B., Martin, J. C., *J. Org. Chem.* **1991**, *56*, 2642–2647.

195 White, K., Feng, J., Ray, A., Lafallme, G., Yu, F., Tsiang, M., Wang, R., McDermott, M., Miller, M., Mackman, R., Cihlar, T., *46th Interscience Conference on Antimicrobial Agents and Chemotherapy Abstract H-251*, San Francisco, USA, **2006**.

196 Cihlar, T., Ray, A. S., Boojamra, C. G., Zhang, L., Hui, H., Laflamme, G.,Vela, J. E., Grant, D., Chen, J., Myrick, F., White, K. L., Gao, Y., Lin, K. Y., Douglas, J. L., Parkin, N. T., Carey, A., Pakdaman, R.,

Mackman, R. L., *Antimicrob. Agents Chemother.* **2008**, *52*, 655–665.

197 Ray, A. S., Vela, J. E., Boojamra, C. G., Zhang, L., Hui, H., Callebaut, C., Stray, K., Lin, K. Y., Gao, Y., Mackman, R. L., Cihlar, T., *Antimicrob. Agents Chemother.* **2008**, *52*, 648–654.

198 Heijtink, R. A., Kruining, J., de Wilde, G. A., Balzarini, J., de Clercq, E., Schalm, S. W., *Antimicrob. Agents Chemother.* **1994**, *38*, 2180–2182.

199 Starrett, J. E. Jr, Tortolani, D. R., Hitchcock, M. J., Martin, J. C., Mansuri, M. M., *Antiviral Res.* **1992**, *19*, 267–273.

200 Andrei, G., Snoeck, R., Goubau, P., Desmyter, J., De Clercq, E., *Eur. J. Clin. Microbiol. Infect. Dis.* **1992**, *11*, 143–151.

25
Clofarabine: From Design to Approval

John A. Secrist III, Jaideep V. Thottassery, and William B. Parker

25.1
Introduction

Many different classes of compounds have been found to have utility in treating a variety of different types of cancers. These compounds act through a variety of different mechanisms, in many cases targeting metabolic differences between normal and cancerous cells, and more recently targeting cancer-specific targets and pathways. An examination of the drugs that have been approved by governments worldwide demonstrates that antimetabolites – compounds that affect the pathways leading to nucleic acids – represent a rich source of anticancer drugs. Almost all of those approved drugs are either nucleosides or compounds that are converted to nucleosides or nucleotides after administration to patients. In the United States, the list of FDA-approved antimetabolites of this type includes 5-fluorouracil (colorectal, breast, stomach, and pancreatic carcinomas), 6-thioguanine (acute non-lymphocytic leukemias), 1-β-D-arabinofuranosylcytosine [acute lymphocytic leukemia (ALL), and acute myelocytic leukemia (AML)], 5-fluoro-2′-deoxyuridine (metastatic colon cancer), fludarabine phosphate [chronic lymphocytic leukemia (CLL)], 2-deoxycoformycin (hairy cell leukemia), cladribine (hairy cell leukemia), gemcitabine [pancreatic cancer, non-small cell lung cancer (NSCLC)], capecitabine (metastatic colorectal and breast cancer), nelarabine (T-cell acute lymphoblastic leukemia and lymphoma), decitabine (myelodysplastic syndrome), and clofarabine (pediatric ALL). Both fludarabine phosphate and clofarabine were discovered and pushed forward preclinically at the Southern Research Institute, and both are products of our preclinical optimization process for nucleoside analogues.

This chapter will focus on the development of clofarabine, and will present that development from our viewpoint as preclinical scientists. A recent review covering many aspects of the development of clofarabine is recommended to the reader for additional details [1].

Modified Nucleosides: in Biochemistry, Biotechnology and Medicine. Edited by Piet Herdewijn

ISBN: 978-3-527-31820-9

25.2 Clofarabine: The Background

Before examining the chronology of the discovery and development of clofarabine, it is important first to identify its current and potential future uses, as well as the companies involved. Clofarabine was approved for the treatment of pediatric ALL in the US in December 2004, and in Europe in May 2006. It has been granted orphan drug status in both the US and Europe. Early efforts toward pushing the drug into clinical trials, beginning in 1992, were initiated at the M. D. Anderson Cancer Center (MDACC) in Houston, Texas, and included clinicians Drs. J. Freireich, M. Keating and H. Kantarjian, as well as pharmacologists Drs. W. Plunkett and V. Gandhi. The initial licensing of the drug by Southern Research Institute was to the Eurobiotech Group in 1998, which utilized MDACC for the clinical trials. The lag time evident in the above dates is truly unfortunate from the standpoint of cancer patients who might have benefited from the drug. The difficulties in licensing clofarabine stemmed from two views prevalent in the pharmaceutical industry at the time: (i) clofarabine was just another fludarabine, and had little chance of making a mark on its own; and (ii) future cancer drugs needed to focus on solid tumors such as colorectal, breast, prostate and lung cancer, and nucleosides were of little interest in that regard. Moreover, even if they had some activity, the market size was too small to be of interest to the larger companies. The CEO of Eurobiotech (most recently called Bioenvision), Dr. Christopher Wood, understood the properties of fludarabine phosphate, and believed that clofarabine had properties that might make it significantly better, and on that basis he was eager to proceed. Similarly, the team at MDACC also believed that clofarabine had the potential to take its own place among anticancer treatments. The eventual approval of clofarabine was aided immeasurably by the commitment of these people.

25.3 The Beginnings

During the early 1980s, two types of nucleoside were found to have very promising selectivity in animal models. These compounds were the 2-halo-2′-deoxyadenosines (**1**), with the halogen being fluorine, chlorine or bromine, and the corresponding β-D-arabinofuranosyl analogues (**2**). The 2′-deoxy compounds had been prepared in several different laboratories [2–4] and examined in various cell lines as potential anticancer drugs. In our laboratories, we examined all three of the 2′-deoxy compounds in a series of experiments in the then-standard L1210 mouse leukemia model system. Interestingly, all three compounds had excellent selectivity, and a summary of that previously unpublished data is shown in Table 25.1 [5]. Although it is easy to see that all three had some selectivity, the chloro and bromo analogues appeared to be the most promising, with some cures seen, and these results warranted further investigation.

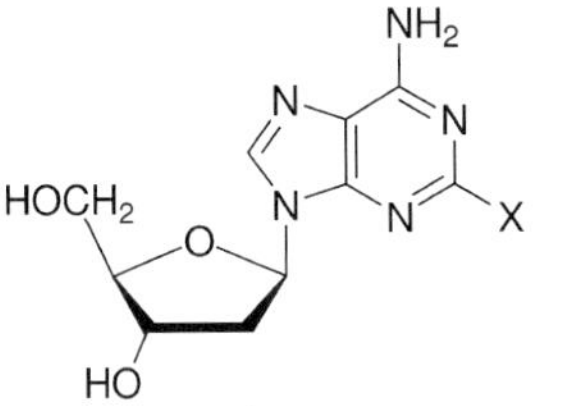
1
X = F, Cl (Cladribine), Br

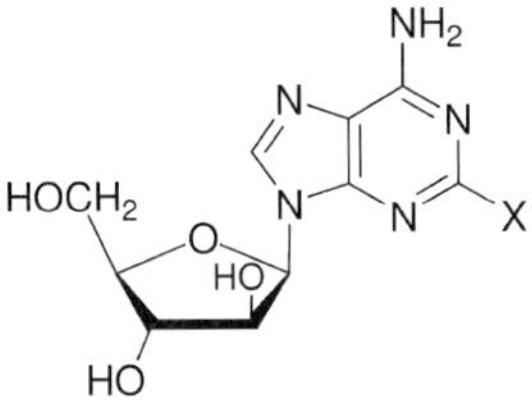
2
X = F (Fludarabine), Cl, Br

Over the next few years, the 2-chloro compound was examined in further detail through a collaboration between John Montgomery at Southern Research Institute and Dennis Carson at the University of California at San Diego, while the 2-bromo compound was further examined by Raymond Blakley and his colleagues at St. Jude Children's Hospital. As events unfolded, the 2-chloro compound (cladribine) was eventually approved in 1992 by the FDA for the treatment of hairy cell leukemia.

In the 2-halo-ara-A series, the fluoro, chloro, and bromo compounds were prepared at Southern Research Institute and elsewhere [2, 6–10], and all three had some activity [11, 12]. The data demonstrated that the fluoro compound was significantly better than the other two, and consequently it was carried forward, eventually being approved in 1991 as the 5′-phosphate (fludarabine phosphate), a prodrug form developed to aid solubility, for the treatment of CLL.

During the first half of the 1980s, some preliminary data were acquired on these compounds which, when combined with previous information regarding the physical properties of the two series, suggested that some improvements in the structures could be made that might have a significant effect on their potential clinical utility. At the time, it was of course not known whether any of these compounds would become approved, and the quest was to prepare compounds that would have enhanced properties that might either achieve approval if the earlier compounds did not, or might be next-generation compounds with more attractive properties than any earlier compounds that did achieve approval.

Table 25.1 Response of intraperitoneally (i.p.) implanted L1210 leukemia to 2-halo-2′-deoxyadenosines.

Compound	Optimal i.p. dose ($\leq LD_{10}$, mg kg^{-1} dose^{-1})[a]	Total dose (mg kg^{-1})	Median % ILS[b] (dying mice only)	Net $\log_{10}$ cell kill[c]	Tumor-free survivors/ total
F-dAdo	25	600	+118	+0.5	0/10
Cl-dAdo	20	480	+150	+2.9	5/10
Br-dAdo	40	960	+125	+1.1	3/6

[a] Treatment schedule was q3 h × 8, Days 1, 5, and 9.
[b] Median day of death of tumored control mice (10^5 cells) was 8 days. ILS, increase in lifespan.
[c] Net $\log_{10}$ reduction in the tumor cell population between the beginning and the end of therapy, based on the median day of death of the mice that died.

Thus, attention was focused on three properties while seeking to improve on the two series presented above. It is easiest to consider the characteristics that might be improved by focusing on the two compounds that were eventually approved, fludarabine and cladribine. In the case of fludarabine phosphate, when administered to an animal it is rapidly cleaved to fludarabine, which enters the cells and is further metabolized [13].

In the case of cladribine, it is well known that 2′-deoxy compounds in the purine series are susceptible to chemical cleavage at low pH, and thus a loss of potency through hydrolytic cleavage is clearly an issue with cladribine. In addition, cleavage of the glycosidic bond by phosphorylases is another means of loss of potency. With both types of cleavage, 2-chloroadenine would be generated, which is a compound with only modest toxicity.

For fludarabine, the chemical hydrolysis of the glycosidic bond is not a significant problem as the presence of the 2′-hydroxyl group provides significant stability, although there is a loss of potency through some phosphorylase cleavage [13–16]. In the case of fludarabine, this process generates 2-fluoroadenine, which is a highly undesirable metabolite. This purine is readily metabolized up to 2-fluoro-ATP, which is an extremely toxic but unselective compound, and so its systemic generation could present a concern. In recent years, attention has been focused on utilizing gene therapy approaches to generate this toxin in tumor cells in a selective manner [17].

The other mechanism of loss of potency that can occur with adenine derivatives is enzymic deamination, which is carried out by adenosine deaminase at the nucleoside level, and by AMP deaminase at the monophosphate level. The incorporation of a 2-halogen into the adenine ring of a nucleoside confers significant resistance to deamination as compared to the parent adenine compounds [11, 12]. Many evaluations have been carried out examining the ability of the 2-haloadenine nucleosides to serve as substrates for adenosine deaminase, and although they are highly resistant to deamination, all have at least some substrate activity. The order of deamination is F > Cl > Br, and within that order the 2-fluoro compounds are significantly more susceptible to deamination than the other two. Thus, in the case of fludarabine there is a minor loss of potency through some deamination [13–16], a pathway that is not a significant problem with cladribine.

The other key issue with regard to nucleoside analogues is their activation to an active metabolite which, in the vast majority of cases, is the nucleoside triphosphate (NTP). In general, nucleosides exert their effects on the biosynthetic pathway leading to DNA, and thus, analogues of 2′-deoxynucleosides are typically of more interest than the building blocks of RNA, although ribonucleoside analogues were also prepared. It was determined that the addition of a 2-halogen did not prevent the phosphorylation of some 2′-deoxynucleoside analogues, and many laboratories have determined that the major enzyme carrying out the initial phosphorylation is generally deoxycytidine kinase [18–20]. Another key observation was that nucleosides with an *arabino* configuration often were also substrates of deoxycytidine kinase. The other enzymes carrying out conversion of monophosphates to their di- and triphosphate metabolites were in general less discriminating, and the majority of nucleosides that could be metabolized to the 5′-monophosphate were converted at a meaningful rate to the triphosphate.

The above-described information relates to the situation during the early to mid-1980s as ways were sought to improve on the activity of this class of potential anticancer nucleosides. The set of simple conclusions drawn from the above information can be summarized as follows:

- A 2-halogen in an adenine ring analogue dramatically reduces deamination, but in general will allow phosphorylation, depending upon the carbohydrate attached.
- A 2-chloro or 2-bromoadenine ring is more desirable than a 2-fluoro, based upon the high toxicity of any 2-fluoroadenine that may be generated, and also based upon its increased ability of 2-fluoroadenine-containing nucleosides to serve as substrates for adenosine deaminase.
- A stabilizing group at C-2′ in the *arabino* configuration – one that will significantly reduce both phosphorylase cleavage and hydrolytic cleavage of the glycosidic bond – is highly desirable.

25.4 The Next Generation of Compounds

Over the years, a highly efficient system was developed for the rapid examination of new compounds in our anticancer drug discovery program, which was strongly supported by the US National Institutes of Health. Whenever a new compound had been prepared and properly characterized, it was submitted for an evaluation of its cytotoxicity in a small series of cancer cell lines, with generally six or seven such lines being derived from various types of human tumors. Typically, the results were available in a few weeks. In all of these cases, the corresponding human tumor xenograft mouse model was available if a compound exhibited significant cytotoxicity. The main challenge was to prepare sufficient material for evaluation in a mouse model, once it had been learned that such an examination was warranted based upon the cytotoxicity profile. When sufficient material became available for such an initial evaluation, the compound was submitted and generally placed into a test system within a month. In parallel, the mechanistic evaluation of compounds of potential interest was started in our biochemistry laboratories. Together, this research effort provided us with the basic information on the activation of new nucleosides to the various phosphorylated metabolites, their effects on DNA, RNA and protein synthesis, and also specific information on key enzymes. Feedback from both the *in-vitro* and *in-vivo* evaluations was thus rapidly available, and we were able quickly to adjust our target structure list based upon this iterative feedback. This simple system prevented us from spending too much time on the synthesis of series of compounds that did not show promise as anticancer drugs.

By utilizing this efficient system, a variety of compounds was evaluated relatively rapidly. The major efforts revolved around carbohydrate modifications with the 2-haloadenines as the bases, and on similar compounds with the nitrogen base somewhat altered.

Examples of ring-altered compounds included 2-fluoro-8-azaadenine nucleosides such as **3** [21], and a ring-fluorinated analogue **4** of formycin A [22]. Unfortunately, however, the two compounds **3** and **4** had characteristics that prevented them from being of therapeutic use. For example, compound **3** was not significantly cytotoxic, presumably because it was not amenable to initial phosphorylation, while **4** was converted to the monophosphate, but not further converted to the di- and triphosphate levels.

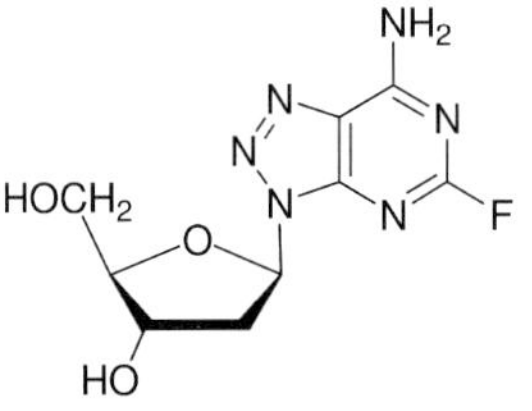

3

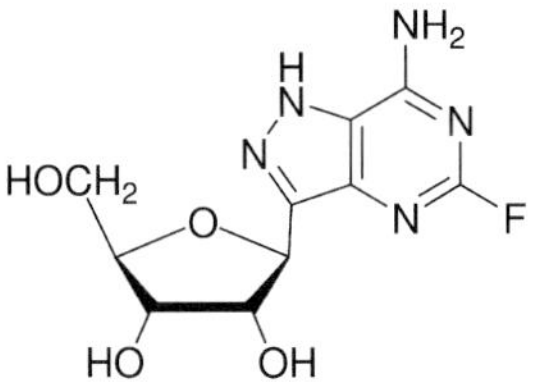

4

In the case of carbohydrate-modified nucleosides, the initial focus – as noted above – was at the 2′-position. It was felt that building in halogen atoms as well as certain other groups with significant electronegativity at the 2′-position would accomplish several goals. First, these compounds should impart significant hydrolytic stability of the glycosidic bond. Second, this alteration might well reduce the ability of these compounds to serve as substrates for phosphorylases. Therefore, attention was redirected towards preparing compound series **5**, which incorporated a 2′-bromine, chlorine, fluorine, azido or amino group. Another related compound, **6**, which also has anticancer activity, has been prepared by the Matsuda group and includes a 2′-cyano group and the *arabino* configuration [23].

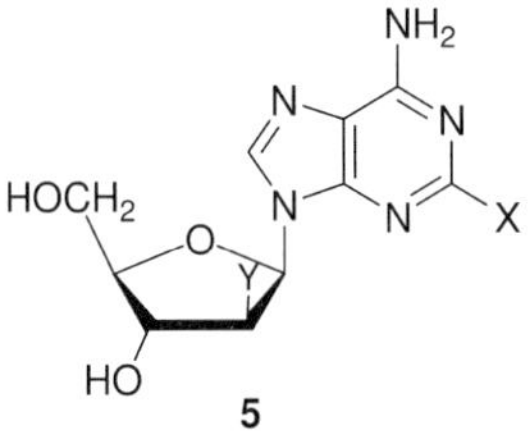

5

X = F, Cl, Br
Y = F, Cl, Br, N_3, NH_2

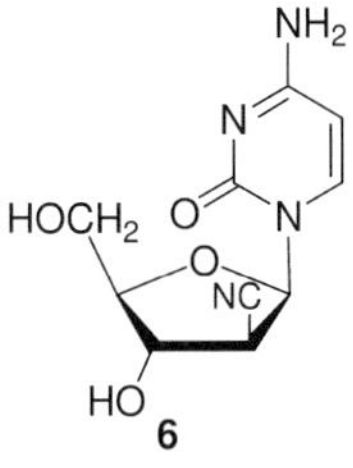

6

Hence, compounds were prepared that incorporated all of these groups at the 2′-position and also contained a 2-haloadenine moiety [24]. At that time, the synthetic routes involved the displacement of a leaving group at C-2′ of a nucleoside with inversion from the *ribo* to the *arabino* configuration, as had been accomplished by Ueda [25]. However, the cytotoxicity results from this series of compounds, with the exception of those incorporating a 2′-fluorine, were not impressive [24], and thus attention was re-focused on compounds with a 2′-fluorine, in turn leading to the preparation of the series of compounds **7**. The incorporation of a fluorine atom has always been attractive, because the size of the atom causes the least disruption, and is the closest to hydrogen. A fluorine also often significantly alters the biochemical properties of a molecule. Initially, the synthetic route was quite laborious because the fluorocarbohydrate precursor required a lengthy synthetic route. However, starting in the 1970s the group of Fox and Watanabe was also preparing nucleosides with a 2′-fluorine [26–28], though with a major focus on pyrimidine nucleosides as antiviral agents, and we were able to take advantage of their synthetic procedures. The initial route (see below) [29, 30] yielded predominantly the desired β nucleoside, but separation of the anomers was necessary. The first synthesis in this target series was completed during the mid-1980s. By utilizing different bases for the coupling, all three compounds **7** were prepared. Based upon the pursuit of one of the pyrimidine nucleosides of antiviral interest, an improved synthesis of the fluorocarbohydrate precursor was developed and published [31], and this route proved to be a very useful advance. Quite recently, another important advance relative to the α/β ratio at C-1′ has been made by utilizing a three-solvent mix for the coupling reaction, which is carried out with 2-chloroadenine rather than 2,6-dichloropurine [32]. This improvement was made during the optimization of the synthesis of clofarabine (7) for manufacturing purposes.

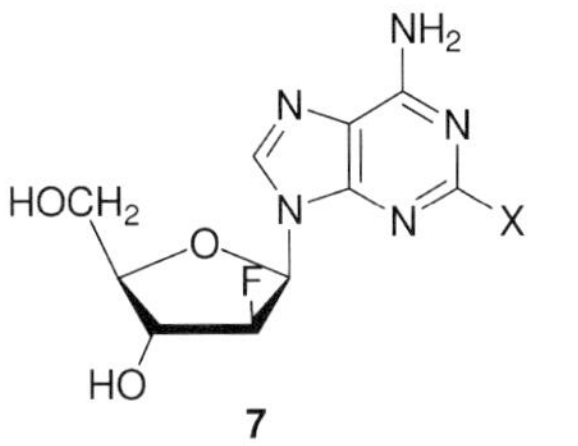

7

X = F, Cl (Clofarabine), Br

Synthetic Route to Clofarabine

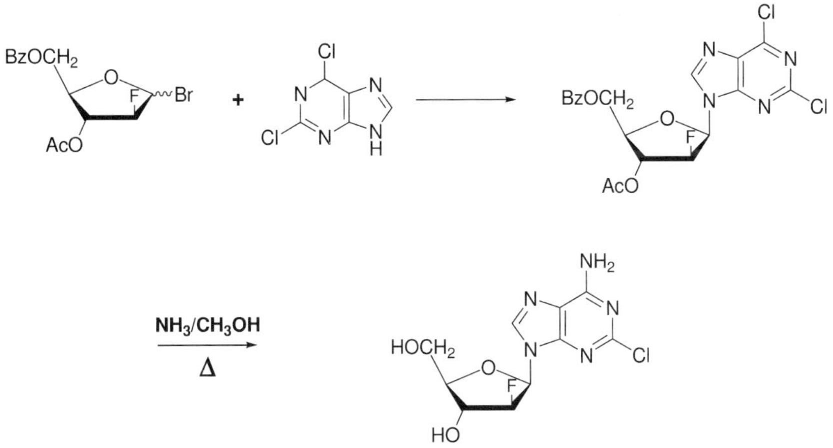

Table 25.2 Anticancer activity of clofarabine against human tumor xenografts in mice.

Tumor	Activity[a)]	Tumor	Activity[a)]
COLO 205 colon	++	A594 lung	+
DLD-1 colon	+	NCI-H23 lung	±
HCC-2998 colon	+++	NCI-H322M lung	±
HCT-15 colon	++	NCI-H460 lung	+
HCT-116 colon	+	DU-145 prostate	+
HT29 colon	+++	LNCAP prostate	+
KM20L2 colon	++	PC-3 prostate	++
SW-620 colon	++	HL-60 leukemia	+++
A498 renal	++	CCRF-CEM leukemia	+
CAKI-1 renal	+++	K-562 leukemia	++
RXF 393 renal	–	MOLT-4 leukemia	–
SN12C renel	++	AS283 lymphoma	±
		RL lymphoma	++

a) – inactive; ± marginal; + minimal; + + good (tumor regressions); + + + excellent (cures).

All three of the 2-haloadenine nucleosides with a 2′-fluorine in the *arabino* configuration were significantly cytotoxic in multiple cancer cell lines [29, 30], in contrast to the other compounds mentioned above, and also in contrast to the 2′-fluoro compound with the *ribo* configuration or the 2′,2′-difluoro analogue [33]. This result encouraged us to move in the direction of human tumor xenograft experiments, and also to embark upon biochemical studies to determine the compound's mechanism of action. The initial animal data strongly suggested that the 2-chloro compound had the most activity, and it was therefore chosen for further experiments. Although the 2-fluoro compound was also of some interest, it was more difficult to synthesize, and even a small amount of glycosidic cleavage yielding 2-fluoroadenine (see above) would be undesirable.

Mechanistic information presenting the unique profile of clofarabine is summarized in the following section. With regard to further animal studies, many different human tumor xenograft models were employed, and much of the data obtained have been presented in one report [34]. A concise summary of clofarabine activity in human tumor models in mice is provided in Table 25.2. The conclusion to be drawn from these animal studies was that clofarabine had significant and often curative activity in a number of systems across a broad spectrum of human tumor types. This type of profile is exactly what is required for a new drug to have a strong chance of success in clinical trials. It is known that there is no direct correlation between outstanding activity in a particular human tumor type in a xenograft model and clinical success with that agent in treating the same type of tumor. In our experience, however, robust, broad-spectrum activity against a variety of human tumors in these models is a good indication that a compound will find clinical utility. Clofarabine clearly had such a profile, and we were eager for it to have the opportunity to move to a clinical trial.

25.5
Mechanism of Action of Clofarabine

Generally speaking, all nucleoside analogues used in the treatment of cancer, including clofarabine, have a similar mechanism of action, namely that they are converted to their respective 5′-triphosphates and inhibit DNA synthesis. However, it is clear from the varying clinical activities of these agents that subtle quantitative and qualitative differences in the metabolism of these agents and their interactions with target enzymes can have a profound impact on their antitumor activity. Thus, a precise understanding of the mechanism of action of each of these compounds is important in order to determine those biochemical actions which are most important to antitumor activity and might aid the rational design and development of new agents.

25.5.1
Transport and Metabolism to Active Metabolites

Clofarabine is efficiently transported into cells [35] by both the human equilibrative nucleoside transporters (hENT1, hENT2) and the human concentrative nucleoside transporters (hCNT2, hCNT3). As noted earlier, the primary enzyme involved in the activation of clofarabine in tumor cells is deoxycytidine kinase [36–38]. Clofarabine is a very good substrate for this enzyme, with K_m and V_{max} values similar to those of deoxycytidine. The structure of deoxycytidine kinase with clofarabine in the active site has recently been determined [39]. The results indicated that the conformation of the enzyme/clofarabine complex was similar to structures of the pyrimidine-bound complexes, and that interactions between the 2-Cl group and its surrounding hydrophobic residues contributed to the high catalytic efficiency of deoxycytidine kinase with clofarabine. Clofarabine is also a good substrate for deoxyguanosine kinase [40], an enzyme that is expressed in mitochondria. The contribution of deoxyguanosine kinase to the phosphorylation of clofarabine in cells is low due to the much higher expression of deoxycytidine kinase activity in most cell types [41], although this may be important in cells that express low activities of deoxycytidine kinase.

Similar intracellular concentrations of clofarabine-5′-monophosphate (the product of the reaction of clofarabine with deoxycytidine kinase) and clofarabine-5′-triphosphate (clofarabine-TP) accumulate in tumor cells treated with clofarabine [37]. This indicates that phosphorylation of the monophosphate of clofarabine is the rate-limiting step in its activation to the triphosphate, and that the monophosphate kinase does not easily tolerate substitutions at the 2-position of a purine nucleoside as large as a chlorine atom. Nucleoside kinases are usually the rate-limiting enzymes in the activation of anticancer nucleoside analogues. Even though clofarabine is a relatively poor substrate for the monophosphate kinase, clofarabine-TP accumulates to high levels in cancer cells treated with clofarabine. Clinical studies have indicated that the concentration of clofarabine-TP in blasts cells is 10-fold greater than the plasma concentration achieved after a 1-h infusion [42].

Phosphorylated metabolites of nucleoside analogues do not readily cross cell membranes, and are therefore trapped in the cell in which they were created; this contrasts with the nucleosides themselves, which will freely distribute across the cell membrane. Consequently, the antitumor activity of these agents can extend well beyond the time that the parent drug circulates in the plasma, because the active metabolite is maintained in tumor cells long after the drug has disappeared from the plasma. The initial half-life for the clearance of clofarabine-TP from cultured tumor cells (CEM) is approximately 2 h [33, 37]. In samples taken from patients treated with clofarabine, more than 50% of the clofarabine-TP was still present in circulating leukemic cells 24 h after the completion of transfusion [42]. The long retention time of clofarabine-TP is believed to be a major contributory factor to the high activity demonstrated by clofarabine in solid tumor xenografts in mice [34].

25.5.2 Inhibition of DNA Synthesis

As with other anticancer nucleoside analogues, the primary activity of clofarabine that is responsible for its antitumor activity is the inhibition of DNA synthesis [33, 36]. RNA and protein synthesis are inhibited by clofarabine only at high concentrations. Clofarabine-TP is a potent inhibitor of ribonucleotide reductase [36, 43], a critical enzyme involved in the *de-novo* synthesis of deoxynucleotides [44]. The activity of ribonucleotide reductase in cells is tightly controlled by the natural deoxynucleoside triphosphates to ensure that the cell has all of the deoxynucleotides needed for DNA synthesis, and in the correct concentrations. dATP is a potent regulator of ribonucleotide reductase activity, and inhibits the reduction of ADP, UDP, and CDP [45]. The effect of clofarabine on intracellular nucleotide pools suggest that clofarabine-TP interacts with ribonucleotide reductase in the allosteric binding site as an analogue of dATP. The inhibition of ribonucleotide reductase activity results in decreases in the levels of deoxynucleotide triphosphates, which are required for the synthesis of DNA. Clearly, decreasing the natural deoxynucleotide pools is sufficient to cause an inhibition of DNA synthesis; this is also the mechanism of action of hydroxyurea, another useful anticancer agent.

Clofarabine is readily incorporated into DNA, although a small amount of drug is also detected in RNA [37]. At low concentrations, most of the clofarabine detected in DNA is incorporated into internal sites in the DNA of CEM cells, which indicates that the elongation of DNA synthesis is not prevented by the drug's incorporation into DNA. The immediate inhibition of DNA synthesis in cells treated with clofarabine (even when most of the clofarabine is in internal positions in the DNA), however, indicates that inhibition of the DNA replication complex is the primary action of clofarabine that results in the death of the tumor cell. The disruption of DNA function due to the incorporation of clofarabine into daughter strands is of secondary importance to the activity of clofarabine.

Clofarabine-TP is a good substrate and inhibitor of DNA polymerases α and ε, two important enzymes involved in the replication of chromosomal DNA [36, 37].

Clofarabine-TP is utilized by DNA polymerase α, with K_m and V_{max} values similar to those of dATP (the natural substrate). Once clofarabine-TP is incorporated into the growing chain, however, the ability of the polymerase to add new nucleotides is significantly less than after dATP incorporation. The inhibition of DNA polymerase α by clofarabine-TP is similar to that of fludarabine-TP [36]. The effect of clofarabine-TP on DNA polymerase ε activity was similar to that seen with DNA polymerase α, except that clofarabine-TP more effectively inhibited chain elongation by DNA polymerase ε [43]. This enhanced inhibition of chain elongation resulted in a significant inhibition of DNA synthesis at clofarabine-TP concentrations that were only 3% of the dATP levels used in the experiment.

The removal of clofarabine from the 3′-end of DNA chains is an important part of the mechanism of action of nucleoside analogues that has not yet been evaluated. If clofarabine can be quickly removed from the 3′-terminus of the DNA chain, then DNA synthesis could continue normally; however, if it is not removed, then it will have a lasting effect on the ability of the DNA polymerase to elongate the DNA chain. The incorporation of two or more clofarabine residues sequentially in the DNA [36] may be harder to repair than single incorporations, and may represent a greater block to DNA synthesis. The continued inhibition of DNA synthesis is another attribute of nucleoside analogues that is a major contributor to their mechanisms of action. A short exposure of tumor cells to circulating drug may have lasting effects on the ability of the tumor cell to replicate its DNA.

The results of the above-mentioned studies concluded that DNA synthesis is inhibited in cells treated with clofarabine by two distinct, but complementary, actions: (i) the inhibition of ribonucleotide reductase; and (ii) the inhibition of DNA polymerase activities. The potent inhibition of ribonucleotide reductase activity by clofarabine-TP enhances its inhibition of the replicative DNA polymerases by decreasing the intracellular concentration of the natural substrate, dATP, which in turn competes with clofarabine-TP for use as a substrate by these enzymes. With respect to the inhibition of ribonucleotide reductase and DNA polymerases, clofarabine combines the features of cladribine (potent inhibition of ribonucleotide reductase) and fludarabine (potent inhibition of DNA polymerase) into one molecule.

25.5.3 Induction of Apoptosis

The inhibition of DNA synthesis by clofarabine is responsible for the induction of the apoptotic response in replicating cells. Inhibition of DNA replication in cells normally leads to the “turning on” of replication checkpoint pathways. Stalled replication forks can threaten DNA replication fidelity, and cells respond to replication blocks by triggering checkpoint pathways that monitor replication fork progression [46]. This monitoring of DNA synthesis operates normally during low-intensity replication stress, and is required for tumor cells to resume cell-cycle progression. However, during chronic or high-intensity replication stress, stalled forks do not restart after removal of the stressor, and this results in irreversible S-phase arrest, possibly mitotic catastrophe, and cell death.

It has been found that replication stress induced by the treatment of CEM cells with clofarabine leads to Chk1 phosphorylation, Chk1 down-regulation, concomitant apoptosis, and cell death (unpublished data). Consistent with our findings on Chk1 activation, it has been found that Cdc25A is down-regulated upon low-dose clofarabine treatment in CEM cells, a consequence of its phosphorylation by Chk1, which is accompanied by accumulation of cells in G_1/S and S [47]. Chk1 and its upstream activator, ATR, phosphorylate a host of substrates involved in the control of the firing of additional replication origins. In addition, it is known that phosphorylation of Chk1 on Ser345 triggers its ubiquitylation, resulting in the proteasomal degradation of Chk1 [48].

The induction of apoptosis in response to replication stress can also be mediated by the tumor suppressor/transcription factor p53. Recently, it was demonstrated that two deoxycytidine analogues – gemcitabine and 4′-thio-araC – can induce the stabilization of p73, a p53 paralogue in both p53 null cancer cells and in cells with wild-type p53 [49]. The p73 protein shares significant sequence similarity with p53, and can transactivate proapoptotic p53 target genes such as Bax and PUMA. In turn, these proteins can induce the release of cytochrome *c* and other apoptogenic molecules from the mitochondrial outer membrane. It has been found that the incubation of CEM cells (in which p53 is inactive) with clofarabine also robustly induced p73 levels (unpublished data).

25.5.4
Activity against Non-Proliferating Cancer Cells

Clofarabine is also active against non-replicating cells [50], and has been shown to interfere with mitochondrial integrity and function in primary CLL cells [51], causing the release of cytochrome *c* and apoptotic inducing factor (AIF-1). This has led to the suggestion that clofarabine induces cell death by initiating the apoptotic cascade in these cells, although the precise mechanism for such activity has not yet been elucidated. Other adenine nucleosides such as cladribine and fludarabine have been shown to induce the intrinsic apoptotic pathway, which causes the release of apoptogenic molecules, such as cytochrome *c* into the cytosol; this leads to the cleavage and activation of caspase 9 (an initiator caspase) and subsequently caspase 3 (an executioner caspase) [52, 53]. The present authors and others have shown that clofarabine induces the depletion of procaspase-9, which is accompanied by the cleavage of caspase-3 and PARP [47] (also unpublished data). It is possible that the cytochrome *c* release is the result of an apoptotic cascade resulting from inhibition of ribonucleotide reductase or DNA polymerase activity.

Clofarabine has also been shown to inhibit DNA repair in quiescent cells [54]. Although clofarabine-TP is a weak inhibitor of DNA polymerase β [36] (a polymerase involved in base-excision repair), as indicated above it is a potent inhibitor of DNA polymerase ε [43], an enzyme which is known to be involved in DNA repair [55]. Clofarabine-TP also inhibits ribonucleotide reductase, in turn causing the reduction of intracellular deoxynucleotides in quiescent cells and the inhibition of DNA repair due to a lack of substrates. Because, due to its spontaneous degradation, DNA is constantly being repaired in quiescent cells, the inhibition of DNA repair may induce the apoptotic response that has been observed in quiescent cells. In addition, clofarabine-TP is able to activate APAF-1-dependent caspase activity in B-CLL cell extracts with kinetic parameters that were identical to dATP [56]. Likewise, clofarabine-TP could contribute to the formation and activation of the apoptosome, which contributes to the activation of the caspase cascade.

25.6 Clofarabine to the Clinic

During the development of clofarabine, none of the preliminary findings was reported publicly until the point had been reached where it was clear that the information was of value. At that time, during the 1990s, patent coverage was filed for and the first results were published. By this time, solid data were available relating to the chemistry, the mechanism of action, and selectivity and potency in animal models. In order to move this compound to the clinic, however, the additional stages of preclinical toxicology, formulation and GMP manufacture had to be addressed. Unfortunately, the resources to fund these aspects were unavailable, and consequently a partner was sought in order to proceed. The search for a licensee began enthusiastically, but it was soon discovered that the interest was very limited (as noted earlier). Consequently, supporting preclinical data continued to be gathered until finally a licensing agreement was signed in 1998. During the 1990s, further studies were conducted with colleagues at the M. D. Anderson Cancer Center in order to push clofarabine towards a clinical trial, in anticipation of a licensing partner being found. The involvement of these colleagues proved to be truly critical for the continuing advancement of the drug during those years.

25.6.1 Clinical Trials and Approval

The first Phase 1 leukemia trial was initiated in 1999, and the first Phase 1 pediatric leukemia trial in 2000. The unmet medical needs in the area of pediatric leukemia propelled clofarabine forward, such that the successes seen in relapsed and refractory patients resulted in an NDA filing early in 2004, and FDA approval at the very end of 2004. Regulatory approval in Europe for pediatric leukemia was granted in mid-2006. In parallel, it has been found that clofarabine appears to have considerable value for adult AML patients, and is being advanced towards additional approvals in this area. At present, many different clinical trials – of single agent and combinations, and including both adult and pediatric leukemias – are being pursued. In addition, expansion in other cancers is also being examined. The recently published review mentioned above provides an excellent summary of some of these directions [1]. As a brief summary, clofarabine has a distinct profile of activity in the clinic, and toxicities that are both manageable and complementary to those of other drugs used to treat the same cancers.

25.7 Summary and Comments

The odyssey of clofarabine began in about 1983, and moved through preclinical and clinical evaluations to the above-mentioned approval over a period of some 21 years. The time lost seeking a licensee was about six to seven years, but once an agreement was in place and clinical trials had commenced, events moved at a fairly rapid pace. The entire process was funded by the NIH through a Program Project Grant, without which there would have been little chance of progressing at the pace ultimately achieved.

As the understanding of clofarabine, its activity and mechanism of action increased over the years, it became clear that this agent was distinct from its predecessors, cladribine and fludarabine. As noted above, clofarabine has multiple mechanisms of action, and it is most likely this multiplicity of effects which is critical to the drug's clinical utility. The present authors have become strong proponents of the view that small changes, at least in nucleosides, can have a significant effect on the clinical and toxicological profile of a drug – as supported by the only structural difference between clofarabine and cladribine being the replacement of a hydrogen atom at the 2′ position with a fluorine atom. Nonetheless, this small difference is sufficient to endow clofarabine with biochemical and clinical activities which differ widely from those of cladribine. This view has been cogently presented in recent reviews focused on nucleoside analogues [57, 58], and is borne out by the results obtained with clofarabine.

In looking back at the chronology of this drug, it was important to identify the correct partner, who would select the correct clinical approach and not be deterred by any adversity encountered along the way. Whilst persistence is critical to success, a degree of luck is equally important. To date, both have contributed to the success of clofarabine.

References

1 Bonate, P. L., Arthaud, L., Cantrell, W. R. Jr., Stephenson, K., Secrist, J. A. III, Weitman, S., *Nat. Rev. Drug Discovery* **2006**, *5*, 855–863.

2 Montgomery, J. A., Hewson, K., *J. Med. Chem.* **1969**, *12*, 498–504.

3 Huang, M.-C., Hatfield, K., Roetker, A. W., Montgomery, J. A., Blakley, R. L., *Biochem. Pharmacol.* **1981**, *30*, 2663–2671.

4 Christensen, L. F., Broom, A. D., Robins, M. J., Bloch, A., *J. Med. Chem.* **1972**, *15*, 735–739.

5 Montgomery, J. A., Trader, M. W., Unpublished results, Southern Research Institute **1981**.

6 Reist, E. J., Goodman, L., *Biochemistry* **1964**, *3*, 15–18.

7 Keller, F., Botvinick, I. J., Bunker, J. E., *J. Org. Chem.* **1967**, *32*, 1644–1646.

8 Montgomery, J. A., Clayton, S. D., Shortnacy, A. T., *J. Heterocyclic Chem.* **1979**, *16*, 157–160.

9 No published synthesis of 2-Br-ara-A exists. Montgomery, J. A., Fowler, A. T., Clayton, S. D., unpublished results, 1-chloro-2,3,5-tri-*O*-benzyl-D-arabinofuranose was coupled with 2,6-dibromopurine in 1,2 dichloroethane in the presence of molecular sieves. This coupled product, after separation of the

β anomer was treated sequentially with ethanolic ammonia and then boron trichloride in dichloromethane to produce 2-Br-Ara-A.

10 Kim, I. Y., Zhang, C. Y., Cantoni, G. L., Montgomery, J. A., Chiang, P. K., *Biochim. Biophys. Acta* **1985**, *829*, 150–155.

11 Montgomery, J. A., *Cancer Res.* **1982**, *42*, 3911–3917.

12 Montgomery, J. A., *Med. Res. Rev.* **1982**, *2*, 271–308.

13 Noker, P. E., Duncan, G. F., El Dareer, S. M., Hill, D. L., *Cancer Treat. Rep.* **1983**, *67*, 445–456.

14 El Dareer, S. M., Struck, R. F., Tillery, K. F., Rose, L. M., Brockman, R. W., Montgomery, J. A., Hill, D. L., *Drug Metab. Dispos.* **1980**, *8*, 60–63.

15 Struck, R. F., Shortnacy, A. T., Kirk, M. C., Thorpe, M. C., Brockman, R. W., Hill, D. L., El Dareer, S. M., Montgomery, J. A., *Biochem. Pharmacol.* **1982**, *31*, 1975–1978.

16 Hersh, M. R., Kuhn, J. G., Phillips, J. L., Clark, G., Ludden, T. M., Von Hoff, D. D., *Cancer Chemother. Pharmacol.* **1986**, *17*, 277–280.

17 Hong, J. S., Waud, W. R., Levasseur, D. N., Townes, T. M., Wen, H., McPherson, S. A., Moore, B. A., Bebok, Z., Allan, P. W., Secrist, J. A. III, Parker, W. B., Sorscher, E. J., *Cancer Res.* **2004**, *64*, 6610–6615.

18 Brockman, R. W., Cheng, Y. C., Schabel, F. M. Jr., Montgomery, J. A., *Cancer Res.* **1980**, *40*, 3610–3615.

19 Carson, D. A., Wasson, D. B., Kaye, J., Ullman, B., Martin, D. W. Jr., Robins, R. K., Montgomery, J. A., *Proc. Natl. Acad. Sci. USA* **1980**, *77*, 6865–6869.

20 Arner, E. S. J., Eriksson, S., *Pharmacol. Ther.* **1995**, *67*, 155–186.

21 Montgomery, J. A., Shortnacy, A. T., Secrist, J. A. III, *J. Med. Chem.* **1983**, *26*, 1483–1489.

22 Secrist, J. A. III, Shortnacy, A. T., Montgomery, J. A., *J. Med. Chem.* **1985**, *28*, 1740–1742.

23 Azuma, A., Nakajima, Y., Nishizono, N., Minakawa, N., Suzuki, M., Hanaoka, K., Kobayashi, T., Tanaka, M., Sasaki, T., Matsuda, A., *J. Med. Chem.* **1993**, *36*, 4183–4189.

24 Secrist, J. A. III, Shortnacy, A. T., Montgomery, J. A., *J. Med. Chem.* **1988**, *31*, 405–410.

25 Fukukawa, K., Uedao, T., Hirano, T., *Chem. Pharm. Bull.* **1983**, *31*, 1582–1592.

26 Wright, J. A., Wilson, D. P., Fox, J. J., *J. Med. Chem.* **1970**, *13*, 269–272.

27 Wright, J. A., Taylor, N. F., Fox, J. J., *J. Org. Chem.* **1969**, *34*, 2632–2636.

28 Watanabe, K. A., Reichman, U., Hirota, K., Lopez, C., Fox, F J. J., *J. Med. Chem.* **1979**, *22*, 21–24.

29 Montgomery, J. A., Shortnacy-Fowler, A. T., Clayton, S. D., Riordan, J. M., Secrist, J. A. III, *J. Med. Chem.* **1992**, *35*, 397–401.

30 Montgomery, J. A., Secrist, J. A. III, US Patent 5,661,136, **1997**.

31 Tann, C. H., Brodfuehrer, P. R., Brundidge, S. P., Sapino, C. Jr., Howell, H. G., *J. Org. Chem.* **1985**, *50*, 3644–3647.

32 Bauta, W. E., Schulmeier, B. E., Burke, B., Puente, J. F., Cantrell, W. R. Jr., Lovett, D., Goebel, J., Anderson, B., Ionescu, D., Guo, R., *Org. Proc. Res. Dev.* **2004**, *8*, 889–896.

33 Parker, W. B., Shaddix, S. C., Rose, L. M., Shewach, D. S., Hertel, L. W., Secrist, J. A. III, Montgomery, J. A., Bennett, L. L. Jr., *Mol. Pharmacol.* **1999**, *55*, 515–520.

34 Waud, W. R., Schmid, S. M., Montgomery, J. A., Secrist, J. A. III, *Nucleosides Nucleotides Nucleic Acids* **2000**, *19*, 447–460.

35 King, K. M., Damaraju, V. L., Vickers, M. F., Yao, S. Y., Lang, T., Tackaberry, T. E., Mowles, D. A., Ng, A. M. L., Young, J. D., Cass, C. E., *Mol. Pharmacol.* **2006**, *69*, 346–353.

36 Parker, W. B., Shaddix, S. C., Chang, C. H., White, E. L., Rose, L. M., Brockman, R. W., Shortnancy, A. T., Montgomery, J. A., Secrist, J. A. III, Bennett, L. L. Jr., *Cancer Res.* **1991**, *51*, 2386–2394.

37 Xie, C., Plunkett, W., *Cancer Res.* **1995**, *55*, 2847–2852.

38 Lotfi, K., Mansson, E., Spasokoukotskaja, T., Pettersson, B., Liliemark, J., Peterson, C., Eriksson, S., Albertioni, F., *Clin. Cancer Res.* **1999**, *5*, 2438–2444.

39 Zhang, Y., Secrist, J. A. III, Ealick, S. E., *Acta Crystallogr. D Biol. Crystallogr.* **2006**, *62*, 133–139.

40 Sjoberg, A. H., Wang, L., Eriksson, S., *Mol. Pharmacol.* **1998**, *53*, 270–273.

41 Arner, E. S. J., Eriksson, S., *Pharmacol. Ther.* **1995**, *67*, 155–186.

42 Gandhi, V., Kantarjian, H., Faderl, S., Bonate, P., Du, M., Ayres, M., Rios, M. B., Keating, M. J., Plunkett, W., *Clin. Cancer Res.* **2003**, *9*, 6335–6342.

43 Xie, K. C., Plunkett, W., *Cancer Res.* **1996**, *56*, 3030–3037.

44 Nordlund, P., Reichard, P., *Annu. Rev. Biochem.* **2006**, *75*, 681–706.

45 Nutter, L. M., Cheng, Y. C., *Pharmacol. Ther.* **1984**, *26*, 191–207.

46 Dimitrova, D. S., Gilbert, D. M., *Nat. Cell Biol.* **2000**, *2*, 686–694.

47 Takahashi, T., Shimizu, M., Akinaga, S., *Cancer Chemother. Pharmacol.* **2002**, *50*, 193–201.

48 Zhang, Y. W., Hunter, T., Abraham, R. T., *Cell Cycle* **2006**, *5*, 125–128.

49 Thottassery, J. V., Westbrook, L., Someya, H., Parker, W. B., *Mol. Cancer Therapeut.* **2006**, *5*, 400–410.

50 Carson, D. A., Wasson, D. B., Esparza, L. M., Carrera, C. J., Kipps, T. J., Cottam, H. B., *Proc. Natl. Acad. Sci. USA* **1992**, *89*, 2970–2974.

51 Genini, D., Adachi, S., Chao, Q., Rose, D. W., Carrera, C. J., Cottam, H. B., Carson, D. A., Leoni, L. M., *Blood* **2000**, *96*, 3537–3543.

52 Klopfer, A., Hasenjager, A., Belka, C., Schulze-Osthoff, K., Dorken, B., Daniel, P. T., *Oncogene* **2004**, *23*, 9408–9418.

53 Leoni, L. M., Chao, Q., Cottam, H. B., Genini, D., Rosenbach, M., Carrera, C. J., Budihardjo, I., Wang, X., Carson, D. A., *Proc. Natl. Acad. Sci. USA* **1998**, *95*, 9567–9571.

54 Yamauchi, T., Nowak, B. J., Keating, M. J., Plunkett, W., *Clin. Cancer Res.* **2001**, *7*, 3580–3589.

55 Miura, S., Izuta, S., *Current Drug Targets* **2004**, *5*, 191–195.

56 Genini, D., Budihardjo, I., Plunkett, W., Wang, X., Carrera, C. J., Cottam, H. B., Carson, D. A., Leoni, L. M., *J. Biol. Chem.* **2000**, *275*, 29–34.

57 Gandhi, V., Plunkett, W., *Clin. Pharmacokinet.* **2002**, *41*, 93–103.

58 Plunkett, W., Gandhi, V., *Cancer Chemother. Biol. Response Modif.* **2001**, *19*, 21–45.

Index

a
ab-initio calculations 57, 158, 166, 309
ab-initio molecular orbital (MO) 157
ACC synthase 234, 237
acetic acid 175
acetic anhydride 175, 404
acetonide-protected amino sugar 567
acquired immunodeficiency syndrome (AIDS) 426
activation enzymes 513
– uridine-cytidine kinase (UCK) 513
acute lymphoblastic leukemia (ALL) 452
acyclic analogues 576
acyladenylate intermediate 375
acyladenylate mimetics 374
– salicyl-AMS 374
acylated acyl carrier protein 227
acyl hydroxamoyl phosphate linker 378
acyl sulfamate linkage 367, 373
addition-elimination reaction 551
adduct formation 97
adefovir dipivoxil 603, 604
adenine heterocycles 385
adenosine kinase 525, 540
adenosine receptors 536
adenine nucleosides 317
– chiral resolution 317
adenosine agonists 434
adenosine analogue(s) 175, 416
adenosine deaminase (ADA) 311, 367
adenosine kinase activity 485
adenosine kinase inhibition 581
adenosine nucleoside 367
– metabolism 367
adenosine receptors 433, 444
– modulators 433
adenosine residues 17
adenosine scaffold 385
adenosine triphosphate (ATP) 223
adenylation domains/enzymes 371
adenylation enzymes 381
– AsbC 381
– DhbE 381
AdoMet 224, 225, 227–231, 234, 236–238, 240
– adenosyl group donor 228
– aminocarboxypropyl group donor 227
– analogues 231, 236
– aziridine analogues 238
– biochemistry 224, 231
– biosynthesis 230
– chemistry 231
– metabolism 230
– methionine side chain 238
– nitrogen analogues 237
– pharmaceutical 240
– radical source 229
– riboswitch 236, 238
– ribosyl group donor 228
– ribosyl moiety 225
– selenium 234
– sulfone analogues 236
– sulfoxide 236
– tellurium analogues 234
AdoMet-dependent enzymes 232, 236, 241
AdoMet-dependent reactions 230, 231
AdoMet-dependent riboswitches 230
AdoMet-dependent transfer reactions 223
AdoMet-initiated radical chemistry 229
AdoMet synthetase(s) 231, 233, 234, 238
affinity chromatography 503
aglycone modifications 571
AHL 228
– metabolism of 228
AHL-mediated quorum sensing systems 227

Modified Nucleosides: in Biochemistry, Biotechnology and Medicine. Edited by Piet Herdewijn

ISBN: 978-3-527-31820-9

AIDS *see* acquired immunodeficiency syndrome
alicyclic groups 456
O-alkyl hydroxylamine 255
alkylated thioadenosines 223
ALL *see* acute lymphoblastic leukemia
N-alkylated nucleosides 311
allylic acetate 550
anti-tumor activities 592
allylic alcohol 328
– deoxygenation 328, 371, 375
aminoacyl tRNA synthetases 371, 375
aminoglycoside-based antibiotics 227
amino group donor 229
amino-substituted side chains 296
AMP deaminase 501
– fluorimetric assay 501
ampullariella regularis 525
anti-AIDS drugs 49
anti-apoptotic enzymatic signal 443
anticancer drug discovery program 635
anti-cancer drugs 587
anticancer nucleoside analogues 639
anti-Epstein-Barr virus (EBV) activity 578
anti-HCMV activity 411
anti-HIV agent 187
anti-parallel triplexes 294
anti-seizure mediator 437
antisense oligonucleotides 134
anti-tumor activities 592
antitumor agent 485
antiviral agents 601
– adefovir 601
– tenofovir 601
apoptosis 641
apoptotic response 643
aqueous solutions 348
aqueous trifluoroacetic acid (TFA) 366, 568
AR-induced vasodilation 439
aristeromycin 532
– modification 532
AR ligands 433
AR proteins 434
area under concentration curve (AUC) 590
arginine residues 226
aryl acid adenylation domains 374
aryl-capped siderophores 388
aryl ring 379
– importance of 379
aspase-dependent apoptosis 519
ataxia telangiectasia mutated (ATM) 519
ATP *see* adenosine triphosphate
ATP binding pocket 385
autoimmune disorders 452
– Crohn's disease 452
– psoriasis 452
autoimmune inflammatory disorders 433
aziridine cofactors 239

b
Bacillus anthracis 370, 381, 492
Bacillus subtilis 121
backbone-protein interactions 338
bacteriophage T4 116
Baeyer–Villiger oxidation 401
base analogues 64
base flipping enzyme 337, 338
base-modified analogues 478
base-pair extension 51
base-pair synthesis 51, 69
base-protected oligomers 164
Baylis–Hillman reaction 326
Benner Hydrogen-Bonding Variants 54
benzimidazole deoxynucleosides 56
– synthesis 56
benzimidazolium triflate 105
benzoate group 572
benzyl chloromethyl ether 465
bicyclic cylopentanons 418
bicyclic intermediate 416
biofilm formation 352
biological systems 6
biomolecule's architecture 6
biotin synthase 229
bipolar depression 241
– anxiety 241
– hypomania 241
– mania 241
bisubstrate inhibitors 374
bone marrow cells 589
bovine rhodopsin 434
butylammonium hydrogen carbonate 346

c
C-5-substituted pyrimidine nucleosides 251
– synthesis 252
Cambridge structural database 309
cancer cells 587
– long-term frequent treatment 587
capecitabine 588, 593
– discovery 588
– efficacy 593
– therapy 594
carbene-mediated intramolecular cyclopropanation 317
carbenoid intermediate 311
carbocyclic–oxetanocin-G (lobucavir) 399
carbocyclic amine 321

carbocyclic analogues 431, 573
carbocyclic nucleosides 309, 393, 403, 416, 526
– activity of 393
– biological activity 393
– mechanism 393
– processing 393
– synthesis 393, 416
carbocyclic oxetanocin analogues 404
carbocyclic oxetanocins 404
carbohydrate-modified nucleosides 636
carbohydrate systems 76
carcinogens 97
cardiac arrhythmias 435
cardiac ischemia 433
cardiovascular system 437
carrrier domain 371
catalytic hydrogenation 458
catechol-O-MTase (COMT) 227
cell-cycle proteins 520
CEM cells 642
chemical shift anisotropy (CSA) 12
chemical shift values 4, 5
chemical synthesis 105
chemokines 440
chiral cyclopentenol 313
chlamydia pneumoniae genotypes 148
p-chlorobenzoate group 569
m-chloroperbenzoic acid (mCPBA) 573
chronic obstructive pulmonary disease (COPD) 439
Chugaev-type reaction 321
circular dichroism (CD)-spectroscopy 139, 282
cis-syn cyclobutane pyrimidine dimer 112
clitocine 567, 575, 578
– aglycone of 578
– analogues 571
– biological activity 567
– cyclopentenyl analogue 575
– isolation 567
– synthesis 567
clofarabine 631, 632, 638, 640
– design 631
– mechanism 638
cobalamin-independent methionine synthetase 231
codon-anticodon interaction 154
colony forming units (CFU) 353
column chromatography 568
compartmentalized self-replication (CSR) 298
complementary DNA 137, 141, 162
– strand 162
condensation domain 371
controlled pore grass (CPG) 153
coproporphyrinogen III oxidase 229
Corey–Bakshi-Shibata (CBS) 547
CPD *see* cyclobutane pyrimidine dimer
– building block 105
– mutagenic properties 113
– photolyases 118
– site 118
– uracil-type 113
CPG-linked oligodeoxynucleotides 163
crude coupling reaction 568
– NMR data 568
cryptochromes 119
cryptosporidium parvum 488
crystallographic analysis 334
crystallographic data 309
cutaneous T-cell lymphoma (CTCL) 452
cyanoacetic acid 458
cyclobutane nucleoside derivatives 398
cyclobutane pyrimidine dimers (CPD) 114
cyclobutane ring 111
cyclohexenyl guanosine-analogue 87
cyclohexenyl nucleic acid (CeNA) 86
cyclohexenyl nucleotides 87
cyclopropanation reaction 316
cyclopropanation strategy 325
cyclopropane derivatives 395
cyclopropane fatty acids (CFAs) 227
cyclopropane ring 397, 414
cyclopropyl homo-nucleosides 397
cyclopropylidene nucleoside 395
cytidine deaminase (CDA) 520
– inhibitor 418
cytomegalovirus retinitis 601
cytosine 100, 103
– hydrate 103, 107
– methylation of 100
– photohydration of 103
cytotoxic drugs 587, 594
cytotoxic processes 434
– apoptosis 434
– calcium overload 434

d

DADMe-immucillin 467, 468
dancing partners 331
daudi host cells 540
de-novo biosynthesis 488
– purine 488, 500
– pyrimidine nucleotides 488
density functional theory (DFT) techniques 100

deoxyribonucleoside analogues 312, 320
Dewar photoproduct 101, 105, 107, 112, 113, 121
– alkaline labile 101
Dewar valence isomer 101, 109, 111
DFCR derivatives 589
diacetylated triol 418
diagnostic coupling constants 309
diaminopurine nucleoside 404
DIBAL reaction 529
dictyostelium discoideum 500
dideoxyribonucleoside analogues 311
Dieckmann condensation 322
Diels–Alder reaction 255, 549
diethylaminosulfur trifluoride (DAST) treatment 198
diethylaniline borane (DEANB) 547
dihydropyrimidine dihydrogenase (DPD) 593
differential scanning calorimetry (DSC) measurements 330
difluorotoluene-containing dTTP analogue (dFTP) 58
difluorotoluene deoxyriboside 56
difluorotoluene isostere of thymine 58
diguanylic acid intermediate 345
diisobutyl aluminum hydride (DIBAH) 184, 411
dimethylbutyl ring 404
dimethylformamide (DMF)-soluble alkynes 253
dinucleoside monophosphate 102, 103, 107, 112
diphenylphosphoryl azide (DPPA) 546
dipole–dipole interactions 6, 284
DNA 75, 76, 97, 103, 337
– base analogues 294
– base structures 66
– blueprint 75
– chain terminator 334
– chips/microarrays 153
– complementary strand 142
– CPG conjugate 163, 164
– cross-hybridization 76
– damaging chemotherapeutic agents 300
– dependent DNA (strand) synthesis 399
– dependent RNA polymerases 164
– detection method 164
– double helix 338
– duplex 14, 58, 60, 109, 157, 160, 264, 284, 289
– fragments 103
– methylation 337
– photolyases 118
– probes 135
– structure 97
– UV irradiation 102
DNA damage 97
– endogenous 97
– environmental 97
DNA–DNA duplex 282
DNA excision repair pathway 121
DNA hybridization 55
DNA-like systems 50
DNA-LNA chimera 135
DNA methyl transferases 282, 337
DNA methylation 224, 226, 337
DNA-modifying enzymes 240
DNA MTase-mediated coupling 238
DNA nucleobases 89
DNA oligomers 134, 168
DNA pairing 57
DNA polymerase 49, 50, 52, 63, 64, 67, 68, 70, 75, 76, 80, 85, 88, 90, 92, 93, 122, 154, 256, 257, 261, 290, 295–297, 429, 609, 642
– importance 49
– mitochondrial 429
– replication 122
– studies 82
– thermostable 257
– TNA-directed 80
– wild-type 80
DNA–protein interactions 292
DNA–RNA duplex 266
DNA precursor 229
DNA primase 299
DNA primer 82, 296
DNA probes 153, 154, 163, 168, 291
DNA repair 642
– enzyme 503
– mechanisms 49
DNA replication 51, 70, 640
– fidelity 50, 641
– processes 89
DNA segments 307
DNA sequences 97, 237
– mutations of 97
DNA strand 161, 270, 285
DNA structures 75, 111
DNA substrates 337
– primers 296
DNA synthesis 64, 75, 155, 291, 299, 511, 640, 641
– inhibition 592, 641
DNA synthesizer 103
DNA template 68, 262
DNA/RNA bases 290
DNA/RNA nucleobase 307

Docking experiments 333
double-stranded DNA 51, 148
Dowex-50WX2 column 360
dose fractionation 592
dose-limiting intestinal toxicity 588
down regulate DPD activity 594
DPD activity 596
DPD down regulators 594
DPD inhibitors 594
drug-efflux pumps 369
drug-receptor interaction 331
drug design 454, 588
– structure-based 454
duck hepatitis B virus (DHBV) polymerase 609

e

Escherichia coli 118, 489
– structures 118
– valyl tRNA synthetase 378
electron-deficient alkenes 321
electroparamagnetic resonance (EPR) methods 293
ELISA assay 575
electrophilic vinyl fluorination 561
enantioselective synthesis 561
endogenous reverse transcriptase 404
enzymatic alkyl transfer rates 238
enzymatic reactions 225
enzyme-catalyzed reactions 223
enzyme-DNA complex 117
– crystallization 117
enzyme inhibitor 373
enzyme-sugar interactions 90
episelenonium intermediate 314
Epstein-Barr virus (EBV) 395, 611
error-free 75
ester groups 419
ethyl chlorofomate 404
ethylene 228
ethylformate 458
ECyd 511
– apoptotic pathway 519
– mechanism 517
– metabolism 517
– synthesis of 517

f

FAAD enzymes 369
FAXS approach 12
FDA-approved antimetabolites 631
flavin adenine dinucleotide (FAD) 118
flavin mononucleotide (FMN)-RNA aptamer 22
fluorescence emission spectra 501
fluorescence resonance energy transfer (FRET) 17, 111
fluorescent tRNAs 502
fluorinated aromatic amino acids 7
fluorinated nucleoside analogues 13
fluorinated reporter group 5
fluorinated ribose units 19
fluorinated TAR 17
fluorine-modified riboses 19
fluorine relaxation 6
fluorine reporter nucleus 6
fluoroaromatic nucleosides 285
fluorophore-derivatized oligonucleotides 263
formamidine acetate 464
formycin phosphates 478
formycins 475, 476, 479, 483, 485, 488, 496, 500–501, 502
– biological activity 485
– modifications 476
– molecular probes 502
– molecular target 488
– sources 483
– spectral properties 479
forodesine HCl 463
– synthesis 463

g

gel-based (PAGE) assay 52, 53
gel mobility data 14
gemcitabine-resistant human pancreatic cancer cells 520
gene silencing 146
genetic switches 230
genomic DNA 68
Giese's method 346
global genome repair (GGR) 121
gluconabacter xylinus 343
glycerol-derived nucleotides 80
glycol-derived nucleoside triphosphates 82
glycosidic bond 99, 113, 385
N-glycosidic linkage 568
glycosyl bond 164
glycosyl donor 545, 559
– mitsunobu condensations 559
glycosyl torsion angle 309
G protein-coupled receptor (GPCR) 433
gram-negative bacteria 227, 369
gram-negative organism 369
gram-negative pathogen 365
– *yersinia pestis* 365
green fluorescent protein 8
– denatured states 8
Grignard reaction 411

ground-state PNP inhibitors 500
Grubb's catalyst 407, 411
guanine analogue 166
guanine substrates 451
guanosine monophosphate (GMP) 526
guanylation 459

h

H-bond acceptor group 67
H-bonding interaction 53
HBV DNA polymerase 608
HBV polymerase inhibition 609
HBV resistance 619
hairpin structure 286
hairy cell leukemia 631
HCMV infections 396
HCV-infected cells 145
HCV replication assay 561
hepatic enzymes 590
hepatitis C virus (HCV) 145
herpes simplex virus (HSV) 256, 570
heteronuclear NMR experiments 102
hexamethyldisilazane (HMDS) 568
hexane nucleosides 309
hexane scaffold 310
hexane system 324
hexane template 318, 324, 326, 327
hexitol nucleic acids (HNA) 134
high-affinity recognition 148
high-fidelity DNA polymerases 92
high-intensity replication stress 641
high-performance liquid chromatography (HPLC) 232, 349, 501
– analysis 105
– tandem mass spectrometry method 101
high-throughput screening 11
highly active antiretroviral therapy (HAART) 425
histone proteins 226
– modification 226
HIV-positive individuals 241
HIV reverse transcriptase (RT) 311, 335
Hoogsteen base pair 123
Hoogsteen hydrogen bonds 138, 293
Horner–Emmons reaction 407
human and calf phosphorylases 496
human cancer colon xenograft models 591
human cancer xenograft models 593
human cytomegalovirus (HCMV) 385, 609
human erythrocytic PNP 452
human fibroblasts 100
human gastric adenocarcinoma cells 520
human genome project 251
human hepatocellular carcinoma (HCC) 241
human immunodeficiency virus (HIV) 534, 601
human liver extract galocitabine 590
human organic anion transporters 607
human osteosarcoma (HOS) 335
human promyelocyticleukemia HL60 cell lines 572
human tumor xenograft experiments 638
hybridization probes 280, 291
hydrazine monohydrate 105
hydrogen bond 111, 167, 455
– donor 491
– energy 166
– formation 111
– system 164
hydrogen-bonded geometry 55
hydrogen-bonding carboxamide group 282
hydrogen-bonding interactions 111, 282
hydrogen-bonding motifs 289
hydrogen-bonding schemes 52
hydrogen-bonding substituents 299
hydrogen-bonding systems 157
hydrophobic base analogues 277
hydrophobic interactions 287
hydrophobic packing 61
hydroxymethyl group 327
hydroxyphosphonomethoxypropyl derivatives 614
hypoxanthine guanine phosphoribosyl-transferase (HGPRT) 526

i

immune system 225
immunostimulatory molecule 353
indole 56
– synthesis 56
infra red (IR) spectroscopy 99
intestinal toxicity 587
intracellular metabolism 606
iodine oxidation 329
ionizing radiation 97, 519
iron chelators 365
iron-sulfur cluster 229
isocarbostyril analogues 286
isopropylidene ketal 366
isothionucleosides 195
– synthesis 195
isotope-labeled AdoMet 232

k

Kaposi sarcoma-associated herpes virus (KSHV) 334
keratinocytes 100
keto-enol tautomers 418

keto ester product 318
ketone-derivatized DNA 255
Klenow enzyme 53
Klenow fragment 58, 92, 286, 296

l
leishmania donovani 503
lethal mutagenesis 298
leukemic myeloblasts 589
Lindlar's catalyst 312
Lineweaver–Burke plot 404
lipid-mediated transfection 145
lithium aluminum hydride 328
lithium tetramethylpiperidide (LiTMP) 461
LNA 139, 140, 143
– analogues 140
– antisense approach 146
– based molecular beacons 148
– containing duplexes 140
– duplex 135
– gapmer antisense oligonucleotide 144
– high-affinity hybridization properties 139
– hybridization 137
– potential therapeutics 143
– probes 147, 148
– use 147
LNA-DNA chimeras 135
LNA-DNA gapmer sequences 137
LNA-DNA mixmer 139
– strand 142
LNA-modified DNA-probes 147
LNA-modified duplexes 142
LNA-modified primers 147
LNA-modified siRNA 146
LNA-nucleotides 143
LNA-phosphorthioate chimeras 146
LNA-specific motif 140
LNA monomers 135, 137–139, 146
LNA nucleotides 145, 148
LNA oligonucleotides 135
– sequences 143
LNA probes 147
LNA sequences 146, 148
locked nucleosides 309
– synthesis 309
low-dose X-irradiation 308
low myelotoxicity 587
lung carcinoma 514
lysine residues 226

m
magnetic fields 12
MbtA 375
– three-dimensional structure 375
– homology model 378, 383
MDL drug data report 10
membrane-damaging protein streptolysin-O 144
metal-adenosyl complexes 229
methanococcus jannaschii 231
methenyltetrahydrofolylpolyglutamate (MTHF) 118
methoxytrityl chloride (MTT-Cl) resin 572
methyl group replacements 238
methyltransferases (MTases) 224
Michael reaction 410
Michaelis–Menten methods 52
micro RNA (miRNA) 227
– double-stranded 227
minor groove hydrogen bond 69
minor groove interactions 64
mitochondrial damage 434
mitochondrial permeability transition pore (MPTP) 443
mitogen-activated protein kinase (MAPK) 433
Mitsunobu condensation 549
Mitsunobu conditions 314, 325
Mitsunobu coupling 317
Mitsunobu esterification reaction 316
Mitsunobu reaction 311, 323
Mitsunobu-type reaction 329
MO *see* molecular orbital
Moffatt oxidation 546
molecular probes 502
moloney murine sarcoma virus (MSV) 610, 612
MOM derivative 561
monoclonal antibodies 101
monophosphate kinase 639
Monte Carlo simulations 109
multidimensional NMR methods 4
multidrug resistance protein (MRP) 607
murine adenosine deaminase (mADA) 9, 10
murine cytomegalovirus (MCMV) 395
murine rectum adenocarcinoma 520
mycobacterium tuberculosis 230, 365, 369, 371
– biosynthesis 371
– potent inhibition 365
myelodysplastic syndrome 631
myocardial ischemia 443

n
natural killer (NK) T cells 439
neplanocin 310, 525, 574
– biological activity 525
– modification 532
– X-ray structure 310

NER 121
– global genome repair (GGR) 121
– pathway 119
– transcription-coupled repair (TCR) 121
neurospora crassa 121
neurotropic retroviruses 612
nicotianamine 228
NMR-based screening techniques 11
– approach 12
NMR spectroscopy 3, 12, 105, 109, 111, 142
NMR study 19, 102
NMR titration experiments 17
non-enzymatic methylation 97
non-polar bases 57
non-polar isosteres 56–59
non-polar nucleoside isosteres 56, 58, 68
non-ribosomal peptide synthetases (NRPSs) 370
non retroviral RNA virus 611
non-small cell lung cancer (NSCLC) 596, 631
non-tumor-selective toxicity 587
– accumulation 587
northern isomers 415
northern thymidine analogue 415
novel antiviral nucleotides 619
NRPS adenylation domains 37, 374, 379, 382
nuclear magnetic resonance (NMR) 3, 99, 135, 266, 282, 330, 436
nuclear overhauser effects (NOEs) 6
nucleic acids 13, 19, 76, 133, 140, 255
– applications 133
– array technology 265
– duplex stability 251
– fluorine modifications 22
– NMR spectroscopy 13
– perspectives 133
– properties 133
– sequences 255
– strands 269
– structure 22
– systems 138
nucleobase 385
– impact 385
– isosteres 56
nucleophilic substitution 176
nucleoside analogues 393
nucleoside antibiotic 365, 475
– metabolism 367
– properties 366
– synthesis 365
– toxicity 367
nucleoside-based drugs 307
– therapeutic range 307
nucleoside diphosphate kinase 607
nucleoside enzymes 331
nucleoside hydrolases 502
nucleoside inhibitors 388
nucleoside kinases 639
nucleoside library 574
nucleoside ligands 436
nucleoside phosphoramidites 253
nucleotide prodrugs 619
nucleoside reverse transcriptase inhibitors (NRTIs) 425
nucleosides 397, 571
– adenosine kinase (AKase) inhibitory effect 571
– cyclobutyl analogues 401
– methylenecyclopropane analogues 397
nucleoside transporters 506
nucleotide-binding enzymes 331
nucleotide excision repair (NER) 119
nuclear magnetic resonance 330
– analysis 330
– experiments 142
– restraints 111
– spectral data 569
– spectroscopic methods 5

o

octamer primers 291
oligonucleotides 103, 105, 107, 135, 153, 155, 159, 162, 164, 166, 253, 290
– chemical synthesis 103, 105
– properties 153, 155, 159
– synthesis 153, 155, 270
– two-step irradiation 107
organopalladium coupling reactions 252
oxetanocin-A 401
oxocarbenium ion 462

p

packing effects 60
PAGE analysis 52
PDC oxidation 534
palladium catalyzed allyclic 554
palladium mediated condensation 552
palladium-mediated coupling reactions 381
pancreas carcinoma 516
PCR amplifications 55
Pd-catalyst 410
Pd-catalyzed allylic substitution 410
Pd-catalyzed cross-coupling reactions 387
Pd-mediated coupling reaction 270
PDP-CPG conjugate 163
PDP strategy 164
peripheral blood mononuclear cells (PBMC) 500

phage vectors 113
phenylalanyl-tRNA synthetase adenylate inhibitor 382
phosphate-binding site 454
phosphodiester backbone conformation 337
phosphodiester backbone-protein interactions 338
phosphodiester group 265, 268
phosphodiester linkage 103, 121
– ATP-independent hydrolysis 121
phosphoramidite approach 329
phosphoramidite building block 105, 162
phosphoramidites 105
phosphoric-carboxylic acid anhydride 373
phosphorodithioate linkage 105
phosphorthioate sequences 146
phosphorylated metabolites 635
photochemical DNA computer 269
photodimers 112
– mutations 112
photolysis-induced nitrogen extrusion 321
photo reactivating enzymes 118
phototautomerism 479
– heterocyclic systems 479
phthalimido group 315
– hydrazinolysis 315
physico-chemical interactions 50
plaque-forming efficiencies 112
plaque reduction assay 575
plasmodium falciparum 492, 500
plasmodium lophurae 489, 500
PNA-containing universal bases 282
PNA-DNA duplexes 282
PNP enzyme 461
PNP inhibition 461
PNP inhibitors 451, 454, 458, 460, 467, 468, 498
– (3-D) structures 489
– chemical structure 475
– DADMe-immucillin-G 467
– first-generation 458
– high-molecular-mass 492
– immucillin-H 467
– low-molecular-mass 490
– role 495
– second-generation 460, 461
– third-generation 467
PNP monomer 490
PNP transition-state inhibitors 460
PNP trimer 457
– first-generation 457
polyamines 228
– biosynthetic pathway 228
polyfluorinated benzenes 61
polyformycin phosphates 478
polymerase behavior 58
polymerase chain reaction (PCR) 257, 287
polymerase enzymes 66
polymerase inhibitors 49
polymerase-mediated extension reaction 264
post-oligonucleotide-synthesis modification 253
post-synthetic modification method 253, 293
primer strand 113
primer-template duplex 58
pro-inflammatory cytokines 439
probe-on-carrier method 163
propynyl-substituted pyrimidine nucleosides 267
protease inhibitors 425
protein-DNA complex 328
protein-ligand interactions 385
protein–protein conjugation 255
protein demethylases 226
protein kinase PKC (PKC) activation 441
protein lysine methylation 226
protein residues 502
proteins 7, 8
– NMR spectroscopy 7
– synthesis 157
prototropic tautomers 475
pseudomonas aeruginosa 343, 369
pseudorotational pathway 309
Pummerer reaction 195
purine-binding pocket 453
purine-metabolizing enzyme 451
– (PNP) 451
purine-pyrimidine scaffold 54
purine nucleoside hydrolases 475
purine nucleoside phosphorylase (PNP) 367, 451, 452, 475, 477, 488
– analogues 180
– inhibitors 488
purine salvage pathway 367
Pyrex-filtered light 107
pyridinium dichromate (PDC) 178, 527
pyrimidine–pyrimidine sites 99
– CC 99
– CT 99
– TC 99
– TT 99
pyrimidine analogues 395
pyrimidine bases 100
pyrimidine dimers 99, 103, 112, 116, 118
– containing DNA 109
– cyclobutane 98
– mutagenesis 109
– mutations 112

– repair 116
– structure 109
– UV-induced 116, 118, 122
pyruvate formate-lyase activating enzyme 229

q
quantum-chemical method 100

r
radical-SAM 229
radioligand binding assay 536
rapid eye movement (REM) 437
real-time PCR 147
receptor-mediated endocytosis 607
replicative enzymes 601
residual dipolar coupling (RDC) 20
resonance energy acceptors 479
restriction endonucleases (REases) 225
restriction sites 264
reverse transcriptases (RT) 49, 51, 59, 399, 608
rheumatoid arthritis 433
rhodamine-modified nucleotide 263
ribofuranose hydroxyls 454
ribofuranose ring oxygen 382
ribonucleic acid 22
– non-native modification 22
ribonucleoside analogues 318, 321
ribose-binding site 454
ribose moieties 13
ribosomal RNA (rRNA) 227
– fragmentations 519
ribosomal synthesis 65
riboswitches 230
ring-altered compounds 636
ring-closing metathesis (RCM) reaction 407, 408, 527, 528, 536
ring-expansion reaction 102
RNA 76
– affinity 6
– cross-hybridization 76
– dependent RNA polymerase (RdRP) 299
– duplex 160, 285
– inhibition 519
– ligand binding 22
– primer 82
– probe 160
– protein complexes 20
– protein interactions 270
– secondary structure equilibria 21
– sequences 141, 148
– substrates 337
– synthesis 511, 525
– transcripts 147
– viruses 298
ROESY experiment 23

s
S-adenosyl-L-methionine 223
SAH hydrolase 525, 537, 543
SAM dependent methyltransferase 525
SAM-riboswitches 230
S-methanocarba AZT analogues 323
– synthesis 323
S-modified nucleotides 330
salmonella typhimurium 343
schiff base-type intermediate 117
schizosaccharomyces pombe 121, 503
selectivity ratio 379
sequence-specific oligonucleotide hybridization 291
severe combined immunodefiecient (SCID) 520
shewanella oneidensis 343
siderophores 370
– background 370
– biosynthesis 365
– inhibitors 365
– producing microorganisms 368
silica gel chromatography 366
simian immunodeficiency virus (SIV) 613
Simmons–Smith cyclopropanation 312, 560
Simmons–Smith reaction 315
single nucleotide polymorphisms (SNPs) 163, 291
single-photon emission computed tomography (SPECT) 439
small interference RNA (siRNA) 227
SNP-specific PCR 147
sodium borohydride reduction 320
sodium cyanoborohydride 467
sodium hydride 458, 465
sodium salt glycosylation method 278
solid-phase reaction technologies 582
solid-phase synthesis 13
spore photoproduct 102
stacking interactions 283
staphylococcus aureus 230, 352
steady-state kinetics 58
steric-based hypothesis 62
steroselective hydroxymethymethylation 536
streptococcus faecium 211
– growth 211
streptomyces citricolor 525
streptomyces griseolus 237
structure-activity relationship (SAR) 331, 433, 526

substrate-bicyclic ester 396
sugar-binding site 454
sugar mimicking linker 610
sugar-modified analogues 478
sugar-phosphate backbone 328, 329
Suzuki reaction 252
Swern oxidation 318

t

Taq DNA polymerase 296
T-cell acute lymphocytic leukemia 463
T-cell lymphoma (CTCL) 457
T-cell-mediated diseases 451
T-cell selective immunosuppression 451
TC sequence 107
T4 DNA polymerase 100
– exonuclease activity 100
template-primer duplex 51, 113, 334
tenofovir disoproxil fumarate 614
– efficacy 614
tenofovir disoproxil fumarate 605
terbium ions 479
tert-butyldimethylsilyl chloride (TMBDS) 103, 403
tert-butylphenoxyacetyl group 105
tetrapropylammonium perruthenate (TPAP) 528
tetra-*n*-butylammonium fluoride (TBAF) 158, 403
therapeutic agents 393
therminator DNA polymerase 82
thermococcus litoralis 88
thermophilic DNA ligase 291
thermophilic DNA polymerases 297
thermus aquaticus DNA polymerase 123, 257
thermus thermophilus 118
thin-layer chromatography (TLC) 185
thioguanine-pyridone 53
thionucleosides 205, 210
– biological activity 210
– synthesis 205
thiol-terminated linkers 256
thioxonucleosides 173, 198
– synthesis 198
thymidine/LNA-thymine probe 147
thymidylate synthase (TS) 331
thymine base 99, 161
thymine-uracil-type CPD 105
TNA libraries 80
toxicity dependent antiviral activities 552
TP gene expression 594
transcription-coupled repair (TCR) 121
transition-state inhibitors 467
transmembrane helices (TMs) 434
tricyclic ring system 567
trifluoromethanesulfonic acid (TfOH) 192
trimethylsilyl trifluoromethanesulfonate (TMSOTf) 568
trimethylsulfonium hydroxide 235
triphosphate derivatives 294
tRNA molecules 154, 371
tRNA synthetase inhibitor 378
T7 RNA polymerase 55
tumor cells 511
tumor mammalian cells 504
tumor necrosis factor (TNF) 355
tumor-selective action 587
turnover number (TON) 235

u

ultraviolet (UV) radiation 97
United States Adopted Names Council (USAN) 463
UUG trinucleotide 20
UV-A photosensitizers 100
UV damage endonuclease (UVDE) 121
UV-induced DNA Damage 97
– cyclobutane pyrimidine dimers (CPDs) 97
– pyrimidine(6-4) pyrimidone photoproducts 97
UV induced DNA lesions 97, 98, 109
UV-irradiated chromatin 100
UV-irradiated DNA 116
UV irradiation 101, 107
UV lesions 113
UV melting study 111
UV radiation 100
uridine cytidine kinase (UCK) 513

v

van der walls contacts 339
varicella zoster virus (VZV) 580
viral polymerases 608
– inhibition 608
vinyl magnesium bromide 210, 529

w

water-ethanol system 285
water-soluble phosphine 252
Watson–Crick base pair 4, 161, 167
Watson–Crick binding site 285
Watson–Crick hydrogen bonds 50, 57, 58, 60
Watson–Crick purine–pyrimidine pair 66
Watson–Crick rule 80
wild-type polymerase 298
Wittig–Horner reaction 408, 412
western blot analysis 522
woodchuck hepatitis virus (WHV) 611

x

x-ray analysis 57
– data 157
x-ray crystallographic 454
– analysis 454, 569
x-ray crystallography 135, 318, 477
x-ray structures 52
x-ray induced cell death 522
xanthomonas oryzae 485
xenograft model 638
XPA–RPA protein 291
XP variant (XPV) cells 122

y

yeast DNA polymerase 123
yersinia pestis 369
yersinia pseudotuberculosis ATCC 369